Ecotoxicology

Ecotoxicology offers a comprehensive overview of the science underpinning the recognition and management of environmental contamination. It describes the toxicology of environmental contaminants, the methods used for assessing their toxicity and ecological impacts, and approaches employed to mitigate pollution and ecological health risks globally. Chapters cover the latest advances in research, including genomics, natural toxins, endocrine disruption and the toxicology of radioactive substances. The second half of the book focuses on applications, such as cradle-to-grave effects of selected industries, legal and economic approaches to environmental regulation, ecological risk assessment, and contaminated site remediation. With short capsules written by invited experts, numerous case studies from around the world and further reading lists online, this textbook is designed for advanced undergraduate and graduate one-semester courses. It is also a valuable reference for graduate students and professionals. Online resources for instructors and students are also available.

Peter G. C. Campbell is Emeritus Professor at the Institut national de la recherche scientifique (INRS) in Québec City, Canada, which he joined in 1968 after completing his PhD. Over the course of his career he has researched elements of analytical chemistry, geochemistry and ecotoxicology, with an emphasis on metal speciation and bioavailability. He co-directed the Metals in the Environment Research Network from 1998–2009 and held a Canada Research Chair in Metal Ecotoxicology from 2002 until his retirement in 2015. He was elected to the Academy of Sciences of the Royal Society of Canada in 2002 and received the SETAC Founders' Award in 2019.

Peter V. Hodson is Professor Emeritus at Queen's University in Kingston, Ontario, Canada. His recent research includes the toxicity of crude oil and oil dispersants to fish embryos and the role of chemicals in the decline of the American eel in Lake Ontario. He has authored numerous peer-reviewed papers, books and technical reports related to the toxicity of chemicals to fish and contamination of the Great Lakes and rivers of Ontario, Quebec, and Alberta. He was President of the Society of Environmental Toxicology and Chemistry (SETAC) (1994–1995) and a member of its Board of Directors and the Board of the SETAC World Council (from 2004 to 2007), serving as the Chair of the World Council Science Committee. He was Program Chair of SETAC's 10th Annual Meeting in Toronto in 1989.

Pamela M. Welbourn is currently an Adjunct Professor in the School of Environmental Studies at Queen's University, Canada. She was on the faculty at the University of Toronto from 1970 to 1990, where she was Director of the Institute for Environmental Studies from 1984 to 1989. Welbourn has published numerous peer-reviewed papers and technical reports, and co-authored the textbook *Environmental Toxicology* (Cambridge University Press, 2002). She has consulted for the private and public sector and also for non-governmental organisations. She holds two teaching awards and currently gives guest lectures and public lectures on ecotoxicology.

David A. Wright is Emeritus Professor of Environmental Toxicology at the University of Maryland, Center for Environmental Science, USA. While advisor to fifteen MS and PhD students, he has conducted numerous studies on the effect of inorganic and organic contaminants on aquatic organisms. He has published numerous articles, books and technical reports on a variety of toxicological and maritime issues such as invasive species, and has co-authored the textbook *Environmental Toxicology* (Cambridge University Press, 2002). He is a Fellow of the Institute for Engineering Science & Technology (IMarEST) and was Chief Scientist at the 2010 Gulf of Mexico Deepwater Horizon spill.

Ecotoxicology

Principal authors

Peter G. C. Campbell
Institut national de la recherche scientifique (INRS),
QC, Canada

Peter V. Hodson
Queen's University, Kingston, ON, Canada

Pamela M. Welbourn
Queen's University, Kingston, ON, Canada

David A. Wright
University of Maryland, Center for Environmental
Science, Cambridge, MD, USA

Contributing authors

Valérie S. Langlois
Institut national de la recherche scientifique (INRS),
QC, Canada

Christopher J. Martyniuk
University of Florida, Gainesville, FL, USA

Christopher D. Metcalfe
Trent University, Peterborough, ON, Canada

Louise M. Winn
Queen's University, Kingston, ON, Canada

CAMBRIDGE
UNIVERSITY PRESS

Guest capsule contributors

Capsule 6.1 *Metal Effects on Fish Olfaction*
Gregory G. Pyle
University of Lethbridge, AB, Canada

Capsule 7.1 *Mobility, Bioavailability and Remediation of PFAS Compounds in Soils*
Michael J. McLaughlin
University of Adelaide, SA, Australia

Capsule 10.1 *Radiological Protection of the Environment*
Nicholas A. Beresford   **David Copplestone**
Centre for Ecology & Hydrology, Lancaster, UK   University of Stirling, Stirling, UK

Capsule 11.1 *Mercury and Silver: A History of Unexpected Environmental Consequences*
Saul Guerrero
Australian National University, Canberra, ACT, Australia

Capsule 12.1 *The Sudbury Soils Study: An Area-wide Ecological Risk Assessment*
Christopher D. Wren   **Glen Watson**   **Marc Butler**
LRG Environmental, Markdale,   Vale Canada Limited,   Glencore Sudbury Nickel Operations, ON,
Canada   Copper Cliff, ON, Canada   Falconbridge, ON, Canada

Capsule 12.2 *Legislation for Chemical Management – Traditional Environmental Knowledge in the Regulation of Chemical Contaminants*
F. Henry Lickers
International Joint Commission, Canada–United States

Capsule 13.1 *The Enduring Legacy of Point-source Mercury Pollution*
John W. M. Rudd   **Carol A. Kelly**
Rudd & Kelly Research, Salt Spring Island, BC, Canada

Capsule 13.2 *Bioremediation of Oil Spills*
Charles W. Greer
National Research Council Canada, Montreal, QC

Capsule 14.1 *Lithium – A Critical Mineral Element: Sources, Extraction and Ecotoxicology*
Heather Jamieson
Queen's University, Kingston, ON, Canada

With additional guest contributions from:

Corinna Dally-Starna   **Case Study 14.1**
Queen's University, Kingston, ON, Canada

Brendan Hickie   **Contribution to Chapter 4**
Trent University, Peterborough, ON, Canada

Adrian Pang   **Case Study 13.2**
Queen's University, Kingston, ON, Canada

Doug Spry   **Contribution to Chapter 12**
Environment and Climate Change Canada, Gatineau, QC, Canada

CAMBRIDGE
UNIVERSITY PRESS

University Printing House, Cambridge CB2 8BS, United Kingdom

One Liberty Plaza, 20th Floor, New York, NY 10006, USA

477 Williamstown Road, Port Melbourne, VIC 3207, Australia

314–321, 3rd Floor, Plot 3, Splendor Forum, Jasola District Centre, New Delhi – 110025, India

103 Penang Road, #05–06/07, Visioncrest Commercial, Singapore 238467

Cambridge University Press is part of the University of Cambridge.

It furthers the University's mission by disseminating knowledge in the pursuit of
education, learning, and research at the highest international levels of excellence.

www.cambridge.org
Information on this title: www.cambridge.org/highereducation/isbn/9781108834698
DOI: 10.1017/9781108819732

© Cambridge University Press 2022

First published 2022

Printed in the United Kingdom by TJ Books Limited, Padstow, Cornwall, 2022

A catalogue record for this publication is available from the British Library.

ISBN 978-1-108-83469-8 Hardback
ISBN 978-1-108-81973-2 Paperback

Additional resources for this publication at www.cambridge.org/ecotoxicology

CONTENTS

Preamble   *page* xv
Preface   xvii
Acknowledgements   xix

PART I APPROACHES AND METHODS   1

1

The History and Emergence of Ecotoxicology as a Science   3

Pamela M. Welbourn and Peter V. Hodson

Learning Objectives   3
1.1   The Science of Ecotoxicology   3
1.2   Historical Landmarks in the Development
       of Ecotoxicology   7
       1.2.1   *Silent Spring* and Pesticides   7
       1.2.2   Mercury   10
       1.2.3   Acidification   10
       1.2.4   Industrial Waste Disposal and Brownfields   11
       1.2.5   Oil Spills   12
       1.2.6   *Our Stolen Future* and
               Endocrine Disruptors   12
1.3   The Emergence of the Science
       of Ecotoxicology   13
1.4   The Turning Point and Formal Regulation of
       Toxic Substances   15
1.5   Solutions That May Lead to New Problems   16
1.6   Conclusions   17
Summary   18
Review Questions and Exercises   18
Abbreviations   19
References   19

2

Measuring Toxicity   23

Peter V. Hodson and David A. Wright

Learning Objectives   23
2.1   The Basics of Environmental Toxicology   23
       2.1.1   Concepts and Definitions   24
               2.1.1.1 What Is Toxicity?   24
               2.1.1.2 Chemical Structure vs Toxicity   24

               2.1.1.3 Nutrients vs Toxicants   25
               2.1.1.4 Expressions of Toxicity   26
2.2   Designing a Toxicity Test: What Is the
       Question?   28
       2.2.1   Test Organisms   29
               2.2.1.1 Laboratory Cultures of
                        Test Organisms   30
               2.2.1.2 Life Stages Tested and
                        Responses Measured   31
       2.2.2   Test Media and Routes of Exposure   32
       2.2.3   Exposure Gradients   33
       2.2.4   Exposure Time   33
       2.2.5   Control Treatments   36
       2.2.6   Other Test Conditions That Affect
               Measured Toxicity   36
       2.2.7   Characterizing Test Conditions and
               Chemical Exposures   38
       2.2.8   Complexities in Toxicity Testing   39
               2.2.8.1 Toxicity Tests for Sparingly
                        Soluble Compounds   39
               2.2.8.2 Sediment and Soil Toxicity Tests   39
               2.2.8.3 Standard vs 'Realistic' Toxicity Tests   41
               2.2.8.4 Surrogate Species for Routine Testing   41
2.3   Statistics for Toxicity Tests   42
       2.3.1   Regression Analyses for
               Computing Toxicity   42
               2.3.1.1 Data Types and Transformations   43
               2.3.1.2 Control Data   44
       2.3.2   Hypothesis Testing: Multiple
               Regression Analyses   44
       2.3.3   Predictive Toxicology: Single Compounds   45
               2.3.3.1 Acute to Chronic Ratios (ACRs)   46
               2.3.3.2 Species Sensitivity Distributions (SSDs) 47
               2.3.3.3 Quantitative Structure–Activity
                        Relationships (QSARs)   47
       2.3.4   Predictive Toxicology: Mixtures   48
               2.3.4.1 Toxic Unit (TU) Model   48
               2.3.4.2 Toxic Equivalent Factor (TEF) Model   49
               2.3.4.3 Target Lipid Model   49
               2.3.4.4 Metal Mixtures   50
               2.3.4.5 Dissecting Complex Mixtures   51
       2.3.5   Moving Away from Traditional
               Toxicity Tests   52
Summary   54
Review Questions and Exercises   54
Abbreviations   55
References   56

3

Contaminant Uptake and Bioaccumulation: Mechanisms, Kinetics and Modelling   61

Peter G. C. Campbell, Peter V. Hodson, Pamela M. Welbourn and David A. Wright

Learning Objectives   61
3.1   General Considerations   61
    3.1.1   Composition and Structure of Biological Membranes   62
    3.1.2   Transport of Solutes Across Cell Membranes   64
        3.1.2.1   Diffusion Through the Lipid Bilayer   65
        3.1.2.2   Diffusion Through Membrane Pores and Channels   65
        3.1.2.3   Carrier-mediated Transport   69
    3.1.3   Endocytosis   71
    3.1.4   Transcellular Transport (e.g., Gill; Intestine; Lung)   71
    3.1.5   Ecotoxicological Perspective on Transmembrane Transport Processes   72
3.2   Uptake Routes   73
    3.2.1   Skin   73
    3.2.2   Lungs   73
    3.2.3   Gills   74
    3.2.4   Digestive System   75
    3.2.5   Olfactory System   76
    3.2.6   Plant Foliage and Roots   76
    3.2.7   Boundary Layers   78
    3.2.8   Uptake by Endocytosis   78
    3.2.9   How Different Exposure Routes Affect the Rates of Toxicant Uptake   79
3.3   Elimination Routes   79
3.4   Bioaccumulation and Uptake–Elimination Kinetics   79
    3.4.1   Toxicant Uptake: Differences Between Lipophilic and Hydrophilic Molecules   81
    3.4.2   Toxicokinetics   83
        3.4.2.1   One-compartment Model   83
        3.4.2.2   Two-compartment Model   84
3.5   Biotransformations   85
    3.5.1   Metals   85
    3.5.2   Organic Xenobiotics   86
3.6   Bioaccumulation and Biomagnification   87
    3.6.1   Metals   87
    3.6.2   Bioaccumulation of Persistent Organic Contaminants   88
        3.6.2.1   Lipophilic Contaminants   88
        3.6.2.2   Interplay Between Bioenergetics and Bioaccumulation of Lipophilic Contaminants   90
        3.6.2.3   Proteinophilic Contaminants   91
Summary   91
Review Questions and Exercises   92
Abbreviations   93
References   93

Appendix 3.1:   Kinetics of a Saturable Transmembrane Carrier System Transporting a Chemical Substrate   95
Appendix 3.2:   Uptake/Loss Kinetics in a Single-compartment System   95

4

Methods in Ecotoxicology   99

Peter V. Hodson and David W. Wright

Learning Objectives   99
4.1   Moving Beyond Environmental Toxicology   99
4.2   Laboratory Versus Field Studies of Ecotoxicology: Strengths and Weaknesses   100
4.3   Surveys, Monitoring and Assessment   102
    4.3.1   Relating Cause and Effect by Surveys and Monitoring   102
    Case Study 4.1   Upstream–Downstream Studies to Assess Whether Pulp-mill Effluents Affect the Sexual Maturation of Fish   104
    4.3.2   Ecoepidemiology: Assessing the Strength of Proposed Cause–Effect Relationships   105
    Case Study 4.2   The Ecoepidemiological Case for Cancer in Fish Caused by Sediment Polycyclic Aromatic Compounds (PACs)   107
    4.3.3   Markers and Indicators of Chemical Exposure and Effects   107
        4.3.3.1   Chemical Markers   108
        4.3.3.2   Biomarkers   108
        4.3.3.3   Bioindicators   109
        4.3.3.4   The Sediment Quality Triad   111
        4.3.3.5   Summary: Markers and Indicators   112
    4.3.4   Palaeo-ecotoxicology: Retrospective Assessment of Contamination and Toxicity   112
    Case Study 4.3   Evidence from Palaeo-ecotoxicology for a Chemical Cause of Reproductive Failure of Lake Trout (*Salvelinus namaycush*) in Lake Ontario   114
    4.3.5   Monitoring the Human Food Supply   115
4.4   Field Experiments   117
    4.4.1   *In Situ* Toxicity Tests   117
    4.4.2   Experimental Plots   118
    Case Study 4.4   The Effectiveness of Fertilizers in Promoting Degradation of Crude Oil Spilled on a Vegetated Wetland   118
    4.4.3   Experimental Ecosystems   119
    Case Study 4.5   Whole-lake Experiment with an Endocrine Disruptor   122
4.5   Modelling Environmental Fate, Behaviour, Distribution and Effects of Chemicals   122
    4.5.1   Chemical Fate Modelling   123
    4.5.2   Bioaccumulation and Effects Modelling   126
    Case Study 4.6   PCB Contamination of the Southern Resident Killer Whale   128
    4.5.3   Integrated Effects Modelling   129
Summary   130

Review Questions and Exercises   131
Abbreviations   131
References   132

5

Ecotoxicogenomics   139

Valérie S. Langlois and Christopher J. Martyniuk

Learning Objectives   139

5.1 Environmental 'Omics': A Role in
    Ecotoxicology Research   139
5.2 Ecotoxicology and Transcriptomics   141
    5.2.1 Application of Ecotoxicogenomics   142
5.3 Ecotoxicology and Proteomics   144
5.4 Ecotoxicology and Metabolomics/
    Lipidomics   146
5.5 Ecotoxicology and Epigenetics   147
5.6 Environmental DNA (eDNA)   149
5.7 Ecotoxicology and the Microbiome
    (Metagenomics)   149
5.8 Ecotoxicology and Bioinformatics   150
5.9 Omics and Adverse Outcome Pathways (AOPs)   152
5.10 Omics in Regulatory Toxicology   153
    5.10.1 Computational Toxicology in
           Regulatory Toxicology   154
    Case Study 5.1 Omics to Reveal Mechanisms
           Underlying Glyphosate Toxicity in Invertebrates
           and Vertebrates   155
    5.10.2 Environmental Omics in
           Regulatory Toxicology   157
    5.10.3 Challenges and Considerations   157
5.11 Emerging Applications for Omics
    in Ecotoxicology   159
    5.11.1 Genome-wide CRISPR Screens
           in Ecotoxicology   160
    5.11.2 Multi-omics, Exposome and Exposomics
           in Ecotoxicology   161
Summary   161
Review Questions and Exercises   162
Abbreviations   162
References   163

PART II TOXICOLOGY OF
INDIVIDUAL SUBSTANCES   169

6

Metals and Metalloids   171

*Peter G. C. Campbell, Pamela M. Welbourn and
Christopher D. Metcalfe*

Learning Objectives   171
6.1 Introduction   171

6.2 Biogeochemistry of Metals and Metalloids   174
    6.2.1 General Properties: Metal Speciation   174
    6.2.2 Mobilization, Binding, Transport and Chemical
           Forms of Metals in the Environment   178
6.3 Biological Availability of Metals in Aquatic and
    Terrestrial Systems   179
    6.3.1 General Considerations   180
    6.3.2 Aquatic Environments: Dissolved Metals   180
    6.3.3 Aquatic Environments: Particulate Metals   184
    6.3.4 Terrestrial Environments   185
    6.3.5 Diet-borne Metals   186
6.4 Mechanisms of Metal Toxicity   187
    6.4.1 Alteration of Enzyme Conformation   188
    6.4.2 Displacement of Essential Cations   188
    6.4.3 Oxidative Stress   189
    6.4.4 Changes to Cellular Differentiation   190
    6.4.5 Behavioural Effects   190
    Capsule 6.1 Metal Effects on Fish Olfaction   191
    (Gregory G. Pyle)
6.5 Metal Detoxification and Tolerance   195
    6.5.1 Metal Speciation Within Cells   195
    6.5.2 Determination of Subcellular
           Metal Partitioning   196
    6.5.3 Links Among Changes in Metal Exposure,
           Changes in Metal Subcellular Distribution
           and the Onset of Deleterious Effects   197
           6.5.3.1 Laboratory Observations   198
           6.5.3.2 Field Observations   199
    Case Study 6.1 Response of Native Freshwater
           Animals to Metals Derived from Base-metal
           Smelter Emissions   199
    6.5.4 Metal Tolerance   202
           6.5.4.1 Occurrence and Origin   202
           6.5.4.2 Approaches   202
           6.5.4.3 Taxonomic Distribution of Tolerance   203
           6.5.4.4 Tolerance Mechanisms   203
           6.5.4.5 Ecotoxicological Implications and
                   Practical Applications   205
6.6 Organometals (Hg, Pb, Sn, As, Sb, Se)   206
6.7 Abiotic Factors Affecting Metal Toxicity   206
    6.7.1 Temperature   206
    6.7.2 pH   207
    6.7.3 Hardness   208
    6.7.4 Salinity   208
    6.7.5 Dissolved Organic Matter   209
6.8 Metal-specific Sections   209
    6.8.1 Mercury   210
           6.8.1.1 Occurrence, Sources and Uses   210
           6.8.1.2 Biogeochemistry   212
           6.8.1.3 Mercury Methylation   214
           6.8.1.4 Biogeochemical Cycle   215
           6.8.1.5 Mercury Biomagnification   217
           6.8.1.6 Environmental Factors Affecting
                   Mercury Bioaccumulation   219
           6.8.1.7 Mercury Bioaccumulation
                   and Monitoring   221
           6.8.1.8 Ecotoxicity   221

6.8.1.9   Detoxification and Tolerance   224
6.8.1.10 Mercury Highlights   225
6.8.2   Cadmium   225
6.8.2.1 Occurrence, Sources and Uses   226
6.8.2.2 Biogeochemistry   226
6.8.2.3 Biochemistry   226
6.8.2.4 Ecotoxicity   227
6.8.2.5 Cadmium Highlights   228
6.8.3   Lead   228
6.8.3.1 Occurrence, Sources and Uses   228
6.8.3.2 Biogeochemistry   230
6.8.3.3 Biochemistry   231
6.8.3.4 Ecotoxicity   232
6.8.3.5 Lead Highlights   234
6.8.4   Copper   234
6.8.4.1 Occurrence, Sources and Uses   234
6.8.4.2 Biogeochemistry   235
6.8.4.3 Biochemistry   236
6.8.4.4 Ecotoxicity   237
6.8.4.5 Copper Highlights   239
6.8.5   Nickel   239
6.8.5.1 Occurrence, Sources and Uses   239
6.8.5.2 Biogeochemistry   240
6.8.5.3 Biochemistry   240
6.8.5.4 Ecotoxicity   241
6.8.5.5 Nickel Highlights   243
6.8.6   Zinc   243
6.8.6.1 Occurrence, Sources and Uses   243
6.8.6.2 Biogeochemistry   244
6.8.6.3 Biochemistry   245
6.8.6.4 Ecotoxicity   246
6.8.6.5 Zinc Highlights   248
6.8.7   Arsenic   248
6.8.7.1 Occurrence, Sources and Uses   248
6.8.7.2 Biogeochemistry   249
6.8.7.3 Biochemistry   251
6.8.7.4 Ecotoxicity   252
6.8.7.5 Arsenic Highlights   254
6.8.8   Selenium   255
6.8.8.1 Occurrence, Sources and Uses   255
6.8.8.2 Biogeochemistry   255
6.8.8.3 Biochemistry   256
6.8.8.4 Ecotoxicity   257
6.8.8.5 Selenium Highlights   259
Summary   259
Review Questions and Exercises   260
Element-specific Questions   261
Abbreviations   262
References   262

7.3   Uptake into Organisms and Bioaccumulation   280
7.4   Cellular Receptors   281
7.5   Metabolism   282
7.5.1   Phase I Reactions   283
7.5.2   Phase II Reactions   287
7.5.3   Phase III Reactions   288
7.5.4   Induction of Metabolism   288
7.6   Compounds of Particular Concern   289
7.6.1   Hydrocarbons: Sources, Applications
        and Concerns   290
7.6.1.1 Polycyclic Aromatic Compounds   290
7.6.1.2 Petroleum Hydrocarbons   292
7.7   Legacy Contaminants   294
7.7.1   Organochlorine Insecticides   294
7.7.2   Polychlorinated Dibenzodioxins
        and Dibenzofurans   296
7.7.3   Polychlorinated Biphenyls   296
7.8   Current Use Pesticides   297
7.8.1   Organophosphate Insecticides   298
Case Study 7.1 Toxicity of Insecticide,
        Monocrotophos, to Swainson's Hawks   299
7.8.2   Carbamate Insecticides   299
7.8.3   Phenylpyrazole Insecticides   301
7.8.4   Pyrethroid Insecticides   301
7.8.5   Neonicotinoid Insecticides   302
7.8.6   Chlorophenoxy Herbicides   302
7.8.7   Bipyridilium Herbicides   303
7.8.8   Glyphosate Herbicide   303
7.8.9   Triazine Herbicides   304
7.8.10 Fungicides   304
7.9   Flame Retardants   305
7.10 Perfluoroalkyl Compounds   307
Capsule 7.1 Mobility, Bioavailability and
        Remediation of PFAS Compounds in Soils   309
(Michael J. McLaughlin)
7.11 Plasticizers   313
7.12 Pharmaceutically Active Compounds   315
Case Study 7.2 Decline of Populations of
        Gyps Vultures in South Asia   317
7.13 Toxicovigilance   318
Summary   319
Review Questions and Exercises   319
Abbreviations   320
References   321

7

Organic Compounds   275

*Christopher D. Metcalfe, David A. Wright and
Peter V. Hodson*

Learning Objectives   275
7.1   Classes of Organic Compounds   275
7.2   Fate in the Environment   277

8

Endocrine Disrupting Chemicals   327

*Christopher D. Metcalfe, Christopher J. Martyniuk,
Valérie S. Langlois and David A. Wright*

Learning Objectives   327
8.1   Endocrine Disruption   327
8.2   The Endocrine System and Its Disruption   328
8.2.1   Neuroendocrine Control   330

8.2.1.1 The Hypothalamic–Pituitary Axis 330
8.2.1.2 Neuroendocrine Disruption 331
8.2.2 Gonadotropins 332
8.2.3 Steroid Hormones 333
8.2.4 Thyroid Hormones 333
8.3 Hormone Receptors 336
8.4 Modes of Action of EDCs 337
8.4.1 Agonists and Antagonists 337
8.4.2 Altered Biosynthesis of Hormones 337
8.4.3 Binding to Hormone Transport Proteins 338
8.4.4 Altered Hormone Receptor Levels and Gene Expression 338
8.5 Examples of EDCs 338
8.5.1 Xenobiotics in Wastewater as Sex Steroid Mimics 341
Case Study 8.1 Gonadal Intersex in Fish 341
8.5.2 Phthalates as EDCs 342
8.5.3 Atrazine as an EDC 343
8.5.4 Flame Retardants as EDCs 343
8.5.5 Legacy Contaminants as EDCs 344
8.5.6 Organotins as EDCs 345
8.6 EDCs as a Human Health Concern 346
8.7 Conclusions 346
Summary 347
Review Questions and Exercises 347
Abbreviations 348
References 349

9

Natural Toxins 355
David A. Wright and Pamela M. Welbourn
Learning Objectives 355
9.1 What Is a Toxin? 355
9.2 Evolutionary Perspective and Role of Natural Toxins 356
9.3 Toxins and Their Mode of Action 356
9.3.1 Toxins Produced by Harmful Algal Blooms 357
9.3.1.1 Domoic Acid 357
9.3.1.2 Saxitoxin 358
9.3.1.3 Brevotoxin 359
9.3.1.4 Okadaic Acid 360
9.3.1.5 Karlotoxin 360
9.3.1.6 Tetrodotoxin 361
9.3.1.7 Microcystins 361
9.3.1.8 Anatoxins 361
9.3.2 Toxins Produced by Vascular Plants 361
9.3.2.1 Naphthoquinones 362
9.3.2.2 Lectins 363
9.3.3 Toxins Produced by Microorganisms: Fungi and Bacteria 363
9.3.3.1 Anthrax Toxin 363
9.3.3.2 Microbial Methylation of Mercury 364
9.3.3.3 Fungal Toxins 364
9.3.4 Toxins Produced by Animals 365
9.3.4.1 Venoms 365

9.4 Defining the Ecological Advantage of Toxin Production 366
9.5 Applications of Natural Toxins 368
9.5.1 Pest-control Products 368
9.5.1.1 Bt insecticide 369
9.5.1.2 Quinones 370
9.5.2 Biological Warfare and Bioterrorism 370
9.6 Conclusions 370
Summary 371
Review Questions and Exercises 372
Abbreviations 372
References 373
Appendix 9.1: Summary of Some Toxins, Their Sources and Effects 375

10

Ionizing Radiation 379
Louise Winn
Learning Objectives 379
10.1 Non-ionizing Versus Ionizing Radiation 379
10.2 Definitions 380
10.2.1 What Is Ionizing Radiation? 380
10.2.2 Units of Measurement 382
10.3 Sources of Ionizing Radiation 383
10.3.1 Background Ionizing Radiation 383
10.3.2 Manufactured Ionizing Radiation for Medical Use 385
10.3.3 Nuclear Weapons 385
10.3.4 Nuclear Power 386
10.3.4.1 Mining and Extraction 386
10.3.4.2 Enrichment, Conversion and Fuel Fabrication 386
10.3.4.3 In-core Fuel Management 386
10.3.4.4 Fuel Reprocessing 387
10.3.5 Nuclear Waste Management 387
10.3.5.1 Short-lived Intermediate and Low-level Waste 388
10.3.5.2 Long-lived Intermediate and High-level Waste 388
10.4 Case Studies 389
Case Study 10.1 The Chernobyl Accident 389
Case Study 10.2 Fukushima Daiichi Nuclear Power Plant 390
10.5 Effects of Ionizing Radiation at the Molecular and Cellular Levels 391
10.5.1 Cell Death 393
10.5.2 DNA Damage 393
10.5.3 Protein Damage 394
10.5.4 Lipid Damage 394
10.5.5 Epigenetic Effects 394
10.5.6 Effects on the Immune System 395
10.6 Risk Assessment of Ionizing Radiation 395
10.7 Ecological Effects of Radiation 398

Capsule 10.1 Radiological Protection of
the Environment   399
(Nicholas A. Beresford and David Copplestone)
10.8 Conclusions   403
Summary   404
Review Questions and Exercises   404
Abbreviations   404
References   405

PART III COMPLEX ISSUES   409

11

**Complex Issues, Multiple Stressors and
Lessons Learned**   411
*Pamela M. Welbourn, Peter G. C. Campbell,
Peter V. Hodson and Christopher D. Metcalfe*
Learning Objectives   411
11.1 Acidification of Freshwater, Terrestrial and
Marine Systems   411
  11.1.1 Freshwater Acidification   412
    11.1.1.1 Chemical Effects   412
    11.1.1.2 Physical Changes   413
    11.1.1.3 Biological Effects and Risks for Sensitive
             Aquatic Systems   413
  11.1.2 The Effects of Acidification on
         Terrestrial Systems   414
  11.1.3 Regulation of Acidic Emissions and
         Recovery of Aquatic and Terrestrial Systems 415
    11.1.3.1 Abatement   416
    11.1.3.2 Treatment   417
  11.1.4 Acidification of Marine Systems: 'The Other
         CO_2 Problem'   417
  11.1.5 Lessons Learned   418
11.2 Metal Mining and Smelting   419
  11.2.1 The Issue   419
Capsule 11.1 Mercury and Silver: A History of
Unexpected Environmental Consequences   420
(Saul Guerrero)
  11.2.2 Processes Involved in the Extraction and
         Purification of Metals   424
  11.2.3 Substances of Concern   426
  11.2.4 Ecotoxicological Impacts of Metal Mining
         and Smelting   427
    11.2.4.1 Rivers   428
    11.2.4.2 Lakes   430
    11.2.4.3 Coastal Marine Environments   430
  11.2.5 Lessons Learned   431
11.3 Engineered Nanomaterials   431
  11.3.1 Routes of Exposure and Environmental Fate 433
  11.3.2 How Do Engineered Nanomaterials Enter
         Living Organisms?   434
  11.3.3 In Search of Nanotoxicity   435
  11.3.4 Lessons Learned   437

Case Study 11.1 Whole-lake Addition
of Nanosilver   438
11.4 Pulp and Paper Production   439
  11.4.1 Evolution of Pulp and Paper
         Environmental Issues   440
    11.4.1.1 Making Paper from Wood   440
    11.4.1.2 Power Dams: Pulp Mills Need Water   442
    11.4.1.3 Oxygen Consuming and Toxic Wastes
             from Wood Pulping   442
    11.4.1.4 Toxic Chemicals from
             Pulp Bleaching   444
  11.4.2 Lessons Learned   445
Summary   446
Review Questions and Exercises   447
Abbreviations   448
References   448

PART IV MANAGEMENT   455

12

**Regulatory Toxicology and Ecological
Risk Assessment**   457
*Peter V. Hodson, Pamela M. Welbourn and
Peter G. C. Campbell*
Learning Objectives   457
12.1 The Need for Chemical Management
and Regulation   457
12.2 Legislation for Chemical Management   458
  12.2.1 The Process of Regulation   459
    12.2.1.1 Policy   459
    12.2.1.2 Legislation   459
    12.2.1.3 Regulations   459
    12.2.1.4 Departmental Responsibilities and
             Options for Chemical Management   460
  12.2.2 International Law and
         Multilateral Agreements   460
  12.2.3 Regulatory Challenges and Disparities   461
    12.2.3.1 Factors That Affect the Development
             and Implementation of
             Chemical Regulations   461
12.3 Applying Ecotoxicology to Support
Chemical Management   462
  12.3.1 Numerical Limits: Criteria, Objectives,
         Standards, Guidelines (contributed by
         Douglas J. Spry)   463
    12.3.1.1 How Numerical Limits
             Are Developed   463
    12.3.1.2 Numerical Limits for Soils, Sediments
             and Biological Tissue   464
    12.3.1.3 Future of Numerical Limits   465
  12.3.2 Ecological Risk Assessment (ERA)   465
    12.3.2.1 The Methodology of Ecological
             Risk Assessment   465

12.3.2.2 Problem Formulation 466
12.3.2.3 Analysis 467
12.3.2.4 Risk Characterization 467
12.3.2.5 Applications of ERA: Specific Chemicals 467
12.3.2.6 Handling Uncertainty: An Integral Part of ERA 468
12.3.3 Regulations for Individual Chemicals and Complex Mixtures in Environmental Media 469
12.3.4 Enforcement of Environmental Regulations 470
Capsule 12.1 The Sudbury Soils Study: An Area-wide Ecological Risk Assessment 471
(Christopher D. Wren, Glen Watson and Marc Butler)
12.3.5 Environmental Surveillance and Monitoring 476
Case Study 12.1 Monitoring Rivers to Assess the Adequacy of Pesticide Regulations 477
12.4 The Future of Environmental Regulation 477
Capsule 12.2 Legislation for Chemical Management – Traditional Environmental Knowledge in the Regulation of Chemical Contaminants? 478
(F. Henry Lickers)
Summary 482
Review Questions and Exercises 482
Abbreviations 483
References 484

13

Recovery of Contaminated Sites 487
Pamela M. Welbourn and Peter V. Hodson
Learning Objectives 487
13.1 Background 487
13.2 Component Disciplines and Goals 488
13.3 Definitions and Concepts 490
13.4 Triggers for Action Towards Recovery 490
13.5 Methods and Approaches for Recovery 491
13.6 Engineering 492
13.6.1 Removal and Off-site Disposal of Contaminated Material 492
13.6.2 On-site Remediation 493
Case Study 13.1 Entombment 493
13.7 Monitored Natural Recovery (MNR) 494
13.7.1 Passive Recovery for Surface Water 494
13.7.2 Passive Recovery and Natural Attenuation for Sediments and Soils 495
Capsule 13.1 The Enduring Legacy of Point-source Mercury Pollution 496
(John W. M. Rudd and Carol A. Kelly)
Case Study 13.2 Recovery of Saglek Bay, Labrador 500

13.8 Bioremediation 503
Capsule 13.2 Bioremediation of Oil Spills 504
(Charles W. Greer)
13.9 Recolonization and Phytoremediation 507
13.9.1 Recolonization by Plants 507
13.9.2 Recolonization by Fish and Other Animals 508
13.9.3 Phytoremediation 509
13.10 Conclusions 510
Summary 511
Review Questions and Exercises 511
Abbreviations 512
References 512

14

Emerging Concerns and Future Visions 515
David A. Wright and Peter G. C. Campbell
Learning Objectives 515
14.1 Climate Change and Its Role in Ecotoxicology 515
14.1.1 Interactions Between Climate Change and Ecotoxicology 517
14.1.1.1 Ecotoxicological Effects of Climate Change on Individual Species 518
14.1.1.2 Interspecific Effects of Climate Change on Ecotoxicology 520
14.1.2 Regional Considerations 521
14.1.3 Future Considerations 522
14.2 Microplastics and Nanoplastics 522
14.2.1 Toxicology of Microplastics 524
14.2.1.1 Adverse Physical Effects Through Tissue Damage and Inhibition of Movement 524
14.2.1.2 Cellular Invasion by Small Particles (Nanospecific Effect) 525
14.2.1.3 Toxicity of Chemical Constituents of Microplastics 525
14.2.1.4 Toxicity of Adsorbed Chemicals 525
14.2.2 Future Considerations 526
14.2.2.1 Establishing Cause and Effect 526
14.2.2.2 Mitigation 526
14.3 Emerging Inorganic Contaminants 528
14.3.1 Trends in Mining Activities 528
14.3.2 Trends in Metal Use 529
Capsule 14.1 Lithium – A Critical Mineral Element: Sources, Extraction and Ecotoxicology 531
(Heather Jamieson)
14.3.3 Future Considerations 534
Case Study 14.1 Deep-sea Mining 534
14.4 Emerging Concerns about Organic Contaminants 539
14.4.1 Monitoring 539
14.4.2 Non-targeted Screening 540
14.4.3 Toxicity Evaluation 541
14.4.4 Predictive Toxicology 542

14.4.5 Applications of Predictive Toxicology in Ecological Risk Assessment 543

14.4.6 Future Considerations 544

Summary 544

Review Questions and Exercises 545

Abbreviations 546

References 547

Epilogue: A Final Perspective 551

Updating Ecotoxicology 551

Ecological Risk Assessments; Environmental Decision-making and Indigenous Rights 551

Reliance on Environmental Modelling in Evaluating New Chemicals 552

Interactions Between Ecotoxicology and Human-induced Environmental Changes 552

Looking to the Future 553

Index 555

PREAMBLE

Almost 20 years ago, we published *Environmental Toxicology*. That book drew on the previous quarter-century, which had seen the gestation and adolescence of the environmental movement, and emphasized the emerging science of environmental toxicology. Students of today will have a different perspective from ours at the end of the twentieth century. Nevertheless, the current ecotoxicology text has been conceived as a follow-up. As a product of the subsequent 20 years of rapidly advancing research, technological progress, scholarly and popular writing, and media communications, it is in no sense a second edition. It is a new book, but it is tempting to review and reflect on the way that we saw the whole topic of toxic substances in the environment at the turn of the century. In our concluding chapter in 2002, we wrote:

At the beginning of the twenty-first century, there is a clear need to reformulate our thinking about the relationship between society and nature. All the preceding viewpoints were initially conceived without any clear evidence of widespread environmental pollution. Well-publicised instances of major contamination events and recent emphases on adverse effects at the population, community, and transboundary ecosystem levels now indicate the pervasiveness of problems. While science and technology have more sophisticated tools than ever before to identify and remediate environmental contamination, global problems will require global solutions and the political will to drive them. Such political decisions can only be made in the light of informed public opinion. This will increasingly involve scientists in the role of communicators and the engagement of politicians, economists, social scientists and philosophers in providing a balanced, realistic framework for ecosystem management.

How much has changed?

Pamela M. Welbourn and David A. Wright

PREFACE

Ecotoxicology is the study of the effects of toxic chemicals on biological organisms, especially at the population, community and ecosystem levels. It falls under the umbrella of environmental science and represents one of the aspects of environmental studies, along with the conservation of species, habitats and ecosystems, the protection of endangered species, and various levels of management for water, soil, wildlife and fisheries.

As is the case for other branches of environmental science, in ecotoxicology, there are reciprocal relationships among scientific investigations and social problems. Although this textbook focuses on the scientific and technological features of ecotoxicology (i.e., the things that we can measure), we think it is important to offer some historical background concerning the social context in which the science has developed. We describe the increasing public awareness and concern about toxic chemicals, and discuss the role that environmental non-governmental actions have played in the development of ecotoxicology. For example, the role of endocrine disrupting chemicals was brought to public attention by the publication of *Our Stolen Future* in 1996 (co-written by Theo Colborn). Research on endocrine disrupting chemicals now constitutes a very significant domain in ecotoxicology.

In choosing the title 'Ecotoxicology', we aim to emphasize the ecosystem and its components at all levels of organization, and – to some extent – downplay the effects of toxicants on humans (often referred to as environmental toxicology). Nevertheless, we acknowledge that on many occasions, studies in environmental toxicology have led to the discovery of unexpected effects on ecosystems. For example, the food chain transfer and magnification of methylmercury, first recognized as causing disease in human populations in the late 1950s, were shown to have similar effects on top-level consumers in the animal kingdom.

Our textbook aims to provide a clear understanding of the broad scope of the discipline of ecotoxicology,

informed by the latest scientific analysis and thinking. Some of the key features are:

- A unique blend of the chemistry, the biology and the regulatory aspects of ecotoxicology.
- The inclusion of chapters on endocrine disruption and ecotoxicogenomics, and considerations of how the recent findings in the field of genomics are beginning to provide tools that may assist our understanding of how chemicals can impact on ecosystem health.
- Consideration of novel contaminants such as engineered nanomaterials, polyfluoroalkyl substances (PFAS) and technology-critical elements (TCEs).
- The coverage of a range of countries in the discussion of regulatory toxicology, including the European Union, the USA, Australia, Canada and New Zealand, as well as international agreements.
- Up-to-date case studies and capsules throughout the text, some written by guest authors, to engage students and provoke interest in topics that touch their daily lives.
- Review questions at the end of each chapter to test the students' knowledge.
- Online resources for instructors (at www.cambridge.org/ecotoxicology), which include solutions to student questions and problems. Online resources for students include the glossary of all the key terms highlighted in brown within the book and additional reading lists.
- The presentation of future visions, emerging concerns, novel contaminants and new technical approaches to understand and mitigate pollution and ecological health risks globally, which can be debated in targeted student discussion sessions or seminars.

This book has been designed for advanced undergraduate and graduate students taking courses on ecotoxicology, environmental toxicology and environmental pollution. It assumes knowledge of some fundamental and widely accepted concepts and biological processes. Students should have some background in basic natural sciences, chemistry, biochemistry and biology. Professional consultants and practitioners may also find this a useful guide in specific areas.

The book was planned with a one-semester senior undergraduate course in mind, but it is also appropriate for graduate students who need to expand their background in ecotoxicology. In designing a particular course, instructors will be able to choose the chapters or chapter sections that fit with the course objectives, knowing that there are frequent cross-references among the chapters that will help the students to make the necessary connections. With few exceptions, the references cited date from 2000 or later. A few earlier references have been included, because we judged these to be seminal.

Book Organization

Throughout, this textbook provides more than a catalogue of toxic chemicals and their effects. It links ecotoxicology to the basic sciences of biology and chemistry that explain why some chemicals are more bioavailable and toxic than others and how chemicals interact with life at the molecular and cellular levels. It links these fundamental interactions to subsequent effects at higher levels of organization, from whole-organism performance to ecological change, with implications for the provision of ecological services such as natural resources.

Within Part I, 'Approaches and Methods', Chapter 1 discusses the history and evolution of this originally hybrid science and provides an overview of the structure of the whole book. Chapters 2, 3, 4 and 5 deal with the 'tools of the trade', some well-established (Chapter 2), some rapidly emerging (Chapter 5). In a number of places in the text, including the Epilogue, we refer to the challenges and pitfalls of relating experimental (lab-based) studies to the real world. We include methods and approaches for determining how potentially toxic

substances can affect living organism at all levels of organization, from the gene to the whole ecosystem, and how these effects can be quantified. In the course of so doing, recent research and technical advances are included, along with selected examples to illustrate the major issues and current approaches to the subject. Recognizing the contribution of technology to the progress of the science of ecotoxicology, significant technological advances are highlighted and evaluated. As such, Chapter 5 is a state-of-the-art overview of the developing science of 'Omics', which underpins the development of Adverse Outcome Pathways, linking effects at the molecular level to successively higher levels of organization, including ecosystems.

Part II, 'Toxicology of Individual Substances', addresses categories of chemicals in classes, with their sources, chemistry and modes of action. This includes Metals and Metalloids (Chapter 6), Organic Compounds (Chapter 7), Endocrine Disrupting Chemicals (Chapter 8), Natural Toxins (Chapter 9) and Ionizing Radiation (Chapter 10). The inclusion of natural toxins is a departure from most ecotoxicology textbooks, but natural plant, animal and microbial metabolites constitute some of the most toxic substances known. Further, there is increasing interest in the use of natural products in pest control and medicine.

Part III, 'Complex Issues', deals with real-world complex issues, covered in Chapter 11. These issues were chosen to illustrate the effects of multiple stressors, the potential implications of 'nanotoxicity', interactions among toxicants and cradle-to-grave effects of industrial processes such as metal extraction/refining and pulp and paper processing. In Part IV, 'Management', Chapter 12 deals with environmental regulation and risk assessment, and Chapter 13 with recovery of ecosystems damaged by chemical contamination. Finally, Chapter 14 looks at emerging issues and anticipated future developments in ecotoxicology. Perhaps you, the readers of this volume, will contribute to an updated edition in another 20 years!

ACKNOWLEDGEMENTS

Many people have contributed to the publication of this textbook and we would like to express our gratitude here for their help. Some of these contributors will be evident to our readers, notably the authors of our capsules and several case studies (see page iv). For more subtle help, for example in chasing down and suggesting relevant references, we thank Charles Driscoll (Syracuse University), Cynthia Gilmour (Smithsonian Environmental Research Center), Bill Keller (Ontario Ministry of the Environment), Gerrit Schüürmann (Helmholtz Centre for Environmental Research) and Shaun Watmough (Trent University). In several cases, data were kindly made available by external colleagues (Matthew Graham, Environment and Climate Change Canada; Thomas Graedel, Yale University).

Various individual chapters benefited from critical review by obliging colleagues (Graeme Batley, CSIRO Land and Water; Claude Fortin and Emilien Pelletier, Université du Québec; Don Mackay, Trent University; Kevin Wilkinson, Université de Montréal;). Philip Rainbow, also a CUP author, provided helpful advice about book publishing. Maame Adai, a Queen's University graduate student, carried out a literature search for the Chapter 6 section on Mercury. Similarly, Jenny Moe, also a Queen's University graduate student, searched the literature for examples of significant case studies to be included in Chapter 1.

We also received essential technical support from Queen's University (Morag Coyne, Library) and the Université du Québec, INRS (Jean-Daniel Bourgault, Service de documentation). Colin Kahn (Queen's University, School of Environment Studies) provided invaluable help in negotiating the ever-changing Internet security requirements. Robert Loney (Trent University) made innumerable contributions to the graphics in our book, not only in converting our rough sketches into publishable figures, but also in patiently providing critical advice about image resolution and fonts. Caroline Doose (an INRS graduate student) provided the careful art work for the biotic ligand model figure in Chapter 6. Anik Giguère (a former INRS graduate student) and Nastassia Urien (a former INRS postdoctoral fellow) also contributed to the figures in Chapter 6. We also thank Myriam Castonguay (coordinator of the Intersectoral Centre for Endocrine Disruptor Analysis, ICEDA) for her graphical skills in assembling figures for Chapter 5. Many of the chemical structures illustrated in the book were obtained from *ChemSpider*, a service provided by the Royal Society of Chemistry. The original structures were then manipulated with *ACD/ChemSketch*, version 2019.1.3, Advanced Chemistry Development, Inc., Toronto, ON, Canada, www.acdlabs.com.

As this book progressed from conception to completion, we have profited from the sage contributions of CUP staff, notably Emma Kiddle, Ilaria Tassistro and Rachel Norridge. In its final stages, the book also benefited from the meticulous analysis of our draft chapters by Lindsay Nightingale in her role as the copy editor.

Finally, we would be seriously remiss if we did not acknowledge the essential support of our marital partners, who had to deal not only with the COVID-19 pandemic but also their preoccupied spouses!

The part and chapter opening images were supplied courtesy of: Sean Justice / The Image Bank / Getty Images (page xx); Science History Institute (page 2); xPACIFICA / Stone / Getty Images (page 22); PASIEKA / Science Photo Library / Getty Images (page 60); Stuart Westmorland / Corbis Documentary / Getty Images (page 98); Andriy Onufriyenko / Moment / Getty Images (page 138); Michel Joffres / 500px / Getty Images (page 168); Robbie Goodall / Moment / Getty Images (page 170); Keenpress / Photodisc / Getty Images (page 294); MedicalRF.com / Getty Images (page 326); Péter Gulyás / EyeEm / Getty Images (page 354); yangna / E+ / Getty Images (page 378); shaunl / E+ / Getty Images (page 408); aydinmutlu / E+ / Getty Images (page 410); the_burtons / Moment / Getty Images (page 454); Robert Brook / Science Photo Library / Getty Images (page 456); Romolo Tavani / iStock / Getty Images Plus (page 486); Paul Souders / Stone / Getty Images (page 518).

I

Approaches and Methods

D. Teniers pinx.ᵗ

THE CHYMIST

T. Major sculp.ᵗ

To Richard Mead. M.D. Physician in Ordinary to his Majesty. F. R. S.
This Print Ingrav'd from an Original Painting of the same Size by David Teniers
is humbly Dedicated by his most Obedient Servant

Thoˢ. Major.

Published May 7ᵗʰ 1760.
accordᵍ to Act of Parliamᵗ   London sold by the Author at the Golden Head   in West Street the upper end of St Martins Lane.

CHAPTER

1 The History and Emergence of Ecotoxicology as a Science

Pamela M. Welbourn and Peter V. Hodson

Learning Objectives

Upon completion of this chapter, the student should be able to:

- Discuss what is meant by 'ecotoxicology' and to place it in context with 'environmental' and 'classical' toxicology.
- Create an historical context for the current issues and practice of ecotoxicology and for the subject matter of this textbook.
- Identify the major links between environmental issues caused by chemical contamination, the public's growing concerns and demands for action, the legislative initiatives to resolve these issues and the evolution of the science of ecotoxicology to support chemical management.
- Explain why many issues of chemical contamination and the science of ecotoxicology are global in nature and require international research and agreements to resolve.

This chapter traces the origin of **Ecotoxicology**, its first definition in 1977 and its connections to related branches of science. Historical landmarks in the development of ecotoxicology are reviewed, including the role of environmental non-government organizations (ENGOs) in increasing the public's awareness of environmental issues and creating pressure for political action. We then address the emergence of the science of ecotoxicology, and finally we look at the way that formal regulation of toxic substances has evolved. As an indication of how some of these ideas are developed in this book, there are numerous cross-references to subsequent chapters.

1.1 The Science of Ecotoxicology

Ecotoxicology is a relatively young science, often seen as a hybrid of ecology and toxicology. The development of many branches of science enhances linkages among intellectual curiosity, technological advances and an awareness of human-related problems, including health, social, political and economic concerns. Ecology and toxicology stand at different points along a spectrum of scientific investigations, from those based solely on intellectual curiosity ('pure' research) to those addressing real-life problems ('applied' research). Ecology can be seen as a pure science, whereas toxicology, by its very nature, is an applied science. However, such distinctions tend to break down when one considers ecotoxicology. Several features of the subject are notable:

- Ecotoxicology seeks to understand the relationships among the structures of chemicals, their environmental behaviour, distribution and effects on species of interest, and the mechanisms linking toxicity to changes in ecosystem structure and function. Although the impetus for an experiment might be a concern for the adverse effects of a specific chemical, i.e., applied science, there is also an intellectual curiosity to know more about causes and consequences, i.e., basic science.

- In the course of ecotoxicological investigations, there may be unanticipated benefits for ecological research, when a new understanding about the basic structure and functioning of an ecosystem is revealed. In other words, new insights are gained about how natural ecosystems operate by studying those perturbed by chemical contamination (including what have been called 'spills of opportunity' (Section 12.3.4; Capsule 13.2)). Research on the mode of action of natural toxins has also provided a deeper understanding of many aspects of basic molecular and cellular biology. For example, the use of the natural tetrodotoxin and other neurotoxins has led to a greater understanding of the structures of membrane sodium channels and various 'gating' mechanisms.

- A critically important aspect of ecotoxicology is the motivation of scientists and technologists to recognize, understand, prevent and mitigate environmental issues caused by chemical contamination. Thus, ecotoxicology is largely an applied science when inspired by real issues of environmental degradation. It contributes to the development of environmental laws and regulatory actions, and spurs the development of technology for waste treatment, environmental monitoring and site remediation. Interactions among toxicologists, ecologists, chemists and environmental engineers make ecotoxicology truly interdisciplinary.

Although the systematic study of the effects of toxic substances on ecosystems is essentially a twentieth-century phenomenon, ecotoxicology has its roots in the study of human or classical toxicology, which has a much longer and somewhat more sinister history. The older science of toxicology dates back to the earliest humans, who used animal venoms and plant extracts for hunting, warfare and assassination. The knowledge of these **poisons** must have pre-dated recorded history (Borgia, 2018), yet elucidating the mechanisms of the toxicity of venoms continues in toxicology today. A review of Shakespeare's plays as well as other historical literature reveals that deliberate poisoning has long been regarded as a means of solving personal and political problems. The practice continues today, exemplified by the poisoning in August 2020 of Alexei Navalny, a Russian opposition figure and anticorruption activist. The nerve agent Novichok was confirmed in his blood samples by

Figure 1.1 Paracelsus, widely regarded as the founding father of modern toxicology. Source: Nastasic.

five laboratories in the Organisation for the Prohibition of Chemical Weapons (OPCW).

The first students of toxicology were physicians and alchemists who often treated cases of poisoning. The science of toxicology, from which ecotoxicology evolved, began with the insights of the Swiss physician Theophrastus Bombastus von Hohenheim (1493–1541) who, in the fashion of the day, Latinized his name to *Philippus Aureolus Theophrastus Paracelsus* (Figure 1.1).

Paracelsus' interest in poisonous substances arose from his recognition that low **doses** of metal salts (Chapter 6) and herbal remedies (to treat human diseases) were therapeutic, whereas higher doses were toxic. When translated, his writings in Latin have many variants in wording, including his most famous declaration: "*What is there that is not poison? All things are poison and nothing is without poison. Solely the dose determines that a thing is not a poison.*" His declaration is somewhat oversimplified and dogmatic, particularly the assertion that '*all things*' (compounds) are toxic. Modern toxicologists have determined that many compounds (e.g., waxes) are too large to be taken up across membranes and to interact with and affect the molecules and cellular processes essential to life. There are also chemicals that cause endocrine disruption at low but not at high doses. Nevertheless, for most compounds that can be accumulated in living tissues, his insights linking chemical exposures to the extent of adverse effects remain fundamental to toxicology and ecotoxicology. The modern expression of his thesis "It is the dose that makes the poison" is a theme revisited many times throughout this textbook. Paracelsus was also responsible for a treatise, *On the Miner's Sickness and Other Diseases of Miners*, published posthumously in 1567 and one of the first works on occupational toxicology. This theme was expanded to a variety of occupational hazards in the classic work of Bernadino Ramazzini (1633–1714), *Discourse on the Diseases of Workers* (1700).

The modern science of toxicology can be traced back to the Spanish physician Mattieu Orfila (1787–1853), who in the nineteenth century published a comprehensive treatise on the toxicity of natural agents. He articulated many of the basic components of the discipline that we know today. These include the relationship between signs (pathology) of toxicity and chemical content of tissues as determined by analysis, mechanisms of eliminating poisons from the body, and treatment with antidotes (Orfila, 1818). The work of physiologist Claude Bernard (1813–1878) introduced a more mechanistic approach to toxicology through controlled experiments on animals in the laboratory.

Beginning in the 1960s, the study of adverse effects of environmental pollutants was primarily concerned with specific events such as discharges and spills of single compounds or mixed effluents, and occupational exposures during the manufacture and use of chemical products. As such, it was regarded as an extension of the earlier 'classical' science of toxicology and came to be called 'environmental toxicology'. Although the initial focus was on chemical toxicity to humans and livestock, public awareness of adverse chemical effects on wildlife escalated as a result of publications such as *Silent Spring* in 1962 (Section 1.2.1). It became increasingly apparent that a more holistic approach was needed to understand how contaminants can harm natural ecosystems.

'Ecotoxicology' was first proposed as a new branch of toxicology by René Truhaut in June 1969 at a meeting of an *ad hoc* Committee of the International Council of Scientific Unions (ICSU) in Stockholm and defined as:

… the branch of Toxicology concerned with the study of toxic effects, caused by natural or synthetic pollutants, to the constituents of ecosystems, animal (including human), vegetable and microbial, in an integral context (Truhaut, 1977).

As defined by Truhaut, ecotoxicology differs from classical or human toxicology in the nature of the target organisms, systems or processes, and the resulting complexity. However, the fundamental scientific approach, including the principles and practice of the scientific method, is entirely similar for both fields.

Along with an increased understanding of how contaminants can affect natural ecosystems came the need for a more proactive approach to regulating environmental pollutants. In the context of the control of toxic substances, the technical terms '**objectives**', '**guidelines**', '**criteria**' and '**standards**' became part of the lexicon of regulations governing the extraction, manufacture, use, disposal and release of chemicals into the environment (Section 12.3.1). All of these terms are subject to non-scientific influences such as risk benefit analysis, public perception and political judgement. Yet all are ultimately based on the fundamental principles of environmental toxicology:

- Measurement of chemical concentrations in different environmental media (hazard identification);
- Assessment of exposure of biota to toxicants;
- Quantitative dose–response relationships associated with chemical exposure.

Figure 1.2 depicts a scheme that encompasses most of the elements comprising the practice of ecotoxicology, including the dissemination of scientific information to the public and to those responsible for formulating and enacting regulatory legislation. The scheme incorporates **ecological risk assessment** (ERA, Section 12.3.2), which considers not only the potential toxicity (hazard) of a substance, but also the potential for exposure of target organisms. This is expressed as 'no exposure, no risk'. This expression is sometimes at odds with what the public perceives as risk, which may be subjective. A well-known example is that many people are more apprehensive about commercial air travel than driving in automobiles. However, the annual fatality rates in the United States for automobile accidents are 50,000, 250-fold higher than the 200 fatalities from commercial airline accidents. Risk perception often has a bearing on how priorities are ranked and regulations enacted.

Definitions of environmental toxicology and ecotoxicology respectively are by no means universally accepted or applied. Wright and Welbourn (2002) was titled *Environmental Toxicology*, but for the current textbook we opted for *Ecotoxicology*. Ecotoxicology signals our emphasis on ecosystems and their components, rather

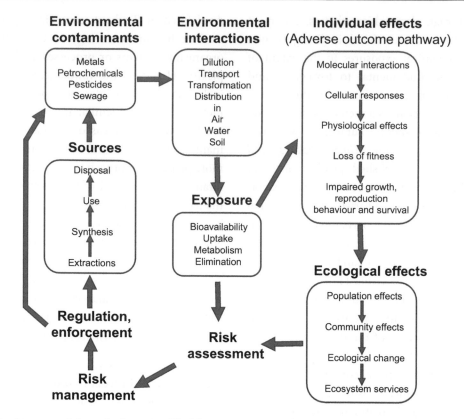

Figure 1.2 The major features of ecotoxicology. Modified from Figure 13.5 of Wright and Welbourn (2002).

than on relations between environmental contamination and the health of individual humans or other species. Chapman (2002) discusses the terms with a somewhat different emphasis and considers that environmental toxicology is a laboratory component of ecotoxicology that is needed for screening chemicals, regulatory testing, and research on new chemicals of concern. As a broader, or 'umbrella' discipline, ecotoxicology considers the full range of chemical effects demonstrated by field and laboratory studies at scales ranging from complete ecosystems to *in vitro* molecular responses. Perhaps more importantly, ecotoxicology investigates the ecological changes that result from chemical emissions and toxic effects on individual species. Ecotoxicology is required for predicting real-world effects and for site-specific risk and impact assessments. Chapman concludes that an "increased emphasis on ecotoxicology represents a shift from reductionist to holistic approaches". In this text, we apply 'environmental toxicology' to activities directed primarily at laboratory toxicity testing and research. Terminology is also blurred for the concepts of contaminant/contamination, pollutant/pollution, respectively, and again we refer to Chapman (2007) (Box 1.1).

Box 1.1 Contamination and pollution

Chapman (2007) defines a contaminant as a substance that is "present where it should not be or at concentrations above background", whereas a pollutant is a contaminant that elicits adverse biological effects. Thus, all pollutants are contaminants, but not all contaminants are pollutants. In this textbook, we adhere to Chapman's definition.

Most authorities would consider that ecotoxicology as a discipline emerged in the mid to late 1960s. As stated by van der Heijden *et al.* (in Finger (1992), pp. 1–40), "What we call environmental problems today were not defined as such before 1965." It is notable that, even though advances in environmental toxicology and ecotoxicology have resulted in part from need, the same time span has seen great advances in technology, providing tools to facilitate the rapid growth of the science. As environmental toxicology and ecotoxicology have evolved, fundamental scientific advances in related fields such as ecology and chemistry have been accelerated, partly in response to the need to better understand the fundamental processes upon which environmental toxicants are acting. In these and other respects, the past few decades have been exceptionally exciting for ecotoxicologists.

It would be misleading to give the impression that there is a clear straight line from scientific research on ecotoxicology to regulation and management. In fact, activists, the media, public opinion and commercial and political turf protection typically play roles and may be in conflict with scientists and regulators. The way in which all these players may be involved has been illustrated in Figure 1.3. In common with other branches of environmental science, public awareness and political pressure stimulate action in the form of regulation or management of ecotoxic substances. The influence of increased social awareness on the development of ecotoxicology is difficult to measure objectively, but the significance of social sciences and politics in ecotoxicology has been pointed out many times. For example, Schüürmann and Markert (1998) stated, "The need to achieve results of practical relevance for politics and society made the development of ecotoxicology rapid indeed." This is not a one-way street. As Harari (2018) points out, "scientists need to be far more engaged with current public debates". He considers that doing academic research and publishing in learned journals are certainly important, but communicating to the general public is equally important. An informed public is essential for bringing about appropriate actions to tackle or remediate environmental problems.

1.2 Historical Landmarks in the Development of Ecotoxicology

In the second half of the twentieth century, a series of dramatic and unanticipated events related to chemicals released into the environment alerted the scientific community to the complexities and unexpected ecological effects of toxic substances (Figure 1.4). In many instances there was public reaction, sometimes stimulated by the writings and actions of ENGOs.

1.2.1 *Silent Spring* and Pesticides

Rachel Carson's 1962 book *Silent Spring* (Carson, 1962) was one of the most influential of such books. Indeed, Carson is credited with spawning the modern environmental movement. She discussed, with dramatic examples and chapter titles such as "And no birds sing",

the risks that synthetic pesticides posed for human and environmental health. She pointed out that many of these harmful chemicals were in our homes and in our foods, and she believed that the public was not properly educated about them, nor protected. In terms of government protection, she emphasized that there was a need for stricter regulations and for more people working to identify the risks and to educate the public.

In Carson's time, most pesticides were neurotoxic organochlorines (Section 7.7.1), widely applied because of their efficacy, persistence in effects, low cost and low toxicity to humans. These insecticides replaced traditional compounds based on arsenic, copper and mercury that were highly toxic to the users. Although there had been great benefits to the control of insects by chlorinated pesticides, notably in agriculture and forestry and for malaria carriers and other vectors of human disease, there were also significant environmental risks. Organochlorines were not only toxic to target species but also ecotoxic, poisoning many non-target organisms. For example, dichlorodiphenyltrichlorethane (DDT) **bioaccumulated** and **biomagnified** in aquatic and terrestrial food webs, often resulting in the death or reproductive failure of consumers at higher trophic levels, most notably predatory birds. A well-known example was the near-extinction of Peregrine falcons (*Falco peregrinus*) in the United Kingdom, due in part to reproductive impairment by DDT in the 1950s and 1960s (Wilson *et al.*, 2018). Even though the dangers of pesticides were known to science in the 1950s, the information had not been translated into any kind of public awareness or regulatory policy prior to the publication of *Silent Spring*. Rosner and Markowitz (2013) described the close relationship between chemical industries and the governmental agencies that were meant to regulate pesticides. In the 1970s, DDT and related industrial compounds such as polychlorinated biphenyls (PCBs) were banned for many uses. It has been said that regulation of DDT was the first example of such policy for a toxic substance based on the risk to non-human targets, but it was certainly not the last. The value was clearly evident in the recovery of populations of predatory birds such as the Peregrine by the 1990s (Wilson *et al.*, 2018).

Another of the troubling properties of organochlorine pesticides was their persistence, with very slow breakdown in the environment. Even with the banning of their many uses, environmental contamination is still a legacy of their past use.

Life History of an Ecotoxicology Issue

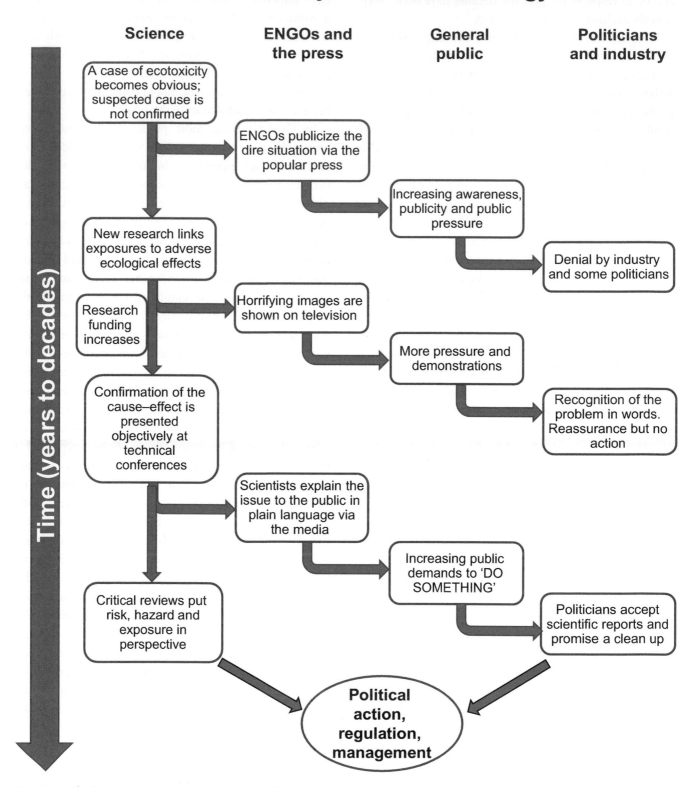

Figure 1.3 The interconnected roles of politics, public opinion and science in the regulatory process for toxic substances.

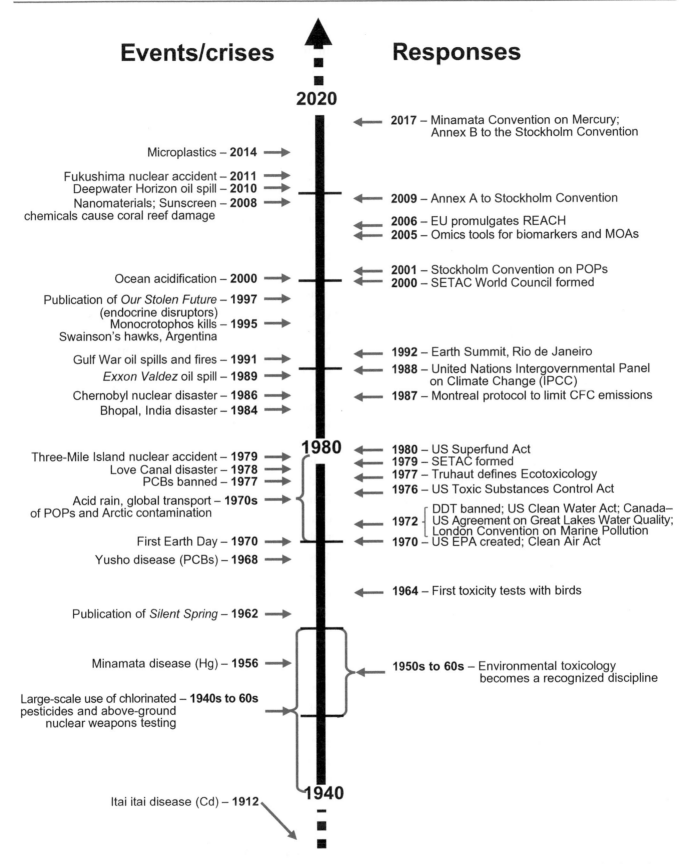

Figure 1.4 Timeline (1940–2020) of environmental issues and regulatory responses associated with chemical contamination. These events are described in detail in Chapters 1, 6, 9, 11 and 12.

1.2.2 Mercury

Carson did not write about mercury (Hg), but its eco-toxicity came to the fore during the same era. The demonstration of food chain transfer of Hg was revealed in human and domestic animal poisonings that occurred in Japan in the 1950s to 1960s, where methylmercury was discharged from a factory to the ocean. There was limited prior understanding of the food web transfer of a toxic substance, and the perpetrator attempted to deny the evidence connecting the mercury emissions to human health. Despite these obstacles, it was proven unequivocally that the ingestion of methylmercury-contaminated fish and shellfish was the cause of many illnesses and a number of deaths in the fishing village of Minamata. The village gave its name to the 'Minamata disease' syndrome, and other areas around the Shiranui Sea in Japan were similarly affected (Hachiya, 2006). At the time, Hg effects on human and domestic animal health were the major concerns, rather than effects on the wider ecosystem. Nevertheless, the case illustrated the importance of the food web transport and **biomagnification** of this metal. While occupational illnesses caused by Hg had long been known, this was the first clearly demonstrated and publicized example of non-occupational mercury toxicity. It stimulated public support for scientific research and regulation of the ecological and toxicological risks related to Hg contamination at numerous sites around the world.

Minamata was not the only example of mercury contamination resulting from industrial activity. Contamination of aquatic ecosystems by the electrochemical production of chlorine for bleached Kraft pulp mills caused Hg poisoning in downstream communities of indigenous people throughout North America (e.g., Section 11.4, Capsule 13.1) and Scandinavia. In the Minamata example, methylmercury was the contaminant, whereas in the chlor-alkali plants it was elemental Hg(0) that, as research showed, was methylated by bacteria (Section 6.8.1.3). It was soon recognized that there were deadly consequences in the form of food web biomagnification of methylmercury: poisoning of human and wildlife consumers of fish followed Hg contamination of aquatic systems where microbial methylation occurred. Research on all aspects of Hg, both ecotoxicological and for human health, has since expanded, culminating in the aptly named Minamata Convention to protect human health and the environment from the adverse effects of mercury emissions.

1.2.3 Acidification

In the 1970s, 'acid rain' became a household phrase. **Acidic deposition** had been described in the nineteenth century but was considered a local industrial issue, with the most recognizable effect being damage to the limestone architecture of historic buildings and public monuments. It was first recognized in Scandinavian countries that fish in soft water and poorly buffered lakes were dying because of acidification (Section 11.1). The impact of acidification on fish and press reports of 'dead lakes' created a sense of drama, bringing the issue from relative obscurity to one of the most prominent environmental problems of the latter half of the twentieth century (Wetstone, 1987). Dealing with the long-range transport of acid precursors such as oxides of nitrogen and sulphur (NO_X and SO_X) required not only a scientific understanding of the sources and effects (e.g., on fish, forests) but also consideration of the transport of acid precursors and their environmental chemistry. The so-called 'target loading' studies (Gorham, 1985; Foster *et al.*, 2001) required modelling to relate the amount of SO_X emitted to the change in pH and adverse effects on **receptors**, sometimes thousands of kilometres from the source.

In the course of tracing the long-range transport of acid precursors, it became evident that effects were not limited to aquatic ecosystems, but that forest dieback could be linked to acid deposition, and that acidic water mobilized toxic metals. One of the most significant aspects of the acid rain crisis was the demonstration that local chemical sources and effects were contributors to regional and global contamination and impacts. It led directly to the recognition that other airborne contaminants, including mercury and volatile organic compounds such as PCBs, were also subject to long-range transport. This dispersion in the atmosphere resulted in significant contamination of ecosystems remote from industrial activity. The more global perspective fostered international cooperation among scientists and the development of international agreements to limit atmospheric emissions from all sources, not only in urban areas.

There has been some indication of ecosystem recovery from acidification following the control of acid rain (Section 11.1), albeit not always to their original states. Acid rain research has provided an enormous amount of ecological and toxicological understanding about the receiving ecosystems: lakes, rivers and forests. Most recently, the global impacts of **ocean acidification** have been recognized, particularly its effects on the growth and survival of many marine invertebrate species, evident as coral reef dieback. Ocean acidification is linked to global warming because a primary causal agent is carbon dioxide, and the formation of carbonic acid in surface waters (Section 14.1).

1.2.4 Industrial Waste Disposal and Brownfields

Liquid, solid and gaseous wastes have always been by-products of industries and other manufacturing activities. When such activities remained small in number or were widely distributed and more or less out of sight, the disposal of industrial waste was not a topic of great concern for the general public. Waste was discharged routinely into water, air and soil, apparently without violating any rules, if indeed any such rules existed at the time. Certainly, the topic of industrial waste did not have the same visibility as human waste disposal. Once the association between disease and human waste was established in the 1800s (Section 4.3.2), the disposal of sewage came under regulation in developed countries. For untreated industrial wastes that often caused visible damage, such as the desertification of land in the vicinity of metal smelters, there seemed to be a more tacit acceptance of the environmental impacts. Perhaps this could be explained in part as an accepted sign of progress, according to the maxim "Where there's muck, there's money", and the attitude that the natural world was created to serve human needs. By the late nineteenth century, this began to change. Activists were writing about the scourge of industrial waste, generally out of concern for human health and aesthetics. Eventually the contamination of water and air by industrial activities did attract public attention when *Silent Spring* triggered the environmental movement in the mid to late 1960s. This change in general attitude also reflected the growing understanding of ecological services, and the real price to be paid for degrading ecosystems.

Burial as a means of disposal was a long-accepted practice and is still used today, but under controlled conditions. A pivotal event concerning buried toxic waste occurred in 1942 at Niagara Falls, NY, USA. Hooker Electrochemical was granted permission by the Niagara Power and Development Company to fill the abandoned Love Canal with waste chemicals from manufacturing chlorinated hydrocarbons and caustics. By 1953, more than 10,000 tons (T) of chemicals had been deposited in the 800 m long canal, at which time the site was closed, capped with clay and soil, and sold to the Niagara Falls Board of Education for a token US$1.00. An elementary school was erected on the site, and developers built 98 homes along the banks of the former canal. During the ensuing 25 years, construction that disturbed the clay seals allowed water to penetrate the dump site, and chemicals migrated into the surrounding community (Paigen, 1982). By 1978, families evacuated the area in response to severe effects on human health, confirmed by epidemiological studies.

Although human health was the major concern in this example, nevertheless there was a lesson to be learned for ecotoxicology concerning the potential for migration of contaminants after burial, and the related risks. The geological and hydrological conditions at a burial site determine the ultimate fate and effects of the buried chemicals.

Colten and Skinner (1995) point out that Love Canal was by no means unique, although no examples were as notorious in the public eye, nor as politicized. Regulatory actions followed rapidly in the enactment in the United States of the Comprehensive Environmental Response, Compensation and Liability Act (CERCLA), often called the Superfund Act. It attributes liability and provides for clean-up of contaminated sites. There is hope that no 'new' Love Canals will be discovered, because waste disposal is now more carefully regulated, and disposal sites are chosen after appropriate geological and hydrological assessment. Nevertheless, there remain an enormous number of identified (and some unknown) contaminated sites that may need future clean-up and restoration (see Chapter 13). Love Canal left legacies beyond the suffering of its residents and the wrangling of the involved parties. It illustrates, as do many other examples, the politicization, attempts at denial and complexities beyond science that impede the acceptance and resolution of many environmental issues.

The term '**brownfields**' is used to describe potentially contaminated old industrial sites that are generally smaller and less expensive to remediate than 'Superfund' sites. Brownfields are typically in urban settings, where industries or smaller operations such as petrol stations have contaminated soils. They are often abandoned but have potential for redevelopment. The basic concepts for regulating and **remediating** brownfields (Chapter 13) involve ERAs with jurisdictional prescriptions for the process, and regulatory agency approvals of clean-up, if required.

1.2.5 Oil Spills

Oil spills have characterized the modern economy since the first commercial oil wells were developed in the 1850s. Although spills were dramatic examples of chemical pollution and environmental degradation, the public's attention to such incidents was relatively brief, because spilled oil rapidly dissipated in the open ocean and access to spills in remote areas was often difficult. By today's standards, the volumes spilled were also relatively small. The damage also appeared to be confined to a limited array of receptors, particularly shorelines near popular beach resorts, and iconic marine species such as seabirds, turtles and seals affected by floating oil. Concerns about oil spills increased considerably on 28 March 1989 when a single-hulled crude oil tanker, the *Exxon Valdez*, ran aground off the coast of Alaska. More than 37,000,000 litres (37,000 t) of Alaska North Slope crude oil (ANSC oil) were spilled into Prince William Sound in the Gulf of Alaska, one of the world's most productive marine ecosystems.

By most standards, this spill was not large, in 2019 ranking 36th globally for all tanker spills. The 1979 *Atlantic Empress* spill (www.itopf.org/knowledge-resources/data-statistics/) in the West Indies (287,000 t) was 7.5 times as large. At the time, the largest known oil spill from any source was from the 1979 Ixtoc-1 oil drilling platform in the Gulf of Mexico (480,000 t). Nevertheless, the *Exxon Valdez* incident completely changed the public's perspective about the environmental importance of oil spills. Much of the public's focus on this spill could be attributed to the increasing importance of television in people's lives, and the access to a remote site facilitated by lightweight camera equipment, aircraft and helicopters, and satellite data transmission. This was the first oil spill presented live on nightly newscasts. It transformed an environmental disaster into public education, providing an immediacy and drama to events that heightened environmental awareness and concerns and sparked intense discussions among the public, regulators, and the shipping and oil industries. Debates centred on marine resources affected or at risk, methods for assigning values to lost 'environmental services' (e.g., fisheries, tourism), human and economic impacts, oil spill clean-up techniques, liabilities for clean-up and the regulation of tanker safety.

The understanding of oil spill impacts was expanded enormously by industry- and government-funded research to bolster court actions that lasted more than 20 years. For the first time, impacts on fish and fisheries appeared as important as impacts on surface resources such as seabirds and shorelines. Oil spills were no longer regarded as transient incidents because oil was found to persist out of sight in intertidal sediments, causing toxic effects on sea otters for decades after the spill. These insights supported new regulatory efforts, including a new Oil Pollution Act (1990) in the United States that required double-hulled tankers for oil shipments in US waters and assigned all damages from oil spills to the companies that owned the spilled oil. A second measure was the creation of an advanced programme for Natural Resources Damage Assessment that mandated studies to identify and assess resource impacts, the degree of 'injury' and measures needed for restoration. The party responsible for the spill could be involved in the assessment and restoration and would be responsible for all costs and for compensating resource users for their losses and expenses. The science developed for the *Exxon Valdez* oil spill also laid a foundation for the unprecedented research effort following the 2010 Deepwater Horizon oil spill from a British Petroleum drilling platform in the Gulf of Mexico (Beyer *et al.*, 2016).

1.2.6 *Our Stolen Future* and Endocrine Disruptors

In 1996, the environmental movement was aroused once again by an alarming publication. Dr Theo Colborn co-wrote a book with Dianne Dumanoski (an environmental journalist) and John Peterson (a leading environmentalist) entitled *Our Stolen Future – Are We Threatening Our Fertility, Intelligence, and Survival? A Scientific Detective Story* (Colborn *et al.*, 1996). *Our Stolen Future* was considered by many to be a sequel to Carson's *Silent Spring* and opened the public's eyes to another significant threat from chemical emissions, that of **endocrine disruptors**. It focused on chemicals that impair the endocrine functions of humans and wildlife with sublethal effects that could determine the fate of entire populations of humans and animals by altering their capacity to have healthy offspring.

In the 1970s, Frederick vom Saal discovered that small changes in hormones in the uterus of mice during pregnancy had lifetime consequences. Alarmed, Dr Colborn reviewed hundreds of related studies and governmental reports. She noted that endocrine-active contaminants were commonly found in humans and wildlife, particularly in blood, placenta and breast milk, as well as in their offspring at birth. To understand the effects of endocrine disruption, it is important to know the history of the pharmaceutical diethylstilboestrol (DES), a synthetic analogue of oestrogen first created for human clinical use and later adopted as a growth promoter for livestock. In the US Midwest during the mid-1960s, mink ranches lost their stock because mink failed to reproduce when fed chickens that had been given DES (Travis and Schaible, 1962). During the same period, women were prescribed DES in some regions to treat problems during pregnancy (e.g., miscarriages and premature births). However, later research linked maternal DES treatments to an array of reproductive and neurological abnormalities in children, including intersexuality, malformed uterus, infertility and depression. Unfortunately, DES crossed the placenta–blood barrier and impaired foetal development.

Subsequent environmental research demonstrated reproductive effects in fish and wildlife exposed to structurally related compounds discharged in sewage and industrial effluents. The suite of known contaminants acting as endocrine disrupting chemicals included many of the widely distributed compounds that are readily bioavailable (e.g., DDT, PCBs, polycyclic aromatic hydrocarbons (PAHs); Chapter 8), transforming a human health issue into an environmental crisis.

Dr Colborn's key messages were:

- To fully appreciate the scope of adverse effects on humans that are caused by environmental toxicants, we must move beyond a focus on cancer. Entire populations (humans, fish and wildlife) experience significant reproductive impairment, characterized by offspring born with an array of sexual, neurological and immune disabilities, often attributed to endocrine disruption.
- The endocrine system in fish and wildlife is closely related to that of humans. Because of this, human epidemiological studies should systematically include animal research in their analyses. Endocrine disruption is far broader than chemical interference with 17β-oestradiol.
- All endocrine axes should be carefully investigated, including thyroid hormones crucial for brain and sexual development, basic metabolism, and many other important physiological functions.

Although Colborn's message emphasized human health, it is nevertheless still relevant to ecotoxicology in the context of a point made earlier, that "the study of human toxicology has sometimes led to the discovery of ecosystem effects, often previously unanticipated".

Endocrine disruption is better integrated into today's governmental vocabulary. Several human health and environmental organizations (e.g., Organisation for Economic Co-operation and Development [OECD], United States Environmental Protection Agency [US EPA]) have developed protocols to screen for and identify endocrine disrupting chemicals. However, despite evidence of a global generalized reduction in human fertility and health of children, the routine assessment of the effects of chemicals on hormonal systems is limited to oestrogens, androgens and thyroid hormones. Few countries have regulated contaminants based on endocrine disrupting capacities.

1.3 The Emergence of the Science of Ecotoxicology

Although many reports about chemical contamination of the environment were published prior to World War II, the post-war era saw a major increase in publications and journals related to what was then called environmental toxicology. Striking examples of environmental degradation, such as those described in Section 1.2, fed growing demands to regulate industrial wastes, to remediate contaminated sites and to limit the environmental risks of new products. The development of legislation to control the production, use and disposal of chemicals, and the creation of agencies to monitor and protect environmental quality, were logical responses to these concerns. However, these initiatives did not immediately solve all environmental problems. Although the laws enabled chemical regulation and management, the tools and science needed to support these actions were still relatively few and undeveloped.

The need for tools and knowledge was not only an issue for governments. Industries that emitted wastes or that marketed products thought to be environmentally damaging were equally concerned. They feared that environmental controls would be overly strict, unevenly applied and cause economic disruption. Not surprisingly, the primary focus of research in the 1970s and 1980s was to develop and standardize accurate and repeatable methods for assessing chemical toxicity to species representative of contaminated environments. The lead agencies were newly created international or government departments, such as the United Nations Environment Programme (UNEP) and the US EPA. Non-governmental agencies such as the American Society for Testing and Materials (ASTM), an international standards association representing about 140 countries, played a significant role by engaging industry and government stakeholders. Committees of the ASTM developed a wide array of consensus-based methods for testing the acute and chronic toxicity of single chemicals to plants, invertebrates, fish and later wildlife. Committee decisions on methods under development were supported by annual technical symposia that provided emerging science related to environmental toxicology and risk assessment. The broad scope of the methods developed by the various national and international initiatives (Box 1.2) was a clear reflection of the breadth and complexity of the environmental issues revealed by the expanding efforts to recognize and regulate chemical emissions.

Although an early focus on developing test methods supported the immediate need to regulate the most severe chemical emissions, it did not foster a broader understanding of the environmental interactions and ecological impacts of existing or ongoing chemical contamination. Committees to develop methods were not a scientific 'home' for those interested in the broader issues of ecotoxicology. The response to this deficiency was the formation of an array of science societies including the International Society of Ecotoxicology and Environmental Safety (SECOTOX; 1972) (www.secotox .uth.gr) and the Canadian Ecotoxicity Workshop (CEW, formerly Aquatic Toxicity Workshop; 1974) (www .ecotoxcan.ca). The largest and most successful has been the Society of Environmental Toxicology and Chemistry (SETAC; 1979) (www.setac.org). The initial membership was balanced among scientists from academia, business and government but later expanded to include ENGOs and graduate and undergraduate students. The society has since grown to a global organization of more than 5,000 members with a World Council to coordinate activities among SETAC Africa, Asia/Pacific, Europe, Latin America and North America. Its inclusivity and openness to a broad ecological perspective are reflected in its publications, annual meetings, technical workshops and special interest groups (Box 1.3), which have fostered many advances in the science and application of ecotoxicology (Table 1.1).

Since 1979, the scope of ecotoxicology has expanded from toxicity test methods to the broad-scale ecological effects of chemical emissions, including the mechanisms that link molecular effects to ecological change. Advances in analytical chemistry have enhanced our capacity to detect the global distribution of contaminants and to identify emerging substances of concern. Because the number of chemicals in commerce far exceeds our capacity to test them all, models have been

Box 1.2 Scope of standard methods

Toxicity test methods

- Specific materials: effluents, industrial products, pesticides
- Duration: acute, chronic
- Species: mammals, fish, invertebrates, plants, algae
- Medium: marine, freshwater, soil, sediments, air
- Responses: acute lethality, reproduction, behaviour
- Statistical analyses of test data

Applying toxicity data

- Product life-cycle assessments
- Product registration
- Product approval (e.g., chemicals safe to use near water)
- Ecological risk assessments (ERAs)

Box 1.3 The breadth of ecotoxicology: selected SETAC interest groups

- Animal Alternatives in Environmental Science
- Bioaccumulation Science
- Chemistry
- Ecological Risk Assessment
- Ecosystem Services
- Effect Modelling
- Ecotoxicology of Amphibians and Reptiles
- Endocrine Disruptor Testing and Risk Assessment
- Environmental Monitoring of Pesticides
- Exposure Modelling
- Indigenous Knowledge and Values
- Metals
- Microplastics
- Nanotechnology
- OMICS
- Pharmaceuticals
- Sediment and Soils
- Wildlife Toxicology

Table 1.1 **Examples of recent advances in methods in ecotoxicology.**

Discipline	Advance	Benefit
Environmental toxicology	Passive dosing	Sustains constant concentrations of hydrophobic test compounds in water by equilibrium partitioning between a sorbent and water (Chapter 2).
	In vitro reporter gene assays	Rapid and sensitive; detect chemicals that induce targeted molecular responses (Chapter 5).
	'Omics' technologies - Toxico-genomics - Toxico-proteomics - Computational toxicology	Identify genes and cell functions activated or suppressed by chemical exposure; provide a mechanistic understanding of chemical effects on physiology, development, behaviour and reproduction; high-throughput screening assays generate large databases to be mined and integrated to reveal complex mechanisms (Chapter 5).
Environmental and analytical chemistry	Passive or biomimetic samplers	Equilibrium partitioning into sorbents integrates variable concentrations of metals and organic compounds over days to weeks of deployment; isolates the 'free' or bioavailable fractions from the total concentrations of chemicals in air, water, sediments or soils (Chapters 2, 4, 6).
	Chemical extraction and analytical techniques and instruments, e.g., single-particle inductively coupled plasma mass spectrometry (among many others)	Continued improvement to the sensitivity, specificity and detection limits for environmental contaminants, including 'non-targeted' analytical approaches; supports research, monitoring, and the identification of compounds of emerging concern; helps to establish mechanistic links between the subcellular partitioning of metals and organics and toxic effects in the lab and field (Chapter 6).
Ecotoxicology	Mesocosm and whole ecosystem experiments	Describe the distribution, concentrations and effects of chemicals in natural or model ecosystems; identify links between molecular effects and changes in the structure and function of populations, communities and ecosystems (Chapter 4).
	Palaeo-ecotoxicology	Analyses of contaminants and micro-fossils in dated sediment cores link trends in environmental contamination to ecosystem change (Chapters 4, 6).
	High-throughput nucleic acid sequencing - eDNA - Metabarcoding	Maps whole populations of organisms in environmental samples; tracking complex taxonomic changes related to chemical contaminants and other environmental stressors (Chapter 5).
	Contaminant modelling	Supports ERAs of chemicals by integrating data on environmental fates, distributions and effects to predict the risks of ecosystem impacts; identifies key variables, data gaps and research needs (Chapters 2, 4).
Hazard and risk assessment	Ecosystem services paradigm	Provides valuable perspective on the social and economic costs of chemical emissions that degrade natural resources (Chapters 4, 10).
	Radiation monitoring	Provides awareness of levels and sources of ionizing radiation (Chapter 10).

developed to integrate this growing knowledge about the ecological behaviour, fate and effects of different types of contaminants. These models are now an integral part of ERAs for new chemicals that are structurally similar to well-studied compounds (Chapters 2, 4, 12), with the added benefit of reducing the numbers of animals used in research.

1.4 The Turning Point and Formal Regulation of Toxic Substances

In general, it is true that until the late 1960s, environmental pollution was considered a minor, if not trivial, offence. However, there are a few exemplary cases indicating that long before the environmental movement, some legal approaches had been made in recognition of the evils of environmental contamination. As early as the thirteenth century, air pollution, typically a result of the increased use of coal as fuel, was recognized when smoke and other airborne materials from incineration were visible and daily irritations. Aesthetic concerns predominated, but some attention was paid to possible human health effects. In the thirteenth century, penalties in England for contravening 'pollution abatement regulations' seem to have been fines or the removal of the offending source (Brimblecombe, 1976).

Prior to the 1970s, legal actions to address environmental contamination relied on common law and existing **statute law** (Section 12.2). In the 1960s and 1970s, many countries responded to the public's concerns about environmental disasters and crises by enacting focused statutes or legislation to regulate chemical emissions (see examples in Figure 1.4). Essential to these initiatives was the creation of ministries of the environment to apply the legislation by developing environmental standards and monitoring programmes, and by initiating research to support these activities. Environment ministries were also formed from components of other existing agencies with responsibilities for resource management (e.g., fisheries, forestry, water).

These developments followed closely and sometimes paralleled scientific and social considerations of environmental toxicology. They surely arose from the work of environmental activists, scientists and the general public's recognition of the risk related to chemicals in the environment. The US Council on Environmental Quality was set up by President Richard Nixon. On the first **Earth Day** in 1970, the Council wrote:

A chorus of concern for the environment is sweeping the country. It reaches to the regional, national and international environmental problems. It embraces pollution of the earth's air and water, noise and waste and the threatened disappearance of whole species of plant and animal life. (Cited in Van Der Heijden (1999))

Environmental legislation grew concurrently. In the modern world, the network of environmental legislation from the level of countries to municipalities is still evolving. It is rich, complex, costly and often controversial, which is not surprising, considering the multi-disciplinary and multi-stakeholder nature of environmental issues. Chapter 12 discusses in more detail the development, context and implementation of some Acts and Regulations from different jurisdictions.

Arguably the most complex aspect of environmental regulation is the international one. We trade and transport produce and goods across international boundaries, with the risk that the originating country may not have the same level of control and regulation of toxic substances (e.g., pesticides) as the recipient. The pathways of chemical emissions from sources to targets, particularly for those that are airborne, do not respect political or geographical boundaries, and many contaminant issues are truly global in nature. There have been increasing levels of international recognition, agreement and research about toxic chemicals. Put together, these are three reasons why international agreements relating to the regulation of toxic substances make sense and have been signed by numerous countries in the decades following the implementation of national laws (Table 1.2). However, it is axiomatic that international law and agreements are fraught with challenges, including assuring the necessary level of commitment to their implementation and enforcement. Note that the dates for the examples in Table 1.2 are considerably later than the dates of related statutes or legislation at the individual country level. These agreements represent landmarks in environmental management, and more details are provided in Chapter 12.

1.5 Solutions That May Lead to New Problems

Recognizing environmental issues does not always lead to their resolution: the solutions may create more problems. For example, we continue to struggle with the conflicts between the positive and negative aspects of pesticide use. The publication of *Silent Spring* was a major impetus for developing the science of ecotoxicology. It highlighted the environmental impacts of pesticides that were highly toxic to fish and wildlife, environmentally persistent and biomagnified in food chains.

As discussed in Section 1.2.1, Rachel Carson's book triggered the development of new and more 'environmentally safe' pesticides, among them the organophosphates, which were highly neurotoxic but degraded rapidly. In spite of the early alerts and the bans on organophosphate pesticides in some jurisdictions, they remain in use in some countries. In Argentina during the summer of 1995/96, forensic analysis indicated that the organophosphate insecticide monocrotophos was responsible for the deaths of 5,095 Swainson's hawks

Table 1.2 **Examples of international agreements that have contributed to the management of global chemical issues.**

Act or convention	Date	Substance(s) of concern	Toxicological issue(s)	Objective	Number of participants
Montreal Protocol	1985	Chlorofluorocarbons (CFCs)	Depletion of stratospheric ozone and increased ultraviolet-B radiation in the troposphere.	Phase out the use of CFCs in aerosol cans and as refrigerants.	179 countries
Stockholm Convention	2001[1]	Persistent organic pollutants (POPs), including the Dirty Dozen: Aldrin, DDT, chlordane, dieldrin, endrin, heptachlor, hexachlorobenzene, mirex, toxaphene, PCBs, chlorinated dioxins & furans	Reproductive failure of predatory fish and wildlife due to high persistence and high rates of bioaccumulation and biomagnification.	Of the Dirty Dozen, nine are banned outright.	184 countries
The European REACH Agreement	2006	All chemicals imported, manufactured or used in amounts ≥1 tonne (t)/yr, per producer or importer	General concerns about chemical safety.	A regulatory initiative for the Registration, Evaluation, Authorisation and Restriction of Chemicals.	European Union
Minamata Convention on Mercury	2013[1]	Mercury and all its compounds	Methylmercury, formed microbially from inorganic mercury, is highly toxic and bioaccumulates and biomagnifies in aquatic food webs. It is a risk for wildlife consumers of fish, and for humans.	Control the anthropogenic releases and toxic effects of mercury by banning or restricting products containing mercury and controlling emissions to air, land and water.	128 signatories

[1] Signed on the dates indicated; both conventions were ratified later by many signatory countries.

(*Buteo swainsoni*) at six separate sites (Goldstein *et al.*, 1999). A detailed account of the events is provided in Case Study 7.1. New farming practices, characterized as 'natural' and 'organic', were also adopted. Unfortunately, 'natural' and 'organic' do not always equal 'environmentally safe'. Research in ecotoxicology continues to address controversies related to the environmental and human health effects of pesticides derived from natural products (e.g., neonicotinoids) and genetically modified crop species tolerant of herbicides or capable of expressing genes for 'natural' insecticides.

Technologies to reduce and treat municipal and industrial wastes evolved more or less in parallel with the science of ecotoxicology and with regulations to limit emissions (Chapter 12). The modern approach to waste treatment corresponds to the three R's: reduce, reuse or recycle. The most important of these is 'reduce'. If less waste is produced, it is less expensive to treat, and there

is less liability for environmental damage. The ultimate solutions are 'closed-loop' processes (recycle), whereby water or air needed to create a product are treated and reused within the industry (Section 11.4.2). For industries that continue to discharge liquid or gaseous emissions, the question must be asked: *If an effluent is not good enough to be reused in the industry, why is it good enough for the environment?*

1.6 Conclusions

Chapter 1 has illustrated the linkages among the growth of human populations, industrial development, and the severity and extent of environmental contamination by chemical wastes. The human health and environmental costs of contamination were often dramatic and extreme, creating a significant public awareness and alarm. Twenty-five years after the publication of *Our Stolen*

Future, most of the recommendations listed in its chapter 12, "Defending Ourselves", are still up-to-date: for example, "Know your water", "Improving Protection", "Redesigning Manufacture and Use of Chemicals", etc. The pressure to resolve these issues forced political and legislative changes in the way resources are extracted, consumed, processed and disposed of. It also highlighted the need for sound science as a technical basis for recognizing, understanding and resolving these issues.

As will become evident in subsequent chapters, increased expenditures on research have provided standard methods for toxicity testing, innovations in analytical chemistry, and a synthesis of knowledge for predictive modelling to improve the efficiency of ERAs. Many advances have been made in the understanding of previously recognized inorganic and organic toxicants. Currently, there are very active research efforts to assess the toxicity and ecological effects of emerging compounds of concern, including pharmaceuticals, endocrine disruptors, perfluorinated compounds and microplastics (Chapter 14). In describing these issues in detail, later chapters review the mechanisms that link the first molecular effects of toxicants to subsequent effects on the physiological responses of individual organisms and changes in ecosystem structure and function.

Summary

This chapter traces the emergence and evolution of the science of ecotoxicology. It also reviews some of the major environmental crises that created an awareness of the human health and ecological costs of chemical contamination. The result was political pressure from the public, ENGOs and the media to regulate chemical emissions and to remediate contaminated sites. Governments globally have responded, creating a need to better understand the environmental behaviour, fate, toxicity and ecological impacts of contaminants. As Chapters 2 to 14 illustrate, the result is the increasing sophistication and rapid development of the science of ecotoxicology, and its application to regulatory toxicology.

Review Questions and Exercises

1. How would you distinguish among human toxicology, environmental toxicology and ecotoxicology?

2. Class discussion: What are some of the important characteristics of the science of ecotoxicology?

3. Trace the historical links between classical (human) toxicology and ecotoxicology.

4. Based on Chapter 1 (or if you have read her book), what do you think Rachel Carson's important messages were in *Silent Spring*?

5. How was the subject of ecotoxicology first described and defined? How did this come about?

6. Discuss the role of politics, public opinion and economics in the evolution of the science of ecotoxicology.

7. What are the main scientific organizations that have responded to the need for tools and knowledge to understand the risks and environmental impacts of chemical contaminants?

8. Why are international agreements on the use and control of certain chemicals important?

9. Provide five examples of international agreements related to environmental contaminants and explain how each was triggered.

10. What is the difference between contamination and pollution?

11. Class discussions: What lessons were learned about chemical sources and effects from the following: (A) the Love Canal; (B) the *Exxon Valdez* oil spill; (C) global mercury cycles; (D) persistent pesticides such as DDT; (E) pharmaceuticals and sewage effluent?

Abbreviations

ASTM	American Society for Testing and Materials	PCB	Polychlorinated biphenyl
CFC	Chlorofluorocarbons	POP	Persistent organic pollutant
DDT	Dichlorodiphenyltrichloroethane	SETAC	Society of Environmental Toxicology and Chemistry
ENGO	Environmental non-governmental organization	US EPA	United States Environmental Protection Agency
ERA	Ecological Risk Assessment		
NGO	Non-governmental organization		

References

Beyer, J., Trannum, H. C., Bakke, T., Hodson, P. V. & Collier, T. K. (2016). Environmental effects of the Deepwater Horizon oil spill: a review. *Marine Pollution Bulletin*, 110, 28–51.

Borgia, V. (2018). The prehistory of poison arrows. In P. Wexler (ed.) *Toxicology in Antiquity*. 2nd ed. London, UK: Elsevier, 1–10.

Brimblecombe, P. (1976). Attitudes and responses towards air pollution in medieval England. *Journal of the Air Pollution Control Association*, 26, 941–945.

Carson, R. (1962). *Silent Spring*. Greenwich, CT, USA: Fawcett Publications.

Chapman, P. M. (2002). Integrating toxicology and ecology: putting the "eco" into ecotoxicology. *Marine Pollution Bulletin*, 44, 7–15.

Chapman, P. M. (2007). Determining when contamination is pollution – weight of evidence determinations for sediments and effluents. *Environment International*, 33, 492–501.

Colborn, T., Dumanoski, D. & Myers, J. P. (1996). *Our Stolen Future – Are We Threatening Our Fertility, Intelligence, and Survival? A Scientific Detective Story*, New York, NY: Dutton.

Colten, C. E. & Skinner, P. N. (1995). *The Road to Love Canal: Managing Industrial Waste Before EPA*, Austin, TX: University of Texas Press.

Finger, M. E. (1992). *The Green Movement Worldwide*, Greenwich, CT: JAI Press.

Foster, K. R., McDonald, K. & Eastlick, K. (2001). Development and application of critical, target and monitoring loads for the management of acid deposition in Alberta, Canada. *Water, Air and Soil Pollution: Focus*, 1, 135–151.

Goldstein, M. I., Lacher, T. E., Woodbridge, B., *et al.* (1999). Monocrotophos-induced mass mortality of Swainson's hawks in Argentina, 1995–96. *Ecotoxicology*, 8, 201–214.

Gorham, E. (1985). Acid rain and aquatic ecosystems: the target loading question. *Intergovernmental Conference on Acid Rain, Quebec City, QC, Canada, 1985*, 91–97; available from University of Minnesota archives.

Hachiya, N. (2006). The history and the present of Minamata disease. *Japan Medical Association Journal*, 49, 112–118.

Harari, Y. N. (2018). *21 Lessons for the 21st Century*. Toronto: Random House.

Orfila, M. J. B. (1818). *Traité des poisons tirés des regnes minéral, végétal et animal, ou toxicologie générale, considérée sous les rapports de la physiologie, de la pathologie et de la médecine légale*. Paris: Crochard.

Paigen, B. (1982). Controversy at Love Canal. *Hastings Center Report*, 12, 29–37.

Rosner, D. & Markowitz, G. (2013). Persistent pollutants: a brief history of the discovery of the widespread toxicity of chlorinated hydrocarbons. *Environmental Research*, 120, 126–133.

Schüürmann, G. & Markert, B. (1998). *Ecotoxicology. Ecological Fundamentals, Chemical Exposure, and Biological Effects.* New York, NY: Wiley.

Travis, H. F. & Schaible, P. J. (1962). Effects of diethylstilbestrol fed periodically during gestation of female mink upon reproductive and kit performance. *American Journal of Veterinary Research*, 23, 359–361.

Truhaut, R. (1977). Ecotoxicology: objectives, principles and perspectives. *Ecotoxicology and Environmental Safety*, 1, 151–173.

Van Der Heijden, H.-A. (1999). Environmental movements, ecological modernisation and political opportunity structures. *Environmental Politics*, 8, 199–221.

Wetstone, G. S. (1987). A history of the acid rain issue. In H. Brooks & C. L. Cooper (eds.) *Science for Public Policy*. Oxford, UK: Pergamon Press, 163–196.

Wilson, M. W., Balmer, D. E., Jones, K., *et al.* (2018). The breeding population of Peregrine Falcon *Falco peregrinus* in the United Kingdom, Isle of Man and Channel Islands in 2014. *Bird Study*, 65, 1–19.

Wright, D. A. & Welbourn, P. (2002). *Environmental Toxicology*, Cambridge, UK: Cambridge University Press.

2 Measuring Toxicity

Peter V. Hodson and David A. Wright

Learning Objectives

Upon completion of this chapter the student should be able to:

- Discuss the principles of toxicology as applied to environmental toxicology.
- Design toxicity tests for microorganisms, plants and animals.
- Understand how experimental factors affect the outcome of toxicity tests.
- Apply modelling tools to move from descriptive to predictive environmental toxicology.

Chapter 1 presented a history of environmental issues related to chemical contamination derived from all aspects of human social and economic development. This history provides us with ample reasons why governments have developed regulatory tools for controlling chemical emissions and their environmental effects. Fundamental to **regulation** is a need to measure the toxicity of chemicals that harm any forms of life, a need that has been met by the evolution of **environmental toxicology** as a scientific discipline (Section 1.1). Environmental toxicology includes standard **toxicity test** methods that support a consistent approach for classifying and regulating contaminants (Chapter 12) and for developing models to predict the toxicity of previously untested compounds and complex mixtures. Equally important, standard

methods reveal the range of chemical sensitivity among test species and the perspective needed to protect the millions of species that have never been tested.

2.1 The Basics of Environmental Toxicology

We begin Chapter 2 with the fundamental concepts of **exposure–response relationships** and define terms that must be used correctly if test data are to be understood, comparable among studies and useful in chemical regulation. The issues of realistic versus standard tests are explored, as well as the essential role of analytical chemistry in toxicity testing, particularly when dealing with sparingly soluble compounds. Statistical methods are presented for calculating toxicity, estimating **thresholds**, comparing toxicity among different species or test conditions, and predicting the toxicity of previously untested compounds from their chemical structure.

Although the principles of toxicology apply in both laboratory and field studies of chemical effects, natural ecosystems are far more complex than laboratory test systems. In Chapter 4, we expand the scope of laboratory studies to address the complexities of **ecotoxicology**, i.e., chemical effects on ecosystem structure and function. We learn how tools such as environmental chemistry, **biomarkers** of exposure and effects, and the principles of **epidemiology** are adapted to assess **cause–effect relationships** in complex ecosystems.

2.1.1 Concepts and Definitions

A toxicity test is the measurement of the adverse effects of chemicals on living organisms, and the amount of chemical and exposure time needed to cause those effects. Underpinning the science of toxicity tests are a variety of definitions and principles (Box 2.1) which will be cited as we learn more about how these principles are translated into test methods.

> ## Box 2.1 Definitions and principles of environmental toxicology
>
> - Toxicity is defined as the inherent capacity of a chemical to cause harm to living organisms.
> - All chemicals are toxic, but some are also nutrients; *it is the dose that makes the poison* (Paracelsus).
> - Every toxic effect begins with an interaction between a chemical and some component of living matter.
> - Exposure is key – if there is no exposure there is no effect.
> - Chemical exposures can be described by environmental concentrations (food, air, water, soil), dose (amount in organisms, tissues, or cellular **receptors**) and exposure time.
> - Chemical structure determines toxic effects and the nature of a chemical's interactions with living matter. Small changes in a chemical's structure may cause large changes in toxicity.
> - Toxicity is characterized by an exposure–response relationship in which effects increase **monotonically** with exposure concentration and time, in proportion to each increment of exposure.
> - Most chemicals exhibit a threshold exposure level, below which toxicity is absent.
> - Toxicity is generally expressed as the response of the average organism, not the most sensitive or most **resistant**.
> - Toxicity is not absolute, but varies with the form of the chemical tested, the organism that is exposed and the conditions of the exposure.
> - The biochemical and physiological consequences of chemical toxicity determine the capacity of an organism to survive, grow, reproduce and cope with other environmental **stressors**.
> - The physiological and behavioural responses of individual organisms to chemical exposure progress to effects on populations, communities and ecosystems.

Measuring toxicity is fundamental to understanding the potential ecological risks of chemical contamination. However, before we can measure toxicity, we must first understand what toxicity is (or is not), its causes and the major factors that control it.

2.1.1.1 What Is Toxicity?

Toxicity is one of several chemical **hazards** (Box 2.2), where hazard refers to the 'inherent capacity to cause harm'. Thus, toxicity is the inherent capacity of chemicals to harm organisms. Nutrient deficiency and suffocation are the exact opposite, i.e., they are the adverse effects caused by inadequate amounts of substances essential for normal metabolism.

Any impairment of normal biochemical and physiological processes will potentially limit the capacity of an organism to cope with its environment, and to survive, grow and reproduce. Widespread effects on individual organisms will also change the population structure of a species. For example, missing or weak year classes occur when reproduction is impaired and few young survive to a reproductive age. If severe, changes in population structure may lead to changes in community structure (e.g., relative abundance of predators or prey), and changes in ecosystem structure and function as energy flow and nutrient distribution change. The chain of conceptual or actual linkages from chemical exposure to effects on organisms and to ecological impacts is referred to as an **adverse outcome pathway** (AOP), an emerging tool for chemical management (Section 4.2).

2.1.1.2 Chemical Structure vs Toxicity

Chemical interactions with **molecular receptors** such as proteins, nucleic acids or fatty acids can block normal physiological processes or trigger otherwise normal processes at an inappropriate time or to an inappropriate extent, with adverse effects on cellular function. The

> ## Box 2.2 Examples of chemical hazards
>
> - Toxicants: chemicals that impair biological functions when accumulated by organisms
> - Nutrient deficiency: lack of vitamins, minerals
> - Suffocation: lack of oxygen
> - Climate change: carbon dioxide; methane
> - Acidification: oxides of carbon, nitrogen or sulphur
> - Fire and explosions: highly reactive compounds

correlations between a chemical's properties (structure, molecular shape, polarity, etc.) and its potency for affecting biochemical processes can be modelled by **quantitative structure–activity relationships (QSARs;** Section 2.3.3.3). These statistical models support the design of drugs or pesticides that target specific biological processes (e.g., neurotransmission) and **ecological risk assessments (ERAs)** of untested compounds (Section 12.3.2). Molecular size also determines apparent toxicity by affecting exposure. Molecules that are very large are not easily **bioaccumulated** because the spaces between membrane lipid and protein molecules are too small to allow rapid diffusion (steric hindrance) (Section 3.1.2.1). Although these chemicals might show biological activity when tested with tissue homogenates, they often appear non-toxic in tests with whole organisms: i.e., *no exposure = no effects.*

The importance of chemical structure in toxicity is illustrated by the highly specific **mode of action (MoA)** of **organophosphate (OP) pesticides** (Box 2.3, Figure 2.1, details in Section 7.8.2).

2.1.1.3 Nutrients vs Toxicants
Some chemicals are both **nutrients** and **toxicants**, exerting biphasic effects on organisms. For example, in

mammals, selenium toxicosis has been known since the 1930s, but the adverse effects of selenium deficiency are also well known. Selenium in trace amounts is an integral component of several enzymes with antioxidant properties and of others involved with thyroid metabolism. Similarly, vitamin D is critical for bone development

---

Box 2.3 Structure, metabolism and toxicity of OP pesticides

Acetylcholine is a chemical messenger at the junction between two neurons (synapse) and is released from the 'upstream' neuron to initiate signal transmission by the 'downstream' neuron. Acetylcholinesterase modulates neural activity at the synapse by destroying acetylcholine after its release. By inhibiting acetylcholinesterase, OP pesticides permit a continuous discharge of the downstream neuron (**tetany**) so that death occurs rapidly because of muscle paralysis and loss of control of respiration. The potency of each OP pesticide varies with small changes in its molecular shape. The more closely it fits the receptor site of the acetylcholinesterase enzyme, the more effectively it blocks acetylcholine degradation and causes tetany.

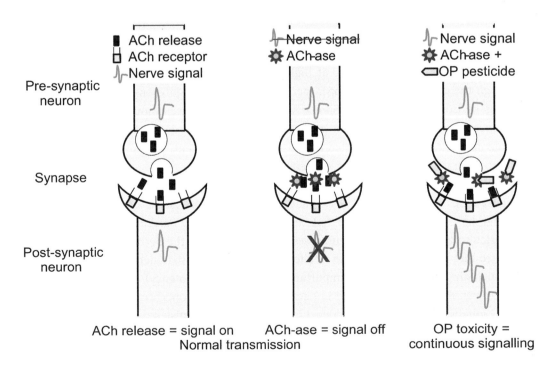

Figure 2.1 Organophosphate pesticides inhibit acetylcholinesterase (AChE), an enzyme that limits neural transmission at synapses by destroying the neurotransmitter acetylcholine (ACh) after its release. If ACh is not destroyed, the post-synaptic neuron fires continuously, causing tetany and death.

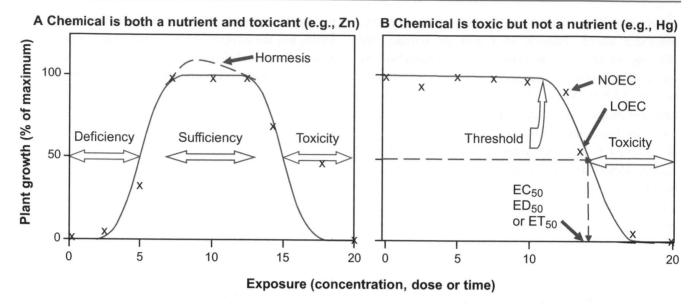

Figure 2.2 The hypothetical effects on plant growth of exposure to chemicals that are both nutrients and toxicants (A; e.g., zinc, Zn) and chemicals that have no role in nutrition (B; e.g., mercury, Hg). Hormesis is a stimulation of an organism's performance at low levels of exposure to chemicals that are toxic at higher levels of exposure (Box 2.4). Acronyms are defined in Box 2.5.

and health, but excess vitamin D causes hypercalcaemia and pathologies in kidneys and other organs. Vitamin A is necessary for the proper development and health of the eye, but an excess causes liver and cardiovascular damage and osteoporosis.

In agriculture, it is essential to know how much fertilizer should be added to soils. Plant growth is limited by very low concentrations of micronutrients such as iron (Fe), which is essential for chlorophyll synthesis and chloroplast function. However, Fe can be toxic at high concentrations through the conversion of hydrogen peroxide into damaging hydroxyl free radicals. Similarly, as soil zinc (Zn) concentrations increase, plant growth will increase to a maximum. Above the threshold for Zn deficiency, plants can tolerate a wide range of concentrations, but at very high soil concentrations, growth will decline to zero (Figure 2.2).

For **pesticides** that have no nutrient value but are applied to sustain agricultural production, we must assess their toxicity to ensure that application rates are safe for non-target species, including humans. Just as important are toxicity tests of industrial wastes to limit their impacts on terrestrial and aquatic ecosystems. Section 12.2 describes the strategies and tools used by governments to control and regulate the environmental fate, distribution and effects of chemical pollutants. The appropriate assessment of toxicity is clearly a crucial component of regulatory toxicology.

The risk or likelihood of toxicity of any chemical depends on the organism that is exposed. For example, in agriculture there is a broad array of **herbicides** that can kill plants considered as weeds on cropland. Some specifically target the grasses, others 'broad-leaved' plants, and still others, thistles, each taking advantage of unique biochemical and physiological sensitivities of the different plant groups. Receptor sensitivity is also a function of the life stage of an organism that is exposed to a chemical. For example, invertebrate and vertebrate embryos are highly sensitive to chemical effects on organogenesis because thousands of genes are expressed in a specific order and at critical times, and each developmental event presents a potentially sensitive target for toxicity.

2.1.1.4 Expressions of Toxicity

As deduced by Paracelsus (Section 1.1), all chemicals are toxic and it is the **dose**, or concentration of a chemical in living tissue, that determines toxic effects. Ideally, the toxicity (or potency) of a compound is best described as the tissue concentration associated with effects, or the amount administered (dose) by injection, diet or some other means of direct exposure. However, ecotoxicology and environmental regulation are concerned with the toxicity of chemicals discharged to different environmental media (air, water, soil). When we measure the toxicity of chemicals in these media, we must recognize that our

Box 2.4 Hormesis

Hormesis describes a biphasic exposure–response relationship for some chemicals; the result is a stimulatory/beneficial effect at low doses but an inhibitory/toxic effect at higher doses (Calabrese, 2008). Hormesis is often described as a balance between a nutritive and a toxic effect, but there are other possible mechanisms underlying hormesis unrelated to nutrition. For example, although arsenic is an established **carcinogen**, populations exposed to low arsenic concentrations may have a reduced cancer risk compared with less exposed populations. Similarly, low levels of ionizing radiation facilitate the prevention and repair of DNA damage resulting from toxicant exposure (Section 10.5).

Hormetic responses have been documented in a broad spectrum of species and for a wide range of organic and inorganic chemicals. At the molecular level, hormesis may represent the activation of cellular stress response pathways. These typically involve receptors and signalling pathways linked to kinases and the activation of transcription factors responsible for the expression of cytoprotective proteins, including antioxidant enzymes, protein **chaperones** and growth factors (Section 7.4). Hormesis has been characterized as a conditioning reaction whereby exposure to a toxic agent triggers a response that mitigates the adverse effect of the toxicant. Such an acclimation response develops over exposure concentrations that are often lower than the toxicity threshold by up to an order of magnitude. Therefore, hormesis may be difficult to discern from existing exposure–response studies that are typically designed to detect only adverse effects.

The exact scope and **aetiology** of hormesis remains a subject of debate. However, its affirmation as a fundamental concept in toxicology may require a reassessment of exposure–response relationships at low toxicant concentrations, including the model of 'no threshold' for carcinogens.

tests provide only an indirect measure of the dose that is toxic. The key link is the extent to which an organism accumulates a chemical from the exposure medium (Chapter 3). *While the tissue dose that is toxic to each species may be constant, the concentrations in environmental media that result in that dose can be highly variable (Section 2.2.7).*

Mixtures of chemicals in industrial effluents or contaminated sediments present further complexities (Section 2.3.4). It is likely that many chemicals in a mixture contribute to toxicity so that observed effects cannot be attributed easily to one specific component. In this case, the effluent is treated as a 'substance', and toxicity is expressed as the per cent dilution associated with effects.

Although organism sensitivity determines toxicity, sensitivity can vary widely among individuals even within populations of the same test organism. To avoid bias because of overly sensitive or highly resistant individuals, toxicity is typically expressed as the response of the average or median (50th percentile) organism in a test population (e.g., LC_{50} – median lethal concentration of a chemical in air, water or food; LD_{50} – median lethal dose injected or administered to test organisms; Box 2.5). These toxicity measures are derived from statistical regressions relating the extent of observed responses to gradients of exposure concentrations or doses (Section 2.3). Toxicity estimates are most precise when based on the 'median' organism because the statistical variance of estimated responses is least at the midpoint of a regression.

The toxicity of a compound also depends on the nature of the response measured and the exposure time (Box 2.1). By definition, **lethal** concentrations of a compound (those causing death) will be higher than those

Box 2.5 Common expressions of toxicity

LC_{50}	Median lethal concentration
LD_{50}	Median lethal dose
EC_{50}	Median effective concentration (for sublethal responses)
ED_{50}	Median effective dose
IC_{50}	Concentration causing 50% inhibition (e.g., of algal growth)
EC_{10}	Concentration causing a 10% response (an estimate of threshold toxicity)
LT_{50}	Median lethal time
ET_{50}	Median effective time
EC	Effect concentration
NEC	No-effect concentration
LOEC	Lowest-observable-effect concentration
NOEC	No-observable-effect concentration
MATC	Maximum acceptable toxicant concentration

causing **sublethal** effects (e.g., reduced growth). Similarly, the concentration or dose needed to cause an effect during a brief (**acute**) exposure will be greater than that needed to cause the same effect during a more prolonged or **chronic** exposure. Therefore, no expression of toxicity is complete without modifiers indicating exposure time and the effects measured. For example, acute lethality tests with animals often measure the 96-hour LC_{50} or LD_{50}, which, not surprisingly, corresponds to a Monday-to-Friday work week. The chronic, sublethal effects of a compound impairing the growth of a test species over 30 days would be expressed as the '30-d EC_{50} for growth'. Note that acute and chronic are relative terms and have no precise definition. A 96-h test would measure the acute toxicity of a compound to a bird species with a life span of months to years, but a test of the same duration would measure chronic toxicity to an insect species living only days to weeks.

An important diagnostic characteristic of toxicity is a monotonic relationship between exposure and effect, i.e., a progressive increase in toxic effects with a progressive increase in exposure (Figure 2.2). Ultimately, there is a threshold concentration or dose below which no toxicity will be observed, no matter how long the exposure. In practice, however, threshold exposures for chronic effects such as cancer can be difficult to define.

Box 2.5 presents several variations of acronyms that express toxicity in terms of exposure time or the extent of inhibition of different physiological processes, such as growth or reproduction. If exposure–response relationships are robust, i.e., based on a very large number of data points with little variance about the regression line, toxicity can be calculated at lower response rates. The concentrations affecting the most sensitive 5% or 10% of a test population (e.g., 96-h EC_5 or EC_{10}) define threshold concentrations below which no effects would be expected during a 96-h exposure. An alternative way to estimate a threshold is to identify the highest concentration that causes no statistically significant effects compared to control organisms (NOEC or no-observable-effect concentration) and the lowest exposure causing significant effects (LOEC or lowest-observable-effect concentration). However, these terms have fallen out of favour because their estimation is easily biased by experimental design (Jager, 2012). The MATC is still used to designate a regulatory limit derived from acute or chronic toxicity tests. Thresholds for specific chemicals can also be derived by plotting **species sensitivity distributions**

(SSDs; Section 2.3.3.2) to calculate the concentration or dose below which a certain percentage (e.g., 95%) of test species would be unaffected by exposure.

In summary, toxicity tests, i.e., the measurement of the adverse effects of chemicals on living organisms and the amount of chemical needed to cause those effects, are fundamental tools for environmental, human health and agricultural protection. Section 2.2 describes the standard test methods needed to support the regulation of single compounds, industrial effluents and other complex mixtures (e.g., crude oil). Innovative research methods are also needed to understand why some chemicals are more toxic than others. For example, Chapter 5 describes the emerging science of high-throughput 'omics' approaches to identify the molecular mechanisms of the toxicity of different compounds.

2.2 Designing a Toxicity Test: What Is the Question?

Toxicity tests are laboratory or field experiments that measure the concentrations, doses or exposure times to a chemical that cause specific biological responses of test organisms. A first step in designing a test is to define its purpose (Box 2.6). For example, if the toxic effects of an insecticide on agricultural insect pests are already well understood, our objective may be to determine how toxicity varies among non-target species. What concentration limits would protect other species outside the spray zone? The known effects on target species will guide the selection of test responses, the range of exposure concentrations or doses, and the route of exposure (e.g., air versus water) for non-target species. Similarly, the biology and habitat requirements of non-target test species will guide the selection of environmental conditions in test chambers.

A common reason for toxicity testing is the need to assess the hazards of newly created industrial chemicals. In this case, we would apply preliminary or exploratory tests with a standard test species to identify the nature of responses caused by a chemical and the range of exposures between no response (e.g., 100% survival) and a complete response (e.g., 100% mortality). These range-finding tests would inform the design of subsequent definitive tests that define toxicity with more **precision**. Toxicity tests may also be applied to routinely monitor industrial effluents to verify that waste treatment has rendered them non-toxic.

Box 2.6 Design considerations for toxicity tests

- Purpose
  - Identify chemical effects
  - **Range-finding** vs **definitive** toxicity tests
  - Assess how toxicity varies with environmental conditions (e.g., temperature)
  - Routine **monitoring** of effluents
- Test substances: pure chemicals; complex mixtures
- Test species, life stage
- Responses measured: lethality, growth, reproduction, physiological or biochemical change
- Test media
  - Water, air, soil, sediment, diet
  - Terrestrial vs aquatic
  - Saline (marine) vs freshwater
- Exposures
  - Exposure routes: media or direct dosing
  - Exposure gradient
  - Time: acute vs chronic
  - Continuous vs episodic (pulsed) exposures
  - Controls: negative and positive
- Environmental conditions
  - Lighting (intensity, wavelengths, photoperiod)
  - Temperature
  - Disturbance
  - Exposure chamber (materials, size)

Knowledge of the MoA of a compound is an important factor in test design to ensure that relevant effects are measured, and with a sufficient degree of sensitivity. For example, pentachlorophenol (PCP), a wood preservative, uncouples oxidative phosphorylation in mitochondria. Uncoupling blocks the conversion of energy stored in carbohydrates to energy in **adenosine triphosphate (ATP)** that is needed to drive biochemical reactions and sustain cell functions. Knowing the MoA of PCP provides ideas for novel tests on energetics and growth that bridge the gap between standard laboratory tests described in this chapter and ecosystem-level effects considered in Chapter 4. Such tests could range from *in vitro* assays of mitochondrial respiration (energetics of cells) to whole-organism tests of aerobic capacity (Section 2.3.5) or to measurements of population change. For example, a relevant population metric could be 'size-at-age', i.e., a measure of growth rates calculated by statistical regressions comparing the weight of organisms to their age.

Once the purpose of a toxicity test has been defined, there are many decisions to make about the details of test design (Box 2.6). These include the species and life stages to be exposed to the test substance, the responses to be measured, and the intensity, duration and route of chemical exposure. The conditions of the test must also be coherent with the question asked. For example, it is logical that tests of a pesticide used in a tropical climate should involve tropical species tested under tropical conditions. As a result, the measured toxicity may not be the same as that observed in tests with temperate species under temperate conditions. Once test conditions have been established, the only test characteristic that should vary is the range of exposure concentrations or doses. Nevertheless, as we shall see in Section 2.2.4, the toxicity of a chemical is not fixed or absolute (Box 2.1). The organism and life stage tested, the responses measured, the test conditions and experimental error all affect our estimates of toxicity.

The following sections describe the principles and experimental conditions that we should consider when designing a toxicity test, with an emphasis on aquatic tests as an example. Many of the recommendations are based on a comprehensive review by Sprague (1969) that defines the methods summarized here, the logic for selecting test conditions, and methods for analysing and presenting toxicity results. Although exciting new methods in toxicology are emerging (e.g., 'Omics', Chapter 5), Sprague's recommendations for valid tests remain fundamental to the acquisition of reliable data.

2.2.1 Test Organisms

The choice of test species, life stages and measured responses in a toxicity test should reflect the nature of the test compound and how it is released to the environment. For example, the organisms most at risk of toxicity from liquid industrial effluents would be aquatic species typical of receiving water ecosystems. Each species has unique adaptations that determine its environmental requirements (e.g., salinity, temperature, oxygen concentrations, light regimes, diet) and its sensitivity to different compounds. A comprehensive programme to assess chemical hazards would include tests with a variety of species to describe the range in toxicity values needed for SSDs (Section 2.3.3.2) used in risk assessments. Box 2.7 lists more than 40 freshwater, marine and terrestrial organisms that are used routinely in toxicity tests.

Box 2.7 Examples of species used in toxicity tests

Aqueous exposures (water only)

- Microorganisms (*Photobacterium phosphoreum* – Microtox™)
- Algae (*Pseudokirchneriella subcapitata*, formerly *Selenastrum capricornutum*)
- Aquatic plants (*Lemna minor*)
- Invertebrates
  - Freshwater (crustaceans: *Daphnia magna*; *Ceriodaphnia dubia*; amphipods: *Hyalella azteca*; rotifers: *Brachionus calyciflorus*)
  - Marine (clams: *Mya arenaria*)
- Fish
  - Freshwater (zebrafish *Danio rerio*; Japanese medaka *Oryzias latipes*; rainbow trout *Oncorhynchus mykiss*; fathead minnow *Pimephales promelas*; guppy *Poecelia reticulata*; goldfish *Carassius auratus*)
  - Marine (Atlantic silversides *Menidia menidia*; Mummichog *Fundulus heteroclitus*; striped bass *Morone saxatilis*)
- Amphibians (common frog *Rana temporaria*; bullfrog *Rana catesbeiana*; woodfrog *Rana silvatica*; Great Plains toad *Bufo cognatus*; African clawed frog *Xenopus laevis*; western clawed frog *Silurana tropicalis*)

Aquatic sediments (water + sediments)

- Invertebrates
  - Freshwater (isopods: *Asellus communis*; infaunal mayflies: *Hexagenia limbata*; amphipods: *Hyalella azteca*, *Rhepoxynius abronius*, *Leptocheirus plumulosus*)
  - Marine (bivalves: *Yoldia limatula*, *Mytilus* sp.; infaunal polychaetes: *Neanthes* sp., *Nereis* sp.; crustaceans: *Mysidopsis* sp., *Callinectes sapidus*, *Ampelisca abdita*)

Air, soil or dietary exposures

- Plants (alfalfa *Medico sativa*; red clover *Trifolium pratense*; lettuce *Latuca sativa*)
- Invertebrates (house fly *Musca domestica*; fruit fly *Drosophila melanogaster*; earthworms *Eisenia fetida*; collembola *Folsomia candida*; insects *Pterostichus oblongpunctatus*)

- Reptiles (western fence lizards *Sceloporus occidentalis*; brown tree snake *Boiga irregularis*; red-eared slider (turtle) *Trachemys scripta elegans*)
- Birds (chicken *Gallus domesticus*; quail *Coturnix coturnix*; domestic ducks *Anas platyrhynchos domesticus*)
- Mammals (rats *Rattus norvegicus*; mice *Mus musculis*)

Although it provides a sense of the diversity of species commonly used, the list represents only a small fraction of those reported by laboratories around the world. For each species, it is critically important to meet their requirements for optimal environmental conditions, social interactions and protection from disturbance during culture and testing. Stress or poor health due to inappropriate culture or test conditions will bias test results and any risk assessments that rely on those results (Section 2.2.6). For novel or non-standard species, preliminary experiments are needed to define the optimal culture and test conditions that support survival, development and growth.

It is important to recognize that most species used in toxicity tests are genetically diverse and not homozygous clones. The result is a wide variation among individuals of the same species in their sensitivity to chemical exposures and corresponding uncertainties in measured toxicities. The precision of test results can be enhanced with identifiable strains of test organisms sustained by inbreeding (e.g., trout from specific hatcheries). However, the data generated might not be accurate predictions of the responses of more diverse populations of wild organisms, which is, after all, one reason why we conduct toxicity tests. The trade-off between precision and **accuracy** highlights the need to define why a test is being conducted. For example, precision and accuracy are emphasized in routine whole-effluent toxicity tests (WET tests) for quality control of industrial effluents but are less important in research experiments to identify ecologically relevant effects.

2.2.1.1 Laboratory Cultures of Test Organisms

Test organisms need to adjust to laboratory and test conditions to avoid the confounding effects of physiological stress caused by confinement and interactions with humans (Conte, 2004). For example, when organisms first arrive in a laboratory, they must recover from

Box 2.8 Developing tolerance to lab conditions

Acclimatization – short-term (minutes to hours) physiological responses to environmental change within the normal limits of tolerance. For example, mammals transferred to a cold environment respond with dermal vasoconstriction to limit blood flow to the skin and heat loss.

Acclimation – longer-term (days to weeks) physiological responses to environmental change. Fish transferred from warm to cold water synthesize fatty acids with a higher proportion of double bonds (polyunsaturated fatty acids) to maintain membrane fluidity and function.

Adaptation – multi-generation genetic responses to environmental change. Selection pressure favours traits that increase rates of survival, growth and reproduction under the new conditions.

transportation stress and undergo a short-term physiological **acclimatization** to the new conditions (e.g., temperature, lighting) (Box 2.8). Before testing, they may need more time to **acclimate** to these new conditions, so that longer-term adjustments do not occur during toxicity tests. For example, if a test temperature will differ from the culture temperature of a fish species (**poikilotherms**), the culture temperature is commonly adjusted by 1 °C/d, with another 5–10 days of holding to allow acclimation to the test temperature.

Genetic adaptations such as unusual **tolerance** to chemical exposure are unlikely to be a factor in laboratory toxicity tests, unless test organisms were collected from a contaminated ecosystem. An increased tolerance to **polycyclic aromatic compounds (PACs)** has been observed in fish populations exposed to industrially contaminated sediments (Di Giulio and Clark, 2015), although the tolerance was lost in subsequent generations. Mosquitofish (*Gambusia affinis*) captured from agricultural drainage systems in the southern United States survived laboratory

exposures to high concentrations of toxaphene and endrin, two chlorinated pesticides applied during the previous 4 years. Compared with unexposed populations, these fish exhibited a very high cross-tolerance to other chlorinated pesticides, such as strobane, heptachlor, chlordane, aldrin, dieldrin and lindane, but not to non-chlorinated OP pesticides. The level of pesticide tolerance was so great that predatory birds were killed by consuming fish that had accumulated high concentrations (Andreasen, 1985). As might be expected, these adaptations typify fish species that reproduce repeatedly within one season, rapidly selecting for traits of individuals that survive frequent or prolonged chemical exposures. The same mechanisms explain the development of pesticide resistance in insects targeted by pesticide application. These tolerant populations would clearly be unsuitable as standard test organisms!

2.2.1.2 Life Stages Tested and Responses Measured

The life stage of a test species is a critical element in our test designs. Measurements of acute lethality are usually made with immature or juvenile organisms, largely for practical reasons. Juveniles are smaller than sexually mature adults so that testing equipment and solution volumes are smaller, and toxicity is unaffected by sex and reproductive condition. For chronic sublethal toxicity, tests that encompass the full life cycle of a species demonstrate which life stages and effects are most sensitive. Although highly informative, full life-cycle tests for most species are uncommon because they are very long and costly. Thus, most sublethal tests focus on shorter experiments with a specific life stage or developmental process. For example, effects on growth are best assessed by measuring changes in the length, weight and condition of rapidly growing juveniles. Reproductive effects are measured when organisms are maturing sexually or actively reproducing; responses could include concentrations of reproductive **hormones**, the appearance of secondary sexual characteristics, and the production and survival of offspring.

Embryos are the most commonly used life stages for assessing sublethal toxicity, largely because those that develop outside the body of the mother (birds, fish) have practical advantages as test organisms compared with older life stages. They are relatively small, immobile (more easily observed), and they derive their nutrients from egg yolk, so that feeding is unnecessary and waste food and excreta do not bias test results (Section 2.2.6). **Teratogenesis**, the generation of structural and functional deformities (**terata**) in developing embryos, can be studied with embryos *in vivo* or embryonic tissues *in vitro*. Teratogenesis is a very sensitive indicator of toxicity because the expression of many genes vital to normal development must be turned on or off in a precise sequence and at precise times. Other useful responses include growth, hatch and development rates. Selecting the duration of chemical exposures when designing an embryo toxicity test can be difficult if the test compound accelerates or delays development. In this case, the test duration could be defined as the time to reach a specific developmental stage, such as 50% hatch of control embryos. Toxicity would be expressed as changes in the per cent hatch of treated embryos within that fixed time. Alternatively, the test could be prolonged until all treated embryos have hatched, with the index of toxicity being the time to 50% hatch at each level of exposure.

2.2.2 Test Media and Routes of Exposure

Terrestrial organisms (and air-breathing aquatic species) can be exposed to chemicals experimentally via the air, diet or drinking water, by dermal application, oral intubation, skin patches or injection, or by maternal transfer during embryo development or lactation. Similar routes are available for aquatic species, except that chemicals are added to water rather than air. Thus, the test medium must match the question being asked: is the concern about chemical contamination of air, water, soil or diet (food web)? For air pollution, most air-breathing species have similar requirements for air quality, although some species that live at higher altitudes may tolerate low oxygen concentrations (**hypoxia**) without physiological stress. For both terrestrial and aquatic species, hypoxia without acclimation to low oxygen concentrations would increase their ventilation and heart rates, thereby increasing the respiratory **uptake** and toxicity of non-polar, low-molecular-weight test chemicals (Section 3.2.9). To avoid confusing toxicity with responses to hypoxia, test solutions containing high concentrations of **biodegradable** organic compounds (e.g., industrial effluents) are typically aerated during testing and oxygen concentrations measured frequently. Alternatively, test species tolerant of hypoxic conditions should be selected.

Aquatic toxicity tests can also be complicated by water supplies that vary in their ionic content, from almost distilled water (e.g., glacier meltwater) to saline ocean water (32 parts per thousand of salts). Inorganic ions in

---

Box 2.9 Terms to describe water solubility

- **Hydrophobic** ('water fearing') – compounds that are relatively insoluble in water and tend to move out of the aqueous phase
- **Hydrophilic** ('water loving') – compounds that are quite soluble in water
- **Lipophobic** ('fat fearing') – compounds that are relatively insoluble in lipids
- **Lipophilic** ('fat loving') – compounds that are soluble in oil and accumulate in fatty tissues and lipid membranes

Hydrophobic and lipophilic are not synonyms. While some substances such as **PCBs** may be both, others are hydrophobic but not lipophilic (e.g., elemental silver). Polar compounds such as phenols are hydrophilic and lipophobic.

---

water interact with ionic forms of elements (e.g., Cu^{2+}; SeO_3^{2-}) and of some organic compounds (e.g., phenolics). The extent of interaction is sensitive to pH (acidity or **alkalinity**), which controls the proportion of metals or organic compounds that are 'free' in solution (i.e., not bound to other ions), which, in turn, determines exposure and toxicity (Section 3.2.7). The **bioavailability** and toxicity of many metal ions and organic compounds with a low water solubility (hydrophobic; Box 2.9) are also decreased by solute **partitioning** to particulate and dissolved organic matter. Biofilms (microbes, algae) that form on hard surfaces in test chambers have a similar effect, emphasizing the need to clean test chambers regularly. Tests of natural soils are even more affected by the high concentrations and diversity of soil chemicals.

A related question is solution renewal. For acute toxicity tests, exposures are often **static**, i.e., fish are exposed to a single batch of test solution in a large tank (e.g., 40–80 l) for 4 days. However, large test chambers are often impractical for tests with several replicates of six or more exposure concentrations. More importantly, static tests increase the risk that concentrations of hydrophobic or unstable test chemicals will decline as they are bioaccumulated, adsorbed to surfaces, degraded or evaporated. The consequence is a non-**steady-state** and often unknown exposure regime, which compromises estimates of toxicity. One alternative is a **semi-static** test, wherein test solutions are replaced periodically with fresh solutions (e.g., 24-h static exposures with daily renewals), reducing requirements for large tanks but at the cost of increased maintenance. The processes that reduce chemical concentrations also occur in semi-static tests, but to a lesser extent over 24 h than during 96 h. Constant exposures can be achieved by the continuous addition of test chemicals to flowing fresh water. A variety of continuous-flow exposure systems have been developed that work well for relatively soluble chemicals (e.g., phenols; some metal salts), but when test chemicals are not very soluble, alternative methods are needed (Section 2.2.8.1).

For some situations such as pesticide applications or oil spills, our objective may be to assess the impacts of periodic or pulse exposures. These are brief exposures to fixed or declining concentrations, followed by an observation period after exposures have ceased. Such tests would typically include concentration gradients with single or multiple pulses of exposure and post-treatment observation periods sufficiently long to allow the expression of delayed effects. For example, fish that survived a 24-h exposure to crude oil without visible harm demonstrated a reduced aerobic capacity and ability to swim when tested 25 days later (Mager *et al.*, 2014). These latent defects have serious implications for feeding, growth and escaping predators, but would not be evident in standard toxicity tests.

All of these factors generate complexities in interpreting the relationships among observed responses of test organisms and 'exposure', i.e., the amount of chemical added to a test system, the form the chemical is in, and the amount actually accumulated by test organisms (Chapter 3). Therefore, the measured toxicity may be unique to each test and not easily comparable among laboratories or transferable to new situations. One solution is to measure the concentrations of test chemicals in the tissues of organisms at the end of each test, providing a measure of dose that links cause and effect, independent of factors affecting bioavailability.

Dietary exposures provide an alternative exposure route with known administered doses, but many factors can still affect measurements of toxicity. To avoid test bias, diets must meet the nutritional requirements of the species and life stages tested, and correspond in size, appearance and palatability to food a species would normally consume in captivity. Proteins, lipids, carbohydrates, fibre and micronutrients may interact with test chemicals to limit their bioavailability and toxicity. Digestive enzymes and acids and the gut **microbiome** can also change the form and concentrations of some toxicants in food, so that the forms and doses accumulated are not the same as the forms and doses administered. As with all toxicity tests, the interpretation and application of toxicity data from dietary exposures depend heavily on a detailed description of exposure methods and characterization of 'dose'.

2.2.3 Exposure Gradients

Toxicity tests typically expose groups of test organisms to gradients of concentrations or doses of the test substance. The objective is to estimate toxicity with the lowest possible variance (greatest precision) and the highest accuracy. Computed toxicities are most precise when gradients range from exposures causing no effects to those causing maximal effects. Ideally, the gradient should include several intermediate exposure levels that cause progressively greater responses and describe a monotonic exposure–response relationship. Exposures are typically spaced at equal intervals on a log scale. Biological systems generally respond to stimuli as a power function, i.e., equal increments of response (e.g., 20% increases in mortality) are caused by proportional increases in exposure (e.g., doubling of concentration). For a previously untested chemical, the gradient of a preliminary or range-finding test would be broad, with exposures increasing stepwise by 10-fold to identify the range bracketing 0 and 100% response (Figure 2.3). The variance for LC_{50}s derived from such tests would be relatively high, reflecting the wide spacing of exposures. Repeating the test with a narrower range of test concentrations (Figure 2.4) generates a gradient of responses, and the confidence limits for the calculated LC_{50}s will be narrower, indicating a more precise estimate of toxicity. Regulatory guidelines and LC_{50}s measured for environmental management of chemical emissions are usually reported in gravimetric units (e.g., µg/l). For research comparing the toxicity of structurally different compounds, molar concentrations are reported (e.g., µmoles/l or µM) to avoid bias created by differences among compounds in molecular weights (e.g., structure–activity models, Section 2.3.3.3).

2.2.4 Exposure Time

Intuitively, time is an essential determinant of toxicity; the longer an organism is exposed to a chemical, the greater will be the observed response. The 'median effect time' (ET_{50}) is the time in which 50% of test organisms respond to a specific chemical exposure (Figure 2.5). To

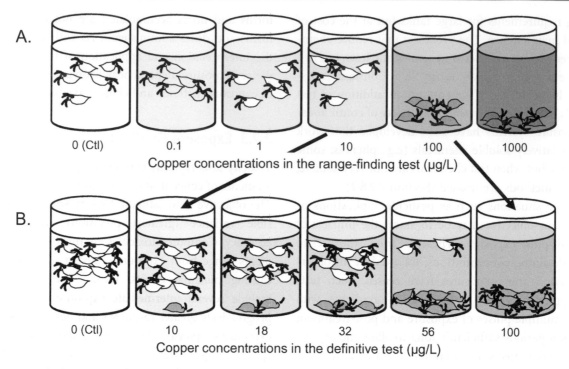

Figure 2.3 Range-finding and definitive toxicity tests to measure survival rates of water fleas (*Daphnia magna*) across a gradient of copper concentrations in water. A: In the range-finding test ($N = 5$/tank), all Daphnia survive at concentrations ≤ 10 µg/l and all die at concentrations ≥ 100 µg/l. B: In the more definitive test ($N = 10$/tank), there is a graded mortality across a narrower range of concentrations, from 10 to 100 µg/l. Ctl, control.

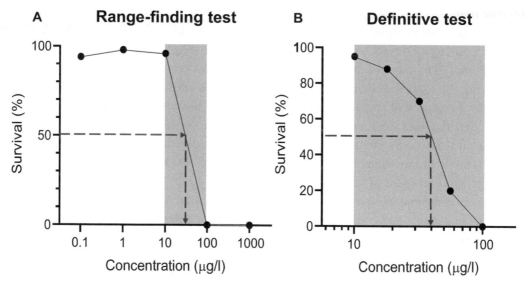

Figure 2.4 Exposure–response relationships showing the survival rates of water fleas exposed to copper. The LC_{50} estimated from the range-finding test is approximately 32 µg/l (logarithmic midpoint of the 10–100 µg/l range). The LC_{50} estimated from the more definitive test with a narrower range of concentrations is 40 µg/l.

measure ET_{50}, time is treated in the same way as concentration or dose, i.e., observations are made at increasing exposure intervals on a log scale (e.g., 1, 2, 4, 8, 16, 32 h). If tests continue beyond one day, observation intervals after the first 24 h can be fixed at increments of 24 h,

corresponding to the daily observations of responses and sampling of exposure media for chemical analyses. At high exposure concentrations, response times are very short (minutes to hours), and as concentrations decrease, response times increase.

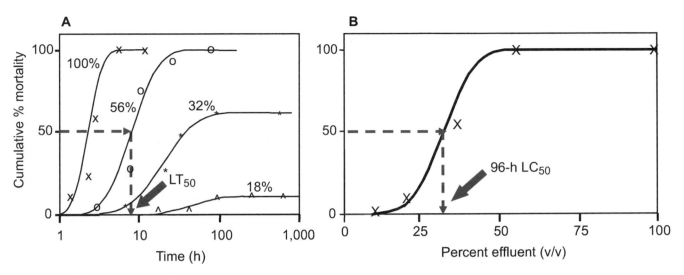

Figure 2.5 A: The effect of exposure time on the mortality of water fleas exposed to increasing dilutions (% v/v) of an industrial effluent. For the 56% (v/v) dilution, the median effective time (LT$_{50}$) is 8.5 h. B: The per cent mortality of water fleas after 96 h of exposure to the effluent dilutions. The 96-h LC$_{50}$ is 30% (v/v) effluent.

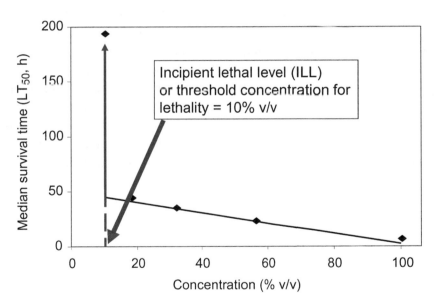

Figure 2.6 The derivation of the threshold lethal concentration of an industrial effluent from median lethal times (LT$_{50}$s) estimated for each dilution.

Test durations depend on the species, the life stage tested and the response measured, which determine whether the test would be considered acute or chronic. The exposure below which no responses are observed, no matter how long the exposure period, is a useful definition of the threshold concentration or dose causing a specific effect. The measurement of thresholds requires that each new exposure time be proportionally longer than the last. For example, in a 96-h lethality test, most mortality may occur within 24–48 h. However, if

one or more organisms die between 48 and 72 h, the test should be extended by 72 h (double the previous exposure time) to ensure that no more mortalities occur; this will lengthen the test from 96 to 144 h. The extended time indicates whether more organisms will die over a longer exposure time and whether the threshold exposure has been defined. The threshold times can be estimated graphically by plotting LT$_{50}$s or ET$_{50}$s (analogous to LC$_{50}$s or EC$_{50}$s) against exposure concentrations (Figure 2.6). The time to 50% response increases as concentrations decrease, until a threshold concentration is reached below which the organisms survive indefinitely. The breakpoint evident in Figure 2.6 will depend to a certain extent on the intervals between exposure concentrations. A similar but more sophisticated approach to defining thresholds is presented in Section 2.3.3.1.

2.2.5 Control Treatments

Quality assurance/quality control (QA/QC) procedures are essential components of any experiment. All toxicity tests must include a negative **control**, i.e., a group of test organisms not exposed to the chemical but held under identical conditions. In tests of acute lethality, results are often considered valid only if the mortality of control organisms is 10% or less during the test (EC, 2007). Control treatments ensure that effects are truly caused by the chemical exposure, not by some unknown test condition (known as a **false positive**). If test chemicals are first dissolved in a solvent before adding to water, an additional 'solvent control' treatment provides assurance that organisms have responded to the chemical, not to the solvent (Section 2.2.8.1). A further check on false positives is to monitor test conditions (e.g., oxygen concentrations; Section 2.2.2) to demonstrate that test conditions are optimal for the test species.

The alternative bias is a **false negative**, i.e., a substance is declared non-toxic under the conditions of the test when in fact it was added to test media at toxic concentrations. This bias typically occurs when compounds with low water solubility (e.g., PCBs) decrease in concentration during the test. False negatives can also occur when a sublethal response such as enzyme inhibition or gene expression is technically difficult to measure. They can be detected by including a **positive control**, i.e., organisms exposed to a well-studied compound at a concentration known to cause the desired effect. If there is no response to the positive control, the methods for measuring responses are at fault.

A critically important procedure for any experiment is to assign test organisms to treatments in a random order and to arrange treatments in a random sequence within an array of exposure chambers. Randomization avoids systematic observer or expectation bias, particularly when the person who measures responses is not informed about treatment distributions.

2.2.6 Other Test Conditions That Affect Measured Toxicity

The results of toxicity tests are clearly dependent on our experimental designs, but they also reflect the influence of many test characteristics that are often overlooked. In experimental science, the **observer effect** states that a phenomenon cannot be measured or examined without perturbing or altering the state of the phenomenon being examined. For toxicity tests, a substance, an organism or a process cannot be subjected to experimentation without intentionally or unintentionally altering the substance, organism or process in ways that affect the measured toxicity. There are two primary ways that toxicity test results can be biased by test methods: changes in the concentration or form of the test substance, and changes in the sensitivity of the test organism (Table 2.1).

In aquatic toxicity tests, test chemicals are assumed to remain at constant concentrations and in a dissolved state, freely available for uptake by aquatic species. However, depending on their properties, they can be lost from solution or change their chemical form. Chemical losses include the evaporation of volatile compounds, the partitioning of hydrophobic compounds to the surfaces of test chambers and to organic matter (e.g., **biofilms**, food or faeces of test organisms), and microbial degradation. The bioaccumulation of chemicals by test organisms also reduces chemical concentrations in water if their biomass loading (ratio of total biomass to solution volume) is excessive. As outlined in Section 2.2.2, the form of a chemical and its bioavailability may also vary with the chemical characteristics of the dilution water. When test compounds interact with dissolved organic matter (e.g., **humic substances**) or inorganic ions (e.g., carbonates) their bioavailability and toxicity will decrease, even though they appear to be dissolved in water. If the dissolved concentrations of test chemicals decrease from **nominal** or added concentrations during the test, the test chemical will appear less toxic than it really is, often to an unknown degree. For example, in static solutions, the concentrations of hydrophobic compounds that partition from water to any surface available can decrease by more than 90% within 24 hours. For this reason, it is essential to estimate toxicity from *measured* concentrations of chemicals in test solutions, not nominal concentrations (Section 2.2.7). These same issues arise when chemicals are added to the diets of test organisms. Interactions of test chemicals with components of the diet, such as lipids

Table 2.1 Experimental conditions that may affect toxicity test results by changing the amount and form of a chemical in test solutions or the organism's sensitivity to chemical exposure.

Condition	Tests affected	Effect on measured toxicity	Cause	Mechanism
Bioavailability and apparent toxicity of test compounds				
Biomass loading rates	Aquatic	Under-estimated	Excess biomass (size or numbers of test organisms)	**Bioaccumulation** reduces the concentrations of test chemicals in test solutions.
Static and semi-static exposures	Aquatic	Under-estimated	Loss of test chemicals of low water solubility from solution	Without continuous solution renewal, concentrations of test chemicals decrease owing to volatilization, absorption to surfaces, and/or biodegradation.
Lighting	All types	Under- or over-estimated	Ultraviolet wavelengths	Photo-oxidation of aromatic compounds in air or water reduces exposure concentrations. In contrast, photo-oxidation of aromatic compounds accumulated by transparent invertebrates and fish embryos releases reactive oxygen species that cause tissue **necrosis** and acute lethality.
Test chamber materials	All types	Under- or over-estimated; may change the nature of effects	Plastics; metals (copper, zinc, cadmium)	Plastics absorb hydrophobic test chemicals and may release toxic plasticizers. Copper, zinc and cadmium can dissolve or desorb into test solutions and are highly toxic.
Feeding	All types	Under-estimated	Accumulation of uneaten food or excreta	Organic wastes sequester toxicants and reduce their bioavailability.
Sensitivity of the test organism				
Disturbance	All types	Over-estimated	Movement in the lab; bright colours; noise; vibration	Organisms stressed by disturbance may be hypo- or hyperactive, changing chemical accumulation and apparent toxicity.
Lighting	All types	Over-estimated	Intensity; wavelengths; photoperiod	Animals: Bright lighting increases sensitivity to disturbance; lack of shade and inappropriate photoperiods cause stress. Plants: Inappropriate wavelengths and low intensity inhibit growth and development.
Temperature	All types	Under- or over-estimated	Differs from culture temperatures or not optimal for the test species	Changes in rates of metabolism, bioaccumulation and toxicity; physiological stress at extremes of a species' thermal tolerance.
Feeding	Aquatic	Over-estimated	Accumulation of uneaten food or excreta	Biodegradation of organic wastes consumes oxygen needed by test organisms and releases toxic ammonia.
Diet quality	All types	Under- or over-estimated	Nutrient deficiencies; impaired metabolism	Lack of essential vitamins; inappropriate amounts of carbohydrates or lipids cause liver damage.

and fibre, can reduce their bioavailability and the dose taken up from the gut.

The sensitivity of a test organism is affected by the **stress** of being handled and confined in test chambers. Confinement increases crowding, limits movement, prevents escape behaviour, and increases competition or dominance by other organisms. A lack of cover and exposure to intense lighting also induce stress in any captive animal, as can sudden noise, chronic or intermittent vibration, or investigators wearing white lab coats (Morgan and Tromborg, 2007). Stress may be evident in startle responses and altered rates of respiration, metabolism, activity, chemical uptake and response to chemical exposures. For example, when molluscs are disturbed, they cease filtering water and accumulating waterborne chemicals (Charifi *et al.*, 2018). Uneven patterns of lighting, movement and noise in a lab reinforce the importance of randomizing the order of treatments to ensure that gradients of chemical exposure are not matched by gradients of disturbance.

Standard test conditions will not eliminate all bias caused by these confounding factors. However, they will improve the consistency and comparability of toxicity estimates among test substances and among laboratories testing the same compounds. Government agencies that regulate environmental contamination such as the Organisation for Economic Co-operation and Development (OECD), Environment and Climate Change Canada (ECCC), and the United States Environmental Protection Agency (US EPA) publish standard toxicity test methods for an array of species to promote consistency in approaches and results (OECD, 2019; EC, 2007; US EPA, 2002). Nevertheless, standardized test methods may not be sufficiently flexible for some research questions, and the data generated may not be universally applicable to site-specific ERAs (Section 2.2.8.3).

Regardless of whether a test follows a standard or a unique method, it is essential that toxicologists report the details of experimental designs so that test results are interpreted correctly by those involved in regulating chemical emissions.

2.2.7 Characterizing Test Conditions and Chemical Exposures

The ideal measure of chemical exposures causing toxicity is the tissue concentration at the site of action (e.g., pesticides in neural tissues). Tissue concentrations integrate variations in exposure and uptake of chemicals, particularly for persistent compounds, and represent the dose that causes observed effects. Unfortunately, if test chemicals are metabolized or excreted rapidly, tissue concentrations will not be reliable indicators of chemical interactions with cellular receptors. Tissue analyses are also more complicated and expensive than analyses of

exposure media, especially in small organisms, and less relevant to regulating industrial emissions.

The most common alternative is the chemical analysis of test solutions. To characterize chemical exposures, we should describe the gradient of test concentrations, the extent and rate of changes in concentrations from the start to the end of the test, and the variance about mean concentrations in each treatment. Control solutions should also be analysed to recognize any background contamination that might affect control responses. For non-steady-state exposures (typically declining concentrations), chemical analyses only at the start and end of an exposure do not represent the actual exposure regime. Such exposures are best characterized by a time series of sampling and analyses to describe temporal trends and to express exposure as the integral of time and concentration.

Exposures to hydrophobic compounds that decline in concentration during toxicity tests can also be estimated with **passive samplers** (Box 2.10) which require less cost, time, effort and analytical complexity than extracting and analysing larger volumes of test solutions containing target chemicals at very low concentrations. However, solution volumes must be sufficient to compensate for the amount of test chemical removed by the sampler; otherwise, it

---

Box 2.10 Passive samplers

Passive samplers are devices coated with a solid matrix such as a silicone polymer that accumulates hydrophobic organic chemicals from aqueous test solutions by **equilibrium partitioning**. When the sampler is immersed in a test solution, hydrophobic compounds diffuse from the water into the siloxane coating until the concentrations in each phase are at equilibrium. If waterborne concentrations decrease, the compound diffuses back into the water until a new equilibrium is reached. At the end of a toxicity test, the accumulated chemical is extracted from the sampler for analysis and its concentration in the test solution calculated from its water–siloxane **partition coefficient**. The calculated concentration represents the integral or 'average' of the variations in the exposure regime experienced by test organisms throughout the test (Redman *et al.*, 2018). Methods for measuring the concentrations of 'free' hydrophobic compounds in toxicity tests with sediments or sediment pore waters are reviewed by Jonker *et al.* (2020).

would bias test results by reducing the exposure of test organisms.

When test chemicals interact with the normal components of a test solution (e.g., metal **cation** interactions with dissolved organic matter in aqueous toxicity tests), the bioavailable fraction should also be measured by analytical methods that differentiate 'free' chemical from the fraction bound to a substrate. Free metal ions can be analysed by specific-ion electrodes or passive samplers developed for metals (Section 6.2.2) or estimated with mathematical models (Section 6.2.1). For unstable organic compounds that degrade due to photolysis or **hydrolysis**, the by-products may be more or less toxic than the parent compound, and specialized analytical methods are needed to discriminate among the different forms.

Characterizing test solutions may require research to create practical sampling and analytical protocols, including QA/QC measures such as analyses of blank, duplicate and certified standard reference samples. Standard curves and their linear range and method detection limits should also be defined and reported (Hodson *et al.*, 2019). Although chemical analyses can be expensive, the cost of under- or over-estimating toxicity can be far greater if the management of chemical discharges is inappropriate, or if toxicity tests must be repeated.

2.2.8 Complexities in Toxicity Testing

Our emphasis so far in describing toxicity test methods has been the need for standard test methods and characteristics to ensure precise (i.e., repeatable) estimates of chemical toxicity. However, we have also encountered characteristics of test substances and test conditions that increase the uncertainty of test results. This section describes variations in test methods to deal with these added complexities.

2.2.8.1 Toxicity Tests for Sparingly Soluble Compounds

Toxicity tests with low solubility compounds can be difficult if these compounds must be mixed in test media at concentrations near their solubility limits. Mixing can be achieved by dissolving test chemicals in suitable solvents and diluting the stock solution in the test medium. For example, metals can be dissolved in acids, organic compounds in water-miscible solvents (e.g., acetone; ethanol), and complex mixtures of hydrophobic compounds, such as crude oil, in oil dispersants. Unfortunately, carrier solvents introduce new compounds to toxicity tests, and their toxicity must be known in advance to ensure that solvent effects are not mistaken for the effects of the test chemical. Even if solvents are at non-toxic concentrations, there is a strong possibility of mixture interactions (Section 2.3.4). This approach must include appropriate solvent controls and a search for signs of toxicity in test organisms that cannot be attributed to the test compound alone.

Solvents used to prepare solutions for static or semi-static tests do not ensure the stability of these solutions. Once dispersed into water, concentrations of hydrophobic compounds may decline, creating the challenge of characterizing non-steady-state exposures (Section 2.2.7). Continuous-flow systems can circumvent this problem. Stock solutions of test compounds in solvent are mixed continuously with flowing water using high-precision pumps. Dilution gradients are created with multiple pumps, or by proportional dilution of the highest concentration.

Recent advances in **passive dosing** technologies provide an alternative to solvent delivery, based on the same principle as passive samplers (Section 2.2.7), but in reverse. Chemicals diffuse into test solutions by equilibrium partitioning from a dosing medium (usually a silicone polymer) that contains a high concentration of the test compound. The highest concentration of a compound in water is determined by its solubility limit. Below this limit, predictable and stable gradients of test concentrations can be established with a gradient of concentrations in the dosing medium (Kiparissis *et al.*, 2003; Butler *et al.*, 2016). Similar approaches are available for metals, where metal ion buffers or ion-exchange resins maintain free metal ions at constant concentrations throughout a toxicity test (Twiss *et al.*, 2001). Although passive dosing methods are somewhat time-consuming to set up, they are far more cost-effective and reliable than pumps and mechanical mixers, and the results are more realistic. Although used in research, they have yet to be adopted for regulatory testing.

2.2.8.2 Sediment and Soil Toxicity Tests

Sediment and soil toxicity tests were first applied in the 1970s to address concerns about contaminated harbours, terrestrial **brownfields**, and the removal and disposal of contaminated materials. Sediment toxicity tests have since been adopted universally for ERAs of the effects of contaminated sediments on **benthic** (bottom-dwelling) aquatic species. Aquatic sediments are the ultimate repository for many pollutants and represent a complicated medium for toxicity testing owing to a diversity of physical and chemical characteristics (e.g., grain size; total organic carbon; pH; oxygen content; water content). These same factors determine the species composition, structure and productivity of communities of benthic species. As sediment toxicity tests have evolved, it has become increasingly clear that the responses of benthic species reflect the physical characteristics of test sediments as well as their chemical contaminants. Thus, in selecting test species, their substrate preference, salinity tolerance, availability and ease of culturing are all critical factors. Species suitable for marine and freshwater sediment tests include invertebrates that live on or just above sediments and those that burrow into sediments (**infaunal** species) (Box 2.7). Infaunal crustaceans are among the phyla most sensitive to sediment contamination, as indicated by changes in growth, fecundity, reproduction and burrowing behaviour. The bioavailability of contaminants from the aqueous and particulate fractions of sediments also depends on the feeding strategies of the test species (Section 6.3.3).

The characteristics of surface and subsurface habitats also affect the bioavailability and measured toxicity of sediment contaminants. For example, in the subsurface, sediments may be **anoxic** due to decaying organic matter. As the chemical environment shifts from an oxidizing state at the surface to a low-oxygen reducing state deeper in the sediments, the form, solubility and bioavailability of chemicals are changed. Some species of **oligochaete** worms create burrows in sediment that they ventilate to introduce fresh water, changing the oxidation state of the burrow environment, diluting the chemicals that diffuse from adjacent sediments and reducing their toxicity. Burrows lined with organic matter also bind compounds diffusing from the sediments. Organisms that dwell on the surface of sediments rich in organic matter, but in contact with oxygen-rich water, can be protected from chemical exposure by the formation of **oxic** surface layers that act as barriers to chemical diffusion from anoxic sediments to the overlying water.

Most standard protocols require that sediment samples be mixed to ensure homogeneity before testing, which means that deeper, anoxic sediments are mixed with oxic surface layers, creating a condition of chemical disequilibrium. Interactions between surface and deeper sediments, and between test chemicals spiked to sediments and

organic matter in oxidized or reduced states, are quite slow, as long as 40 days (Vandegehuchte *et al.*, 2013). In view of this complexity, it is often recommended that sufficient storage time be allowed for test chemicals to equilibrate with all the components of the test environment.

Alternatives to whole-sediment toxicity tests include tests of sediment leachates and **porewater** extracts obtained through squeezing or centrifugation of whole sediments or by the installation of 'peepers' to sample pore water without disturbing sediment structure. Peepers are cells containing a fluid that is separated from the sediments by a dialysis membrane, allowing dissolved chemicals to diffuse into the cell without contamination by particulates. When the concentrations of target chemicals in the cell are estimated to be in equilibrium with the concentrations in sediment pore waters, the peeper is removed and the cell contents analysed. These manipulations provide a measure of the bioavailability of sediment metals (Section 6.3.3) and non-ionic organic compounds, as estimated by models of equilibrium partitioning (Di Toro *et al.*, 1991). Although sediment mixing and pore-water extraction simplify test methods and promote standardization of test methods, they may be less relevant to ERAs if they disrupt the physical structure that influences the survival, growth and reproduction of test species.

Water overlying sediments can also be tested to measure the toxicity of chemicals diffusing from contaminated sediments. For example, tests with frog embryos held in mesh cups in water overlying sediments contaminated with an azo dye demonstrated its release to water at genotoxic and carcinogenic concentrations (Balakrishnan *et al.*, 2016). Another theme in sediment toxicity testing is that measures of ecosystem integrity should include indicators at the population and community level. Consequently, sediment toxicity tests are an integral part of the **sediment quality triad** (Chapman, 1990), a 'weight-of-evidence' approach used in ERAs for sediment-borne contaminants (Section 4.3.3.4). Simpson and Batley (2016) provide a comprehensive treatise on the practical aspects of sediment toxicity testing.

A review of soil toxicity test methods (Alves and Cardoso, 2016) touched on many of the same issues of test design underlying sediment and aquatic toxicity tests. Test species (Box 2.7) must be selected carefully to ensure relevance for site-specific risk assessments, ease of culture and recognition of traits that affect the extent of chemical exposures (e.g., dietary requirements). The optimal test conditions are selected for each test species, and measured responses include lethality, behaviour (soil invertebrates), growth and reproduction. Exposures are acute or chronic, and chronic tests with invertebrates include feeding, usually in the form of organic matter added to the soil. Test media are either artificial soils (mixtures of sand, clay and peat) or natural soils representative of a given region. Natural soils are usually frozen before testing to kill soil species that are not part of the test, and are analysed to describe characteristics such as pH, texture and moisture that may affect the toxicity of added chemicals. Similar considerations apply to tests of soils from contaminated sites, but with the added challenge of finding a matching uncontaminated soil as a control treatment. All soils are physically and chemically complex. As with aquatic sediments, the bioavailability of chemicals added to soil decreases with time due to interactions with soil water and organic and inorganic matter. For stable measures of toxicity, soils spiked with test chemicals may require weeks to months of equilibration before testing.

2.2.8.3 Standard vs 'Realistic' Toxicity Tests

Our emphasis in this chapter has been on learning the basic principles and methods of toxicity testing, illustrated by standard methods that define clear exposure–response relationships and consistent estimates of toxicity. Standard tests underpin chemical management programmes that classify and manage individual compounds and industrial emissions by their effects and toxicity (Section 12.3.1). They also provide the starting point for ERAs of chemical emissions and contaminated sites (Section 12.3.2). However, to understand and predict the ecological impacts of a given compound or mixture of compounds, we must account for changes in toxicity due to the environmental characteristics of the ecosystems we wish to protect. Toxicity is likely to vary widely in natural ecosystems where physical, chemical, biological and ecological interactions affect the form and concentration of chemicals, their bioavailability to endemic species and the sensitivity of those species to chemical exposures.

The potential for ecosystem-specific toxicity highlights a need for 'realistic' tests to measure toxicity under environmental conditions typical of different ecosystems. These could include standard protocols modified to assess the effect on toxicity of one condition at a time, or the interacting effects of several conditions (e.g., temperature and salinity; Section 2.3.2). The purpose would be to identify which factors cause the greatest changes in toxicity and the direction of change. Alternatively, tests could be conducted in experimental ecosystems where many conditions vary simultaneously, as might be expected in real ecosystems. The ultimate in realism is the experimental addition of chemicals to real ecosystems, e.g., in Canada's Experimental Lakes Area (ELA) in northwestern Ontario (Section 4.4.3). The ELA experiments have provided invaluable insights about ecological effects and chemical interactions that could not have been acquired in any other way. However, the drawbacks are those of cost and complexity of design and execution, and the challenge of finding suitable reference ecosystems. There are also obvious limitations to extrapolating results from one unique ecosystem to another that differs markedly in its physical, chemical and biological characteristics.

These issues are discussed in detail in Chapter 4, "Methods in Ecotoxicology", where we explore the challenges of relating cause and effect in ecosystems affected by chemical emissions and emerging tools that help us to understand chemical effects in the world beyond the laboratory.

2.2.8.4 Surrogate Species for Routine Testing

The routine testing of industrial effluents consumes large numbers of plants and animals, with an estimated annual global consumption of approximately 5,000,000 vertebrates, mostly fish species (Norberg-King et al., 2018). This number raises legitimate moral and ethical concerns about animal welfare, and many alternatives to fish have been proposed (Box 2.11). The selection of substitute species is based on a definition of 'sentient beings', i.e., species capable of feeling pain and emotions. Non-sentient

Box 2.11 Alternative species and test methods for routine whole-effluent toxicity tests

- Non-sentient life stages (e.g., fish embryo/larval tests)
- Non-sentient species (e.g., invertebrates, microbes)
- *In vitro* tests (e.g., fish cells in culture; microbial mutagenicity tests)
- Non-destructive biomarkers of sublethal effects developed from AOPs (Section 4.2)
- Chemical analyses to compare the concentrations of effluent components to known thresholds of toxicity
- Computer modelling of mixture toxicity based on chemical analyses and QSARs (Section 2.3.3.3)

organisms are considered to include microorganisms, plants and invertebrates, as well as vertebrate embryos prior to the onset of active feeding. However, the identification of non-sentient species depends on the criteria by which sentience is defined, a subject of much research and philosophical debate. A comparison of the lethal and sublethal acute toxicity values for 2,581 effluent samples from 20 oil and gas facilities demonstrated that marine and freshwater invertebrates were more sensitive than fish in more than 90% of cases (Hughes *et al.*, 2021). Although the study supports substituting invertebrate species for fish in routine monitoring, most effluents tested had a low toxicity. Conclusions about the relative sensitivity of invertebrates and fish might be very different if effluents contain compounds affecting physiological processes unique to vertebrates. Nevertheless, the adoption of surrogate tests for routine effluent testing is worthwhile, if only to reduce costs. However, it increases the importance of environmental monitoring as a quality-control measure on effluent regulations to ensure that ecosystems have been protected from the effects of chemicals that are not acutely toxic in short-term tests (Section 12.3.5).

2.3 Statistics for Toxicity Tests

Statistics play three principal roles in environmental toxicology. In regulatory testing, statistical methods are needed to determine whether toxicity exceeds a legal threshold, indicated by either the size of a response to a fixed exposure, or a measured LC_{50} or EC_{50} compared to a legal limit. Published LC_{50}s and EC_{50}s for an array of different test species also form the technical basis for establishing environmental quality **guidelines** and **standards** for specific compounds (Section 12.3.1). A second role is hypothesis testing, e.g., determining the difference between two populations with some predetermined level of confidence, particularly in research. The third role is predictive (e.g., estimating the outcome of chemical–biological interactions) and often involves the influence of one or more independent variables on one or more measures of effect (dependent variables). Some predictive models address the most important obstacle in chemical regulation, i.e., that the number of chemicals in commerce (e.g., >140,000 in the European Union; UNEP, 2013) far exceeds our capacity to test them all. Other ecological models may predict the toxicity of chemicals under the unique environmental conditions of specific ecosystems (Section 4.5).

2.3.1 Regression Analyses for Computing Toxicity

Regression analysis is a statistical method that estimates linear or curvilinear relationships between a treatment (independent variable, e.g., chemical concentrations) and a response (dependent variable, e.g., the proportions of test animals affected) (Section 2.3.1.1; Figure 2.7). Statistical methods for toxicity tests assess differences among a few very narrowly defined test populations, e.g., those exposed to different chemical concentrations (treatments) and differences between treatments and controls. They also assess the relationship between test populations and the larger population from which they were taken. Measurements of treatment/treatment and treatment/control differences require calculations that account for the variance associated with observed effects (e.g., mortality). Conventional **endpoints** (e.g., EC_{50}s) provide reproducible and quantitative relationships between chemical exposure and some measure of effects on test organisms. As previously discussed, calculating the midpoint of the dose–response curve generates the smallest statistical error.

Differences between or among treatments can be measured with varying degrees of precision, depending on sample sizes, i.e., the numbers of organisms and replicates used. Larger sample sizes decrease the variance associated with endpoint measurements and increase the precision with which treatment differences can be determined. Precision for population (treatment) differences is characterized by a target error rate, usually 5% and expressed as 95% confidence limits about the measured toxicity value. This means there is one chance in 20 of reaching the wrong conclusion based on the data obtained.

Even though such errors can be reduced by increased replication of each treatment, there are theoretical and practical limits to the number of replicates in any test. Most toxicity data are derived from controlled laboratory tests and involve single chemicals and very small populations of test organisms. Even standard tests attempt to simulate what would happen in a large population by observing as few as 10 to 30 organisms (e.g., 10×3 replicates) per exposure or treatment. In acute lethality tests, the number of organisms per exposure treatment is often 10, and each organism represents 10% of the exposed population. This number is sufficient to create relatively smooth regressions comparing per cent

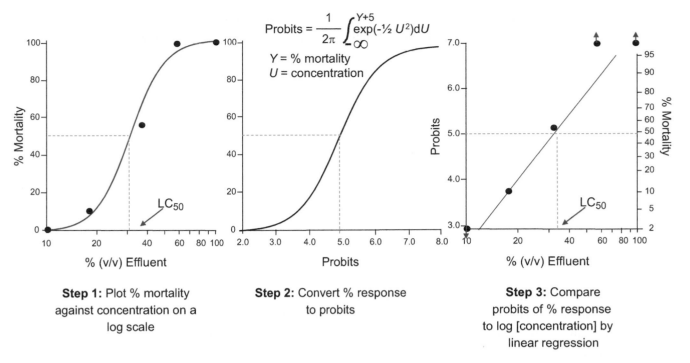

Figure 2.7 Data transformations for probit analysis of quantal mortality data to calculate LC$_{50}$s from a linear relationship. Adapted from Finney (1971) with permission from Cambridge University Press.

response to a gradient of chemical exposures, without the need for larger tests. The size of an experiment can also be minimized by testing only one replicate of each treatment, i.e., each treatment generates one estimate of per cent response for a given level of chemical exposure. Although this might appear statistically inadequate, statistical regressions represent running means, with each observed response equivalent to a replicate of the regression. In effect, a toxicity test with five exposure levels has been replicated five times. Adding multiple treatments at the same exposure level or running the entire test multiple times with test organisms drawn from a single population or batch is not replication. Instead, this would be considered test repetition, and the results would reflect the precision of the test method, i.e., the variance among repeated measures of the same endpoint. Toxicity tests would be replicated if run with different populations of the same species, under different conditions (e.g., alternative sources of test media), or by different laboratories.

2.3.1.1 Data Types and Transformations

Median effective exposures (e.g., LC$_{50}$s) are defined graphically as the midpoints of exposure–response curves relating some measure of response to the logarithm of exposure concentrations or dose. Responses are often characterized as **quantal** or **binomial**, i.e., existing in one of two states, such as alive or dead. For example, 3 of 10 fish responding to a chemical exposure represents a 30% response. These exposure–response relationships are usually non-linear **sigmoidal** or S-shaped curves because there is a wide range of sensitivity among test organisms (e.g., Figure 2.7, Step 1). A few highly sensitive organisms create the lower left-hand tail of the curve at low-level exposures while a few 'tough old survivors' tolerate high concentrations and create the upper right-hand tail. In a concentration gradient with 10 organisms per treatment, it is unlikely that one exposure level will cause an exact 50% response. Therefore, an LC$_{50}$ must be predicted or interpolated from a statistical regression relating response (dependent variable) to exposure (independent variable).

Because of cost, toxicity tests are usually limited to five or six exposure concentrations, plus suitable controls. However, LC$_{50}$s predicted from sigmoidal or curvilinear regressions of only five or six data points would include very large error limits. The solution is to linearize the sigmoidal curves by plotting response data on a probability or **probit** scale. There is also a sigmoidal relationship between per cent mortality and calculated probits (Figure 2.7, Step 2). Probits refer to the number of standard deviations of each observed response from the median or 50% response. Thus, it includes values above and below the mean % response, and 5.0 is added to each probit value to eliminate negative numbers. As a result, when probits of per cent mortality are compared with the log of concentration, the relationship becomes linear (Figure 2.7, Step 3), providing the maximum power of linear regression statistics to calculate the minimum error limits about predicted LC_{50}s. Step 3 also includes a per cent response scale. It corresponds to the probit scale and shows that the probit transformation stretches the sigmoidal distribution in opposite directions, centred on the 50% or median response.

Many sublethal responses to chemical exposure cannot be considered quantal or binomial. For example, growth expressed as organism weight or the activity of a specific enzyme corresponds to a **continuous variable**, i.e., data that can have any value within biological limits, and is not restricted to 'either–or'. The relationship between chemical exposures and a continuous response may also be curvilinear, but continuous data cannot be transformed into a probabilistic or probit function. In this case, EC_{50}s or ED_{50}s must be derived from curvilinear polynomial regressions. When the relationship between exposure concentrations and response data corresponds to a sigmoidal curve, logistic analyses can be applied to compute EC_{50}s (Turcotte *et al.*, 2011). Logistic analyses were first developed in ecology to describe exponential growth rates. Their application to toxicology allows the specification of a non-zero background response (i.e., elevated baselines representing background control responses). Many computer-based statistical packages include probit and logistic analyses of exposure–response data in spreadsheet formats.

Chronic or sublethal tests, particularly those with bird or fish embryos, may also require larger sample sizes, often 20–50 embryos per treatment. Larger test populations support the measurement of multiple responses during a single test, such as chemical analyses of tissues, molecular assays of 'omic' responses (Chapter 5), histopathology, or analysis of proteins and lipids as an index of growth or caloric content. By including more organisms per treatment, more questions can be asked about the nature of toxic effects and more value obtained for a given cost and effort. Groups of embryos sampled at intervals can also describe the rate of onset of responses, although subsampling may cause **selection bias** (e.g., sampling the slowest, most intoxicated animal) unless each subsample is exposed as a separate randomized treatment.

2.3.1.2 Control Data

When computing lethal and sublethal endpoints, we must consider control responses. Where control mortality is low, simple subtraction of control values from treatment values will suffice. However, species that spawn large numbers of offspring (e.g., sea urchin, *Arbacia*) often sustain high rates of embryo and larval mortality, even under control conditions, perhaps 30% over a 96-h period. In such cases, adjustment of a treatment value relative to a control is made with Abbott's formula (Abbott, 1925):

$$\text{Adjusted \% response}$$
$$= \frac{\text{Test \% response} - \text{Control \% response}}{100 - \text{Control \% response}} \times 100$$

$$(2.1)$$

For a 70% test response and a 30% control response, the adjusted test response becomes:

$$\frac{70 - 30}{100 - 30} \times 100 = 57\% \qquad (2.2)$$

Often, sublethal (chronic) data are not normally distributed (as determined by a chi-square or Shapiro–Wilk test) or homogeneous (e.g., Bartlett's test). In such cases, results may be analysed using a non-parametric procedure such as Steel's multiple rank comparison test (for equal size replicates) or Wilcoxon's rank sum test (replicates of unequal size). Normally distributed data can be analysed using Dunnett's test. In certain instances, non-normally distributed data may be transformed to a normal distribution using a mathematical device that has the effect of lessening differences between data points (e.g., arcsine transformations). One case is multiple regression analysis, which is applied when responses are compared to more than one independent variable.

2.3.2 Hypothesis Testing: Multiple Regression Analyses

Toxicity tests usually focus on a single toxic agent and its effect on a lethal or sublethal endpoint over a defined exposure period, but we are often concerned with multiple variables that contribute to multiple toxic effects. These observed responses may represent a suite of dependent variables that might be examined collectively or independently. In such cases we can use multiple regression analysis to examine and quantify the relationships among complex toxic factors and their biological effects.

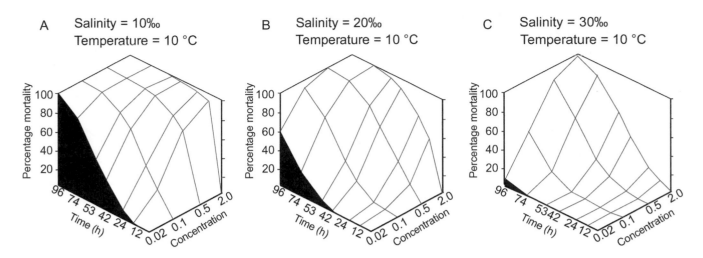

Figure 2.8 Response surfaces describing the effect of salinity (parts per thousand, ‰) on the toxicity of trimethyltin (mg/l) to hermit crab (*Uca pugilator*) larvae. Adapted from Wright and Roosenburg (1982) with permission from Springer-Verlag.

For example, if metal toxicity is examined at different concentrations, exposure times, temperatures and salinities, each test condition is regarded as an independent variable and a toxic response such as mortality is the dependent variable. In such an instance, the regression analysis is performed on arcsine transformations of % mortality, which are back-transformed for presentation in a regression equation, or as response surfaces. The statistical transformation creates a normal distribution of % mortality data as a requirement for the regression calculation.

Figure 2.8 shows response surfaces describing the mortality of hermit crab larvae (*Uca pugilator*) as a function of trimethyltin concentration and exposure time at different salinities (Wright and Roosenburg, 1982). These response surfaces describe clear relationships among a matrix of three independent variables (trimethyltin concentration; exposure time; salinity) and a single dependent variable, mortality. Toxicity increased with trimethyltin concentration and exposure time but decreased with increasing salinity. Although this represents a more complicated experimental design than a standard acute toxicity test, it is an order of magnitude simpler than the complex cause–effect relationships encountered in field experiments. As we move from the laboratory to the field, we are confronted with a huge range of biological effects (i.e., dependent variables) not included in laboratory tests (Section 4.2). Beyond the single-species/single-chemical exposure–response relationships, we must consider detrimental effects at higher levels of biological organization.

These include interactive effects at the population and multi-species (community) levels, where the impact on one group may result in significant changes to other components of the food web.

2.3.3 Predictive Toxicology: Single Compounds

The objective of models that predict chemical toxicity is to provide regulatory tools that afford the maximum protection for the largest number of species without the high costs of testing the universe of chemicals and organisms. The models reviewed below include those that relate chronic toxicity to less expensive measures of acute lethality (e.g., acute to chronic ratios, ACRs). Others describe the relationships between toxicity and the physical or chemical characteristics of previously tested chemicals to predict the toxicity of untested chemicals. For example, QSARs correlate the toxicity of organic compounds to structural properties such as molecular size or the presence and distribution of specific substituents that influence interactions with cellular receptors. Equivalent models for metals (quantitative ion character–activity relationships or QICARs) are discussed in Section 6.2.1. Other models relate toxicity to physical properties such as lipid solubility, expressed as K_{OW} (**octanol–water partition coefficients**; the ratio of chemical concentrations in n-octanol to concentrations in water at equilibrium). One such model, the **target lipid model** (TLM), is presented in Section 2.3.4.3 as one of several tools to predict mixture toxicity. Although models can rapidly assess chemical hazards, a fundamental principle should not be ignored: *the quality and accuracy of model predictions are limited by the quality and accuracy of the data used to build and run the model.* For many classes of compounds, another important limitation is our incomplete understanding of cause–effect relationships. For all models, the uncertainty in model predictions due to errors inherent in the input data should be recognized by estimating the error limits (e.g., 95% confidence limits) for predicted toxicity.

2.3.3.1 Acute to Chronic Ratios (ACRs)

Because of the relative ease and economy with which acute toxicity can be measured, acute data are often the most common, and sometimes the only, measures of toxicity for chemicals and complex mixtures. To maximize environmental protection, regulatory agencies are increasingly focused on more sensitive endpoints. Part of this information is derived from abbreviated chronic tests (shorter than a full life-cycle test) with early life stages of test species, and *in vitro* screening tests for specific responses such as mutagenicity (Section 7.5.1).

When chronic toxicity data are unavailable, extrapolations are often made from existing datasets for known chemicals and species. One example is the ACR, i.e., the ratio of an acute LC_{50} to a measure of chronic toxicity, such as the MATC,

$$ACR = Acute\,LC_{50}/MATC \qquad (2.3)$$

This ratio has been used for regulatory purposes to estimate chronic toxicity from LC_{50} values where gaps exist in empirical chronic data. The ACR is a somewhat crude means of estimating chronic toxicity because it assumes a constant ratio, despite observations that the MoA for chronic effects often differs considerably from that for acute lethality. Refinements for regulatory purposes include the use of data from the most sensitive test species. The ACRs usually represent the mean ratios for a variety of known compounds, and predicted chronic toxicities should include 95% confidence limits calculated from the variance of the mean ACR.

Other models have been developed to estimate threshold toxic concentrations to support risk assessments. A three-dimensional approach published by the US EPA extrapolates an NEC (Box 2.5) from acute toxicity data (Ellersieck *et al.*, 2003). The NEC is operationally defined as the $LC_{0.01}$ from a probit curve after an infinite exposure

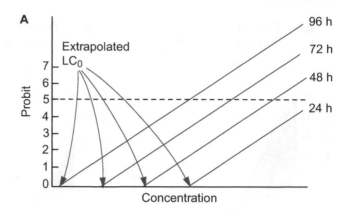

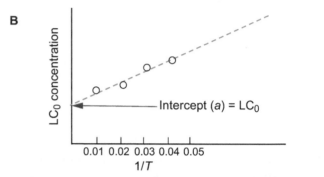

Figure 2.9 A: Derivation of a 'no-effect' or threshold concentration (LC_0) by regression of dose–response data at different times to extrapolate to an $LC_{0.01}$, considered to be equivalent to the LC_0. B: Plot of LC_0 vs 1/time (T) to estimate the LC_0 at an infinite exposure time, equivalent to a time-independent threshold. Reprinted with permission from Springer Nature from Figures 5 and 8 of Giesy and Graney (1989).

time. In other words, $LC_{0.01}$ is used as a surrogate for LC_0, a hypothetical value. The programme performs two separate regressions, initially determining the $LC_{0.01}$ at each time interval and then plotting these $LC_{0.01}$ values against reciprocal time to extrapolate to a hypothetical infinite exposure time (Figure 2.9). Although NEC values derived from acute toxicity data for several fish species are strongly correlated with corresponding MATCs from more time-consuming chronic tests, some caution is needed in accepting the predictions. Threshold concentrations predicted by the NEC model represent a 0.01% (equivalent to 0%) response rate, but the model makes this prediction by analysing data from acute lethality tests with only 10 organisms per treatment. Because each observed mortality represents a 10% response rate, the input data are 1,000-fold less precise than the model prediction. In practical terms, the estimation of thresholds from the NEC model is similar to assessing the threshold dilution required to render harmless a toxic effluent (Figure 2.6).

In that case, serial dilutions were tested empirically until chronic exposures no longer caused mortalities of test organisms, no matter how long the exposure.

2.3.3.2 Species Sensitivity Distributions (SSDs)

One goal of ERAs is to define chemical concentrations in environmental media that will be protective of most species in an ecosystem. The proportion of species protected is, to a certain extent, an arbitrary value that reflects the degree of conservatism applied by regulatory authorities in each nation (Section 12.3.1.1). Some agencies hope to protect all species by selecting a concentration that will protect the most sensitive species tested, while others estimate the concentration that will protect 95% or more of test species, without having to test every species. The range of sensitivity to a chemical can be illustrated by SSDs based on toxicity data for nine or more species, a sample size that ensures a minimum statistical power of the SSD regression. However, the number required for the regulatory application of SSDs varies among jurisdictions. The SSD approach assumes that the range of sensitivities for a small number of species is a reasonable approximation of the range for all related species (e.g., all marine fish).

To construct an SSD, species are ranked by their sensitivity to a given chemical, starting with the most sensitive species (i.e., with the lowest LC_{50} or EC_{50}). Ranks are expressed as proportions or percentiles of the species tested; for 10 species, the most sensitive would represent the 10th percentile. The **HC_5, or hazardous concentration** toxic to the 5th percentile species, is estimated by regression analyses relating percentile to log LC_{50} or EC_{50} (Figure 2.10). The SSD model can represent many taxa (Bejarano and Mearns, 2015) or a narrow taxonomic grouping (e.g., fish species; Lee *et al.*, 2015). Figure 2.10 illustrates that oil toxicity to fish embryos varies by almost three orders of magnitude among species tested, and that most fish species should be protected from chronic embryotoxicity at oil concentrations <0.6 µg/l (expressed as the summed or total concentrations of measured polynuclear aromatic hydrocarbons).

A predicted HC_5 for any chemical is not absolute but includes error limits derived from the regressions comparing toxicity endpoints to exposure concentrations (Belanger and Carr, 2019). The size of error limits will depend on the quality of the data used to develop the SSD, and these data should be screened for the quality of test methods before they are included in the analysis.

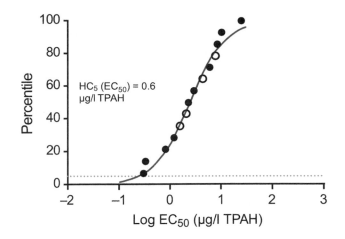

Figure 2.10 Species sensitivity distribution (SSD) for the sublethal toxicity of crude oil to freshwater (open symbols) and marine (filled symbols) species of fish. The EC_{50}s are reported as concentrations of total polynuclear aromatic hydrocarbons (TPAH). The HC_5 was estimated from the intersection of a 5th percentile line (dotted line) with a non-linear regression relating toxicity to oil concentrations. Adapted from Lee *et al.* (2015) with permission from the Royal Society of Canada.

Uncertainty in estimated HC_5s can be caused by grouping species that respond to a chemical in different ways. For example, the MoA for the toxicity of an herbicide to a plant species is likely to be quite different from the MoA for its toxicity to a vertebrate species. Restricting an SSD to a narrow range of taxa (e.g., class or phylum) will increase the reliability and confidence in the estimated HC_5. For ecological risk assessments, it must also be recognized that SSDs apply only to the response measured, and not to other effects such as avoidance or a reduction in population size.

2.3.3.3 Quantitative Structure–Activity Relationships (QSARs)

Models that predict the biological activity of organic chemicals based on their physicochemical properties are widely used by pharmaceutical companies to design compounds with known beneficial effects and minimal side effects. In ecotoxicology, the same approaches are used for ERAs and to prioritize the regulation of new compounds, including pharmaceuticals, personal care products, and agricultural and industrial chemicals. These QSARs assume that each substructure of a compound makes a specific contribution to the properties of the compound. Thus, similar compounds having the same functional groups in similar spatial arrays should have similar MoAs in target organisms and predictable toxicities. Several electronic and spatial properties of chemicals may influence biological effects, including three-dimensional components and interactions between and among subcomponents.

Computerized **principal component analysis** software are often employed to identify which molecular characteristics are of primary importance in producing a biological effect. Simple or multiple regressions can relate the biological effects of chemicals to one or more physicochemical parameters associated with these substructures. Mathematical expressions used to describe the relationships are:

$$Y = b + aX \text{ for simple linear regressions} \qquad (2.4)$$

or

$$\begin{array}{l} Y = b + a_1 X_1 + a_2 X_2 + a_3 X_3 + \cdots \\ \text{for multiple regressions} \end{array} \qquad (2.5)$$

where Y is the estimate of toxicity (dependent variable), X (or X_1) is the known physicochemical parameter(s) or descriptor(s) (independent variables), and a and b represent the slope and intercept of the regression equation, respectively. If the value of R^2 (coefficient of determination) is high (e.g., 0.99), as much as 99% of the variability of Y can be explained by the independent variables being considered. When few empirical data are available to establish regulations, the predictive capabilities of QSARs (organic compounds) and QICARs (metal species, Section 6.2.1) make them highly attractive. Software versions have become increasingly sophisticated, with both empirical and computer-based components.

2.3.4 Predictive Toxicology: Mixtures

Chemical contaminants are almost always found in mixtures; single compounds in the environment are the exception, not the rule. However, most of our knowledge about chemical toxicity is derived from testing one compound at a time. Risk assessments can be very difficult for mixtures that contain hundreds or thousands of components, particularly when they are diverse in their structure, behaviour, environmental fate and toxic effects. The measured toxicity of such mixtures may represent the contributions of many constituents, or only a few high-potency or high-concentration components acting independently. Similarly, the concentrations of mixture components may vary widely among industries, treatment technologies and sampling times. In this case, the simplest approach is empirical, i.e., measure the toxicity of the actual mixture to integrate the exposure and effects of all components at once (Section 12.3.3). Nevertheless, it can be difficult to transfer the observed exposure–response relationships from one risk assessment to another.

2.3.4.1 Toxic Unit (TU) Model
The toxicity of mixtures represents the combined effects of components of a mixture that have similar or opposite interactions with cellular receptors. The interactions can be described as:

- **Additive** – Two or more chemicals mixed together cause effects that equal the sum of their toxicities, characterized as a simple addition of their effects (response addition) or their partial contributions to a toxic concentration (concentration addition).
- **Synergistic (potentiation)** – More-than-additive toxicity. Two or more chemicals mixed together cause effects that exceed the sum of their toxicities.
- **Antagonistic** – Less-than-additive toxicity. Two or more chemicals mixed together cause effects that are less than the sum of their individual toxicities because the presence of one compound interferes with the action of another.
- **Independent action** – No interactions. One chemical at any concentration has no effect on the toxicity of another.

For mixtures that have been characterized chemically, toxicity is often predicted from mathematical models, including the **toxic unit (TU)**, **toxic equivalent factor (TEF)** (Section 2.3.4.2) and target lipid (TLM) (Section 2.3.4.3) models. For the TEF and TLM models, the toxicity of each component of a mixture is expressed in molar units (e.g., μmol/l) rather than gravimetric units (μg/l) because each compound has a unique chemical structure and molecular weight. This is not needed for the TU model, because TUs are the ratios of the measured concentrations of each chemical in a mixture to their toxic concentrations.

The TU approach helps us to identify different types of mixture interactions. If the sum of TUs (ΣTUs) calculated from the LC_{50}s of two compounds in any test solution equals 1.0 (Figure 2.11), 50% mortality would be expected. If the ΣTUs < 1.0, less mortality would be expected; when the ΣTUs > 1.0, more mortality would be expected. Mixtures of A and B in a test solution exhibit independent action (i.e., no interactive effects) when 50% mortality occurs when $TU_A = 1$, no matter how many TUs of chemical B are present (or *vice versa*). Joint action would occur if 50% mortality occurs when the sum of TU_A and $TU_B < 2.0$, represented by a data point within the square. The joint action would be:

- 'Strictly additive' if 50% mortality occurred when the sum of TU_A and $TU_B = 1.0$ (diagonal line);
- 'More-than-additive' if 50% mortality occurred when the sum of TU_A and $TU_B < 1.0$;
- 'Less-than-additive' if 50% mortality occurred when the sum of TU_A and $TU_B > 1.0$ but < 2.0.

'Synergism' or 'potentiation' are synonyms for 'more-than-additive' toxicity; 'antagonism' occurs if 50% mortality is evident only when TU_A or $TU_B > 1.0$, represented by a data point anywhere outside the square.

2.3.4.2 Toxic Equivalent Factor (TEF) Model

For compounds that share a common MoA (e.g., **dioxin-like compounds (DLCs)**: chlorinated dioxins, furans and biphenyls; Section 7.7.3), the TEF for a single compound A can be calculated as the ratio of its toxicity to that of a reference compound, often the one considered most toxic (e.g., **2,3,7,8-tetrachlorodibenzo-p-dioxin, or TCDD**):

$$TEF_A = (LD_{50A}/LD_{50TCDD}) \qquad (2.6)$$

For example, a chlorinated dioxin with a potency about 1/10 that of TCDD would have a TEF of 0.1. To predict the toxicity of a mixture of DLCs, toxic equivalent quantities (TEQs) for each component (e.g., compound A) are calculated from their TEFs multiplied by their measured concentrations (C) in the mixture:

$$TEQ_A = C_A \times TEF_A \qquad (2.7)$$

As with TUs, about 50% of test organisms should exhibit toxicity when the sum of TEQs for all components of a mixture exceeds the LC_{50} for the reference compound (Table 2.2).

2.3.4.3 Target Lipid Model

The TLM combines models of bioaccumulation, structure–activity relationships and toxic units, and is

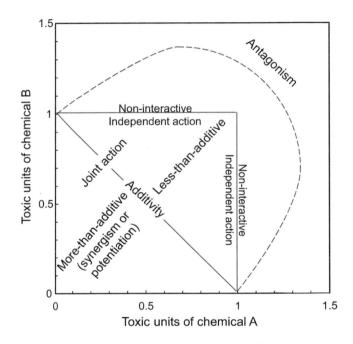

Figure 2.11 A graphical representation of possible interactive effects of two chemicals, A and B. The X and Y axes show the relative contributions of chemicals A and B to the toxicity of a mixture of the two. Adapted from Sprague (1970) with permission from Elsevier.

Table 2.2 **The toxic equivalent quantities (TEQs) of five structurally similar compounds sharing a common mode of action, calculated as the product of their toxic equivalent factors (TEFs) and their measured concentrations in a mixture.**
Because the sum of TEQs equals 1.73 mg/l and exceeds the LD_{50} of the reference compound, the mixture would be expected to cause more than 50% mortality of exposed organisms.

	LD_{50} (mg/l)	TEF	Concentration in a mixture (mg/l)	TEQ (mg/l)
Reference compound	1.0	1	0.2	0.2
Congener A	10	0.1	8	0.8
Congener B	50	0.02	5	0.1
Congener C	200	0.005	62	0.31
Congener D	1,000	0.001	320	0.32
TOTAL				1.73

often used to estimate the acute lethality to aquatic species of mixtures of narcotic organic compounds with a wide range of K_{OW}s (Di Toro *et al.*, 2000). The TLM assumes that **narcosis** or baseline toxicity is the common MoA among compounds, i.e., the bioconcentration of hydrophobic compounds in tissue lipids leads to concentrations that disrupt membrane function and cause death. The median lethal dose (LD_{50}, μmol/g lipid) is termed the **critical body burden (CBB)**. The waterborne concentrations of hydrophobic compounds corresponding to their CBBs (i.e., the LC_{50}) can be calculated from strong linear relationships between log K_{OW} and the log of lipid-normalized **bioconcentration factors (BCF**; ratio of tissue concentration [μmol/g lipid] to water concentration [μmol/ml]). In effect, the K_{OW} of each compound is a physical model of its bioconcentration by aquatic species, and K_{OW} can be estimated from chemical structures by web-based models (www.vcclab.org/lab/alogps/). To estimate the toxicity of a mixture of chemicals in water, the TLM calculates the LC_{50} of each component of the mixture from its unique K_{OW} and a single universal molar value of the CBB for acute lethality (Di Toro and McGrath, 2000). The contribution of a single compound to the toxicity of the mixture is expressed as a TU (ratio of the concentration in water to the LC_{50}; Section 2.3.4.1), and if the sum of TUs across all compounds in a mixture is equal to or greater than 1.0, 50% or more mortality is expected.

The TLM can successfully predict the acute lethality of mixtures of a wide array of organic compounds. Most importantly, it highlights the very strong role of water–lipid partitioning in determining chemical exposure, bioaccumulation and tissue dose, which are major determinants of toxicity, regardless of MoA. The CBB for nonpolar organic chemicals is assumed to be about 2.5 μmol/g lipid (Di Toro *et al.*, 2000) and varies little among the chemicals considered (e.g., Type I polar vs Type II nonpolar narcotics). However, it may vary considerably among species, perhaps reflecting species-dependent differences in rates of bioaccumulation (McCarty and Mackay, 1993).

The drawbacks to the TLM model are that error limits on predicted toxicities are often wide, and the model was developed for a single MoA, narcosis. Error limits on predicted toxicity increase when log $K_{OW} > 6.5$ because the K_{OW}–toxicity relationship deviates from linearity, likely because steric hindrance to diffusion slows the rate of transfer of large molecules across lipid membranes (Section 3.1.2.1). Models relating toxicity to K_{OW} can also

include significant errors because LC_{50}s for highly hydrophobic compounds are difficult to measure accurately (Section 2.2.8.1). Deviations from toxicity–K_{OW} regressions may also reflect alternative or multiple MoAs of compounds with unique molecular structures and interactions with different cell receptors. Although there are many MoAs for the sublethal effects of organic compounds (McCarty and Mackay, 1993), there are often too few data to support model development for each MoA. For example, a correction factor must be applied for petroleum-derived polycyclic aromatic compounds (PACs; which include PAH) to account for 'additional' toxicity compared to baseline narcotics (Di Toro *et al.*, 2007). The application of the TLM to the chronic effects of PACs on cardiac development of fish embryos has been challenged because narcosis does not include critical PAC interactions with ion channels and membrane function of cardiac myocytes (Incardona, 2017). As with TEQs, the models will be most accurate if applied to organic compounds that interact with the same cellular receptors and share closely related structures (e.g., hydrocarbons) (McGrath *et al.*, 2018).

2.3.4.4 Metal Mixtures

A useful introduction to modelling the toxicity of metal mixtures can be found in Meyer *et al.* (2015) who provide a guide to the terminology, regulatory framework, qualitative and quantitative concepts, and experimental designs for studying chemical mixtures. Meyer *et al.* (2015) provide methods for data analysis and visualization, with an emphasis on bioavailability and metal–metal interactions in mixtures of waterborne metals. This paper also introduces a series of 11 studies published in a special section of the *Environmental Toxicology and Chemistry* journal as the Metal Mixtures Modelling Exercise (MMME). Other reports have examined the literature on metal mixtures to evaluate the prevalence of 'more-than-additive' or 'less-than-additive' toxicity (Liu *et al.*, 2017) (Section 2.3.4.1).

Current metal-mixture toxicity models incorporate the basic principles of the **biotic ligand model (BLM)** (Section 6.3.2), using chemical **speciation** calculations to compute free ion activities of metals and major cations. The models evaluate their competitive binding to one or more binding sites on an organism using conventional competitive equilibrium chemistry. A single binding site implies that the different metals have the same mode of entry to the organism, whereas multiple binding sites suggest independent modes of entry. These chemical considerations are linked to metal toxicity by correlating accumulated metal concentrations to toxicity using potency factors or toxicity-response functions.

One of the important conclusions from the MMME exercise was that models calibrated from single-metal exposures provide reasonable predictions of metal mixture toxicity, assuming concentration-addition or independent action (Van Genderen *et al.*, 2015). Indeed, there is a consensus that additivity best describes metal mixture toxicity, and that examples of synergistic or more-than-additive responses are rare.

2.3.4.5 Dissecting Complex Mixtures

Bioassays are specialized toxicity tests used to measure the concentrations of a substance in a mixture that causes a specific effect, even when the identity of the substance is unknown or there is no appropriate analytical method to measure it (Box 2.12). **Toxicity identification and evaluation (TIE)** and **toxicity reduction and evaluation (TRE)** refer to the application of bioassays to first identify the toxic effects and potency of whole industrial effluents.

Box 2.12 Bioassays vs toxicity tests

'Bioassay' is a term often substituted incorrectly for 'toxicity test'. Toxicity tests measure the potency of a compound for causing adverse effects. In contrast, bioassays measure the concentration of a substance of known potency by the extent of a response in living tissues (cells, tissues, whole organisms) known to react in a predictable way to that substance.

Bioassays were first developed to measure the concentrations of compounds when suitable methods for chemical analyses were unavailable. For example, the heart rate of some vertebrates responds in a predictable way to intravenous injections of vasoactive compounds such as adrenaline. By injecting these species with blood serum from experimental animals, adrenaline concentrations in the serum can be measured by the extent to which heart rate increases in the bioassay species. Thus, factors controlling adrenaline secretion can be investigated in the experimental species, providing an early understanding of endocrinology. Bioassays are used in ecotoxicology to identify components of complex mixtures that cause specific toxic effects or the source of those components within an industry or a watershed.

Subsequent bioassays of effluent treated to remove or concentrate compounds with specific characteristics (e.g., solubility in acids or base) demonstrate whether toxicity changes with treatment and whether these changes suggest the identity of the toxic agent. For example, waterborne ammonia exists in toxic (NH_3, un-ionized) and non-toxic (NH_4^+, **ionized**) forms, and the percentage of total ammonia in the toxic form increases with water pH. An increase in effluent toxicity when pH is increased would point to ammonia as the agent to target with effluent treatment. Bioassays of each effluent stream within an industry can also identify the source of specific toxicants in an industrial process. By this means, highly concentrated effluent streams can be eliminated by process control or treated before they are diluted in a much larger volume of final effluent that is more difficult and expensive to treat.

A related analytical approach is called **effects-driven chemical fractionation (EDCF)** of complex mixtures. An EDCF first identifies the chemical composition, toxic effects and potency of the whole mixture. The components are then separated into fractions according to their chemical characteristics, e.g., by distillation (volatility), extraction into specific solvents, or **chromatographic** separation by molecular weight (**size-exclusion chromatography**) or polarity (**ion chromatography**) (Figure 2.12). Each fraction is re-tested by bioassays to identify which retained the original toxic effects and which did not, and whether the potency increased relative to the whole mixture. In subsequent steps, non-toxic subfractions are discarded, toxic subfractions are further fractionated with more precise extraction methods to isolate and concentrate compounds with closely related structures, and the new fractions are tested. At each stage in fractionation, the toxic components become more concentrated and chemical analyses can more easily identify which compounds are associated with toxicity and which are not. For heavy fuel oil, the fractions richest in two- and >six-ringed PACs were less toxic than those dominated by three-to-four- or four-to-six-ringed PACs (Figure 2.12). The data indicate that molecular structures of PACs that controlled their separation into different fractions also determined their toxicity, and that some PAC structures are far more toxic than others.

An EDCF applied to an industrial effluent can identify novel chemicals of concern, prioritize which chemicals should be monitored in environmental media, and guide industrial waste management and ERAs of related

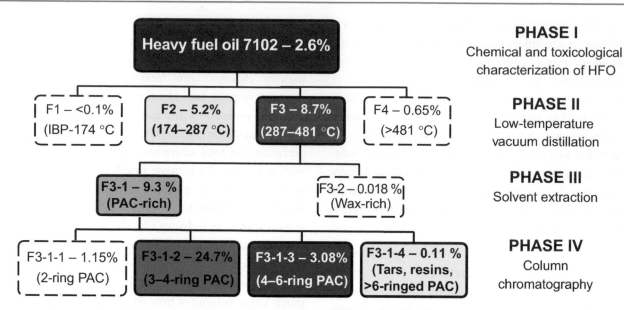

Figure 2.12 Effects-driven chemical fractionation of Heavy Fuel Oil (HFO) 7102, a complex mixture. Phase I tests demonstrated that HFO was toxic to rainbow trout (*Oncorhynchus mykiss*) embryos and contained 2.6% by weight of polycyclic aromatic compounds (PACs). Fractionation separated embryotoxic compounds into fractions rich in three- to six-ringed PACs (% by weight shown for each fraction F). Fractions in the darker red boxes were most toxic, while fractions in white boxes were non-toxic. IBP = incipient boiling point. Adapted from Bornstein *et al.* (2014) with permission from John Wiley and Sons. The toxicity of each fraction was reported by Adams *et al.* (2014).

mixtures. The challenges with this approach include the large volumes of a mixture needed at the start of the process, the resources needed to dissect, chemically analyse and test each fraction, and the potential for error if compounds that acted synergistically to cause mixture effects are separated into different fractions. The separation and identification of toxic compounds into distinct fractions can also be difficult if compounds with similar physical and chemical characteristics are not equally toxic. This would be typical of compounds for which the location of a specific side chain determines receptor interactions and toxicity without affecting key properties such as molecular size, hydrophobicity and solvent interactions.

2.3.5 Moving Away from Traditional Toxicity Tests

Acute lethality tests have the advantage of a straightforward experimental design enabling the standardized comparison of toxicity data among species and chemicals. While mortality is the basis for many hazard assessment models, these models have several drawbacks. For example, a focus on mortality provides no information on the MoA of the chemical and its potential chronic and sublethal effects. Test sensitivity and relevance can be improved with longer tests at real-world exposure levels

and measurements of sublethal responses. Historically the two most common sublethal toxicity endpoints were somatic growth and reproductive performance, but recently, developmental toxicity has become more commonly studied (Section 2.2.1).

Measures of growth and reproduction support integrated approaches such as the **scope for growth (SFG)** and **dynamic energy budget (DEB)** methods. SFG represents the difference between the energy absorbed from food on the one hand, and the energy needed for maintenance of normal functions and expended through respiration and lost through excretion on the other. The remaining energy is assumed to be available for the production of tissue (growth), the development of gametes (reproduction) and detoxification. Thus:

$$SFG = Ab - (R + U) \qquad (2.8)$$

where SFG is the scope for growth (J/h), Ab is the absorbed ration (J/h), R is energy lost in respiration (J/h), and U is the energy lost in excretion (J/h).

The energy absorption rate (Ab) is calculated as:

$$Ab = \text{Ingestion rate } (J/h) \\ \times \text{ Absorption efficiency} (\%) \qquad (2.9)$$

Although the DEB model adopts the same general approach, it differs in its allotment of assimilated energy and, unlike the SFG model, makes provision for energy storage capacity. Both models are subject to refinements related to food availability, food quality and assimilation efficiency (Filgueira *et al.*, 2011).

Measurements of SFG have been correlated with measures of biodiversity in benthic communities and inversely related to pollution gradients in San Diego Bay (USA), the North and Irish Sea coastlines (UK), Bermuda and Iceland (Crowe *et al.*, 2004). Energy budget studies related to environmental contamination have largely been carried out on suspension-feeding bivalve molluscs, usually *Mytilus edulis*. Molluscs are sedentary, facilitating calculations of energy balance unaffected by motility. Measuring SFG in sedentary organisms has advantages in terms of the relative ease of conducting tests, but this approach does not readily integrate changes in the behaviour of the exposed animals, particularly motile organisms.

The assessment of chemical effects on behaviour has generally lagged behind other aspects of ecotoxicology. An inherent attraction of this approach is that behaviour should be a good integrator of all biochemical and physiological changes following chemical exposure (including olfactory detection). However, potential complications include experimental artefacts caused by behavioural responses to confinement of experimental organisms and the presence of monitoring equipment and lab personnel. Behavioural responses to chemical exposure can lead to a variety of sublethal effects that may have significant effects on **fitness** and survival, often at chemical exposures lower than those causing acute or chronic toxicity. Consider a toxicant-induced change in phototaxis that increases an organism's visibility to predators, or subtle changes in reproductive behaviour that diminish or eliminate fertilization. Chemical effects may cause either hypo- or hyperactivity, either of which could affect prey capture or predator escape, depending on whether the respective strategies involve stealth, speed or concealment. Behavioural changes may result from direct neurotoxicity, such as narcosis or neuroreceptor inhibition (Box 2.3). Alternatively, reduced mobility or stamina may follow respiratory distress caused by metabolic inhibition (e.g., PCP, Section 2.2) or chemical damage to respiratory epithelia. Chemoreception is a key factor in the way that many organisms perceive their environment. Inhibition of chemoreception by chemical exposure may significantly reduce survivability through impaired detection of food, predators, potential partners and, in the case of migrating animals, home territory and spawning grounds. Table 2.3 presents examples of various potentially deleterious behavioural endpoints resulting from chemical exposure.

One of the most interesting new approaches to identifying toxic effects and MoAs is the rapidly evolving field of 'omics' (**genomics**, **proteomics**, **metabolomics**, etc.) described in detail in Chapter 5. By analysing the molecular responses of organisms exposed to a previously untested compound, strong inferences can be drawn about which physiological processes and which cells and tissues will be affected. Molecular responses can direct the design of subsequent whole-organism tests to assess chemical effects on measures of performance including aerobic capacity, sexual maturation, reproduction, embryo development and behaviour. A second major

Table 2.3 **Examples of behavioural toxicity.**

Toxicant	Affected organism	Behavioural effect	Reference
Metals	Juvenile coho salmon (*Oncorhynchus kisutch*)	3-h exposure to 5 µg/l Cu impairs avoidance behaviour and increases detection by predatory cutthroat trout (*Oncorhynchus clarkii clarkii*).	McIntyre *et al.* (2012)
Pesticides	Amphipod (*Gammarus fossarium*)	Feeding and locomotor activity impaired at 126 µg/l chlorpyrifos (organophosphate).	Xuereb *et al.* (2009)
	Chinook salmon (*Oncorhynchus tshawytscha*)	Olfactory-mediated alarm response impaired on exposure to >1 µg/l diazinon. Homing behaviour impaired at 10 µg/l.	Scholz *et al.* (2000)
	Atlantic salmon (*Salmo salar*)	Exposure to 1.0 µg/l simazine and atrazine impaired olfactory response of males to female priming hormone prostaglandin $F_{2\alpha}$.	Moore and Lower (2001)
PACs	Seabass (*Dicentrarchus labrax*)	28-d benzo(a)pyrene exposure ($\leq$8 µg/l) reduced swimming activity at or below concentrations causing biomarker responses.	Gravato and Guilhermino (2009)

benefit is the application of molecular responses of wild populations of plants and animals as biomarkers of chemical exposure and toxicity. Response patterns among multiple biomarkers will provide **forensic** tools to diagnose transient or continuous exposures to specific chemicals, and to predict or detect effects at the population level caused by impairment of reproduction or growth. As Chapter 4 shows us, linking cause and effect in contaminated ecosystems is often fraught with uncertainty. Molecular biomarkers will greatly improve our capacity to identify chemical causes of ecological impacts and AOPs that link exposure to important effects.

Summary

In Chapter 2, we explore the principles of environmental toxicology, the various ways chemical toxicity is measured, and the importance of standard test conditions for reliable and reproducible results to support the regulation of chemical emissions. Toxicity tests measure chemical exposures or concentrations causing lethality, sublethal effects on growth, reproduction and behaviour, and indirect measures of fitness such as energy budgets. However, the numbers of chemicals and species at risk of chemical exposure exceed our capacity for testing. Recognizing this issue, Chapter 2 reviews models that predict the toxicity of untested chemicals from the toxicity of compounds with closely related structures or MoAs, either singly or in mixtures. The results of toxicity tests are very sensitive to test conditions that affect chemical exposures. Chapter 3 illustrates how the biology of test species and the chemical characteristics of water influence the rates and routes of chemical uptake by microorganisms, plants and animals.

Review Questions and Exercises

1. What is a toxicity test?

2. What is an exposure–response relationship?

3. What are the essential starting points for chemical effects on natural ecosystems?

4. Explain why estimates of toxicity, such as the LC_{50}, are based on the median organism, not the most sensitive or most resistant.

5. Explain the following phrases:
   a. "Toxicity is not absolute."
   b. "If there is no exposure, there is no effect."
   c. "It is the dose that makes the poison."

6. What is the role or importance of time in measurements of toxicity? How is time represented in the expressions of adverse effects?

7. What are the differences between an LC_{50} and an LT_{50}? How is response time used to estimate threshold concentrations causing toxicity?

8. What is the difference between an LC_{50} and an LD_{50}? Which gives the most accurate and reproducible measure of toxicity? Which is the most useful for managing the industrial emissions of chemicals, and why?

9. Define 'threshold toxicity' and explain how it can be estimated from response times.

10. What is the 'observer effect' and why is it important in toxicity testing?

11. What are the advantages and disadvantages of toxicity tests with embryos rather than juvenile or adult organisms?

12. In aquatic toxicity tests, why are test solutions renewed periodically or continually? What alternative methods are available for maintaining constant concentrations of hydrophobic test compounds in water?

13. Why are toxicity response data 'linearized' by the calculation of probits of per cent response prior to the estimation of $LC_{50}s$?

14. What is the difference between synergism and antagonism?

15. In a spreadsheet, construct a species sensitivity distribution (SSD) for the acute lethality of a new pesticide (X) to the following marine species and estimate the HC_5.

	Species	Pesticide X LC_{50} (mg/l)
1	Atlantic cod	5.6
2	Atlantic sole	23
3	Striped bass	2.5
4	Atlantic salmon	15
5	Mummichog	75
6	Atlantic silversides	150
7	Atlantic herring	32
8	Capelin	60
9	Pollock	11
10	Mackerel	25
11	Gudgeon	38

Abbreviations

Ab	Absorbed ration	ET_{50}	Median effective time
ACh	Acetylcholine	HC	Hazard concentration
AChE	Acetylcholinesterase	IC_{50}	Concentration causing 50% inhibition (e.g., of algal growth)
ACR	Acute-to-chronic ratio		
AOP	Adverse outcome pathway	J	Energy lost in excretion
ATP	Adenosine triphosphate	K_{OW}	Octanol–water partition coefficient
BCF	Bioconcentration factor	LC_{50}	Median lethal concentration
BLM	Biotic ligand model	LD_{50}	Median lethal dose
CBB	Critical body burden	LOEC	Lowest-observable-effect concentration
DEB	Dynamic energy budget	LT	Median lethal time
DLC	Dioxin-like compound	MATC	Maximum acceptable toxicant concentration
EC	Effect concentration		
ECCC	Environment and Climate Change Canada	MMME	Metal Mixtures Modelling Exercise
EC_{10}	Concentration causing a 10% response	MoA	Mode of action
EC_{50}	Median effective concentration	NEC	No-effect concentration
ED_{50}	Median effective dose	NOEC	No-observable-effect concentration
EDCF	Effects-driven chemical fractionation	OECD	Organisation for Economic Co-operation and Development
ELA	Experimental Lakes Area		
ERA	Ecological risk assessment	PAC	Polycyclic aromatic compound

PAH	Polycyclic aromatic hydrocarbon	TLM	Target lipid model
PCBs	Polychlorinated biphenyls	TPAH	Total or sum of concentrations of all PAH in a sample
PCP	Pentachlorophenol		
QA/QC	Quality assurance/quality control	TRE	Toxicity reduction and evaluation
QSAR	Quantitative structure–activity relationship	TU	Toxic unit
R	Energy lost in respiration	UNEP	United Nations Environment Programme
SSD	Species sensitivity distribution	US EPA	United States Environmental Protection Agency
TCDD	2,3,7,8-tetrachlorodibenzo-p-dioxin		
TEF	Toxic equivalent factor	v/v	Ratio of volumes = dilution
TEQ	Toxic equivalent quantity	WET	Whole-effluent toxicity test
TIE	Toxicity identification and evaluation	WQC	Water quality criterion

References

Abbott, W. S. (1925). A method of computing the effectiveness of an insecticide. *Journal of Economic Entomology*, 18, 265–267.

Adams, J., Bornstein, J. M., Munno, K., *et al.* (2014). Identification of compounds in heavy fuel oil that are chronically toxic to rainbow trout embryos by effects-driven chemical fractionation. *Environmental Toxicology and Chemistry*, 33, 825–835.

Alves, P. R. L. & Cardoso, E. J. B. N. (2016). Overview of the Standard Methods for Soil Ecotoxicology Testing. In M. L. Larramendy & S. Soloneski (eds.) *Invertebrates – Experimental Models in Toxicity Screening*. Rijeka, Croatia: InTech, 35–56.

Andreasen, J. K. (1985). Insecticide resistance in mosquitofish of the Lower Rio Grande valley of Texas – an ecological hazard? *Archives of Environmental Contamination and Toxicology*, 14, 573–577.

Balakrishnan, V. K., Shirin, S., Aman, A. M. *et al.* (2016). Genotoxic and carcinogenic products arising from reductive transformations of the azo dye, Disperse Yellow 7. *Chemosphere*, 146, 206–215.

Bejarano, A. C. & Mearns, A. J. (2015). Improving environmental assessments by integrating species sensitivity distributions into environmental modeling: examples with two hypothetical oil spills. *Marine Pollution Bulletin*, 93, 172–182.

Belanger, S. E. & Carr, G. J. (2019). SSDs revisited: part II – practical considerations in the development and use of application factors applied to species sensitivity distributions. *Environmental Toxicology and Chemistry*, 38, 1526–1541.

Bornstein, J. M., Adams, J., Hollebone, B., *et al.* (2014). Effects-driven chemical fractionation of heavy fuel oil to isolate compounds toxic to trout embryos. *Environmental Toxicology and Chemistry*, 33, 677–687.

Butler, J. D., Parkerton, T. F., Redman, A. D., Letinski, D. J. & Cooper, K. R. (2016). Assessing aromatic-hydrocarbon toxicity to fish early life stages using passive-dosing methods and target-lipid and chemical-activity models. *Environmental Science & Technology*, 50, 8305–8315.

Calabrese, E. J. (2008). Hormesis: why it is important to toxicology and toxicologists. *Environmental Toxicology and Chemistry*, 27, 1451–1474.

Chapman, P. M. (1990). The sediment quality triad approach to determining pollution-induced degradation. *Science of the Total Environment*, 97/98, 815–825.

Charifi, M., Miserazzi, A., Sow, M., *et al.* (2018). Noise pollution limits metal bioaccumulation and growth rate in a filter feeder, the Pacific oyster *Magallana gigas*. *PLoS ONE*, 13, e0194174.

Conte, F. S. (2004). Stress and the welfare of cultured fish. *Applied Animal Behaviour Science*, 86, 205–223.

Crowe, T. P., Smith, E. L., Donkin, P., Barnaby, D. L. & Rowland, S. J. (2004). Measurements of sublethal effects on individual organisms indicate community-level impacts of pollution. *Journal of Applied Ecology*, 41, 114–123.

Di Giulio, R. T. & Clark, B. W. (2015). The Elizabeth River story: a case study in evolutionary toxicology. *Journal of Toxicology and Environmental Health B*, 18, 259–298.

Di Toro, D., McGrath, J. & Hansen, D. (2000). Technical basis for narcotic chemicals and polycyclic aromatic hydrocarbon criteria. I. Water and tissue. *Environmental Toxicology and Chemistry*, 19, 1951–1970.

Di Toro, D. M. & McGrath, J. A. (2000). Technical basis for narcotic chemicals and polycyclic aromatic hydrocarbon criteria. II. Mixtures and sediments. *Environmental Toxicology and Chemistry* 19, 1971–1982.

Di Toro, D. M., McGrath, J. A. & Stubblefield, W. A. (2007). Predicting the toxicity of neat and weathered crude oil: toxic potential and the toxicity of saturated mixtures. *Environmental Toxicology and Chemistry*, 26, 24–36.

Di Toro, D. M., Zarba, C. S., Hansen, D. J., *et al.* (1991). Technical basis for establishing sediment quality criteria for nonionic organic chemicals using equilibrium partitioning. *Environmental Toxicology and Chemistry*, 10, 1541–1583.

EC (2007). *Biological Test Method: Acute Lethality Test Using Rainbow Trout*. Ottawa, ON: Environment Canada, Method Development and Applications Section (MDAS), Biological Methods Division.

Ellersieck, M. R., Asfaw, A., Mayer, F. L., *et al.* (2003). *Acute-to-Chronic Estimation (ACE v 2.0) with Time-Concentration-Effect Models*. User Manual and Software.

Filgueira, R., Rosland, R. & Grant, J. (2011). A comparison of scope for growth (SFG) and dynamic energy budget (DEB) models applied to the blue mussel (*Mytilus edulis*). *Journal of Sea Research*, 66, 403–410.

Finney, D. J. 1971. *Probit Analysis*. Cambridge UK: Cambridge University Press.

Giesy, J. P. & Graney, R. L. (1989). Recent developments in and intercomparisons of acute and chronic bioassays and bioindicators. *Hydrobiologia*, 188/189, 21–60.

Gravato, C. & Guilhermino, L. (2009). Effects of benzo(a)pyrene on seabass (*Dicentrarchus labrax* L.): biomarkers, growth and behavior. *Human and Ecological Risk Assessment: An International Journal*, 15, 121–137.

Hodson, P. V., Adams, J. & Brown, R. S. (2019). Oil toxicity test methods must be improved. *Environmental Toxicology and Chemistry*, 38, 302–311.

Hughes, S. A., Maloney, E. M. & Bejarano, A. C. (2021). Are vertebrates still needed in routine whole effluent toxicity testing for oil and gas discharges? *Environmental Toxicology and Chemistry*, 40, 1255–1265.

Incardona, J. P. (2017). Molecular mechanisms of crude oil developmental toxicity in fish. *Archives of Environmental Contamination and Toxicology*, 73, 19–32.

Jager, T. (2012). Bad habits die hard: the NOEC's persistence reflects poorly on ecotoxicology. *Environmental Toxicology and Chemistry*, 31, 228–229.

Jonker, M. T. O., Burgess, R. M., Ghosh, U., *et al.* (2020). *Ex situ* determination of freely dissolved concentrations of hydrophobic organic chemicals in sediments and soils: basis for interpreting toxicity and assessing bioavailability, risks and remediation necessity. *Nature Protocols*, 15, 1800–1828.

Kiparissis, Y., Akhtar, P., Hodson, P. V. & Brown, R. S. (2003). Partition-controlled delivery of toxicants: a novel *in vivo* approach for embryo toxicity testing. *Environmental Science & Technology*, 37, 2262–2266.

Lee, K., Boufadel, M., Chen, B., *et al.* (2015). *Expert Panel Report on the Behaviour and Environmental Impacts of Crude Oil Released into Aqueous Environments*. Ottawa, ON: Royal Society of Canada.

Liu, Y., Vijver, M. G., Pan, B. & Peijnenburg, W. J. G. M. (2017). Toxicity models of metal mixtures established on the basis of "additivity" and "interactions". *Frontiers of Environmental Science & Engineering*, 11, e10.

Mager, E. M., Esbaugh, A. J., Stieglitz, J. D., *et al.* (2014). Acute embryonic or juvenile exposure to Deepwater Horizon Crude Oil impairs the swimming performance of Mahi-Mahi (*Coryphaena hippurus*). *Environmental Science & Technology*, 48, 7053–7061.

McCarty, L. S. & Mackay, D. (1993). Enhancing ecotoxicological modeling and assessment. Body residues and modes of toxic action. *Environmental Science & Technology*, 27, 1718–1728.

McGrath, J. A., Fanelli, C. J., Di Toro, D. M., *et al.* (2018). Re-evaluation of target lipid model–derived HC5 predictions for hydrocarbons. *Environmental Toxicology and Chemistry*, 37, 1579–1593.

McIntyre, J. K., Baldwin, D. H., Beauchamp, D. A. & Scholz, N. L. (2012). Low-level copper exposures increase visibility and vulnerability of juvenile coho salmon to cutthroat trout predators. *Ecological Applications*, 22, 1460–1471.

Meyer, J. S., Farley, R. A. & Garman, E. R. (2015). Metal mixture modeling evaluation project: 1. Background. *Environmental Toxicology and Chemistry*, 34, 716–740.

Moore, A. & Lower, N. (2001). The impact of two pesticides on olfactory-mediated endocrine function in mature male Atlantic salmon (*Salmo salar* L.) parr. *Comparative Biochemistry and Physiology Part B: Biochemistry and Molecular Biology*, 129, 269–276.

Morgan, K. N. & Tromborg, C. T. (2007). Sources of stress in captivity. *Applied Animal Behaviour Science*, 102, 262–302.

Norberg-King, T. J., Embry, M. R., Belanger, S. E., *et al.* (2018). An international perspective on the tools and concepts for effluent toxicity assessments in the context of animal alternatives: reduction in vertebrate use. *Environmental Toxicology and Chemistry*, 37, 2745–2757.

OECD (2019). *OECD Guidelines for the Testing of Chemicals. Effects on Biotic Systems*. Paris, France: Organisation for Economic Co-operation and Development, Section 2, www.oecd-ilibrary.org/environment/oecd-guidelines-for-the-testing-of-chemicals-section-2-effects-on-biotic-systems_20745761

Redman, A. D., Butler, J. D., Letinski, D. J., *et al.* (2018). Technical basis for using passive sampling as a biomimetic extraction procedure to assess bioavailability and predict toxicity of petroleum substances. *Chemosphere*, 199, 585–594.

Scholz, N. L., Truelove, N. K., French, B. L., *et al.* (2000). Diazinon disrupts antipredator and homing behaviors in chinook salmon (*Oncorhynchus tshawytscha*). *Canadian Journal of Fisheries and Aquatic Sciences*, 57, 1911–1918.

Simpson, S. L. & Batley, G. E. (2016). *Sediment Quality Assessment – A Practical Guide*, Clayton South, Victoria, Australia: CSIRO Publishing.

Sprague, J. B. (1969). Measurement of pollutant toxicity to fish. I. Bioassay methods for acute toxicity. *Water Research*, 3, 793–821.

Sprague, J. B. (1970). Measurement of pollutant toxicity to fish. II. Utilizing and applying bioassay results. *Water Research*, 4, 3–32.

Turcotte, D., Akhtar, P., Bowerman, M., *et al.* (2011). Measuring the toxicity of alkyl-phenanthrenes to early life stages of medaka (*Oryzias latipes*) using partition-controlled delivery. *Environmental Toxicology and Chemistry*, 30, 487–495.

Twiss, M. R., Errécalde, O., Fortin, C., *et al.* (2001). Coupling the use of computer speciation models and culture techniques in laboratory investigations of trace metal toxicity. *Chemical Speciation and Bioavailability*, 13, 9–24.

UNEP (2013). *Global Chemicals Outlook - Towards Sound Management of Chemicals*. Washington, DC: United Nations Environment Programme.

US EPA (2002). *Methods for Measuring the Acute Toxicity of Effluents and Receiving Waters to Freshwater and Marine Organisms*. Washington DC: Office of Water.

Van Genderen, E., Adams, W., Dwyer, R., Garman, E. & Gorsuch, J. (2015). Modeling and interpreting biological effects of mixtures in the environment: Introduction to the metal mixture modeling evaluation project. *Environmental Toxicology and Chemistry*, 34, 721–725.

Vandegehuchte, M. B., Nguyen, L. T. H., De Laender, F., Muyssen, B. T. A. & Janssen, C. R. (2013). Whole sediment toxicity tests for metal risk assessments: On the importance of equilibration and test design to increase ecological relevance. *Environmental Toxicology and Chemistry*, 32, 1048–1059.

Wright, D. A. & Roosenburg, W. H. (1982). Trimethyltin toxicity to larval *Uca pugilator*. Effects of temperature and salinity. *Archives of Environmental Contamination and Toxicology*, 11, 491–495.

Xuereb, B., Lefèvre, E., Garric, J. & Geffard, O. (2009). Acetylcholinesterase activity in *Gammarus fossarum* (Crustacea Amphipoda): linking AChE inhibition and behavioural alteration. *Aquatic Toxicology*, 94, 114–122.

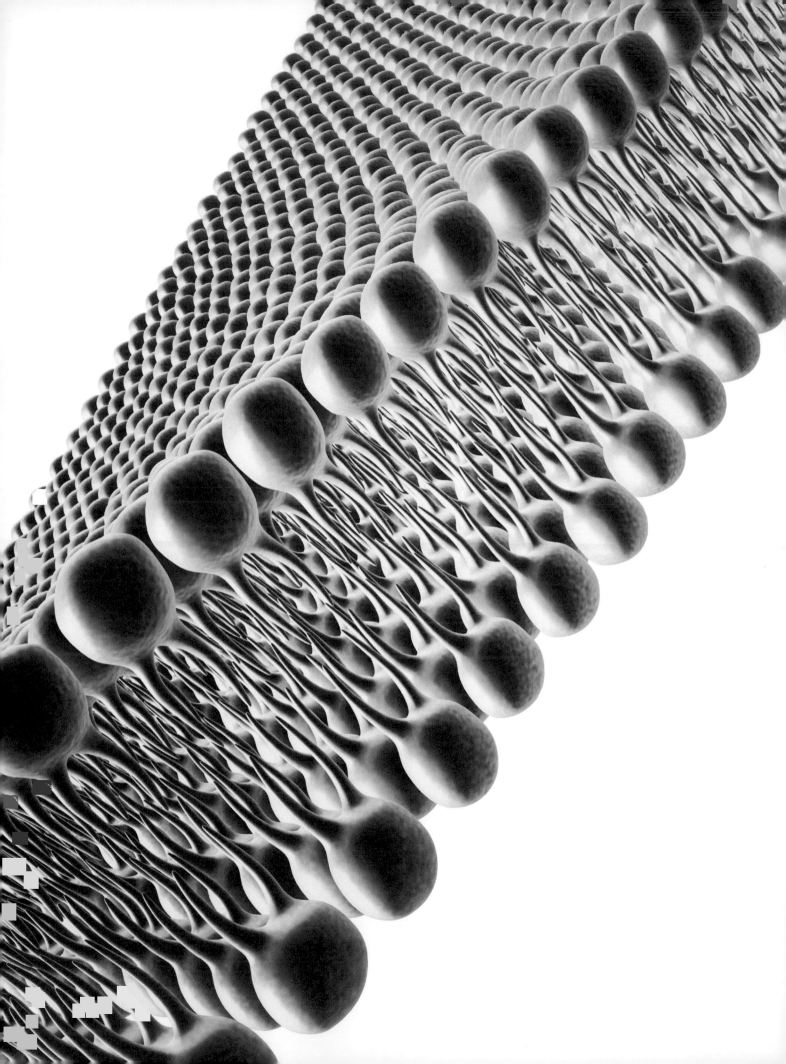

CHAPTER

3

Contaminant Uptake and Bioaccumulation: Mechanisms, Kinetics and Modelling

Peter G. C. Campbell, Peter V. Hodson, Pamela M. Welbourn and David A. Wright

Learning Objectives

Upon completion of this chapter, the student should be able to:

- Describe the abiotic–biotic interface and explain how toxicants can interact with and traverse biological membranes (including both aquatic and terrestrial biota).
- Understand the differences between active and passive transport, and between mediated and unmediated transport across membranes.
- Appreciate the importance of kinetics (i.e., the rates of uptake, transformation and elimination of toxicants) and understand the factors that govern these rates.
- Discuss the idea of cell membranes as a possible target of toxicant action (i.e., how toxicants may perturb membrane function).
- Understand the concepts of bioavailability, bioconcentration, biomagnification and biotransformation.

Chapter 3 initially focuses on the biological membranes that separate living organisms from the external environment. These membranes play an essential role in controlling the uptake of nutrients and potential toxicants, as well as the elimination of the waste products produced by cellular metabolism. Our goal in the initial section is to review membrane structure and function, and to consider how toxicants interact with cell membranes and enter (or leave) living cells. This section leads naturally to a consideration of possible sites of uptake and elimination. These sites include respiratory tissues (lungs, gills), digestive organs and excretory structures, as well as plant roots or leaves. We then briefly consider the intracellular fate of toxicants, introducing some of the metabolic concepts that are developed more fully in Chapters 6, 7 and 8. The final section of the chapter deals with the net result of the preceding processes, i.e., the tendency of a chemical to bioaccumulate or biomagnify in living organisms and food webs. This final section provides a link to the examples of environmental modelling presented in the next chapter.

3.1 General Considerations

In Chapter 2 we emphasized that toxicity is initiated by the interaction of a **toxicant** with a sensitive site, most often a site located within the organism. To gain access to an organism and reach such sensitive intracellular sites, the toxicant must first cross a **plasma membrane**. To reach the plasma membrane, the toxicant may have to pass through a protective barrier (e.g., a plant cell wall or an egg chorion or shell) – see Section 3.2. However, in the present section we focus on the cell membrane itself as the interface between the living (biotic) and non-living (abiotic) realms. To predict whether a given toxicant will be able to enter an organism, we clearly need to understand the nature and permeability of plasma membranes, in particular those that separate the organism from its surrounding environment. Our aims in this first section are to describe the various pathways by which a toxicant can enter cells, and to identify the molecular properties of the toxicant that influence its ability to exploit these pathways.

In addition, readers should keep in mind that controlled transmembrane transport of inorganic ions and metabolites is of crucial importance if normal cell functioning is to be maintained. In other words, any substance that interacts with ion channels and carriers and interferes with their normal functioning may induce toxicity. Note also that the outer layer of most plants and animals is relatively impermeable, because of the need to maintain osmotic balance in aquatic or terrestrial environments. However, there are specific tissues that are quite permeable, including respiratory structures (e.g., gill and lung epithelia in animals, stomata in plants) and structures for the uptake of nutrients and water (e.g., gastrointestinal endothelia in animals, roots in plants). These are also the tissues that are most involved in transmembrane transport of toxic substances.

3.1.1 Composition and Structure of Biological Membranes

The chemical composition of biological membranes, i.e., their lipid, protein and carbohydrate content, has been known since the 1920s. The relative proportions of these components vary among different types of biological membranes, but lipids normally dominate for all levels of organization, i.e., protists, plants and animals. Unlike proteins and carbohydrates, which can be defined in terms of their molecular composition, lipids are a much more diverse class of organic molecules, defined largely by their hydrophobicity (immiscibility in water).

Examples of membrane lipids include phospholipids, glycolipids and cholesterol. The common feature of these lipids is that they are **amphiphilic**, i.e., that they possess both a polar 'head-group' (hydrophilic) and a non-polar hydrocarbon 'tail' (hydrophobic). For example, in a representative phospholipid such as phosphatidylcholine, the partially esterified phosphate moiety is the polar head-group and the lengthy fatty acid side chains are the non-polar tail (Figure 3.1). This juxtaposition of polar and non-polar components defines a surfactant, i.e., a 'surface-active' molecule that tends to concentrate at the air–water interface or at solid surfaces. To illustrate this, if you isolate the lipids from a biological membrane and add them to water, they will collect at the air–water interface. The polar head-group will be immersed in the water, with the non-polar part of the molecule extending above the water's surface.

The non-polar groups of individual lipid molecules are attracted to each other and coalesce to form a lipidic microlayer at the water surface. If the suspension is agitated, the lipids tend to self-aggregate and form liposomes, in which they are arranged in a lipid bilayer, surrounding a central aqueous core (Figure 3.2; Box 3.1).

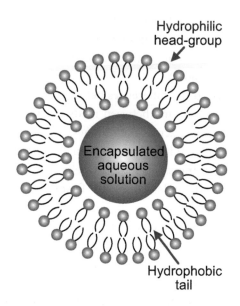

Figure 3.2 Unilamellar liposome.

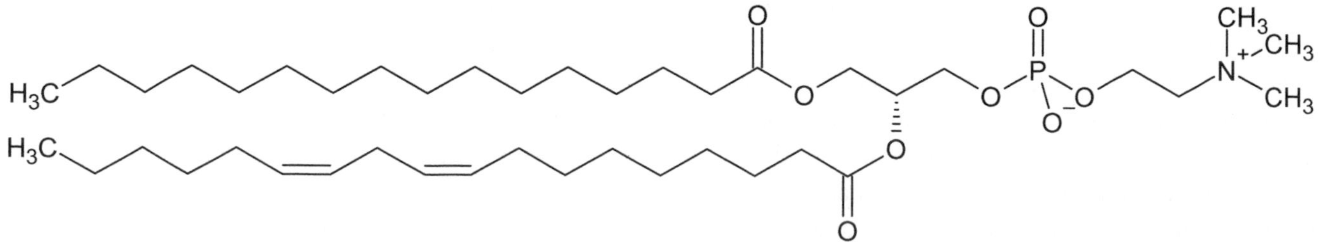

Figure 3.1 Structure of a typical phospholipid, phosphatidylcholine.

Box 3.1 Usefulness of liposomes as models of biological membranes

Liposomes are useful but incomplete models of biological membranes, since they lack the proteins that play key functional and structural roles in true biological membranes. However, techniques exist to introduce membrane proteins into liposomes, and such preparations have been used to study the transport properties of the proteins.

The fundamental structural feature of a cell membrane is the lipid bilayer depicted in Figure 3.2 – the outer or **apical** surface is in contact with the external environment, whereas the inner or **basolateral** surface interacts with the cytoplasm. To this structure we must now add the other membrane components. The constituent proteins may span the bilayer, in which case they are referred to as **integral proteins**, or they may be located on the membrane inner or outer surface (**extrinsic proteins**). In the context of movement of toxicants across cell membranes, integral proteins are of particular importance. They span the membrane from one aqueous face to the other, sometimes only once, but often with many consecutive spanning sequences. Those that have multiple spanning regions are of particular relevance in this chapter as they can be configured to offer a protein-lined hydrophilic pathway through which polar substrates can pass into and out of cells (Section 3.1.2).

The generic biological membrane that we have just described is commonly referred to as the 'fluid mosaic model' (Figure 3.3). The term 'fluid' merits some explanation. To function properly and to allow the protein conformational changes that underpin mediated transport processes, the biological membrane must be in a fluid state, not ordered. Within the membrane, the polar head-groups are quite immobile, whereas the lipid tails are normally in constant motion. However, these hydrophobic tails are attracted to each other, and if the temperature is lowered, they will assume a more ordered or gel state. Such a phase transition from a fluid state to an ordered arrangement would adversely affect the membrane's essential functions.

Living organisms can respond to changes in the ambient temperature by changing the composition of their membrane lipids (Section 2.2.2.1; Box 2.8). For example, the introduction of more unsaturated fatty acids into the hydrophobic tails produces 'kinks' in the tails and results in less interaction between adjacent tails (i.e., less tightly packed and more fluid). This dynamic character of biological membranes extends to individual lipid molecules that can switch position from the outer face of the membrane to the inner (cytosolic) side, and also to some of the integral proteins, which can move from the cell membrane to one of the internal membranes (e.g., those enclosing vacuoles) and back again.

To help the reader put the fluid mosaic model into context and appreciate the role of transmembrane transport, Figure 3.4 illustrates where biological membranes

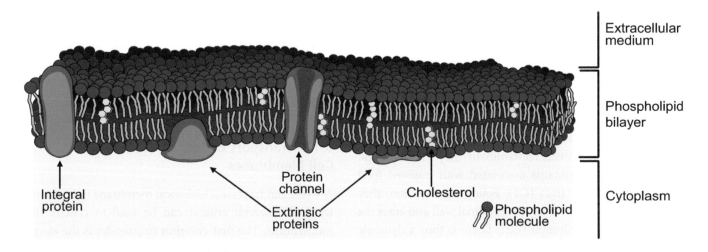

Figure 3.3 Fluid mosaic model for a biological membrane, typically about 80 Å thick. Note the phospholipid bilayer and the presence of integral proteins spanning the outer (apical) surface and the inner (basolateral) surface. These proteins can act as pores, channels and transporters. Adapted from a figure released into the public domain (Wikimedia Commons) by its author, LadyofHats.

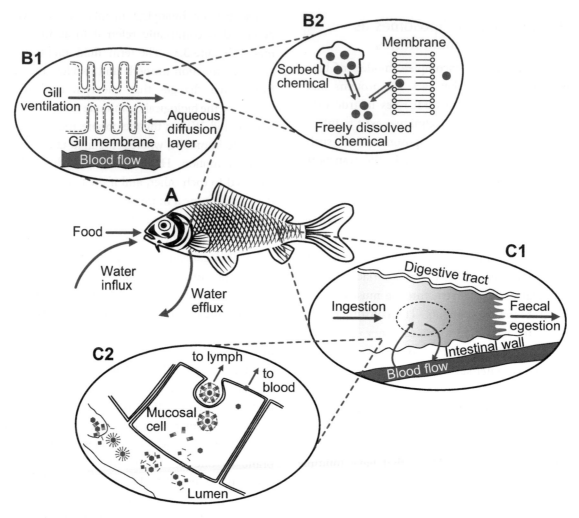

Figure 3.4 Simplified scheme illustrating the principal routes by which a contaminant can enter a fish (A), taken as a representative aquatic animal. The membrane barriers are shown for the fish gills (B1 and B2) and for the digestive tract (C1 and C2). Figure adapted from Gobas and Morrison (2000). Copyright (2000) by Don Mackay. Reproduced with the permission of Taylor and Francis Group, LLC, a division of Informa plc.

intervene in the uptake of a contaminant by a fish, either from the ambient water or from the ingested food. The figure depicts the movement of water across the fish gill (B1, B2) and the transport of dissolved contaminants across the gill membrane to the bloodstream. Adsorbed contaminants that enter the gill cavity on the surface of suspended particles may be subject to desorption. Once desorbed, the dissolved contaminant may cross the gill membrane. Contaminants associated with ingested food enter the intestinal tract (C1). Following digestion, they may leave the lumen, cross the intestinal wall and enter the bloodstream or the lymph (C2). There is thus a dynamic balance between uptake of the contaminant from water (via the gills) and food (via the intestinal tract), and its loss via the gills or by faecal egestion. We revisit the uptake of contaminants by these two routes in Section 3.6 and in

Section 4.5.2. Note too that the epithelial membranes illustrated in the figure are potential targets for toxicity, notably by narcotic pollutants (Section 2.3.4.3) and by other substances (e.g., cadmium, Cd; lead, Pb) that can disrupt the normal functioning of the membranes (Section 6.4).

3.1.2 Transport of Solutes Across Cell Membranes

A solute can traverse a biological membrane in a variety of ways, and several criteria can be used to classify these mechanisms. The first criterion to consider is the **electrochemical gradient** across the membrane (Box 3.2). Does transport occur from a region of high concentration to one of low concentration? In such cases, the movement across the membrane can occur without requiring any energy

Box 3.2 Electrochemical gradient

An electrochemical gradient consists of two components, the chemical gradient, or difference in solute concentration across a membrane, and the electrical gradient, or difference in electrical charges across a membrane. When the solute is present at different concentrations on either side of a permeable membrane, it will tend to move across the membrane from the compartment of higher concentration to the compartment of lower concentration. If the solute is charged (i.e., a cation or anion), it will carry an electrical charge across the membrane. In such cases, if the distribution of charges across the membrane is unequal, the resulting difference in electric potential drives ion diffusion in both directions across the membrane until the charges are balanced in both compartments.

and is considered **passive transport**. Only in cases where the movement occurs in the opposite direction, from low to high concentration, does the movement require metabolic energy and fall in the category of **active transport**. The second criterion to examine is if the solute crosses the membrane 'unaided', i.e., without interacting with a specific receptor embedded in the membrane.

An example of what is referred to as **unmediated transport** would be the passage of a polar solute through an aqueous pore in the membrane. In contrast, if the transport of the solute involves an interaction with a carrier embedded in the membrane, it would be classified as **mediated transport**. Examples of these various modes of transmembrane transport are presented in Figure 3.5. For convenience, transport in this figure is shown as proceeding from left to right (uptake), but similar modes of transport also operate in the opposite direction (elimination). In the following subsections, we describe these modes of transport in order of increasing complexity, proceeding from the top of Figure 3.5 to the bottom. As a complement to this figure, Table 3.1 gives examples of solutes that are transported across cell membranes by one or more of these mechanisms.

3.1.2.1 Diffusion Through the Lipid Bilayer

The simplest form of transport of solutes across biological membranes is by diffusion through the lipidic regions of the membrane, a pathway that is limited to non-polar lipophilic molecules (Figure 3.5, pathway 1). To enter the central lipophilic portion of the membrane, the solute molecule must first shed any associated water molecules (i.e., lose its hydration sheath) and then 'squeeze' between the head-groups of the phospholipids. Because of this size constraint, this pathway is limited to molecules of low to moderate molecular weight (e.g., <500 Da), and their shape may also affect their uptake. For relatively small molecules, the **octanol–water partition coefficient** (K_{OW}) of the molecule can be used to predict how readily it will traverse the membrane (see Chapter 7). This relationship can be explained by the solubility–diffusion model. According to this model (Figure 3.6), the solute must first dissolve in the membrane, cross it by simple diffusion and then partition into the internal aqueous phase. The rate of diffusion across the membrane will depend upon two factors: (i) the difference in solute concentrations within the membrane (i.e., between the outer and inner faces of the membrane, $C_{o,m}$ and $C_{i,m}$) as illustrated in Figure 3.6, and (ii) the resistance that the lipid barrier offers to the movement of the solute across the membrane. Resistance varies according to the packing of the phospholipids (tightly or loosely), the strength of the interactions between the diffusing solute and the phospholipids, and the thickness of the membrane (d). **Diffusion coefficients** within the membrane fall off very steeply with molecular size; for very large organic molecules such as polychlorinated biphenyls (PCBs), there is resistance to diffusion such that K_{OW} is no longer a good predictor of the rate of movement across the membrane.

3.1.2.2 Diffusion Through Membrane Pores and Channels

Simple diffusion of low molecular weight (LMW) and neutral solutes across the membrane may also occur through membrane pores. The most common example of such a process is the movement of the water molecule itself across the membrane through pores known as **aquaporins** (Figure 3.5, mechanism 2). This movement can be tracked by adding radiolabelled water (i.e., tritiated, ^3H_2O) to one side of the membrane and monitoring the rate at which equilibrium is established between the inner and outer solutions. This movement is much faster for biological membranes than for synthetic phospholipid membranes, an observation that puzzled biophysicists for many years, until the discovery of aquaporins in 1992. These protein-lined pores are very narrow, discriminating against ions but allowing the passage of water molecules in

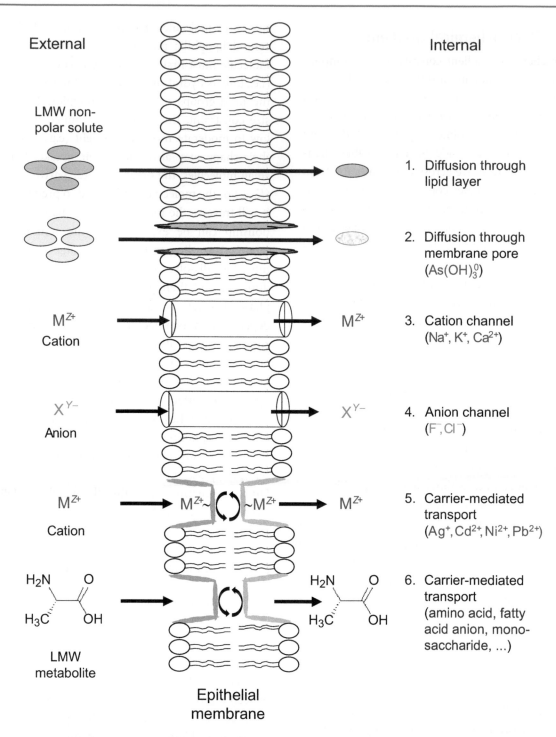

External

LMW non-
polar solute

1. Diffusion through
   lipid layer

2. Diffusion through
   membrane pore
   $(As(OH)_3^0)$

M^{Z+}
Cation

M^{Z+}

3. Cation channel
   (Na^+, K^+, Ca^{2+})

X^{Y-}
Anion

X^{Y-}

4. Anion channel
   (F^-, Cl^-)

M^{Z+}
Cation

$M^{Z+}\sim$ $\sim M^{Z+}$ M^{Z+}

5. Carrier-mediated
   transport
   $(Ag^+, Cd^{2+}, Ni^{2+}, Pb^{2+})$

H_2N O
H_3C OH

LMW
metabolite

H_2N O
H_3C OH

6. Carrier-mediated
   transport
   (amino acid, fatty
   acid anion, mono-
   saccharide, ...)

Internal

Epithelial
membrane

Figure 3.5 Summary of different modes of transport of solutes across biological membranes (LMW = low molecular weight; M^{Z+} = a cation with a charge of $+Z$; X^{Y-} = an anion with a charge of $-Y$). The ellipses with their two side chains correspond to phospholipids, the ellipse being the polar head-group and the side chains being the non-polar tails (cf. Figure 3.1). The ⟳ symbol represents a transmembrane carrier.

single file, at a rate that nevertheless far exceeds the diffusion of water across a protein-free synthetic membrane. In addition to water, other small and uncharged substrates can move into and out of cells through aquaporin channels; examples include ammonia, carbon dioxide, urea, glycerol, antimonite ($Sb(OH)_3^0$) and arsenite ($As(OH)_3^0$), the latter two molecules being of ecotoxicological interest (Section 6.8.7.3).

The range of solutes that are transported across biological membranes includes many charged ions and

Table 3.1 **Examples of solutes that are transported across cell membranes.**

Substance transported	Means of transport	Examples of normal solutes	Examples of potentially toxic solutes
Small non-polar (lipophilic) molecules	Simple, unmediated diffusion across the lipidic portion of the membrane	Hormones – thyroxine – cortisol 	Organochlorine pesticides & polychlorinated biphenyls (PCBs) Polycyclic aromatic hydrocarbons Hydrogen cyanide (HCN) Elemental mercury (Hg(0)) Some organometallic compounds – **methylmercury** chloride (CH_3HgCl) Neutral metal complexes with negligible polarity – $HgCl_2^0$ – $Cd(diethyldithiocarbamate)_2^0$ Okadaic acid, diarrhoeal shellfish poisoning toxin (Section 9.3.1.4)
Small neutral polyhydroxo species	Diffusion through a membrane pore	Urea Glycerol	Arsenite, $As(OH)_3^0$ Antimonite, $Sb(OH)_3^0$
Cations	Channels	Sodium, Na^+ Potassium, K^+ Calcium, Ca^{2+}	Thallium, Tl^+ Cadmium, Cd^{2+}
Anions	Channels	Chloride, Cl^- Phosphate, PO_4^{3-}	Fluoride, F^- Arsenate, AsO_4^{3-}
Cations	Carriers	Cuprous copper, Cu^+ Ferrous iron, Fe^{2+}	Silver, Ag^+ Lead, Pb^{2+}
Small polar molecules	Carriers	Amino acids, peptides Monosaccharides Vitamins Fatty acid anions	Pharmaceuticals – codeine, ibuprofen, paracetamol (acetaminophen), Prozac® Plant alkaloids – cocaine, morphine, nicotine, quinine Assorted xenobiotics – polyfluorinated alkyl acids – conjugates of compounds metabolized by Phase I, Phase II and Phase III reactions
Neurotransmitters	Ion-coupled (H^+, Na^+, Na^+/Cl^-, K^+) carriers	Monoamines Gamma amino butyric acid (GABA) Glycine, L-glutamate	L-glutamate agonists, e.g., domoic acid (Section 9.3.1.1)

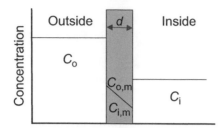

Figure 3.6 Dissolution–diffusion model for the movement of lipophilic substrates across biological membranes. The shaded rectangle in the middle of the figure represents the biological membrane, of thickness d. The solute on the outside is present at a concentration (C_o) higher than that in the internal solution (C_i), and passive diffusion will occur from left to right. Within the membrane itself, the solute concentration at the external surface ($C_{o,m}$) is greater than that at the internal surface ($C_{i,m}$).

metabolites, for which neither diffusion across the lipid portion of the membrane nor diffusion along an aquaporin channel can suffice to maintain homeostasis. Transmembrane transport of these polar substrates involves the membrane's integral proteins, which may operate as channels, carriers or pumps. Ion channels (Figure 3.5, pathways 3 and 4) are similar to aquaporins, in that they involve integral proteins, but they are configured to allow the passage of cations or anions. Since these are *charged* solutes, the rate of transport depends not only on the concentration difference between the outer and inner solution, but also on the difference in electrical potential across the membrane. Measurements of the transmembrane potential gradient in living cells generally fall in the 60–140 millivolt (mV) range, with the inside of the cell being negative with respect to the external solution. Potential differences in this range strongly favour the accumulation of cations in the **cytosol** and the exclusion of anions. For example, for a gradient of 60 mV, the intracellular concentration of a monovalent cation such as Na^+ or K^+ at equilibrium would be about tenfold higher than the external concentration.

The selectivity of ion channels is based in part on the electrical charges of the amino acids that encircle the entry into the channel. For example, negative charges at the mouth of the channel will repel anions and attract cations, whereas the opposite is true for positive charges. One of the first such channels to be identified was the gramicidin channel. Gramicidin is a mixture of six small peptides produced by a soil bacterium, *Bacillus brevis*; collectively, these peptides possess antibiotic properties. They insert themselves into lipid bilayers and adopt a shape that creates a narrow hole in the middle with an internal

diameter equivalent to about two adjacent water molecules. The gramicidin channel is selective for monovalent cations such as Na^+ or K^+. When the cation concentration in the external solution is increased, the channel exhibits a saturation-type response, with the flux through the channel levelling off and tending towards a plateau. This response reflects the limited transport capacity of the channel – it is thought that when one cation enters the channel, a second cation cannot follow it until the first cation has diffused the length of the channel and exited. In other words, the saturation response reflects the existence of a finite number of channels per unit surface area of the membrane.

The gramicidin channel is always open, but other ion channels can be 'gated', meaning that they can exist in an open or closed state. The acetylcholine receptor (AChR) channel is a well-studied example of a **ligand-gated ion channel**, where the on–off switch for neural activity is controlled by the binding or release of a 'ligand' molecule to a receptor site in the channel protein (Box 3.3). The binding of the ligand molecule causes a change in the conformation of the channel and causes it to open; such 'allosteric' changes in conformation are a common feature of proteins, particularly for enzymes. In this case, acetylcholine is the ligand molecule that binds to the channel protein. Other examples are discussed in

Box 3.3 Very different definitions of the word 'ligand'

The word 'ligand' is used here to mean a molecule, in this case acetylcholine, that binds to a receptor site in the channel protein and in so doing promotes an allosteric change in the shape (conformation) of the protein that causes the channel to open. This coupling between ligand binding and channel opening is referred to as 'gating'. A ligand can be an extracellular mediator, such as acetylcholine, or an intracellular mediator such as an ion or nucleotide.

Elsewhere in this book (notably in Chapter 6), 'ligand' is used to define an ion or a molecule that has the ability to form coordinate bonds with metal cations, forming a complex. In this case, the 'ligand' is the receptor, whereas in the ion channel case the channel protein is the receptor!

In still other parts of the book (notably in Chapters 7 and 8), 'ligand' refers to molecules that bind to specific cellular receptors and thereby induce a physiological response.

Section 7.8.2. In **voltage-gated ion channels**, the opening and closing of the channel are controlled by the electrical potential of the membrane.

The AChR channel is impermeable to anions and at high cation concentrations, it saturates in the same manner as the gramicidin channel. It is of low specificity, allowing many different cations to pass through it. Other gated ion channels (e.g., for Na^+, K^+, Ca^{2+}) include what are known as 'selectivity filters' and demonstrate higher selectivity. These filters discriminate among ions on the basis of their effective size by creating narrow restrictions in the channel, which force the ion into intimate contact with the channel walls. In aqueous solution, cations all exist as hydrated ions, meaning that their effective radius is *greater* than the radius of the ion itself. To pass through the narrow restrictions in the channel, the cation must shed most or all of its associated water molecules. To ease the shedding process, which would otherwise be energetically very costly, the channel is lined with hydrophilic electron-rich donor groups that interact with the cation to replace the water molecules and allow the cation to pass the restriction. For example, the potassium channel is the narrowest of the ion channels, and it achieves its selectivity for the K^+ ion in this manner. The hydrophilic binding sites in the channel are relatively weak, but they are perfectly placed to interact with the K^+ ion and bind strongly enough to replace the water molecules in the ion's primary hydration sheath. However, for the Na^+ ion (which is smaller and holds onto its water molecules more tightly), these water molecules remain associated with the central cation, and the hydrated Na^+ ion remains too large to pass through the potassium channel.

$$H_2O$$
$$H_2O \cdots \underset{\underset{H_2O}{|}}{\overset{\overset{|}{H_2O}}{M^{z+}}} \cdots OH_2$$

Hydrated cation

Before leaving the subject of ion channels, let us summarize what we have learned in this section. Cell membranes contain an array of channels, which are selective for the passage of essential cations (Na^+, K^+, Ca^{2+}) or anions (Cl^-). These channels involve integral proteins

that span the membrane. The ions move through the channels by simple diffusion, and this movement is not coupled to an energy source, meaning that channels can only move solutes 'downhill' (passive transport). Some channels are always open, but most can be open or closed, with gating implemented by a change in the conformation of the membrane protein. Passage of a cation through a channel requires the stripping of most if not all of the water molecules that are bound to the ion, and this requirement allows channels to discriminate among different cations. Finally, channels are able to move ions across biological membranes up to 10,000 times *faster* than membrane carriers, the transport mechanism discussed in the next section. Any chemical that interferes with these normal cellular processes will induce toxicity.

3.1.2.3 Carrier-mediated Transport

The transport mechanisms discussed to this point work well for small to moderately sized lipophilic molecules and for ions. However, to move larger hydrophilic molecules (e.g., amino acids, monosaccharides) across a biological membrane or to perform active transport against an electrochemical gradient, carrier-mediated transport is required (Figure 3.5, mechanisms 5 and 6; Box 3.4). As was the case for the ion channels, these transporters are integral, membrane-spanning proteins that are members of the Major Facilitator Superfamily (MFS), but in their transport activity they resemble enzymes rather than pores. In the carrier model, it is assumed that the protein

Box 3.4 Examples of transmembrane carriers

Carriers used for essential ions and 'borrowed' by non-essential elements (in **bold**)

- ATP-driven pumps: H^+ Na^+ K^+ **Tl^+** Ca^{2+}
- Divalent metal transporter, DMT: Fe^{2+} Mn^{2+} **Cd^{2+}**
- Zrt- & Irt-like protein, ZIP: Fe^{2+} Mn^{2+} Ni^{2+} Zn^{2+} **Pb^{2+}**
- Cu-transporter, CTR: Cu^+ **Ag^+**
- Cation diffusion facilitator, CDF: Fe^{2+} Mn^{2+} Zn^{2+}
- Anion transporter: PO_4^{3-} **AsO_4^{3-}**

ATP-binding-cassette transporters (ABC transporters)

- Multidrug resistance (MDR) protein (also known P-glycoprotein); extremely wide substrate selection, able to transport hundreds of structurally unrelated compounds out of cells (Section 7.5.3).

can adopt two conformations, one in which it is open to the outside solution and the other in which it is open to the cytosol. Once the substrate binds to the carrier protein, a conformational change exposes the substrate on the opposite face of the membrane, from which it can be released. This change in conformation is represented by the circular arrows for pathways 5 and 6 in Figure 3.5.

The protein carrier molecules behave like enzymes, but unlike enzymes they do not 'transform' their substrate; they simply deliver it to the other side of the membrane, unchanged. Carriers can distinguish between stereoisomers (e.g., between L- and D-amino acids, or between D- and L-glucose), with the 'wrong' stereoisomer acting as a competitive inhibitor. Similarly, carriers exhibit saturation (Michaelis–Menten) kinetics, and the substrate flux across the membrane tends towards a plateau as the substrate concentration is increased (Section 3.4.1). Examples 5 and 6 in Figure 3.5 show a single substrate moving across the membrane (i.e., a **uniporter**). However, the transport of one substrate may be coupled with that of a second substrate. With a **symporter**, the substrates cross the membrane in the same direction, whereas with an **antiporter** they move in opposite directions (Figure 3.7).

In contrast to transmembrane pores and channels, carriers can be linked to an energy source, and thus are able to move substrates 'uphill', against the electrochemical gradient; such carriers are referred to as **pumps**. The energy needed to drive this **active transport** can come from a variety of sources:

- *Coupled transporters* draw upon the energy stored in concentration gradients to link the *downhill* transport of one substrate, typically an inorganic ion, to the uphill transport of another solute (this approach works in both directions: Figure 3.7, symport and antiport).
- *ATP-driven transporters* link uphill transport to the hydrolysis of ATP → ADP. These pumps fall into three separate categories: P-driven pumps, which establish key electrochemical gradients for such ions as Na^+, K^+, H^+ and Ca^{2+}; V-type transporters, which pump H^+ ions into intracellular vesicles and acidify them; and ABC-transporters (ATP-binding-cassette transporters), which primarily move polar metabolites into and out of cells.
- *Light- or redox-driven pumps* couple the uphill transport of the substrate with energy derived from incident light or from a redox reaction.

In animal cells, the Na^+ ion is often the inorganic ion involved in **cotransport**, since its external concentration is much higher than its internal concentration and it thus provides a positive electrochemical gradient into the cell. The Na^+ is subsequently pumped back out of the cell by the Na^+–K^+ pump. By maintaining the Na^+ gradient, the pump indirectly supports the cotransport of other solutes. The energy-requiring Na^+–K^+ pump is an antiporter, with Na^+ and K^+ travelling in different directions with a specific stoichiometry (3:2); the mismatch in charges is responsible for generating the transmembrane potential. In microorganisms and plants, and in internal membranes that enclose vacuoles, the proton (H^+) is usually the ion that is cotransported, instead of Na^+. The necessary proton gradient is maintained by the action of the H^+-pump.

As a guide to the transport of solutes across cell membranes, Table 3.1 provides some examples of the 'normal' traffic of ions and low-molecular-weight solutes into and out of living cells. More importantly, the table also includes examples of toxic substances that can 'borrow' these mechanisms and enter cells in the same manner as essential nutrients. Lipophilic toxicants such as organochlorine molecules (Section 7.7.1) cross the plasma membrane by the relatively indiscriminate process of partitioning from the external environment into the lipid bilayer. Accordingly, it is more difficult for the cell to regulate their movement than is the case for ions and LMW metabolites. In the next section, we shift our focus

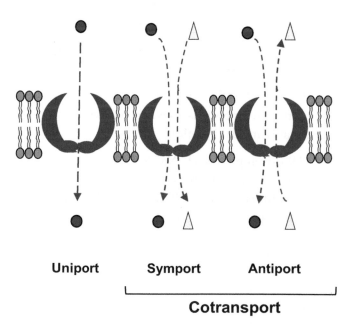

Uniport   Symport   Antiport

Cotransport

Figure 3.7 Types of carrier-mediated transport. The transported molecule is shown as a red filled circle and the cotransported ion as a triangle.

from individual solutes to consider the transmembrane transport of fluids and small particles (colloids).

3.1.3 Endocytosis

In discussing the composition and structure of biological membranes (Section 3.1.1), we referred to the fact that membrane proteins can sometimes move from the plasma membrane to internal (organelle) membranes. Another example of the dynamic nature of plasma membranes is the ability of cells to invaginate portions of their plasma membranes and form small membrane-enclosed vesicles in the cytosol, a process known as **endocytosis**. In addition to acting as a mechanism for recycling membrane phospholipids and proteins, the vesicles formed in this manner may enclose substances that were originally present in the external medium or that were previously adsorbed onto the cell's surface. In this manner, the endocytotic process can be harnessed to acquire nutrients by moving fluids, large molecules and colloidal particles into the cell; contaminants may be inadvertently imported in this manner too, including those that would not typically cross intact membranes. The process can also be used in the reverse sense (**exocytosis**) to eliminate waste products, isolate the metabolic products of toxicants that have entered the cell or transport large molecules into the circulatory system. For example, in vertebrates thyroid hormones are transported from the thecal cells of the thyroid gland into the bloodstream by exocytosis (Section 8.2.4).

There are two types of endocytotic activity, defined by the size of the resulting vesicle. In **pinocytosis** or 'cell drinking', a widespread process in eukaryotic cells, the invagination results in the ingestion of the extracellular fluid (with its solutes and colloids) and the formation of small intracellular vesicles (~150 nm). In some cases, a specific receptor on the cell surface binds to a recognized substrate, and this binding triggers the invagination process. In this receptor-mediated mechanism, both the receptor site and the substrate are internalized and then dissociate. Molecules that are known to trigger receptor-mediated pinocytosis include low-density lipoproteins, some hormones, and transferrin, an iron-binding protein; some toxins may also enter cells by this route. In **phagocytosis**, the invagination process is initiated by contact with a small particle and results in the formation of a somewhat larger intracellular vesicle (>250 nm) that contains the engulfed particle. Physical contact with the particle surface is necessary but not sufficient of itself to trigger phagocytosis, because here too some form of recognition is necessary.

In both these examples of endocytosis, one could argue that neither the liquid nor the particle has truly entered the cytosol, since each is still held within a membrane-bound vesicle. However, the vesicles are short-lived and typically fuse with organelles known as **endosomes**. These vesicles serve as 'sorting stations' and determine the fate of the ingested material (e.g., recycling in the case of membrane components, or digestion in the case of nutrient material).

It is clear that endocytosis plays an important nutritive role in single-celled organisms such as protists, and in this way it may be an entry route for environmental contaminants that are associated with small particles. Similarly, there is microscopic evidence showing the presence of metals in membrane-bound vesicles in the gills and digestive cells of some deposit- or filter-feeding molluscs (Marigómez *et al.*, 2002). These organisms are known to bioaccumulate metals to a high degree, and it has been suggested they are able to take up particle-bound metals by endocytosis followed by **intracellular digestion**. However, it is also possible that dissolved metals are taken up and then shunted into the membrane-bound vesicles for storage or detoxification. In other words, the role of endocytosis as a vector for exposure to environmental contaminants remains controversial and understudied. We revisit this question in the section on uptake routes (Section 3.2.8).

3.1.4 Transcellular Transport (e.g., Gill; Intestine; Lung)

To this point we have considered the passage of solutes, colloidal particles and extracellular fluids across plasma membranes. What we have learned is directly applicable to single-celled organisms such as bacteria or microalgae, but for animals and higher plants we must also consider transport across epithelial barriers. At such living surfaces, transport can occur in between or across cells (Figure 3.8). Normally the junctions between epithelial cells are 'tight' and there is minimal passage of solutes or solution via this route. It follows that to cross the epithelium, ions and molecules must cross a plasma membrane twice, at the apical (outward-facing) surface and at the basolateral (inward-facing) surface that interacts with the internal circulatory network (blood, lymph; phloem, xylem).

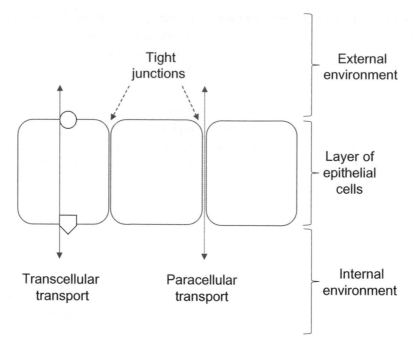

Figure 3.8 Transport routes for solutes crossing epithelial barriers.

Configuration of cells into complex membranes or epithelia often confers 'orientation' to those cells. Cells within an epithelium may move chemicals in a specific direction. For example, ions that enter a cell through transporters at the outer or apical surface of an epithelium may be transported from the cell to the blood by a pump located at the inner basolateral membrane. Because the electrochemical gradients across the apical and basolateral domains will differ, the nature of the transporters will also differ. In Section 3.2 we consider the epithelial barriers that are encountered in different plant and animal structures (roots, foliage; skin, gills, gut, lungs, olfactory organs).

3.1.5 Ecotoxicological Perspective on Transmembrane Transport Processes

To conclude this section on transmembrane transport, let us consider the ecotoxicological ramifications of what we have learned. We set out to describe the various pathways that toxicants can use to enter cells and wreak intracellular havoc (Table 3.1). Based on what we have seen, we can now predict the properties that will be of particular importance with respect to the toxicant's ability to enter cells. For organic and **organometallic** toxicants, these properties will include water solubility, hydrophilicity, lipophilicity and molecular size. To a lesser extent, these same properties will also affect the uptake of inorganic contaminants such as individual metals and metalloids.

However, as we discuss in Chapter 6, for non-essential elements such as cadmium, lead and mercury, it will also be important to consider 'ion mimicry'. To enter cells via pathways designed for essential elements such as copper, iron and zinc, the non-essential metal will have to masquerade as an essential metal and 'fool' a pore or a carrier that normally transports essential elements.

Linked to this idea of mimicry, there are also examples where the non-essential metal can 'hitch a ride' on a low-molecular-weight (LMW) metabolite such as citrate, histidine or thiosulphate (Figure 3.9). Each of these substrates is recognized as a legitimate substrate by various plasma membranes, notably in bacteria, algae and **enterocytes**. However, as is discussed in Chapter 6, each substrate can also form complexes with metals such as cadmium. It

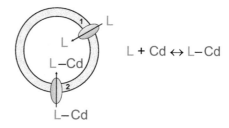

$$L + Cd \leftrightarrow L{-}Cd$$

Figure 3.9 Piggy-back transport of a toxicant (cadmium) across a biological membrane. The normal role of the membrane carrier is to transport 'L' (a low-molecular-weight metabolite) into the cell, but it can be fooled into transporting L along with Cd as the L–Cd complex.

turns out that some membrane transporters are not very selective, and they internalize not only the free substrate but also the substrate–metal complex. In other words, the metal can be piggy-backed into the cell in association with LMW metabolite.

In addition, we have alluded on several occasions to the key role of the cell membrane in maintaining cellular homeostasis. Clearly the integrity of the cell membrane is essential, since any substance that interacts with the membrane and changes its properties (fluidity; permeability; leakiness) may induce deleterious effects. For a more complete description, see Section 2.3.4.3 on narcosis and the target lipid model. Similarly, given the importance of controlled transmembrane transport of inorganic ions and metabolites, any substance that interacts with ion channels and carriers and interferes with their normal functioning may induce toxicity. In other words, in describing the critical role of transmembrane transport, we have also identified a number of potential toxicity 'targets', the normal functioning of which can be impeded by interaction with toxicants.

3.2 Uptake Routes

In terrestrial animals, the three principal routes of uptake of xenobiotic chemicals are the skin, the respiratory system and the gastrointestinal tract. Respiratory and digestive epithelia are specialized to permit selective molecular uptake, and their very large surface areas and extensive vascular perfusion evolved to maximize gas and nutrient exchange, respectively. It follows that these tissues are more vulnerable to the entry of toxic chemicals than is the skin. In aquatic animals, gills are modified for both respiratory exchange and the regulation of electrolytes, and this may permit entry of dissolved substances through diffusion or ionoregulatory pathways. In plants, too, root surfaces are specialized for the uptake of water and electrolytes and present a mode of chemical accumulation that is in marked contrast to foliar uptake.

3.2.1 Skin

The skin of terrestrial animals and amphibians acts as a protective barrier to isolate them from their environment, and it was originally thought to be impervious to environmental toxicants. However, it is now regarded as an important route of entry for several xenobiotics, although uptake is slower than through other epithelia.

In mammals, the skin consists of three layers: the outer, non-vascularized layer, the epidermis; the middle, highly vascularized layer, the dermis; and the inner layer, hypodermis, which consists of connective and adipose tissue. Birds, reptiles and fish lack the hypodermis. The primary barrier to chemical absorption is the outer layer of the epidermis, which contains a dense lipid and keratin matrix. Organic solvents may cross this layer by solubilizing constituent lipids. Particularly effective are small molecules with some lipid and aqueous solubility such as dimethyl sulphoxide (DMSO), chloroform and methanol. These may increase the permeability of the skin to other compounds, which may be absorbed in solvent-aided transport.

In some aquatic animals, the skin is protected by a layer of **mucus** that is continually being produced and sloughed off. Since the mucopolysaccharides that make up the mucus can interact with ionic and hydrophobic solutes, this sloughing-off process helps limit an animal's exposure to waterborne contaminants. For embryos, there are multiple added layers, including exterior layers of jelly in some fish and amphibian species, the egg chorion in all fish species (multi-layered), hard eggshells in birds, and egg 'purses' in some sharks (e.g., rays). Presumably the hair of mammals, the feathers of birds and the scales of reptiles also play a role in reducing the contact of contaminants with skin. Indeed, in some species contaminants such as mercury are detoxified by being moved from the circulatory system into the hair or feathers, which are then shed.

3.2.2 Lungs

In air-breathing vertebrates, the respiratory system consists of three regions: the nasopharyngeal canal, the trachea-bronchial region and the pulmonary region. The first two regions are protective in nature and are responsible for filtering large particles (e.g. >10 μm). These particles are trapped in mucus secreted by goblet cells and transported by ciliated epithelial cells to the oral or nasal cavity, where they are swallowed or expelled.

Dense particles have greater momentum in the upper respiratory tract and tend to collide with the walls of the tract, particularly where the airways curve. Smaller particles (i.e., those less than 1 μm) are carried to the lower respiratory tract, the pulmonary region. This region consists of increasingly subdivided bifurcating lobes ending in terminal bronchioles, which are small tubes of approximately 0.5 to 1 mm in diameter. These further subdivide into alveolar ducts and alveoli, which are tiny sacs (150–350 μm). Retention of respirable particles drops markedly between diameters of 10 and 1 μm, remains low from 1 to 0.1 μm, and increases again from 0.1 to 0.01 μm. Toxicants may enter lungs on this finely divided particulate matter, as aerosols (e.g., sprayed pesticide formulations) or as gases (e.g., NO_X, ozone, peroxyacyl nitrates). The absorption of any toxicants entering the lungs in this manner is highly dependent on their water solubility.

The presence of many foreign chemicals and particles in the lungs stimulates chemotactic and phagocytic activity in a variety of specialized protective cells such as macrophages. Phagocytic cells activated in this way release potent oxidants, notably hydrogen peroxide, which while aiding the phagocytic process may exacerbate the potential for tissue damage through elevated oxidant activity.

3.2.3 Gills

Gills are the primary respiratory organs of aquatic animals and are typified by a thin epithelium, which serves to minimize the distance between the oxygenated external medium and the internal body fluids. Gills range from simple external epithelial appendages to elaborate structures containing numerous folds or lamellae (Figure 3.10) enclosed in a branchial chamber that is ventilated by a continuous stream of water. In most aquatic organisms, the remainder of the external epithelium is relatively impermeable.

Like lungs, gills are designed to present a large, highly permeable surface area for gas exchange and are the primary sites for the exchange of dissolved gases and chemicals. The ratio of the gill surface area to body mass is particularly large because of the severe limits on oxygen availability from water. In contrast to air (21% oxygen by volume), oxygen concentrations in water that has been equilibrated with the atmosphere are quite low, ranging from 15 mg/l at 0 °C to 8 mg/l at 30 °C. Although the large gill surface area for gas exchange compensates for the low concentration of dissolved oxygen, it increases the exposure of gill-breathing animals to waterborne toxicants. The destruction of gill epithelial cells and death by suffocation is a known mechanism for the acute lethality of many waterborne chemicals.

Gill epithelia also contribute to important functions such as ionic and osmotic regulation. In freshwater animals, electrolytes are actively taken up by specialized ion-transporting cells in gill epithelia. In fish, these are called chloride cells. Excess water accumulated through osmosis is excreted through a kidney or kidney-like organ. Marine fish continually 'drink' seawater to compensate for osmotic water loss, and the gut epithelium is responsible for the excretion of excess electrolytes.

In gills, the exchange of gases and other solutes is facilitated by a countercurrent flow wherein the circulation of blood through the lamellae runs counter to the flow of water past the gill surface. The laminar flow of water creates a boundary layer immediately adjacent to the gill surface, which is further modified by the presence of mucus. Many of the chemical interactions occurring at the gill surface are dictated by the chemistry of this boundary layer.

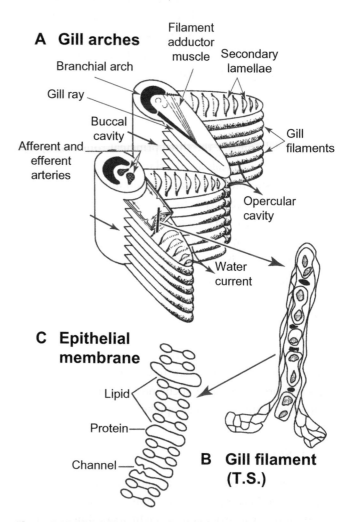

Figure 3.10 Fish gill structure from the tissue to the molecular level. T.S., transverse section. Reproduced from Wright & Welbourn (2002) with permission from Cambridge University Press.

Gill surfaces are composed of one or more layers of epithelial cells joined by tight junctions, which may still allow limited extracellular solvent and solute movement. The outer membrane of these cells is host to a variety of ion channels or carriers, which assure the uptake and excretion of physiological electrolytes and trace metals. The hydrogen ion (H^+) and several metals (Cd^{2+}, Ni^{2+}, Zn^{2+}) have been shown to compete with calcium (Ca^{2+}) for transporters at gill surfaces. Calcium is responsible for maintaining the integrity of cellular junctions in gill epithelia, and low ambient calcium concentrations may increase leakiness of these normally tight junctions. The gills of many organisms also contain cells that excrete mucus continuously to protect the external surface of the epithelium. The mucus limits the accumulation of particulate and dissolved material that binds to the mucus and is lost as the mucus is sloughed off. In fish, the mucus has a rich glycoprotein content and forms a polyanionic matrix capable of binding many cationic trace metals. The rate of mucus excretion increases dramatically when fish respire water containing high concentrations of chemicals that can irritate or damage the gill epithelium.

3.2.4 Digestive System

The absorption of toxicants through the digestive tract represents a major uptake route in multicellular animals. Molecular uptake is, after all, the major function of the gastrointestinal system, which is superbly adapted for this purpose. For example, the apical surface of the intestinal epithelium is covered with microvilli, which serve to increase the surface area available for transmembrane transport. In addition, the apical membrane hosts a very high density of the transporters needed for the uptake of the nutrients liberated by the digestive processes in the intestinal lumen.

The digestive tract may be considered as an extension of the external medium, and all its contents may be regarded as being outside the body. Uptake of ions and nutrients can occur along the whole length of the digestive system, including the mouth and rectum, although in mammals most absorption occurs in the small intestine. The gastrointestinal tract possesses specialized transport systems for certain nutrients such as amino acids, fatty acids, carbohydrates, iron, sodium and calcium, although many lipophilic chemicals are absorbed by simple diffusion (see Section 3.6.2.1).

In the gastrointestinal tract, chemicals are in an aqueous medium with variable pH, and their entry may be strongly influenced by the prevailing pH. If a chemical is an organic acid or base, it is most readily absorbed as the non-polar or non-ionized form of the acid or base conjugate pair. Organic acids, HA, dissociate to produce anions A^-, and organic bases combine with hydrogen ions (H^+) to form cations BH^+. The extent of dissociation is determined by the pH and the acid dissociation constant, K_a, according to the Henderson–Hasselbach equations (Box 3.5).

Box 3.5 Henderson–Hasselbach equations: the relationships among pH, pK_a and the dissociation of weak acids and bases

For weak acids (HA)	For weak bases (HB^+)
$HA \rightleftharpoons H^+ + A^-$	$HB^+ \rightleftharpoons H^+ + B$
$K_a = \frac{[H^+][A^-]}{[HA]}$	$K_a = \frac{[H^+][B]}{[HB^+]}$
$\log K_a = \log\left[\frac{[H^+][A^-]}{[HA]}\right]$	$\log K_a = \log\left[\frac{[H^+][B]}{[HB^+]}\right]$
$\log K_a = \log[H^+] + \log\frac{[A^-]}{[HA]}$	$\log K_a = \log[H^+] + \log\frac{[B]}{[HB^+]}$
$-\log[H^+] = -\log K_a + \log\frac{[A^-]}{[HA]}$	$-\log[H^+] = -\log K_a + \log\frac{[B]}{[HB^+]}$
$pH = pK_a + \log\frac{[A^-]}{[HA]}$	$pH = pK_a + \log\frac{[B]}{[HB^+]}$

As defined here, pK_a is the negative $\log_{10}$ value of the dissociation constant, K_a.

For a weak acid such as benzoic acid, with a $pK_a = 4$, the non-ionized (neutral) form will predominate in the mammalian stomach (pH 2) and this will favour absorption by diffusion (right-hand side of equation (3.1)). The ionized form of benzoic acid will predominate in the small intestine (pH 6) and absorption will not be favoured (left-hand side of equation (3.1)). Nevertheless, the enormous surface area of the small intestine still facilitates the absorption of substantial quantities of the benzoate anion.

$$\tag{3.1}$$

3.2.5 Olfactory System

By their very nature, olfactory organs are designed to respond to substances in the ambient air and water. Indeed, the epithelial surfaces of these organs are densely populated with receptor cells that are sensitive to proteins, small peptides, and derivatives of amino acids and fatty acids. These molecules are normally detected by G-protein-coupled receptors (GPCRs), the largest family of cell-surface receptors. These receptors play a key role in such chemoreception phenomena as social and reproductive behaviour, prey detection and predator avoidance (Scott and Sloman, 2004). The receptors are also potential targets for environmental toxicants. Examples of the impairment of chemoreception are discussed in Chapter 6 for metals (e.g., copper, cadmium).

Unlike other organs that are exposed to the ambient environment (e.g., gills, gastrointestinal tracts), olfactory organs are generally not considered as 'routes of entry' for toxicants. In terrestrial organisms, these organs are exposed to air. Airborne concentrations of toxicants may be of concern for terrestrial life, particularly humans in their workplace and wildlife inhabiting densely industrialized environments. Examples would include the exposure of wildlife to point sources of gases such as sulphur dioxide (SO_2) or fine particulate material (dust) and their short-term exposure to sprayed pesticides. For fish, however, there is evidence from laboratory experiments that inorganic Hg(II), Cd(II) and Mn(II) can enter the nerves enervating the water-exposed **olfactory rosette**

and migrate towards the brain by axonal transport. This transport route has also been suggested for some organometals (e.g., tributyltin). In both cases, axonal transport could potentially circumvent the blood–brain barrier that prevents many toxicants from reaching cerebral tissue.

3.2.6 Plant Foliage and Roots

Toxicants enter plants by the same pathways as nutrients, through foliar or root uptake. In the former case, the extensive leaf canopy provides a vast surface area for the uptake of gases, liquids and particulates. Stomata are the major routes for foliar uptake of CO_2 and release of O_2, as well as being the route of uptake of gaseous contaminants such as SO_2. They are usually located on the lower (abaxial) leaf surface, whereas the upper (adaxial) surface is covered for most plant species with a wax-like cuticle. The deposition of particles onto leaves brings contaminants that are sorbed to the particles into contact with the plant surface. These airborne contaminants include metals, radionuclides, polycyclic aromatic compounds (PACs) and sulphur (often associated with metals), all of which become bound to leaf cuticles. Rainfall also interacts with vegetation surfaces, leaching elements from the particles and favouring their stomatal uptake. Lipophilic compounds mobilized in this manner can also enter the foliage by diffusion through the cuticle. Examples of toxicants known to enter plants in this manner are compiled in Box 3.6.

In addition to washing off powders and aerosols deposited on leaf surfaces, rainfall (particularly if acidic) also leaches substances released by the plant onto the leaf surface. Together, these substances may be transferred to the soil where some of them (e.g., metals) become available to the root systems of plants The throughfall and stemflow resulting from rainfall are important

---

Box 3.6 Examples of toxicants known to enter plants via stomata

Gases: Sulphur dioxide (SO_2)
             Nitrogen oxides (NO_X)
             Ozone (O_3)
             Elemental mercury (Hg^0)
Solutes: Hydrogen fluoride (HF)
             Metals (Cd, Cu, Pb, Zn)
             Metalloids (As, Sb)
             Herbicides (atrazine, glyphosate)
             Polycyclic aromatic compounds (PACs)

vectors moving contaminants and nutrients from the atmosphere to the soil (Section 11.1).

Let us now consider the role of plant roots. Root epidermal cells including their root hairs are in direct contact with the soil or other nutrient media, and they control most of the uptake of essential nutrients, through highly regulated processes. Water and nutrients (and dissolved contaminants) move radially from the epidermis through the root cortex via the **symplast**; dissolved material crosses the cortex and encounters its innermost layer, termed the **endodermis**. The endodermis surrounds the **stele**, the central cylinder of the plant. Parts of the cell walls of the stele have a layer of **suberin**, termed the Casparian strip, which has been considered as a partial or selective barrier for the diffusion of solutes from the soil to the stele. It may also provide the rest of the plant with some protection from potentially toxic substances. However, there is still no complete understanding of the role of the endodermis with its Casparian strip.

Three types of transport from the soil to the root have been described (Barberon and Geldner, 2014). The first is the so-called apoplastic (passive, mass flow) route that can occur across the cell walls and in the intercellular space, and the second is through cellular connections, called the symplastic or active pathway, involving **plasmodesmata**, which provide cytoplasmic continuity between individual adjacent cells. A third route, a coupled transcellular pathway, has been described relatively recently, with influx and efflux transporters. These transporters are distributed at opposite ends, or poles, within the cell, not only in the endodermis but also in cortical and epidermal cells of the root. Figure 3.11 illustrates the three types of pathway.

Once water, nutrients or contaminants have entered a vascular plant, their 'long-distance' transport is accomplished through the **xylem**, a series of connected non-living cells that occupy the **stele**, the central cylinder of the plant. A second long-distance transport system, the **phloem**, also in the stele, is responsible for distributing the products of photosynthesis to the rest of the plant. In the xylem, water and dissolved substances are moved up through the stems and into the leaves or other plant

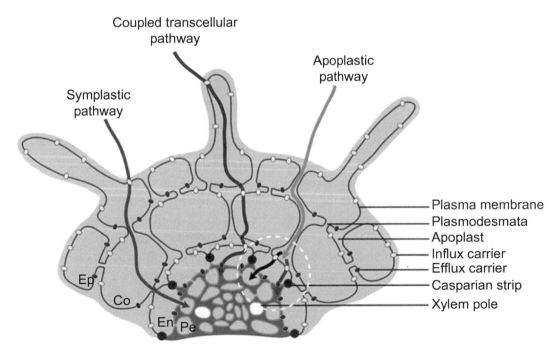

Figure 3.11 Illustration of the three different pathways involved in the transport of solutes from the soil to the plant root endodermis. The symplastic pathway (indicated in grey) requires at first a selective uptake into a cell and then transport from one cell to the other through plasmodesmata. The coupled transcellular pathway (in red) involves influx (in yellow) and efflux (in purple) carriers to transport solutes from one cell to the other. The apoplastic pathway (in blue) corresponds to a passive transport in the extracellular space and is blocked by the Casparian strip at the level of the endodermis. Co = Cortex; En = endodermis; Ep = epidermis; Pe = pericycle. Adapted from Barberon and Geldner (2014) with permission from Oxford University Press on behalf of the American Society of Plant Biologists.

outgrowths by three processes, which are all passive: transpiration (based on the negative pressure in the leaves), capillarity (based on adhesion and cohesion) and root pressure (based on negative water potential in the root cells). The rates of movement of toxicants across plant surfaces and within the plant, as well as their target sites, depend on the physicochemical properties of the toxicant (e.g., its water solubility and its partitioning between the aqueous and gas phases). Environmental conditions such as temperature, humidity, exposure to ultraviolet light and the height of the leaf boundary layer also play a role.

3.2.7 Boundary Layers

In models used to predict the uptake of a dissolved toxicant, it is normally assumed that the solute concentrations at the epithelial surface are the same as those that are measured in the bulk exposure solution. At a surface where solutes are being moved back and forth across the epithelium, the chemical conditions within the boundary layer close to the surface may actually differ from those in the bulk solution. Such 'microenvironments' have been described for fish gills, for root surfaces (the '**rhizosphere**') and for the boundary layer surrounding algal cells (the 'phycosphere'). For example, measurements of the pH of water entering and leaving a fish gill show that water passing through the gill chamber may be acidified by respired CO_2 or alkalinized by release of ammonia, depending upon the pH of the inspired water. Similarly, respiration by plant roots increases the CO_2 concentration in the soil and lowers the pH in the soil water close to the root surface. In addition, some plant species excrete low-molecular-weight carboxylic acids (e.g., malic acid; citric acid) and in this manner acidify the rhizosphere and favour the mobilization of otherwise inaccessible soil nutrients. The opposite trend can be seen in the boundary layer around a photosynthesizing algal cell, where the pH is normally higher than that in the bulk solution.

You may be wondering why we are mentioning these microenvironments. These changes in the chemical composition of the boundary layer will only be important for 'sensitive' solutes. For example, in Section 3.2.4 we describe how a change in pH could influence the chemical form of a weak carboxylic acid such as benzoic acid. Similarly, in Section 6.7.2 we discuss how a change in pH can affect the chemical form ('speciation) of a metal or metalloid. Metal speciation may also be sensitive to the presence of metal-complexing molecules that are excreted into these microenvironments. In Section 7.2 (Figure 7.8), we describe how changes in pH affect the speciation of an oestrogenic compound, triclosan, that is used as an antibacterial compound in various personal care products. The take-home message here is two-fold: (i) the chemical conditions close to an epithelial surface may differ from those in the bulk solution, and (ii) these differences may affect the uptake rate of 'sensitive' solutes.

3.2.8 Uptake by Endocytosis

At most of the entry points described previously in this section, uptake is dominated by dissolved solutes. However, having introduced the phenomenon of endocytosis (Section 3.1.3), we should briefly consider its possible role in toxicant uptake. As a starting point, consider known examples of nutrient uptake by endocytosis. Some species of **mixotrophic** algae (e.g., *Dinobryon* sp.) can assimilate particulate phosphate (in the form of bacteria) by phagocytosis, and it would be reasonable to assume that other bacteria and their cellular contents could also gain entry by this route. A second example involves iron as the nutrient and receptor-mediated pinocytosis as the mechanism. Iron in mammalian blood is transported as a complex with transferrin (two Fe^{3+} ions per molecule of the protein). To leave the bloodstream and enter a cell, the Fe-transferrin complex binds to a receptor site on the plasma membrane. The receptor site–transferrin complex enters the cell by pinocytosis, and the vesicle fuses with an endosome. The resulting vesicle is acidified, leading to the release of Fe^{3+}, which is reduced to Fe^{2+} and then moved into the cytosol via a divalent metal transporter. Other metals are known to bind to transferrin and circulate in the bloodstream (e.g., 80–90% of aluminium in blood plasma is present as the aluminium-transferrin complex). This observation has led to the suggestion that the receptor-mediated endocytotic pathway of cellular iron uptake may be available for other (non-essential) metals (Taylor and Simkiss, 2004). A third example comes from a recent study of the involvement of endocytosis in the uptake of silver nanoparticles by a marine mussel, *Mytilus galloprovincialis*. Blocking clathrin-mediated endocytosis protected against silver toxicity, a result that suggests that this uptake pathway facilitates toxicity (Bouallegui *et al.*, 2018).

These examples illustrate the potential role that endocytosis could play in toxicant uptake, notably in the case of engineered nanoparticles and microplastics (Sections 11.3 and 14.2 respectively), but a clear indication of its quantitative importance remains elusive. It is fair to say that ecotoxicologists often pay lip service to the possible role of endocytosis, but that they rarely try to quantify its importance. In contrast, the role of exocytosis is better defined, because it is an essential step in the movement of lipids across the basolateral membrane in **enterocytes**. Lipids are hydrolysed in the intestinal lumen and the hydrophobic moieties (fatty acids) diffuse across the apical membrane into the enterocytes. Within the enterocytes, the fatty acids are re-esterified and interact with cholesterol and proteins to form **chylomicrons**, which are expelled across the basolateral membrane by exocytosis into the lymphatic system, from which they enter the bloodstream via the thoracic duct. Lipophilic contaminants that have also crossed the apical membrane may associate with the lipids in the chylomicrons and in this manner can be cotransported in the exocytosis step (Kelly *et al.*, 2004) – see Section 3.6.2.1.

3.2.9 How Different Exposure Routes Affect the Rates of Toxicant Uptake

Before leaving this section on exposure routes, we should provide some perspective on how different exposure routes may influence an animal's overall exposure to a toxicant. Uptake by any route will be limited by the volume of material to which the organism is exposed, the concentration and form of the toxicant in the ingested or inspired material, and special features that control bioavailability (e.g., degree of ionization). For animals, the lungs or gills are primary exposure routes because the animal must take up oxygen continuously, and high volumes of air or water are needed to provide sufficient oxygen. Gill-breathing organisms are particularly at risk because water is the 'universal solvent', and larger numbers of waterborne chemicals are accessible than is the case for air-breathing animals.

Much of the information on water–gill exchanges has been derived from studies of the relationship between oxygen exchange rates and gill ventilation rates. For chemicals that behave like oxygen, a non-polar solute that rapidly enters the gill bloodstream by simple diffusion, the ultimate limit to uptake is blood flow, i.e., the rate at which blood carries oxygen away from the gills. Thus, uptake rates of oxygen, or similar solutes (non-polar low-molecular-weight compounds that also rapidly

enter the gill bloodstream), will increase to a maximum as ventilation rates increase. In contrast, if the apical gill membrane limits solute uptake (e.g., in the case of metals that interact with ion channels or large hydrophobic compounds that diffuse slowly through membranes), increased water flow through the gill opercular cavity will have little effect on uptake rates.

The diet is often the primary route of exposure for chemicals that do not readily dissolve in water (e.g., chlorinated pesticides) or which biomagnify to high levels in the ingested material. However, diet consumption is episodic and the amount of material ingested is limited by stomach size. Nevertheless, some compounds may be present in very high concentrations in diets compared to concentrations in water or air and compounds that are persistent and not readily excreted (e.g., PCBs) may accumulate to tissue concentrations that are toxic.

3.3 Elimination Routes

To this point we have emphasized routes by which toxicants can enter living organisms. As a complement to this information on uptake routes, it seems appropriate to add a few words about routes of elimination. They too involve the passage of the original toxicant, or often its transformation or conjugated product, across a biological membrane and back into the ambient environment. For small compounds typical of normal metabolism, breakdown products may represent their complete catabolism to carbon dioxide. Routes of elimination will differ among different organisms, such as unicellular organisms, vascular plants, invertebrates (aquatic, terrestrial) and vertebrates (aquatic, terrestrial). Examples of known routes of elimination can be found in Table 3.2.

For different metals, metalloids or organic molecules, the routes of elimination will vary as a function of size, hydrophilicity, lipophilicity, affinity for proteins, and vapour pressure. These are factors similar to those that affected the uptake of the parent compound.

3.4 Bioaccumulation and Uptake–Elimination Kinetics

In this section, we first define some key terms that will be used here and in subsequent chapters to describe the uptake, transformation and elimination of potential toxicants. We then consider these processes sequentially,

Table 3.2 **Routes by which toxicants or their metabolites can be eliminated from a living organism.**

Organ	Mechanism	Toxicant examples
Vertebrates		
Gill	Respiration. Blood–water exchange. Nitrogenous waste excretion and ionic regulation via ATPase-based transporters.	Low-molecular-weight, water-soluble compounds (polyfluorinated alkyl acids).
Lung	Exhalation and blood–air exchange.	Low-molecular-weight, volatile compounds (peroxyacyl nitrates).
Kidney	Filtration; blood → urine exchange. Ionic regulation. Organic anion and nitrogenous excretion via transporters.	Low-molecular-weight water-soluble compounds and metabolites (<300 Da); cadmium, nickel, zinc. Endogenous anions (uric acid); exogenous anions (aflatoxin).
Liver/intestine	Filtration; blood exchange → bile → duodenum/intestine → egestion → faeces (urine is combined with faeces in amphibians, birds and reptiles).	Phase I and Phase II metabolites of non-polar molecules (generally >300 Da but species-specific); Phase III metabolism conjugates.
Mammary gland	Lactation → suckling juveniles.	Lipophilic compounds (PCBs) or calcium analogues (Sr).
Placenta	Nourishment → foetus.	Methylmercury. Hydrophilic compounds (phthalate metabolites) and proteinophilic compounds (perfluorinated compounds).
Ovary (egg-laying vertebrates)	Ovulation → egg deposition in oviparous vertebrates.	Methylmercury, selenium. Organic compounds with affinity for the lipoprotein moiety of vitellogenin (many persistent organic pollutants or POPs).
Feathers	Blood → feather exchange → moulting.	Lead, mercury, nickel.
Antlers	Blood → antler → shedding.	Fluoride
Salt-glands (seabirds, turtles)	Blood → salt-gland secretion.	Cadmium, lead, mercury.
Invertebrates.		
Gill (bivalve molluscs, crustaceans)	Respiration. Haemolymph → water exchange. Rapid elimination of low-molecular-weight hydrocarbons.	Naphthalene, oil hydrocarbons.
Nephridia, Malpighian tubules	Filtration. Excess salt elimination, water conservation in terrestrial animals and high salinities. Nitrogenous waste excretion.	Urea, uric acid, endogenous and exogenous organic anions.
Trachaea and/or lungs (terrestrial invertebrates)	Exhalation. Haemolymph → air exchange.	Low-molecular-weight, volatile compounds.
Hepatopancreas	Exocytosis of granules.	Type B (sulphur-rich) granules containing metals such as cadmium, copper, lead, silver.
Exoskeleton	Moulting	Various metals eliminated in exuviae.
Plants.		
Foliage	Evaporation; volatilization from leaf surface.	Compounds of arsenic, selenium and mercury.
	Exudation from **trichomes** (outgrowths of epidermal cells, analogous to root hairs).	Calcium/cadmium crystals extruded through trichomes.
	Salt excretion by halophytes.	Sodium and chloride; can be accompanied by mercury.

Table 3.2 (*cont.*)

Organ	Mechanism	Toxicant examples
	Guttation in the transpiration stream through leaf pores.	High concentrations of insecticides (e.g., neonicotinoids) measured in guttation drops during early stages of plant development.
	Leaf abscission.	Shedding of leaves and other plant parts results in the release of any substances that were stored in vacuoles or other cell components.
Roots	Exudation	Root exudates are known to influence the conditions in the rhizosphere, but the role of roots in eliminating toxicants is unclear.

Table 3.3 **Summary of terms used to describe contaminant accumulation in living organisms.**

Bioaccumulation	Overall process by which the organism accumulates the contaminant, by all routes of exposure (air, water, diet).
Bioconcentration factor BCF	Ratio of concentrations of a contaminant in the organism (C_B) to the concentration in an exposure medium such as water (C_W), measured at steady state, but determined in the absence of any dietary uptake. BCF = C_B/C_W.
Bioaccumulation factor BAF	Ratio of concentrations of a contaminant in the organism (C_B) to the concentration in the ambient environment (C_{env}), measured at steady state. BAF = C_B/C_{env}. The denominator may be the total aqueous concentration, the dissolved aqueous concentration or sometimes the total sediment concentration (BSAF = C_B/C_{sed}; biota to sediment accumulation factor). In all cases, the organism is exposed (normally in the field) to all routes of exposure.
Biomagnification	Process by which the organism accumulates diet-borne contaminants to concentrations that exceed those in the diet. Progressive increase with trophic level in the tissue concentrations of compounds acquired from the diet; a property typical of persistent compounds, particularly those that are hydrophobic and accumulate in lipids. Biomagnification factor (BMF) = C_B/C_{diet}, measured at steady state.
Octanol–water partition coefficient K_{OW}	Ratio of the concentration of a contaminant in n-octanol (a surrogate for biological lipids) to its concentration in water, at equilibrium. Usually reported as its logarithm, $\log_{10} K_{OW}$.
Elimination	Process by which an organism rids itself of an accumulated contaminant. Elimination includes excretion across a cell surface – for unicellular organisms – or via a specialized organ in multicellular organisms (liver, kidney, salt-gland, Malpighian tubules, lungs, gills). In some cases, the term 'elimination' is also erroneously used to describe the metabolic degradation of the contaminant.
Depuration	Loss of a substance from an organism owing to elimination and/or degradation.

showing how they can be modelled and how their rates can be estimated.

To predict the effect of an environmental contaminant on an organism living in a natural environment, we need to know how much of the contaminant reaches a sensitive site within the organism or on its surface. In Table 3.3 we have compiled terms that are widely (and sometimes erroneously!) used to describe the adsorptive or absorptive processes that deliver the contaminant to the sensitive site.

3.4.1 Toxicant Uptake: Differences Between Lipophilic and Hydrophilic Molecules

Lipophilic toxicants that enter cells through the lipidic portion of the plasma membrane do so by a process of passive diffusion. Their passage through the lipid portion of the membrane is principally limited by the molecular size and lipid solubility of the compound (Section 3.1.2.1). Other factors that affect the rate of diffusion include the thickness and surface area of the membrane.

The overall diffusive process is defined by Fick's law, equation (3.2):

$$J_{o \to i} = \frac{(C_o - C_i) \times A \times D}{d} \qquad (3.2)$$

where J is the diffusion rate (moles/s); $C_o - C_i$ is the concentration gradient between the outer and inner faces of the membrane (mol/cm^3); A is the membrane surface area (cm^2); d is its thickness (cm); and D is the diffusion coefficient of the diffusing molecule (cm^2/s) within the membrane.

The initial rate of simple diffusion will increase linearly with an increase in the external concentration of the toxicant. As the toxicant concentration builds up within the organism, the reverse process (diffusion back into the exposure medium) will become increasingly important, and the net uptake rate will decrease. However, if the toxicant is sequestered within the organism (or distributed internally to a different compartment), effectively lowering its concentration in the cytosol, the net uptake rate across the membrane will continue to increase linearly as the external concentration is raised.

The tendency of a chemical to move from the aqueous to the lipid phase (hydrophobicity) can be reasonably well predicted as a function of its K_{OW}, usually expressed on a logarithmic scale. A higher log K_{OW} value, relative to those for other compounds, means that the chemical has a greater affinity for the n-octanol phase than do the other compounds. Uptake across the lipid portion of biological membranes tends to increase over the log K_{OW} range from 0.5 to 4, plateaus for values from 4 to 6, and then decreases for log K_{OW} values greater than 6 (Figure 3.12). The downward inflection point is variable, somewhere between 6 and 7, depending on the nature (e.g., shape, composition) of the compounds included in the analysis. Note that increasing log K_{OW} values also imply that the substance will become less and less soluble in water, and this eventually becomes a practical limitation on bioaccumulation, contributing to decreased uptake for highly hydrophobic molecules (log K_{OW} > 6). In addition, highly hydrophobic compounds usually have a high molecular weight and, as discussed earlier in this chapter and in Chapter 7, large molecular size hinders uptake.

In contrast to lipophilic molecules, polar and ionic toxicants cross epithelial barriers via pores, channels and carriers, and their uptake is subject to saturation. In other words, their rate of uptake does not increase linearly as a function of their aqueous concentration, but

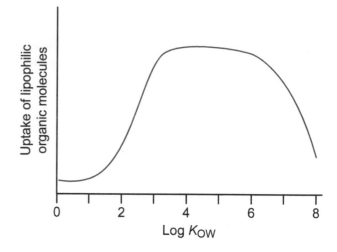

Figure 3.12 General relationship between the uptake rate of a lipophilic organic molecule and its log K_{OW} value. Adapted from McKim *et al.* (1985) with permission from Elsevier. The original study followed the uptake of a suite of organic chemicals into rainbow trout gills.

rather tends towards a plateau. In the case of channels (Section 3.1.2.2), this saturation effect is an indication that the maximum transport capacity of the channel has been reached. For mediated transport, however, the saturation indicates that the toxicant is reversibly bound to a finite number of carrier proteins in the membrane.

The kinetics of a saturable transmembrane carrier system can be explained by analogy to enzyme substrate dynamics (Appendix 3.1). We assume that the system has a finite number of carrier proteins available for chemical (substrate) transport. Carrier-mediated transport may be inhibited in either a competitive or non-competitive manner. Competitive inhibition results in a change in the value of K_m (substrate affinity – see Appendix 3.1) with respect to the normal uptake of the substrate in the absence of a competitor. Such a situation would cause a reversion from curve (a) (Figure 3.13) to curve (b); in the latter, the total number of carrier sites P_0 remains unaltered (i.e., the V_{max} value remains constant) but it takes a higher concentration of the normal substrate to reach the V_{max} value (i.e., a higher K_m value). Non-competitive inhibition, on the other hand, is characterized by a reduction in the overall number of available or functional carriers. This reduction results in a decreased value for the upper asymptote of the saturation curve (i.e., a lower maximum flux rate), although K_m is not affected (curve (c), Figure 3.13).

In Figure 3.13, the Y-axis corresponds to the concentration of the substrate–carrier complex (PS) and thus

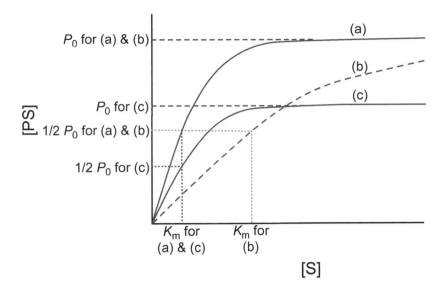

Figure 3.13 Michaelis–Menten plots of saturable, carrier-mediated transmembrane transport. Transport rate varies as a function of [PS], the concentration of the substrate–carrier complex. [S] = substrate concentration. Plot (a) = absence of inhibitor; plot (b) = competitive inhibition; plot (c) = non-competitive inhibition. Reproduced from Wright & Welbourn (2002) with permission from Cambridge University Press.

represents the uptake rate for three systems, (a), (b) and (c). Systems (a) and (b) have the same carrying capacity (i.e., the same V_{max} value), whereas (a) and (c) have the same substrate affinity (K_m value).

3.4.2 Toxicokinetics

Toxicant uptake can occur through direct absorption from ambient media such as air or water or through the ingestion of contaminated food. After uptake, toxic chemicals follow a variety of pathways. They may enter short-term storage (i.e., tissues where they have a limited residence time), enter long-term tissue storage, or undergo rapid elimination or metabolism. In egg-laying species, toxicants such as POPs or selenium may be transferred maternally into the yolk of developing eggs; the developing embryo is subsequently exposed as the yolk is absorbed. Here we describe some of the models that have been developed to describe the kinetics of chemical bioaccumulation by an individual organism. Models tend to be either empirical or mechanistic in nature, although many integrate these two approaches. Typical are the compartmental models developed initially by pharmacologists.

3.4.2.1 One-compartment Model
In the simplest of these models, the organism is considered to be a single homogeneous compartment interacting with the external environment (Figure 3.14), where

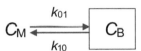

Figure 3.14 Toxicant uptake with the organism considered as a single homogeneous compartment.

C_B is the toxicant concentration in the organism, C_M is the concentration in the ambient medium, and k_{01} and k_{10} are rate constants for uptake and loss, respectively.

If the organism is exposed to a constant concentration of the toxicant, C_M, the concentration within the organism will increase towards a plateau (Figure 3.15A), where the rate of uptake equals the loss rate and a steady-state condition is reached. Following termination of toxicant exposure, the rate of uptake, $k_1 C_M$, falls to zero, and the elimination of the toxicant is conventionally expressed as the biological half-life ($t_{1/2}$). This is the time taken for an organism to clear half of the toxicant content of its body. This term is derived from a logarithmic plot of toxicant clearance (Figure 3.15B). Mass balance equations describing the uptake/loss kinetics of a single-compartment system are shown in Appendix 3.2.

A variety of uptake processes may be conveniently studied using unicellular organisms such as algal cells, and much has been learned about the kinetics of chemical bioaccumulation from such simple systems. However, in multicellular organisms, where epithelial surfaces consist

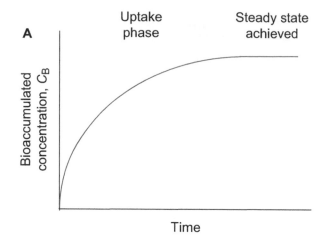

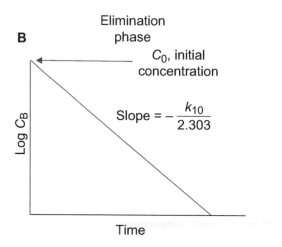

Figure 3.15 A: Increase in the bioaccumulated concentration (C_B) over time until steady state is achieved. B: Loss of C_B over time in the absence of exposure (note the natural log scale of the Y-axis for panel B).

of many cells arranged in complex membranes, the situation rapidly becomes more complicated. In such cases, trans-epithelial chemical movement can be studied by comparing the external medium with a defined bodily fluid, such as blood, which serves as a vector to transport the toxicant to internal tissues where it may be stored or metabolized. For such organisms, the simple one-compartment model described above would still be applicable provided that (a) the toxicant remains unchanged in the circulatory fluid and is not (or only very slowly) taken up by the tissues, or (b) the toxicant freely diffuses throughout the blood and tissues without any rate-limiting diffusional barrier. Such conditions resemble those that are stipulated for compounds considered as narcotics in the target lipid model (Section 2.3.4.3; Di Toro *et al.* (2000)). This toxicant category encompasses a very large number of small bioavailable compounds with

environmental significance (e.g., halogenated and non-halogenated aliphatic and aromatic hydrocarbons, alcohols, ethers, furans and ketones).

3.4.2.2 Two-compartment Model

Often the time-course of toxicant concentration in blood plasma following its rapid introduction (e.g., by injection) is a curve rather than a straight line. Such a curve is the result of toxicant distribution into more than one compartment. The simplest case is when the toxicant distributes rapidly into the plasma and tissues, but its excretion proceeds more slowly.

In such a model, for the simple case of a pulse or bolus exposure to a contaminant, the concentration in the plasma (C_P) would increase very rapidly (e.g., following an injection) and then decline in two distinct phases. The first phase corresponds to the rapid distribution of the contaminant into the tissues (k_{12}) and its slow elimination from the plasma (k_{10} in Figure 3.16). This is followed by a slow phase as the contaminant is cleared progressively from the tissues back into the plasma (k_{21}) and is excreted. If C_P is plotted over time, the resulting curve can be deconvoluted to obtain estimates of the rates of movement into and out of the tissues, and the rate of excretion (Figure 3.17).

The ability of a chemical to exchange between the blood compartment and one or more tissue compartments owes much to its physical and chemical characteristics. The 'blood compartment' in this case is treated as a homogeneous liquid, distinct from tissues served by blood vessels. A large, relatively hydrophilic molecule is much

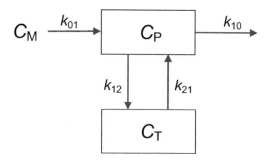

Figure 3.16 Toxicant uptake with the organism considered as two homogeneous compartments. C_M = toxicant concentration in the exposure medium; C_P = toxicant concentration in the plasma; C_T = toxicant concentration in the tissue. The rate constants are labelled 01 for uptake from the medium to the plasma, 10 for elimination from the plasma, 12 for transfer from the plasma to the tissue, and 21 for return from the tissue to the plasma.

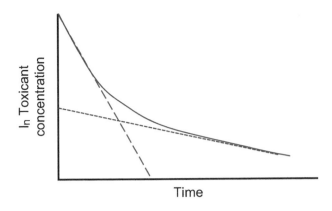

Figure 3.17 Biphasic kinetics of elimination of a toxicant from a two-compartment system (as represented in Figure 3.16) after a pulse or bolus exposure. Dashed line = distribution from plasma to tissue. Dotted line = elimination from plasma. Solid curved line = total body concentration (plasma + tissue).

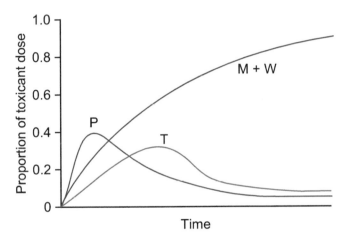

Figure 3.18 Relative changes in a single toxicant dose in different body compartments over time. P = plasma compartment; T = tissue compartment; M + W = sum of metabolism and waste compartments. Reproduced from Wright & Welbourn (2002), with permission from Cambridge University Press.

more likely to remain in the bloodstream than small, hydrophobic molecules, which will be rapidly distributed to tissue lipids. In some cases, several different tissues (each with different rate constants) might be substituted for the single-tissue compartment described above. If, for any tissue, the rate constant from blood to tissue (k_{12}) = the rate constant from tissue to blood (k_{21}), that tissue can be functionally combined with blood as a single compartment.

To this point we have considered the toxicant to be 'persistent' within the organism, once it has crossed the epithelial barrier. In some cases, this is a reasonable assumption, but often the toxicant will be subject to

biotransformation (see Section 3.5). In such cases, C_B will decline both by excretion and by metabolic transformation. Beginning with a single source of toxicant available to the organism, the relative changes in toxicant dose occurring in compartments P (plasma), T (tissues), M (metabolized toxicant) and W (excreted toxicant) are shown qualitatively in Figure 3.18. Various metabolic or biotransformation reactions are described in the next section.

3.5 Biotransformations

To this point in this chapter, we have treated toxicants as unreactive solutes. However, at the end of the previous section we introduced the idea that some toxicants can be metabolized, with resulting changes in their chemical properties and their fate within the host organism. Both metals and organic xenobiotics are subject to transformation once they have traversed the cell membrane and entered the intracellular environment. In this new environment, they are confronted by conditions very different from those encountered in the extracellular environment, notably with respect to such factors as pH and redox potential, but also given the vast array of enzymatically catalysed reactions available within living cells. Here we discuss what happens to both classes of contaminant, in general terms, and provide links to more detailed information about the metabolic fate of specific contaminants, as discussed in Chapter 6 (Metals) and Chapter 7 (Organic contaminants).

3.5.1 Metals

Metals are naturally present in the environment, and their presence pre-dates the origins of biological life. Indeed, many of them are essential for life, and accordingly living cells have developed ways to shuttle them across the plasma membrane, transport them to where they are needed in the host organism, and deliver them to the appropriate target sites (Section 6.5.1). However, when confronted with excessively high concentrations of an essential metal, or with modest to high concentrations of a non-essential metal, living cells need to limit the rate of entry of the metal, to accelerate its rate of elimination, or to 'transform' the metal into a relatively unavailable form. Because a given metal cannot be 'degraded' or made to change into another element (despite the alchemists' best efforts!), the word transform is used here to mean a change in the state, or **speciation**, of the metal. Examples of such changes in 'state' can be

Table 3.4 **Examples of metal biotransformations (essential metals are in bold).**

Metal transformation	Example
- Oxidation	$As(III)(OH)_3 \rightarrow As(V)O(OH)_3$
- Reduction	$Hg(II) \rightarrow Hg(0)$
- Formation of carbon–methyl bonds (methylation)	$As(III)(OH)_3 \rightarrow CH_3\text{-}As(O)(OH)_2$ $Hg(II) \rightarrow CH_3\text{-}Hg\text{-}X$ **Se**(VI) $\rightarrow CH_3SeCH_3$
- Formation of other carbon–metal bonds	$As(III)(OH)_3 \rightarrow$ arsenosugars, arsenolipids **Se**(VI) $\rightarrow$ selenocysteine
- Complexation with detoxifying intracellular ligands	$Cd(II) \rightarrow$ Cd-metallothionein $Cd(II) \rightarrow$ Cd-phytochelatin **Fe**(III) $\rightarrow$ Fe-ferritin
- Formation of inorganic concretions (intracellular)	Al(III), **Mn**(II), **Zn**(II) $\rightarrow$ pyrophosphates Ag(I), Cd(II), Hg(II); **Cu**(II), **Zn**(II) $\rightarrow$ sulphides
- Formation of inorganic concretions (extracellular)	**Ca**(II); Cd(II) $\rightarrow$ carbonates
- Sequestration in structural components	Pb(II) $\rightarrow$ bones **Ca**(II) $\rightarrow$ shells

found in Table 3.4 and they are discussed in more detail in the metal-specific sections of Chapter 6.

The principle behind most of these transformations is to form a strong bond between the metal and either a dissolved ligand or a solid phase, and thus render the metal 'unavailable', at least temporarily.

3.5.2 Organic Xenobiotics

The intracellular fate of organic molecules differs markedly from that of the metals. The molecule can be considered to be entering a biochemical 'factory', within which it will be subject to an extensive array of potential reactions. The key word in the preceding sentence is 'biochemical', because the vast majority of reactions going on within living organisms are enzymatically catalysed. Accordingly, if our contaminant/toxicant is to be transformed by an enzymatic reaction, it must be able to 'fool' one or more enzyme systems into accepting it as a substrate.

Throughout their evolution, both bacteria and archaea (prokaryotes) and complex organisms (eukaryotes) have had to deal with the problem of handling a variety of organic chemicals accumulated from the environment (exogenous) or produced by the organism itself (endogenous). Foreign, exogenous compounds, termed xenobiotics, have multiplied many thousandfold as a result of manufacturing activities. It can be considered as a tribute to the versatility and efficacy of the enzyme systems involved

that the great majority of these new compounds are successfully metabolized and detoxified. The primary reason is that species have evolved with enzymes that interact with structural features common to biologically active molecules such as amino and nucleic acids, proteins, lipids, carbohydrates, etc. These features include amine, hydroxyl, thiol, phenolic and carboxylic acid functional groups, branched and linear alkane and alkene side chains, as well as aromatic and heterocyclic rings. Not surprisingly, many of these same molecular features are found in xenobiotics, so that normal metabolic pathways can remove or degrade them. The exceptions would be persistent xenobiotics that contain molecular components for which enzyme systems have not evolved. Examples of such components include aromatic rings in which halogen atoms have replaced most of the hydrogen atoms (e.g., PCBs) or fatty acids where multiple fluorine atoms have been introduced (e.g., perfluoroalkanoic acids).

As is discussed in detail in Section 7.4, the transformation of non-polar organic chemicals occurs sequentially through three different types of reactions referred to as **Phase I**, **Phase II** and **Phase III**. The conjugated metabolites resulting from **Phase I** and **Phase II** reactions are eliminated via **Phase III** reactions involving ABC-transmembrane transporters, multidrug resistance (MDR) proteins and P-glycoprotein (Box 3.4). Their role is to move these products, most of which are organic anions, out of the cell. The anionic groups act as 'tags', which facilitate the identification of a metabolite as a substrate for membrane transporters in the OAT (organic anion transporter) and MDR families. Specific examples of this elimination step are discussed in Sections 7.5.3 and 7.5.4.

In vertebrates, Phase I enzymes (**cytochrome P450** or **CYP1A**) are synthesized at high concentrations in endothelial cells, i.e., those that line blood vessels. Therefore, the highest activities of these enzymes occur in highly vascularized liver, kidney and respiratory tissues (lungs, gills) where the enzymes can provide protection against respiratory exposure to organic compounds and facilitate the biliary and urinary excretion of the by-products of CYP1A metabolism. Their functions are not limited to metabolism of xenobiotics but include many critically important metabolic functions, such as the synthesis of steroid hormones.

3.6 Bioaccumulation and Biomagnification

As developed in Section 3.4.2 (Toxicokinetics; equation (3.9) in Appendix 3.2), the net bioaccumulation of a stable contaminant, i.e., one that is not subject to metabolic transformation, will reflect a balance between its rate of uptake and its rate of elimination. The actual body concentration of the contaminant may be affected by **growth dilution** (sometimes termed biodilution), but the body burden (expressed not as a concentration but as a mass) will continue to increase over time and will only be influenced by the rates of uptake and elimination. An example of this behaviour is presented in Case Study 4.6 for PCB accumulation in killer whales (*Orcinus orca*). In the following discussion, we first consider the bioaccumulation of metals and then persistent bioaccumulative organic contaminants.

3.6.1 Metals

Considerable success has been achieved in predicting steady-state metal concentrations in exposed (aquatic) organisms with the dynamic multi-pathway bioaccumulation (DYM-BAM) model (Luoma and Rainbow, 2005). This model takes into account metal uptake from both water and ingested solids (food), and it also considers metal elimination and growth dilution. At steady state, the bioaccumulated metal concentration, $[M]_{organism}$, can be predicted according to equation (3.3):

$$\frac{d[M]_{organism}}{dt} = k_u [M^{z+}] + (AE \times IR \times [M]_{food}) - k_e [M]_{organism} - k_g [M]_{organism} \quad (3.3)$$

The terms in this equation are defined as follows:

- k_u, k_e and k_g are rate constants for M-influx from water (expressed in units per litre per gram per day), M-efflux (1/d) and growth (1/d), respectively;
- $[M^{z+}]$ is the free metal ion concentration in water (nmol/l);
- $[M]_{organism}$ is the metal concentration in the organism of interest (nmol/g dry weight);
- $[M]_{food}$ is the metal concentration in the food consumed by the organism (nmol/g dry weight);
- AE is the efficiency (%) with which the organism assimilates metal M from its food; and
- IR is the rate at which food is ingested by the organism (g prey/g animal/d).

In other words, the model integrates information from the animal's habitat (metal concentrations in water and food) and physiological data (influx rates from water, influx rates from food, rate constants of loss, and growth rates), to yield estimates of the steady-state metal concentration in the target organism.

Luoma and Rainbow (2005) reviewed the use of biodynamic modelling to predict species-specific metal bioaccumulation in the field. In a comparison of model predictions with field observations, for 6 metals and 1 oxyanion (Ag, Cd, Co, Cu, Cr, Zn; Se) and 14 animal species, whole-organism metal concentrations varied by over seven orders of magnitude. To accommodate this very wide range of body concentrations, the data were log-transformed, and results for the same animal species but from different studies were pooled. On a log–log scale, most predictions agreed with the observations within a factor of two, with an overall coefficient of determination (r^2) between predictions and observations of 0.98 (Figure 3.19).

In principle, the dynamic modelling approach could be refined to estimate the internal dose of a given metal in a particular organ (rather than in the whole body), and then relate this dose to a metal-induced effect (Sappington *et al.*, 2011). However, experiments in the laboratory and in the field have shown that if the internal dose is reached slowly, the effects on the organism's function are reduced and its chances of survival are greater than if the internal dose is attained more rapidly.

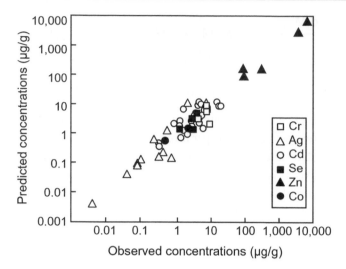

Figure 3.19 Trace metal bioaccumulation as predicted by biodynamic modelling (*Y*-axis) and compared with independent observations of bioaccumulation in the species of interest (*X*-axis). Figure reproduced from Rainbow (2018) with permission from Cambridge University Press.

Stated differently, if the internalization of the metal occurs reasonably slowly, some degree of acclimation can be inferred (e.g., by turning on its metal-detoxification pathways and sequestering the metal in a non-available form, or by accelerating the metal efflux rate). On the other hand, if the internalization flux is high, the exposed animal is unable to detoxify all of the incoming metal, a critical internal dose is attained, function is reduced and the organism eventually succumbs or some physiological function is impaired (Andrès *et al.*, 1999; Baudrimont *et al.*, 1999). It follows that to use the DYM-BAM approach to predict metal toxicity (rather than metal bioaccumulation), one should consider not only the calculated steady-state metal concentration, but also the rate at which this internal dose is reached.

Note too that the biological parameters in DYM-BAM models are modelling 'constants', which are normally determined in the laboratory where a **metal-naïve** test organism is exposed to a range of metal concentrations. Initial metal uptake rates are determined as a function of the metal exposure concentration. Following exposure, the organisms are placed in a metal-free environment and the metal elimination rates determined during the depuration phase. However, this approach may be too simplistic. Specifically, the key epithelial properties of aquatic organisms that govern metal accumulation and toxicity may well be affected by any antecedent exposure to the metal. Organisms that are chronically exposed to

metals can alter the characteristics of their epithelial membranes and influence both the affinity of the metal-binding sites and the rate at which the metal is internalized (Niyogi *et al.*, 2004). Given these two factors (i.e., the need to consider metal influx rates and the differences between metal-naïve organisms and field-exposed organisms), and despite its acceptance in a research environment, the DYM-BAM model has not yet been widely applied in a regulatory context.

3.6.2 Bioaccumulation of Persistent Organic Contaminants

The accumulation of organic contaminants can be modelled in much the same way as described for metals, except that the first two terms in equation (3.3), i.e., uptake from water or from the diet, require some modification. In the following two sections, we consider two distinct classes of organic contaminants: the classic POPs, such as lipophilic organochlorine pesticides and PCBs, and proteinophilic perfluoroalkyl substances that have recently been discovered to be biomagnified in some food webs. The environmental behaviour and toxicology of these compounds are discussed in detail in Chapter 7; here we consider their uptake and bioaccumulation.

3.6.2.1 Lipophilic Contaminants

For lipophilic organic contaminants, the uptake from water or from food will normally involve passive diffusion through the lipid bilayer (whereas for metals in their ionic form, uptake involves channels or carriers). An example of the application of such a model to lipophilic contaminants and freshwater fish is developed in Section 4.5.2.

A lipophilic contaminant will cross the epithelial membrane and partition into the organism's lipid pools, leading to bioconcentration from the external medium, either water or air. For herbivores and carnivores, their diet will also contribute to the overall bioaccumulation of the contaminant. For some organic contaminants such as PCBs and dichlorodiphenyltrichloroethane (DDT), the resulting chemical concentration (on a lipid-weight basis) in the consumer (C_B, as defined earlier) may exceed those in the consumed prey (C_{prey}). When this increase occurs successively over at least three links in a food chain, it is no longer simply bioaccumulation but rather **biomagnification**. Using the fugacity concept, as defined in Section 4.5.1, food-chain biomagnification corresponds to an increase in fugacity with increasing trophic level.

This phenomenon is of crucial importance for regulators, since it can result in contaminant concentrations in top consumers that exceed toxicity thresholds. For example, past declines of fish-eating birds such as eagles in the Laurentian Great Lakes in North America were traced to reproductive impairment caused by the biomagnification of pesticides (now banned – see Sections 1.2 and 7.7.1). Food chain dynamics, biotransformation and growth also have a significant influence on chemical biomagnification, as reviewed by Gobas *et al.* (2009). Note that biomagnification is virtually absent for metals, with the notable exception of mercury, where the organometallic compound methylmercury behaves more like an organic molecule than a metal ion (Section 6.8.1).

As defined in the preceding paragraph, biomagnification necessarily requires efficient dietary uptake. It follows that we should understand how lipids and lipophilic chemicals behave during the digestive process. Experimental studies have shown that when a mixture of natural lipids and lipophilic xenobiotics is present in an animal's diet, the lipids and the xenobiotics show up simultaneously in the gastrointestinal tract (GIT), in the lymphatic flow or in both. Digestion of natural lipids in the intestinal lumen involves the following steps:

- Triglycerides are hydrolysed by lipases, yielding monoglycerides, glycerol and fatty acids;
- The monoglycerides plus cholesterol, fat-soluble vitamins and lipophilic contaminants are emulsified with the aid of bile salts and enclosed in mixed **micelles**;
- The micelles ferry the hydrolysis products to the intestinal mucosa, where they are released and diffuse passively across the apical surface of the enterocytes;
- Within the enterocytes, the fatty acids are esterified with glycerol, reforming triglycerides;
- These fatty globules combine with proteins to form **chylomicrons**, and these small particles are expelled from the enterocyte by exocytosis across the basolateral membrane;
- Finally, the exocytosed compounds can either enter into the circulation via the hepatic portal vein or into the lymphatic system, where they are transported as chylomicrons or as 'free agents'.

An unstirred water layer exists at the surface of the enterocyte, and there is some debate as to whether the micelles actually collide with the membrane and release their 'cargo' directly or whether they disintegrate in the vicinity of the membrane. In this latter case, the cargo molecules would have to diffuse across the unstirred layer, and this would probably become the step that limits the rate of absorption. The pH in the unstirred layer is lower than that in the lumen, and this would favour the protonation and absorption of any ionized molecules (Section 3.2.4).

How can the generic digestive process that we have just explained produce biomagnification, and what are the properties of the contaminant that would favour this phenomenon? In their thoughtful analysis of these questions, Kelly *et al.* (2004) describe several ways in which the combination of digestion + passive diffusion could result in biomagnification.

1. In the *digestion model*, it is assumed that digestion and absorption of the food diminish and simplify the matrix in which the contaminant was originally ingested. This simplification results in an increase in the concentration and the fugacity of the contaminant in the GIT and creates a positive gradient between the GIT and the organism. Provided that elimination by excretion or metabolism is negligible, the chemical concentration and fugacity in the organism will increase to match that in the GIT. The degree of magnification (i.e., C_B/C_{prey}) will be a function of two factors: (i) the extent to which the faecal expulsion rate drops below the ingestion rate, and (ii) the extent to which the fugacity capacity of the food is lowered by the absorption of lipids, proteins and other dietary components.

2. In the *micelle-mediated diffusion model*, intestinal absorption of the contaminant molecules (enhanced by mixed-micelle facilitation) is assumed to occur in the upper GIT in association with dietary lipid absorption. Elimination of the chemical is decoupled in time and space from absorption and occurs in the lower digestive tract. Despite a fugacity gradient in the lower digestive tract that favours elimination, the rate of elimination is slow. It has been suggested that the absence of micelles in the lower digestive tract results in slower mass transfer across the brush border membrane into the lumen of the intestine.

3. In the *fat flush diffusion model*, the absorption of dietary lipids in the upper intestine results simultaneously in a decrease of Z (the fugacity capacity) in the material remaining in the lumen and a (temporary) increase of Z in the intestinal tissue. Stated differently, the contaminant fugacity increases in the lumen and decreases in the intestinal tissue, creating a positive fugacity gradient from the lumen into intestinal tissue.

These processes of lipid and chemical uptake from the GIT have been described elegantly by MacDonald *et al.* (2002) as a combination of steps involving both solvent depletion and solvent switching, with biomagnification ultimately arising from the efficient absorption of lipids and lipophilic chemicals from the gut and the depletion of absorbed lipid as it is metabolized for energy (Section 3.6.2.2).

The influence of the K_{OW} on dietary absorption efficiency has been studied experimentally in fish, birds and mammals. For chemicals with log K_{OW} values <7, absorption efficiency is relatively constant, but it falls off for log K_{OW} values above this threshold. This trend is consistent with a sequence of micellar transport, diffusion of the contaminant through the unstirred water layer, followed by diffusion through the phospholipid bilayer; in such a sequence, the slowest step controls the overall rate of gastrointestinal uptake of the contaminant. For substances with log K_{OW} values <6, uptake is controlled by micellar transport or diffusion across the apical membrane of the enterocyte, whereas for substances with very high log K_{OW} values and correspondingly very low water solubility, their low concentrations in the unstirred water layer will limit the overall absorption process (Kelly *et al.*, 2004).

The three models described above probe the important influence of the digestive process on biomagnification and offer different perspectives on how the uptake of a chemical compound can exceed its elimination by respiration, urinary excretion and metabolism. Ultimately, the biomagnification factor (BMF) will reflect both organism physiology and the physicochemical properties of the ingested compound.

To illustrate the influence of animal physiology, consider fish as an example of aquatic poikilotherms (i.e., cold-blooded organisms that do not regulate their internal temperature). For such organisms, loss of contaminants by gill ventilation is an important elimination route. Elimination to the surrounding water declines as a function of the contaminant's K_{OW}, leading to greater retention of the contaminant. For log K_{OW} values >5, biomagnification will occur unless the contaminant is rapidly metabolized. Respiratory elimination of low-molecular-weight volatile contaminants also occurs for air-breathing homeotherms (i.e., warm-blooded organisms), but to air rather than to water. The relevant physicochemical property of the contaminant is thus its octanol–air partition coefficient (K_{OA}). As values of K_{OA} increase, the efficiency of respiratory elimination declines and the potential for biomagnification increases. Kelly *et al.* (2004) suggest that

if log K_{OA} > 5, respiratory elimination is ineffective, and biomagnification will occur unless log K_{OW} is less than ~2 (which favours rapid excretion via the urine) or the contaminant is rapidly metabolized.

3.6.2.2 Interplay Between Bioenergetics and Bioaccumulation of Lipophilic Contaminants

As we have learned above, the uptake of chemicals from water or air can be predicted from chemical properties such as partition coefficients (K_{OW}; K_{OA}) and the extent to which they are metabolized can be estimated. Nevertheless, BCFs and BMFs of highly persistent compounds are not constant but vary considerably with the biology of each species. The extent of bioaccumulation and biomagnification differs between homeotherms and poikilotherms and between males and females, and with their rates of growth and reproduction, life span and trophic level. The common feature among these variables is the influence of energetics on the accumulation and retention of persistent compounds derived from the diet.

In any animal, energy obtained from food is allocated to maintaining vital functions (basal metabolism), supporting activities such as foraging or locomotion (active metabolism), and enabling growth and reproduction. The ratio between energy allocated to growth and total energy consumed (food conversion efficiency or FCE) is higher in young animals than in adults; in young animals it can be up to 10%. For adult male homeotherms (mammals, birds), growth has ceased and their FCEs are 0%, i.e., all energy accumulated from their diets is expended on maintenance and activity. Among the poikilotherms, fish are unique in that they grow continuously. If food is abundant, FCEs range from >5% for rapidly growing young to almost 0% for older fish. In female fish, FCEs vary widely according to the stage of their reproductive cycle, but over their lifetime, FCEs still exceed 0%.

In contrast to energy allocation, the retention of POPs accumulated from the diet is far greater, with contaminant conversion efficiencies (CCEs) as high as 95%; the missing 5% represents amounts excreted in faeces or metabolized by CYP1A enzymes. In a growing animal with an FCE of 5% and a CCE of 95%, the whole-body *amounts* of POPs will increase 19 times faster than biomass (i.e., energy stored as new tissue). The actual tissue *concentration* of POPs represents the integral of the amount consumed and the increasing biomass of the growing organism. The rate of increase in tissue concentrations will be least in rapidly growing young animals and greatest in adults as growth slows. Concentrations will also increase in adult females, but decrease somewhat each time they give birth (or lay eggs), owing to the amounts transferred to their young; over their lifetime, females will thus accumulate lower amounts of POPs than males. The young may experience brief spikes of contamination during their early growth phase due to the high concentrations of lipids transferred during lactation (mammals; Case Study 4.6) or absorption of egg yolk (fish, birds, reptiles, amphibians).

The interactions between FCEs and CCEs are repeated at each trophic level so that the amounts of POPs consumed by high-trophic-level predators represent the cumulative total of the POPs passed on from lower trophic levels. The extent of biomagnification also increases with the size and age of predators because of their greater lifetime consumption of contaminated prey.

3.6.2.3 Proteinophilic Contaminants

In the preceding discussion of persistent bioaccumulative contaminants, we have emphasized lipophilic molecules and their propensity for biomagnification. However, in recent years a new class of organic molecules has been shown to biomagnify, as exemplified by perfluoroalkyl contaminants (PFCs) (Section 7.10). Prime examples of these compounds are perfluorinated alkyl acids (PFAAs) such as perfluorooctane sulphonic acid (PFOS – now banned) and C_7–C_{14} perfluorocarboxylic acids (PFCAs). They are similar to lipophilic contaminants in that they are readily taken up, both from water (gills) and the diet (gastrointestinal tract), and are persistent (i.e., only slowly degraded). However, unlike lipophilic POPs, which tend to biomagnify in all food webs, biomagnification of PFCs is principally limited to the upper trophic levels in mammalian food webs. This difference in behaviour can be linked to the chemical properties of the PFCs. The latter differ from lipophilic contaminants in that they are amphiphilic, possessing both polar and non-polar moieties. Because of their relatively high aqueous solubility (>500 mg/l) and hydrophilic nature, the PFCs are readily eliminated from gills (water-respiring animals) but *not* from lungs (air-breathing animals). As a result, biomagnification occurs in mammalian food webs, notably in marine Arctic animals, but not in food webs where the top-level consumer is a fish (Kelly *et al.*, 2009). Biomagnification of PFCs has also been documented for terrestrial Arctic food webs (e.g., lichens–herbivores–carnivores), but the trophic magnification factors are lower than in the marine environment.

In exposed organisms, internal PFC concentrations tend to be higher in protein-rich compartments such as the blood and the liver than in other tissues or fluids (milk, blubber, muscle). *In vitro* studies have shown that PFCs bind quite strongly to specific proteins, such as plasma proteins (e.g., albumin) or liver fatty acid binding protein (L-FABP). Comparisons of bioaccumulated concentrations among different organisms are thus normalized with respect to protein content, rather than on the basis of body lipids. The PFC concentrations correlate positively with protein content in organism tissues/fluids, rather than lipid content, and these compounds are referred to as being 'proteinophilic'. Curvilinear relationships have been observed between bioaccumulation factors (BAFs) or bioconcentration factors (BCFs) and perfluorinated chain length, both in laboratory exposures and in field studies (Ng and Hungerbuhler, 2014).

Since PFCs exhibit appreciable hydrophilicity, their behaviour at epithelial surfaces would be expected to differ from that of lipophilic molecules. There has been considerable debate as to how PFCs cross epithelial barriers. To date, specific transporters have not been identified for gill or gut membranes, but organic anion transporters are involved in transmembrane transport of PFCs in liver and kidney tissues. This model for transmembrane transport of anionic PFCs is analogous to the known movement of endogenous fatty acids across cell membranes as carboxylate anions (Table 3.1).

Summary

Having considered how to measure the toxicity of environmental contaminants in the laboratory (Chapter 2), in the present chapter we shifted the focus to the contaminant itself and considered how it enters and accumulates in a target organism. To exert toxicity, an environmental contaminant must first interact with a biological (epithelial) membrane. This interaction may simply involve crossing the membrane and entering the intracellular environment, but in some cases the contaminant may contribute to toxicity by perturbing the normal functioning of the membrane. The ability of a contaminant to cross the cell membrane depends on its intrinsic properties (e.g., size, lipophilicity, affinity for membrane proteins). In some cases, the contaminant is able to masquerade as an essential nutrient and 'fool' the transporters that are embedded in the membrane (molecular or ionic mimicry). Once the contaminant has entered the intracellular environment, it is subject to various transformations, which may reduce or, in some cases, actually enhance its toxicity. Persistent contaminants are subject to bioaccumulation, and the trophic transfer of these bioaccumulated contaminants may lead to biomagnification. These laboratory observations are revisited in Chapter 4, where we consider the environmental fate of contaminants at the ecosystem level.

Review Questions and Exercises

1. Why should ecotoxicologists understand the role and functioning of cell membranes?

2. Which cell membranes are likely of most importance in ecotoxicology?

3. What drives the movement of toxicant molecules and ions across cell membranes?

4. Why is the movement of water molecules across biological membranes much faster than their movement across synthetic membranes (phospholipid bilayers)?

5. Explain the relevance and importance of a membrane channel and a membrane carrier.

6. Explain the difference between active and passive transport.

7. How would you determine the octanol–water partition constant (K_{OW}) of an organic molecule?

8. Explain why the octanol–water partition constant can be used to predict the uptake rate of organic molecules (cf. Figure 3.12).

9. Why does the relationship illustrated in Figure 3.12 resemble a bell-shaped curve?

10. How can non-essential metals and metalloids enter living cells?

11. Is the octanol–water partition constant a useful predictor of the uptake and toxicity of metals and metalloids? Explain your reasoning. What about organometals?

12. Normally a toxicant must enter a cell to induce toxicity, but there are exceptions to this rule. Explain how a toxicant could exert toxicity without crossing a cell membrane.

13. How does one differentiate among bioaccumulation, biomagnification and trophic transfer?

14. What are the properties of organic contaminants that favour biomagnification?

15. Why might an organic contaminant be subject to biomagnification in an aquatic food chain but not in a terrestrial food chain?

16. For what types of toxicants might endocytosis play an important role in their uptake?

17. Give specific examples of competitive and non-competitive inhibition of transmembrane carriers and explain how such competition might adversely affect some normal physiological functions.

18. Describe an experiment designed to track the route of uptake of a chemical by an organism, its compartmentalization within the organism and its biological half-life.

19. Explain the difference between apoplastic and symplastic transport in plant roots (use a diagram).

Abbreviations

··

ABC transporter	ATP-binding-cassette transporter	Irt	Iron transporter protein
AChR	Acetylcholine receptor	K_{OA}	n-octanol–air partition coefficient
ADP	Adenosine diphosphate	K_{OW}	n-octanol–water partition coefficient
ATP	Adenosine triphosphate	LMW	Low molecular weight
BAF	Bioaccumulation factor	MDR	Multidrug resistance
BCF	Bioconcentration factor	PACs	Polycyclic aromatic compounds
BMF	Biomagnification factor	PCB	Polychlorinated biphenyl
CDF	Cation diffusion facilitator	PFC	Perfluoroalkyl contaminant
Da	Dalton	POP	Persistent organic pollutant
DYM-BAM	Dynamic multi-pathway bioaccumulation model	Z	Fugacity capacity
		ZIP	Zrt- & Irt-like protein
GIT	Gastrointestinal tract	Zrt	Zinc transporter protein

References

··

Andrès, S., Baudrimont, M., Lapaquellerie, Y., *et al.* (1999). Field transplantation of the freshwater bivalve *Corbicula fluminea* along a polymetallic contamination gradient (River Lot, France): I. Geochemical characteristics of the sampling sites and cadmium and zinc bioaccumulation kinetics. *Environmental Toxicology and Chemistry*, 18, 2462–2471.

Barberon, M. & Geldner, N. (2014). Radial transport of nutrients: The plant root as a polarized epithelium. *Plant Physiology*, 166, 528–537.

Baudrimont, M., Andrès, S., Metivaud, J., *et al.* (1999). Field transplantation of the freshwater bivalve *Corbicula fluminea* along a polymetallic contamination gradient (River Lot, France): II. Metallothionein response to metal exposure. *Environmental Toxicology and Chemistry*, 18, 2472–2477.

Bouallegui, Y., Ben Younes, R., Oueslati, R. & Sheehan, D. (2018). Redox proteomic insights into involvement of clathrin-mediated endocytosis in silver nanoparticles toxicity to *Mytilus galloprovincialis*. *Plos ONE*, 13, e0205765.

Di Toro, D. M., McGrath, J. A. & Hansen, D. J. (2000). Technical basis for narcotic chemicals and polycyclic aromatic hydrocarbon criteria. I. Water and tissue. *Environmental Toxicology and Chemistry*, 19, 1951–1970.

Gobas, F. A., de Wolf, W., Burkhard, L. P., Verbruggen, E. & Plotzke, K. (2009). Revisiting bioaccumulation criteria for POPs and PBT assessments. *Integrated Environmental Assessment and Management*, 5, 624–637.

Gobas, F. A. P. C. & Morrison, H. A. (2000). Bioconcentration and biomagnification in the aquatic environment. In R. S. Boethling & D. Mackay (eds.) *Handbook of Property Estimation Methods for Chemicals – Environmental and Health Sciences*. Boca Raton, FL: Lewis Publishers, 189–231.

Kelly, B. C., Gobas, F. A. P. C. & McLachlan, M. S. (2004). Intestinal absorption and biomagnification of organic contaminants in fish, wildlife, and humans. *Environmental Toxicology and Chemistry*, 23, 2324–2336.

Kelly, B. C., Ikonomou, M. G., Blair, J. D., *et al.* (2009). Perfluoroalkyl contaminants in an Arctic marine food web: trophic magnification and wildlife exposure. *Environmental Science & Technology*, 43, 4037–4043.

Luoma, S. N. & Rainbow, P. S. (2005). Why is metal bioaccumulation so variable? Biodynamics as a unifying principle. *Environmental Science & Technology*, 39, 1921–1931.

MacDonald, R., MacKay, D. & Hickie, B. (2002). Contaminant amplification in the environment. *Environmental Science & Technology*, 36, 456A–462A.

Marigómez, I., Soto, M., Cajaraville, M. P., Angulo, E. & Giamberini, L. (2002). Cellular and subcellular distribution of metals in molluscs. *Microscopy Research and Technique*, 56, 358–392.

McKim, J., Schmieder, P. & Veith, G. (1985). Absorption dynamics of organic chemical transport across trout gills as related to octanol–water partition coefficient. *Toxicology and Applied Pharmacology*, 77, 1–10.

Ng, C. A. & Hungerbuhler, K. (2014). Bioaccumulation of perfluorinated alkyl acids: observations and models. *Environmental Science & Technology*, 48, 4637–4648.

Niyogi, S., Couture, P., Pyle, G., McDonald, D. G. & Wood, C. M. (2004). Acute cadmium biotic ligand model characteristics of laboratory-reared and wild yellow perch (*Perca flavescens*) relative to rainbow trout (*Oncorhynchus mykiss*). *Canadian Journal of Fisheries and Aquatic Sciences*, 61, 942–953.

Rainbow, P. S. (2018). *Trace Metals in the Environment and Living Organisms: The British Isles as a Case Study*, Cambridge, UK: Cambridge University Press, 106.

Sappington, K. G., Bridges, T. S., Bradbury, S. P., *et al.* (2011). Application of the tissue residue approach in ecological risk assessment. *Integrated Environmental Assessment and Management*, 7, 116–140.

Scott, G. R. & Sloman, K. A. (2004). The effects of environmental pollutants on complex fish behaviour: integrating behavioural and physiological indicators of toxicity. *Aquatic Toxicology*, 68, 369–392.

Taylor, M. G. & Simkiss, K. (2004). Transport of colloids and particles across biological membranes. In H. P. Van Leeuwen & W. Koster (eds.) *Physicochemical Kinetics and Transport at Biointerfaces*. Chichester, UK: John Wiley & Sons Ltd, 357–400.

Wright, D. A. & Welbourn, P. (2002). *Environmental Toxicology*, New York: Cambridge University Press.

APPENDICES

Appendix 3.1 Kinetics of a Saturable Transmembrane Carrier System Transporting a Chemical Substrate

It is assumed that the system has a finite number of carrier proteins available for chemical (substrate) transport. The terms used in the equations below are defined as follows: total molar concentration of protein carrier, $[P]_0$; molar concentration of unoccupied carrier, $[P]$; molar concentration of free chemical substrate, $[S]$; molar concentration of substrate carrier complex, $[PS]$.

The dissociation constant for substrate $\leftrightarrow$ carrier interaction is the Michaelis constant, K_m.

$$K_m = \frac{[P] \times [S]}{[PS]} = \frac{([P]_0 - [PS]) \times [S]}{[PS]} \tag{3.4}$$

$$K_m \times [PS] = ([P]_0 - [PS]) \times [S] \tag{3.5}$$

and

$$[PS] = \frac{[P]_0 \times [S]}{K_m + [S]} \tag{3.6}$$

The rate of uptake, V, is a proportional to the concentration of the substrate–carrier complex, i.e., $V = k \times [PS]$, and the rate is given by the expression

$$v = \frac{k \times [P]_0 \times [S]}{K_m + [S]} \tag{3.7}$$

When the carrier reaches saturation, $V = k \times [P]_0$. This can be described as the maximum rate, V_{max}, at which the system will operate and

$$v = \frac{V_{max} \times [S]}{K_m + [S]} \tag{3.8}$$

where V_{max}, $[S]$ and V may all be determined experimentally, and K_m is the substrate concentration at which the carrier is half-saturated. This value may be derived graphically from plots such as that in Figure 3.13. The K_m value derived from these plots provides a measure of the affinity of the carrier for the chemical (substrate); a smaller K_m value indicates a higher carrier–substrate affinity.

Appendix 3.2 Uptake/Loss Kinetics in a Single-compartment System

In a one-compartment model, the bioaccumulation of a toxicant by an organism can be written as a mass balance equation wherein the net rate of accumulation is the difference between the uptake and loss (elimination) rates:

$$\frac{dC_B}{dt} = k_{01}C_M - k_{10}C_B \tag{3.9}$$

Integration of equation (3.9) gives

$$C_B = \frac{k_{01}}{k_{10}} C_M \times \left(1 - e^{k_{10}t}\right) \tag{3.10}$$

where C_B is the toxicant concentration in the organism at time t, and C_M is the concentration of the toxicant in the ambient medium.

At this point,

$$\frac{dC_B}{dt} = k_{01}C_M - k_{10}C_B = 0 \tag{3.11}$$

and

$$\frac{C_B}{C_M} = \frac{k_{01}}{k_{10}} \tag{3.12}$$

Following termination of toxicant exposure, the rate of uptake, k_1C_M, falls to zero, and the elimination term becomes

$$\frac{dC_B}{dt} = -k_{10}C_B \tag{3.13}$$

Integration of equation (3.13) gives

$$C_B = C_0 e^{-k_{10}t} \tag{3.14}$$

where C_0 is the concentration in the organism at the beginning of the elimination process. The logarithmic form of this equation,

$$\log C_B = \log C_0 - \frac{k_{10}t}{2.303} \tag{3.15}$$

plots as a straight line with a slope of $-k_{10}/2.303$ (Figure 3.15B).

For a single-compartment system, the time taken to clear half of the bioaccumulated toxicant concentration, C_B (C_0 in Figure 3.15B) is defined as the biological half-life ($t_{1/2}$). Thus,

$$\ln \frac{C_0}{2} = \ln C_0 - k_{10} \times t_{1/2} \tag{3.16}$$

and

$$t_{1/2} = \frac{\ln 2}{k_{10}} = \frac{0.693}{k_{10}} \tag{3.17}$$

4

Methods in Ecotoxicology

Peter V. Hodson and David W. Wright

Learning Objectives

Upon completion of this chapter, the student should be able to:

- Discuss the comparative strengths and weaknesses of laboratory and field studies of chemical contamination and effects.
- Explain why standard toxicity tests do not provide a sufficient understanding of chemical effects on populations, communities and ecosystems.
- Identify different approaches for ecological surveys and monitoring to assess the distribution and extent of chemical contamination and to link contamination to ecological effects.
- Explain weight-of-evidence (WoE) approaches to direct the management of chemical emissions and remediation of contaminated sites, including ecoepidemiology as a framework for assessing the strength of proposed cause–effect relationships.
- Design field experiments that provide a broader understanding of ecological effects than laboratory tests.
- Explain the role of modelling in predicting the environmental behaviour, fate and effects of different contaminants.

Ecotoxicology investigates and assesses the distribution and effects of chemical contaminants in ecosystems, distinct from laboratory measurements of **bioaccumulation** and toxicity to single species (Chapters 2 and 3). The overall goal of this chapter is to introduce tools needed to recognize the extent of chemical contamination, adverse effects on exposed species, and the linkages between toxic effects on organisms and ecological change. The tools include **ecoepidemiology**, **biomarkers** and **bioindicators** of chemical effects, **palaeo-ecotoxicology**, chemical and effects monitoring, field experiments and modelling. The application of these tools supports the better management and control of chemical emissions and chemical contamination, and the mitigation of chemical effects on natural ecosystems.

4.1 Moving Beyond Environmental Toxicology

In Chapter 2 we explored how the toxicity of individual chemicals can be measured with laboratory tests of single species, an approach that characterizes **environmental toxicology** (Chapter 1). Here, we pose the question: can toxicity data collected under tightly controlled conditions reliably predict chemical effects in natural ecosystems? For example, in the laboratory we can measure the toxicity of crude oil to a model species such as the domestic duck. Nevertheless, our results may have no relevance to the responses of marine seabirds exposed to spilled oil under winter storm conditions when food is scarce. Even if our lab tests provide a reasonable estimate of toxicity in the field, do they predict the ecological impacts of losing large numbers of seabirds?

The primary roadblocks to a broader understanding of the ecological effects of contaminants are the diversity and complexity of ecosystems and how those characteristics interact with chemical exposures and effects (Gessner and Tlili, 2016). Globally, there are an estimated 8,700,000 microbial, plant and animal species (Mora *et al.*, 2011). These species occupy ecological niches as different as the canopy of tropical rainforests, the sediments of deep-sea trenches, Arctic and Antarctic ice packs, high-altitude deserts and the **hyporheic zones** of deep geological formations. This enormous range of environmental conditions is paralleled by an equally diverse array of genetic, biochemical, physiological, behavioural and morphological traits of species that have adapted to these environments. These traits determine each species' habitat selection, food web interactions and responses to chemical exposures (Poteat and Buchwalter, 2014).

It is unlikely that the sensitivity to chemical toxicity of even 0.01% of Earth's species will ever be tested (examples, Box 2.7). Most species are difficult to obtain in reasonable numbers, are of a size that precludes testing, are difficult to culture in the laboratory, or have a high ecological, social or economic importance (e.g., **keystone species**; iconic species; species at risk). Similarly, it is unlikely that the diverse environmental conditions typical of most natural ecosystems can be replicated in laboratory experiments. For the more than 350,000 chemicals in commerce (Wang *et al.*, 2020), this puzzle is further complicated by the wide range in properties such as solubility and persistence that determine their distribution, concentration and **bioavailability**.

From the perspective of ecotoxicology, how can we recognize the ecological effects of chemical toxicity when they occur, and what effects should we anticipate? If we see ecological change at the level of individuals, populations or communities, can we be certain they were caused by chemical contamination and not by other environmental factors? If multiple chemicals contaminate a given ecosystem, either in sequence or concurrently, how do we identify those that cause effects and those that are benign? Many of these challenges are evident in the complex issues presented in Chapter 11.

4.2 Laboratory Versus Field Studies of Ecotoxicology: Strengths and Weaknesses

As we learned in Chapters 2 and 3, toxicity begins with the route and rate of exposure of an organism to a chemical and subsequent interactions between the chemical and a **molecular receptor**. This interaction becomes evident as an impairment of physiological and developmental processes, up to and including death, a sequence of interactions known as an **adverse outcome pathway** (AOP) (e.g., Figures 4.1, 4.13, 14.2). An AOP is a *conceptual* model that summarizes the state of knowledge about the **mode of action** (MoA) and sequence of events connecting chemical exposures to effects on whole organisms, and ultimately on populations, communities and ecosystems. For any species and chemical compound, AOPs summarize the known linkages among:

- chemical exposure;
- **pharmacokinetics** (bioaccumulation/**bioconcentration**, metabolism, tissue distribution, excretion);
- **pharmacodynamics** (molecular, biochemical and physiological interactions or responses);
- rates of survival, growth and reproduction;
- subsequent changes to populations of the same species and communities of multiple species; and
- ecosystem structure and function.

At their simplest, AOPs are linear representations of **cause–effect** linkages. With increasing knowledge, AOPs become more complex and non-linear as molecular interactions are linked to cellular and physiological responses (Figure 5.7), the health and life stage of exposed organisms, and environmental variables that influence chemical exposure and toxicity. This 'bottom-up' environmental toxicology approach is essential for understanding why chemicals are toxic and how toxicity compares among chemicals. It provides the knowledge and data that underpin virtually all **environmental standards** and **regulations** for chemical emissions. However, it has little capacity to predict ecological effects because the scale and complexity of interactions increase with each level of organization (molecules to ecosystems). *The effects of chemical exposure at any level of organization (e.g., a community of species) includes functions and emergent properties that cannot be predicted from the next lowest level of organization (e.g., populations of a single species).* Effects on growth, reproduction and mortality have obvious implications for populations of affected species, but what happens to ecosystem structure and function when one or more species are so affected?

To understand and manage the ecological effects of chemical toxicity, field studies are needed to describe spatial and temporal patterns of contamination and

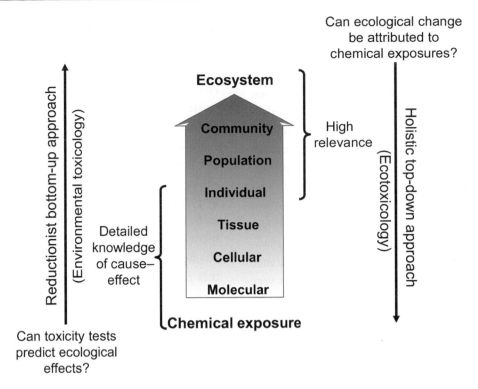

Figure 4.1 Trade-offs: 'Bottom-up' (reductionist) laboratory studies of toxicity and mode of action (adverse outcome pathways – central arrow) propose specific cause–effect relationships but have limited capacity to predict ecological impacts. 'Top-down' (holistic) field studies of ecological change are highly relevant to ecosystem management but provide little understanding of possible causes.

related ecological change. When patterns coincide in time or space, cause–effect relationships can be proposed and assessed by the strength of statistical correlations (Thomas *et al.*, 2021). Such 'top-down' or holistic approaches (Figure 4.1) include studies of sites contaminated by acute chemical spills, ongoing industrial emissions, and field experiments.

Ecological change is inferred from measures of ecosystem structure and function. For example, ecosystems under **stress** may be characterized by trends in the abundance and distribution of species across landscapes (e.g., marine reef communities) and the presence, absence or abundance of keystone species. Similarly, changes in primary and secondary production, respiration, energy flow through food webs and predator–prey relationships indicate effects on ecosystem function. Population-level effects may first be evident as changes in **ecosystem services**, such as the quality or abundance of species harvested by fisheries or forestry and reflected in changes in the **demographics** of local populations (Box 4.1). Unexpected changes in fish or wildlife populations observed during natural resource assessments may be the first **signs** of ecotoxicity when they closely follow industrial developments or other new sources of

contaminants. Inferences about chemical cause can be derived by subsequent surveys of spatial and temporal patterns of contamination and effects. However, caution is always advised: *correlation does not equal causation.*

A common theme throughout this chapter is the challenge (and opportunity) of inferring cause–effect relationships from field studies of contaminated sites or

Box 4.1 Population demographics that follow chemical toxicity

- Mortality
  - age distribution (year-class strength)
- Growth
  - **size-at-age** (growth rate)
- Reproduction
  - sex ratio
  - age-at-first maturity
  - fecundity (# of eggs/g biomass of females)
  - reproductive life span
  - number and size of offspring
- Behaviour
  - timing, location and size of annual migrations

ecosystems that have been disturbed by known or unknown factors. Although most top-down approaches provide information highly relevant to ecological impacts (Figure 4.1), *descriptive field studies are not experiments*. For the most part, we have no control over which species, life stages, populations or communities are exposed to chemical contaminants, the extent to which they are exposed, or their movements into or out of contaminated areas. Unlike experiments, chemical emissions to ecosystems are not distributed randomly to control for bias caused by unrecognized factors that affect exposures and effects. Similar ecosystems can be examined as **reference** systems to infer ecological conditions when contaminants are substantially absent (Section 4.3), but reference sites are not the same as experimental controls. It is very uncommon to find ecosystems that are contaminated by only a few chemicals of interest, or that are unaffected by non-chemical **stressors** including urban and industrial development, agriculture, resource extraction and climate change (Posthuma *et al.*, 2016).

One effective way to deal with differences in scale between laboratory and field assessments is to integrate information from field surveys of contaminated ecosystems with laboratory toxicity data by ecological modelling and comparisons to reference ecosystems. In Sections 4.3 to 4.5 we will learn how top-down approaches and modelling of chemical exposure and toxicity can be combined to help us recognize, measure and understand how toxic effects on individual species translate into ecological impacts.

4.3 Surveys, Monitoring and Assessment

For many chemicals, **objectives**, **guidelines** and **standards** (Section 12.3.1) have been developed for air, water, sediments and soils. Their purpose is to define concentration limits or benchmarks below which no toxicity is expected and above which there is an increasing likelihood of toxic effects and ecological damage. Similar limits have been developed to ensure the safety of food, whether it is produced by agriculture or by a natural resource. However, setting a standard for environmental protection does not guarantee that there are no ecological effects. **Surveys** and **monitoring** (Box 4.2; Section 12.3.5) of chemical concentrations in environmental media and of signs of ecological damage are essential to verify that standards and their enforcement are successful, and that new chemical issues are identified before they become

---

Box 4.2 Surveys versus monitoring

Survey – an examination of a system to determine its status. An **ecological survey** is the assessment of an ecosystem to determine its status, and whether the existence, degree and location of a condition warrant further research, management or monitoring.

Ecological monitoring – the repeated assessment of an ecosystem to determine whether a condition has changed in response to specific management policies or practices. For ecotoxicology, it may include **chemical monitoring** of contamination of air, water, sediment, soils or biota, and **biomonitoring** of the characteristics of biota (e.g., abundance, deformities) that reflect the effects of chemical contamination.

---

critical. At contaminated sites, surveys and monitoring are also needed to direct the planning and implementation of site **remediation** (Case Study 13.1).

Although surveys and monitoring may generate clear patterns of chemical contamination and ecological change (Section 4.3.1), contaminated sites include natural gradients of conditions that determine the abundance and diversity of species. Therefore, survey tools are needed to build the case that it is chemical contamination that causes ecological effects and to justify the effort and cost of remediation. These tools include ecoepidemiology (Section 4.3.2), specific biomarkers and indicators of chemical exposure and effects (Section 4.3.3), and palaeo-ecotoxicology to establish time order of contamination and effects (Section 4.3.4).

4.3.1 Relating Cause and Effect by Surveys and Monitoring

An ecological survey is descriptive and does not involve experimentation or other manipulation. The results may indicate the need for *ecological monitoring*, i.e., surveys repeated at regular intervals, usually following a fixed spatial pattern, and designed to reflect the findings of the initial survey. Monitoring demonstrates whether a problem is being resolved or is growing in size or intensity, and may forecast future conditions or trigger more stringent chemical controls if conditions do not improve or if they exceed some predefined **threshold**. Such information can also be used to support court cases concerning the sources, timing and effects of chemical emissions.

Survey tools for aquatic and terrestrial ecosystems are well-developed and are applied routinely in **ecological risk assessments** (ERAs; Section 12.3.2) of proposed urban or industrial developments. For example, new mining projects often require roads, residential facilities, airports, rail lines and power plants in addition to the mine, processing plants and **tailings** ponds. Ecological surveys can inform the planning, location and design of these new developments. They describe what resources are at risk, the likely environmental effects of expected emissions, and the indicators of impacts that should be included in subsequent monitoring programmes. Surveys and monitoring related to new developments correspond to classic 'before-and-after' studies in which chemical concentrations and ecological state post-construction are compared to pre-construction baselines to establish whether contamination and impacts have occurred. Because the same ecosystem is examined before and after the development, comparisons are conceptually and statistically robust and support inferences about cause and effect (Section 4.3.2).

For ERAs of existing developments, there is often no detailed knowledge of the ecological conditions that existed prior to the development, and no temporal patterns of effects to suggest chemical toxicity. To judge whether an ecosystem has been chemically degraded, nearby ecosystems having similar characteristics of climate, topography, elevation, etc. are examined as reference sites where conditions should be typical of an uncontaminated system (Reynoldson *et al.*, 2014). For rivers receiving a point-source effluent discharge, or for diffuse inputs from urban development, these surveys might correspond to an 'upstream–downstream' design (Figure 4.2). The abundance, diversity, health and productivity of aquatic species upstream of a source of contaminants are treated as baseline or reference conditions to judge whether toxic effects occur downstream. Related data, such as periodic fish kills, tissue concentrations of specific compounds, and **sublethal** responses unique to those compounds (Section 4.3.3) can reinforce conclusions about the cause of downstream effects.

Upstream–downstream surveys, or related comparisons between contaminated and 'clean' sites, are not as conceptually or statistically strong as before-and-after studies. The reason is that most ecosystems include multiple **ecotones** or gradients of environmental conditions called a **river continuum** (Doretto *et al.*, 2020) (Box 4.3). As might be expected, these natural gradients determine the biodiversity, species composition and productivity of aquatic communities at any point along the river's length and may mask gradients of ecological change caused by chemical toxicity. If gradients of natural factors are not recognized and measured in upstream–downstream studies, conclusions about the ecological effects of chemical contamination may be incorrect. The important influence of local environmental conditions on ecological responses

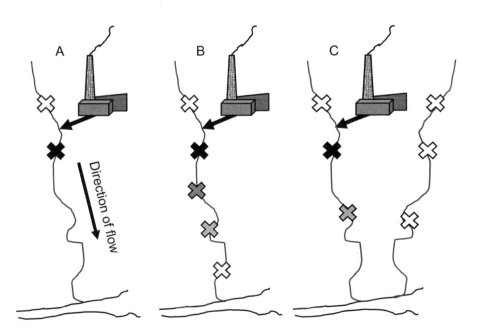

Figure 4.2 Upstream–downstream study designs to assess the effects of an industrial effluent on river ecology. A: Two-site design. B: Gradient design. C: Gradient design plus an uncontaminated reference river (Case Study 4.1).

Box 4.3 River continuum

A river continuum refers to natural gradients of ecosystem characteristics and conditions in a river, from its origin as a small headwater stream to its mouth. With distance downstream, rivers grow in volume and complexity as water, nutrients, sediments and organic matter are added progressively from tributaries and the surrounding watershed. Physical gradients of sediment characteristics and water quality (e.g., pH; ion content) are also created by changes in a river's slope, width, depth, current velocity, shading and substrates (e.g., limestone). The changes in physical and chemical characteristics of rivers determine gradients of species abundance and ecosystem structure and function. Understandably, gradients of any of these conditions may be altered by inputs of municipal sewage, agricultural run-off or industrial effluents.

to chemical contamination is termed **ecological context** (Clements *et al.*, 2012).

An extended version of the upstream–downstream design is one with multiple up- and downstream sites (Figure 4.2B). This design assumes an **assimilation capacity** of contaminated ecosystems, i.e., a capacity to recover from toxicity as chemicals become less bioavailable due to dilution, evaporation, adsorption to particulates, **complexation** with dissolved organic matter, binding to sediments or microbial degradation (Sections 11.4.1.3; 13.7). Multiple upstream sites reveal the extent of natural gradients associated with changing river flow and ecological conditions. At downstream sites close to chemical emissions, contamination and effects may be obvious, but further downstream, we may see a progressive recovery in species abundance, diversity and productivity as contaminant concentrations and bioavailability decrease. As with simple upstream–downstream studies, this design can be adapted to studies of lakes, ocean coasts and terrestrial sites if gradients of chemical concentration radiate from a point source. Parallel gradients of contamination and effects present a more compelling case for cause–effect relationships than a simple two-site comparison and provide greater statistical power due to larger sample sizes.

A more powerful but less common version of the upstream–downstream survey is to add a reference river with a similar upstream–downstream pattern of sampling sites (Figure 4.2C). Where ecotones obscure potential effects of chemicals, comparisons of ecotones between reference and contaminated rivers can highlight effects that were not obvious when only the contaminated river was considered. An interesting example is a study of the effects of a pulp-mill effluent on a **benthic** fish species in the Saint-Maurice River, Quebec, Canada (Case Study 4.1).

The obstacles to all gradient designs are the greater resources needed to sample more sites and to characterize non-chemical gradients. These designs are also limited by the length of a contaminated river and the number of downstream sites accessible for sampling; many industries are located near the mouths of rivers, where there is no 'downstream'. These same principles apply to surveys of terrestrial as well as aquatic ecosystems. For example, the survival, pollinator activity, physiological performance and gene expression indicative of stress in the giant Asian honey bee (*Apis dorsata*) were correlated to a gradient of air pollution in Bangalore, India (Thimmegowda *et al.*, 2020). Causation was reinforced by similar responses in fruit flies (*Drosophila melanogaster*) caged along the same gradient, although the specific toxic constituents of polluted air were not identified.

Case Study 4.1 Upstream–Downstream Studies to Assess Whether Pulp-mill Effluents Affect the Sexual Maturation of Fish (Gagnon *et al.*, 1995)

The discharge of industrial effluents provides many examples of upstream–downstream effects on aquatic species. However, in the Saint-Maurice River, Quebec, the effects of a pulp-mill effluent on the sexual maturation of white suckers (*Catostomus commersoni*) were confounded by **nutrient** enrichment (eutrophication) from a nearby discharge of municipal sewage. Compared to those upstream, fish sampled at sites up to 95 km downstream of the pulp mill were contaminated by chlorophenols and by chlorinated dioxins and furans. They responded with an increased activity of **cytochrome P450 (CYP1A)** enzymes (Figure 4.3) that facilitate contaminant excretion (Sections 3.5.2, 7.5). They also grew faster than fish at upstream sites, likely in response to nutrient stimulation of algal growth and ecosystem productivity. However, their regulation of reproductive steroid **hormones** (oestradiol, testosterone) appeared abnormal. Effluent-exposed fish were also older and larger at first sexual maturity than upstream reference fish and the relationship between

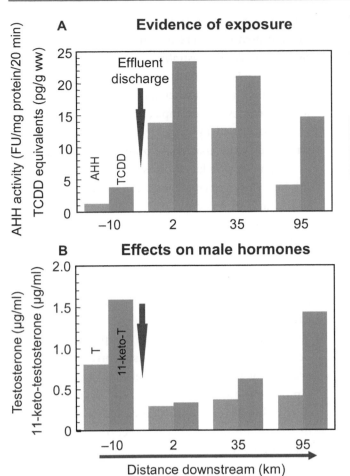

A Evidence of exposure

B Effects on male hormones

Distance downstream (km)

Figure 4.3 A: Induction of liver aryl hydrocarbon hydroxylase activity (AHH; CYP1A enzyme) and contamination by dioxin equivalents of white sucker sampled from the Saint-Maurice River upstream and downstream of a bleached kraft pulp mill. **B:** Plasma sex steroid concentrations in male white suckers. FU = fluorescence units (measuring enzyme activity by amount of fluorescence product created per mg protein per 20 minutes). Figure based on data from Hodson et al. (1992).

Table 4.1 The relative severity of chemical contamination and reproductive responses of white sucker sampled downstream of the discharge of pulp-mill and municipal effluents to the Saint-Maurice, Quebec, and downstream of a municipal discharge to the Gatineau River. The data were derived from Gagnon et al. (1995).

Response + = increase 0 = no change − = decrease	Saint-Maurice River: Pulp-mill + municipal effluents	Gatineau River: Municipal effluent only
Tissue contamination: chlorinated phenols, guaiacols, veratrols	+++	Trace
Liver cytochrome P450 enzyme activity	+++	0
Growth (size-at-age)	+++	+++
Reproductive impacts		
Age at sexual maturity	0	−
Size at sexual maturity	++	−
Fecundity–age relationship	Disrupted	+
Gonad size – males	+	++
Gonad size – females	−	++

fecundity (number of eggs/g female fish) and body weight was disrupted. Taken together, these responses suggested mild effluent effects because increased growth rates obscured reproductive impairment.

To separate nutrient effects from pulp-mill effects, the Gatineau River was sampled as a reference ecosystem. It closely resembles the Saint-Maurice, with similar upstream–downstream gradients in nutrient enrichment from a single source of municipal sewage, but no pulp-mill effluent. White suckers in the reference river were not contaminated with chlorinated compounds and showed no increase in activity of detoxification enzymes (Table 4.1). Compared to upstream, fish downstream of the sewage discharge grew faster in response to the

increase in available energy and invested more energy in reproduction. They had the same patterns of fecundity–age relationships as fish upstream, but they matured sexually at a younger age and at a lower weight. In contrast, the fish exposed to pulp-mill effluent failed to take advantage of the energy available from nutrient enrichment and invested much less in reproduction than fish in the reference river. The physiological responses of suckers from the two rivers are strikingly different and illustrate the critical importance of sampling designs in recognizing the ecological responses to industrial discharges.

4.3.2 Ecoepidemiology: Assessing the Strength of Proposed Cause–Effect Relationships

Ecological complexity and the co-occurrence of multiple environmental stressors are major obstacles to concluding that ecological change can be attributed to chemical contamination. Establishing the strength of proposed cause–effect relationships is a critical factor in justifying management actions to correct a problem. A similar challenge often hinders our ability to correctly identify

Box 4.4 Epidemiology adapted to ecotoxicology

... is *"the study of the distribution and determinants of diseases in human populations"*

- **Epizootiology** – epidemiology applied to diseases in animal populations.
- **Ecoepidemiology** – the application of epidemiology to identify the distribution and determinants of ecological change, and the links and pathways between them.

For each, the purpose is to relate cause to effect and to strengthen the case for corrective action.

and treat the causes of human disease. This challenge gave rise to the science of **epidemiology** (Box 4.4), pioneered in the mid-1800s by a London (UK) doctor, John Snow. Snow examined geographic patterns of cholera epidemics in London slums and used logic and a simple experiment (removing the handle from a water pump) to deduce that the cause was a discrete source of sewage-contaminated drinking water.

Fox (1991) applied the principles, methods and logic of epidemiology to an ecological context. He listed seven criteria for judging the strength of possible cause–effect relationships, and Adams (2003) demonstrated their application to field studies of contamination. These criteria include:

1. **Strength of association**: Cause and effect must coincide, expressed as the statistical likelihood that the occurrence of a putative cause increases with the occurrence of an effect and that a diminution of cause is associated with a reduction in occurrence (e.g., Case Study 4.1). Is the change in a characteristic of a chemically exposed species, population, community or ecosystem large (**relative risk**) compared to changes in the same characteristic of an unexposed system? Although statistical significance bolsters the case for cause and effect, statistical significance is not the same as biological or ecological significance, which remain judgement calls.

2. **Consistency of association**: Has a proposed cause–effect relationship been observed in other circumstances, by other observers, and in other ecosystems? Consistency is strong support for that association. For example, the observation of pulp-mill effluent effects

on fish reproduction (Case Study 4.1) was not unique but repeated elsewhere in Canada, the United States, Sweden, Finland, Australia and New Zealand.

3. **Specificity**: Is the effect diagnostic of exposure? This criterion is strongest when an effect has only a limited array of causes. For example, hypotheses that bird kills might be caused by **organophosphate pesticides** are strengthened by analyses of brain acetylcholinesterase activity in freshly killed birds. Few other compounds inhibit this enzyme and cause immediate and dramatic neurotoxicity.

4. **Time order**: Does the cause always precede the effect, and does removal of the cause diminish the effect? Time order corresponds to before-and-after studies (Section 4.3.1; Case Study 4.2) and is the strongest argument for concluding that chemical emissions are causing ecological effects.

5. **Biological gradient**: Is there a spatial correlation between exposure and response, corresponding to a gradient of relative risk parallel to a gradient of contamination? In Case Study 4.1, the degree of reproductive impairment of fish downstream of an effluent discharge diminished in parallel to a decrease in exposure to chlorinated compounds.

6. **Experimental evidence (predictive performance)**: Can an experimentally observed association between cause and effect predict the response when a cause occurs in nature? For example, pharmaceuticals that cause physiological effects in lab animals should logically cause similar effects in aquatic species when discharged in human sewage. Male fish in sewage-contaminated rivers in England exhibited signs of **feminization** and reproductive impairment coincident with the presence of synthetic oestrogen compounds from birth control pills. Similarly, the behaviour of fish is affected when psycho-active drugs such as antidepressants are discharged in sewage (Table 8.2).

7. **Biological plausibility (coherence)**: Is a proposed cause–effect relationship plausible and consistent with the biology of a species, population, community or ecosystem and with the MoA of a chemical stressor? Coherence is a 'make-or-break' criterion. For example, if groundwater contamination by a solvent is the proposed cause of fewer bird species that inhabit a forest canopy, the cause–effect relationship would not seem coherent. What exposure route would link the birds to groundwater chemicals? Nevertheless, a lack of coherence may reflect a lack of knowledge of the specialized

Box 4.5 Weight-of-evidence (WoE)

A WoE decision integrates multiple lines of evidence, including the role of natural gradients of environmental conditions, to assess whether observed ecological changes are caused by contaminants (Stevenson and Chapman, 2017).

The WoE approach is part of the **precautionary principle**, whereby decisions are made on the management of environmental issues in the absence of absolute certainty about cause and effect to ensure that environments are protected in a timely fashion. It is the opposite of the 'wait-for-evidence' approach that risks further and more severe damage.

biology of a species, unique exposure routes or MoAs, or the influence of natural gradients such as the river continuum. Coherence can also be weakened if chemicals are not bioavailable because they bind to inorganic or organic matter.

The criteria for judging the strength of proposed cause–effect relationships provide the logic needed to decide whether a source of chemical emissions should be managed, based on the WoE (Box 4.5). However, as Case Study 4.2 illustrates, decisions are not straightforward. To apply these criteria systematically, Cormier *et al.* (2010) propose an operational framework to guide the systematic collection and analysis of evidence, although the terminology is somewhat different from that presented by earlier authors.

Case Study 4.2 The Ecoepidemiological Case for Cancer in Fish Caused by Sediment Polycyclic Aromatic Compounds (PACs)

As an example of the power of ecoepidemiology, Rafferty *et al.* (2009) examined 18 different studies of putative skin and liver cancers in brown bullheads (*Ictalurus nebulosus*). Bullheads are a freshwater benthic fish species of eastern North America. Near steel mills where sediments are contaminated with **carcinogenic** PACs (e.g., benzo[a]pyrene; BaP), bullheads exhibit cancer-like lesions on their skin, lips and barbels, and pre-cancerous and cancerous lesions of the liver and bile ducts commonly associated with BaP metabolites. The prevalence and severity of these conditions were greater at contaminated sites compared to reference sites and increased with age, coherent with the

aetiology of cancer derived from laboratory experiments. At one site, the prevalence of liver lesions decreased following the closure of a steel mill, the proposed source of carcinogens, and increased again when the contaminated sediments were disturbed by dredging, declining once more thereafter. This temporal pattern provided strong support for the proposed cause–effect relationship. However, Rafferty *et al.* (2009) concluded that the case for a chemical cause of skin neoplasms was relatively weak, because many other stressors, including bacteria and parasites, caused similar skin lesions. In contrast, the case for liver neoplasms was:

- strong when the biological and technical plausibility and strength of association were considered;
- moderate for consistency of association, time order, dose–response and experimental evidence; and
- weakest for specificity of the relationship.

Given the number of chemicals in contaminated sediments and the relatively small proportion tested for carcinogenicity in fish, the lack of specificity was not surprising.

The review by Rafferty *et al.* (2009) is an excellent example of a WoE approach to making decisions, supported by the systematic application of epidemiological criteria. It directly addresses the dilemma facing environmental managers, i.e., the need to make decisions about environmental enforcement and remediation when associations of cause and effect include uncertainty. Rafferty *et al.* (2009) discuss these challenges and present several arguments in support of the precautionary principle, i.e., the need to act on protecting the environment when the stakes are high even though the scientific case linking cause and effect is not black and white.

4.3.3 Markers and Indicators of Chemical Exposure and Effects

The challenges of linking observed ecological change with exposure to chemicals can also be addressed by including chemical markers, biomarkers and **bioindicators** in surveys and monitoring. Chemical markers include tissue concentrations of chemicals or their metabolites to assess whether they exceed a known effects threshold, such as the **critical body burden** for **narcosis** (Chapter 2). Biomarkers are responses of individual organisms that are often diagnostic of exposure to specific chemicals or groups of chemicals. They may include changes in behaviour, enzyme activity, physiology, gene expression or the prevalence of specific cancers. Bioindicators represent responses of populations, communities or ecosystems to chemicals. Typical indicators of their state or function include abundance, growth rates, productivity, energy flow, species composition, etc. To demonstrate pulp-mill effluent effects on fish, Case Study 4.1 used chemical markers of exposure (tissue concentrations of chlorinated compounds), biomarkers of exposure (CYP1A induction) and toxicity (serum hormone concentrations) and bioindicators of toxicity (age-at-maturity).

4.3.3.1 Chemical Markers

Many chemicals including metals and chlorinated organics persist in organisms, often in specific tissues (Box 4.6), perhaps as a form of detoxification. For example, **poly-chlorinated biphenyls (PCBs)** accumulate in lipids, and their concentrations represent the sum of current and past exposures, with only minor amounts excreted. Metals such as lead are incorporated into the structure of growing bone in the place of calcium. Their distribution in a cross-section of bone can be measured with specialized techniques such as **energy dispersive X-ray analysis (EDXA)** to demonstrate whether exposure occurred in the past or is ongoing. For example, fish grow continuously, and each year of life adds an annual growth ring to their **otoliths** (bones in the semicircular canals used for balance). Counting the annual rings under a microscope and measuring lead concentration in each by EDXA provides an accurate history of the lifetime lead exposure of that fish. In studies of fish populations with high rates of deformities, this history of exposure can support a plausible hypothesis that episodic lead exposures caused these deformities.

The same principle applies to any species that grows continuously, assuming that the contaminant does not move. The growth rings of trees provide decade- to century-long records of the years (ring count) and climate conditions (size of rings) associated with chemical exposures (wood analyses) (Saint-Laurent *et al.*, 2011). Analyses of deer antlers that are shed annually demonstrate the history of exposure and accumulation of fluoride (Jelenko and Pokorny, 2010), an element that weakens bone and may kill ungulates by starvation if their jaws break. As a bonus, tree growth rings, fish fin rays, and mammalian antlers or teeth can be sampled without killing the specimen.

Box 4.6 Biological matrices used to date the history of chemical exposures

Recent

- Blood – lead
- Hair, feathers – mercury
- **Scat** – metals and organic compounds
- Lipids – lipophilic contaminants

Long-term

- Mammalian teeth, bones – lead, **radionuclides**, fluoride
- Whale baleen plates – metals
- Tree growth rings – metals
- Mussel shell growth rings – metals

4.3.3.2 Biomarkers

Biomarkers can be derived from known mechanisms of toxicity, such as **endocrine** disruption. Changes in the expression of specific genes related to chemical metabolism and toxicity constitute a class of biomarkers that are sensitive indicators of exposure and potential effects. For example, in most vertebrates, increased concentrations of cytochrome P450 (*cyp1a*) mRNA are clear signs of exposure to **dioxin-like compounds** (DLCs; chlorinated dioxins, furans and PCBs) or to PACs. Changes in gene regulation are usually followed by measurable biochemical responses, in this case the induction or increased synthesis and enzyme activity of CYP1A proteins that oxygenate PACs. Similarly, in animals exposed to some metals, there is an increased synthesis of metallothionein proteins (or **phytochelatins** in plants) that bind to metals and render them non-toxic (Section 6.5.2). Measures of the subcellular **partitioning** of a metal between metal-binding proteins and other cell constituents can discriminate between the amount that is bound and detoxified and the amount that is still freely available to cause toxicity. Other biomarkers include increased concentrations of metabolites in excretory pathways, such as the hydroxylated or conjugated derivatives of PACs in bile.

Biomarkers can provide specific and sensitive signs of chemical exposure and biological effects but are often transient. For example, the induction of CYP1A enzyme activity by brief exposures to PACs persists only as long as they remain in tissues. Because PACs are metabolized by CYP1A enzymes to accelerate their excretion, CYP1A activity declines quickly once PAC exposures cease. The value of measuring induction is the ability to recognize current exposures and to forecast **chronic** or delayed effects, e.g., the liver cancer that may follow CYP1A metabolism of BaP to carcinogenic derivatives (Case Study 4.2). In contrast, the bioaccumulation of persistent inducers of CYP1A activity (e.g., DLCs) causes prolonged induction.

Lead provides an example in which chemical and biological markers can be combined in vertebrates to discriminate between current and past exposures. The concentration of lead in the blood and its inhibition of erythrocyte **δ-aminolevulinic acid dehydratase (ALA-D)**, a blood enzyme that is uniquely sensitive to lead, demonstrate current exposures. In contrast, lead concentrations in calcified growth rings of bone (Section 4.3.3.1) show detailed histories of past lead exposures.

Even though some biomarkers are sensitive early responses to chemical exposure, responses are not necessarily followed by adverse effects. Brief exposures to BaP rapidly induce CYP1A enzymes, BaP oxygenation and metabolite excretion without toxic effects. In this case, CYP1A induction is a defence mechanism that shortens BaP exposures and reduces the risk of toxicity. In contrast, chronic exposures and prolonged CYP1A induction enable the continuous production of highly reactive dihydrodiol and **epoxide** metabolites of BaP. In this case, CYP1A induction increases the risk that BAP metabolites will react with nucleic acids to form **DNA adducts** that are characteristic biomarkers of genetic damage and cancerous lesions. Other examples of biomarkers include the inhibition of acetylcholinesterase by organophosphate pesticides, or the increased concentrations of **vitellogenin** (VtG) in male vertebrates exposed to oestrogenic compounds (Section 8.4). At a whole organism level, biomarkers include changes in behaviour (Section 2.3.5; e.g., avoidance of contaminated areas) and reduced aerobic capacity (e.g., reductions in maximum activity levels). Although these signs of chemical effects are not unique to a specific chemical, they can provide insight about potential effects at higher levels of organization, particularly when combined with biochemical and chemical markers (Section 5.9).

4.3.3.3 Bioindicators

Chemical markers and biomarkers provide strong evidence that organisms have been exposed to chemicals and are responding to that exposure, but how can we know whether these exposures and responses were sufficient to cause ecological change? The answers may be obtained through the measurement of bioindicators, i.e., measures of the structural and functional characteristics of an ecosystem that reflect its state or function.

Population Indices

Population demographics (Box 4.1) provide many of the **endpoints** used to determine chemical effects on populations (Case Study 4.1). Relatively simple populations of algae may be quantified in terms of cell numbers, overall biomass or biomass surrogates, such as chlorophyll-a assays. More dynamic, functional measurements include doubling times of cell densities, rates of photosynthesis, and respiration rates. For species with more complex life cycles, demographics include structural indicators such as the size and age distributions of a population, and functional indicators such as fecundity, a measure of reproductive potential. They may involve precise counts of offspring numbers and survival or estimates of the size of wildlife cohorts by repetitive surveys throughout their life cycle.

Initially developed as a tool for the study of population dynamics, the **life table** provides a more detailed evaluation of reproductive success through the determination of the intrinsic rate of natural population increase, r. This parameter is derived from the equation

$$\sum l_x m_x e^{-rx} = 1 \qquad (4.1)$$

where l_x = number of living females on day x/number of females at the start of the life table, x = time (d), and m_x = number of female offspring produced on day x/number of living females on day x. In view of the cost of collecting such detailed data, this approach has rarely been used to assess chemical toxicity.

More comprehensive demographic models, such as Leslie matrix models, encompass specific developmental stages defined as egg → larva → juvenile → adult. Such stage-based models can project population matrices for different species under various levels of toxic stress. When designing a field sampling programme, it is important to have information on the normal population dynamics of a species including the seasonality and rate of development of different life stages. For example, increased nutrient availability may accelerate the life cycle of a particular species, in which case more frequent sampling would be needed to account for the more rapid transition from one developmental stage to the next. This is not an issue for species that reproduce only once per year. The relative toxicity of a chemical to different life stages must also be accounted for. Therefore, the season, duration and intensity of a chemical exposure are critical in determining its effect, and determined by when the chemical is discharged, how much is discharged, where it is distributed and how persistent it is in the environment. These factors are all determined by the unique characteristics of each chemical and ecosystem.

Community Indices

The simplest structural indicators of community change include the presence or absence of indicator species (e.g., those tolerant of metal exposures) and types of species (e.g., K- vs r-strategy of reproduction, Box 4.7). More complex indicators are community biomass size spectra and indices of biodiversity. For example, the Benthic Assessment of Sediment model (BEAST) (Reynoldson *et al.*, 1995) relates benthic invertebrate species composition to the physical and chemical characteristics of sediments. Functional indicators are used less commonly than structural indicators, but would include rates of reproduction, growth and survival, grazing rates of herbivores, and species interactions (e.g., predator/prey relationships). Methods to measure changes at the landscape and ecosystem level include nutrient and energy flow and habitat suitability indices. Even economic indicators such as fisheries or forestry yields can indicate ecosystems under stress.

From a purely quantitative perspective, many well-known indices of community diversity can indicate

chemical effects in polluted areas; several are summarized in Table 4.2, although many variants are in use. **Diversity indices** are measures that combine evenness and richness of organisms within a community, with a particular weighting for each. **Species richness** is defined as the number of species counted in a particular sample. **Species evenness** indicates how uniformly the numbers of organisms in a sample are distributed among species. Diversity indices have often been used as measures of community perturbation by chemical pollutants. However, from a qualitative perspective, strict adherence to species abundance and numbers of organisms may overlook other important chemical effects, such as the nature of the species surveyed. Pollution may reduce the abundance of native species sensitive to toxic stress, but their decline may be offset by concurrent increases in opportunistic or invasive species more tolerant of chemical exposure. Although chemical toxicity may radically change the species composition of an exposed community relative to a reference community, these shifts may not be matched by changes in diversity indices.

Table 4.2 **Examples of community diversity indices used in pollution assessment. Different authors define similar indices with different symbols and give different weights to the data collected.**

Margalef $d = \frac{s-1}{\ln N}$	Indicator of specific diversity (d) of a community where s = number of species, N = number of individuals, and $\ln$ = natural logarithm.
Pielou $E = \frac{H}{H_{max}}$	Index of Evenness (E) measures how equally numbers of organisms are distributed among species in a sample. H_{max} is a value for H when individuals are spread evenly among the species present.
Simpson $D = 1 - \frac{\sum n(n-1)}{N(N-1)}$	Diversity index (D) where n = number of organisms of a particular species and N = total number of organisms in the community.
Brillouin $H_b = \frac{1}{N} \log \frac{N!}{n_1!\, n_2!...n_i!}$	Species richness index (H_b) where N = number of species in a community and n_i = number of individuals of the ith species.
Shannon–Weaver $H' = -\sum_{i=1}^{S} \frac{n_i}{N} \ln\left(\frac{n_i}{N}\right)$	Diversity index (H') where N = number of individuals in the sample, S = number of species in the sample and n_i = number of individuals of the ith species.

Box 4.7 Species reproductive strategies

r (reproduction)-selected species produce many offspring and invest little energy into each to ensure their survival. Many species of plants, invertebrates and fish scatter thousands of seeds or eggs, but with no parental care to ensure their survival. The offspring develop rapidly and the number that survive is haphazard, depending on environmental conditions and predator abundance.

K (carrying capacity)-selected species produce only a few slowly developing young and invest a significant amount of energy and time to ensure their survival. An extreme example is the 95-week gestation period of elephants (*Loxodonta africana*), followed by 3–5 years of lactation plus 5 years of juvenile development.

However, numbers do not tell the whole story: ecological function is critically important. Although diversity indices depend on dominant versus rare species, those that incorporate only richness and evenness in the same metric overlook changes in the functional aspects of community structure, i.e., the role of species adversely affected by chemical pollution. Rare or obscure species often occupy key roles in the food chain, such as predators or prey. Their loss can trigger ecological change well out of proportion to their original abundance by changing trophic efficiency, including nutrient transfer through a food web. For example, the elimination or shifts in trophic hierarchy of two top predators (Gulf killifish, *Fundulus grandis*; Stingray, *Dasyatis pastinaca*) in a Florida saltmarsh community followed heat stress from a power plant. Combined with compensatory changes in trophic transfer, the result was a 20% reduction in trophic carbon flow (i.e., productivity) (Ulanowicz, 2004). Similarly, Baird *et al.* (2004) reported significant changes in food web structure and energy flow in the Neuse River, North Carolina, due to **hypoxia**. Important driving functions were declines in filter-feeding benthic invertebrates, such as oysters, with subsequent increases in the abundance of planktonic algae and free-living bacteria.

To better recognize the roles of different ecosystem components, **indices of biotic integrity (IBI)**, were developed to describe freshwater fish stocks according to the number and identity of native/exotic fish species, number and identity of pelagic/benthic species, and percentage of omnivores, insectivores and top carnivores. Biotic indices were expanded to include pollution-tolerant and intolerant species and indices widely used as pollution indicators. The approach is highly versatile, with bioindicators of pollution drawn from a variety of taxonomic groups (Hering *et al.*, 2006). However, there are no fixed baselines, and indices must often be recalibrated region-by-region to recognize the unique sensitivity to specific chemicals of different indicator species, environments or ecozones.

For regulatory purposes, some approaches have been integrated. In a study of chemical pollution in three French estuaries, Blanchet *et al.* (2008) combined five methods of biotic assessment adopted by the European Water Framework Directive. The indices, based on benthic fauna, covered a range of ecological roles and chemical sensitivities of key groups/species. This WoE approach (Box 4.5) is derived from composite IBI scores

and statistical analyses. It facilitates regulatory decisions on pollution control yet highlights the influence of water quality and sediment characteristics independent of chemical pollution.

4.3.3.4 The Sediment Quality Triad

The **sediment quality triad** was developed to support decisions on remediation of contaminated sediments (Chapman, 1990). The triad includes three unique approaches to measuring the effects on communities of benthic invertebrates of chemicals in marine and freshwater sediments. Their purpose is to answer the following questions:

1. *Have thresholds of toxicity been exceeded?* In this approach, sediments collected from suspected 'hotspots' and uncontaminated reference sites are analysed chemically to map the geographic extent of contamination and to identify areas where concentrations exceed established guidelines or standards.
2. *Are the sediments toxic to benthic invertebrates representative of contaminated and reference sites?* Laboratory **toxicity tests** with invertebrate species typical of the sites surveyed demonstrate whether contaminated samples cause lethal or sublethal effects at the measured ambient chemical concentrations.
3. *Are there ecological effects that correspond to the spatial pattern of sediment contamination and toxicity?* Field surveys measure the abundance, diversity and community structure of sediment-dwelling invertebrates at the contaminated and reference sites.

All biological and toxicity data for contaminated sites are expressed as a ratio to the values measured for reference sites, because sediment toxicity and benthic community structure vary with ecological context (Clements *et al.*, 2012). For example, in the absence of contaminants, gradients of sediment grain size, organic carbon content and oxygen concentration determine the distribution of benthic species in the same way as the river continuum. If so, the observed gradients of species abundance may not reflect gradients of contaminant concentration. Therefore, judgements about the risks of leaving contaminated sediments in place or of dredging them for site remediation (Section 13.6) must be based on the WoE approach. The strengths and weaknesses of the triad are linked to the experience and professional judgements of scientists who design the studies, interpret the data and develop an opinion on the relative risks of contamination.

The triad was first developed over concerns about the acute lethality of very high concentrations of sediment-bound metals and organic compounds in harbours near industrial development. As these problems were addressed, it was recognized that many **hydrophobic** and lipophilic compounds (e.g., PCBs, **methylmercury**) could **biomagnify** to higher concentrations in food webs. Because species of fish and birds that consume benthos may range beyond the immediate site of contamination, the potential zone of influence would be much larger than the site itself. In response, surveys of tissue contaminant concentrations of benthic species have been added as the 'fourth leg' of the triad, and other indicators of the ecological impacts of contaminated sediments have also been considered (Chapman and Hollert, 2006).

The sediment quality triad is valued because it creates a perspective on the size of degraded areas, the severity of degradation, priorities for remediation, and the potential environmental consequences of remedial actions. The patterns of response also provide insights about the relative contribution of different contaminants to the biological characteristics of each site (Table 4.3) and support decisions based on the WoE approach.

4.3.3.5 Summary: Markers and Indicators

Chemical markers, biomarkers and bioindicators of chemical toxicity increase our capacity to interpret the results of surveys and monitoring, to recognize the ecological impacts of chemical contamination and to relate ecosystem-level effects to specific chemical exposures.

Biomarkers provide early warning of emerging changes in populations, communities and ecosystems before damage becomes irreversible. In all studies of contaminated ecosystems, it is critically important to characterize the distribution and severity of chemical contamination with the same level of detail and effort as is spent in measuring ecological responses. This requires integrated programmes for biological and chemical sampling, and the chemical analysis of target and non-target contaminants with a high degree of sensitivity. Analytical results that are 'less than the detection limit' provide no useful information if detection limits exceed thresholds for toxicity.

4.3.4 Palaeo-ecotoxicology: Retrospective Assessment of Contamination and Toxicity

For ecoepidemiology (Section 4.3.2), time order (temporal change) is among the most powerful criteria for linking chemical causes to ecological effects. However, our capacity to define temporal trends in exposure and effects is weak if industrial activities began long before the recognition of contaminant issues and the implementation of monitoring. This is also a problem for substances that were discharged before sensitive analytical tools were developed for environmental monitoring. Some of these issues can be resolved by analysing samples from tissue and sediment archives derived from ongoing monitoring programmes, but such archives are uncommon and not easily accessed.

Table 4.3 **Interpreting the results of the sediment quality triad.**

Case	Chemical contamination	Toxicity	Benthic community response	Possible interpretation
1	Yes	Yes	Yes	Strong evidence for pollution-induced degradation.
2	No	No	No	Strong evidence for no pollution-induced degradation.
3	Yes	No	No	Contaminants present but not sufficiently bioavailable to cause toxicity.
4	No	Yes	No	Unknown chemicals and/or conditions may cause degradation.
5	No	No	Yes	Impacts not due to chemicals.
6	Yes	Yes	No	Chemicals are likely stressing the system.
7	No	Yes	Yes	Unmeasured chemicals are causing degradation.
8	Yes	No	Yes	Chemicals are present but not bioavailable. Impacts may be due to indirect effects of chemicals.

'Yes' indicates an impact relative to reference site sediments. 'No' indicates no difference from the reference site.
Modified from Chapman (1990).

One ingenious solution is to study 'natural archives' of persistent markers and indicators stored in substrates that accumulate over time, such as glacial ice, terrestrial soils and aquatic sediments. Chemicals often persist in these media because of their inherent stability (e.g., PCBs, lead), the absence of oxygen needed for biodegradation, and/or low temperatures. Reconstructing past ecological conditions (**palaeo-ecology**) involves the collection of ice, sediment, soil or rock cores. The surface layers of each core represent the most recently deposited materials while successively deeper layers represent progressively older deposits.

Palaeolimnology is the analysis and interpretation of lake sediment cores to infer long-term environmental change and is a well-developed example of palaeo-ecology that relies on the slow accumulation of sediments in the deeper (profundal) parts of lakes, known as settling zones. A typical analysis starts with the collection of vertical cores of sediment (usually 5–10 cm in diameter, 30–200 cm long) sampled from a lake. The cylindrical cores are sliced like a sausage into evenly spaced sections. The key to linking ecosystem structure to past environmental conditions is the accurate dating of successively deeper layers of each core. Dates can be estimated from the activities of **radioisotopes** that decay at known rates

(e.g., ^{210}Pb; ^{137}Cs from atomic bomb testing) or from the concentrations of anthropogenic contaminants used during a specific era. For example, sediment concentrations of stable lead increased from 1921 when tetraethyl-lead was first added to gasoline as an anti-knock agent, to its phase-out, beginning in the 1970s (Figure 6.20). The age of each core section indicates sedimentation rates and the rate of change in chemical concentrations throughout the core (Figure 4.4).

Effects on lake ecosystems are assessed by identifying and counting the seeds and spores of plants and fungi (**palynology**) and the micro- and macrofossils of phytoplankton and zooplankton in each slice of the sediment core. Fossils are the parts of each species that do not decay, such as the siliceous cell walls of diatoms and the exoskeletons of zooplankton. Because the shapes of these structures are unique to each species, the species composition, relative abundance and community structure of a lake can be described for each slice of sediment. Similarly, the nature of adjacent terrestrial ecosystems can be inferred from the seeds and spores of terrestrial plants. Combined with the dating of each slice, we now have an historical record of the lake's plankton community. Changes in community structure provide a bioindicator of stresses on the lake, when the stress first occurred,

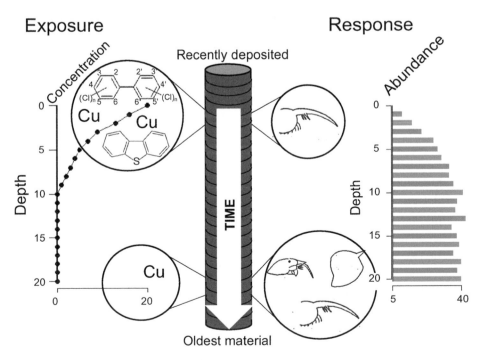

Figure 4.4 Palaeo-ecotoxicology methods. Measurements of contaminant concentrations in sediment cores are combined with counts of plankton microfossils and core dating to infer the history of chemical effects on aquatic communities. Adapted with permission from Korosi *et al.* (2017). © 2017 American Chemical Society.

how much conditions have changed, how rapidly they changed, and whether there are signs of subsequent recovery.

Palaeolimnology has been used to assess the effects of nutrient enrichment in lakes (eutrophication) and pH shifts caused by acid rain. For acid rain, planktonic species uniquely sensitive to acidification disappeared in the core in parallel with increasing numbers of more acid-tolerant species as predators and competitors were eliminated. For example, during periods of known ecological stress, the appearance of zooplankton species that are easily captured by fish signals the **extirpation** of predatory fish species. Thus, the history and importance of factors that affect the structure and function of lake ecosystems can be deduced from changes in species composition of successively older slices from sediment cores.

The application of palaeolimnology to describe the history of chemical contamination and effects is called palaeo-ecotoxicology. It is most applicable to chemicals that persist in **anoxic** lake sediments and that are immobile in sediments (e.g., PCBs, Pb, Cd). In contrast, low-molecular-weight degradable organics (e.g., phenols) may not persist, and some redox-sensitive elements (e.g., As, Se) can be **biotransformed** to derivatives that move among sediment layers and obscure relationships between time and concentration (Sections 6.8.7, 6.8.8). Taken together, the temporal sequence of contaminant accumulation in cores and changes in the community composition and relative abundance of indicator species can strengthen hypotheses linking chemical causes to ecological effects (Figure 4.4). When the sediment record of chemical contamination is combined with fisheries data, the role of PCBs in the reproductive impairment and population decline of fish species can also be deduced (Case Study 4.3). Similarly, analyses of cores from remote Arctic lakes have demonstrated the role of atmospheric transport (**global distillation hypothesis**; Section 7.2) and **biovectors** (migrating fish and wildlife) in transferring PCBs from industrial zones in temperate latitudes to remote lakes in the Arctic (Korosi *et al.*, 2017).

As methods develop, and more data on temporal patterns of sediment contamination become available, palaeo-ecotoxicology will provide valuable historical perspectives on environmental contamination and ecological impacts. The application of **eDNA** technologies (Chapter 5) to identify species present at specific times may also increase the power of the method to document ecological change.

Case Study 4.3 Evidence from Palaeo-ecotoxicology for a Chemical Cause of Reproductive Failure of Lake Trout (*Salvelinus namaycush*) in Lake Ontario

Palaeo-ecotoxicology was an essential tool in defining the role of DLCs in the reproductive failure of lake trout in Lake Ontario (Cook *et al.*, 2003). The abundance of lake trout decreased progressively from 1930 to 1950, apparently owing to overfishing and predation by the invasive sea lamprey (*Petromyzon marinus*). After 1950, there was evidence of reproductive failure, but the cause was unknown, and by 1960, natural populations were extirpated. To sustain populations, stocking with hatchery-reared trout was initiated in 1971 in conjunction with sea lamprey controls and limits on commercial fishing. However, there was no successful spawning until 1986, when embryonic and juvenile trout were first observed and interpreted as the progeny of stocked fish. The loss of naturally reproduced lake trout coincided with the loss of deepwater sculpins (*Myoxocephalus quadricornis*), a species not subject to lamprey attacks or commercial fishing, suggesting that a common factor prevented reproduction of both species.

One proposed cause of reproductive failure was exposure to DLCs, based on their propensity to biomagnify in food webs, their transfer from female fish to developing ova, and their high toxicity to embryos. Among the DLCs, **2,3,7,8-tetrachlorodibenzo-p-dioxin (TCDD)** is the most toxic, and the embryotoxicity of a mixture of DLCs can be estimated from **toxic equivalent factor (TEF)** models (Section 2.3.4.2). For lake trout embryos, the threshold tissue concentration for mortality is 30 pg/g TCDD equivalents; 100% mortality occurs at 100 pg/g or more.

To test the hypothesis of DLC-induced reproductive failure, two sediment cores from deep water zones of Lake Ontario were sectioned and dated (^{137}Cs; ^{210}Pb), and sediment concentrations of DLCs were measured and normalized to total organic carbon. Concentrations of DLCs expressed as TCDD 'toxic equivalent concentrations' (TECs) increased from the 1940s to the 1970s in parallel with industrial emissions, declining thereafter with emission controls. For each year represented by core sections, the TECs in adult trout were estimated from known **biota–sediment accumulation factors (BSAFs)**, i.e., the ratios of tissue concentrations of DLCs to sediment concentrations. Concentrations of DLCs in trout eggs were calculated from observed egg/adult ratios (0.7) to create a temporal trend of TCDD equivalents in trout eggs in parallel to sediment concentrations (Figure 4.5). To validate this model, the predicted egg concentrations were compared to egg concentrations derived from measured TCDD equivalents in adult trout between 1979 and 1990, and to measured concentrations in herring gull eggs sampled from Lake Ontario between 1970–1990. In both cases, there was a close correspondence between measured and predicted TCDD equivalents, confirming that calculated egg concentrations between 1930 and 1990 were suitable for estimating the extent of embryotoxicity.

Overall, the temporal trends of sediment concentrations of DLCs predicted that TCDD equivalents in trout eggs increased from <10 pg/g before 1932, peaked at 270 pg/g by the mid-1960s and declined thereafter to about 20 pg/g by 1988 (Figure 4.5). Concentrations

exceeded the threshold for 100% mortality of eggs between 1948 and 1976 and the thresholds for partial mortality from 1932 to 1948, and 1976 to 1991. Most importantly, the pattern of embryo mortality predicted from tissue concentrations of TCDD equivalents matched the temporal pattern of lake trout harvest derived from Lake Ontario fisheries data (Figure 4.6). The close correspondence between long-term trends in DLC contamination of sediments and the reproductive success of lake trout was strong time-order evidence for the ecological effects of DLC emissions.

4.3.5 Monitoring the Human Food Supply

Surveys and monitoring of the chemical contamination of wild organisms, particularly fish, are required to protect human consumers from the effects of persistent and

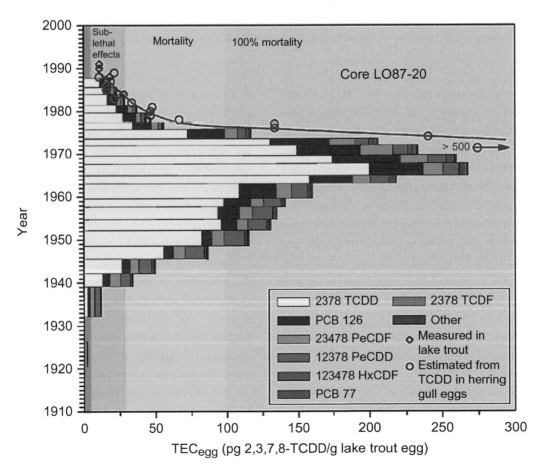

Figure 4.5 Concentrations of dioxin-like compounds (DLCs) in trout eggs, expressed as TCDD equivalent concentrations (TECs) calculated from concentrations of DLCs in dated 1-cm sections of Lake Ontario sediment cores. Concentrations expected to cause 0%, 0–100% and 100% mortality are indicated by pink, blue and grey zones, respectively. The TECs in eggs predicted from sediment DLCs are matched by TECs measured in lake trout (red circles) and estimated from DLCs measured in herring gull eggs (blue circles). Pe = penta; Hx = hexa; T = tetra; CDD = chlorinated dibenzodioxins; CDF = chlorinated dibenzofurans. Adapted with permission from Cook *et al.* (2003). Copyright 2003 American Chemical Society.

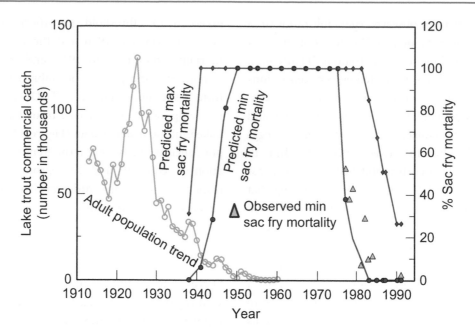

Figure 4.6 Temporal changes in the commercial catch of Lake Ontario lake trout compared to trout sac fry mortality predicted from sediment core analyses of TCDD equivalents. Triangles represent the observed mortality of sac fry raised from eggs collected from adult lake trout stocked to Lake Ontario. Adapted from Cook *et al.* (2003) with permission from the American Chemical Society.

bioaccumulative compounds (Section 12.3.3). If concentrations exceed the recommended limits for human consumption, specific batches of food may be banned for sale. In extreme cases, the sale of certain foods is banned entirely, and highly contaminated lakes, rivers or coastal waters are closed to commercial and sports fishing. In less extreme cases, public health advisories recommend that the consumption of certain species be limited to smaller size classes (Section 6.8.1). Fish grow continuously, and concentrations of persistent and bioaccumulative compounds (e.g., methylmercury) increase progressively with age and size. Consumption advisories often recommend limits location by location, species by species, and size class by size class (e.g., Guide to Eating Ontario Fish 2017–18; www.ontario.ca/page/eating-ontario-fish-2017-18). Bon appetit!

Increasing the public's awareness of the risks of chemical contamination of fish and wildlife often generates considerable support for environmental research, protection and remediation to limit the loss of valued ecosystem services. In Canadian Provinces and American States bordering the North American Great Lakes, public support has led to the successful management and remediation of contaminated sites and chemical emissions. Analyses of edible fish collected since 1970 illustrate long-term trends of declining concentrations

of contaminants such as PCBs (Bhavsar *et al.*, 2010) and the successful remediation of industrial harbours. Success is evident in the increased number of species and size classes of fish that are now safe to consume. Nevertheless, as analytical methods have improved, monitoring has also identified sites where more remedial work is needed or where there is contamination by previously unrecognized chemicals of concern, including brominated and fluorinated compounds (Sections 7.9; 7.10).

Monitoring fish and wildlife to protect human health also provides extensive databases that support research on chemical distribution and food web accumulation. However, one limitation to this form of monitoring is the focus on the 'edible portion'. For example, lead concentrates in calcified structures such as bone, but not in muscle, so that spatial patterns of lead contamination are difficult to discern from analyses of edible tissues. A similar issue applies to species that have a low fat content and do not accumulate high concentrations of hydrophobic organic compounds. To reduce these sources of bias, some monitoring programmes measure whole-body concentrations of contaminants in small minnows that live close to shore. A 30-year retrospective analysis of PCBs in Great Lakes spottail shiners (*Notropis hudsonius*) demonstrated clear trends of declining concentrations (French *et al.*, 2011), suggesting that control strategies have been successful. Correlations among PCB concentrations in shiners, warmer summer temperatures and an increased abundance of invasive mussel species (*Dreissenids*) also indicated that ecological interactions play important roles in food web contamination.

4.4 Field Experiments

Sections 4.2 and 4.3 describe many of the limitations of observational field studies commonly used to link ecological change to chemical contamination. A less common approach is the experimental application of chemicals to existing ecosystems, isolated parts of ecosystems or constructed model ecosystems. Field experiments often provide insights about ecotoxicology that cannot be obtained in any other way. The more that experimental conditions mimic those of natural ecosystems, the more likely it is that measured effects can be related to observed changes in ecosystem structure and function following chemical emissions.

4.4.1 *In Situ* Toxicity Tests

Toxicity can be assessed *in situ* under site-specific conditions by measuring chemical accumulation and effects on organisms caged in effluents, groundwater from contaminated sites, and contaminated soils and sediments (Oikari, 2006). This approach integrates many of the local conditions that affect chemical exposure and toxicity to native organisms. The test organism could be a cultured species transferred to the field or a native species collected from a reference site. Installing cages along gradients of contamination allows continuous chemical exposures without having to collect, store or dilute effluents or soil samples. Repeated observations of caged organisms and analyses of water or soils sampled adjacent to the cages provide data on effects, the rate at which effects occur, and chemical concentrations associated with those effects. After a predetermined exposure interval, tissues of test organisms can also be sampled to measure chemical and biological markers of exposure and effects.

Although caging studies avoid some of the artificial constraints of laboratory tests, they are best suited for acute exposures to minimize bias due to caging stress, particularly for species that are not typically held in captivity (Oikari, 2006). Stress may be induced by social interactions among test organisms that typically disperse, and by exposure to parasites, predators and natural gradients of environmental conditions. Starvation may also be an issue, although planktivorous species would have access to prey drifting through the cages. These confounding factors can be assessed by cages installed at uncontaminated reference sites, analogous to an upstream–downstream survey design (Section 4.3). For vertebrate embryos, more prolonged experiments are possible because feeding is not an issue, although naturally high mortality rates of fish embryos require large sample sizes to support statistical comparisons. It is also difficult to describe **exposure–response relationships** unless integrated samplers are placed adjacent to cages, an approach applied successfully to demonstrate the chronic toxicity of oil-contaminated sediments to lake whitefish (*Coregonus clupeaformis*) (Debruyn *et al.*, 2007).

A variation on this design is the exposure of assemblages of native aquatic species to dilution gradients downstream of chemical sources. Substrates such as ceramic tiles or glass slides are caged in reference areas where bacteria, fungi, algae, protists and invertebrate species colonize the slides to form communities representative of the local environment. After colonization, the substrates are transferred to cages distributed along a dilution gradient of an industrial effluent. The responses measured are community-level bioindicators such as the presence or absence of indicator species, the species diversity and community structure of each substrate, and the total weight or biomass of organisms. If there is sufficient biomass, tissue analyses of suspected contaminants and histopathology could also be performed. The primary drawback to these designs is that mobile vertebrate and invertebrate species are not included in colonized substrates, and cages must be designed to exclude grazing species (e.g., snails) that could consume substantial portions of a colonized community.

The toxicity of single compounds or mixtures of compounds can also be measured *in situ* by adding them to containers of natural communities of phytoplankton and zooplankton. Marshall *et al.* (1983) conducted toxicity tests in the open waters of southern Lake Michigan, USA by filling 18-l polyethylene carboys with unfiltered lake water containing native phytoplankton and zooplankton species at densities typical of natural communities. The carboys were spiked with a gradient of zinc and cadmium concentrations, plus traces of their radioisotopes (^{65}Zn, ^{109}Cd), and suspended below the lake's surface. After two weeks, algal productivity and the survival rates, community structure and tissue metal concentrations of 16 species of zooplankton were measured. The two metals clearly impaired primary productivity and zooplankton survival at concentrations well below those causing toxicity in laboratory tests. Toxicity was greater in the field, in part because the densities of organisms in lake water were much lower than in laboratory experiments. In small-scale laboratory tests, a large biomass of test organisms can reduce the concentrations of bioavailable metals through respiratory **uptake** and adsorption to exterior surfaces. In this field experiment, there were fewer organisms per litre of solution so that a higher proportion of each metal was in a form that was bioavailable and toxic. As well, there was a wide variation among species in their sensitivity to metals, with many more sensitive than typical laboratory test species. Most importantly, the measured toxicities were highly relevant to Great Lakes conditions.

4.4.2 Experimental Plots

Agricultural research has demonstrated the power of field experiments in the development of crops and cropping practices. Different experimental treatments (e.g., seed type, fertilizer concentrations) are applied to blocks of agricultural lands with different but relatively homogeneous characteristics (e.g., sloping fields facing north, south, east and west). These blocks may be further divided into subplots of land according to moisture gradients, proximity to forests, or changes in soil characteristics. Randomization of treatments among subplots within the larger blocks avoids the systematic influence of environmental factors on the treatments applied. With sufficient replication, these experiments have significant experimental power and can reliably identify effects on crop production caused by treatments (main effects), environmental variables (subeffects) and effects stemming from interactions between the two. These experimental designs and statistical methods have also been applied in ecotoxicology, for example to assess whether oil biodegradation on contaminated beaches can be enhanced by nutrient additions (Case Study 4.4).

Case Study 4.4 The Effectiveness of Fertilizers in Promoting Degradation of Crude Oil Spilled on a Vegetated Wetland

Experimental plots have been applied to assess whether ammonium nitrate fertilizer would stimulate plant growth, oxygen and nutrient transport to sediments, and natural microbial degradation of crude oil applied to a freshwater wetland (Venosa et al., 2002) (Capsule 13.2). Four replicates of five nutrient treatments were assigned to 4×5 m plots of marsh grasses along a gently sloping shore of the St. Lawrence River, Quebec. The plots were flooded twice daily by freshwater tides driven by ocean tides at the river's mouth. The treatments included controls (nutrients, no oil), natural attenuation (oil only), oil plus nutrients (plants intact), oil plus nutrients (plants cut), and an alternative source of nitrogen (oil plus sodium nitrate) (Figure 4.7). A curtain of oil-absorbing material around each plot prevented cross-contamination to adjacent plots and the surrounding environment. Sediment samples were collected from each plot at regular intervals for more than a year to measure oil concentrations, microbial activity, and the bioavailability of petroleum hydrocarbons to fish (Hodson et al., 2002). The toxicity of sediments was assessed with two snail species in laboratory tests and by caging snails of one species within each plot following the application of oil (Lee et al., 2002).

Over 15 months, the once-only application of fertilizers enhanced plant growth. Growth in the oil-only treatments was equivalent to that of wetland plants outside the plots, indicating that the enclosures did not disturb normal growth and that the oil treatments were not phytotoxic. Ammonia concentrations in sediments increased with fertilization and were highest in plots where the vegetation was cut, reflecting measured increases in the growth and activity of petroleum-degrading bacteria. Although concentrations of petroleum hydrocarbons in sediments decreased over time, loss rates were higher in the more aerobic surface sediments than in anoxic deeper sediments, and there were no marked differences in overall degradation rates among nutrient treatments. The stimulation of microbial growth without a corresponding acceleration of oil degradation was attributed to a lack of oxygen in fine-grained wetland soils. The oxygen contributed by the more rapidly growing plants was insufficient to enhance the oxidation reactions needed for oil degradation.

The bioavailability of PACs to fish in laboratory sediment tests was directly related to PAC concentrations in each sample. The induction of CYP1A enzymes in fish exposed to oiled sediments decreased in parallel with the progressive loss of oil from treated plots over the 15-month experiment. The loss reflected oil

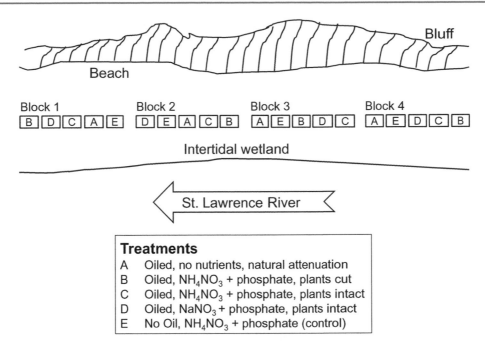

Treatments
A Oiled, no nutrients, natural attenuation
B Oiled, NH_4NO_3 + phosphate, plants cut
C Oiled, NH_4NO_3 + phosphate, plants intact
D Oiled, $NaNO_3$ + phosphate, plants intact
E No Oil, NH_4NO_3 + phosphate (control)

Figure 4.7 The distributions of oil and fertilizer treatments among four blocks of five naturally vegetated experimental beach plots (N = 20). Each 4 × 5 m plot received one of the five fertilizer and oil treatments.

biodegradation, weathering of oil exposed to the air and diurnal tides that diluted the oil in treated sediments with clean sediments washed into the plots.

After one month, the mortality of caged snails increased with plant cutting (loss of cover), nutrient treatments (ammonia toxicity), and confinement and overcrowding in cages; survival was greatest in the oil-only treatment. In laboratory tests, oiled sediments sampled one year after the last fertilizer treatment were much less toxic to snails than those containing fresh oil. Chronic sublethal effects of oil were lowest in tests of sediments treated with fertilizers compared to natural attenuation. The herbivore snail species was also less affected than the detritivore species, consistent with observations of more oil in the gut of the detritivores.

This study illustrates some of the challenges and unexpected and valuable findings of long-term field experiments. The chemical analyses and bioavailability studies demonstrated the anticipated loss of oil from wetland sediments over a year, but not by the expected mechanism of nutrient-stimulated microbial degradation. Although plant growth increased with soil fertilization, the fine wetland sediments limited interstitial oxygen concentrations and prevented the enhanced degradation of oil by aerobic microbes. Sediment oil concentrations were apparently affected more by tidal dissolution and dilution than by microbial metabolism. Snail survival

in situ demonstrated that cage design prevented conclusions about oil effects because many snails escaped. Nevertheless, the study highlighted the importance of sediment oxygen concentrations in microbial degradation, the relative persistence of oil, and the greater risks of oil exposure and toxicity to detritivore snails than to their herbivorous cousins.

4.4.3 Experimental Ecosystems

An understanding of the ecological impacts of chemical contamination can also be achieved with model ecosystems or components of ecosystems of different sizes and complexities (Table 4.4). Bench-top aquaria or terraria (**microcosms**) are constructed 'systems' that contain only a few species of plants and animals, but allow simultaneous multi-species testing and some limited interactions among species. Microcosms are not true ecosystems because they are not self-sustaining and experiments are generally brief (days to weeks). Large indoor or outdoor tanks or enclosures (**mesocosms**) are also constructed ecosystems and include more species, more complex species interactions, and some community interactions typical of real ecosystems; test conditions may be stable for weeks to months. Very large enclosures of water or land (**macrocosms**; akin to experimental plots, Section 4.4.2) start with portions of natural ecosystems and

Table 4.4 **Examples of experimental ecosystems applied to studies of chemical fate and effects.**

Type	Experimental system, size and duration	Test organisms and effects measured	Substances	Reference
Aquatic				
Microcosms	200-ml flasks, 25 d	Community metabolism and survival of algae, ciliate protozoa, rotifers, **oligochaetes**, bacteria	Cu; **herbicide**; chlorinated pesticides	Sugiura (1992)
Mesocosms	Tanks 2 m diam × 0.5 m deep; 1,400 l lake water, 160 l lake sediment; 11 d	Abundance, biodiversity and community structure of bacteria, **cyanobacteria**, phytoplankton and zooplankton	Diluted bitumen	Cederwall et al. (2020)
Macrocosms	Experimental stream, 20 m long × 1 m wide × 0.3–0.7 m deep, volume = 10.9 m³; 18 mo	Abundance, taxa richness, and community structure of macrophytes, phytoplankton, periphyton, benthic and pelagic invertebrates, microbes, fish	Cu	Joachim et al. (2017)
Whole ecosystems	Whole lake, 7.2 × 10⁵ m³; 18 mo	Silver accumulation by fish	Ag nanoparticles	Martin et al. (2018)
Terrestrial				
Microcosms	Glass tanks 20 × 13.5 × 15 cm; 500 g of homogenized grassland surface soil; 45 d	Abundance, genus richness, species evenness and richness of a mixed species assemblage of 100 nematodes added to each soil microcosm	Inorganic Hg	Martinez et al. (2019)
Mesocosms	3 × 3 m cages on contaminated soils at an oil refinery; 4–8 wk	Morphology, immune responses of cotton rats (Sigmodon hispidus) caged for up to 8 weeks	Petroleum hydrocarbons	Propst et al. (1999)
Macrocosms	20 × 20 m plots of mature spruce forest (Picea abies); 4 y	Al concentrations in soil, ground vegetation, spruce needles and leaf fall; tree crown condition, root growth, magnesium uptake	AlCl₃ added to soils	De Wit et al. (2001)
Whole ecosystems	Hubbard Brook experimental forest; 12–13-ha plots; 19 y	Accumulation of biomass of living trees	Acid rain; Ca added to soil	Battles et al. (2014)
Terrestrial–aquatic coupling				
Mesocosms	Terrestrial: 51 × 25 × 51 cm high; 81 kg of soil; seeded with seven species of pasture crops; 1 y Wetlands: 3.66 × 1.22 × 0.81 m high; partially filled with sand; three zones: partially, periodically, rarely flooded; 9 mo	Terrestrial and wetland communities: Vegetation growth; soil and sediment biodiversity and compositional shifts of bacterial, fungal, algal communities	Cu nanopesticide	Carley et al. (2020)
Whole ecosystems	METAALICUSᵃ experimental watershed; 6 y	Fate and transport of mercury deposited on boreal forest soils (simulated atmospheric deposition)	Stable and isotopic Hg	Oswald et al. (2014)

ᵃ METAALICUS = Mercury Experiment To Assess Atmospheric Loading in Canada and the United States.

support functioning food webs, predator–prey interactions and sustainable ecological processes that permit long-term experiments (months to years). For example, 'limnocorrals' are freshwater macrocosms that consist of large plastic bags or enclosures (e.g., 10 m diameter × 10 m deep). The bags may be closed at the bottom to simulate open-water processes, or open to sediments with the side walls anchored to minimize water exchange. The inclusion of sediments allows studies of sediment–water interactions, the responses of benthic species and the fate of hydrophobic chemicals.

Definitions of the size and complexity of micro-, meso- and macrocosms are not absolute, and often overlap. Nevertheless, a critical issue is that *size matters*. Micro- and mesocosms offer more flexibility than macrocosms or whole-ecosystem experiments, so that a wider array of chemicals and environmental conditions can be assessed with more replication and statistical power. These smaller systems have proven invaluable in comparisons among compounds of specific effects, such as the propensity for pesticides to persist, bioaccumulate or biomagnify through food webs. Based on the experience gained with small experimental systems, the persistence and distribution of chemicals in ecosystems, including biomagnification, can now be predicted by computer models (Section 4.5).

The advantage of larger model ecosystems is the increased number of ecological processes and effects that can be studied, and the confidence in the relevance of such effects. Schindler (1998) compared studies of aquatic bench-scale (bottle), mesocosm-scale and whole-lake experiments and concluded that the gains in statistical power of highly replicated small-scale studies are often outweighed by spurious conclusions about ecological effects. The stability of whole-lake ecosystems allows prolonged treatment and recovery times that reveal slow biogeochemical processes not evident in short-term experiments. As experimental systems increase in size, 'realism' increases owing to corresponding increases in ecological complexity, productivity, stability and longevity. Larger systems include all primary, secondary and tertiary production processes, along with degradation and nutrient recycling. Longer experiments with stable systems also enable the completion of life cycles of long-lived, predatory species and unique insights about ecological impacts.

The gold standards of field experiments are those in which complete ecosystems (e.g., a series of similar forest plots or lakes) are treated as experimental units, each with a natural assemblage of species and food web interactions (Table 4.4). By definition, these systems are realistic, with diverse and stable communities of indigenous species, and the natural exchanges of species, nutrients and energy with adjacent ecosystems provide a broader perspective on chemical effects. With greater size and complexity, larger systems increase the likelihood that important interactions among chemicals, species and ecological processes will be discovered. Nevertheless, experiments with natural systems include some obstacles:

- The legal and ethical issues of adding pollutants to natural systems, including the need for decontamination and mitigation of any experimental effects;
- A scarcity of multiple but similar and unperturbed ecosystems available for experiments;
- The need for remote and secure sites to avoid the confounding effects of other activities or sources of chemicals;
- The practical limitations of replicating large-scale multiple treatments; and
- The risk of experimental bias caused by oversampling upper trophic levels and perturbing ecosystem structure and function.

One very important benefit of experimental ecosystems is the discovery of the unexpected. For example, the acidification of a whole lake in Canada's Experimental Lakes Area (ELA) (www.iisd.org/ela/) first impaired the reproduction of minnow species at pH 6, well above the pH causing toxicity in lab tests; surprisingly, predatory lake trout seemed unaffected. However, after several years of acidification, the trout began to starve (Figure 4.8), an indirect effect of the declining abundance of minnows and other prey species. Only the analysis of large-scale, whole-lake linkages among nutrient cycling, food web energy transfer, species diversity, and fish reproduction and growth provided the detailed understanding needed to connect acidification to ecological change and to impacts on trout. Most importantly, the detailed understanding of acid effects allowed accurate predictions of effects in whole-lake experiments in Wisconsin (Section 11.1.1) and in lakes 'naturally' acidified by industrial emissions (Schindler *et al.*, 1990). Similar insights were gained from whole-lake studies of an endocrine disruptor (Case Study 4.5).

Figure 4.8 The effects of lake acidification on the growth and condition of lake trout. A: Appearance in 1979 at pH 5.4. B: Appearance in 1982 at pH 5.1. Reprinted from Schindler *et al.* (1985) with permission from the American Association for the Advancement of Science.

Case Study 4.5 Whole-lake Experiment with an Endocrine Disruptor

The addition of 5–6 ng/l of synthetic oestrogen (**17α-ethinyloestradiol or EE2**) to a whole lake over 5 years provided an ecological perspective on endocrine disruption of fish. Effects measured included intersex conditions in males (gonads had both male and female characteristics) and altered **oogenesis** in females (Kidd *et al.*, 2007). Among species, the response to EE2 was greatest in fathead minnows and pearl dace (*Margariscus margarita*), with increased synthesis (induction) of VtG (egg-yolk protein) by male fish; VtG is typically found only in female fish (Palace *et al.*, 2009). In addition to the increased incidence of intersex condition, there were histopathological changes in livers, kidneys and gonads. Lake trout and white sucker were less affected, with VtG induction the primary effect. At a population level, impacts were dramatic. Within 5 years, fathead minnows and pearl dace were almost extirpated from the lake owing to reproductive failure. A similar but smaller effect was observed in lake trout, a predator of minnows, but

their lower rates of reproduction could have been due to a loss of condition caused by a lack of prey. These very large changes in fish community structure would also have major impacts on the lake's food web, biodiversity and energy flow. The results corresponded very well with effects on fish observed in English rivers receiving high volumes of treated domestic sewage containing measurable concentrations of EE2 (Jobling *et al.*, 1998). In that case, the cause–effect relationship was less certain because sewage contains many pharmaceuticals and other organic compounds. These whole-lake experiments resolved many controversies about the cause and consequences of endocrine disruption.

4.5 Modelling Environmental Fate, Behaviour, Distribution and Effects of Chemicals[1]

A main theme of this chapter is the challenge of extrapolating laboratory toxicity data to real ecosystems. This task is complicated by the sheer number of chemicals in the environment, their concentrations and movements among different ecosystem compartments, and the uncertainty about which forms are bioavailable and toxic. An understanding of ecological effects is also limited by the large number of species in various developmental stages that interact within and among trophic levels. Their chemical exposures and sensitivity will vary spatially and temporally according to environmental conditions that change with season and climate. It is clear that very large amounts of chemical and biological data are needed to successfully predict chemical toxicity in highly complex natural ecosystems.

As one solution to these complexities, mathematical modelling has become an integral part of ecotoxicology. Fate models estimate the distribution of chemicals in or among various environmental compartments such as air, water, sediment, soil and biota. Effects models describe known phenomena and can be used to test or confirm that our understanding of mechanisms or processes are consistent with observations at specific sites. Models may also help us to identify errors in hypotheses, highlight

[1] Dr Brendan Hickie, Trent University, Peterborough, Ontario, Canada, contributed Section 4.5 on modelling, including text, figures, equations and a Case Study.

knowledge gaps, and support the regulation and management of environmental contamination.

Steady-state models describe systems in which chemical concentrations are expected to be stable (i.e., what the system will tend towards given stable conditions and sufficient time). In contrast, dynamic models track changes in chemical concentrations over time, driven by factors that are not stable, such as varying rates of chemical input, environmental or biological conditions, or factors that control chemical movement among different environmental compartments.

The process of constructing a model requires a focus on specific processes to constrain some of the complexity, simplify the picture and help us consolidate ideas and develop hypotheses. Building a mathematical model is an iterative process starting with a conceptual model that outlines the known components and processes expected in the system and how they are connected. Conceptual models may be simple box and arrow diagrams, formal or informal, and deliberate or intuitive, but always mindful of the model's intended use or purpose. Transforming a conceptual model to a quantitative model involves finding numerical measures to characterize a system (state variables) and the mechanistic or empirical equations that describe the model's processes and how they are linked. It is prudent to start with a relatively simple model and to add additional terms and processes as needed, being mindful of the adage that a good model should be "as simple as possible, as complex as necessary". A successful model that simulates an ecological reality can be invaluable, but simple first-stage models only give an approximate simulation, and their results must be treated with appropriate caution.

Some aspects of models used in ecotoxicology can be traced to functional elements of ecological models from the 1960s. With the advent of more powerful computers in the 1970s, ecological models evolved rapidly from being largely descriptive to much more quantitative. These models increased our comprehension of ecosystem dynamics, including complex feedback mechanisms that characterize ecosystems, and addressed the inherent variability in natural systems and uncertainty in measurements. For ecotoxicology, our understanding of the distribution of chemicals among environmental media has been improved by incorporating physicochemical properties such as **partition coefficients** (e.g., octanol/water, K_{OW}; octanol/air, K_{OA}; air/water, K_{AW} or **Henry's Law** constant) into models. These improvements include

their capacity to predict accumulation in organisms, both by direct uptake and by trophic transfer through food webs (Section 7.2).

There are three types of models commonly used in ecotoxicology: chemical fate models, bioaccumulation models, and integrated risk and effects models. Any of these models can be developed with generic environmental or biological characteristics for broad evaluative and regulatory use or for exploring differences among chemicals in environmental behaviour. Models can also be focused to fit site-specific characteristics, to examine past or potential risks of chemical contamination, or to explore which remediation scenarios reduce environmental risk. The output of fate models may supply both direct and indirect information on chemical exposures and inferences about potential effects. In 'effects models', this approach may be further extended to include information about the actual nature of the biological effect.

4.5.1 Chemical Fate Modelling

Chemical fate refers to the environmental distribution and disposition of chemicals following their emissions (Box 4.8) and is an essential component of ERAs. Fate models that describe the distribution and concentrations of a chemical in two or more environmental compartments are called multimedia fate models (MMFMs). They range from simple closed-box chemical equilibrium

Box 4.8 Chemical fate

… is "where and how a chemical is distributed after it is introduced into the environment"

Fugacity (f) – the tendency of a chemical to escape from a medium or phase, expressed in units of pressure (Pa).

Equilibrium – a state in which there is no net movement of chemical among two or more environmental compartments, and where chemical potential or fugacity is equal among the compartments (i.e., $f_1 = f_2 = f_3\ldots$).

Steady state – a state in which there is no change in concentrations over time in a system across two or more compartments. Steady state can be achieved without the system being at equilibrium.

models to open steady-state (non-equilibrium) models and complex dynamic models that track the global inputs, distribution and deposition of POPs such as lindane (MacLeod *et al.*, 2011). Fate models provide a comprehensive picture of the environmental behaviour of specific chemicals, transport processes and an overall mass balance that predicts concentrations in specific environmental compartments. The groundbreaking work by Donald Mackay (Parnis and Mackay, 2020) developed a series of MMFMs of varying complexity for organic chemicals that rely on the concept of fugacity (f). Fugacity expresses the tendency of a chemical to escape from one compartment or phase to another, and is expressed in units of pressure (Pascals, Pa). Fugacity is linearly related to concentration by equation (4.2):

$$C = Zf \qquad (4.2)$$

where C = chemical concentration (mol/m^3), Z = the fugacity capacity constant of the phase [mol/(m^{3}Pa)] and f = fugacity of the chemical (Pa). The fugacity capacity expresses the tendency of a phase to absorb the chemical, such that the compound will tend to move to compartments with high capacities. The Z values for various phases are determined by the properties of the chemical and of environmental media such as the organic carbon content of soils, sediments and particulate matter.

The dissolved oxygen concentration in an aquarium provides a simple illustration of the concept of fugacity when the concentration in freshwater at 15 °C is kept at saturation (10 mg/l) by bubbling air through the water:

1. About 20% of air is oxygen; thus, the oxygen in air has a *partial pressure* (i.e., fugacity, *f*) of about 0.2 atmospheres or *20,220* Pa (1 atmosphere = 101,100 Pa).
2. At this point, we cannot see that the water is saturated with oxygen owing to the difference in units, but by definition we know that for the water to be saturated with oxygen, that oxygen too must have a fugacity of 20,220 Pa.
3. This can be shown to be true by converting the oxygen concentration (*C*) of 10 mg/l to a molar concentration of 0.31 mol/m^3 and dividing by the Z value for oxygen in water:
    ◦ Z_w = 1/Henry's Law constant for oxygen = C (mol/m^3) = Z_{water} (mol/m^3 Pa) × f (Pa); Z acts as a conversion factor between concentration and fugacity.
4. Reversing equation (4.2), $f = C/Z$ = (0.31 mol/m^3)/ (1.53×10^{-5} mol/m^3 Pa) = *20,261* Pa.

If the oxygen content is below saturation, oxygen will diffuse from the air bubbles into the water (i.e., from high to low fugacity) until it becomes saturated and reaches equilibrium (i.e., $f_{air} = f_{water}$), illustrating the utility of fugacity in determining the direction that a chemical will move between environmental phases.

The versatility of the fugacity concept is illustrated by its scalability from transmembrane chemical transport in a single organism to chemical dispersion on a global scale. For example, fugacity-based models are used to evaluate the long-range atmospheric transport of POPs, including the characterization of the multi-stage global distillation (evaporation/condensation) process known as the '**grasshopper effect**' (Gouin *et al.*, 2004) (Section 7.2).

The simplest Level I fugacity-based MMFM describes the distribution of a chemical in a closed system among air, soil, water, sediment and aquatic biota at equilibrium (i.e., equal fugacity in all compartments, Figure 4.9). The fate of a compound in this simple model is determined by the size and fugacity capacities of each compartment and by the physical and chemical properties of the compound including water solubility, vapour pressure, K_{OW}, K_{OA} and K_{AW} (Figure 7.5). By excluding many environmental processes, this model provides a simple view of the environmental fate of a chemical, i.e., where concentrations will be highest due to partitioning, and a useful introduction to the value of models in ecotoxicology.

Level II and III fugacity models provide progressively more realistic descriptions of the fate of organic chemicals, moving from equilibrium to non-equilibrium steady-state conditions. These models consider many important environmental processes (Figure 4.9), including:

- chemical emission rates from a point source;
- chemical fluxes into and out of the model system with flows of water and air (advection); and
- rates of chemical removal from circulation by degradation, burial in sediments (chemical sinks), and transport among environmental compartments.

In these models the rate of transport (N, mol/h) between media is calculated as the product of the rate constant D (mol/h × Pa) and the difference in fugacities between the two compartments under investigation:

$$N = D[f_1 - f_2] \qquad (4.3)$$

The fugacity rate constant D can be applied to a wide range of processes, both abiotic and biotic, and to both steady-state and dynamic models. A system will remain

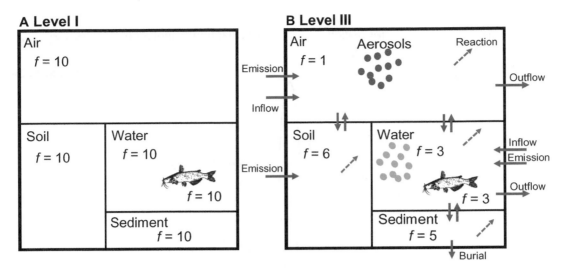

Figure 4.9 A: A Level I multimedia fate model shows the relative equilibrium (equal fugacity) partitioning of a conserved (i.e., non-reacting) chemical in a multimedia setting. Equilibrium and steady-state conditions are assumed to apply in this closed system. B: A more complex Level III model assumes steady state, but compartments are not at equilibrium and different fugacities apply to each medium. Rates of inter-media transport are calculated between the various compartments, including rates of chemical reactions and inputs and outputs from the open model.

at or close to an equilibrium distribution of a chemical if the set of D values describing chemical transport into and out of compartments are similar in magnitude. A non-equilibrium distribution develops if the D values are markedly different, i.e., when chemical transport becomes rate-limited, leading to a difference in fugacities between compartments.

Level IV multimedia fate models add further complexity by calculating a dynamic solution to the non-equilibrium chemical mass balance. Level IV models show how chemical concentrations in various media change over time in the model system in response to temporal changes in chemical loadings, as might occur following the regulation of production or release of a chemical. It is noteworthy that progressing to higher-order models places greater demands on finding appropriate parameter values needed to calibrate the model and to accurately predict chemical fate and temporal trends.

Several processes described by these models are of particular interest to ecotoxicologists. These models can estimate the total and bioavailable concentrations of chemicals in air, water or food, which are starting points for modelling bioaccumulation by individual organisms or transfer through food webs (Section 4.5.2). Generic MMFMs provide the basic structure from which other models can be developed to focus on particular environmental compartments or to combine them into multiseg-mented models that address the site-specific dispersal of

chemicals, regionally or even globally. For example, fugacity-based multisegmented models address the transport, deposition and fate of POPs at scales ranging from regions (Prevedouros *et al.*, 2004) to continents (MacLeod *et al.*, 2008) and to the entire globe (MacLeod *et al.*, 2011). A recent example integrated a fugacity-based MMFM with a 'hydrobiogeochemical' model to provide a detailed understanding of the seasonal behaviour of POPs in a catchment such as the River Thames in England (Lu *et al.*, 2016). Global-scale models have been particularly useful in highlighting the potential for the grasshopper effect to transfer POPs from mid-latitude sources to the Arctic. These global-scale models contributed to the science behind the 2001 Stockholm Convention and the international regulation of the production and use of over 30 POPs or chemical groups (www.pops.int; Table 1.2).

Several MMFMs that focus specifically on contaminants in aquatic ecosystems have been developed, with many based on the Quantitative Water, Air, Sediment Interaction (QWASI) model (Mackay *et al.*, 1983). This family of models provides a more detailed examination of the many processes that control the fate of chemicals in aquatic systems and can predict concentrations in water and sediment that are of particular interest to ecotoxicologists and chemical regulators. It has been applied to organic contaminants in the North American Great Lakes (Mackay, 1989), and adapted for metals and other non-volatile ionic chemicals (Csiszar *et al.*, 2011). The

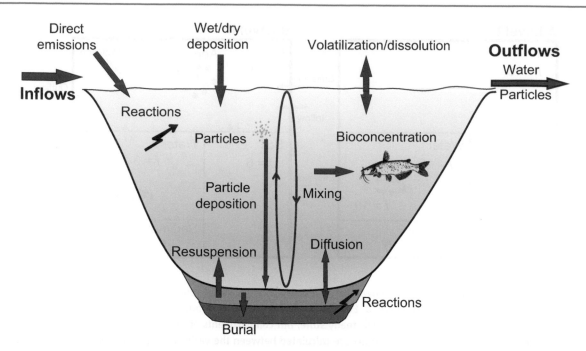

Figure 4.10 The compartments and processes represented in a steady-state mass balance model of chemicals in a lake ecosystem. The model requires data on masses (e.g., biota, particulates, sediments), flows (amounts entering or leaving a lake), reaction rates and concentrations. Mixing is accounted for in the calculated residence time of water in the lake. The amounts of chemical in each compartment are adjusted according to the reaction rate of the chemical in the water column, biota or sediments. The characteristics of the chemical that must be known include its air–water, biota–water and sediment–water partition coefficients. Adapted from Wright and Welbourn (2002) with permission from Cambridge University Press.

QWASI model has also been the basis for multisegmented aquatic models for lakes (Mackay and Hickie, 2000); Figure 4.10), river catchments (Warren *et al.*, 2007) and marine fjords (Lun *et al.*, 1998).

4.5.2 Bioaccumulation and Effects Modelling

The process of bioaccumulation (definitions, Table 3.3) connects environmental concentrations to accumulated dose, and effects may be estimated from established critical body burdens. Early bioaccumulation models adapted well-established concepts and equations from pharmacology and focused on a steady-state solution for an individual organism treated as a single compartment, such as a fish (Figure 4.11). This model considers chemical uptake from water (respiration) and food and uptake is driven by mass-specific bioenergetic demands for the oxygen and nutrients needed for resting and active metabolism, growth and reproduction. **Elimination** processes include chemical losses via the gills, faeces and/or urine, biotransformation to more readily eliminated metabolites, and a pseudo-loss resulting from organism growth (**biodilution**). **Bioamplification**, on the other hand, is the increase in chemical concentration in an

organism resulting from a loss of body mass. Rate constants for uptake, tissue distribution and elimination of neutral organic chemicals can generally be estimated from K_{OW} and K_{OA}. Although enzyme-mediated biotransformations can significantly reduce the bioaccumulation of many organic chemicals, rate constants for metabolism (k_M) are difficult to predict. They are determined by specific molecular structures that are often unrelated to K_{OW} or other physical or chemical properties (Arnot *et al.*, 2009). Similarly, uptake and elimination rates of metals are difficult to predict and must be determined experimentally (Poteat and Buchwalter, 2014).

These equations can be adapted to air-breathing organisms by changing the terms for exchange across gills to those for lungs and, for birds and mammals, changing the underlying bioenergetics from those for **poikilotherms** to those for homeotherms. Most bioaccumulation models assume that chemical concentrations are in equilibrium among tissues within an organism (i.e., equal fugacities or equal lipid-normalized concentrations of hydrophobic organic chemicals). In some instances, multicompartment 'pharmacokinetic' bioaccumulation models may be needed to describe non-equilibrium distributions of chemicals among tissues. Examples would be models to

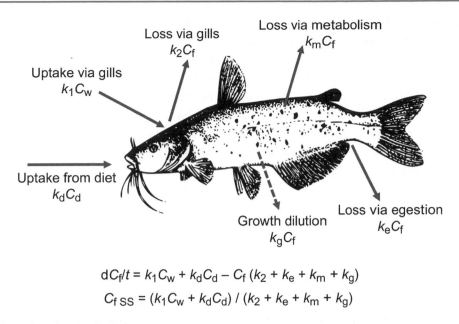

$$dC_f/t = k_1C_w + k_dC_d - C_f(k_2 + k_e + k_m + k_g)$$

$$C_{f\,SS} = (k_1C_w + k_dC_d) / (k_2 + k_e + k_m + k_g)$$

Figure 4.11 Fish bioaccumulation model. Rate constants (k) for uptake and loss include: k_1 = uptake (water); k_2 = depuration (gills); k_m = metabolism; k_d = diet; k_e = egestion (faeces); and k_g = growth dilution. Concentrations (C) include: C_w = water; C_f = whole fish; and C_d = diet. The differential equation describes the time-course of chemical bioaccumulation considering both uptake and loss processes. The lower equation calculates the concentration in the fish at steady state. These equations can also be expressed in terms of fugacities (f) and fugacity rate constants (D) (Parnis and Mackay, 2020).

characterize the critical role of gastrointestinal uptake in the biomagnification of chemicals (Drouillard *et al.*, 2012) or complex elimination processes such as maternal transfers of chemicals to an egg or foetus. Modelling non-equilibrium processes is often better addressed with a dynamic model that tracks changes in chemical concentrations over an organism's lifetime. Dynamic models can be solved readily by breaking down the calculations into repeated finite time-steps using spreadsheets or customized computer programs. Dynamic models have the added benefit of incorporating important aspects of life history (e.g., growth; reproduction; lactation), season (e.g., temperature; food availability or quality; gain or loss of lipid or mass) and time-varying exposure concentrations that affect bioaccumulation. Examples of such models can be found for fish (Daley *et al.*, 2014), birds (Norstrom *et al.*, 2007) and mammals (Case Study 4.6).

To model bioaccumulation and biomagnification through food webs, models for individual organisms (described above) are linked together for a set of species representative of different trophic levels. Their defined feeding preferences determine the chemical concentrations in their diet. Food web models can be written following conventional rate constant form (Arnot and Gobas, 2004) or in fugacity form (Campfens and MacKay, 1997). The organisms included in the food web can be generic in nature, such as phytoplankton → zooplankton → forage fish → piscivore, if the model is used to evaluate chemical bioaccumulation or trophic magnification potential. Food webs can also include endemic species for site-specific applications such as interpreting data from chemical monitoring programmes, conducting ERAs, or estimating bioaccumulation or biomagnification potential for species that are difficult to sample directly.

Most food web models that provide a steady-state solution are relatively simple to develop and calibrate and provide adequate solutions for most species and situations. The exceptions are long-lived species with complex life histories such as killer whales (*Orcinus orca*) that may never reach steady-state concentrations over their lifetime (Case Study 4.6). This problem can be overcome in part by including different life stages of such species in a steady-state model (Alava *et al.*, 2012).

Although most bioaccumulation models have been developed for aquatic food webs, Gobas *et al.* (2016) reviewed an array of terrestrial models that fell into two groups. Models of agricultural food webs characterize the transfer of chemicals from air and soil to plants, herbivores (e.g., cattle) and humans that consume milk and meat (McLachlan, 1996). Wildlife models describe chemical transfers from air and soil to herbivores and carnivores, such as a lichen → caribou → wolf model

(Kelly and Gobas, 2003) and a soil → earthworm → shrew model (Armitage and Gobas, 2007). Models that include humans as the end receptor with a diverse diet need to include both aquatic and terrestrial food chains (Czub and McLachlan, 2004). Case Study 4.6 presents an example of the application of several different bioaccumulation models to address the complexities of PCB bioaccumulation by the endangered Northeast Pacific southern resident killer whale (SRKW) population.

Case Study 4.6 PCB Contamination of the Southern Resident Killer Whale

The SRKW population is endangered, numbering less than 90 individuals. Its critical habitat includes the highly urbanized waters of the Salish Sea and Puget Sound (Figure 4.12A), bordered by Vancouver and Victoria (British Columbia, Canada) and Seattle (Washington, USA). The principal threats to the

viability of the population are food limitations owing to the decline in abundance of chinook salmon (*Oncorhynchus tshawytscha*, their primary prey), noise and disturbance from shipping, and bioaccumulative POPs, such as PCBs (DFO, 2018). Unfortunately, killer whales are the ideal organism for accumulating high concentrations of POPs. They are large (2,700 to over 5,000 kg) apex predators that have a long life span (often 50–90 years), large stores of lipid in their blubber, and a limited capacity to eliminate POPs. These traits were evident from the study by Ross *et al.* (2000) of PCB concentrations in the blubber of the fish-eating 'resident' and marine-mammal eating 'transient' populations from these coastal waters: they were among the most contaminated cetaceans in the world.

There are practical and ethical limitations on working with and sampling an endangered population of marine mammals. One solution has been the application of models to improve our understanding of the bioaccumulation and biomagnification of PCBs by SRKWs and

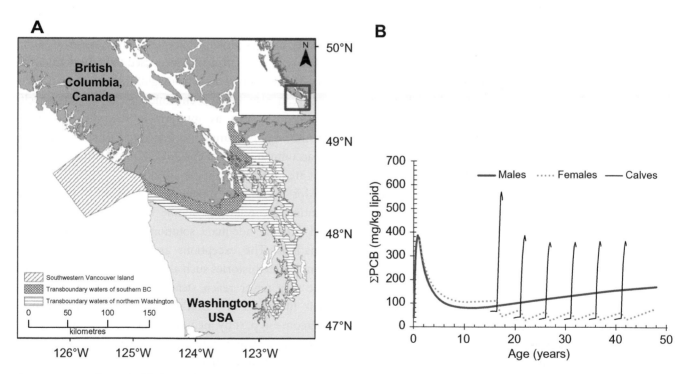

Figure 4.12 A: The critical habitats for southern resident killer whales (SRKWs). The hatched areas in the transboundary waters of southern British Columbia and off southwestern Vancouver Island are designated under Canada's Species at Risk Act (https://laws.justice.gc.ca/eng/acts/S-15.3/). The hatched areas in the transboundary waters of northern Washington State are designated under the United States Endangered Species Act (www.fws.gov/endangered/laws-policies/). Reprinted from DFO (2018). B: Lifetime changes in the concentrations of PCBs (mg/kg) in the blubber of orcas, predicted from a model by Hickie *et al.* (2007). Solid lines show the rapid accumulation of PCBs from birth until weaning for six successive progeny. The source is maternal transfer by lactation, and the first calf receives the highest dose. After weaning, juvenile growth rates exceed the rate of uptake of dietary PCBs, so that tissue concentrations decrease through biodilution until growth slows at maturity. For females, PCB concentrations decline with each birth (dotted line), but for males, concentrations continue to climb through dietary intake.

other populations, and to evaluate their potential for health effects. Hickie *et al.* (2007) developed an individual-based model to explain the effects of age, sex and reproductive activity on PCB concentrations in blubber over the lifetime of males and females (Figure 4.12B). The model showed that contaminant 'off-loading' during nursing could explain why adult females have lower contaminant burdens than males of similar age and why nursing calves should have some of the highest PCB concentrations in the population. Results from this model were incorporated into a dynamic population-based bioaccumulation model to reconstruct the history of PCB concentrations in the SRKW population from the 1930s, when PCBs were first manufactured, and the potential for future declines in contamination. The dynamic model revealed that concentrations likely peaked in the late 1960s to 1970s at levels 2.0- to 2.6-fold higher than concentrations measured in biopsy samples in the 1990s, and that future declines would likely be gradual. Ross *et al.* (1996) estimated the threshold PCB concentration causing health effects in captive harbour seals (*Phoca vitulina*) of 17 mg/kg (lipid weight). By comparison, the predicted PCB concentrations in SRKW exceed this threshold in all members of the population, and concentrations in most of the population will remain above this threshold for several more decades.

One of the challenges in understanding PCB contamination in the SRKW population is that they are a mobile predator, as are their primary prey, chinook salmon. Thus, the PCBs accumulated by SRKWs reflect a diverse exposure via atmospheric deposition to the Pacific Ocean as well as inputs from coastal sources. This presents considerable challenges to assessing whether PCBs in dredged sediments dumped in parts of the SRKW critical habitat could increase the contamination of the population. To address this question and to define water and sediment PCB concentrations that would not harm killer whales, Alava *et al.* (2012) developed a steady-state bioaccumulation model with unique food webs for each of seven areas spanning the SRKW range. For each food web, the model predicted PCB concentrations in adults and calves based on measured concentrations in sediment, water and air, as well as predictions that integrated the movement of whales among the areas. In these scenarios, ambient PCB concentrations in adult whales were within the range of (or exceeded) health-effect tissue concentrations (1.3 to 17 mg/kg lipid) in some individual

areas, or when averaged over all areas. Additional model scenarios demonstrated that existing sediment quality guidelines aimed primarily at protecting benthic invertebrates were vastly under-protective of whales. The model was then used to estimate sediment guideline concentrations that would protect the health of long-lived apex predators such as killer whales.

More recently, Desforges *et al.* (2018) developed a novel individual-based dynamic model that combines population growth and PCB bioaccumulation models and integrates concentration–response relationships for whale calf mortality and mortality due to immunotoxicity. The model forecast the impacts of PCBs over the next 100 years on the viability of 17 globally distributed killer whale populations that were divided into seven groups defined by their known PCB concentrations. The results predicted that four groups of whales with PCB concentrations equal to or greater than 40–113 mg/kg are at a high risk of population collapse. Interestingly, the SRKW population fell into the group just below these four. Forecasts from such models are not free of controversy, owing to the challenges of parameterizing the models, as shown by the critique by Witting (2019) and the subsequent response by Desforges *et al.* (2019).

4.5.3 Integrated Effects Modelling

In Section 2.3.3, we saw how models can relate molecular shape and properties such as hydrophobicity to their bioaccumulation potential and MoAs. The physicochemical properties of organic compounds have been used to formulate QSARs, and similar models, QICARs, have been applied to inorganic substances (Section 6.2.1). Recent developments in rapid nucleic acid sequencing have transformed our perception of molecular events governing the metabolism of chemicals and possible acclimation to chemical exposure (Chapter 5). While the primary focus of such effects models relates to the toxic MoA, they can provide important information on the bioavailability of specific classes of compounds and the nature of their effects on organisms. As such, they may play an important role in ERAs and chemical regulation. Ashauer and Jager (2018) make the point that AOPs have potential as toxic effects models at different levels of biological organization but would need further development for application beyond the cellular → physiological → individual pathway. Such a transition would require the identification and validation of

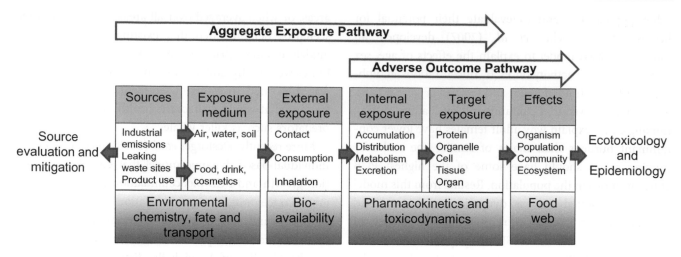

Figure 4.13 An aggregate exposure pathway (AEP) coupled with an adverse outcome pathway (AOP) linking sources of chemicals to target molecular sites, and effects from molecular sites to ecological responses. Each box represents a key event or measurable change in a chemical state, concentration and biological interaction that is essential, but not necessarily sufficient, for the movement of a chemical from a source to a target site. AEPs and AOPs can be used to accumulate information for source mitigation, predictions of toxic effects and ecoepidemiology. Figure and caption adapted with permission from Teeguarden *et al.* (2016). © 2016 American Chemical Society.

AOPs that are common to different taxonomic groups, or at least have common elements that could be reconciled. These AOPS could be integrated into an aggregate exposure pathway (AEP) framework (Figure 4.13) to provide a complete analysis from chemical source to outcome for risk-, **hazard**- or exposure-based assessments (Teeguarden *et al.*, 2016).

A similar argument might apply to energy-based models. In Section 2.3.5 we considered dynamic energy budget models that characterize toxic effects in an organism, using input parameters such as rates of feeding and reproduction. While energy budgets can span the life cycle of an organism (Jager *et al.*, 2010), we have yet to link this kind of model with changes that may occur among different species at different trophic levels. A broader perspective should also consider the overarching effects of environmental factors such as climate change (Figure 14.2).

Summary

In Chapter 4, we have learned about an array of tools that we can use to identify and measure the ecological effects of chemical contamination. These tools include ecoepidemiology to test the strengths and weaknesses of proposed cause–effect relationships and to assess the need for remedial actions. We have seen how environmental monitoring programmes can be designed to separate effects due to chemical contamination from ecological change linked to gradients of natural environmental conditions. Biomarkers and bioindicators further connect chemical exposures to responses of individual organisms and increase the specificity of cause–effect relationships, and palaeo-ecotoxicology correlates long-term ecological change with the timing of chemical contamination. Hypotheses about chemical effects on complex food webs can be assessed with experimental ecosystems and mathematical models integrate our detailed understanding of chemical fate and effects to help us understand the ecological risks of contamination. However, much remains to be investigated.

Review Questions and Exercises

1. Why is it difficult to predict community-level effects of chemical contamination from observed effects on a population of a single species?

2. Explain the 'weight-of-evidence' approach to chemical management. Why might it be used instead of a 'wait-for-evidence' approach?

3. What are the strengths and weaknesses of the 'top-down' and 'bottom-up' approaches to linking chemical exposures to ecological effects?

4. In surveys of a population of birds that live in the forest edge adjacent to agricultural fields and are exposed through their diet to a pesticide, what would you measure to demonstrate that there had been effects on reproduction? How would you design a field sampling programme to ensure that observed effects were not due to some other factor? What might those factors be?

5. Define coherence and explain why it is an important criterion for assessing cause and effect in ecoepidemiology.

6. Explain what a 'river continuum' is. Why is it important in studies of the effects of point sources of contaminants on aquatic species? Are there equivalent 'continuums' in terrestrial ecosystems?

7. Why is 'time order' a strong criterion for linking chemical exposures to ecological effects? Explain with an example.

8. What is the difference between a biomarker and a bioindicator?

9. What does a diversity index measure? What data would you collect to calculate a diversity index?

10. What are the four elements of the sediment quality triad? Why was the fourth element added?

11. What samples and measurements are essential for a palaeo-ecotoxicological assessment of ecological effects in lakes contaminated by a POP?

12. What is the ecological importance of monitoring the chemical contamination of the human food chain?

13. If you were asked to design an experiment to assess the biomagnification potential of a newly discovered POP, what experimental approaches are available, which would you choose, and why?

14. Define fugacity. Why is it an important property of chemicals for ERAs?

15. What is the difference between an aggregate exposure pathway and an adverse outcome pathway?

Abbreviations

AEP	Aggregate exposure pathway	DNA	Deoxyribonucleic acid
AHH	Aryl hydrocarbon hydroxylase	EDXA	Energy dispersive X-ray analysis
ALA-D	Aminolevulinic acid dehydratase	EE2	17α-ethinyloestradiol
AOP	Adverse outcome pathway	ELA	Experimental Lakes Area
BaP	Benzo[a]pyrene	ERA	Ecological risk assessment
BSAF	Biota–sediment accumulation factor	f	Fugacity
CDD	Chlorinated dibenzodioxins	IBI	Index of Biotic Integrity
CDF	Chlorinated dibenzofurans	K_{OA}	Octanol–air partition coefficient
CYP1A	Cytochrome P450 enzyme	K_{OW}	Octanol–water partition coefficient
cyp1a	Cytochrome P450 gene	MMFM	Multimedia fate models
DLC	Dioxin-like compound	MoA	Mode of action

mRNA	Messenger ribonucleic acid	TEC	Toxic equivalent concentration
PAC	Polycyclic aromatic compound	TEF	Toxic equivalent factor
PCBs	Polychlorinated biphenyls	VtG	Vitellogenin
POPs	Persistent organic pollutants	WoE	Weight-of-evidence
SRKW	Southern resident killer whale	Z	Fugacity capacity constant
TCDD	2,3,7,8-tetrachlorodibenzo-p-dioxin		

References

Adams, S. M. (2003). Establishing causality between environmental stressors and effects on aquatic ecosystems. *Human and Ecological Risk Assessment: An International Journal*, 9, 17–35.

Alava, J. J., Ross, P. S., Lachmuth, C., *et al.* (2012). Habitat-based PCB environmental quality criteria for the protection of endangered killer whales (*Orcinus orca*). *Environmental Science & Technology*, 46, 12655–12663.

Armitage, J. M. & Gobas, F. A. P. C. (2007). A terrestrial food-chain bioaccumulation model for POPs. *Environmental Science & Technology*, 41, 4019–4025.

Arnot, J. A. & Gobas, F. A. P. C. (2004). A food web bioaccumulation model for organic chemicals in aquatic ecosystems. *Environmental Toxicology and Chemistry*, 23, 2343–2355.

Arnot, J. A., Meylan, W., Tunkel, J., *et al.* (2009). A quantitative structure-activity relationship for predicting metabolic biotransformation rates for organic chemicals in fish. *Environmental Toxicology and Chemistry*, 28, 1168–1177.

Ashauer, R. & Jager, T. (2018). Physiological modes of action across species and toxicants: the key to predictive ecotoxicology. *Environmental Science: Processes & Impacts*, 20, 48–57.

Baird, D., Christian, R. R., Peterson, C. H. & Johnson, G. A. (2004). Consequences of hypoxia on estuarine ecosystem function: energy diversion from consumers to microbes. *Ecological Applications*, 14, 805–822.

Battles, J. J., Fahey, T. J., Driscoll, C. T., Blum, J. D. & Johnson, C. E. (2014). Restoring soil calcium reverses forest decline. *Environmental Science & Technology Letters*, 1, 15–19.

Bhavsar, S. P., Gewurtz, S. B., McGoldrick, D. J., Keir, M. J. & Backus, S. M. (2010). Changes in mercury levels in Great Lakes fish between 1970s and 2007. *Environmental Science & Technology*, 44, 3273–3279.

Blanchet, H., Lavesque, N., Ruellet, T., *et al.* (2008). Use of biotic indices in semi-enclosed coastal ecosystems and transitional waters habitats—implications for the implementation of the European Water Framework Directive. *Ecological Indicators*, 8, 360–372.

Campfens, J. & MacKay, D. (1997). Fugacity-based model of PCB bioaccumulation in complex aquatic food webs. *Environmental Science & Technology*, 31, 577–583.

Carley, L. N., Panchagavi, R., Song, X., *et al.* (2020). Long-term effects of copper nanopesticides on soil and sediment community diversity in two outdoor mesocosm experiments. *Environmental Science & Technology*, 54, 8878–8889.

Cederwall, J., Black, T. A., Blais, J. M *et al.* (2020). Life under an oil slick: response of a freshwater food web to simulated spills of diluted bitumen in field mesocosms. *Canadian Journal of Fisheries and Aquatic Sciences*, 77, 779–788.

Chapman, P. M. (1990). The sediment quality triad approach to determining pollution-induced degradation. *Science of the Total Environment*, 97/98, 815–825.

Chapman, P. M. & Hollert, H. (2006). Should the Sediment Quality Triad become a tetrad, a pentad, or possibly even a hexad? *Journal of Soils and Sediments*, 6, 4–8.

Clements, W. H., Hickey, C. W. & Kidd, K. A. (2012). How do aquatic communities respond to contaminants? It depends on the ecological context. *Environmental Toxicology and Chemistry*, 31, 1932–1940.

Cook, P. M., Robbins, J. A., Endicott, D. D., *et al.* (2003). Effects of aryl hydrocarbon receptor-mediated early life stage toxicity on lake trout populations in Lake Ontario during the 20th century. *Environmental Science & Technology*, 37, 3864–3877.

Cormier, S. M., Suter, G. W. & Norton, S. B. (2010). Causal characteristics for ecoepidemiology. *Human and Ecological Risk Assessment: An International Journal*, 16, 53–73.

Csiszar, S. A., Gandhi, N., Alexy, R., *et al.* (2011). Aquivalence revisited — New model formulation and application to assess environmental fate of ionic pharmaceuticals in Hamilton Harbour, Lake Ontario. *Environment International*, 37, 821–828.

Czub, G. & McLachlan, M. S. (2004). A food chain model to predict the levels of lipophilic organic contaminants in humans. *Environmental Toxicology and Chemistry*, 23, 2356–2366.

Daley, J. M., Paterson, G. & Drouillard, K. G. (2014). Bioamplification as a bioaccumulation mechanism for persistent organic pollutants (POPs) in wildlife. In Whitacre, D. (ed.) *Reviews of Environmental Contamination and Toxicology*. Springer: Cham, 107–155.

De Wit, H. A., Mulder, J., Nygaard, P. H., *et al.* (2001). Aluminium: the need for a re-evaluation of its toxicity and solubility in mature forest stands. *Water, Air and Soil Pollution: Focus*, 1, 103–118.

Debruyn, A. M. H., Wernick, B. G., Stefura, C., *et al.* (2007). *In situ* experimental assessment of lake whitefish development following a freshwater oil spill. *Environmental Science & Technology*, 41, 6983–6989.

Desforges, J.-P., Hall, A., McConnell, B., *et al.* (2018). Predicting global killer whale population collapse from PCB pollution. *Science*, 361, 1373.

Desforges, J.-P., Hall, A., McConnell, B., *et al.* (2019). Response to L. Witting: PCBs still a major risk for global killer whale populations. *Marine Mammal Science*, 35, 1201–1206.

DFO (2018). *Recovery Strategy for the Northern and Southern Resident Killer Whales (*Orcinus orca*) in Canada [Proposed]*. Ottawa, ON: Fisheries & Oceans Canada.

Doretto, A., Piano, E. & Larson, C. E. (2020). The River Continuum Concept: lessons from the past and perspectives for the future. *Canadian Journal of Fisheries and Aquatic Sciences*, 77, 1853–1864.

Drouillard, K. G., Paterson, G., Liu, J. & Haffner, G. D. (2012). Calibration of the gastrointestinal magnification model to predict maximum biomagnification potentials of polychlorinated biphenyls in a bird and fish. *Environmental Science & Technology*, 46, 10279–10286.

Fox, G. A. (1991). Practical causal inference for ecoepidemiologists. *Journal of Toxicology and Environmental Health*, 33, 359–373.

French, T. D., Petro, S., Reiner, E. J., Bhavsar, S. P. & Jackson, D. A. (2011). Thirty-year time series of PCB concentrations in a small invertivorous fish (*Notropis hudsonius*): An examination of post-1990 trajectory shifts in the Lower Great Lakes. *Ecosystems*, 14, 415–429.

Gagnon, M. M., Bussieres, D., Dodson, J. J. & Hodson, P. V. (1995). White Sucker (*Catostomus commersoni*) growth and sexual maturation in pulp mill-contaminated and reference rivers. *Environmental Toxicology and Chemistry*, 14, 317–327.

Gessner, M. O. & Tlili, A. (2016). Fostering integration of freshwater ecology with ecotoxicology. *Freshwater Biology*, 61, 1991–2001.

Gobas, F. A. P. C., Burkhard, L. P., Doucette, W. J., *et al.* (2016). Review of existing terrestrial bioaccumulation models and terrestrial bioaccumulation modeling needs for organic chemicals. *Integrated Environmental Assessment and Management*, 12, 123–134.

Gouin, T., Mackay, D., Jones, K. C., Harner, T. & Meijer, S. N. (2004). Evidence for the "grasshopper" effect and fractionation during long-range atmospheric transport of organic contaminants. *Environmental Pollution*, 128, 139–148.

Hering, D., Johnson, R. K., Kramm, S., *et al.* (2006). Assessment of European streams with diatoms, macrophytes, macroinvertebrates and fish. A comparative metric-based analysis of organism response to stress. *Freshwater Biology*, 51, 1757–1785.

Hickie, B. E., Ross, P. S., Macdonald, R. W. & Ford, J. K. B. (2007). Killer whales (*Orcinus orca*) face protracted health risks associated with lifetime exposure to PCBs. *Environmental Science & Technology*, 41, 6613–6619.

Hodson, P. V., Ibrahim, I., Zambon, S., Ewert, A. & Lee, K. (2002). Bioavailability to fish of sediment PAH as an indicator of the success of *in situ* remediation treatments at an experimental oil spill. *Bioremediation Journal*, 6, 297–313.

Hodson, P. V., McWhirter, M., Ralph, K., *et al.* (1992). Effects of bleached kraft mill effluent on fish in the St. Maurice River, Quebec. *Environmental Toxicology and Chemistry*, 11, 1635–1651.

Jager, T., Vandenbrouck, T., Baas, J., De Coen, W. M. & Kooijman, S. A. L. M. (2010). A biology-based approach for mixture toxicity of multiple endpoints over the life cycle. *Ecotoxicology*, 19, 351–361.

Jelenko, I. & Pokorny, B. (2010). Historical biomonitoring of fluoride pollution by determining fluoride contents in roe deer (*Capreolus capreolus L.*) antlers and mandibles in the vicinity of the largest Slovene thermal power plant. *Science of the Total Environment*, 409, 430–438.

Joachim, S., Roussel, H., Bonzom, J.-M., *et al.* (2017). A long-term copper exposure in a freshwater ecosystem using lotic mesocosms: invertebrate community responses. *Environmental Toxicology and Chemistry*, 36, 2698–2714.

Jobling, S., Nolan, M., Tyler, C. R., Brighty, G. & Sumpter, J. P. (1998). Widespread sexual disruption in wild fish. *Environmental Science & Technology*, 32, 2498–2506.

Kelly, B. C. & Gobas, F. A. P. C. (2003). An Arctic terrestrial food-chain bioaccumulation model for persistent organic pollutants. *Environmental Science & Technology*, 37, 2966–2974.

Kidd, K. A., Blanchfield, P. J., Mills, K. H., *et al.* (2007). Collapse of a fish population after exposure to a synthetic estrogen. *PNAS Environmental Sciences*, 104, 8897–8901.

Korosi, J. B., Thienpont, J. R., Smol, J. P. & Blais, J. M. (2017). Paleo-ecotoxicology: What can lake sediments tell us about ecosystem responses to environmental pollutants? *Environmental Science & Technology*, 51, 9446–9457.

Lee, L. E. J., Stassen, J., McDonald, A., *et al.* (2002). Snails as biomonitors of oil-spill and bioremediation strategies. *Bioremediation Journal*, 6, 373–386.

Lu, Q., Futter, M. N., Nizzetto, L., *et al.* (2016). Fate and transport of polychlorinated biphenyls (PCBs) in the River Thames catchment – insights from a coupled multimedia fate and hydrobiogeochemical transport model. *Science of the Total Environment*, 572, 1461–1470.

Lun, R., Lee, K., De Marco, L., Nalewajko, C. & Mackay, D. (1998). A model of the fate of polycyclic aromatic hydrocarbons in the Saguenay Fjord, Canada. *Environmental Toxicology and Chemistry*, 17, 333–341.

Mackay, D. (1989). Toxic substances modelling: fugacity concepts. In M. G. Singh (ed.) *Systems and Control Encyclopedia*. Oxford: Pergamon Press, 4897–4901.

Mackay, D. & Hickie, B. (2000). Mass balance model of source apportionment, transport and fate of PAHs in Lac Saint Louis, Quebec. *Chemosphere*, 41, 681–692.

Mackay, D., Joy, M. & Paterson, S. (1983). A quantitative water, air, sediment interaction (QWASI) fugacity model for describing the fate of chemicals in lakes. *Chemosphere*, 12, 981–997.

MacLeod, M., von Waldow, H., Tay, P., *et al.* (2011). BETR global – a geographically-explicit global-scale multimedia contaminant fate model. *Environmental Pollution*, 159, 1442–1445.

MacLeod, M., Woodfine, D. G., Mackay, D., *et al.* (2008). BETR North America: a regionally segmented multimedia contaminant fate model for North America. *Environmental Science and Pollution Research*, 8, 156–163.

Marshall, J. S., Parker, J. I., Mellinger, D. L. & Lei, C. (1983). Bioaccumulation and effects of cadmium and zinc in a Lake Michigan plankton community. *Canadian Journal of Fisheries and Aquatic Sciences*, 40, 1469–1479.

Martin, J. D., Frost, P. C., Hintelmann, H., *et al.* (2018). Accumulation of silver in Yellow Perch (*Perca flavescens*) and Northern Pike (*Esox lucius*) from a lake dosed with nanosilver. *Environmental Science & Technology*, 52, 11114–11122.

Martinez, J. G., Quiobe, S. P. & Moens, T. (2019). Effects of mercury (Hg) on soil nematodes: a microcosm approach. *Archives of Environmental Contamination and Toxicology*, 77, 421–431.

McLachlan, M. S. (1996). Bioaccumulation of hydrophobic chemicals in agricultural food chains. *Environmental Science & Technology*, 30, 252–259.

Mora, C., Tittensor, D. P., Adl, S., Simpson, A. G. B. & Worm, B. (2011). How many species are there on earth and in the ocean? *PLOS Biology*, 9, e1001127.

Norstrom, R. J., Clark, T. P., Enright, M., *et al.* (2007). ABAM, a model for bioaccumulation of POPs in birds: validation for adult herring gulls and their eggs in Lake Ontario. *Environmental Science & Technology*, 41, 4339–4347.

Oikari, A. (2006). Caging techniques for field exposure of fish to chemical contaminants. *Aquatic Toxicology*, 78, 370–381.

Oswald, C. J., Heyes, A. & Branfireun, B. A. (2014). Fate and transport of ambient mercury and applied mercury isotope in terrestrial upland soils: insights from the METAALICUS watershed. *Environmental Science & Technology*, 48, 1023–1031.

Palace, V. P., Evans, R. E., Wautier, K. G., *et al.* (2009). Interspecies differences in biochemical, histopathological, and population responses in four wild fish species exposed to ethynylestradiol added to a whole lake. *Canadian Journal of Fisheries and Aquatic Sciences*, 66, 1920–1935.

Parnis, J. M. & Mackay, D. (2020). *Multimedia Environmental Models – The Fugacity Approach*, Boca Raton, FL: CRC Press.

Posthuma, L., Dyer, S. D., de Zwart, D., *et al.* (2016). Eco-epidemiology of aquatic ecosystems: separating chemicals from multiple stressors. *Science of the Total Environment*, 573, 1303–1319.

Poteat, M. D. & Buchwalter, D. B. (2014). Phylogeny and size differentially influence dissolved Cd and Zn bioaccumulation parameters among closely related aquatic insects. *Environmental Science & Technology*, 48, 5274–5281.

Prevedouros, K., MacLeod, M., Jones, K. C. & Sweetman, A. J. (2004). Modelling the fate of persistent organic pollutants in Europe: parameterisation of a gridded distribution model. *Environmental Pollution*, 128, 251–261.

Propst, T. L., Lochmiller, R. L., Qualls, C. W. & McBee, K. (1999). *In situ* (mesocosm) assessment of immunotoxicity risks to small mammals inhabiting petrochemical waste sites. *Chemosphere*, 38, 1049–1067.

Rafferty, S. D., Blazer, V. S., Pinkney, A. E., *et al.* (2009). A historical perspective on the "fish tumors or other deformities" beneficial use impairment at Great Lakes Areas of Concern. *Journal of Great Lakes Research*, 35, 496–506.

Reynoldson, T. B., Day, K. E. & Norris, R. H. (1995). Biological guidelines for freshwater sediment based on Benthic Assessment of Sediment (the BEAST) using a multivariate approach for predicting biological state. *Australian Journal of Ecology*, 20, 198–219.

Reynoldson, T. B., Strachan, S. & Bailey, J. L. (2014). A tiered method for discriminant function analysis models for the Reference Condition Approach: model performance and assessment. *Freshwater Science*, 33, 1238–1248.

Ross, P., De Swart, R., Addison, R., *et al.* (1996). Contaminant-induced immunotoxicity in harbour seals: wildlife at risk? *Toxicology*, 112, 157–169.

Ross, P. S., Ellis, G. M., Ikonomou, M. G., Barrett-Lennard, L. G. & Addison, R. F. (2000). High PCB concentrations in free-ranging Pacific killer whales, *Orcinus orca*: effects of age, sex and dietary preference. *Marine Pollution Bulletin*, 40, 504–515.

Saint-Laurent, D., Duplessis, P., St-Laurent, J. & Lavoie, L. (2011). Reconstructing contamination events on riverbanks in southern Québec using dendrochronology and dendrochemical methods. *Dendrochronologia*, 29, 31–40.

Schindler, D. W. (1998). Whole-ecosystem experiments: replication versus realism: the need for ecosystem-scale experiments. *Ecosystems*, 1, 323–334.

Schindler, D. W., Frost, T. M., Mills, K. H., *et al.* (1990). Comparisons between experimentally-acidified and atmospherically-acidified lakes during stress and recovery. *Proceedings of the Royal Society Edinburgh*, 97 B, 193–226.

Schindler, D. W., Mills, K. H., Malley, D. F., *et al.* (1985). Long-term ecosystem stress: the effects of years of experimental acidification on a small lake. *Science*, 228, 1395–1401.

Stevenson, R. W. & Chapman, P. M. (2017). Integrating causation in investigative ecological weight of evidence assessments. *Integrated Environmental Assessment and Management*, 13, 702–713.

Sugiura, K. (1992). A multispecies laboratory microcosm for screening ecotoxicological impacts of chemicals. *Environmental Toxicology and Chemistry*, 11, 1217–1226.

Teeguarden, J. G., Tan, Y.-M., Edwards, S. W., *et al.* (2016). Completing the link between exposure science and toxicology for improved environmental health decision making: the aggregate exposure pathway framework. *Environmental Science & Technology*, 50, 4579–4586.

Thimmegowda, G. G., Mullen, S., Sottilare, K., *et al.* (2020). A field-based quantitative analysis of sublethal effects of air pollution on pollinators. *Proceedings of the National Academy of Sciences USA*, 117, 20653–20661.

Thomas, P. J., Newell, E. E., Eccles, K., *et al.* (2021). Co-exposures to trace elements and polycyclic aromatic compounds (PACs) impacts North American river otter (*Lontra canadensis*) baculum. *Chemosphere*, 265, e128920.

Ulanowicz, R. E. (2004). Quantitative methods for ecological network analysis. *Computational Biology and Chemistry*, 28, 321–339.

Venosa, A. D., Lee, K., Suidan, M. T., *et al.* (2002). Bioremediation and biorestoration of a crude oil-contaminated freshwater wetland on the St. Lawrence River. *Bioremediation Journal*, 6, 261–281.

Wang, Z., Walker, G. W., Muir, D. C. G. & Nagatani-Yoshida, K. (2020). Toward a global understanding of chemical pollution: a first comprehensive analysis of national and regional chemical inventories. *Environmental Science & Technology*, 54, 2575–2584.

Warren, C., Mackay, D., Whelan, M. & Fox, K. (2007). Mass balance modelling of contaminants in river basins: application of the flexible matrix approach. *Chemosphere*, 68, 1232–1244.

Witting, L. (2019). Rebuttal against Desforges *et al.* (2018): No evidence for a global killer whale population collapse. *Marine Mammal Science*, 35, 1197–1200.

Wright, D. A. & Welbourn, P. (2002). *Environmental Toxicology*, New York: Cambridge University Press.

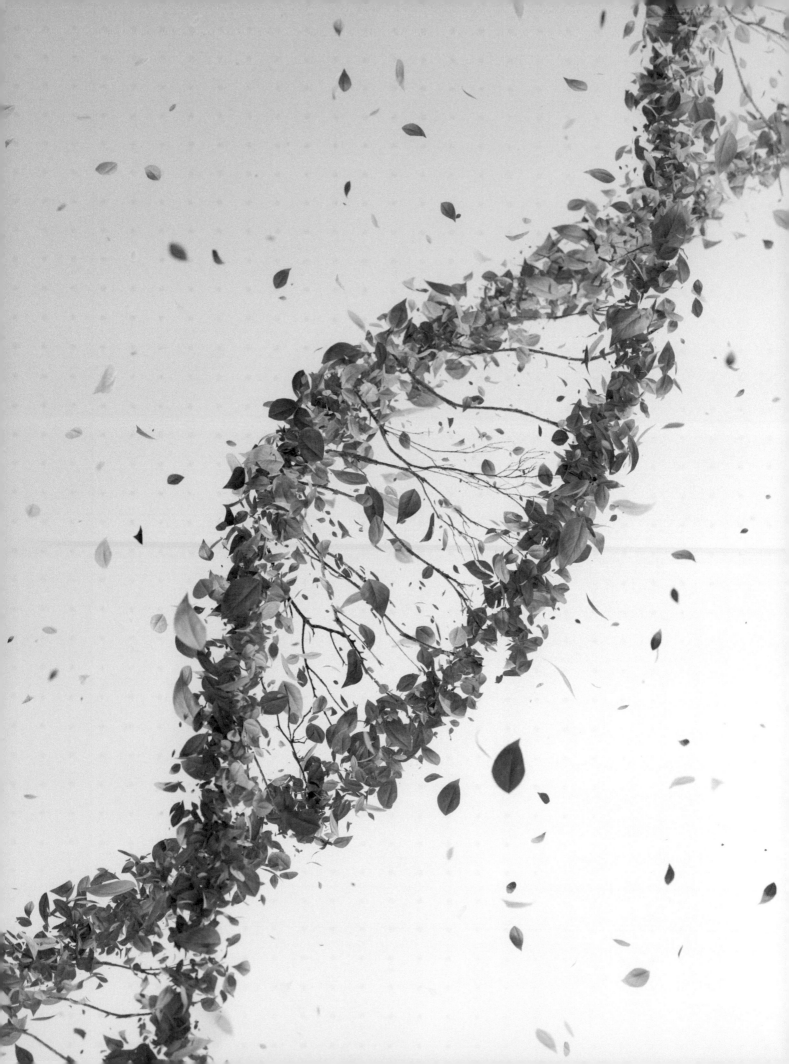

CHAPTER

5

Ecotoxicogenomics

Valérie S. Langlois and Christopher J. Martyniuk

Learning Objectives

Upon completion of this chapter, the student should be able to:

- Distinguish among the different omics technologies and discuss strengths and weaknesses of each approach.
- Describe how omics data can be incorporated into an adverse outcome pathway framework.
- Understand the function of high-throughput cell-based reporter assays and explain how they are utilized to determine molecular initiating events.
- Discuss how new technologies (e.g., CRISPR/Cas9 gene editing) can be used to evaluate chemical toxicity and mode of action or MoA.

This chapter introduces the reader to 'omics' and discusses advantages and limitations of different omics approaches in ecotoxicology. Technological advances in molecular biology now allow for the simultaneous assessment of the totality of genes, proteins, metabolites and lipids in a tissue or organism (i.e., omics). While comparatively new, these approaches are yielding novel insights about the molecular mechanisms underlying responses to exposures to environmental contaminants. Chapter 5 begins by describing the use of ecotoxicogenomics from a historical perspective, moving towards the objective of using such data to predict adverse outcomes and to prioritize risk assessment strategies. We draw on examples from the ecotoxicology literature and highlight approaches used to integrate large datasets to better discern organismal responses to environmental contaminants. Novel concepts such as the exposome, environmental DNA technology, computational biology, adverse outcome pathways and gene-editing tools are introduced.

5.1 Environmental 'Omics': A Role in Ecotoxicology Research

Advances in analytical and molecular methods have considerably improved our ability to detect chemicals in the environment. Next-generation sequencing, instruments that can sequence multiple DNA strands in parallel, and mass spectrometry, an analytical method that can measure the mass-to-charge ratio of ions, now allow us to measure biological responses to chemicals on a fine scale. Using these techniques, scientists can quantify hundreds to thousands of DNA fragments, RNA molecules (e.g., microRNA, messenger RNA, long non-coding RNA), proteins, or metabolites in a single experiment. These powerful **omics** approaches increase the sensitivity and reliability of biomarker detection and interpretation. They can also generate new mechanistic information about a chemical that interacts with organisms in our environment.

The word 'omics' refers to the large-scale generation and interpretation of molecular data, and it aims to describe the totality of molecular responses in the cell or organism. Hence, **proteomics** is the study of all the proteins in a cell or tissue, whereas **metabolomics** is the study of all the metabolites in a cell or tissue. Hundreds of terms have adopted the suffix: for example terms such as cellomics, tissueomics, immunomics, reproductomics and microRNAomics have appeared at one time in scientific literature. In the context of ecotoxicology, we discuss only the major omics here. These technologies can reveal new pathways of toxicity. At the turn of the twenty-first century, environmental toxicologists began to leverage these high-throughput technologies (technologies that allow precise and simultaneous measurement of thousands of genes, proteins or metabolites) to study chemicals. However, there were many questions and some hesitation in using such technologies in ecotoxicology.

- How would data generated from these technologies be used for chemical risk assessment?
- How reliable and reproducible were these types of data?
- How would academics, industry, government agencies and environmental managers interpret these data for practical applications to protect wildlife and promote safer chemicals?
- How much expertise or training would be required to interpret these complex datasets?
- What computational tools were required to analyse the data?
- Could we associate a molecular response to a phenotypic endpoint that could be regulated?

Today, we still do not have complete answers to these questions, but as a scientific community over the past several decades, we have made significant strides in addressing these concerns. With any new technology, a great deal of research and effort goes into optimization, validation and data collection prior to the technology becoming widely adopted by a scientific discipline.

As a result of these technological advances, environmental toxicologists can now measure chemical-induced molecular responses in organisms on an unprecedented scale. This has generated terabytes of molecular data, such that computational tools are required to integrate multi-omics data. In this chapter, we begin with a brief history of omics technologies in ecotoxicology (**transcriptomics**, **proteomics** and **metabolomics**), and end with the most recent advances (environmental DNA, epigenetics and microbiome analysis), which are expected to propel a paradigm shift in how we measure the impacts of environmental pollutants (Figure 5.1).

New chemicals continue to be produced and released into the environment. High-throughput assays are now used in toxicology to better annotate **adverse outcome pathways** (AOPs; Sections 4.2, 5.9). Omics is expected to become increasingly integrated into the AOP framework to assist regulatory agencies with chemical risk management and best practices. Furthermore, on the horizon are new genetic tools that can be used in ecotoxicology to determine mechanisms of chemical action. These technologies and

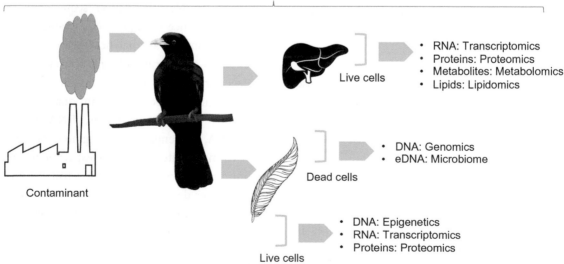

Figure 5.1 An example of omics applications in ecotoxicology. A bird is exposed to aerosols from a contaminated plume of smoke which affects the organism in multiple ways. Different molecular endpoints can be measured in its liver and feathers (both in dead and living cells) using omics technologies.

concepts (e.g., **CRISPR/Cas9** gene editing) will be discussed later in this chapter (Section 5.11).

5.2 Ecotoxicology and Transcriptomics

At the beginning of the 'genome revolution' (early to mid-1990s), researchers recognized the value of obtaining molecular information in individuals following chemical exposures in the context of environmental toxicology. The idea that genetic variability, gene expression, protein activity and metabolites could be measured following an exposure to a toxicant was an exciting prospect. It was recognized that these molecular responses could provide novel insights about how chemicals affected the homeostasis, physiology and health of organisms. Such data could also reveal toxicity pathways that potentially lead to cell and tissue damage as well as disease. Intriguing was the possibility of using these molecules (e.g., transcripts, proteins, metabolites) as **biomarkers** of exposure (Section 4.3.3) to predict adverse outcomes in wildlife as well as humans. By characterizing molecular responses in organisms following chemical exposures, we are better able to predict adverse responses based on a 'molecular signature' (i.e., the molecular profile produced by a chemical).

The techniques used to measure all transcripts (i.e., **transcriptomics**) simultaneously were developed early in the genome revolution. A transcript is defined as an RNA molecule transcribed from a DNA template. When a chemical interacts with a receptor, there is a 'downstream' response at the DNA and protein level. At the level of DNA, this involves the expression of specific responsive genes to protect against the chemical exposure (such as the antioxidant system), or an induction of genes needed to produce the enzymes that metabolize the chemical (such as members of the cytochrome P450 enzyme family). In the case of endocrine disrupting compounds (Chapter 8), the cell can respond inappropriately to an interaction of a chemical with a ligand (for example, a chemical mimicking a hormone), and a transcriptomic response can ensue that is typical of the hormone being mimicked. This can be detrimental if the signalling is initiated at the wrong time in the reproductive season or at the wrong stage of development.

Transcriptomics measures gene expression responses in individuals exposed to a chemical relative to a control group (i.e., unexposed). Such experiments are conducted to determine how the transcriptome (i.e., the entire collection of transcripts in a cell or tissue) responds to the exposure. By so doing, one gains insight into the mechanism of action of the chemical, as well as the cellular response to the exposure (i.e., compensatory, or protective mechanisms to avoid further damage). The analysis of gene expression captures information about messenger RNA (mRNA), which is eventually translated into proteins, to generate insight about the biological functions affected by exposure to the contaminant. It is important to emphasize that the term 'transcriptomics' refers to whole transcriptome analysis – measuring thousands of mRNAs in a single sample. However, we often refer to the term 'gene expression' when fewer select genes are being analysed using other approaches such as **real-time reverse transcriptase-polymerase chain reaction** (RT RT-PCR; also known as quantitative PCR or **qPCR**).

Examples of the utility of qPCR in the field of ecotoxicogenomics are numerous, but here is an interesting study performed on birds. Using chicken eggs as a model system, researchers assessed the toxicity of water leachates from artificial turf used in playgrounds and in replacement of grass fields. Eggs were exposed to leachates by either coating or dipping, or by injecting the leachates into the air space or yolk sac. Using qPCR, the scientists demonstrated that several genes were dysregulated by the leachate, including genes involved in the stress response and thyroid hormone pathways (Xu *et al.*, 2019). This molecular impairment, along with other toxic effects noted by the research team, raise questions about the ecological risks of artificial turf leachates.

Some of the earliest omics studies in ecotoxicology utilized a reverse northern blot (macroarray) (Larkin *et al.*, 2002; Brown *et al.*, 2004) to study chemical exposures in standard test species. DNA from a target gene was amplified through PCR and 'spotted' in excess onto specialized membranes. The DNA strands were denatured into single strands and fixed onto the membrane. In parallel, mRNAs from the samples were isolated, reverse transcribed into complementary DNA (cDNA is complementary to the DNA strand used to initially transcribe the mRNA) using radioactive nucleotides, and subsequently hybridized to their complementary DNA strand counterpart on the membrane. This yielded a double-stranded radioactive molecule on the membrane that could be quantified. Membranes were subsequently exposed to radiation-sensitive film, and the more intense the radioactivity observed on the film, the more abundant the transcript.

The use of macroarrays prevailed for a relatively short time before new technology allowed micro-printing of the DNA spots. These new high-density arrays or microarrays involve printing ('spotting') of DNA fragments or shorter oligonucleotides onto glass slides. The advantages of microarrays over macroarrays included a higher density of gene arrays, less DNA material required for spotting, and less-harmful fluorescent dyes could be used instead of radioactive compounds. At the onset of their use in toxicology, microarrays were utilized to investigate metals, by-products of industrial chemicals, and oestrogenic compounds (Nuwaysir et al., 1999; Bartosiewicz et al., 2001; Iguchi et al., 2002). Although microarrays generated abundant information at the transcript level, there were obstacles that required attention prior to widespread adoption of the technology. These concerns related to cost, technical expertise, data interpretation and 'added value' for environmental monitoring and ecological risk assessment (ERA).

Despite initial challenges, the science of genomics has made significant gains in improving interpretation, increasing sensitivity and reliability, and reducing costs associated with DNA sequencing. Since the turn of this century, there has been a huge increase in the use of transcriptomics in ecotoxicology. The platforms now preferred for whole transcriptome analysis (or **RNA-seq**) vary from sequencers that generate read lengths shorter than 100 base pairs (bp) (e.g., Illumina MiSeq) to high-throughput sequencers that generate reads greater than 100 bp (e.g., Illumina HiSeq or TruSeq). 'Read length' refers to the number of base pairs sequenced while 'numbers of reads' refer to the number of times a transcript is sequenced. **RNA-seq** refers to the sequencing of all the transcripts present in one sample. As mentioned above, microarrays can also measure all the significant transcripts expressed from a genome. The main distinction between RNA-seq and microarrays is as follows: RNA-seq will sequence all the RNA (mRNAs, microRNAs and others), and thus it is referred to as a *de novo* approach. One does not need to know the DNA sequence of each gene *a priori*. The molecules in the sample are first sequenced, then compared to existing databases of sequences to determine the identity of each molecule, using homology or sequence similarity. In contrast, microarray analysis requires that the DNA sequence for a gene on the platform is known, as gene-specific sequences are printed on the slides. As one can see, RNA-seq is a significant advantage for toxicologists working with non-model species for which little genomic information is available. Another advantage of RNA-seq over microarrays is that other types of RNA molecules, such as **miRNAs** or long non-coding RNA, can be measured. Given the amount of data generated, new fields of research called bioinformatics and computational toxicology have emerged to analyse these massive datasets (Section 5.8).

5.2.1 Application of Ecotoxicogenomics

Transcriptomics has been used to determine the scope of adverse effects for industrial compounds, metals, pesticides and various pharmaceuticals. More than 1,000 peer-reviewed studies have been published on transcriptome responses of organisms following chemical exposure, i.e., that relate to ecotoxicology. Thus, in some cases, we can now predict which transcripts will most likely be altered in expression by exposure to specific chemicals. For example, when a fish is exposed to **17α-ethinyloestradiol** (EE2), the active ingredient in birth control pills, transcript levels of aromatase (aromatizes testosterone to 17β-oestradiol) and vitellogenin (egg yolk precursor protein expressed in multiple egg-producing taxa) increase in abundance. This is because EE2 is lipophilic and readily crosses plasma membranes in cells, acting to bind to the oestrogen receptors that are present in the cytosol (Chapter 7). These EE2–receptor complexes migrate to the nucleus and bind to specific DNA regions (oestrogen-response elements or EREs) known to be controlled by activated oestrogen receptors, such as the aromatase gene. Once a receptor–ligand complex is bound to the promotor region of an oestrogen-responsive gene such as aromatase or vitellogenin, transcription begins following recruitment of cofactor proteins, and the transcribed messages are transported from the nucleus to the cytosol. Increased activation of these 'EE2–receptor complexes' can promote higher levels of transcriptional responses that are sustained over prolonged periods of time. Therefore, vitellogenin and aromatase mRNAs are reliable biomarkers of exposure to exogenous chemicals that have oestrogenic activity. In fact, more than 1,000 studies have used the biomarker vitellogenin in some way to measure oestrogenic chemicals. Vitellogenin has been adapted into cell-based reporter assays, and measured in laboratory experiments with fish, as well as in surveys of wild fish to assess the oestrogenic potential of the water (Table 5.1). The gene can also be induced prematurely in early life stages of many egg-bearing vertebrates following exposure to oestrogens. There are many more examples of transcriptomics applications in ecotoxicology. These studies not only address oestrogenic pharmaceuticals, but also include an array of other chemicals that affect animals. Here we present two additional examples of toxicogenomic applications in ecotoxicology (Box 5.1).

Table 5.1 **Examples of applications of the transcriptomic approach to freshwater and marine fish.**

Species	Approach	Known contaminants	References
Laboratory-raised animals exposed to field-collected water			
Fathead minnow *Pimephales promelas*	Fish, caged in the field (48 h); microarray	Endocrine disruptors	Denslow and Sabo-Attwood (2015)
	Fish, exposed in the laboratory (48 h); RNA-seq	Oil sands process water	Wiseman *et al.* (2013)
Rainbow trout *Oncorhynchus mykiss*	Fish (juvenile), caged in the field (14 d), up- and downstream from an outfall; microarray	Municipal wastewater effluent	Ings *et al.* (2011)
Japanese medaka *Oryzias latipes*	Fish, exposed in the laboratory (14 d); microarray	Diluted bitumen	Madison *et al.* (2020)
Wild-caught animals chronically exposed to metals and other contaminants in field situations			
Yellow perch *Perca flavescens*	Fish, caught along metal contamination gradient; RNA-seq; liver	Metals (Cd, Cu, Zn)	Pierron *et al.* (2011)
	Fish, from reference lake, transferred to and caged in contaminated lake (4 wk); microarray; liver	Metals (Cd, Cu, Zn)	Defo *et al.* (2015); Bougas *et al.* (2016)
	Fish caught in St. Lawrence River, up- and downstream from Montreal; microarray; liver	Municipal and industrial effluents (Montreal)	Houde *et al.* (2014)
Brown trout *Salmo trutta*	Fish, caught along metal contamination gradient; RNA-seq; gill, gut, kidney, liver	Metals (Cd, Cu, Ni, Pb, Zn)	Webster *et al.* (2013)

Box 5.1 Examples of transcriptomics and ecotoxicology

Example 1

Iron sulphide nanoparticles are proposed to be useful in remediation efforts. These nanoparticles, owing to their high adsorption capacity, are efficient in immobilizing a broad spectrum of contaminants (e.g., metals, radionuclides, chlorinated organic compounds and polychlorinated biphenyls) in soil and water. However, prior to their use, one needs to ensure that nanoparticles exhibit little to no toxicity to organisms. Chinese researchers exposed adult zebrafish to 10 mg/l of an iron sulphide nanoparticle mixture for 96 h and conducted transcriptomic analyses on the fish livers. Over 3,200 genes were downregulated in fish from the treated group compared to the control group, and the molecular signature suggested a stress response. Gene ontology (GO) analysis, a method used to group genes by their function, indicated that genes involved in 'response to stimulus' and 'metabolic and cellular processes' were significantly enriched after the exposure. More specifically,

transcripts involved in the inflammatory response, detoxification, oxidative stress and DNA damage were altered by the treatment. This transcriptomic analysis revealed the molecular toxicological pathways associated with exposure to the engineered iron sulphide nanoparticles, raising questions about potential risks to aquatic organisms when using this remediation technology (Zheng *et al.*, 2018).

Example 2

In a second example, researchers measured the effects of arsenate water contamination on developing amphibians with a custom microarray pre-designed for their frog species of interest (western clawed frog, *Silurana tropicalis*). Several clusters of genes involved in **cell signal transduction**, cell survival, development and **histone remodelling** were altered after the frog embryos were exposed to arsenate. Based on the activation of specific molecular pathways, the authors concluded that amphibians exposed to arsenate at concentrations found in wells supplying drinking water were coping with arsenate toxicity during early frog development (Zhang *et al.*, 2015).

5.3 Ecotoxicology and Proteomics

Proteins are gene products that can serve as biomarkers of the effects of contaminants in an organism or as biomarkers of exposure. The analysis of protein abundance can involve traditional fluorescence two-dimensional (2D) differential **gel electrophoresis** or mass spectrometry (Figure 5.2). Proteomics is the study of the entire proteome complement within a cell, tissue or organism. Different proteomic methods have been used to determine how exposures to environmental toxicants regulate, directly or indirectly, the abundance of proteins. Proteomics also includes the study of protein translation, protein modification (protein adduct formation, phosphorylation, acetylation, **ubiquitination**) and protein–protein interactions. Studying the proteome is important because proteins are the primary functional molecules of the cell and include essential enzymes, structural proteins, cell signalling molecules, and transcription factors that are required for cell homeostasis. An altered abundance of these proteins because of exposure to environmental contaminants can have significant detrimental impacts on cell physiology and organismal performance. However, difficult choices must be made when using an omics technology to study toxicology, because every method used to evaluate chemicals has both strengths and weaknesses (Box 5.2)

Some of the first studies investigating proteins in environmental toxicology applied western blotting approaches, a method that is still heavily relied upon today. Proteins are first separated by one-dimensional electrophoresis. Proteins within the complex environmental mixture (e.g., soil, water, tissues of organisms) become separated based on their charge and their size. Denatured and separated proteins are transferred to a nitrocellulose or PVDF (polyvinylidene difluoride) membrane, and antibodies that are specific for the detection of the denatured protein are incubated with the synthetic cell membrane. After the blot is incubated with a secondary antibody

Box 5.2 Strengths and weaknesses of proteomics

Proteins are advantageous as biomarkers of chemical exposure and effect because they are longer-lived than many transcripts, which can be transient in expression. Proteins are arguably more biologically meaningful than an expressed gene because gene expression does not always correspond to protein translation. The disadvantages of proteomics include a relatively high cost, a requirement for significant amounts of biological material, and fewer molecules quantified compared to transcripts. For example, one might measure 50,000 transcripts with sequencing, but a strong proteomics experiment may yield information for fewer than 5,000 molecules. This is because the technology has limitations, and many proteins are difficult to detect.

containing a chemical for detection (horseradish peroxidase, for example), one can visualize the protein bands using densitometry analysis. The objective is to compare protein abundance among experimental groups to determine if the protein has increased or decreased in abundance relative to the experimental control group.

There are many examples of proteomics applications in ecotoxicology. A Norwegian team measured the expression levels of two proteins isolated from the livers of Arctic seabird chicks (the black-legged kittiwake, *Rissa tridactyla*, and the northern fulmar, *Fulmarus glacialis*). These chicks were exposed to airborne halogenated organic contaminants (measured hydroxy-polychlorinated biphenyls [PCBs], methyl sulphone and dichlorodiphenyldichloroethylene [DDE] among others) in Svalbard, Norway. Using western blot analysis, the cytochrome P450 (CYP) 1A and 3A proteins were measured. These enzymes are involved in Phase I oxygenating reactions (Section 7.5.1). The authors concluded that the upregulation of these proteins, along with the presence of various hydroxylated PCBs in chick

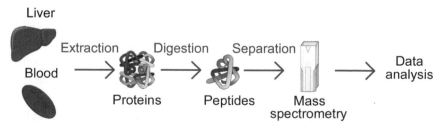

Figure 5.2 General pipeline, from tissue processing to data analysis, in proteomics research.

tissues and complementary gene expression and enzymatic activity data, indicated that the chicks were biotransforming the contaminants (Helgason *et al.*, 2010).

One limitation of western blotting is that specific antibodies are required to detect the protein. For many environmentally relevant species and non-model organisms, species-specific antibodies may not yet exist for the proteins of interest. This was the case in the previous example in Arctic birds: the authors stated that their western blot data should be carefully interpreted, as the anti-fish antibodies used in the study may also bind to other proteins in seabirds. Antibodies from closely related species can be tested through trial and error for binding efficiency but having a specific antibody for the species under study is the gold standard. Another limitation of western blotting is that proteins are studied one at a time, which is not ideal when one is interested in how the entire proteome responds to environmental contaminant exposure. For these types of questions, gel-based electrophoresis and analytical chemistry with a mass spectrometer can provide higher throughput.

Two-dimensional differential gel electrophoresis (2D-DIGE) separates complex mixtures of proteins based on their isoelectric point and size. Isoelectric focusing refers to the movement of a protein along a medium with a pH gradient in one dimension under the influence of an electric current. It stops moving at a point where the overall net charge of the protein is 0 (isoelectric point or point at which the protein carries a neutral charge). Proteins are further separated in a second dimension on denaturing gels, and specific fluorescent dyes are used to label protein samples from different experimental groups. Zebrafish (*Danio rerio*) were exposed for 180 d to perfluorononanoic acid, a synthetic perfluorinated carboxylic acid and fluorosurfactant, and were sampled for differential protein expression analysis using 2D-DIGE. Over 50 proteins were differentially expressed compared to the control group, revealing that proteins involved in cell metabolism, structure, motility, immunity, signal transduction and cell communication are involved in the complex MoA of perfluorononanoic acid in fish (Zhang *et al.*, 2012). Proteomics has also been used to study protein changes in response to perfluorinated chemicals in peripheral blood mononuclear cells of European eels in contaminated environments (Roland *et al.*, 2014).

Some of the latest advances in ecotoxicoproteomics have come from advances in liquid chromatography (LC) and MS. Improvements to the analytical methods and instrumentation are related to improved sensitivity, reproducibility, and capacity for environmental proteomics. When coupled to LC, **tandem mass spectrometry** (LC-MS-MS) can detect thousands of proteins in a single sample. Proteins are first cleaved into peptides by enzymes or other chemicals, and the peptides are separated based on their size and charge. Liquid chromatography is used to further separate molecules prior to injection into a mass spectrometer, typically using electrospray ionization. In this approach, molecules are charged in the instrument and travel through the spectrometer at different rates. A detector at the other end records the 'time of flight' and intensity of the signal. Ultimately, the mass spectra generated are compared to a database of information on characteristics of specific peptide sequences (known mass, time of flight) and an identification can be made. Powerful computer algorithms search through these databases at an extremely fast rate, comparing each signal to those accumulated in the database being searched. Mascot (Matrix Science Ltd, London, UK) is one such example of a software search engine that is commonly used in proteomics and, combined with LC-MS-MS, is a powerful biomarker discovery tool for environmental toxicology, not only for proteomics but also for metabolomics (Section 5.4).

There are two major quantitative proteomics methods (label-free and labelling approaches). Label-free proteomics includes spectral counting (i.e., counting the number of mass spectra matching with a specific peptide derived from a protein) and absolute quantitation (i.e., using a heavy label peptide as an internal standard to quantify a target peptide). Label-based methods utilize a 'tag' that is chemically attached to a specific amino acid or peptide. This tag is ultimately used to quantify changes in protein abundance. Labelling methods include stable isotope labelling, isotope coded affinity tags, isobaric tagging for relative and absolute quantitation (or iTRAQ), and tandem mass tags (or TMTs). The power of the iTRAQ and TMT labels is that peptides from multiple samples can be labelled with tags and measured all together by MS. This improves one's capacity to compare protein changes among experimental groups, because each sample is tagged with a different small molecule identifier. This approach also increases accuracy and reliability because each sample is processed within a single experiment, reducing variability due to batch effects (i.e., errors in comparisons among samples that are processed at different times).

Proteomics has been used to identify proteins that are differentially expressed in animals exposed to environmental contaminants. Some earlier experiments applied 2D gel electrophoresis to separate proteins in species exposed to contaminants. These included juvenile cod (*Gadus morhua*) exposed to polycyclic aromatic hydrocarbons (PAHs) and North Sea crude oil, the freshwater amphipod *Gammarus pulex* exposed to PCBs, and the marine alga *Scytosiphon gracilis* exposed to copper (Bohne-Kjersem *et al.*, 2009; Contreras *et al.*, 2010; Leroy *et al.*, 2010). In another study, livers from zebrafish treated with the organochlorine pesticide dieldrin were analysed by both label-free (spectral counting) and label-based methods (iTRAQ; LC-MS-MS) (Simmons *et al.*, 2019). Each approach quantified different proteome responses in the fish. The authors proposed that differences in protein identification were likely due to variability in sample preparation and analytical instruments used. This is not to say that the methods are faulty or unreliable. Unique proteins in a cell can number in the 100,000s if one considers protein modification (Box 5.3) and various splice variants (i.e., a transcript that has been post-transcriptionally modified). Any given proteomics technique may identify a few thousand proteins in general.

Another example of proteomics comes from studies assessing **triclosan** exposure in chicken eggs. Using the iTRAQ technique, the researchers showed that this antimicrobial agent altered the expression of proteins involved in lipid and energy metabolism in chicken offspring. Only male chickens were sensitive to triclosan based upon increased response of enzymes involved in Phase I and Phase II detoxification (Guo *et al.*, 2018). Another example of proteomics and toxicology comes from research in plants. Cadmium (Cd) is a common

environmental pollutant in soils (Section 6.8.2), and it is phytotoxic. Wheat cultivars exposed to Cd showed changes in multiple proteins based upon a tandem mass tag approach, and several proteins were related to glycerolipid metabolism, protein export, DNA replication, ribosome structure and cyanoamino acid metabolism (Jian *et al.*, 2020). The proteomics study revealed that certain strains of wheat are more apt than others at utilizing glutathione for detoxification and protection from metals.

So far, we have reviewed only a few examples of how the proteome responds to pollutants. Different proteomics techniques will reveal different aspects of the 'protein-universe' in a tissue. An advantage of using multiple proteomics tools is that one can increase the view of the proteome, but a downside is that challenges arise in comparing data across laboratories due to differences in technical preference and equipment. This is because the characterization of the proteome is dependent upon technique and instrumentation, and standardized operating protocols for the community are not always available (nor completely agreed upon!).

5.4 Ecotoxicology and Metabolomics/Lipidomics

In a similar manner to transcripts and proteins, metabolites can be measured to describe adverse effects in organisms. All small-molecule metabolites generated internally by a living organism at a given time can be determined with a **metabolomics** approach. In human medicine, we have been using metabolites as biomarkers of disease for a long time; for example, blood glucose is a biomarker for diabetes. It is important to remember that genes, proteins and metabolites are interconnected, because genes code for proteins that act as enzymes to produce metabolites. Glycolysis, for example, involves several enzymatic reactions that convert precursor molecules into metabolites required for ATP production during oxidative phosphorylation. Quantifying alterations in metabolites provides significant information about physiological responses to chemical exposures. While metabolites are highly relevant to physiology, they can also be very sensitive to change and rapid degradation. Therefore, sample preparation and analysis must be done with great care, especially if the metabolite has a short half-life.

When conducting metabolomic analyses, gas chromatograph/MS, high-performance liquid chromatography/MS, or liquid chromatography–quadrupole time

---

Box 5.3 Modified proteins can act as biomarkers of exposure

Proteomics also involves the study of post-translational modifications (PTMs). Proteins can be modified as a result of acetylation, glycosylation, nitrosylation, **ubiquitination** and **methylation**. The function or activity level of proteins can thus be altered with post-translational modifications of their structure. Environmental contaminants can alter the patterns, frequency or locations of these PTMs, which in turn could impair protein function.

of flight MS (LC-QTOF) can be used to separate metabolites and characterize profiles. For LC-QTOF, metabolites are extracted and separated under different analytical conditions (e.g., C18 columns and positive and negative ion modes). Data obtained in this manner can be analysed to determine what metabolites were present in the samples (i.e., amino acids, lipids, etc.) and also to determine whether there were any that were different in concentration relative to controls. The method uses the retention time on the column to identify each molecule. This method can also be called 'untargeted data analysis', because the goal is to identify as many molecules as possible in the sample. Conversely, one can apply a targeted approach to quantify molecules of interest. As an example, Irish scientists were able to detect subtle metabolite differences in the plasma of rodents exposed to dietary dioxins. Twenty-four ions were used to establish a prediction model, which included lysophosphatidylcholine and tyrosine metabolites (O'Kane *et al.*, 2013). An example of metabolic applications specifically for ecotoxicology comes from plants. Environmental metabolomics can be used to determine changes in a plant's metabolism, yielding unique signatures or 'fingerprints' that indicate stress due to herbicides or metal contamination. The common duckweed (*Lemna minor* L.) is an aquatic freshwater plant often used to monitor environmental contamination. Kostopoulou *et al.* (2020) exposed duckweed to metribuzin, a broad-spectrum herbicide that affects photosystem II, and to glyphosate, another broad-spectrum systemic herbicide and crop desiccant. Metabolites related to the tricarboxylic acid cycle (i.e., maleate and fumarate), carbohydrates and sucrose, and the fatty acids palmitate and oleate changed in abundance following exposure to one or both herbicides, yielding new insights into chemical toxicity in plants.

Another rapidly emerging tool related to metabolomics is **lipidomics**. Lipids can be short- or long-chained organic compounds that serve as structural components or signalling molecules in numerous biochemical pathways. They can also be branched chained (alkylated) with varying numbers of double bonds (saturated versus unsaturated). In general, there are three major groups of lipids: triglycerides (fats and oils), diglycerides (phospholipids) and steroids. These categories can be further divided into diverse classes, such as sphingolipids, ceramides, saccharolipids, glycerolipids and polyketides. The integrity of the plasma membrane is dependent on phospholipids, and disruptions in their biosynthesis or stability by chemicals

can affect membrane structure and function (Section 3.1.1). Lipids play an important role in many biological functions, such as metabolism, immunity, endocrinology, growth and reproduction. For example, egg-laying animals require specific lipoproteins, such as vitellogenin, for normal oocyte development and egg viability. In fact, lipidomics studies have identified lipid components for a healthy fish egg versus one that is of poor quality (Schaefer *et al.*, 2018). Chemicals that induce poor-quality eggs through altered lipid biosynthesis can potentially impact species at a population level (Case Study 8.1). Another important aspect of lipidomics that is relevant to chemical exposures is that of lipid peroxidation which is a result of oxidative stress. A specific branch of lipidomics seeks to identify oxidized lipids to verify novel biomarkers of exposure and to increase knowledge about the lipid classes that are perturbed by environmental chemical exposure. Lipids are regulated by the **peroxisome proliferator-activated receptors** (PPARs), and a significant number of chemicals modulate these receptors (Dong *et al.*, 2016). In particular phthalates (e.g., di(2-ethylhexyl)phthalate, or DEHP), which are added to plastics to increase resilience and flexibility, interact with these receptors to induce downstream changes in lipid deposition and PPAR expression (Ito *et al.*, 2019).

Lipidomics is an emerging field related to metabolomics, and one that is gaining traction in ecotoxicology. One significant obstacle for lipidomics is that many lipids remain uncharacterized in terms of their biological activity and role. Nevertheless, it is anticipated that lipidomics will improve our understanding of the links between chemical exposure and adverse outcomes, as reviewed by Dreier *et al.* (2019).

5.5 Ecotoxicology and Epigenetics

What if the antidepressant drug your grandmother took years ago has affected your ability to cope with stress? The answer may surprise you. Researchers found that fish embryos exposed for only 6 d to fluoxetine, a selective **serotonin reuptake inhibitor** drug, showed diminished stress hormone levels at adulthood. This was observed in the offspring over three generations (Vera-Chang *et al.*, 2018). This phenomenon is explained by the ability of some drugs or chemicals to alter the local cellular environment surrounding DNA, which is called the epigenome. In this example for antidepressants, enzymes associated with DNA methylation and histone

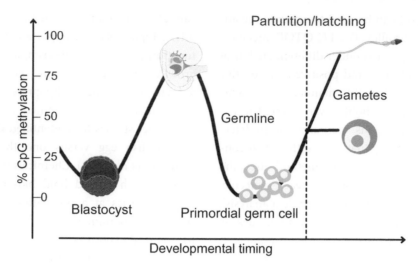

Figure 5.3 DNA methylation, measured as % CpG content, can vary over embryonic mammalian development. In this figure, there is lower methylation present in the blastocyst stage and higher methylation at hatching/birth. Methylation is reduced again in the primordial germ cells but increases in frequency in gametes when they are needed for reproduction. Adapted from a figure released into the public domain (Wikimedia Commons) by its author, Mariuswalter (Creative Commons Attribution –Share Alike 4.0 International licence).

modifications (essential in allowing gene expression to proceed) were altered in expression. Other ecotoxicological studies also demonstrate multi-generational effects in animals. For example, zebrafish (*Danio rerio*) exposed to the fungicide azoxystrobin produced offspring (F1 generation) that showed lower survival and impaired development compared to offspring from unexposed parents (Cao *et al.*, 2019). In addition, genes with important roles in the endocrine system, oxidative damage response and apoptosis showed altered expression in offspring that had never been exposed to the fungicide (in contrast to their parents). This suggests that chemical exposure in the parental generation can affect their offspring. Similarly, inorganic contaminants, such as Cd and Pb, caused significant DNA methylation in the testes of rats that had been exposed for 1 year to metal-contaminated soils (Nakayama *et al.*, 2019). Aberrant DNA methylation was associated with sperm defects in these rodents. The role of **epigenetics** in ecotoxicology has been reviewed by Chatterjee and colleagues (Chatterjee *et al.*, 2018).

Epigenetics is also the reason why some genes are only activated during specific developmental stages (e.g., genes related to organ formation) or only in one sex (Figure 5.3). Epigenetics includes several mechanisms of action, including DNA methylation, histone modification and miRNA silencing, which can be measured by methods such as **enzyme-linked immunosorbent assays** (ELISA), bisulphite sequencing, chromatin immunoprecipitation (ChIP), and others.

DNA methylation occurs mainly in DNA regions rich in cytosines (C) and guanines (G); these regions are also known as CpG islands. Once the CpG island of a gene or a small DNA section is methylated, transcription can no longer occur. One of the key enzyme classes involved in DNA methylation includes DNA methyltransferases. Their role is to catalyse the reaction to methylate cytosines. ELISA methods can be used to quantify resulting changes in protein level and are widely applied in biochemistry for protein and hormone quantification. Alternatively, the expression of genes coding for DNA methyltransferases can be measured using transcriptomics. In addition, bisulphite sequencing is increasingly used to detect methylated CpG regions. The advantage of this technique compared to ELISA is that it screens the entire epigenome rather than only a few proteins at a time. Treatment with bisulphite converts all cytosines into uracil, except for methylated-cytosines. Complementary sequencing analysis identifies the methylated genes or DNA regions. DNA compaction through histone modification can also prevent transcription and can be measured by chromatin immunoprecipitation, where histone proteins are immunoprecipitated, purified and sequenced to determine which DNA regions were highly compacted after exposure to a given contaminant.

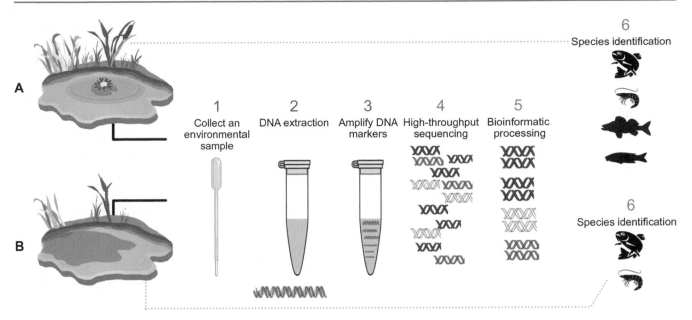

Figure 5.4 The application of eDNA technology could be used to compare the biodiversity between different habitats. For example, a sample of water taken in a pristine site (A) could show high diversity once analysed for eDNA (i.e., DNA sequences of many different organisms are found), wheras less diversity could be found in a highly contaminated environment (B) (i.e., DNA sequences of few organisms are detected).

5.6 Environmental DNA (eDNA)

Environmental surveys using **environmental DNA**, or eDNA methodology, are increasing in popularity for research and for environmental regulation. Its robustness, low cost, lack of animal sacrifice and high sensitivity explain this success (Wang *et al.*, 2019). The science of eDNA aims to describe the macrobiotic DNA present in an environmental sample, so it is important to distinguish eDNA from **metagenomics** (further described in Section 5.7) as these methods target two different groups of living organisms: macrobiota versus microbiota.

Mostly used in conservation biology, the principle of eDNA is to sequence either the DNA of all species or the DNA of a target species present in a water, air, sediment or soil sample. The analysis of eDNA allows one to detect the macrobiota living near the sample collection point, as organisms will shed cells into their surrounding environment. Most often, eDNA probes are developed to detect mitochondrial DNA instead of nuclear DNA because mitochondrial DNA has over a hundred to a thousand copies per cell, compared to nuclear DNA, of which there are often only two copies. This highly sensitive technique provides accurate information on biodiversity but can also identify the presence of invasive or endangered species (Veldhoen *et al.*, 2016). Profiles of eDNA have demonstrated the impacts of anthropogenic activities on

biological communities and ecosystem functions, and this is a rapidly emerging field of research. For example, eDNA profiles can be described before and after an oil spill to quantify the extent of an impact, and to monitor aquatic species recovery with time (Figure 5.4). One ongoing challenge is to use this technique in such a way as to quantify species abundance, rather than simply the presence or absence of the species in the environment.

Although eDNA is a persistent molecule, environmental factors, such as temperature, solar radiation, pH and bacteria, can influence its detection. This is also true of some environmental contaminants. Organophosphorus pesticides, for example, contribute to eDNA degradation, decreasing its persistence in the environment and increasing false-negative type I errors (Yang *et al.*, 2019). New related research opportunities continue to be developed, such as eRNA and eProtein analysis.

5.7 Ecotoxicology and the Microbiome (Metagenomics)

Since 2015, there has been an increasing interest in the role of the **microbiome** in organismal responses to chemical exposure. **Microbial 16S sequencing** and **metatranscriptomics** have added exciting new dimensions to the study of chemical exposures and animal health, and these approaches have been described conceptually (Adamovsky *et al.*, 2018). Currently, the most common approach is bacterial 16S rRNA gene sequencing, whereby the composition of the microbial communities associated with the skin, gastrointestinal tract, nasal cavity and other tissues is described. This approach provides information about the relative abundance and diversity of bacterial species. Bacterial 16S rRNA gene sequencing amplifies variable regions of bacterial genomes using a primer-based approach to generate operational taxonomical units or OTUs. OTUs are important for assessing microbial communities, especially in terms of richness, as well as measures of microbial diversity. Reductions in microbial community diversity and richness can be detrimental to the health of the host individual because pathogenic bacteria may proliferate under some scenarios.

Chemicals that have been investigated from a microbiome perspective in ecotoxicology include methylmercury and triclosan in species such as fathead minnow and mice (Narrowe *et al.*, 2015; Bridges *et al.*, 2018). It is important to keep in mind that the interaction between chemicals in our environment and the microbiome is complex, and that the host will respond to chemical exposure dynamically with the microbiome (Box 5.4). During the process of gut dysbiosis, defined here as an imbalance in microbial homeostasis with its host, the gastrointestinal system can experience significant inflammation. This can lead to disrupted gastrointestinal cell communication, loss of tissue integrity (i.e., leaky gut), inability to obtain nutrients, and ulceration and haemorrhage. Another important point to make is that the microbiome can be a significant ally in the detoxification and metabolism of environmental contaminants within the gut. For example, aryl hydrocarbon receptors (AhRs), involved in detoxification, can be activated in the gut, and microbial species can metabolize the chemicals ingested by animals, rendering them potentially less toxic to the individual (reviewed in Adamovsky *et al.* (2018)). Conversely, microbial communities may also activate chemicals into more harmful compounds. We have much to learn regarding the *in situ* microbial degradation of chemicals by the various microbiomes present in organisms.

Understanding the significance of the microbiome in ecotoxicology is a new area of research, and there is much to learn about the role of tissue-specific microbiomes in chemical detoxification or bio-activation. Emerging computational tools allow one to predict how changes in microbial communities lead to altered metabolite production in the gastrointestinal system (e.g., software such as PICRUSt; Langille *et al.* (2013)). While it may be clear that chemicals can decrease richness and abundance of beneficial microbes, it is less clear how chemicals perturb the intricate community networks that microbial communities share with their host. Symbionts and pathogenic species, as well as the host immune system, strike a delicate balance to maintain the overall health of the gastrointestinal system. If community structure is fragmented in any way as a result of chemical exposures, there could be severe consequences for the individual and adverse health outcomes. This has been demonstrated with plasticizers, such as bis(2-ethylhexyl)phthalate in zebrafish. Feeding the plasticizer to fish induced fragmentation of microbial communities within the gastrointestinal system (Figure 5.5). Researchers are now using multi-omics approaches to integrate microbial community changes with molecular responses in the host gastrointestinal tract, generating novel insights into how a chemical exposure is mitigated by microbial-host defences. Another challenge is that individuals can differ in the dominant species of microbe present in the microbiome. This may have implications for toxicity if a species lacks microbes for metabolizing ingested chemicals. Microbiome research promises to have an exciting future in ecotoxicology.

Box 5.4 The microbiome is complex and abundant

The array of species in the microbiome is vast, composed of bacteria, protists, viruses and fungi. These organisms are critical for animal and human health, and synthesize many vitamins, nutrients and signalling molecules within the body (i.e., microbial metabolites). There are more microbes in the human body than there are cells, and this is estimated in humans to be at a ratio of 10 to 1. There are also networks of soil fungi that support forest ecosystems.

5.8 Ecotoxicology and Bioinformatics

There are intrinsic obstacles to using omics data in toxicology, and perhaps the most significant include data management, integration and interpretation. The field of **bioinformatics** has addressed many of these challenges. Extremely large databases are generated with omics, and

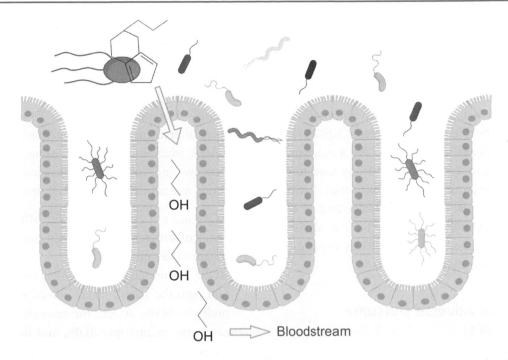

Figure 5.5 Schematic of the gut microbiome. Different bacteria in the lumen can metabolize chemicals. These chemicals can be absorbed through the gut epithelium and can make their way into the bloodstream (indicated by arrows leading into bloodstream). Chemicals in turn can affect microbial community abundance and richness. 16S sequencing measures the abundance and richness of the microbiota by amplifying variable DNA regions.

knowledgeable and highly trained bioinformaticians are needed along with standardization of methods for data analysis and interpretation. As such, computational toxicology will continue to play a central role in the assessment of chemicals. Integration of different omics datasets is addressed in Section 5.9.

Computational algorithms and graphical capacity continue to improve. Interpretation of omics data was difficult in the beginning and could not be integrated into a broader biological picture. Researchers were left to manually connect the dots among thousands of genes, proteins or metabolites. This was an overwhelming task. Early on, it was recognized that more sophisticated computational methods were required to categorize molecules into functional groups. To address this problem, bioinformaticians developed unbiased methods to computationally interrogate omics data. These analyses use network analysis (e.g., connecting groups of genes together) (Figure 5.6) and multi-omics tools for data integration (Section 5.11.2).

Several excellent online and commercial resources are available that can be used to gain a broader understanding of omics data. Data mining is becoming more routine, and now omics data from different studies can be used to obtain a global molecular perspective on the physiological responses that follow a chemical exposure. Each of the

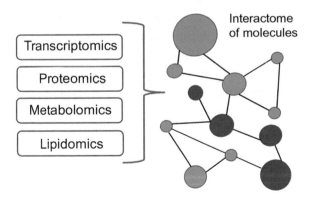

Figure 5.6 Omics data (e.g., transcriptomics, proteomics, metabolomics) can be integrated to build interaction networks or pathways. Red entities represent the upregulation of a molecule, and the green entities represent the downregulation of a molecule in response to a chemical exposure. The size of the circle can represent magnitude of change in the molecule, with larger circles representing a larger response relative to the smaller circles. Computational programs can deduce specific pathways affected by chemical exposures.

omics disciplines has unique resources available for mapping and comparing toxicological data among studies. For example, transcriptomics has the Comparative Toxicogenomics Database, a resource managed by the MDI Biological Laboratory and North Carolina State University. This database reports gene–chemical relationships for thousands of environmental contaminants and can rapidly compare molecular signatures for different chemicals. The database is a significant advance for toxicology in general because researchers can obtain a list of the genes that are responsive to a specific chemical. It can also be used to predict how chemicals from the same class may affect biological responses. KEGG (Kyoto Encyclopedia of Genes and Genomes) has a rich database of biochemical pathways that can be freely used to illustrate complex responses to environmental toxicants. Other open-source platforms are available (e.g., R codes, Cytoscape, Gene Ontology Resource) for analysis, and students are encouraged to independently explore these different avenues for data informatics.

5.9 Omics and Adverse Outcome Pathways (AOPs)

One approach to strengthen chemical risk assessment is the **adverse outcome pathway** framework or **AOP** (Sections 4.2, 4.5.3, 6.4, 14.1.3, 14.4.3, 14.4.4). Molecular data are limited in use in ecotoxicology when they are not generated within the context of the organism's biology. If molecules are to be useful predictors of adverse effect, biomarkers of effect must be functionally linked by biochemical responses to physiological functions. Phenotypic endpoints that have been associated with molecular responses include teratogenicity, tissue pathology, growth, fecundity and egg quality. The AOP concept was developed as a framework for linking molecular responses to population-level outcomes, with high relevance for ERAs.

More specifically, an AOP describes the relationship between **molecular initiating events** (MIEs) and subsequent key events (KEs) within an AOP. MIEs can be defined as a direct interaction between a chemical and a molecular receptor. Omics data can be important for elucidating what the MIEs are, especially if transcriptomics identifies a suite of genes that are activated following a ligand–receptor interaction (e.g., the induction of oestrogen responsive genes leads to the hypothesis that the chemical binds to the oestrogen receptor as a MIE). This interaction can initiate a molecular cascade and subsequent physiological changes in the organism such as an increased metabolic rate, release of a hormone or maturation of gonads. In turn, these events may lead to adverse outcomes if the animal cannot compensate or maintain physiological homeostasis. AOPs are typically described in linear fashion, beginning with the MIE

linked to a KE at the cellular, organ and organism levels. The KEs are subsequent steps that can ultimately lead to an adverse outcome (a regulatory-relevant population-level response) (Figures 4.13, 5.7, 14.2). It is important to note that AOPs are not necessarily linear and can comprise a complex network of KEs. Key events can include altered metabolites or inhibition of enzymes in the AOP, which can be revealed using omics approaches. Villeneuve and colleagues (Villeneuve *et al.*, 2014a, b) describe five principles of the AOP framework:

- AOPs are not chemical-specific. MIEs can occur as a result of different chemical activators, each of which initiates the same AOP.
- AOPs are modular, meaning they are flexible and contain 'specific parts'. These modules can be used in multiple AOPs. A KE, for example, within an AOP can occur in multiple AOPs, and thus AOPs are not always independent from one another.
- An AOP is a pragmatic unit of development and evaluation. AOPs strengthen the rationale for measuring a specific endpoint in regulatory toxicology.
- AOP networks are the functional unit of prediction for most real-world applications.
- AOPs are living documents. As additional experimental data are collected, KEs within an AOP become more strongly supported (or not) over time, and AOPs can be updated and modified accordingly.

Each AOP includes an MIE, i.e., an interaction between a chemical toxicant and biological molecule within the organism that spurs a response. MIEs can be broadly described as non-endocrine and endocrine. MIEs defined as non-endocrine include AhR activation (although this pathway interacts significantly with endocrine pathways such as oestrogen receptor signalling) and mitochondrial dysfunction. Chemicals can bind proteins within the mitochondria and affect oxidative respiration. Rotenone, for example, is a pesticide that binds to complex I of the electron transport chain to inhibit oxidative respiration. Several environmental toxicants have been reported to inhibit proteins in the electron transport chain. Inhibition of oxidative respiration can lead to altered mitochondrial membrane potential and ATP production. This is significant because cells that do not have a sufficient supply of energy are unable to conduct basic cellular processes, in turn leading to detrimental effects on the organism. Metals can also bind specific components of the cell membrane, leading to adverse outcomes.

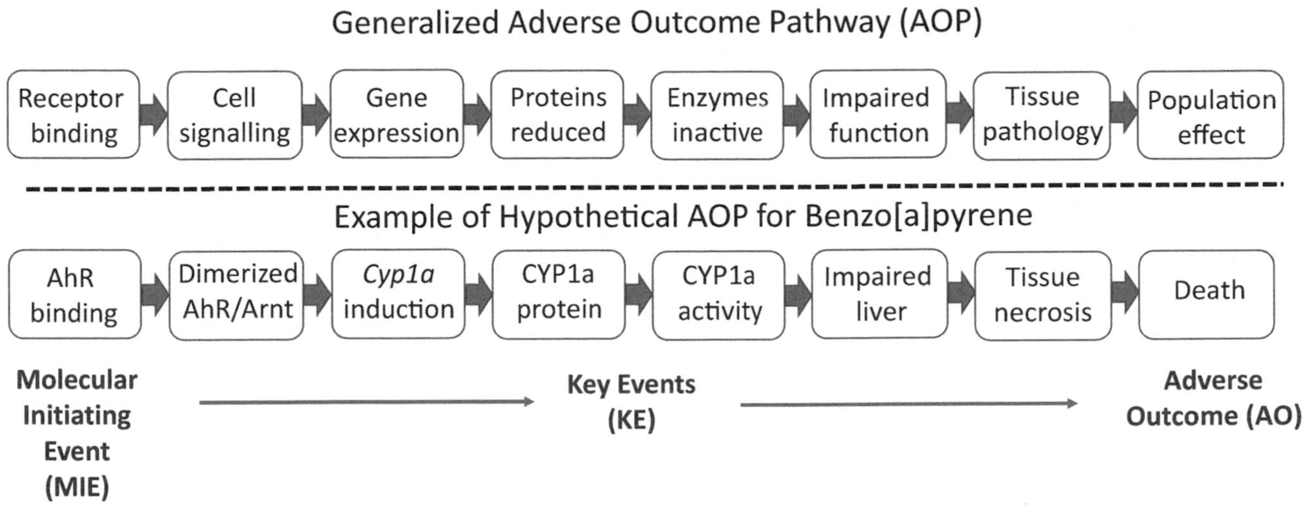

Generalized Adverse Outcome Pathway (AOP)

Receptor binding → Cell signalling → Gene expression → Proteins reduced → Enzymes inactive → Impaired function → Tissue pathology → Population effect

Example of Hypothetical AOP for Benzo[a]pyrene

AhR binding → Dimerized AhR/Arnt → *Cyp1a* induction → CYP1a protein → CYP1a activity → Impaired liver → Tissue necrosis → Death

Molecular Initiating Event (MIE) ——→ **Key Events (KE)** ——→ **Adverse Outcome (AO)**

Figure 5.7 The adverse outcome framework. The pathway on the top is a generic adverse outcome, highlighting the major components, i.e., molecular initiating event (MIE), key event (KE) and adverse outcome (AO). The pathway on the bottom is one example for an AO for a well-studied chemical, benzo[a]pyrene (BaP). AhR is the aryl hydrocarbon receptor, ARNT is the aryl hydrocarbon receptor nuclear translocator, and CYP1A is the cytochrome P450 1a involved in phase I of detoxification. Over-activation of AhR signalling may result in detrimental effects in tissues over time. Omics approaches such as transcriptomics and proteomics can play an important role in identifying MIEs and characterizing KEs.

MIEs defined as endocrine-related include androgen receptor antagonism, aromatase inhibition, thyroid receptor agonism and oestrogen receptor activation, among others. Oestrogen receptor activation has been broadly studied in ecotoxicology from the molecular to the population level. Different chemicals such as 17α-ethinyloestradiol, nonylphenol, methoxychlor, bisphenol A, diethylstilbestrol and an array of other industrial chemicals, pesticides, pharmaceuticals and phyto-oestrogens can act as initiators of the MIE of oestrogen receptor activation. This array of activators highlights the first point above, i.e., that AOPs are not chemical-specific. Enzyme inhibition could be either a molecular interaction (e.g., polar or non-polar binding) that causes a reversible change in enzyme function (e.g., aromatase inhibition by fadrazole), or a chemical reaction in which the structure of the enzyme is permanently changed, resulting in loss of function (e.g., chronic exposure to organophosphates may permanently inhibit acetylcholinesterase).

The KE is typically a description of a biological process. Examples would include decreased 17β-oestradiol synthesis, altered cortisol levels or decreased uptake of vitellogenin into the oocyte. Each of these KEs can potentially lead to reduced fecundity and subsequent population-level declines. An important point about KEs in an AOP is that the KE must be measurable and causally linked to the AO. Each KE can occur in multiple AOPs and can provide a more comprehensive view of risks associated with exposure. This is important for predictive ecotoxicology, as one can imagine how different MIEs can converge on a common KE, eventually becoming an AO.

Several resources are available for exploring and developing AOPs for environmental toxicology (see Figure 14.2 as an example incorporating climate change). An AOP Wiki is a collaborative development of AOP descriptions and information based on experimental evidence for linkages between KEs and AOs. The Wiki is a collaborative effort by several global organizations and agencies that include the European Commission, US EPA, US Army, OECD and SAAOP (Society for the Advancement of Adverse Outcome Pathways). There is also an AOP knowledge base that presents chemical, biological and toxicological ontology that can be used to develop AOPs. To assist with graphical interpretation and development, AOP Explorer (https://aopwiki.org/) and Effectopedia (https://norecopa.no/3rset-resources/introduction-to-effectopedia) are useful resources for researchers.

5.10 Omics in Regulatory Toxicology

Regulatory toxicology covers the legal aspects of toxicological applications and is used by governments to protect the environment as well as animal and human health (Section 12.2). In fact, the phrase used to describe the intersect between environment, animal and human health is "One Health". There is currently much debate in regulatory toxicology about how best to utilize omics data. Regulators and environmental managers require that the molecular response be strongly associated with an adverse effect. For example, it is difficult to demonstrate to a risk assessor that, if a specific gene, protein or metabolite has changed in individuals from a population, immediate protection of the population is required. For molecular data, it is still not well understood what threshold of response warrants action. This is discussed in more detail in Section 5.10.2. Moreover, one must demonstrate that these molecular responses are strongly linked to an adverse outcome. Several principles must be satisfied if a cause-and-effect relationship is to be proposed, including strength of the relationship, consistency of response, plausibility, temporality, experimental evidence, evidence for a dose response or biological gradient, and specificity as examples (Section 4.3.2). As one can imagine, it becomes extremely difficult to satisfy these criteria for a molecular response. Different laboratories must reliably and consistently measure the molecular response under an array of conditions to demonstrate an adverse outcome. Nevertheless, the conversation remains open among agencies, regulators, managers and scientists on how best to incorporate molecular knowledge into risk assessment. Below, we highlight some new approaches for high-throughput methods to assist with generating evidence for cause-and-effect relationships and discuss some of the challenges in using omics technologies in the field (i.e., environmental omics). While laboratory experiments provide rigorous conditions for experimental manipulation, the natural environment is much more uncertain, and variables other than chemicals can influence the collection and interpretation of omics data (Section 4.3).

5.10.1 Computational Toxicology in Regulatory Toxicology

The sheer number of chemicals in the environment that can pose risks to ecosystem health presents immense challenges for toxicological screening and testing. These chemicals include not only the parent compounds but also the breakdown products from microbial degradation. Chemicals enter aquatic and terrestrial environments from multiple sources (e.g., municipal wastewater treatment and sewage, gas and oil industry, mining activities,

automobile exhaust, electronic waste, plastic degradation). One can quickly conclude that it is an overwhelming task to evaluate every chemical in every species potentially exposed to that chemical. Novel chemistry is being developed for new chemicals and environmental toxicologists are now turning more than ever to high-throughput *in vitro* testing platforms (e.g., cell-based, ligand binding assays) to discover those chemicals that pose a high risk to wildlife and humans for endocrine disruption, mitochondrial dysfunction, impairments to cell proliferation and growth, and cytotoxicity.

The document entitled "*Toxicity Testing in the 21st Century*" outlined a vision and a strategy for how to conduct toxicity testing in the future (US National Research Council, 2007). The overarching goal is to increase the capacity of chemical testing by reducing the cost and time associated with such intense activities. Using whole-animal models to study chemicals one by one, while highly informative, can be extremely expensive. The objectives of high-throughput toxicity testing are to cover a diverse range of chemicals and mixtures, as well as a broad range of potential adverse effects. *In vitro* toxicity testing also addresses concerns about the use of animals in toxicity assays and can offer mechanistic insights that may not be readily attainable with animal models. These platforms can identify MIEs and contribute experimental evidence for cause and effect.

The **ToxCast** programme was developed to identify new mechanisms of chemical activity and to prioritize chemical testing (Figure 5.8). The idea is to use high-throughput *in vitro* toxicity testing as well as computational approaches to predict which chemicals may be most harmful to wildlife and humans. These chemicals are further tested in animal models to better discern the relationship between exposure and adverse health effects. In addition, computational biology can be used to predict how toxicants with similar moieties or chemical structures alter biological endpoints. For example, chemicals that contain a benzene ring have been identified using ToxCast data as those most likely to affect mitochondrial membrane potential (Dreier *et al.*, 2019). QSARs and QICARs are examples of another approach to understand what properties of a given chemical are related to its activity (Sections 2.3.3.3; 6.2.1). The ToxCast programme can identify chemicals for further study (i.e., prioritization) and can screen new emerging chemical classes.

There are limitations to high-throughput screening, and they can be subject to false positives and negatives.

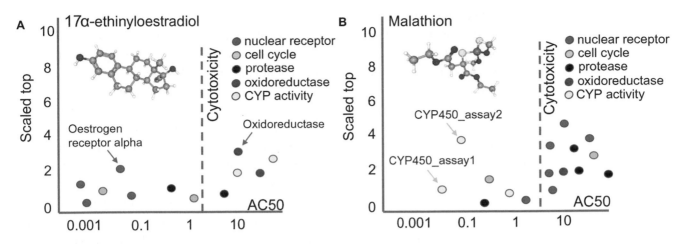

Figure 5.8 High-throughput receptor activation assays implemented under the Tox21 and ToxCast Programmes generate toxicity data that can be helpful in ecotoxicology. These graphs are example summaries for two environmental contaminants: A, 17α-ethinyloestradiol (a pharmaceutical) and B, malathion (a pesticide). The x-axis is a log scale of the chemical concentrations tested in the assays. The dashed vertical line indicates cytotoxicity of the chemical. Cell assays activated to the left of the vertical line are considered active or positive while those to the right of the dashed line are considered inactive, likely because of cytotoxicity. In A, 17α-ethinyloestradiol is positive for many steroid receptor assays (blue dots), and this occurs at very low concentrations. In B, malathion is positive for assays related to the cytochrome P450 family (yellow dots) [TOXCAST: EPA ToxCast Screening Library (comptox.epa.gov)]. AC50 refers to the concentration at 50% of maximum activity. In the label for the vertical axis, 'scaled top' refers to the relative response activity for all the tests conducted with the chemical.

In addition, it is important to differentiate cytotoxicity from the MoA. Data are collected to ensure reporter responses are not due to cytotoxicity. A reporter is a gene that is attached to a regulatory sequence of another gene that induce visible (fluorescent and luminescent proteins) characteristics that indicate a certain molecular event (e.g., ligand binding or gene expression) in the cell or animal has taken place. Data are rigorously evaluated prior to becoming publicly available, and the assays have several QA/QC checkpoints to ensure scientific rigour. A long-term goal of the ToxCast programme is to prioritize the chemicals that merit additional studies. If problematic, these chemicals may be replaced in time with safer alternatives. Chemicals are first assessed for purity prior to being tested in *in vitro* assays, and multiple doses of the chemical are used. Cell-based *in vitro* assays primarily utilize human and rat cell lines. These receptor-based or biochemical assays determine the potential for a chemical to bind to a receptor, enzyme or transporter. It is important to recognize that not all data are transferable to all species, and receptors in wildlife may respond in a more sensitive manner, highlighting the importance of developing and comparing different *in vitro* assays for several vertebrate and invertebrate representatives. Nevertheless, valuable mechanistic information can still be obtained. The ToxCast dashboard hosts the processed data and

can be accessed by external users to explore specific questions about chemicals of interest.

Case Study 5.1 Omics to Reveal Mechanisms Underlying Glyphosate Toxicity in Invertebrates and Vertebrates

Agrochemicals and plant protection products (PPPs) are used globally to reduce damage to crops and increase the world's food supply. Beginning in the 1940s, synthetic pesticides have had a long history of successes and controversy (Section 1.2.1); environmental protection agencies must carefully assess chemicals used on crops to prevent wildlife (and human) exposures that can adversely affect both short- and long-term health outcomes. However, food security is a top priority for all countries, and the use of agrochemicals is often a necessity. Glyphosate in its commercial form (i.e., as the active ingredient of RoundUp®) is a broad-spectrum herbicide that was first commercially produced by the Monsanto Company in the 1970s. The herbicide acts on the plant enzyme 5-enolpyruvylshikimate-3-phosphate synthase, a part of the shikimate pathway, which is responsible for the biosynthesis of aromatic compounds. Disruptions in this pathway ultimately impair plant metabolism.

Glyphosate is the most widely used herbicide in North America (Section 7.8.8), and estimates are staggering, at 1.6×10^9 kg of glyphosate used as the active ingredient in US markets since it was first registered for use in 1974 (Benbrook, 2016). In North America, both the US Environmental Protection Agency and Health Canada have reviewed data and reassessed the safety and uses of glyphosate, concluding that proper use of the herbicide at label-recommended concentrations does not lead to significant risk to wildlife or the environment. The European Food Safety Authority has also expressed the position that there is a low likelihood of risk to animal and human health from current uses of glyphosate in terms of cancer and endocrine disruption. However, several bioassays used to test toxicity pathways do not capture all molecular pathways. In addition, formulations and degradation products can be more toxic than the parent chemical.

Such a scenario can warrant omics approaches to uncover new potential avenues of toxicity in non-target organisms. In a study by Simões *et al.* (2018), Glyphosate-based formulations (Montana®) were added to soil, and the transcriptomics and proteomics responses of the ecotoxicological model organism *Folsomia candida* were studied. *F. candida* is a species of springtail (soil-dwelling invertebrate) and one widely used to study soil contamination. Transcriptome and proteome responses revealed cellular stress effects, disturbance of Ca^{2+} homeostasis, and changes in reproductive proteins (i.e., vitellogenin-1 and vitellogenin-2). The study used both gene and protein data to identify new targets of toxicity, such as calcium regulation. In another study, in fish, juvenile brown trout were exposed to one of three concentrations of glyphosate or RoundUp for 2 weeks, and transcriptome responses were measured in the liver (Webster and Santos, 2015). Hepatic gene expression profiles were related to oxidative stress, cellular stress response and apoptosis. The authors suggested (1) oxidative stress is probably a central mechanism of glyphosate and RoundUp toxicity, and (2) environmentally relevant concentrations of glyphosate or RoundUp may exert adverse health effects in wild fish populations. These are two ecotoxicological examples of how omics approaches can reveal new avenues for herbicide effects in non-target organisms (Figure 5.9).

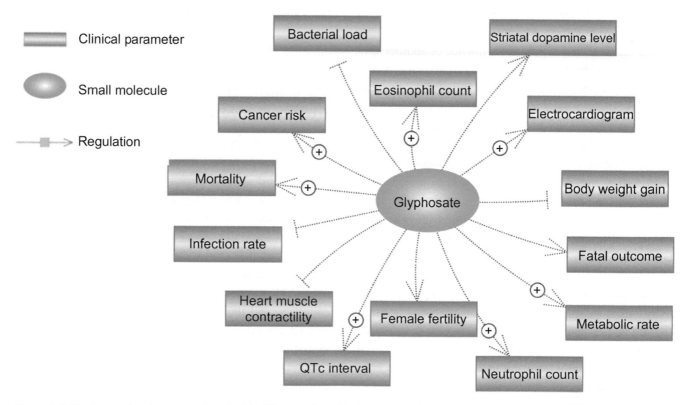

Figure 5.9 Glyphosate has been associated with different clinical indicators of adverse health in organisms. These relationships are built using a weight-of-evidence approach in Pathway Studio® (Elsevier) to reveal adverse outcomes associated with glyphosate exposure. Pointing arrows (with plus symbol) indicate a stimulation of the clinical sign, while the "T" arrows represent inhibition of the clinical sign.

Complementary information on glyphosate can be found in Section 7.8.8.

5.10.2 Environmental Omics in Regulatory Toxicology

Transcriptomics, proteomics, metabolomics, lipidomics and epigenetics have been used to study wild populations of organisms to determine the impact of environmental contamination. Molecular responses have been linked to chemical burden and phenotypic endpoints of relevance, such as the case of vitellogenin and egg production (Section 5.2.1). These phenotypic endpoints are important for environmental managers and regulators; examples of key phenotypes important for environmental regulation are an individual's rate of growth and reproductive output (i.e., fecundity). These are penultimate metrics that can be assessed in environmental monitoring programmes, and there is active debate as to whether omics-based data can assist in improving biomonitoring in the natural environment.

An example of a well-designed and successful programme is the Environmental Effects Monitoring (EEM) Programme in Canada (Section 12.3.5). This programme acts as a quality control measure to ensure that the emissions regulations (chemical composition, effluent toxicity, volumes discharged) governing specific industrial sectors (pulp mills; mines) are sufficient to protect the environment. It relies on the assessment of both aquatic invertebrates (Canadian Aquatic Biomonitoring Network, known as CABIN) and vertebrates (primarily fish species). For fish, the Canadian EEM programme records the gonadosomatic index (GSI), a measure of the gonad weight relative to the body weight of the individual, in a diverse array of species. Decades of research have established mean GSI values for fish throughout Canada, determining an expected range for GSI for each species. If the GSI, considered to be a reliable surrogate measure for reproductive output, falls outside an established 95% confidence interval determined from historical data, further investigation is initiated into the cause of the impact. Other indicators of biological impacts such as sex steroids (17β-oestradiol and 11-ketotestosterone) have been investigated for utility in environmental monitoring programmes. These values have been determined for fish from several pristine and contaminated sites. However, although they are highly relevant to reproductive success and population stability, reproductive steroids can be variable within individuals, and government environmental monitoring programmes have not yet adopted the use of hormone concentrations.

The concept of genomics in ecotoxicology-based assessments has been around for some time (~25 years); however, omics data have not yet been widely accepted by ecosystem managers and environmental risk assessors for making decisions. There is currently much debate about which molecular endpoints are most useful for environmental impact assessments. For example, should regulators make decisions on a single gene, a protein or a specific metabolite? It is important to keep in mind that any data used to make an environmental impact decision must be agreed upon by regulators and industry partners, and must be defensible in a court of law (i.e., a black or white indicator of adverse effect with little room for uncertainty). There must be sound reasons for basing economic and environmental decisions on a molecular endpoint. The endpoint must show reasonable sensitivity and demonstrate a predictable response under various environmental scenarios. It must be reliably measured in multiple laboratories, with a small margin of measurement error.

There have been some efforts to incorporate omics into effects-based environmental monitoring programmes. For example, the Laboratory Watchfrog from the French National Natural History Museum uses amphibian models combined with toxicogenomics approaches to study impacts of contaminants on neurodevelopment, to transfer this approach into future regulatory toolkits. Thus far, however, there remains no clear consensus regarding the incorporation of omics endpoints into industry and government regulations. Complementary information on regulations can be found in Section 12.2.

5.10.3 Challenges and Considerations

While there has been significant discussion on the incorporation of omics data into ERAs (Bahamonde et al., 2016), major challenges remain. Both intrinsic and extrinsic variability in molecular responses can be problematic and limit the use of omics data in risk assessment. In the case of GSI as a phenotypical endpoint, significant effort has been made to understand the natural variability of GSI in species. Variability is associated with time of year, habitats, temperature, sexual maturation and reproduction, migrations, feeding cycles, evolution of prey items with growth, metamorphosis and senescence. These efforts have been conducted in both Canada and in Europe (Munkittrick *et al.*, 2002; Cogne *et al.*, 2019). This is necessary, as variable endpoints require a large sampling size to detect small differences among populations. Cowie *et al.* (2015) measured 15 transcripts in the fathead minnow ovary to determine variability of genes in a tissue. Some transcripts, such as the oestrogen receptor isoforms, showed low relative variability among individuals. In contrast, other transcripts, such as the 5α-reductase enzymes, showed high endogenous variability among individuals. If these transcripts were to be used in an environmental-monitoring scenario, they would require large sample sizes, which might not be feasible in field collections. Ideally, one would desire a molecular biomarker that was sensitive to a chemical exposure but showed low natural variability within individuals.

To address questions about the variability and the detection of consistent responses across laboratories, inter-laboratory comparisons have been conducted with transcriptomics. Laboratories were tasked with measuring molecular responses in fathead minnows following a low dose exposure to 17α-ethinyloestradiol (Feswick *et al.*, 2017). While a subset of common transcripts was identified by each of the six laboratories, there were also unique responses observed by each independent research group. This study highlights the need for further standardization of the methodology.

Another limitation that hinders the incorporation of molecular endpoints into environmental monitoring programmes is that the molecular biomarker (biomarker of effect) must in some way be 'responsible' for the physiological change, which is difficult to demonstrate. This type of evidence is required for industry and lawmakers. For example, experimental and computational studies support the observation that decreases in vitellogenin protein can lead to reduced fecundity in fish (Miller *et al.*, 2007). However, it is less clear whether changes in other transcripts and proteins persist and can be unequivocally associated with population-level effects. The AOP framework is one such tool that can assist in making these required linkages. One must also remember that the abundance of transcripts is time-dependent, which confounds the interpretation of the molecular marker as an indicator of an adverse effect. Homeostatic mechanisms following a chemical insult can restore transcripts to preexposure levels, and thus timing is important.

To avoid depending on a single gene, interacting genes (or a gene network) may be useful for monitoring populations at risk. Interactome networks, or the relationships among molecular entities, can be quantified to improve interpretation (Martyniuk, 2018). Baseline data would be required for these interactomes, and admittedly this is not a trivial task. To evaluate molecular endpoints is technically complicated compared to morphometrics, and there are more steps required to generate a measurable value. Moreover, there is a need for rapid processing and ultra-low-temperature storage of samples, which can be difficult in laboratory conditions, and even more so in the field. Often trade-offs are needed between time and effort devoted to measures of phenotypic responses (e.g., GSI) and collection and preservation of samples for molecular work. In this regard, metabolites may be more straightforward to measure in animals within an environmental monitoring programme compared to genes or proteins; however, metabolite data can also be complicated to interpret and are influenced by an animal's diet, age, sex, reproductive state or overall health status.

As with any environmental monitoring study, site selection is of paramount importance when using molecular data in environmental assessments (Section 4.3.1). Typically, upstream sites are selected as a reference site for areas impacted by sewage facilities and municipal wastewater treatment facilities, pulp and paper mills, or other industrial activities. However, it may be necessary to select multiple reference sites to improve omics data interpretation. The effect of reference site on omics data was explored in a study of the rainbow darter in the Grand River, Ontario, Canada. It was revealed that transcriptome data vary by as much as 30% based upon relative comparisons to single control sites (Marjan *et al.*, 2017). To circumvent this issue, historical data within a site may be useful for transcriptome data, and the frequency and types of genes (i.e., based on their functions) that are altered can be assessed over time. Again, scientists, legislators and industry partners must first decide on what is acceptable or tolerated when it comes to a change in a molecular response. The inability to agree on what kind of a change in a transcript or protein is unacceptable has hindered the incorporation of molecular endpoints into environmental monitoring programmes.

A final consideration to be mentioned here is the type of omics approach to be used in field-based studies (Table 5.2). Each approach has limitations and provides different perspectives of a toxicant exposure, and thus these tools are complementary. Transcriptomics has disadvantages because data are highly time-dependent, and

Table 5.2 **Advantages and disadvantages of using omics approaches in regulatory toxicology.**

Omics approach	Advantages	Disadvantages
eDNA	Rapidly identify species present in the environment without exhaustive field surveys.	Protocol standardization lacking; type I and type II errors. A type I error rejects a true null hypothesis, while a type II error accepts a false null hypothesis.
Transcriptomics	Thousands of potential biomarkers assessed.	Transient expression; type I and type II errors. Relevance to biological function uncertain. High variability among individuals.
Proteomics	Proteins are functional units of the cell. More long-lived than transcripts (days versus hours).	Can be expensive to obtain data. Limited number of proteins identified; often those most abundant in tissue.
Metabolomics	Metabolites (nutrition-derived, synthesized in body) are shared and ubiquitous across species (e.g., glucose, cholesterol). Targeted versus untargeted approaches do not require *a priori* knowledge of molecular composition of species (e.g., their genes). Detection of metabolites derived from a specific chemical are a 'smoking gun' for exposure.	Sensitive to diet and other environmental perturbation.
Lipidomics	High relevance to metabolism and oxidative stress.	Uncertain what changes mean as the function of many lipids unknown.
Epigenomics	Multi-generational perspective.	Difficult to conduct studies in wild populations.

linkages between a change in a transcript and a functional impact in the individual are difficult to discern. Proteomics studies can be limited if the focus is on the relative abundance of proteins, rather than on protein function, as the abundance and activity of a protein are not always correlated. For example, phosphorylation through kinase activity is often required for enzyme function. In this case, measuring the abundance of the protein may be less important than determining its activity because assessing the activity of the protein may be more relevant to understanding the biological response. Phospho-proteomics, the study of phosphorylation patterns of proteins, can provide additional insight into which proteins are likely to be active or inactive, but such studies are technically demanding. A major disadvantage of metabolomics is that metabolites can also be transient and highly dependent upon diet and environmental factors. These variables are difficult to control or tease apart when sampling wild populations.

Other variables such as individual genetics can confound the interpretation of omics data following exposure to chemicals. Genomic variability from individual to individual, across populations and species is a natural and critically important phenomenon. Research employing genomics and transcriptomics on the Atlantic killifish

(*Fundulus heteroclitus*) along the Atlantic coast of the United States has pinpointed molecular variability among individuals in subpopulations. This variability can explain tolerance to high pollution loads, revealing that the AhR pathway acts as a 'molecular checkpoint' for sensitivity/resistance to environmental contaminants (Reid *et al.*, 2016). Individual and species differences in genetics can subsequently make it challenging to interpret molecular data from multiple species in a polluted environment. This can be partially addressed by focusing on the more abundant and ubiquitous species in each polluted habitat, but this approach still has limitations as species sensitive to the chemical are overlooked. Metabolites, on the other hand, are structurally conserved from species to species (glucose is glucose, whether it is measured in an aquatic invertebrate, fish or bird). However, here we are referring to metabolites that are the normal products of metabolism.

5.11 Emerging Applications for Omics in Ecotoxicology

To this point, we have discussed some of the more common omics technologies used in ecotoxicology. Technology continues to improve, and new methodology is being used to evaluate chemical–molecule interactions and biological responses. In this section, we discuss the **CRISPR/Cas9** system and how it can be used to identify genes that confer 'resistance or susceptibility' to environmental pollutants. We also discuss the growing concept of the exposome, integrating multi-omics data to understand lifetime exposures to chemicals and how this relates to toxicity and wildlife disease.

5.11.1 Genome-wide CRISPR Screens in Ecotoxicology

CRISPR is part of the bacterial immune system that has been adapted as a gene-editing tool in the medical and biological sciences (Figure 5.10). CRISPR/Cas9 stands for Clustered Regularly Interspaced Short Palindromic Repeats/Associated Protein 9. A protospacer adjacent motif or PAM sequence in double-stranded DNA is cut at nucleotides just prior to the site using the Cas9 enzyme. Non-homologous end-joining or homology-directed repair is typically used to generate the gene knockouts. CRISPR/Cas9 is revolutionizing the medical sciences, and

CRISPR-Cas9 screen

Step 1
sgRNA pooled library constructed

Step 2
Cell line of choice is transduced with lentivirus

Step 3
Grown in culture

Step 4
Treated with a toxicant

Control   Chemical

Step 5
RNA-seq counts reads that confer resistance or sensitivity to toxicant

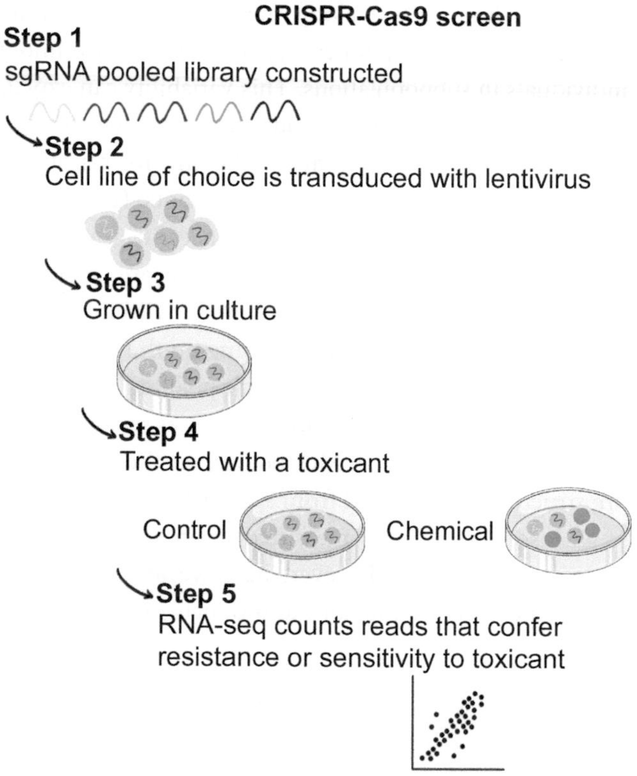

Figure 5.10 General workflow of a CRISPR-Cas9 screen to identify genes that are sensitive or resistant to the chemical. RNA-seq counts the number of sequencing reads to obtain information on abundance.

ecotoxicology may not be far behind. The technology can determine the relative importance of a gene to a toxicity pathway, or it can uncover new genes and pathways that metabolize a toxicant. The advantage of CRISPR gene editing over other approaches (i.e., zinc finger nucleases, transcription activator-like effector nucleases, viral systems and transposon-mediated gene editing) is that CRISPR/Cas9 offers high sensitivity and target specificity, and is less labour-intensive than pre-existing methods.

It is now possible to **knock out** (or **knock in**) specific genes of interest to toxicologists such as detoxification enzymes (cytochrome P450 enzymes) or transporters such as multidrug resistance protein 1 (MDR1). By doing so, ecotoxicologists can quantify physiological or biochemical responses that occur due to the loss or gain of gene function, providing insight about the relative importance of enzymes, transporters, receptors and signalling molecules in a chemical-induced response. Knock-in models for zebrafish and cell lines already exist to study environmental chemicals. For example, zebrafish have been gene-edited with CRISPR/Cas9 so that the promoter for vitellogenin, the egg yolk precursor, is tagged with a green fluorescent protein (Abdelmoneim *et al.*, 2020). Upon exposure to oestrogenic chemicals, the gene is turned on and the fish will glow green when exposed to water containing oestrogenic compounds. This is a powerful tool for assessing water quality and chemicals that bind oestrogen receptors (i.e., xeno-oestrogens). Gene-editing methods are expected to reveal new understanding at the level of the genome: which genes are involved in a biological response to a toxicant, which are the most sensitive genes, and which confer sensitivity or resistance to a chemical exposure.

CRISPR screens have been used to study molecular responses to chemical exposures and have been used to reveal genes that confer resistance or sensitivity to pesticides and metalloids such as arsenic. The workflow is such that the single guide (sg)RNAs (i.e., a nucleotide that targets a specific region of the gene) are designed using web-based algorithms and packaged as vectors into a lentivirus that infects a specific cell type (hepatocytes, immune cells, etc). Once the unique sgRNA has been delivered into each cell, the cells are cultured in the presence of a toxicant of interest, and selection occurs in the 'knockout pool' of cells. Genetic disruption by the CRISPR approach can reveal those genes that confer resistance to the chemical compared to those that render the cell more sensitive. After the selection period, RNAseq is used to count the final number of transcripts that are present in the pool. Genome-wide CRISPR screens have been conducted for halogenated aromatic hydrocarbons and PAHs. For example, a mouse genome-wide CRISPR library was used to identify putative transcripts important in the AhR signalling pathway (Sundberg and Hankinson, 2019). Other studies have used CRISPR screens to determine genes important for sensitivity and resistance to chemical exposure. Genome-wide CRISPR screening was used to identify genes that determine tolerance to acetaldehyde toxicity and arsenic toxicity.

5.11.2 Multi-omics, Exposome and Exposomics in Ecotoxicology

Multi-omics (or exposomics) involves the integration of the different levels of molecular data that we have discussed in this chapter. This is important because each of the omics offers unique insights into the molecular responses that are occurring within the cell or tissue following exposure to a toxicant. For example, transcriptome data coupled to metabolomics can generate insight into the relationship between enzyme expression and intermediate products during metabolism. Data on protein abundance and post-translational modifications can also reveal how a chemical affects enzyme expression/ function in relation to metabolic processes. Multi-omics is proving useful in ecotoxicology and offers a comprehensive view of key events that lead to adverse outcomes induced by environmental pollutants. Computational approaches using multi-omics datasets are used to understand not only animal and human diseases, but also organism responses to environmental toxicants. Integrated knowledge about the actions of pharmaceuticals, pesticides, industrial compounds, or other potential toxicants is expected to facilitate biomarker discovery and improve environmental monitoring strategies in the future. However, a significant challenge is computational ability; there are several software tools available for systems biology and multi-omics integration such as Cytoscape and Mixomics among others. Major challenges for multi-omics datasets include data visualization (e.g., how does one graphically portray such data?) and interpretation (e.g., if protein abundance for an enzyme decreases, but the metabolite concentrations increase, how will this be interpreted?). Nevertheless, as the costs associated with next-generation sequencing and metabolomics continue to decrease and the amount of information gathered continues to increase, integration of data using computational tools will play more of a role in ecotoxicology.

First proposed in 2005, the exposome represents an attempt to consider the complex exposures that organisms face. This concept integrates lifelong exposures to all external environmental factors (e.g., contaminants, radiation, diseases, diet) and internal biological factors (e.g., hormones, inflammation, oxidative stress) from *in utero* to old age. Exposome analysis can reveal an improved picture of how toxicants affect animal health by accurately measuring exposures along with the effects of these exposures. For example, *in vivo* solid-phase micro-extraction was used to extract toxicants and various endogenous metabolites in white suckers (*Catostomus commersoni*) living in or outside the contaminated oil sands region of the Athabasca River (Alberta, Canada) (Roszkowska *et al.*, 2019). The authors measured significant changes in the biochemical metabolite signatures in the fish muscle between contaminated and non-contaminated sites, together with highly different concentrations of petroleum-related compounds. This exposome study confirmed that several metabolic pathways in fish are altered in response to mixtures of environmental pollutants. Fish and wildlife exposome studies are starting to emerge, but the field remains at an early stage of development. Exposome testing can also be assisted by exposomics analysis, which combines genomics, metabolomics, lipidomics, transcriptomics and proteomics into a multi-omics approach to understanding toxicity.

Summary

Ecotoxicogenomics has advanced tremendously over the past several decades, and the amount of molecular data amassed about chemicals and their biological effects in wildlife is staggering. High-throughput omics technologies have generated novel insights into chemical MoAs and the consequences of exogenous chemical–ligand interactions. Elucidating MIEs for AOPs and characterizing the diversity of molecular responses that occur in a cell or tissue are critical to understanding physiological responses that manifest during or after a chemical exposure. Omics technologies expand our capacity to detect, regulate and prioritize chemicals in the environment, acting as biomarkers of exposure. Continued diligence is required to set safe limits on air and water quality and to monitor industrial activity and municipal wastewater treatment. Omics science is anticipated to play a continuing role in environmental protection for decades to come.

Review Questions and Exercises

1. Compare and contrast the advantages and disadvantages of transcriptomics, proteomics and metabolomics in the study of ecotoxicology.

2. Which omics techniques would you use if you were to study the effects of an oil spill in a lake? Discuss the limitations of your approach and some considerations you would have for your study.

3. Describe how an omics approach can be used to identify a molecular initiating event in an adverse outcome pathway.

4. Describe the role of environmental DNA in an environmental monitoring programme.

5. What is the difference between metagenomics and microbiome 16S sequencing?

6. How would you design an experiment if you aimed to study the effects of a new pesticide on the progeny of an exposed frog?

7. Draw a theoretical adverse outcome pathway for bisphenol A. Discuss why some governments have banned this contaminant from baby bottles.

8. Define what is meant by the following terms: exposome; lipodomics; eDNA.

9. What are the differences between label-free and labelling methods in proteomics? Describe how proteomics data can assist you in determining a chemical's mode of action.

Abbreviations

2D-DIGE	Two-dimensional differential gel electrophoresis		DEHP	Di(2-ethylhexyl)phthalate
AO	Adverse outcome		eDNA	Environmental DNA
AOP	Adverse outcome pathway		EE2	17α-ethinyloestradiol
C18	HPLC column with a stationary phase containing 18 carbon atoms		EEM	Environmental effects monitoring
			ELISA	Enzyme-linked immunosorbent assay
CABIN	Canadian Aquatic Biomonitoring Network		GO	Gene ontology
			GSI	Gonadosomatic index
CpG island	DNA regions rich in cytosines, C, and guanines, G		HPLC	High-performance liquid chromatography
CRISPR/Cas9	Clustered Regularly Interspaced Short Palindromic Repeats/Associated Protein 9		iTRAQ	Isobaric tagging for relative and absolute quantitation
			KE	Key event
CYP	Cytochrome P450		KEGG	Kyoto Encyclopedia of Genes and Genomes

LC	Liquid chromatography	PPAR	Peroxisome proliferator-activated receptor
LC-QTOF	Liquid chromatography–quadrupole time of flight	PTMs	Post-translational modifications
LC-MS-MS	Liquid chromatography/tandem mass spectrometry	PVDF	Polyvinylidene fluoride or polyvinylidene difluoride
MIE	Molecular initiating event	qPCR	Quantitative polymerase chain reaction
miRNA	MicroRNA	RNA-seq	RNA-sequencing
mRNA	Messenger RNA	RT RT-PCR	Real-time reverse transcriptase-polymerase chain reaction
MS	Mass spectrometry		
OTUs	Operational taxonomical units	(sg)RNA	Single guide RNA
PCBs	Polychlorinated biphenyls	TMT	Tandem mass tag

References

Abdelmoneim, A., Clark, C. L. & Mukai, M. (2020). Fluorescent reporter zebrafish line for estrogenic compound screening generated using a CRISPR/Cas9-mediated knock-in system. *Toxicological Sciences*, 173, 336–346.

Adamovsky, O., Buerger, A. N., Wormington, A. M., *et al.* (2018). The gut microbiome and aquatic toxicology: an emerging concept for environmental health. *Environmental Toxicology and Chemistry*, 37, 2758–2775.

Bahamonde, P. A., Feswick, A., Isaacs, M. A., Munkittrick, K. R. & Martyniuk, C. J. (2016). Defining the role of omics in assessing ecosystem health: perspectives from the Canadian environmental monitoring program. *Environmental Toxicology and Chemistry*, 35, 20–35.

Bartosiewicz, M., Penn, S. & Buckpitt, A. (2001). Applications of gene arrays in environmental toxicology: fingerprints of gene regulation associated with cadmium chloride, benzo (a) pyrene, and trichloroethylene. *Environmental Health Perspectives*, 109, 71–74.

Benbrook, C. M. (2016). Trends in glyphosate herbicide use in the United States and globally. *Environmental Sciences Europe*, 28, 1–15.

Bohne-Kjersem, A., Skadsheim, A., Goksøyr, A. & Grøsvik, B. E. (2009). Candidate biomarker discovery in plasma of juvenile cod (*Gadus morhua*) exposed to crude North Sea oil, alkyl phenols and polycyclic aromatic hydrocarbons (PAHs). *Marine Environmental Research*, 68, 268–277.

Bougas, B., Normandeau, E., Grasset, J., *et al.* (2016). Transcriptional response of yellow perch to changes in ambient metal concentrations—a reciprocal field transplantation experiment. *Aquatic Toxicology*, 173, 132–142.

Bridges, K. N., Zhang, Y., Curran, T. E., *et al.* (2018). Alterations to the intestinal microbiome and metabolome of *Pimephales promelas* and *Mus musculus* following exposure to dietary methylmercury. *Environmental Science and Technology*, 52, 8774–8784.

Brown, M., Davies, I. M., Moffat, C. F., *et al.* (2004). Identification of transcriptional effects of ethynyl oestradiol in male plaice (*Pleuronectes platessa*) by suppression subtractive hybridisation and a nylon macroarray. *Marine Environmental Research*, 58, 559–563.

Cao, F., Li, H., Zhao, F., *et al.* (2019). Parental exposure to azoxystrobin causes developmental effects and disrupts gene expression in F1 embryonic zebrafish (*Danio rerio*). *Science of the Total Environment*, 646, 595–605.

Chatterjee, N., Gim, J. & Choi, J. (2018). Epigenetic profiling to environmental stressors in model and non-model organisms: ecotoxicology perspective. *Environmental Health and Toxicology*, 33, e2018015–0.

Cogne, Y., Almunia, C., Gouveia, D., *et al.* (2019). Comparative proteomics in the wild: accounting for intrapopulation variability improves describing proteome response in a *Gammarus pulex* field population exposed to cadmium. *Aquatic Toxicology*, 214, e105244.

Contreras, L., Moenne, A., Gaillard, F., Potin, P. & Correa, J. A. (2010). Proteomic analysis and identification of copper stress-regulated proteins in the marine alga *Scytosiphon gracilis* (Phaeophyceae). *Aquatic Toxicology*, 96, 85–89.

Cowie, A. M., Wood, R. K., Chishti, Y., *et al.* (2015). Transcript variability and physiological correlates in the fathead minnow ovary: implications for sample size, and experimental power. *Comparative Biochemistry and Physiology Part B: Biochemistry and Molecular Biology*, 187, 22–30.

Defo, M. A., Bernatchez, L., Campbell, P. G. C. & Couture, P. (2015). Transcriptional and biochemical markers in transplanted *Perca flavescens* to characterize cadmium- and copper-induced oxidative stress in the field. *Aquatic Toxicology*, 162, 39–53.

Denslow, N. D. & Sabo-Attwood, T. (2015). Molecular bioindicators of pollution in fish. In R. H. Armon & O. Hänninen (eds.) *Environmental Indicators*. Dordrecht, The Netherlands: Springer Netherlands, 695–720.

Dong, X., Li, Y., Zhang, L., *et al.* (2016). Influence of difenoconazole on lipid metabolism in marine medaka (*Oryzias melastigma*). *Ecotoxicology*, 25, 982–990.

Dreier, D. A., Denslow, N. D. & Martyniuk, C. J. (2019). Computational in vitro toxicology uncovers chemical structures impairing mitochondrial membrane potential. *Journal of Chemical Information and Modeling*, 59, 702–712.

Feswick, A., Isaacs, M., Biales, A., *et al.* (2017). How consistent are we? Interlaboratory comparison study in fathead minnows using the model estrogen 17alpha-ethinylestradiol to develop recommendations for environmental transcriptomics. *Environmental Toxicology and Chemistry*, 36, 2614–2623.

Guo, J. H., Nguyen, H. T., Ito, S., *et al.* (2018). In ovo exposure to triclosan alters the hepatic proteome in chicken embryos. *Ecotoxicology and Environmental Safety*, 165, 495–504.

Helgason, L. B., Arukwe, A., Gabrielsen, G. W., *et al.* (2010). Biotransformation of PCBs in Arctic seabirds: characterization of phase I and II pathways at transcriptional, translational and activity levels. *Comparative Biochemistry and Physiology Part C: Toxicology and Pharmacology*, 152, 34–41.

Houde, M., Giraudo, M., Douville, M., *et al.* (2014). A multi-level biological approach to evaluate impacts of a major municipal effluent in wild St. Lawrence River yellow perch (*Perca flavescens*). *Science of the Total Environment*, 497, 307–318.

Iguchi, T., Watanabe, H., Katsu, Y., *et al.* (2002). Developmental toxicity of estrogenic chemicals on rodents and other species. *Congenital Anomalies*, 42, 94–105.

Ings, J. S., Servos, M. R. & Vijayan, M. M. (2011). Hepatic transcriptomics and protein expression in rainbow trout exposed to municipal wastewater effluent. *Environmental Science and Technology*, 45, 2368–2376.

Ito, Y., Kamijima, M. & Nakajima, T. (2019). Di (2-ethylhexyl) phthalate-induced toxicity and peroxisome proliferator-activated receptor alpha: a review. *Environmental Health and Preventive Medicine*, 24, e47.

Jian, M., Zhang, D., Wang, X., *et al.* (2020). Differential expression pattern of the proteome in response to cadmium stress based on proteomics analysis of wheat roots. *BMC Genomics*, 21, e343.

Kostopoulou, S., Ntatsi, G., Arapis, G. & Aliferis, K. A. (2020). Assessment of the effects of Metribuzin, glyphosate, and their mixtures on the metabolism of the model plant *Lemna minor* L. applying metabolomics. *Chemosphere*, 239, e124582.

Langille, M. G., Zaneveld, J., Caporaso, J. G., *et al.* (2013). Predictive functional profiling of microbial communities using 16S rRNA marker gene sequences. *Nature Biotechnology*, 31, 814–821.

Larkin, P., Folmar, L. C., Hemmer, M. J., *et al.* (2002). Array technology as a tool to monitor exposure of fish to xenoestrogens. *Marine Environmental Research*, 54, 395–399.

Leroy, D., Haubruge, E., De Pauw, E., Thomé, J. & Francis, F. (2010). Development of ecotoxicoproteomics on the freshwater amphipod *Gammarus pulex*: identification of PCB biomarkers in glycolysis and glutamate pathways. *Ecotoxicology and Environmental Safety*, 73, 343–352.

Madison, B. N., Wallace, S. J., Zhang, J., Hodson, P. V. & Langlois, V. S. (2020). Transcriptional responses in newly-hatched Japanese medaka (*Oryzias latipes*) associated with developmental malformations following diluted bitumen exposure. *Comparative Biochemistry and Physiology Part D: Genomics Proteomics*, 35, e100685.

Marjan, P., Bragg, L. M., MacLatchy, D. L., Servos, M. R. & Martyniuk, C. J. (2017). How does reference site selection influence interpretation of omics data? Evaluating liver transcriptome responses in male rainbow darter (*Etheostoma caeruleum*) across an urban environment. *Environmental Science and Technology*, 51, 6470–6479.

Martyniuk, C. J. (2018). Are we closer to the vision? A proposed framework for incorporating omics into environmental assessments. *Environmental Toxicology and Pharmacology*, 59, 87–93.

Miller, D. H., Jensen, K. M., Villeneuve, D. L., *et al.* (2007). Linkage of biochemical responses to population-level effects: a case study with vitellogenin in the fathead minnow (*Pimephales promelas*). *Environmental Toxicology and Chemistry* 26, 521–527.

Munkittrick, K. R., McGeachy, S. A., McMaster, M. E. & Courtenay, S. C. (2002). Overview of freshwater fish studies from the pulp and paper environmental effects monitoring program. *Water Quality Research Journal*, 37, 49–77.

Nakayama, S. M. M., Nakata, H., Ikenaka, Y., *et al.* (2019). One year exposure to Cd- and Pb-contaminated soil causes metal accumulation and alteration of global DNA methylation in rats. *Environmental Pollution*, 252, 1267–1276.

Narrowe, A. B., Albuthi-Lantz, M., Smith, E. P., *et al.* (2015). Perturbation and restoration of the fathead minnow gut microbiome after low-level triclosan exposure. *Microbiome*, 3, e6.

Nuwaysir, E. F., Bittner, M., Trent, J., Barrett, J. C. & Afshari, C. A. (1999). Microarrays and toxicology: the advent of toxicogenomics. *Molecular Carcinogenesis: Published in cooperation with the University of Texas MD Anderson Cancer Center*, 24, 153–159.

O'Kane, A. A., Chevglier, O. P., Graham, S. F., Elliott, C. T. & Mooney, M. H. (2013). Metabolomic profiling of in vivo plasma responses to dioxin-associated dietary contaminant exposure in rats: implications for identification of sources of animal and human exposure. *Environmental Science and Technology*, 47, 5409–5418.

Pierron, F., Normandeau, E., Defo, M. A., *et al.* (2011). Effects of chronic metal exposure on wild fish populations revealed by high-throughput cDNA sequencing. *Ecotoxicology*, 20, 1388–1399.

Reid, N. M., Proestou, D. A., Clark, B. W., *et al.* (2016). The genomic landscape of rapid repeated evolutionary adaptation to toxic pollution in wild fish. *Science*, 354, 1305–1308.

Roland, K., Kestemont, P., Loos, R., *et al.* (2014). Looking for protein expression signatures in European eel peripheral blood mononuclear cells after in vivo exposure to perfluorooctane sulfonate and a real world field study. *Science of the Total Environment*, 468–469, 958–967.

Roszkowska, A., Yu, M., Bessonneau, V., *et al.* (2019). In vivo solid-phase microextraction sampling combined with metabolomics and toxicological studies for the non-lethal monitoring of the exposome in fish tissue. *Environmental Pollution*, 249, 109–115.

Schaefer, F., Overton, J., Krüger, A., Kloas, W. & Wuertz, S. (2018). Influence of batch-specific biochemical egg characteristics on embryogenesis and hatching success in farmed pikeperch. *Animal*, 12, 2327–2334.

Simmons, D. B. D., Cowie, A. M., Koh, J., Sherry, J. P. & Martyniuk, C. J. (2019). Label-free and iTRAQ proteomics analysis in the liver of zebrafish (*Danio rerio*) following dietary exposure to the organochlorine pesticide dieldrin. *Journal of Proteomics*, 202, 103362.

Simões, T., Novais, S. C., Natal-da-Luz, T., *et al.* (2018). An integrative omics approach to unravel toxicity mechanisms of environmental chemicals: effects of a formulated herbicide. *Scientific Reports*, 8, e11376.

Sundberg, C. D. & Hankinson, O. (2019). A CRISPR/Cas9 whole-genome screen identifies genes required for aryl hydrocarbon receptor-dependent induction of functional CYP1A1. *Toxicological Sciences*, 170, 310–319.

US National Research Council (2007). *Toxicity Testing in the 21st Century: A Vision and a Strategy*. Washington, DC: US National Research Council.

Veldhoen, N., Hobbs, J., Ikonomou, G., *et al.* (2016). Implementation of novel design features for qPCR-based eDNA assessment. *PLoS ONE*, 11, e0164907.

Vera-Chang, M. N., St-Jacques, A. D., Gagne, R., *et al.* (2018). Transgenerational hypocortisolism and behavioral disruption are induced by the antidepressant fluoxetine in male zebrafish *Danio rerio*. *Proceedings of the National Academy of Sciences USA*, 115, E12435–E12442.

Villeneuve, D. L., Crump, D., Garcia-Reyero, N., *et al.* (2014a). Adverse outcome pathway (AOP) development I: strategies and principles. *Toxicological Sciences*, 142, 312–320.

Villeneuve, D. L., Crump, D., Garcia-Reyero, N., *et al.* (2014b). Adverse outcome pathway development II: best practices. *Toxicological Sciences*, 142, 321–330.

Wang, P. Y., Yan, Z. G., Yang, S. W., *et al.* (2019). Environmental DNA: an emerging tool in ecological assessment. *Bulletin of Environmental Contamination and Toxicology*, 103, 651–656.

Webster, T. M. U., Bury, N., van Aerle, R. & Santos, E. M. (2013). Global transcriptome profiling reveals molecular mechanisms of metal tolerance in a chronically exposed wild population of brown trout. *Environmental Science and Technology*, 47, 8869–8877.

Webster, T. M. U. & Santos, E. M. (2015). Global transcriptomic profiling demonstrates induction of oxidative stress and of compensatory cellular stress responses in brown trout exposed to glyphosate and RoundUp. *BMC Genomics*, 16, e32.

Wiseman, S. B., He, Y., Gamal-El Din, M., *et al.* (2013). Transcriptional responses of male fathead minnows exposed to oil sands process-affected water. *Comparative Biochemistry and Physiology, Part C: Toxicology and Pharmacology*, 157, 227–235.

Xu, E. G., Lin, N., Cheong, R. S., *et al.* (2019). Artificial turf infill associated with systematic toxicity in an amniote vertebrate. *Proceedings of the National Academy of Sciences USA*, 116, 25156–25161.

Yang, B., Qin, C., Hu, X. J., *et al.* (2019). Enzymatic degradation of extracellular DNA exposed to chlorpyrifos and chlorpyrifos-methyl in an aqueous system. *Environment International*, 132, e105087.

Zhang, J., Koch, I., Gibson, L. A., *et al.* (2015). Transcriptomic responses during early development following arsenic exposure in western clawed frogs, *Silurana tropicalis*. *Toxicological Sciences*, 148, 603–617.

Zhang, W., Liu, Y., Zhang, H. & Dai, J. (2012). Proteomic analysis of male zebrafish livers chronically exposed to perfluorononanoic acid. *Environment International*, 42, 20–30.

Zheng, M., Lu, J. & Zhao, D. (2018). Toxicity and transcriptome sequencing (RNA-seq) analyses of adult zebrafish in response to exposure carboxymethyl cellulose stabilized iron sulfide nanoparticles. *Scientific Reports*, 8, e8083.

PART

Toxicology of Individual Substances

6

Metals and Metalloids

Peter G. C. Campbell, Pamela M. Welbourn and Christopher D. Metcalfe

Learning Objectives

Upon completion of this chapter, the student should be able to:

- Distinguish between inorganic and organic contaminants, and explain how the environmental behaviour and toxicology of metals and metalloids differ from those exhibited by organic chemicals.
- Illustrate the differences in environmental behaviour and toxicity within the metals and metalloids class of contaminants and distinguish between essential and non-essential elements.
- Develop criteria that can be used to identify the metals and metalloids that are of the highest priority from an ecotoxicological perspective.
- Describe the links among mechanisms of metal toxicity, cellular strategies for metal detoxification and metal tolerance.

In the preceding chapters, we have considered how plants and animals can encounter toxic substances in their environment, how these substances may be assimilated and how they may cause toxicity. The present chapter is the first of a series of five chapters in which we consider in detail specific classes of toxic substances, the focus here being on metals and **metalloids**. Subsequent chapters will address organic chemicals, both synthetic (Chapters 7 and 8) and natural (Chapter 9), as well as ionizing radiation (Chapter 10).

This chapter deals with metals collectively, considering such factors as their geological origins, their uses, their mobilization and behaviour in the environment, and their interactions with living organisms. In the final section, we examine in detail eight metals or metalloids that are recognized as potential environmental stressors (arsenic, cadmium, copper, lead, mercury, nickel, selenium and zinc). Among this group, readers encounter elements that behave as cations in the receiving environment or as oxyanions; they will also note elements that are biologically essential and others that have no known biological role. The aim here is to help readers appreciate the diversity of behaviours and toxicities within the metals and metalloids group.

6.1 Introduction

Inorganic substances can become environmentally problematic as a result of their mobilization or chemical modification through human activities. In contrast to organic contaminants, most of which are synthetic chemicals, the majority of inorganic contaminants occur naturally. Table 6.1 compares organic contaminants and metals, highlighting differences in their origin, their environmental fate and their interactions with living organisms. As a result of these differences, the strategies and approaches used for management and regulation of inorganic contaminants will differ in many respects from those applied to organic contaminants (Section 12.3.2.5).

Some metals and metalloids form organometallic compounds (e.g. arsenic, selenium, mercury; Section 6.6), and their regulation will call upon concepts that are relevant for both inorganic and organic contaminants.

The primary reference for inorganic contaminants is the Periodic Table of the elements, an extract from which is reproduced in Figure 6.1. Although metals and metalloids dominate the Periodic Table, nevertheless their total number is finite (<90) and very much lower than the number of known organic contaminants. Another point of contrast between inorganic and organic contaminants is that a number of inorganic elements are not only potentially toxic but are also required as nutrients (Section 2.1.1.3; Figure 2.2A). For these elements, depending upon whether they are required in trace amounts, in large amounts or not required at all, the dose–response patterns vary. This concept has been captured as three basic patterns of dose-response, as illustrated in Figure 6.2. In this figure, stimulation is shown as a positive response and inhibition is shown as a negative response, both on the y-axis. The type A pattern, exemplified by a macronutrient such as calcium (Ca), is of little concern in an ecotoxicological context and is not discussed in any further detail in this chapter. The type B pattern, for metals that are required in small amounts but are toxic at higher concentrations, is exemplified by copper (Cu) and selenium (Se) (Sections 6.8.4 and 6.8.8, respectively). For example, very low doses of Cu can result in nutrient deficiency, resulting in suboptimal growth; as the supply increases, up to a certain point, there is a positive response. This is the range over which Cu is required as a micronutrient. After the optimum concentration is reached, there is no further positive response and normally there is a plateau, as shown in the type B pattern. At increasingly higher doses, Cu is in excess and begins to have harmful effects, which cause the growth rate to decrease (i.e., it has reached concentrations that are toxic). The logical result of a further increase in Cu dose beyond that shown in pattern B in Figure 6.2 would be death. The type C pattern, where there is no positive response, is exemplified by mercury (Hg), cadmium (Cd) and lead (Pb) (Sections 6.8.1, 6.8.2 and 6.8.3, respectively).

In the real world, the threshold concentrations in Figure 6.2 are often difficult to determine. The first reason for this hurdle is that the range of concentrations in the medium (water, soil, etc.) that represents the optimum for the biological system is sometimes rather narrow (e.g., for Se). Secondly, and more importantly, the response depends on the chemical form of the element, which affects its biological availability, a term discussed in detail in Section 6.3. A third reason is the inherent variation among organisms of the same species in their relative sensitivity to deficiency or toxicity.

Life on Earth evolved in the presence of all the elements in the Periodic Table, but only a subset was subsumed into living cells (Frausto da Silva and Williams, 2001). A partial list of essential elements would include macronutrients such as Ca, Mg, Na and K, as well as micronutrients such as Co, Cu, Fe, Mn, Mo, Se and Zn. These essential elements are used in living organisms for oxygen transport, for maintaining the required conformation of enzymes and other macromolecules, and as partners in cellular redox chemistry, storing or giving up electrons. The listed essential micronutrients, as well as the much more numerous non-essential metals shown in Figure 6.1, can exert toxicity by a variety of mechanisms (Section 6.4). These mechanisms include displacement of essential metals from enzymes as well as interaction at active sites in enzymes, both of which would inhibit enzymatic activity (through **competitive inhibition**). Metals may also bind to enzymes at sites away from the active site, affecting the enzyme's conformation, again inhibiting its activity (via **non-competitive inhibition**). They may also interfere with gene transcription and in this manner affect enzyme regulation and synthesis. In addition, redox-sensitive metals such as Fe and Cu can exert toxicity by promoting the generation of reactive oxygen species (ROS) within the cell, resulting in oxidative stress.

Much of the work on the ecotoxicology of metals has been carried out in the laboratory, not always under conditions that approach those encountered in the real world. Of particular concern are experiments that have been performed at metal concentrations far in excess of those encountered even in the most contaminated environments. Given the non-linearity of exposure–response relationships, the use of the results of such experiments to predict the effects of metals in the aquatic and terrestrial environments is fraught with difficulty. That said, there are some well-documented field results that do demonstrate the effects of metals in such environments. Examples include the effects of metals on the olfaction of native fish, on the **benthic** communities in lakes downwind from smelters, and on cattle, sheep and horses that consume seleniferous plants. These examples and others are presented in Capsule 6.1 and in the metal-specific sections (6.8) at the end of the present chapter. For

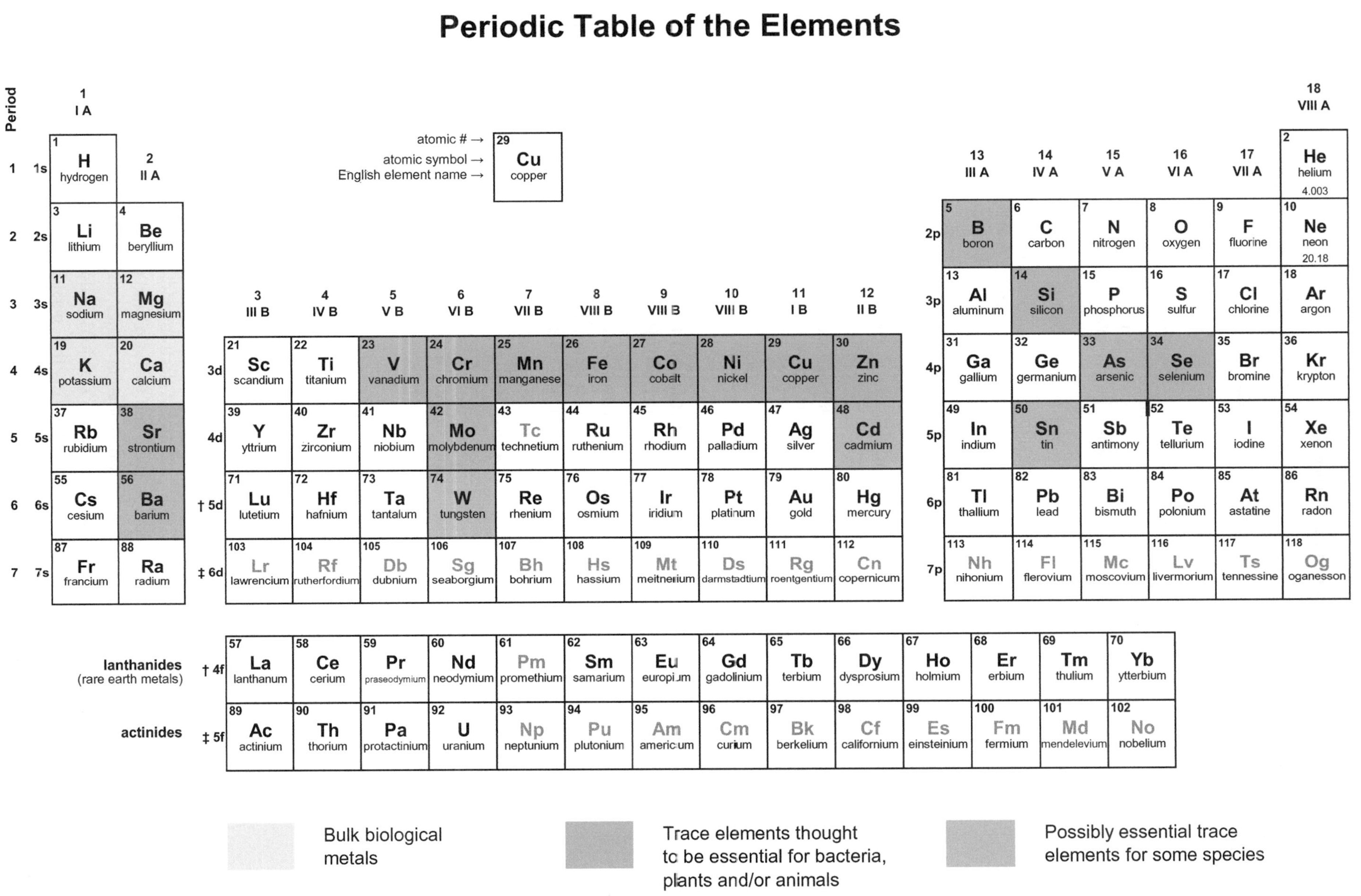

Periodic Table of the Elements

Figure 6.1 Extract from the Periodic Table of the elements, showing the metals and metalloids that are considered to be essential for life. The figure was prepared using an interactive version of the Periodic Table available from MrBigley@MrBigler.com. The designations as 'Bulk biological elements', 'Trace elements thought to be essential' and 'Possibly essential trace elements' are derived from Williams and Rickaby (2012) and Sigel and Sigel (2019).

Table 6.1 Comparison of metals and anthropogenic organic contaminants.

Characteristics	Metals/metalloids	Organic contaminants[a]
Origin	Geogenic; naturally present in the Earth's crust; number limited by the Periodic Table. Background concentration variable.	Absent from the Earth's crust; number without limit and increasing. Background concentration approaches 0.
Environmental fate See Sections 3.5 and 3.6	Not subject to degradation; infinitely persistent. 'Half-life' inapplicable, except for radioactive elements. Speciation changes (reversible). Biomagnification rare (except for organo-metals).	Subject to various degradation processes (hydrolysis, photolysis, biodegradation). 'Half-life' concept useful. Parent compound → derivatives (irreversible change). Biomagnification common.
Interactions with living cells See Chapter 3	Uptake normally takes place by facilitated transport, across a lipid bilayer. Toxicity specific to the metal.	Uptake often takes place by simple diffusion across a lipid bilayer. Narcotic effects common; metabolites may be toxic. See Chapters 7 and 8.
Biological role *(needed to support life)*	Often biologically essential. Bioaccumulation and homeostasis are natural phenomena.	None; complex defence mechanisms against the accumulation of toxic compounds.

[a] As used in this table, the term 'contaminants' refers to organic compounds of anthropogenic origin; it does not include toxins, which are the subject of Chapter 9. The great majority of organic contaminants will be synthetic organic chemicals, but some will fall in a grey area where both anthropogenic and natural sources contribute to the concentrations found in the environment (e.g., polycyclic aromatic compounds, Section 7.6.1).

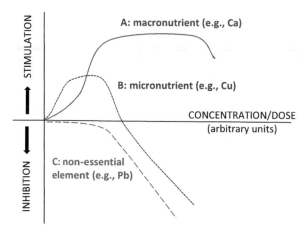

Figure 6.2 Typical dose–response relationships for a required macronutrient, for a required micronutrient and for an element with no known biological function in the test organism.

inclusion in these sections, we have chosen metals and metalloids that are of major concern in ecotoxicology (Hg, Cd, Pb, Cu, Ni, Zn; As, Se). This selection was made not only to illustrate some of the common features among these elements, but also to demonstrate how the biogeochemistry and ecotoxicity of apparently similar metals and metalloids may differ.

6.2 Biogeochemistry of Metals and Metalloids

Prior to discussing the effects of metals on living systems, we address how metals behave in the environment and consider the factors that may affect this behaviour, i.e., their biogeochemistry. In the following two sections, we first introduce the concept of chemical **speciation** and then consider how metals and metalloids interact with the non-living components of aquatic and terrestrial environments.

6.2.1 General Properties: Metal Speciation

Most metallic elements tend to lose their outer electrons, forming positively charged ions (cations). Some elements lose all of their outer electrons, generating monovalent (Na^+, K^+), divalent (Cd^{2+}, Zn^{2+}) or trivalent (Al^{3+}) cations, whereas other elements may lose a variable number of electrons and exist in different oxidation states (Fe^{2+} and Fe^{3+}, Cu^+ and Cu^{2+}). Transformations between two oxidation states are referred to as reduction and oxidation reactions (redox) and are sometimes very slow (leading to the persistence of **metastable species**). A minority of the metallic elements, on the other hand, react with water and form neutral polyhydroxo-species or oxyanions (e.g. As, Mo, Se, Sb, V). These latter elements are often lumped together as 'metalloids' and they share a number of properties: absence of cationic forms; existence as oxyanions; existence of multiple oxidation states. Another commonality among this group of elements is the presence of methylated forms (for As, Sb and Se). As a result, their geochemical behaviour differs markedly from that of the other (cationic) metallic elements. A few elements straddle these two groups (e.g., the chromium cation, Cr^{3+}, and the chromate anion, CrO_4^{2-}).

Metal cations exist in aqueous solution as the free hydrated cation (**aquo ion**), $M^{z+}(H_2O)_n$, and as various complexes; the $z+$ designates the charge on the cation, and n refers to the number of water molecules in the primary hydration sphere surrounding the cation. The formation of these complexes, referred to as metal 'complexation', involves the interaction of the cation with a '**ligand**' present in solution

Hydrated cation or aquo ion

$$M^{z+} + L^{n-} \underset{k_d}{\overset{k_f}{\rightleftharpoons}} ML^{(z-n)+} \tag{6.1}$$

where M^{z+} = the free metal ion with its associated water molecules; L^{n-} = the ligand; k_f = the rate constant for the formation of the complex; k_d = the rate constant for the dissociation of the complex. The formation of the complex in equation (6.1) is shown as an equilibrium reaction. For most complexes of importance in natural waters, the rates of formation and dissociation of the complex are reasonably rapid and the assumption that equilibrium is attained is valid. The relative importance of the free metal ion and the complex will be determined by the ratio of $k_f / k_d = K_C$, the complexation constant, and by the concentrations of the free metal ion and the ligand; higher values of K_C correspond to a greater degree of complexation.

The association of the cation and the ligand may simply result in the formation of an ion pair in solution (referred to as an outer-sphere complex). Alternatively, it may involve the formation of a coordinate bond between the ligand (as an electron donor) and the cation (as an electron acceptor); this more intimate association is referred to as an inner-sphere complex. It is important to distinguish here between metal complexes (with relatively weak *coordinate* bonds linking the metal and the ligand) and organometallic forms (with much stronger *covalent* metal–carbon bonds that are found, for example, in methylmercury, CH_3HgCl).

Examples of inorganic ligands commonly present in natural waters include the anions chloride (Cl^-), fluoride (F^-), sulphate (SO_4^{2-}), thiosulphate ($S_2O_3^-$) and carbonate (CO_3^-) (Figure 6.3). Organic ligands are often present too, derived from natural sources such as phytoplankton or soil leachates and also from anthropogenic sources such as municipal sewage and industrial wastewaters. If the organic ligand is multi-dentate, i.e., if it has two or more donor groups, it may form multiple coordinate bonds with a single central metal cation; the resulting complex is called a **chelate** and the complexation is termed chelation. Chelates tend to be more stable than complexes formed with monodentate ligands. Table 6.2 gives examples of different metal forms that are found in natural surface waters or in the pore waters present in soils and sediments.

Table 6.2 **Examples of metal complexes found in natural surface waters or in soil pore waters.**

Species	Examples[a]
A Free metal ions	$Al^{3+}(H_2O)_6$ $Cu^{2+}(H_2O)_6$
B Hydroxo-complexes	$AlOH^{2+}$, $Al(OH)_2^+$, $Al(OH)_3^0$, $Al(OH)_4^-$ $FeOH^{2+}$, $Fe(OH)_2^+$, $Fe(OH)_3^0$, $Fe(OH)_4^-$ $Cu(OH)_2^0$, $Hg(OH)_2^0$
C Simple inorganic complexes	AlF^{2+}, AlF_2^+ $CdCl^+$, $CdCl_2^0$, $CdCl_3^-$ $HgCl_2^0$, $HgOHCl^0$ $CuCO_3^0$ $CdSO_4^0$
D Simple organic complexes	
(i) synthetic	$Cu–EDTA^{2-}$ $Cd–NTA^-$
(ii) natural	Cd-alanine, Cd-citrate Fe–**siderophore**
E Metals associated with natural dissolved organic matter	Al, Fe, Cu, Pb or Hg – fulvic or humic acids

[a] EDTA = ethylenediaminetetraacetic acid; NTA = nitrilotriacetic acid.

Environmental chemists refer to the distribution of a metal among its various forms as its 'speciation', a term borrowed from biologists (Templeton *et al.*, 2000). In this chapter we adopt the chemists' terminology.

Chemical equilibrium models can be used to calculate how different metals are distributed among their various possible forms. These models take into account the reactions in which metal cations can participate. Examples include complexation (equation (6.1)), oxidation and reduction, and precipitation reactions. In addition, the models can be set up to take into account exchanges of gases such as O_2 and CO_2 between the atmosphere and the aqueous solution. The models are not limited to a single cation, but rather are able to solve the equilibrium equations simultaneously for all the cations and anions in solution. The equilibrium constant for each of the reactions is required to perform these calculations. For many of the common metals, these constants are readily available and are supplied as part of a default thermodynamic database that comes with the software package. Normally these constants are given for dilute solutions and a reference temperature (25 °C). The model then 'adjusts' the constants to account for the true **ionic strength** and temperature of the solution; these two variables affect the activity coefficients of the cations and anions present in solution, and changes in temperature can affect the equilibrium constant itself. To run the model, the total

concentrations of each cation and anion are required, as well as the partial pressure of CO_2 in the gas phase in equilibrium with the aqueous solution. The required input data include those ions that do not interact with the metal of interest but nevertheless affect the overall ionic strength of the solution (e.g., Ca, Mg, Na, K); for most applications, the pH is fixed by the modeller. An example of the output from such a calculation is shown in Figure 6.4 for cadmium, for solutions where the salinity was increased progressively; note that the relative contribution of the free Cd^{2+} ion decreases from about 80% to less than 5% as the salinity increases.

The approach described in the preceding paragraph works well for aqueous solutions of defined composition that contain the common inorganic cations and anions, as well as simple (monomeric) organic ligands (Table 6.2 and Figure 6.3, classes B, C and D). However, systems that contain **natural organic matter** (NOM; class E) pose a major challenge for such models, largely because NOM does not contain a single type of binding site with a well-defined complexation constant (K_C, equation (6.1)). Instead, NOM offers a wide range of metal binding sites, some of which have a high affinity for metals whereas others may bind the metal only weakly (Section 6.7.5). At low concentrations, the metals will seek out and bind to the strongest sites (and the apparent complexation constant will be high). On the other hand, at high metal

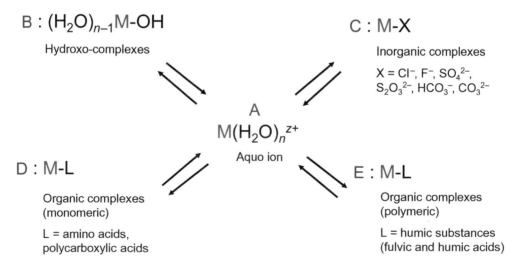

Figure 6.3 Speciation of cationic metals (M) – classes of metal forms found in natural waters. A, B and C = inorganic complexes; D and E = organic complexes.

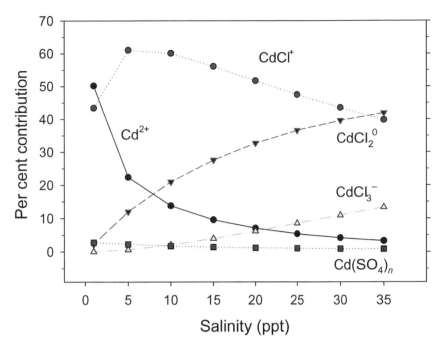

Figure 6.4 Example of the output from a chemical equilibrium model, showing how the speciation of cadmium changes as the salinity (and thus the chloride concentration) increases. Salinity is expressed in parts per thousand (ppt).

concentrations, these high-affinity sites will already be occupied; only low-affinity sites will be available for any additional metal, and thus the apparent complexation constant will be low. Fortunately, this problem was recognized in the 1990s, and several models have been developed and calibrated that are able to predict what proportion of a metal such as Cu^{2+} or Cd^{2+} will be bound to NOM at a given pH and specific metal and NOM concentrations. Examples of such models include:

- Windermere Humic Acid Model, WHAM (Tipping *et al.*, 2011);
- Non-Ideal Competitive Adsorption (NICA)-Donnan model (Milne *et al.*, 2003; Gustafsson, 2019).

Both of these models are widely used in metal ecotoxicology. These models are not yet as reliable as those that consider only synthetic organic ligands such as EDTA (Table 6.2). Nevertheless, even given this caveat, this type of modelling has resulted in respectable progress in providing a chemical basis for understanding the behaviour of metals in a variety of environments.

For a given aqueous solution, each chemical equilibrium model should in principle yield the same results. When this is not the case, the differences in predictions from one model to another can normally be traced to differences in the underlying thermodynamic data (e.g.,

the complexation constant for a particular metal–ligand complex in the database for model 'A' may differ from that in the database for model 'B'). It is important to emphasize that these models are by definition *equilibrium* models (i.e., they do not consider the kinetics of the various reactions). This constraint is usually reasonable for complexation reactions, which occur rapidly for most monovalent and divalent metals. For more information about the kinetics of metal complexation reactions and some of the chemical equilibrium models that are available and their use, please see the Further Reading section online.

The tendency of a metal to form coordinate complexes with various ligands is of importance not only for predicting its speciation in natural surface and pore waters, but also for predicting its **intrinsic toxicity**. Pioneering work in the early 1980s showed that it was possible to predict the toxicity of many metal ions on the basis of their ion characteristics. Among the characteristics that were tested, the relative preference of a given metal for binding to ligands offering oxygen as the donor atom, or to ligands with reduced sulphur as the donor atom, proved very useful. **Thiols** are a good example of this latter class of ligand (R-SH), where R represents an alkyl or other organic constituent and SH is a sulphydryl functional group. Metals with a high affinity for oxygen donor atoms are commonly referred to as 'hard cations', whereas those with a high affinity for thiols, such as Ag^+, Cd^{2+} and Hg^{2+}, are often referred to as 'soft' metals and are characterized by high intrinsic toxicity. A useful predictor of a metal's toxicity is its 'covalent index' or $\chi_m^2 r$, where χ_m is the metal's electronegativity and r is its ionic radius. For updates on this application of quantitative ion character–activity relationships (QICARs), please see the online Further Reading section.

6.2.2 Mobilization, Binding, Transport and Chemical Forms of Metals in the Environment

Metals are released naturally from the Earth's crust by weathering, a term used to describe the action of atmospheric precipitation (rain, snow and their constituents) on rocks and secondary minerals. Some minerals are more prone to weathering than others; for example, under similar weathering conditions, sedimentary rocks such as sandstone, limestone and shale yield greater quantities of solutes than do rocks of igneous origin such as granite or basalt. For a given mineral, under natural conditions, pH, natural dissolved organic matter (DOM) and temperature are the main factors that affect the rate of metal release by weathering. Cationic metal release will normally be favoured by low pH values, warm average temperatures and the presence of natural organic matter. In addition to weathering, metals can also be released by such natural phenomena as wind erosion of arid areas, volcanic emissions and hydrothermal vents on the ocean floor.

Metals are also released as a result of human activities, with evidence of metal release from metallurgical activity as early as 6000 BCE (Smol, 2002). The knowledge that metal concentrations in our rivers, lakes and oceans are influenced by inputs from both natural and anthropogenic sources suggests that the ratio of anthropogenic to natural fluxes of various metals (i.e., the **mobilization** factor) could be used as an indicator of the metals for which the biogeochemical cycle has been the most perturbed by human activities. This approach was pioneered by Nriagu (1979), updated by Klee and Graedel (2004) and used by the United Nations Environment Programme (UNEP) to identify problematic metals – Figure 6.5 (UNEP, 2013). For these data to be useful in an ecotoxicological context, one would need to link the 'mobilization factor' to the 'intrinsic toxicity' of the metal and also take into account the actual metal

concentrations that are present in the Earth's surface environment (Campbell et al., 1985).

Once metals have been released to the aquatic environment, their concentrations in oxygenated natural waters may sometimes be controlled by **precipitation** reactions. For example, at pH values close to neutrality, Al^{3+} and Fe^{3+} tend to form very insoluble metal hydroxides, $Al(OH)_3$ and $Fe(OH)_3$. However, the concentrations of most other metals in such waters are controlled not by precipitation but rather by sorption reactions onto solid surfaces. In the absence of oxygen (e.g., in the anoxic hypolimnion of a stratified lake or in the interstitial water of an organic-rich lake sediment), the solubility of many metals is controlled by their precipitation as sulphides (Hansen et al., 1996).

As a result of the affinity of cationic metals for suspended particles, river, lake and ocean sediments are the ultimate sink for most metals released into natural waters. In other words, metals that reach the sediments tend to stay there, especially in water bodies where O_2 is present in the bottom waters. This tendency proves to be useful in the field of **palaeolimnology** (Smol, 2002), where sediment cores are often used as chronological records of past releases of metals and other contaminants to natural waters (Section 4.3.4). Profundal lake sediment cores, with their stratigraphy maintained, can provide evidence of the mobilization and distribution of metals in the past, up to the present day. An example of this use of sediment cores is discussed in Section 6.8.3, where we show how lead concentrations increased as a result of the industrial revolution and then declined in response to the adoption of environmental regulations, the most dramatic of which was arguably the phasing out of lead in petrol (gasoline) in the mid-1980s.

Most metals released into the terrestrial environment from anthropogenic sources tend to stay there, given the abundance of solid surfaces; a notable exception to this generalization is mercury, which is subject to reduction to Hg^0, a volatile form that can be degassed into the atmosphere (Section 6.8.1.1). Physical erosion of soils will introduce metals into the aquatic environment in particulate form, but the particles introduced in this manner will behave similarly to other **autochthonous** particles and tend to settle to the bottom sediments.

As can be deduced from the preceding discussion, in natural waters metals tend to partition between dissolved and particulate forms. Filtration of the water sample is often used to distinguish between these forms. A sample of

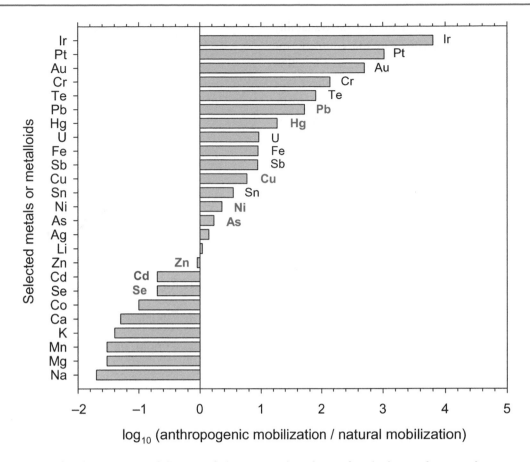

Figure 6.5 Comparison of anthropogenic mobilization of elements to the release of each element by natural processes. The symbols for the metals that are discussed in Section 6.8 are shown in colour (green = essential; blue = nonessential). The figure was prepared using the data compiled by Klee and Graedel (2004) and used with the authors' permission.

water that has been filtered is chemically different from 'whole' water. Even the definition of 'filtered' is not absolute, although most water samples that are described as filtered have been passed through a membrane filter with 0.45-μm pores. This operation sounds simple, but in fact the effect of filtration has been shown to be far more complicated than first appreciated (Horowitz *et al.*, 1996), particularly for trace metals. Known problems include inadvertent contamination of the filtrate (Box 6.1), progressive clogging of the filter pores (which changes their effective pore size), and absorption and retention of water in the filter pores. These phenomena pose problems not only for understanding the true nature of the metal forms present in a particular water body, but also from a regulatory perspective (where permissible metal concentrations in the receiving environment may be expressed as those measured in filtered samples). Note that *in situ* diffusion samplers can be used to minimize the potential contamination and clogging problems associated with filtration (Fortin *et al.*, 2010).

Box 6.1 Beware of pre-2000 measurements of trace metal concentrations in lakes and oceans!

Trace metals often occur in natural waters at microgram per litre (μg/l) concentrations or less. A microgram per litre is equivalent to 1 μg of metal per 10^9 μg of water; a ratio of $1:10^9$ corresponds approximately to three seconds in a century.

In other words, it is very easy to contaminate a water sample inadvertently. Many of the data obtained before 2000 were affected by contamination and should be regarded with scepticism.

6.3 Biological Availability of Metals in Aquatic and Terrestrial Systems

The biological availability of a metal will obviously play a key role in determining whether it causes a beneficial or deleterious effect. However, a metal's bioavailability is not an inherent characteristic. On the contrary, the bioavailability of a particular metal will vary as a function of the ambient conditions (which will affect the metal's speciation) and according to the target organism. In this section we discuss the interplay between these two factors.

6.3.1 General Considerations

In most cases, for a metal to be 'bioavailable' it must be able to cross the epithelial barrier and enter the intracellular environment; the only exceptions would be when there is a metal-sensitive site on the apical surface of an epithelial membrane, i.e., a site that is accessible to the metal from the extracellular environment. Examples of such 'sensitive surface sites' are discussed in the metal-specific section for copper (Section 6.8.4). As indicated in the preceding sections, metals tend to exist in natural waters, sediments and soils as charged species (either cations or anions). Such species cannot traverse a phospholipid bilayer by simple diffusion (Section 3.1). It follows that the biological availability of most metals will depend upon their ability to enter biological cells by **facilitated uptake**, i.e., uptake involving transport proteins embedded in the phospholipid membrane, or via pores or channels (Figure 6.6, mechanism 1). Such transport proteins or channels are designed for the uptake of essential (nutrient) metals; for a non-essential metal such as Ag, Cd, Pb or Hg to enter a cell, it must 'fool' the cell and masquerade as an essential metal (e.g., Cd^{2+} as Zn^{2+}; Cd^{2+} and Pb^{2+} as Ca^{2+}; Ag^+ as Cu^+; see Bridges and Zalups (2005)).

Some organometallic forms such as methylmercury chloride (CH_3HgCl^0) are **lipophilic** and can 'cheat' and diffuse across the phospholipid barrier directly (no carrier involved; Figure 6.6, mechanism 3). Similarly, some neutral metal–ligand complexes can also enter cells in this manner (e.g., ML_n^0 complexes with L = oxine, diethyldithiocarbamate or xanthates). The superscript zero in these chemical formulae indicates that the net charge of the metal form is zero. This is a necessary condition that must be satisfied for a metal form to be lipophilic, but it is insufficient on its own; some metal complexes have no net electrical charge but are still polar owing to a partial separation of charge within the complex (e.g., CH_3HgOH^0), and they remain hydrophilic despite their neutral charge.

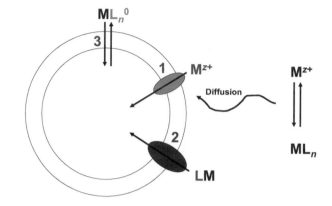

Metal–organism interactions

1. Facilitated transport M^{z+}

2. Facilitated transport LM

3. Simple diffusion ML_n^0

Figure 6.6 Metal–organism interactions (M^{z+} = metal cation; ML_n = metal–ligand complex; ML_n^0 = neutral lipophilic metal–ligand complex; LM = metal bound to an assimilable ligand).

In addition to considering the net electrical charge borne by the metal form, it is also important to take into account its physical state (dissolved versus particulate). Plants (phytoplankton, attached algae, rooted aquatic plants and terrestrial plants) can only assimilate dissolved metals (Section 6.3.4). Foliar uptake of particles is known to occur (Schreck *et al.*, 2012), notably via stomata, but metals associated with such particles must first dissolve if they are to exert any toxicity. Filter-feeders (zooplankton) and deposit-feeders (benthic invertebrates), on the other hand, can ingest particulate metals; as long as the metals remain in the animal's gut, they are not strictly speaking 'assimilated', but if the conditions in the gut favour metal dissolution, they can become bioavailable during their passage through the gastrointestinal tract. The same argument holds for higher organisms (e.g., fish, birds and mammals) (Meyer *et al.*, 2005).

6.3.2 Aquatic Environments: Dissolved Metals

Prior to 1970, studies on the toxicity of metals to aquatic organisms consisted of exposing the test organism to added metal salts in its normal growth medium. Not surprisingly, the results of these experiments tended to be difficult to reproduce. Starting in the 1970s, researchers began to use defined exposure media and to manipulate the experimental conditions (pH, **hardness**, presence of metal-binding ligands, etc.). At the same time, the first chemical equilibrium speciation models were computerized, and it became easier to calculate what chemical forms of the metal were present in the exposure media. François Morel and William Sunda were pioneers in this area.

From these experiments, it became clear that the best predictor of the bioavailability of a dissolved metal was the free metal ion activity, $[M^{z+}]$ (Campbell, 1995). An early example of this type of result is given in Figure 6.7. Copper accumulation by a unicellular marine alga was determined in the presence of different concentrations of TRIS (tris (hydroxymethyl)-aminomethane, $(HOCH_2)_3CNH_2$, a synthetic pH buffer that also complexes metals such as Cu); comparison of panels A and B in Figure 6.7 clearly shows that the free Cu^{2+} concentration was a much better predictor of Cu accumulation than was the total Cu concentration in the exposure medium.

Why is the biological response related to the free metal ion concentration? The explanation of this relationship requires a brief exploration of the chemical equilibria involved (Campbell and Fortin, 2013). To elicit a biological response from a target organism and/or to accumulate within this organism, a metal must first interact with a cell membrane. For hydrophilic metal species, this interaction with the cell surface can be represented in terms of the formation of M–X-cell surface complexes, where X-cell is a cellular ligand present at the cell surface.

In the simplest case, where the hydrated free metal ion is the species reacting at the cell surface, one can envisage the following reactions:

Solution equilibrium

$$M^{z+} + L \underset{\longleftarrow}{\overset{K_C}{\rightleftharpoons}} M-L \qquad (6.2)$$

$$K_C = \frac{[ML]}{[M^{z+}][L]}$$

Surface reaction of M^{z+}

$$M^{z+} + X\text{-cell} \underset{\longleftarrow}{\overset{K_2}{\rightleftharpoons}} M-X\text{-cell} \qquad (6.3)$$

$$[M-X\text{-cell} = K_2[X\text{-cell}][M^{z+}] \qquad (6.4)$$

where K_C and K_2 are conditional equilibrium constants; L is a ligand in solution; charges on complex species are omitted for simplicity; molecules of water held in the metal's coordination sphere are not indicated. It is assumed that the biological response elicited by the metal will be proportional to the concentration of M–X-cell complexes at the cell surface, i.e. [M–X-cell]. This assumption implies that the biological effect occurs at the cell surface or that concentrations of M–X-cell at the cell surface are correlated to the flux of metals that enter the cell and cause toxicity. If the concentration of free X-cell sites does not change appreciably as the metal concentration in solution is varied, equation (6.4) indicates that the biological response will vary directly as a function of $[M^{z+}]$. This response corresponds to the linear

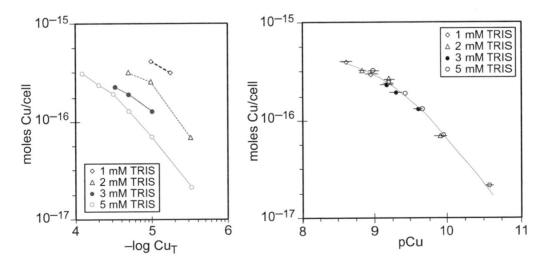

Figure 6.7 Copper bioaccumulation by a marine alga. A: Moles of Cu per algal cell expressed as a function of the negative log of the total Cu concentration in the exposure medium. B: Moles of Cu per algal cell expressed as a function of the negative log of the free Cu^{2+} concentration (pCu) in the exposure medium. Figure redrawn from Sunda and Guillard (1976), with permission from the authors.

portion of the Michaelis–Menten uptake curve described in Figure 3.13. A similar situation prevails if a metal complex (ML) is the species reacting at the cell surface, provided (i) that the reaction proceeds by ligand exchange and (ii) that the metal in the resulting surface complex is bound only to X-cell (i.e., M–X-cell) and water molecules (Campbell, 1995).

In the first formal presentation of this conceptual model of metal–organism interactions, known as the free ion activity model or FIAM, Morel (1983) suggested that "The most important result to emerge is the universal importance of the free-metal ion activities in determining the uptake, nutrition and toxicity of all cationic trace metals." Subsequent results, summarized below, have indicated a need to attenuate this statement, but nevertheless the importance of the free metal ion activity as a predictor of metal bioavailability remains indisputable.

The applicability of the FIAM has been demonstrated for a variety of epithelial surfaces, for both plants (e.g., unicellular algae; aquatic and terrestrial rooted plants) and animals (e.g., fish and invertebrate gills). All the early studies were carried out on marine species in seawater media, where the pH and the concentrations of Ca and Mg were reasonably constant. When the model was subsequently tested on species living in fresh waters, where there is a much greater variability of these key parameters, its performance was less satisfactory. This problem was resolved when it was realized that the metal-binding site on the epithelial surface, i.e., X-cell, could also bind to the hardness cations, Ca^{2+} and Mg^{2+}. In other words, high hardness could afford some protection against the binding of the metal ion, M^{z+}, to the epithelial surface. Indeed, long before the formulation of the FIAM it was known that many cationic metals were less toxic when tested in hard waters. The FIAM was reformulated during the 1990s to account for the protective effect of Ca^{2+} and Mg^{2+}. This evolution corresponds to its transition to the **biotic ligand model** (BLM), where the focus shifted from the free metal ion in solution to include the binding site on the epithelial membrane. In its present incarnation, the BLM accounts for the roles of the hardness cations, the pH (concentration of the proton, H^+) and metal cation binding by organic ligands such as natural DOM (Paquin et al., 2002).

As depicted in Figure 6.8, the BLM focuses on the interface between a living organism and its environment. The left-hand side of the figure depicts the various forms of metal M that are present in the ambient solution (Section 6.2.1). Note that we have attributed a positive charge of z to the cation, where z could be 1, 2 or 3; the charges of the inorganic complexes will depend upon

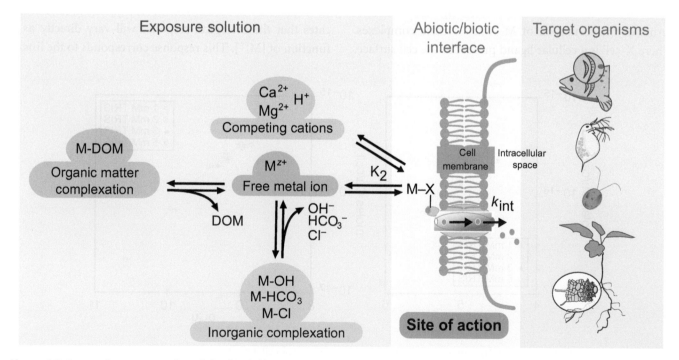

Figure 6.8 A generic representation of the biotic ligand model depicting the speciation of the metal in the ambient solution, its interaction with a cell membrane (forming the M–X-cell surface complex) and its internalization (k_{int}).

the value of z, but for simplification these charges have been omitted. The various reactions are assumed to be at equilibrium, including the reaction of the free metal ion with the cell membrane (equilibrium constant K_2, as defined in equation (6.3)). Both the free metal ion and metal ion complexes can react with the binding site labelled X on the apical surface of the cell membrane (X being the 'biotic ligand'), but only the metal cation can cross the membrane and enter the cytosol (k_{int}, which represents the internalization rate constant for the metal of interest). The proton (H^+) and the hardness cations (Ca^{2+} or Mg^{2+}) present in solution can compete with the metal for binding to the biotic ligand and in this manner reduce metal uptake. The right-hand side of Figure 6.8 portrays a selection of organisms for which the BLM has been shown to apply (fish, zooplankton, algae, rooted plants).

There are a number of implicit assumptions in the BLM construct:

1. The metal is present in the exposure solution as the free cation and its complexes.
2. An equilibrium is established between metal species in the bulk solution and those at the biological surface (equation (6.3)). In other words, the organism exposed to metals (or at least its membranes in contact with water) is treated as another ligand for metal binding, with its own metal binding characteristics.
3. Metal transport in solution towards the epithelial membrane and the subsequent surface complexation reaction, yielding M–X-cell, occur rapidly.
4. The biological response (e.g., metal uptake, metal-stimulated growth or metal-induced toxicity) is dependent on the concentration of the M–X-cell surface complex, [M–X-cell].
5. Variations of [M–X-cell] as a function of [M^{z+}] in solution follow a Michaelis–Menten relationship; provided the concentration of free sites, [X-cell], remains relatively constant in the range of metal concentrations of interest, variations in [M–X-cell] will follow those of [M^{z+}] in solution (equation (6.4)).
6. During exposure to the metal of interest, the nature of the biological surface remains constant (i.e., the metal does not induce any changes in the nature of the plasma membrane or its ion transporters).

If these assumptions are met, then at constant pH and constant hardness, the biological response of the exposed organism should vary as a function of the free metal ion activity in the exposure solution. Indeed, the vast majority of experiments that have been conducted to verify the applicability of the BLM have shown that this is the case. There are, however, some exceptions to this rule, as explained in the next paragraph.

Known cases where BLM predictions of metal uptake or toxicity have proved to be inaccurate have been grouped into a limited number of classes (Campbell and Fortin, 2013). These are cases where one of the following applies:

1. The rate of diffusive metal transport from the bulk solution, towards the cell membrane, is slower than the rate at which the cell can assimilate the metal (Degryse and Smolders, 2012); in such cases (which are rare), labile metals other than the free metal ion can contribute to the rate-limiting flux.
2. Metal uptake occurs by simple diffusion of intact lipophilic metal complexes across the cell membrane (Phinney and Bruland (1994) – Figure 6.6, mechanism 3).
3. The metal forms a stable complex with an assimilable ligand, and metal uptake involves the transmembrane transport of the intact metal complex rather than the metal cation (Fortin and Campbell (2001) – Figure 6.6, mechanism 2).
4. The metal exerts its toxic action by binding to sites present at the membrane surface (i.e., membrane transport is not a necessary condition for metal-induced toxic effects to occur) (Wilkinson et al., 1990).

As outlined up to this point, the BLM is a chemical equilibrium construct with limited biological content. Many of the original experiments that contributed to the development of the BLM were carried out at quite high but still reasonable metal concentrations and for relatively short exposure periods; under such conditions, the assumption that the biological surface does not change as a result of metal exposure is probably realistic. However, for longer-term and non-lethal exposures, there is good evidence that exposed organisms can acclimate to metal exposure (Section 6.5.4). In some cases, this acclimation involves changes to the epithelial membrane and to the metal-binding sites, X-cell (Niyogi et al., 2004); such changes are not accounted for in current versions of the BLM. Similarly, but at the other end of the metal exposure scale, the BLM does not apply to scenarios where the metal is present at such high concentrations that toxicity is characterized by destruction of (gill) epithelial cells, and the organism dies because O_2 and/or CO_2 transfer are impaired.

In natural systems, metals are often present as components of metal mixtures. When more than one metal is present in an exposure scenario, they may interact in solution (e.g., competing for the same dissolved ligands), at the biological surface (e.g., competing for binding at the epithelial surface) or in the intracellular environment. The BLM is inherently well suited to deal with such interactions in the exposure solution and at the biological surface. In a real-world example of metals competing for the same ligands in solution, Hoppe *et al.* (2015) added aluminium (Al) to a natural water sample and showed that this addition increased Cu toxicity towards the water flea, *Daphnia magna*; at the tested concentrations, Al alone was not toxic to the test invertebrate. The authors concluded that the added Al reacted strongly with the natural ligands present in the water sample, displaced some of the previously bound Cu and increased the free Cu^{2+} concentration in solution.

At the biological surface, the BLM can be set up with one or several binding sites. A single binding site implies that the different metals will compete for a common route of entry to the organism; at low metal concentrations, this competition will be negligible, and the metal toxicity will be additive. However, as the metal concentrations are increased, competition at the point of entry will result in toxicity becoming less-than-additive. On the other hand, if the metals in the mixture have separate routes of entry, competition no longer exists, and toxicity will be additive regardless of the metal concentrations. Recent research on the toxicity of metal mixtures to individual organisms suggests that additivity best describes the toxicity of most metal mixtures; examples of synergistic or more-than-additive responses are rare (Van Genderen *et al.*, 2015; Liu *et al.*, 2017).

6.3.3 Aquatic Environments: Particulate Metals

To this point in the discussion of the bioavailability of metals in the environment, we have focused on the aquatic environment and the water column, but what do we know about metals that have reached aquatic sediments? The sediment–water interface is important in ecology; it plays an important role in controlling fluxes of nutrients and contaminants into and out of the sediment, and it supports a very diverse benthic fauna. As mentioned earlier (Section 6.2.2), metals tend to end up in sediments. The total metal concentration in a lake sediment, for example, can serve as an indication of the degree to which the lake has been 'contaminated' by metals, irrespective of the source: atmospheric deposition, weathering of rocks in the lake watershed or industrial effluents. However, the total metal concentration in a lake sediment is not a good predictor of whether the benthic organisms living there are likely to be affected by the sediment-associated metals. Healthy benthic communities with high species richness are sometimes found at extremely contaminated sites (Section 4.3.3.4).

The disconnect between the total concentration of a metal in a sediment and its bioavailability stems largely from the fact that metals in sediments are partitioned between the sediment pore water (also referred to as sediment interstitial water) and the various solid phases present in the sediment. The biological availability of a dissolved metal is inherently much higher than that of a metal bound to or within a solid particle. It follows, unsurprisingly, that metal concentrations in sediment interstitial water ($[M]_i$) are a good predictor of metal bioavailability to sediment-dwelling organisms. In effect, porewater concentrations reflect the metal's chemical potential at the sediment–water interface; changes in this chemical potential will affect the metal's bioavailability.

Two approaches can be used to estimate $[M]_i$: one applies to *oxic* conditions and assigns control of $[M]_i$ to sorption reactions on such sorbents as Fe-, Mn-oxyhydroxides or sediment organic matter (Campbell and Tessier, 1996); the second applies to *anoxic* conditions and assumes that $[M]_i$ is controlled by precipitation-dissolution reactions with reactive amorphous sulphides (acid-volatile sulphides, AVS). This second approach is often referred to as the AVS-SEM model, where SEM stands for simultaneously extracted metals. Provided that the molar value for SEM is less than that for AVS, the model predicts negligible toxicity (Hansen *et al.*, 1996). The two approaches differ in their choice of reactions controlling metal solubility in sediment pore waters (Box 6.2). This divergence stems from different concepts of what constitutes a biologically important sediment. In fully oxidized, surficial sediments, where amorphous sulphide levels should be vanishingly low, $[M]_i$ should be controlled by sorption reactions. On the other hand, in partially oxidized, suboxic sediments, where significant AVS levels would be expected to persist, exchange reactions with amorphous sulphides would control metal partitioning between dissolved and solid phases.

Box 6.2 Equilibrium partitioning models in sediments: sorption or precipitation?

Since the turn of the century, views regarding the relative importance of oxic versus anoxic control of porewater metal concentrations have oscillated. However, recently the pendulum has settled on a more nuanced central vision of what controls metal cation concentrations in sediment pore water, with an appreciation that both sorption and precipitation reactions can intervene as the oxygenation of the surface sediment changes (e.g., seasonally or as a function of drought) and as the sediment ages.

Note that an equilibrium partitioning approach is also commonly used to evaluate the concentration of organic contaminants in sediment pore water (Burgess et al., 2013).

Because of small-scale spatial heterogeneity in natural sediments, the distinction between oxic and anoxic sediments is often blurred, and choosing between the two approaches is not at all straightforward. Most aerobic benthic organisms survive in sediments that are underlain or even surrounded by anaerobic material, which constitutes a potential source of AVS. Note too that seasonal changes in lake oxygen concentrations associated with thermal stratification (summer) or under winter ice cover may also affect the redox conditions in the bottom sediments. Finally, in a given sediment, metal bioavailability may also vary from one sediment-dwelling organism to another, as a function particularly of their respective burrowing or feeding behaviour. Additional information about the toxicity of metals associated with aquatic sediments can be found in Section 2.2.8.2.

6.3.4 Terrestrial Environments

Moving from the aquatic environment onto the land surface, we will now consider the biological availability of metals in the terrestrial environment. Metals in soils may be present as dissolved metals in soil pore water, as metals adsorbed on soil particles (clays, iron and manganese oxides, organic matter) and as metals held within soil particles. For microorganisms and rooted plants, metals will be taken up from the soil pore water. In contrast to the situation for an aquatic sediment, pore water in soils does not form a continuous medium, so

that diffusive and convective transport to the organism surface may limit metal uptake. For plants, root growth (leading to the root encountering fresh surfaces) and the creation of root microenvironments (e.g., exudates present in the **rhizosphere**) also have to be considered. For soil invertebrates, metal uptake may occur from soil pore water, through their external epithelium, or from ingested soil particles (in which case digestive processes within the gut and gut passage time may be important – Section 3.2.4).

In common with the situation in sediments, metal concentrations in soil interstitial water are recognized as good indicators of the metal's bioavailability. Intuitively, one might expect that an equilibrium partitioning approach, analogous to that described earlier for estimating porewater metal concentrations in sediments, might be applicable to soils (Weng et al., 2001). However, often this is not the case, presumably because the assumption of equilibrium is invalid. In soils, the resupply of a metal from the solid phase (via kinetically controlled desorption and/or dissolution reactions) can greatly influence its bioavailability. In addition, in contrast to the situation for aquatic sediments and wetlands, terrestrial soils are only rarely saturated with water, and their water content is typically characterized by an important temporal variation (linked to antecedent rainfall) and spatial heterogeneity (linked to changes in topography and soil composition).

Largely for these reasons, soil scientists often use various chemical extraction techniques to estimate the concentrations of metals that are readily exchangeable and thus the most likely to enter the soil interstitial water and be 'phytoavailable' or 'bioavailable'. These extractions were originally used to assess the (micro)nutrient status of soils and to determine the need for addition of fertilizer. Indeed, the use of partial extractants to improve predictions of the availability of metals to organisms has been an active research area in soil science for more than 80 years, with much of the early work aiming to predict the availability of essential metals for plants. Several reviews detailing the development of these methods and their use in predicting availability of metals in soils can be found in the online Further Reading list. Generally, much more information about metal bioavailability in soils is available for plants than for soil microorganisms or invertebrates.

Typical extractants are shown in Box 6.3. The weaker salt extractants (e.g., those based on chloride or nitrate

Box 6.3 Typical solutions used for the partial extraction of metals from soils

- Neutral salts (e.g., $CaCl_2$, $NaNO_3$), designed to remove metals present in the soil pore water and to desorb electrostatically adsorbed metals.
- Dilute acids (e.g., HCl), to desorb metals specifically adsorbed to Fe and Mn oxides.
- Reducing agents (e.g., hydroxylamine, dithionite), to solubilize metals held within Fe and Mn oxides.
- Chelating agents (e.g., EDTA, DTPA), to compete with metal-binding sites on the surface of soil solid phases and extract the metal.
- Oxidizing agents (e.g., hydrogen peroxide), to solubilize metals bound within solid organic matter.

salts) and diffusion-based tests (e.g., diffusive gradients in thin films, DGT; Marrugo-Madrid *et al.* (2021)) tend to be better correlated with biological responses than are the fractions extracted with stronger chelating extractants (e.g., diethylenetriaminepentaacetic acid, DTPA). Some success has been achieved with the partial extraction approach, usually within narrow ranges of soil types and for specific biological targets. However, these relations between 'extractible metal' and 'bioavailable metal' tend to break down when the protocols are applied to a wide range of soils.

In recent years, the extraction approach has been extended to higher metal concentrations and the investigation of metal toxicity thresholds in soils. In such studies, toxicity tests are carried out by spiking clean soils with a single metal, allowing the metal to equilibrate with the soil and then determining the effects on soil microbial processes, rooted plants and soil invertebrates. Increases in soil pH, clay content, organic matter content or **cation exchange capacity** (CEC) have all been found to mitigate toxicity of cationic metals to soil organisms, presumably because all such increases will promote stronger metal binding to the soil solids.

Metals found in smelter- and mine-impacted soils often originate from smelter emissions, from the aerial deposition of windblown tailings, or from the transformation of solid mine wastes (Section 11.2). Such soils often contain distinctive metal forms that are unlike those encountered in natural soils, or even in soils that have been amended with soluble metal salts or biosolids from sewage treatment operations. For example, work on soils around the copper and nickel smelters in Sudbury, Ontario, Canada has demonstrated the presence of distinctive metal-containing particles, the morphology of which clearly suggests a pyrometallurgical origin. Metals within such smelter-derived particles tend to be quite stable in the soil environment, and their contribution to overall metal bioavailability will normally be low. For example, Degryse *et al.* (2004) reported that exchangeable Zn was about 2-fold lower in such soils than in soils where Zn^{2+} salts were added and then aged. Similar results were demonstrated for exchangeable ^{63}Ni in a collection of 100 soil samples representing a worldwide range of conditions and contrasting Ni sources (Echevarria *et al.*, 2006). To assess metal bioavailability in these cases, i.e., for metals of pyrometallurgical origin, it is useful to compare the 'production' function (the rate at which the metal in a smelter-derived recalcitrant form is solubilized) with the 'capacity' function (the ability of the soil to bind the solubilized metal and keep it out of the soil pore water). Metals that enter soil solution as a result of chemical transformation of the original contaminant source in the soil are subject to interactions with other components of soil, the most important being the soil solid phases. In the specific case of smelters dealing with metal sulphide ores (Section 11.2), the acidic emissions from the smelting process will favour mobilization of the smelter-derived metals that contaminate the surrounding soils.

6.3.5 Diet-borne Metals

The preceding discussion of particulate metals leads naturally to a consideration of the uptake of metals from food. Note here that for organic contaminants, researchers tend to distinguish between **bioconcentration** (i.e., uptake from water or air) and **bioaccumulation** (i.e., simultaneous uptake from both food and water or air; Section 3.4). However, this distinction is not rigorously respected by researchers in the metals sector, where it is more common to distinguish between 'waterborne' and 'diet-borne' metals.

Until fairly recently, the uptake of diet-borne metals was neglected by researchers (with the notable exception of Hg and Se, where the importance of the diet was recognized early on; e.g., Adams *et al.* (2005)). An important catalyst for the expanded interest in diet-borne metals was a publication by Fisher and Hook (2002), who reported that when marine copepods were exposed to metals (Ag, Cd, Hg, Se, Zn) through their diet, their reproductive capacity decreased by up to 75%; fewer eggs were produced, and the hatching success was lower for the eggs that were produced. These sublethal effects occurred at metal concentrations that were two to three orders of magnitude lower than dissolved metal concentrations that were acutely toxic. The authors suggested that metals accumulated in copepods by trophic transfer reach internal tissues, where only modest increases above background levels might depress reproductive capability. In their experiments, metals that accumulated from the dissolved phase were deposited primarily on external surfaces, where they had negligible biological effects at environmentally realistic concentrations. This distinction between the accumulation of waterborne and diet-borne metals is not surprising for Hg and Se, but is atypical for Ag, Cd and Zn.

Digestive physiology varies considerably among different animals, both in terms of the prevailing chemical conditions in the digestive tract (e.g., pH, redox potential) and the length of time the food spends in the tract (Campbell et al., 2005a). For example, conditions in the digestive tract of zooplankton are not greatly different from those in the ambient water, and the literature suggests that zooplankton only access the metals that are present within their prey in dissolved form. For deposit-feeding benthic invertebrates, most of the available information comes from studies on marine organisms (e.g., polychaetes). Measurements of *in vivo* fluid properties (redox potential, pH, dissolved oxygen) in the guts of such deposit feeders suggest that the pH is close to neutrality and that the fluid in the gut is anoxic (Weston et al., 2004). In contrast, digestion in higher animals is much more destructive, and the passage time in their gastrointestinal tract is longer than in zooplankton or small deposit feeders. Accordingly, the assimilation efficiency for the metals present in a prey item will vary from one prey item to another, as well as from one consumer to another. Negative effects of ingested metals may occur in the digestive tract itself (e.g., effects on the intestinal membrane or on the intestinal microbiome; Clearwater et al. (2005)). Alternatively, the metals may enter the animal's circulatory system after traversing the intestinal membrane barrier and thereby reach other internal organs.

It is beyond dispute that metals are subject to trophic transfer and that they can move up aquatic and terrestrial food webs as a result of uptake from the diet. However, what is of particular interest in an ecotoxicological context is whether this trophic transfer results in increasing metal concentrations as one moves up the food web. This phenomenon is commonly referred to as **biomagnification** and among metals is known to occur for methylmercury (Sections 3.5 and 6.8.1.5).

Trophic transfer factors (TTFs) are used to describe the relationship between the steady-state metal concentrations in an organism and the metal concentrations in its ambient environment (i.e., in water and food). They can be calculated for single or multiple transfers (steps) in a food chain, with each step representing a consumption event (grazing or predation). Biomagnification occurs when the TTF is >1.0 through three or more trophic levels, i.e., as a result of at least two trophic transfers (Cardwell et al., 2013). Using both laboratory and field data, these authors reviewed published data on single and multiple trophic transfers for five metals, three that are biologically essential (Cu, Ni, Zn) and two non-essential elements (Cd, Pb). Not surprisingly, trophic transfer factors in a given food chain proved to be metal-specific, and furthermore the TTF for a given metal varied from one food chain to another. For all the studied metals, however, there was an inverse relationship between the TTF and the actual concentration of the metal in the exposure medium (water, sediment, soil). Most importantly, Cardwell et al. (2013) concluded that Cd, Cu, Ni, Pb and Zn "generally do not biomagnify in (freshwater) food chains consisting of primary producers, macroinvertebrate consumers and fish". In fact, the tendency was rather for 'biominification', with TTFs < 1. In other words, although some trophic transfer occurs for all metals and metalloids, biomagnification only occurs for a very limited number of elements (e.g., Hg, in its methylated form).

6.4 Mechanisms of Metal Toxicity

In this section, we focus primarily on the mechanisms of toxicity of non-essential metals such as cadmium, lead and mercury. Essential metals such as copper, nickel and zinc can also exert toxicity if their concentrations exceed the levels that are required for normal metabolism (an example of the "dose makes the poison" rule mentioned in Chapter 2). However, as discussed in Section 6.8, cellular regulation and detoxification mechanisms for these essential cations are normally able to control their intracellular availability and limit but not remove the potential for toxicity.

A review of the literature on the mechanisms by which non-essential cations induce toxicity in organisms can be bewildering, given the many ways that metals can be harmful to plants and animals (Wu *et al.*, 2016). However, in this section we are concentrating on the action of metals at the molecular level. This idea is presented in Chapter 5, where we introduce the **adverse outcome pathway** (AOP) approach and define what are known as **molecular initiating events** (MIEs). For metals, the generic MIEs can be summarized under the following headings:

- alteration of enzyme conformations;
- displacement of essential cations;
- generation of oxidative stress;
- effects on cellular differentiation.

Some examples of these different MIEs are presented below. Later in this chapter (Section 6.8), where we consider and compare individual metals, we have included a text box for each metal to summarize the adverse outcomes (i.e., symptoms or signs of toxicity) that characterize the toxicity induced by each metal.

6.4.1 Alteration of Enzyme Conformation

The three-dimensional structure of an enzyme, i.e., its conformation, is maintained by intramolecular interactions within the polypeptide backbone of the protein and between its side chains. These interactions may include covalent bonds between sulphydryl groups (-SH) associated with the amino acid cysteine. Soft metals and some metalloids can bind to these sulphur groups and thus alter the enzyme's conformation (shape) so that it no longer retains its structure and so loses its enzymatic activity. This type of interaction, shown schematically in Figure 6.9, is known as allosteric inhibition. Metals can also inhibit enzyme activity by binding directly to functional groups located within the enzyme's active site.

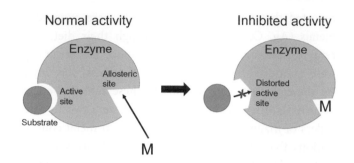

Figure 6.9 Illustration of how the binding of a metal (M) far from an enzyme's active site can induce a conformational change and render the enzyme inactive.

Various toxic metals preferentially affect different enzyme systems in various organs, thereby inducing unique toxicological effects. For example, in aquatic animals that rely upon their gills for respiration, ion regulation is maintained primarily through the activity of Na^+/K^+-ATPase enzymes associated with the gill membrane. Disruption of the activity of these enzymes through exposure to toxic cations, such as Ag^+, Cd^{2+} and Pb^{2+}, can lead to osmoregulatory imbalance and **acidosis** in such organisms (Henry *et al.*, 2012).

6.4.2 Displacement of Essential Cations

Zinc is an important enzyme 'cofactor' that helps to maintain the activity of a very wide range of cellular enzymes. For example, Zn is a cofactor that is required for the activity of delta-aminolaevulinic acid dehydratase (ALAD). Exposure to lead inhibits the activity of this enzyme by displacing the Zn. The inhibition of ALAD or the resulting increase in the levels of delta-aminolaevulinic acid (ALA) can be used as a **biomarker** of Pb exposure in organisms (Section 6.8.3). Since ALAD catalyses an essential step in the synthesis of haem and also the synthesis of chlorophyll, inhibition of this enzyme can lead to anaemia in animals that produce haemoglobin as a respiratory metalloprotein and reduced photosynthesis in all green plants.

Essential cations are also constituents of many biological molecules. In some examples, non-essential cations have a higher affinity for the binding sites within these biomolecules than do the essential cations. For example, the structure of chlorophyll includes a chlorin ring with a central 'core' of four nitrogen atoms to which

Cysteine

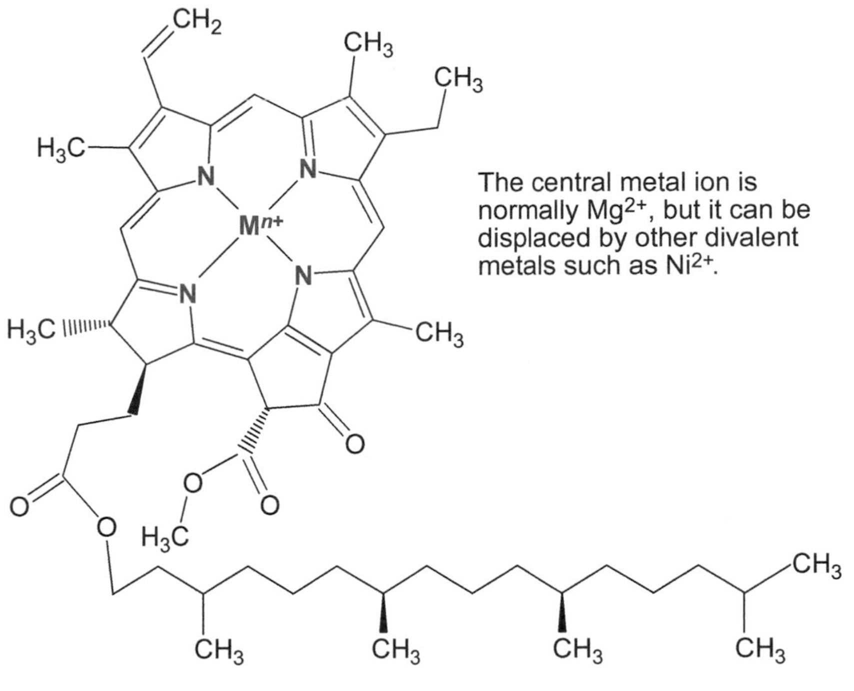

Figure 6.10 The structure of chlorophyll-*a*, which includes a chlorin ring, with four nitrogen atoms surrounding a central magnesium atom.

is bound the magnesium cation (Figure 6.10). The Ni^{2+} ion has a higher affinity for nitrogen than Mg^{2+} and so can displace the magnesium ion and reduce the photosynthetic capacity of exposed plants (Section 6.8.5).

6.4.3 Oxidative Stress

All aerobic organisms are subject to oxidative stress because of the cellular generation of reactive oxygen species (ROS), such as hydrogen peroxide (H_2O_2), the superoxide radical ($O_2^{\bullet}$) and the hydroxyl radical ($HO^{\bullet}$). For example, the process of oxidative phosphorylation is driven by the transfer of free electron pairs along a series of respiratory cytochromes. However, electrons sometimes 'escape' and are released into the cytoplasm, where they can react with water to produce ROS. Although ROS play an essential role in various cellular processes, an imbalance in endogenous ROS levels adversely affects the oxidation of lipids and membranes and can cause RNA and DNA lesions. Oxidative DNA and RNA damage can interfere with cellular replication and gene transcription, and lead to genetic mutations and chromosomal instability. Aerobic organisms have developed enzymatic mechanisms and rely on cellular antioxidants (e.g., Vitamin E) to

counteract this natural process. However, if the production of ROS overwhelms the capacity of the cell to respond, the resultant 'oxidative stress' can lead to oxidative damage to lipid membranes, enzymes and nucleic acids, leading to cell death. Oxidative damage to DNA may be a mechanism by which some metals cause cancers.

Metals can induce oxidative stress through a variety of mechanisms. Redox-sensitive transition metals such as Fe(III) or Cu(II) can be reduced to ferrous or cuprous ions, respectively, which results in the release of electrons that generate $^{\bullet}OH$ free radicals by the **Fenton reaction** (equations (6.5) and (6.6)).

$$Fe^{2+} + H_2O_2 \rightarrow Fe^{3+} + HO^{\bullet} + HO^- \qquad (6.5)$$

$$Cu^+ + H_2O_2 \rightarrow Cu^{2+} + HO^{\bullet} + HO^- \qquad (6.6)$$

Since the cellular levels of iron and copper are usually well regulated, this typically occurs only under very high exposure scenarios. Cadmium itself cannot participate in redox reactions, but nevertheless exposure to cadmium consistently leads to signs of oxidative stress (Section 6.8.2). It has been suggested that Cd induces oxidative stress not by getting involved in the Fenton reaction, but rather by inhibiting various enzymes that are involved in the cellular defence against oxidative stress (e.g., superoxide dismutase and the catalase enzyme that destroys H_2O_2). Another mechanism that has been put forward involves the displacement of Cu and Fe from various cytoplasmic and membrane proteins by Cd (Waisberg *et al.*, 2003); the release of ferric and cupric ions in unbound form into the cytoplasm could promote oxidative stress.

Metals can also induce oxidative stress by disrupting the capacity of the peptide glutathione to counteract the generation of ROS. The cycling of glutathione between the reduced form (GSH) and oxidized form (GSSG) is catalysed by the activity of glutathione reductase and glutathione peroxidase, in tandem with the activity of glucose-6-phosphate dehydrogenase (Figure 6.11). The resultant scavenging of electrons through the reduction of H_2O_2 and other reactive species controls oxidative stress. However, inorganic contaminants can disrupt this process through various mechanisms. For example, As(III) induces oxidative stress because it is a potent inhibitor of glutathione reductase (Section 6.8.7), whereas methylmercury can interact directly with glutathione, deplete cellular levels of GSH and perturb the GSH-based antioxidant system (Farina and Aschner, 2019) – Section 6.8.1.

As the reader can judge from the foregoing discussion, metal-induced oxidative stress is a common occurrence. For example, Szivak *et al.* (2009) exposed the freshwater alga *Chlamydomonas reinhardtii* to a suite of seven redox active and redox inactive metals (Ag, As, Cd, Cr, Cu, Fe, Pb, Zn), five of which are discussed in Section 6.8. The exposure concentrations were environmentally realistic and did not affect algal photosynthesis. However, intracellular ROS concentrations did increase, with the maximal ROS induction decreasing in the order Pb(II) > Fe(III) > Cd(II) > Ag(I) > Cu(II) > As(V) > Cr(VI) > Zn(II); As(III) and Cr(III) had no detectable effect. This demonstrated ability of many different metals to induce oxidative stress has an unfortunate consequence, in that the detection of this sign of toxicity cannot be used as a metal-specific biomarker of exposure or effects.

6.4.4 Changes to Cellular Differentiation

Epidemiological studies involving occupational exposures to arsenic, nickel, cadmium and hexavalent chromium have demonstrated that these metals are carcinogens (Chen *et al.*, 2019). However, these elements do not cause mutations and therefore do not cause cancers through the classical carcinogenesis model of accumulated mutations to oncogenes and tumour suppressor genes. There is increasing evidence that chronic metal exposure can alter the differentiation of stem cells to produce cancer stem cells that have the capacity to proliferate and develop into malignant tumours (Wang and Yang, 2019). The molecular mechanisms for altered differentiation of stem cells may be dysregulation of gene expression. Epigenetic alterations in gene expression are changes in gene expression that occur regardless of the sequence of nitrogenous bases in DNA. These alterations include histone modifications, methylation of nitrogenous bases and microRNA regulation (Wu *et al.*, 2016) (Section 5.5).

6.4.5 Behavioural Effects

To conclude this section on mechanisms of metal toxicity, we expand our horizon to include behavioural effects. Capsule 6.1 focuses on the effects of metals on fish olfaction, this being one of the better examples of how metals can interfere with the chemical signals that enable aquatic animals to detect their prey and to identify an appropriate reproduction partner.

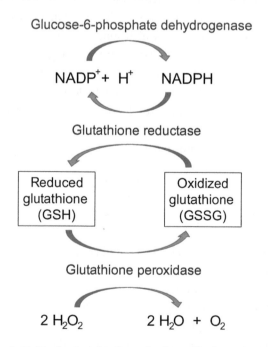

Glucose-6-phosphate dehydrogenase

$NADP^+ + H^+$ NADPH

Glutathione reductase

Reduced glutathione (GSH)   Oxidized glutathione (GSSG)

Glutathione peroxidase

$2 H_2O_2$ $2 H_2O + O_2$

Figure 6.11 Mechanism for the reduction of hydrogen peroxide through the cycling of glutathione from its reduced form (GSH) to its oxidized form (GSSG) in association with various catalysing enzymes.

Capsule 6.1 Metal Effects on Fish Olfaction

Gregory G. Pyle

Department of Biological Sciences, University of Lethbridge, Alberta, Canada

Ecology of Fish Olfaction

Fish perceive their environment in a much different way than humans. Humans rely more on vision to perceive environmental cues at a distance than fish do because light can more readily transmit through air than water. Where visual acuity is inhibited, a more reliable way for aquatic animals to perceive their environment is to detect dissolved chemicals. The sensory perception of chemicals – chemosensation – includes olfaction (the sense of smell) and gustation (the sense of taste). Here, we will focus on olfaction.

Water can dissolve more substances than any other naturally occurring liquid. In natural aquatic ecosystems, some of the dissolved substances include organic matter from dead and decaying biomass, ions that contribute to water hardness or salinity, or other information-containing molecules (or 'infomolecules') that convey important information about an animal's immediate surroundings. The amount of information conveyed to an aquatic animal by these infomolecules is impressive. Simply on the basis of their sense of smell, fish can detect predators in their immediate environment, locate potential food sources, navigate migration routes, and evaluate the fitness and appropriateness of potential mates (Figure 6C.1). Anything that can impair a fish's ability to perceive these important cues has the potential to cause significant ecological effects.

Effects of Metals on Fish Olfaction

Dissolved metals are well known as powerful neurotoxicants. In many early studies that examined metal effects on fish olfaction, researchers used metal concentrations much higher than those that might be observed under natural conditions. More recently, however, several studies have exposed fish to environmentally relevant concentrations to determine whether metals are likely to induce olfactory

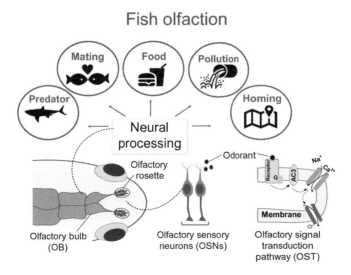

Figure 6C.1 Fish perceive olfactory cues that alert them to the presence of predators, the identity of appropriate mates, the location of food, the presence of contaminants, or migratory routes. Odorant molecules bind to receptors on epithelial sensory neurons. The resulting signal transduction cascade propagates an electrical impulse to the brain for information processing, which allows the animal to mount an appropriate response to the stimulus. Figure used with permission from Parastoo Razmara (Razmara, 2021).

dysfunction at concentrations that occur in real contaminated systems. In some cases, metals have been shown to affect olfaction at concentrations that were previously thought to be 'safe'.

Copper is one of the best-studied metals in aquatic ecotoxicology, including analyses of its effects on fish olfaction (Pyle and Mirza, 2007). One of the earliest studies to establish the underlying mechanism of Cu effects on fish olfaction investigated the histopathology of olfactory structures in two estuarine species after a waterborne exposure to Cu or a Cu injection (Gardner and LaRoche, 1973). Fish in the waterborne exposure showed evidence of vasodilation, necrosis of olfactory sensory neurons (OSNs), and hyperplasia of supporting cells, which effectively reduced the surface area of the sensory epithelium. Injected fish did not show the same type of damage, which suggested that the lesions were induced by direct exposure to waterborne Cu.

More recent studies have corroborated these early observations and have provided a deeper mechanistic understanding of the effect. Copper does not appear to accumulate in OSNs, the olfactory nerve or the olfactory bulb, but rather accumulates in melanosomes of melanophores located in the *lamina propria* of the olfactory epithelium, effectively sequestering Cu from the OSNs. However, once the available Cu exceeds the melanosome capacity for sequestration, Cu induces apoptosis (i.e., programmed cell death) in OSNs (Bettini *et al.*, 2006). At relatively high Cu concentrations, severe, non-specific epithelial lesions occur that contribute to olfactory dysfunction. At lower concentrations, apoptosis is restricted to relatively mature OSNs coupled with an increase in basal cell proliferation and differentiation. Consequently, the controlled loss of mature OSNs from the apical surface of the olfactory epithelium may be an adaptive response that breaks the connection between an epithelial surface in direct contact with potentially toxic environmental contaminants and the sensitive underlying brain. OSNs are one of only a few types of neurons in the vertebrate nervous system capable of replacement.

The increased proliferation and differentiation of pluripotent basal cells suggest that olfaction impaired by Cu-induced apoptosis is capable of recovery. Under relatively low Cu concentrations, fish are able to at least partially recover olfactory function even under continuous exposure conditions (Dew *et al.*, 2012). After an olfactory-impaired fish has been removed from Cu-contaminated water and is allowed time for recovery, olfactory function is usually restored within 10 days to 2 weeks. Higher exposure concentrations require longer recovery times. However, when fish are exposed during sensitive development periods (e.g., as a developing embryo), the olfactory system may become permanently impaired, regardless of any subsequent exposure to Cu later in its life (Carreau-Green *et al.*, 2008).

Interestingly, Cu nanoparticles (CuNPs) induce olfactory dysfunction by a completely different mechanism than for dissolved Cu ions (Razmara *et al.*, 2020). Fish exposed to CuNPs do not show any recovery during continuous exposure. Instead, the longer the exposure to CuNPs, the more severe the olfactory dysfunction. A recent study examined gene expression profiles (using RNA sequencing) of olfactory tissues from rainbow trout (*Oncorhynchus mykiss*) exposed to equitoxic concentrations of dissolved Cu and CuNPs (Razmara *et al.*, 2021). Of approximately 51,000 sequenced genes, 1,000 unique genes were differentially expressed in each exposure with only 156 overlapping targets. This observation clearly highlights important differences between exposing fish to dissolved Cu and CuNPs.

Like Cu, cadmium can also impair olfactory function in fish. Unlike Cu, Cd can accumulate in OSNs and induce cellular necrosis. Cadmium does not appear to be transported to secondary neurons, nor does it appear to accumulate in the brain. In other words, the blood–brain barrier is effective for protecting the brain against Cd accumulation and toxicity (Scott *et al.*, 2003). At high exposure concentrations, Cd can inhibit the EOG response, but not at lower, environmentally relevant concentrations. Other metals, such as Zn or Ni in the exposure water may compete with Cd for epithelial binding sites and reduce Cd-induced olfactory toxicity (Dew *et al.*, 2016). Copper does not reduce Cd-induced olfactory toxicity, which suggests that Cu and Cd target different apical receptors on the olfactory epithelium.

Under natural conditions, Hg can exist in both inorganic and organic forms. Inorganic Hg can be converted into methyl Hg (CH_3Hg) (Section 6.8.1), either through natural biotic or abiotic processes in the environment or in the fish gut. At lower exposure concentrations, olfaction is initially impaired by inorganic Hg, but can recover after 96 hours of continuous exposure; this is similar to observations by Dew *et al.* (2012) for continuous exposure to relatively low waterborne Cu concentrations. However, unlike Cu (but like Cd), Hg accumulates in the olfactory epithelium, and in particular the OSNs and olfactory nerves. Fish exposed to either water- or diet-borne CH_3Hg showed Hg accumulation in olfactory rosettes. Fish accumulated 10-fold more Hg when the exposure was from a waterborne source than from a dietary source.

Fish olfaction can be measured directly from the olfactory epithelium using a neurophysiological technique called electro-olfactography (EOG). An electro-olfactogram measures odour-evoked extracellular field potentials from the surface of the olfactory epithelium. Fine electrodes are used to measure the loss of cations from the extracellular environment when the olfactory tissue is stimulated by an odour resulting in a bulk depolarization of OSNs. The amplitude of the EOG response reflects olfactory acuity towards the chemosensory stimulus. More Hg is required to inhibit an olfactory electrophysiological response in fish than is required for Cu; however, the EOG waveform is different when fish are exposed to Hg than when they are exposed to Cu. This observation suggests that waterborne Hg may affect slow-acting Ca^{2+} channels at the apical surface of the OSN.

Zinc is also known as an olfactory toxicant in fish, but it has received considerably less research attention. Zinc appears to target OSNs directly, even at relatively high exposure concentrations, while leaving accessory or supporting cells intact. Longer-term exposure can induce more severe and widespread damage; however, recovery is still possible given enough time. Mammalian research has demonstrated that Zn taken up through the olfactory epithelium can be transported to the olfactory bulb and elsewhere in the brain, including synaptic termini. The mechanism of Zn uptake into these termini was via a Zn-specific transporter. Zinc antagonizes N-methyl-D-aspartate (NMDA) receptors and inhibits gamma-amino butyric acid (GABA) release, both of which affect neurological function (Kay and Tóth, 2008). Consequently, olfactory inhibition by Zn may not occur at the olfactory epithelium, but rather downstream in secondary neurons or other neurological structures.

Several other metals have been shown to affect olfaction in fishes. Nickel accumulates in OSNs and is transported by anterograde axoplasmic flow to the olfactory bulb, but never crosses the blood–brain barrier. It preferentially targets microvillous over ciliated OSNs, which can impair a fish's ability to perceive important food cues while leaving its ability to perceive social cues intact (Dew *et al.*, 2014). Aluminium can also impair fish olfaction, especially under low pH conditions. At relatively low Al concentrations, damage to the olfactory epithelium was limited to non-sensory cilia associated with supporting cells and an increase in necrotic cells and cellular debris, whereas OSNs remained largely intact. However, severe damage involving OSNs did occur when concentrations were significantly elevated. Chromium, Mn, Co, Pb and Ag have all been shown to impair fish olfaction; however, much more research is required to understand the mechanisms by which olfactory dysfunction is achieved.

Many of the effects observed under controlled laboratory conditions appear to translate to wild fish populations. In one early study, spawning Atlantic salmon avoided migrating up rivers with high concentrations of Cu and Zn (Saunders and Sprague, 1967). A similar avoidance response was observed in migrating Atlantic salmon (*Salmo salar*) in low-pH streams contaminated with Al (Åtland and Barlaup, 1995). Iowa darters (*Etheostoma exile*) failed to avoid traps treated with a conspecific alarm cue in a metal-contaminated lake, whereas in a clean lake the avoidance response remained intact (McPherson *et al.*, 2004). Yellow perch (*Perca flavescens*) collected from metal-contaminated lakes in a Ni and Cu mining district in northern Ontario, Canada, showed impaired EOG responses to conspecific cues and food cues, relative to those from clean lakes (Azizishirazi and Pyle, 2015). This observation suggests that under natural conditions and continuous metal exposure, fish olfaction does not recover, contrary to laboratory observations from fish continuously exposed to Hg or Cu discussed above. However, when these fish were transferred into clean water, olfactory function was restored.

Conclusions

Metals are well-known neurotoxicants that can have profound effects on a fish's olfactory system and perception of its natural surroundings. Effects of metals that occur in fish olfactory epithelium differ markedly from those that occur in the gill (Lari *et al.*, 2019). Moreover, metals are not the only contaminants that can affect fish chemosensory function. Contaminants as diverse as low-molecular-weight hydrocarbons associated with bitumen extraction (Lari and Pyle, 2017), pharmaceuticals and personal care products, pesticides and other environmental contaminants have been shown to perturb chemoreception in fish (Tierney *et al.*, 2010).

Because fish mediate vital life processes through olfaction, it is important to understand how different metals might interfere with olfaction and its associated chemical communication systems. Metal concentrations do not need to be sufficiently high to affect survival, growth or reproduction directly to have an impact on fish populations. Fish that cannot find food, detect predators or identify an appropriate mate because of metal-induced olfactory dysfunction fail to perform in an ecological context and suffer from 'ecological death' (Scott and Sloman, 2004). Much more research is necessary to understand the broader ecological implications of these effects as our freshwater ecosystems continue to receive metal-contaminated inputs from human activities.

Capsule References

Åtland, Å. & Barlaup, B. T. (1995). Avoidance of toxic mixing zones by Atlantic salmon (*Salmo salar L.*) and brown trout (*Salmo trutta L.*) in the limed River Audna, southern Norway. *Environmental Pollution*, 90, 203–208.

Azizishirazi, A. & Pyle, G. G. (2015). Recovery of olfactory mediated behaviours of fish from metal contaminated lakes. *Bulletin of Environmental Contamination and Toxicology*, 95, 1–5.

Bettini, S., Ciani, F. & Franceschini, V. (2006). Recovery of the olfactory receptor neurons in the African *Tilapia mariae* following exposure to low copper level. *Aquatic Toxicology*, 76, 321–328.

Carreau-Green, N. D., Mirza, R. S., Martínez, M. L. & Pyle, G. G. (2008). The ontogeny of chemically mediated antipredator responses of fathead minnows *Pimephales promelas*. *Journal of Fish Biology*, 73, 2390–2401.

Dew, W. A., Azizishirazi, A. & Pyle, G. G. (2014). Contaminant-specific targeting of olfactory sensory neuron classes: connecting neuron class impairment with behavioural deficits. *Chemosphere*, 112, 519–525.

Dew, W. A., Veldhoen, N., Carew, A. C., Helbing, C. C. & Pyle, G. G. (2016). Cadmium-induced olfactory dysfunction in rainbow trout: effects of binary and quaternary metal mixtures. *Aquatic Toxicology*, 172, 86–94.

Dew, W. A., Wood, C. M. & Pyle, G. G. (2012). Effects of continuous copper exposure and calcium on the olfactory response of fathead minnows. *Environmental Science & Technology*, 46, 9019–9026.

Gardner, G. R. & LaRoche, G. (1973). Copper induced lesions in estuarine teleosts. *Journal of the Fisheries Research Board of Canada*, 30, 363–368.

Kay, A. R. & Tóth, K. (2008). Is zinc a neuromodulator? *Science Signaling*, 1, 1–6.

Lari, E. & Pyle, G. G. (2017). Rainbow trout (*Oncorhynchus mykiss*) detection, avoidance, and chemosensory effects of oil sands process-affected water. *Environmental Pollution*, 225, 40–46.

Lari, E., Razmara, P., Bogart, S. J., Azizishirazi, A. & Pyle, G. G. (2019). An epithelium is not just an epithelium: effects of Na, Cl, and pH on olfaction and/or copper-induced olfactory deficits. *Chemosphere*, 216, 117–123.

McPherson, T. D., Mirza, R. S. & Pyle, G. G. (2004). Responses of wild fishes to alarm chemicals in pristine and metal-contaminated lakes. *Canadian Journal of Zoology*, 82, 694–700.

Pyle, G. G. & Mirza, R. S. (2007). Copper-impaired chemosensory function and behavior in aquatic animals. *Human and Ecological Risk Assessment*, 13, 492–505.

Razmara, P. (2021). *Effect of Copper Nanoparticles on Rainbow Trout Olfaction and Recovery*. PhD Dissertation, Lethbridge, AB, Canada: Dept. of Biological Sciences, University of Lethbridge.

Razmara, P., Sharpe, J. & Pyle, G. G. (2020). Rainbow trout (Oncorhynchus mykiss) chemosensory detection of and reactions to copper nanoparticles and copper ions. *Environmental Pollution*, e113925.

Razmara, P., Imbery, J. J., Koide, E., *et al.* (2021). Mechanism of copper nanoparticle toxicity in rainbow trout olfactory mucosa. *Environmental Pollution*, 284, e117141.

Saunders, R. L. & Sprague, J. B. (1967). Effects of copper–zinc mining pollution on a spawning migration of Atlantic salmon. *Water Research*, 1, 419–432.

Scott, G. R. & Sloman, K. A. (2004). The effects of environmental pollutants on complex fish behaviour: integrating behavioural and physiological indicators of toxicity. *Aquatic Toxicology*, 68, 369–392.

Scott, G. R., Sloman, K. A., Rouleau, C. & Wood, C. M. (2003). Cadmium disrupts behavioural and physiological responses to alarm substance in juvenile rainbow trout (*Oncorhynchus mykiss*). *Journal of Experimental Biology*, 206, 1779–1790.

Tierney, K. B., Baldwin, D. H., Hara, T. J., *et al.* (2010). Olfactory toxicity in fishes. *Aquatic Toxicology*, 96, 2–26.

6.5 Metal Detoxification and Tolerance

To this point in Chapters 3 and 6, we have considered how metals behave in the aquatic and terrestrial environments, how they interact with living organisms, how they are able to traverse biological membranes and enter the intracellular environment, and how they may cause toxicity. In the present section we will explore what happens to metals once they enter living cells, including how their intracellular fate can result in detoxification and the acquisition of metal tolerance.

6.5.1 Metal Speciation Within Cells

The intracellular environment differs markedly from the exposure environment that has been described earlier in this chapter. For example, the intracellular pH is tightly controlled close to neutrality, but the external pH may fluctuate widely (Section 6.7.2). Similarly, the concentration of metal-binding organic ligands within a cell is much higher than that in a natural surface water or in a sediment or soil pore water. In addition, the redox potential within cells is lower than that in oxic external environments. Given these different chemical conditions in the extracellular and intracellular environments, we can expect the metal's speciation to change too. One of the most dramatic changes is that free metal ion concentrations within the cell are exceedingly low; accordingly, if rapid metal transfers are to occur within a cell (a necessary condition for essential metals), they must involve the direct exchange of metal ions between two ligands. Dissociation of the free metal ion from the donor ligand and its recombination with the acceptor ligand would be prohibitively slow (Finney and O'Halloran, 2003).

Intracellular metal-binding ligands can be grouped into three classes, based on the consequences of the metal–ligand linkage (Mason and Jenkins, 1995). The first class consists of ligands that benefit from the metal binding (e.g., an **apo-metalloenzyme** that requires a specific metal cofactor to function properly), whereas the second class includes metal-sensitive ligands for which the binding of the metal produces a negative effect (e.g., enzymes that are inactivated when bound to an inappropriate metal). The goal of metal detoxification is thus to limit the formation of metal–ligand complexes of this second type. Finally, ligands that demonstrate neither beneficial nor negative effects when bound to the metal would fall into a metal-insensitive third class, corresponding to metals that are sequestered in non-labile intracellular pools (e.g., metal-rich granules).

When an essential metal such as Cu or Zn enters a cell, it is normally captured by a metal-**chaperone** (typically a small protein) and transported within the cell to a waiting partner (typically an enzyme) (Figure 6.12). This intracellular metal trafficking is particularly elegant from a metal complexation perspective (Finney and O'Halloran, 2003). First the metal-chaperone must bind the incoming nutrient metal very tightly (e.g., in an atypical coordination environment), so as to protect the metal and prevent its indiscriminate binding to inappropriate ligands within the cell. Then, when the metal-chaperone complex encounters the waiting partner, the interaction between the partner and chaperone results in a distortion of the chaperone and a weakening of its hold on the nutrient metal, which allows the metal to transfer to the waiting partner. Examples of 'partners' might include **apoenzymes** that need the metal as a cofactor, or a **finger-protein** that needs the metal to maintain its proper conformation.

If the incoming metal is not an essential metal, or if the influx of a nutrient metal exceeds biological needs, the organism can deploy various protection mechanisms. These detoxification mechanisms may include:

a. exclusion of the incoming metal (i.e., shutting down the transport across the epithelial membrane);
b. isolation of the metal internally (i.e., moving the metal into a vacuole within the cell);

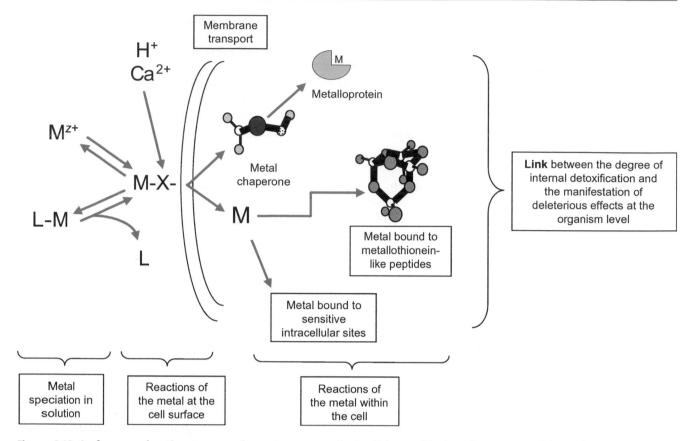

Figure 6.12 Surface complexation, transmembrane transport and subcellular partitioning of a non-essential metal M. Adapted from Campbell and Couillard (2004) with permission from the Presses de l'Université du Québec.

c. eliminating the metal from the cell (i.e., acceleration of metal efflux); or

d. binding the offending metal within the cell, either as a non-labile complex or as a particle.

Mechanisms (a) and (c) will lead to a decrease in the cellular metal concentration, whereas mechanisms (b) and (d) will result in a change in the metal's partitioning within the cell. Methods that have been used to determine the subcellular partitioning of metals are described briefly in the next section.

6.5.2 Determination of Subcellular Metal Partitioning

Why should we be interested in subcellular metal partitioning? Earlier in this chapter, we saw how knowledge of the speciation of a metal in water, or its partitioning in a sediment or soil, can help us to understand how 'bioavailable' the metal will be and to evaluate the risk the metal will pose for aquatic or terrestrial life. Now that we have moved into the subcellular environment, we can

logically ask if there is also a link between the subcellular partitioning of a metal and its internal bioavailability. In other words, does the bioaccumulated metal pose a risk, either to the plant or animal that harbours the metal, or to the consumer that feeds on the plant or animal?

The answer to this question is not straightforward, because it will depend on how the metal behaves once assimilated, i.e., on its subcellular partitioning (Figure 6.13). As indicated in the preceding section, the bioaccumulated metal may have been detoxified, for example by sequestration in insoluble granules, by isolation in membrane-bound vesicles (lysosomes), or by complexation with cytosolic ligands such as glutathione (GSH), phytochelatins (PCs) or metallothioneins (MTs) (Box 6.4). If this is the case, the plant or animal harbouring the metal will not be affected by the accumulated metal (except, perhaps, indirectly, if there is an appreciable metabolic cost to the detoxification). The form in which the bioaccumulated metal exists may also affect its fate if the plant or animal is consumed as food. For example, metals that have been detoxified and are present as insoluble metal-rich granules tend to pass right through the digestive system

Box 6.4 Metallothioneins and phytochelatins

Metallothioneins, first discovered in the equine kidney in 1957, are now recognized as ubiquitous metal-binding peptides found in a very wide range of taxonomic groups. They contain multiple cysteine groups and have a high affinity for soft metals. Within living cells they play a dual role, as a reservoir of labile Cu and Zn (i.e., as a 'buffer' for these essential metals) and as a detoxifying ligand for soft non-essential metals such as Ag, Cd and Hg.

Initially discovered in yeasts and plants, **phytochelatins** (PCs) are glutathione-derived metal-binding peptides that play a major role in the detoxification of Cd and As in plants. Unlike metallothionein, the synthesis of which is under ribosomal control, the synthesis of PCs involves an enzyme, PC-synthase, that is not synthesized in the ribosome but rather by a cytosolic transpeptidase. PCs were originally thought to be present only in plants and yeast, but recently the gene coding for PC-synthase has been discovered in invertebrate genomes. This finding raises the possibility that PCs may play a wider role in metal detoxification than was first suspected (Bundy *et al.*, 2014).

of the consumer without being assimilated, whereas metals present in soluble forms in the cytosol are normally more available for assimilation.

To determine the subcellular distribution of metals in living organisms, one must first isolate the different metal-binding ligands and then determine the amount of metal associated with each ligand pool. A typical example of this approach is shown in Figure 6.14.

To complete the differential centrifugation approach, and to determine the speciation of the dissolved cytosolic metals, various chromatographic approaches can be used to separate the different cytosolic molecules involved in metal binding. Mounicou *et al.* (2009) provide a useful summary of the various 'hyphenated' techniques that can be used to separate and quantify metal complexes with biological macromolecules.

6.5.3 Links Among Changes in Metal Exposure, Changes in Metal Subcellular Distribution and the Onset of Deleterious Effects

In the preceding sections, we have laid out the basis for what has been referred to in the ecotoxicological literature as the 'spillover' hypothesis, as it applies to the prediction of metal toxicity. According to this hypothesis, when

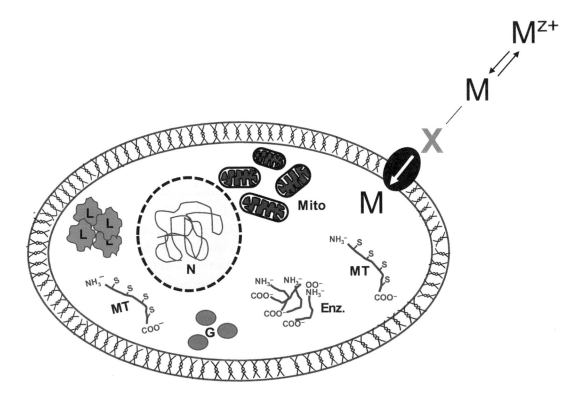

Figure 6.13 Intracellular fate of metals showing possible metal-binding sites. Metal-sensitive sites = mitochondria (Mito), nucleus (N), enzymes (Enz.). Metal-insensitive sites (detoxified metal) = metal-rich granules (G), metallothionein (MT)-like peptides, lysosomes (L).

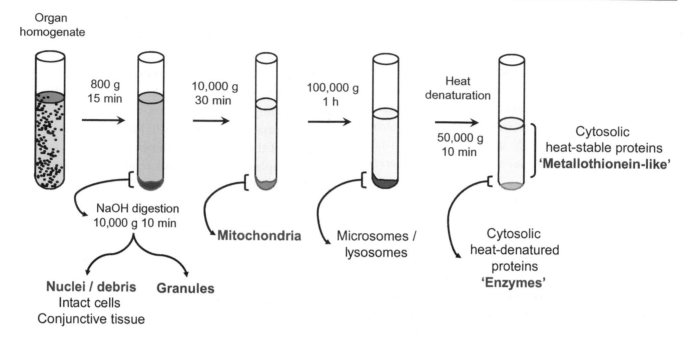

Figure 6.14 Typical experimental protocol used to determine the subcellular partitioning of metals, involving organ homogenization, differential centrifugation and heat denaturation.

the influx of a (non-essential) metal into an organism is reasonably slow, the exposed organism will be able to activate one or more of its metal detoxification mechanisms (Section 6.5.1, mechanisms (a)–(d)) and avoid any deleterious effects. However, if the metal influx is too high, the detoxification mechanisms may prove to be inadequate, and as a result, some of the metal will bind to inappropriate metal-sensitive sites within the cell and toxicity will ensue. The corollary of this hypothesis is that there will be a threshold exposure concentration below which an organism will be able to detoxify the incoming metal and avoid metal-induced negative effects. A similar pattern of 'on/off detoxification' is also observed for organic contaminants (Chapter 7). In the present section, we consider whether laboratory and field data support this hypothesis.

6.5.3.1 Laboratory Observations

A convincing demonstration of the sequence of events described in the preceding paragraph is illustrated in Figure 6.15 (Sanders and Jenkins, 1984). Newly hatched larvae from the marine crab *Rhithropanopeus harrisii* were exposed to free Cu^{2+} concentrations ranging from $10^{-12.4}$ M to $10^{-7.9}$ M until they had moulted. At the end of the experiment the researchers measured the dry weight of the larvae (i.e., their growth), homogenized the larvae, isolated the cytosol by ultracentrifugation and determined how the bioaccumulated Cu was distributed within the cytosol. Three fractions were separated by size-exclusion chromatography: a high-molecular-weight fraction (HMW, >20 kDa, including Cu-metalloenzymes + haemocyanin); a metallothionein-like peptide fraction (MTLP, 10–12 kDa); and a low-molecular-weight fraction (LMW, <5 kDa). At exposure concentrations less than $10^{-10.6}$ M Cu^{2+}, cytosolic Cu was associated largely with metallothionein-like and HMW ligands (Figure 6.15A). Over this range of $[Cu^{2+}]$, the MTLP fraction increased, the HMW fraction decreased, and the LMW fraction remained very low. As the Cu^{2+} exposure concentration increased above $10^{-10.6}$ M, Cu appeared in the LMW fraction (<5 kDa) and accumulated in this pool (Figure 6.15A), as well as in the MTLP pool. Crab growth rate fluctuated at the lowest Cu concentrations but slowed at $[Cu^{2+}] > 2 \times 10^{-11}$ M; this reduction in larval growth coincided with Cu accumulation in the MTLP and LMW pools (Figure 6.15B). Sanders and Jenkins suggested that above this threshold Cu^{2+} concentration, the induction of MT-like proteins was insufficient to prevent Cu from spilling over into the LMW fraction, even though MTLP concentrations continued to increase. They suggested that accumulation of Cu in the LMW pool was the best predictor of toxic effects on the larvae.

It would be tempting to generalize from the preceding example, but several subsequent laboratory studies with a similar experimental design but different test organisms

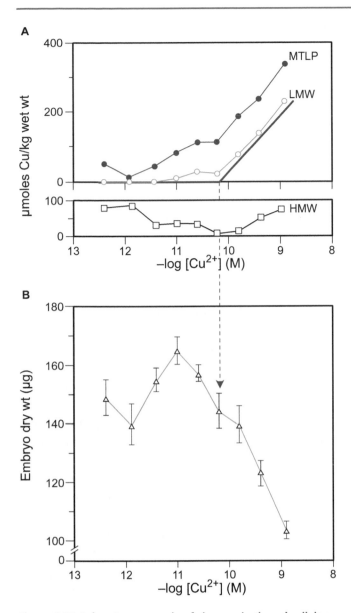

Cd-sensitive fractions, suggesting that detoxification of Cd was ineffective. Similarly, Kamunde and MacPhail (2008) exposed juvenile rainbow trout (*Oncorhynchus mykiss*) to a constant sublethal waterborne Cu concentration ($10^{-6.2}$ M) for 21 d and followed Cu accumulation and partitioning in the fish liver. Although they noted a time-dependent accumulation of Cu in the liver, it accumulated concurrently in metabolically active and detoxified pools; they concluded that the spillover hypothesis did not hold under the exposure conditions that prevailed in their experiment.

6.5.3.2 Field Observations

The preceding laboratory observations suggest that there is a spectrum of possible subcellular responses to metal stress, ranging from effective detoxification at low to moderate exposures, to ineffective detoxification even for very low exposures. Note too that these results were generated in a laboratory setting with metal-naïve test organisms, i.e., with organisms that had not previously been exposed to elevated metal concentrations. In the following Case Study, we will consider the results of studies that were designed to test the applicability of the threshold or spillover model with native aquatic animals living along a metal contamination gradient.

Case Study 6.1 Response of Native Freshwater Animals to Metals Derived from Base-metal Smelter Emissions

To establish a possible link between laboratory manipulations and field observations, it is instructive to compare the collection of native animals along a metal concentration gradient in the field with the experimental manipulation of metal exposure concentrations in the laboratory. In such comparisons, we are trying to mimic a controlled lab experiment by choosing field sites that offer a gradient of metal concentrations.

Base-metal smelters emit airborne metals from their stacks, and as a result, lakes located downwind from such smelters often exhibit a marked concentration gradient for some metals in the water column, in the sediments and indeed in the resident biota (Section 11.2). By carefully choosing a series of lakes upwind and downwind from such a smelter, one can often find habitats that were similar originally (before the smelter was constructed) but now

Figure 6.15 Laboratory example of changes in the subcellular partitioning of Cu as a function of exposure (marine crab, *Rhithropanopeus harrisii*). The free Cu^{2+} concentration increases from left to right along the *x*-axis (logarithmic scale). Figure reproduced from Sanders and Jenkins (1984), with permission from the University of Chicago Press.

and different metals have failed to show the threshold response indicated for the LMW fraction in Figure 6.15A. For example, in a study with the marine polychaete *Neanthes arenaceodentata* and Cd, Jenkins and Mason (1988) found that the subcellular distribution of Cd did not change when the Cd concentration in the exposure medium was increased; Cd was present in the LMW pool even for the lowest exposure concentration. There was no evidence of a threshold below which the polychaete could prevent the accumulation of Cd in potentially differ markedly with respect to metal exposure. Although not a true experiment (Section 4.3.1), a spatial study along such a metal concentration gradient is conceptually analogous to laboratory exposures carried out at different metal concentrations (as though the lakes had been 'titrated' over time with different quantities of metal).

The foregoing approach can be used to determine whether animals living along a metal concentration gradient in the field behave according to the threshold model. If this were the case, then in the least contaminated lakes they would be able to effectively detoxify the metal(s) and prevent them from binding to putatively metal-sensitive sites, whereas at some point in the progression to the more contaminated end of the gradient, a threshold would be exceeded and 'spillover' would occur into the metal-sensitive sites. On the other hand, if the partial or ineffective detoxification model applies, as described in Section 6.5.3.1 for the marine polychaete (Jenkins and Mason, 1988) and rainbow trout (Kamunde and MacPhail, 2008), then we would expect to see metal binding to inappropriate ligands even at the low end of the metal exposure gradient.

In the design of a field study of this nature, several criteria must be met. First and foremost, of course, there must be a marked metal exposure gradient, but all other characteristics of the lakes and their catchments must be as similar as possible. Ideally the concentration gradient would represent legacy contamination, rather than the result of present-day (current) emissions. The native aquatic animals chosen as biomonitors should be sedentary (non-migratory), such that they integrate the conditions in their lake. They must also be metal-tolerant, meaning that they can be found all along the metal contamination gradient, even at the high end. In the example described below, the smelters were located in northern Canada, on the Canadian Precambrian Shield, and accordingly the chosen lakes were relatively isolated and unaffected by other industrial or urban contaminants. Three different types of aquatic animals were collected: juvenile yellow perch (*Perca flavescens*; target organ = liver), the floater mollusc (*Pyganodon grandis*; target organs = digestive gland and gills) and the phantom midge (*Chaoborus* sp.; whole animal) (Campbell *et al.*, 2005b; Rosabal *et al.*, 2012).

The responses of the biomonitor organisms to the metal exposure gradient were studied at the chemical (total metal accumulation; metal subcellular partitioning) and biochemical (metallothionein induction) levels. Since the biomonitor organisms had lived their entire lives in the lakes from which they were collected, it was assumed that their bioaccumulated metal concentrations represented a steady state of uptake, storage and elimination. Results showed that for a given biomonitor species, metal concentrations varied among its organs; the liver was an important site for metal accumulation in the yellow perch, whereas the gills and digestive gland were important binding sites for the bivalve. As expected, the metal concentrations in the animals did reflect their varying exposure along the contamination gradients. For example, ratios of the maximum tissue Cd concentration to the lowest tissue Cd concentration ($[Cd]_{max}/[Cd]_{min}$) were 18 for the bivalve digestive gland, 28 for the bivalve gills, 14 for the perch liver and 29 for the whole body of the midge, with the highest values being for the most contaminated lake. Comparable ratios for Ni were 36 for the perch liver and 89 for the midge.

Despite this considerable range in tissue metal concentrations, the subcellular distribution of the metals did not vary appreciably along the exposure gradient (see Figure 6.16 for an example of this behaviour, for Cd in yellow perch liver and the bivalve digestive gland). Although metallothionein induction increased as a function of metal exposure (as indicated by the increase in the metals bound to the heat-stable protein (HSP) fraction), the resulting steady-state concentrations of metallothionein were insufficient to prevent Cd and Ni from binding to potentially metal-sensitive sites within the subcellular environment (mitochondria and heat-denaturable proteins). There was no evidence of a threshold exposure concentration below which Cd or Ni were absent from the putative metal-sensitive fractions; in other words, detoxification of Cd and Ni was incomplete even in the least contaminated lakes. For all of the animals examined, the amount of Cd or Ni found in a given subcellular fraction was directly proportional to the total amount of the metal in the organ. This proportionality is similar to the results described in the preceding section for the laboratory exposures of the marine polychaete, *N. arenaceodentata*, and rainbow trout, *O. mykiss*. Biochemical and population-level measurements were generally consistent with this demonstration of incomplete metal detoxification. For example, a wide range of population-level parameters (littoral density, live weight, dry viscera biomass, production and cumulative fecundity) decreased in bivalves living in lakes along the Cd exposure gradient, without any apparent threshold (Perceval *et al.*, 2004). Similarly, the normal cortisol stress response of juvenile

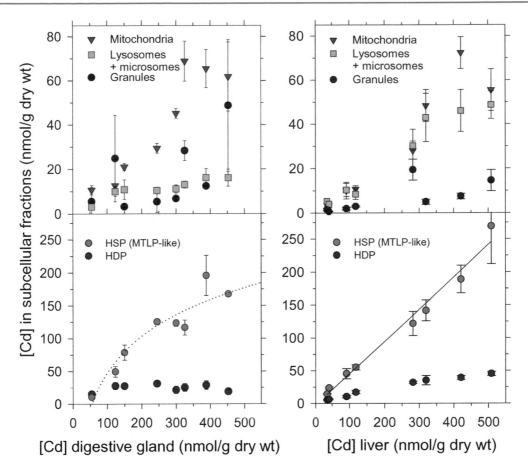

Figure 6.16 Field studies of the subcellular partitioning of cadmium in two biomonitor species, a bivalve (*Pyganodon grandis*) on the left and a freshwater fish (*Perca flavescens*) on the right. The biomonitors were collected along a metal contamination gradient. HDP = heat-denatured protein; HSP = heat-stable protein; dw = dry wt. Figure derived from data presented in Campbell *et al.* (2005b) with permission from Elsevier.

yellow perch was progressively suppressed as Cd exposure and bioaccumulation increased (Couture *et al.*, 2015).

In another field example from the freshwater environment, Cain *et al.* (2004) collected benthic invertebrates along a metal concentration gradient in the Clark Fork River, Montana, USA, and determined the subcellular partitioning of cytosolic metals in five insect taxa. As in the lake studies described in Case Study 6.1, Cd concentrations increased progressively along the metal exposure gradient. Although subcellular partitioning patterns varied among the different taxa, in all cases Cd accumulated in both the 'metal-detoxified' and the 'metal-sensitive' fractions, without any suggestion of a lower threshold below which the metal-sensitive fractions were protected.

These field results pose an interesting problem. On the one hand, there is compelling evidence for the incomplete detoxification of Cd and Ni, even in animals living in relatively uncontaminated environments. On the other

hand, there is no evidence that the capacity of the bivalves or the fish to produce metallothionein is limited. On the contrary, comparison of animals collected from lakes in the middle of the contamination gradient with those at the high end shows that steady-state concentrations of metallothionein continue to increase along the entire gradient. In other words, metallothionein was synthesized in proportion to the amount of metal in the tissues, but not in sufficient quantity to bind all the Cd or Ni. In Case Study 6.1 of the lakes on the Canadian Precambrian Shield, the smelters had been in operation for 70 or more years, meaning that many generations of bivalves, fish and other fauna would have been produced since the onset of contamination. Given the long duration of the metal exposure (>70 years), there has been ample time for adaptive (genetic) change to have occurred in response to metal-specific selective pressures (Section 6.5.4).

In summary, it would appear that metallothionein synthesis and metal detoxification incur an appreciable metabolic cost. Under chronic exposure conditions, aquatic animals must establish a trade-off between the 'cost' of detoxifying a non-essential metal such as Cd and the 'cost' of allowing some of the incoming Cd to bind to metal-sensitive sites. Indeed, there is some evidence for an inverse response, where detoxification is less effective at the lower end of the contamination gradient and becomes more effective when the animal 'wakes up' to the fact that it is being poisoned (Rosabal *et al.*, 2012).

6.5.4 Metal Tolerance

In the preceding section, we referred to sites on the Canadian Precambrian Shield that have been affected for over 70 years by smelter emissions, noting that there has been ample time for adaptation to have occurred. We now examine the phenomenon of metal tolerance, reviewing the range of organisms and situations in which it occurs and examining what research has revealed concerning mechanisms of tolerance. In the course of reviewing the occurrence of metal tolerance, we look for insights into the basic biological concepts of adaptation and evolution; in examining mechanisms, we determine whether tolerant organisms operate by using the same homeostatic mechanisms that are used by their non-tolerant congeners for maintaining intracellular metals at a non-toxic level, or whether they have developed and use novel or unique mechanisms.

Various meanings and definitions have been attached to the term tolerance (and likewise to resistance). We use the term tolerance to mean the ability to withstand conditions of exposure to abnormally high concentrations of substances (elements or compounds) that would otherwise cause adverse biological effects. However, all definitions and examples of tolerance are, in fact, relative, not absolute, and this fact sometimes leads to confusion in the literature.

6.5.4.1 Occurrence and Origin
Metal tolerance is heritable. Many well-documented examples occur within a species; **ecotypes** or strains or races of a given species exhibit an ability to function at concentrations of a given element that are toxic to the original strain (sometimes called the wild type). This is believed to result from the selection of pre-adapted individuals in the original population, which then reproduce under conditions that are adverse to the wild type,

forming a distinct population. When tests have been performed, the tolerant individuals are typically still sufficiently genetically similar to the wild type to breed with it.

Examples of plants from mine sites and other recently contaminated sites have provided some convincing examples of the rapid development of tolerance. In contrast, for some species of plants, sometimes referred to as 'edaphic endemics', entire species appear to be tolerant to certain potentially toxic chemicals. Such endemics are found in distinctive and often chemically unusual situations, e.g., the flora of **serpentine** and other soils on **ultramafic** rocks. These soils have very high concentrations of Ni, Cr and Co (in addition to Mg and Fe) and support a distinctive flora, many species of which have been shown to be **hyperaccumulators** of some metals or metalloids.

The evolution of edaphic endemics is a long-standing question that remains largely unresolved (Armbruster, 2014). Some researchers consider that such species possess a constitutional tolerance to elements. Serpentine flora have been established over more than a million years, an evolutionary history very different from the rather rapid selection of metal-tolerant ecotypes on mine sites. Ernst (2006) has suggested that "based on recent taxonomic revisions", some of these endemics are in fact **ecotypes** of more widespread species. El Mehdawi and Pilon-Smits (2012) consider that since Se hyperaccumulation occurs in relatively few genera and species, the hyperaccumulators probably evolved from non-hyperaccumulators. However, they also point out that Se hyperaccumulation, being an ancient trait, may be constitutive; seleniferous soils may once have been more widespread, so that the trait may have been lost when soil Se concentrations decreased. Whatever their evolutionary origin, for ecotoxicologists these hyperaccumulators provide excellent material for the study of tolerance mechanisms.

6.5.4.2 Approaches
Field observations of organisms surviving and functioning under conditions that are toxic to most other organisms provide some indication of tolerance. However, these 'survivors' are frequently subject to multiple stressors, such as low nutrient status, low moisture and extreme pH. It is clearly necessary to test for metal tolerance in the laboratory under controlled conditions. Such tests can also be designed to establish cause–effect relationships between the metals in the source environments and relative tolerance. Some taxonomic groups lend themselves more than others to such experimental

approaches. Unicellular organisms such as Protista, fungi, yeasts and microalgae, with their relative ease of culturing and short generation times, are obvious candidates. For herbaceous vascular plants, root elongation tests have been developed for rapid screening to compare metal-exposed seedlings with controls. Invertebrates similarly lend themselves to such controlled testing.

Situations where an entire species is tolerant to a specific element and frequently hyperaccumulates this element have also proven to be useful for investigating mechanisms of tolerance. Since plants are immobile and have only limited capacities for modifying adverse conditions in their rhizosphere, in colonizing such challenging soils they have had to develop internal tolerance mechanisms or metal exclusion strategies. Definitive information has been obtained by comparing closely related plant species that differ in their respective metal tolerance. Specific examples of hyperaccumulating plants can be found in Sections 6.5.4.4 (Cd, Se), 6.8.5 (Ni) and 6.8.6 (Zn).

6.5.4.3 Taxonomic Distribution of Tolerance

Reported examples of tolerance among taxa may reflect the convenience of testing, rather than any true quantitative distribution. Nevertheless, an overview of reports of metal tolerance does reveal that the phenomenon has been recognized in a wide range of autotrophic and heterotrophic organisms, living in terrestrial and aquatic ecosystems. The phenomenon is believed to have arisen independently, many times; there is little evidence for phylogenetic relationships. Indeed, within one genus, *Arabidopsis*, *A. halleri*, a Zn/Cu-tolerant hyperaccumulator, and *A. lyrata sp. petraea*, a non-tolerant non-hyperaccumulator, are sufficiently closely related to have been experimentally crossed, yet are distinct in their respective tolerance to Cd and Zn (Courbot *et al.*, 2007). Box 6.5 shows the range of taxonomic categories for which tolerance to metals has been demonstrated, but it is not meant to provide a comprehensive review. The apparent prevalence of Cu, Cd and Zn in the box is probably a function of the researchers' focus, rather than a true picture of the frequency of elements for which tolerance exists.

6.5.4.4 Tolerance Mechanisms

Reference was made earlier in this chapter (6.5.3.1) to the 'spillover' hypothesis, as it applies to the prediction of metal toxicity. A number of detoxification mechanisms have been demonstrated, and some apply across a wide

---

Box 6.5 Examples of taxonomic categories where metal or metalloid tolerance has been documented

Bacteria	Ag, Cd, Cu, Hg, Pb, Zn; Mg
Cyanobacteria	Cu
Microalgae (freshwater)	Cd, Co, Cu, Ni, Zn
Microalgae (marine)	Cd, Cu, Pb, Zn
Macroalgae (marine)	Cu
Yeasts	Cd, Hg, As, Sb, Se
Fungi	Cd, Cu, Cr, Pb, Zn, Mo, V
Aquatic angiosperms	Cd, Cu, Pb, Zn
Terrestrial angiosperms	Cd, Cu, Ni, Pb, Zn, Se
Aquatic invertebrates	Cd, Cu, Zn
Terrestrial invertebrates	Cd
Fish (freshwater)	Cd, Cu, Ni, Zn

---

range of taxonomic groups. We now examine recent research to determine whether metal-tolerant organisms use the same types of detoxification mechanisms as sensitive forms, and whether there are common mechanisms across taxonomic groups.

Metals that are commonly present at high concentrations on mine sites and other industrially contaminated sites, and to which tolerance has been demonstrated, include Cd, Cu, Pb and Zn. Some demonstrated or proposed tolerance mechanisms for these elements are common to a number of eukaryotic organisms (Wysocki and Tamas, 2010). These include binding of the metal by glutathione (GSH), metallothionein (MT) or phytochelatins (PCs), which would limit its internal bioavailability, and the activation of metal transporters designed either to promote efflux of the excess metal to the external environment or to promote its transfer into vacuoles. Genomic studies of metal-tolerant and non-tolerant plants suggest that tolerance is normally not related to a single dominant mechanism, but rather involves an array of complementary responses. For example, Chiang *et al.* (2006) found that two plant metallothioneins (MT2b and MT3), ascorbic acid peroxidase (APX) and certain efflux and uptake metal transporters were all expressed at higher levels in the Zn/Cu-tolerant hyperaccumulator *Arabidopsis halleri* than in the non-tolerant *A. thaliana*. Enzymatic measurements confirmed that peroxidase activity was indeed higher in *A. halleri*, leading the authors to conclude that its tolerance was largely attributable to heightened protection against metal-induced oxidative stress.

Currently, it is not possible to make any generalized statements about mechanisms across the taxonomic groups that display metal tolerance, but some patterns are beginning to emerge. For example, in a study of Cd tolerance "from bacteria to humans", Prévéral *et al.* (2009) describe how Cd detoxification is achieved by complexation with an endogenous thiol (e.g., GSH or a PC), followed by export of the Cd-thiol complex by a membrane transporter in the ATP-binding cassette family (SpHMT1). Cadmium tolerance required both the thiol complexation step and the presence of a functional transporter; to confer tolerance, the transporter required the presence of an adequate supply of GSH and ATP. Overexpression of the SpHMT1 transporter enhanced tolerance to Cd (but not to Ag, As(III), As(V), Sb(III) or Hg(II)). These results led the authors to conclude that for Cd tolerance, "a common highly conserved mechanism has been selected during the evolution from bacteria to humans".

As is the case for MTs, the demonstration of the synthesis of PCs in the presence of a metal does not automatically prove that PCs are involved in tolerance mechanisms. For the Zn- and Cd-tolerant hyperaccumulator plant *Thlaspi caerulescens*, PCs do account for Zn tolerance, through the stimulation of the synthesis of PCs by Zn and the subsequent sequestration of Zn in leaf vacuoles. However, Ebbs *et al.* (2002) compared the tolerant *T. caerulescens* with the related but non-tolerant *T. arvense*. Although both species produced PCs in response to Cd, and although leaf and root PC levels for both species were positively correlated with tissue Cd, total PC levels in the hyperaccumulator were generally lower than in the non-tolerant species. They concluded that PCs do not appear to be involved in Cd tolerance in this hyperaccumulator and discussed other possibilities including the role of LMW organic ligands such as citrate and histidine. These ligands would have the general ability to bind with Cd and render it less available, but would not account for any metal specificity in tolerance.

To this point, we have considered tolerance mechanisms that operate for metal cations. Examples of tolerance to metalloid oxyanions are also well documented. For example, the genus *Astragalus* contains a high number of Se-hyperaccumulating species, with some of these species accumulating up to 0.6% of their foliar dry weight as Se, whereas other *Astragalus* species are non-tolerant and do not hyperaccumulate Se. Selenium-tolerant species of *Astragalus* synthesize methyl-selenocysteine, and it had been suggested by several authors that synthesis of this seleno-amino acid might contribute to Se tolerance. Neuhierl and Böck (1996) compared *Astragalus cicer* and *A. bisculatus*, a non-tolerant and tolerant species, respectively. Through labelling and cell culture experiments, they demonstrated that the tolerant species produced an enzyme (SecMT) that catalyses the methylation of selenocysteine to methyl-selenocysteine (MeSeCys) but is much less effective in methylating normal L-cysteine. In other words, Se tolerance is achieved by a shunting mechanism, in which selenocysteine is intercepted and prevented from intruding on the downstream sulphur pathways that cysteine normally follows. In non-accumulators and non-tolerant species, the SecMT enzyme is absent, and the selenocysteine that is produced by the substitution of Se for S in normal sulphur metabolism is either incorporated into proteins or serves as a substrate elsewhere in the sulphur biochemical cycle, interfering with sulphur metabolism and causing toxicity.

Methyl-selenocysteine

Returning to our initial question about mechanisms, we see that in most of the examples for which mechanisms have been demonstrated, tolerant organisms utilize the same types of homeostatic mechanisms as their less-tolerant congeners, namely:

- shutting down metal influx across the epithelial membrane;
- accelerating metal efflux;
- isolating the metal internally (vacuole; granule);
- binding the metal within the cytoplasm as a non-labile complex;
- shunting the metal away from sensitive metabolic targets;
- upregulating antioxidant enzymes;
- altering the environmental conditions at the abiotic–biotic interface (e.g., the rhizosphere around plant roots).

The only truly novel mechanism is the interception and shunting aside of Se.

Clearly we cannot make generalizations about mechanisms of metal tolerance. For most metals, whether essential or non-essential, there is still a need to clarify whether detoxification mechanisms, many of which are quite well known, are in fact consistently applicable to explaining tolerance. Genomics and the development of adverse outcome pathways (AOPs) promise to provide useful tools in this respect.

The question of specificity remains. We know that certain organisms are tolerant to more than one metal, but it is not yet clear in the case of co-tolerance whether a single mechanism is in operation. In fact, for a number of examples the prevailing opinion is that tolerance is metal-specific. For example, some grasses growing on mine sites and exposed to Cu and Cd demonstrated tolerance only to Cu, not Cd. Even where there is tolerance to more than one metal, conclusive proof that one mechanism confers tolerance to another remains in question. In other words, co-tolerance does occur, but the respective mechanisms may well be metal-specific.

6.5.4.5 Ecotoxicological Implications and Practical Applications

For individual ecotypes or species, one may question if there is any ecological advantage to the development of tolerance, beyond the obvious advantage of surviving and reproducing in contaminated environments. The idea that accumulated metals may protect a plant from herbivores has been termed the 'elemental defence hypothesis'. Evidence comes from Se-tolerant accumulator plants, on seleniferous soils. Proposals have been made that for tolerant plants that are hyperaccumulators, protection from herbivory and disease may be conferred. For example, accumulation of Se may give plants a selective advantage by protecting them from invertebrate herbivores (Hanson et al., 2004) and from microbial infection (Hanson et al., 2003). In these two experimental studies, brown mustard (Brassica juncea) was dosed with Se. The tissue concentration and chemical form of Se applied in the experimental plants were comparable to those in plants in the field, such as A. bisculatus. The treatment with Se protected the plants from invertebrate grazing (e.g., by aphids and moths) and from fungal damage. There is some evidence that a similar ecological advantage is conferred by hyperaccumulation of Cd and Ni (Stolpe et al. 2017). More widespread observations, ideally under field conditions, are necessary to substantiate such a claim as an ecological advantage.

To this point in our discussion, we have emphasized the metal tolerance of particular species, but it is also possible to consider the tolerance of a whole community. The development of community tolerance may involve the acclimation of the indigenous species that make up the community, or it may result from the elimination of sensitive species and selection for their tolerant congeners. In principle, the development of community tolerance could be used as a diagnostic tool, to determine whether a given community was currently exposed to a particular toxicant or toxicant class (or had been in the recent past). To quote Blanck (2002), "An increase in community tolerance compared to the baseline tolerance at reference sites suggests that the community has been adversely affected by toxicants." This approach has earned an acronym, PICT ('pollution-induced community tolerance') and has been tested in a number of site-specific risk assessments.

The PICT approach has been applied in aquatic and terrestrial environments, usually for reasonably sedentary communities. Examples of communities that have been studied include soil microbes and invertebrates, periphyton in streams, and phytoplankton and benthic invertebrates in small lakes. Community responses have included changes in photosynthesis, respiration, and incorporation of thymidine or leucine as an indicator of microbial growth. The application of the PICT approach in the field involves collecting water, soil, sediment or biofilm samples at uncontaminated reference sites and at sites where contamination (past or present) is suspected. The samples are then tested under controlled conditions, usually in the laboratory and with a single toxicant (Chapter 2), and the community response is determined. The observation that communities at the contaminated sites are more tolerant to the toxicant than those at the reference sites constitutes a 'positive' response, i.e., that community tolerance has developed at the contaminated site. Such responses have been documented for stream periphyton (Cd, Cu, Zn), lake phytoplankton (As), soil microbes (Cd, Cu, Ni, Pb, Zn) and sediment invertebrates (Cd, Cu, Zn) (Blanck, 2002; Faburé et al., 2015; Oguma and Klerks, 2017).

Finally, there are practical applications of the capacity of tolerant organisms, particularly hyperaccumulators, in the restoration and recovery of contaminated sites (Sections 13.9.1, 13.9.3). Metal-tolerant grasses, for example, have been cultured and used extensively for revegetation of mine or roadside sites. Direct use of tolerant plant ecotypes has already been in use for years, with some seed now being commercially available. More complex intervention has been proposed for the clean-up of such sites, utilizing the accumulative capacity of terrestrial and aquatic microorganisms and primary producers, to remove contaminants from water or soils, with the possibility of 'mining' the elements from the accumulator organisms. This area of applied research may develop into engineering procedures for the clean-up of contaminated sites, clean-up being legally required in many jurisdictions.

Engineering of the genes responsible for tolerance holds promise for the refining of such approaches. For example, over the past two decades, researchers have begun to identify the genetic factors influencing Se acquisition and accumulation by plants. Application of this knowledge has allowed the genetic manipulation of Se metabolism to increase Se accumulation in harvested tissues, with potential benefit for **phytoremediation** strategies. Similar studies are on record for other toxic elements (Section 13.8).

6.6 Organometals (Hg, Pb, Sn, As, Sb, Se)

To this point in Chapter 6, we have dealt exclusively with cationic metals and their complexes. A small number of metals can also exist as 'organometals' or 'organometallic compounds'. In these forms, the metal is bound to one or more carbon atoms by covalent bonds, as is the case in the molecules shown below.

Methylmercury Tributyltin
$H_3C-Hg-X$ $(CH_3CH_2CH_2CH_2)_3-Sn-X$
(X = Cl, OH) (X = Cl, OH, O-)

Tetraethyllead Arsenobetaine
$(H_3C-H_2C)_4-Pb$ $(H_3C)_3-As^+-CH_2C(O)O^-$

Covalent bonds are much stronger than the coordinate bonds that are present in the metal cation–ligand complexes that were described in Section 6.2.1. Some of the organometals, such as tributyltin (a biocide) and tetraethyllead (an anti-knock agent formerly added to gasoline), are manufactured industrially and can be considered to be environmental contaminants, whereas others are produced biochemically (e.g., methylmercury, by microorganisms; arsenobetaine, by crabs, shrimp and other marine animals). In addition, some of the oxyanions (e.g., As and Se) can be incorporated into natural biomolecules such as arsenosugars and selenoproteins.

Given the presence of covalent metal–carbon (M–C) or metal–oxygen–carbon (M–O–C) bonds, organometals tend to exhibit a hybrid behaviour in the environment, approaching that of organic molecules (Chapter 7). Many organometals are less polar than metal–ligand complexes and exhibit a higher lipid solubility, leading to a higher bioavailability. As discussed earlier (Table 6.1), the terms 'degradation' and 'half-life' have little or no significance for cationic metals, but for organometals they are relevant. The covalent bonds binding the metal to the organic moiety are subject to chemical and photochemical degradation (e.g., oxidation; hydrolysis), and the organic moiety itself may be subject to biodegradation. The regulatory approach for organometals necessarily differs from that used for cationic metals and is largely based on the rules that apply to organic contaminants (e.g., water solubility, octanol–water partitioning coefficient, half-lives in the atmosphere, in surface waters and in soils).

Additional details about organometals can be found in the metal-specific sections at the end of the present chapter (Hg, Section 6.8.1; Pb, Section 6.8.3; As, Section 6.8.7; Se, Section 6.8.8) and in Section 8.5.6 (organotin compounds).

6.7 Abiotic Factors Affecting Metal Toxicity

In the preceding sections in this chapter, we have seen how metals behave in the aquatic and terrestrial environments, how they interact with living organisms, and how they are employed and handled within living cells. Now we can step back, integrate what has come before, think about the whole organism and consider in a more general fashion how various abiotic factors may affect metal toxicity. Environmental regulators often refer to these variables as 'toxicity modifying factors' (which can justify different water quality criteria for different receiving waters, as is discussed in Section 12.3.1). These factors include those that primarily affect the receptor and its bioaccumulation and response to a metal (e.g., temperature) and those that primarily influence the chemistry or form of the metal and its bioavailability (e.g., pH).

6.7.1 Temperature

Ectothermic metabolism increases approximately 2-fold for every 10 °C change in temperature (i.e., the $Q_{10} = 2$).

Changes in metabolism may be reflected by changes in respiratory rate, chemical absorption, detoxification and excretory rates, all of which may affect chemical toxicity. A rise in temperature may thus increase the uptake of metals, but at the same time can increase detoxification and elimination rates; the final outcome will depend on the relative effects of temperature on these different processes. At the cellular level, a rise in temperature causes an increase in lipid fluidity resulting in increased membrane permeability, although this effect may only be temporary. Enzymes may exhibit a generalized increase in activity, but some may move beyond their optimal functional range. We might expect the influence of temperature changes on homeotherms (organisms that maintain a constant internal temperature) to differ from its influence on **poikilotherms** (organisms whose internal temperature varies as a function of the ambient conditions – all species except birds and mammals), but this remains speculative.

In considering the influence of temperature on metal toxicity, two parameters are of overriding importance: acclimation temperature and experimental temperature. Acclimation temperature has different meanings in the literature. Within the context of a toxicity test, it is defined as the temperature at which organisms are held prior to toxicant exposure (Section 2.2.1.1; Box 2.8) and should be the same as the test temperature. For example, Hodson and Sprague (1975) demonstrated a 4- to 5-fold lower Zn toxicity in Atlantic salmon (*Salmo salar*) at $3\,^{\circ}C$ compared with $19\,^{\circ}C$ when determined as a 24-h LC_{50}. Paradoxically, when the assay was extended to 2 weeks, the temperature effect disappeared; toxicity remained unchanged at $19\,^{\circ}C$ but the $3\,^{\circ}C$ LC_{50} decreased and was virtually the same as at $19\,^{\circ}C$. In effect, temperature did not change the concentration of Zn causing toxicity. Instead, it changed the rate at which Zn toxicity was expressed, likely because the rates of metabolism, respiration and accumulation of Zn in gills, the target of acute effects, were slower at $3\,^{\circ}C$ than at $19\,^{\circ}C$. In contrast, if salmon acclimated to $11\,^{\circ}C$ were exposed to Zn at $3\,^{\circ}C$, toxicity increased two-fold compared to toxicity at $11\,^{\circ}C$. However, at $19\,^{\circ}C$ Zn toxicity was unchanged, demonstrating an influence of cold stress but not heat stress. Clearly, caution must be exercised when interpreting temperature effects from acute toxicity tests. When carried out on indigenous polar organisms at cold temperatures, these tests should be prolonged to account for possible delays in the manifestation of toxicity (Chapman, 2016 – Section 14.1).

6.7.2 pH

The pH of natural waters varies spatially as a function of the buffering capacity of the surface soils and the nature of the underlying geology. At a given location, it may also vary on a daily cycle in response to the day–night cycle of photosynthesis, or because of industrial discharge of acidic or alkaline effluents. Within the construct of the biotic ligand model (BLM), changes in pH can influence the toxicity of cationic trace metals in two ways. For example, a decrease in pH will tend to decrease the extent of metal complexation (which would be expected to increase the metal's bioavailability). On the other hand, a decrease in pH will also increase the competition between the H^+ ion and the metal for binding at the epithelial surface (which would be expected to decrease the metal's bioavailability). These effects operate in opposing directions, and the net effect will vary from one metal to another and from one biological species to another (Wang *et al.*, 2016). In some cases, particularly for very low pH waters, the H^+ ion itself may perturb the ionoregulatory functions of the epithelial membrane, leading to additional stress and an increase in overall toxicity (Section 11.1).

An interesting field example of the sometimes counter-intuitive effect of pH was provided by temporal monitoring of the bioavailability of Cd in lakes near metal smelters in northern Ontario and Quebec, Canada (Croteau *et al.*, 2002). When the smelters introduced measures to reduce their airborne emissions of SO_2 and particulate metals, the total concentrations of Cd in the lake waters decreased. However, over the same time period, Cd concentrations increased in the phantom midge (*Chaoborus punctipennis*, a recognized biomonitor). In effect, the acidity of the lakes also decreased during this same period, i.e., their pH rose, and thus there was less competition between the H^+ ion and Cd^{2+}; net uptake of Cd increased despite its lower aqueous concentration.

Changes in pH may also affect the bioavailability of lipophilic metal species present in solution, which may traverse the plasma membrane without forming a surface complex (Section 6.3.2). For example, acid conditions tend to favour the predominance of mercuric chloride ($HgCl_2$) for inorganic mercury and monomethylmercury chloride (CH_3HgCl) for methylmercury. Such neutral chloro-complexes exhibit appreciable lipid solubility and can cross membranes by simple diffusion, whereas the hydroxo-complexes (e.g., $Hg(OH)_2$), which predominate at higher pH values, cannot. The octanol–water partitioning coefficient (K_{OW}) for $HgCl_2^0$ is about 70 times higher than that of $Hg(OH)_2^0$.

6.7.3 Hardness

The principal components of water **hardness** are the divalent cations calcium and magnesium. Hardness is traditionally defined in terms of $CaCO_3$ equivalents: 0–75 mg/l for soft water, 75–150 mg/l for moderately hard water, 150–300 mg/l for hard water and >300 mg/l for very hard water. Hardness alone has a negligible effect on the speciation of other cations, although it is usually correlated with pH and **alkalinity**, two factors that can affect cation speciation. The correlation among hardness, alkalinity and pH often makes it difficult to differentiate among the effects of these factors. However, experimental manipulations have demonstrated that, at a fixed combination of pH and alkalinity, increasing water hardness is often associated with a reduction in metal toxicity.

Notwithstanding disputes concerning the relative influence of the correlates mentioned in the preceding paragraph, there is a large body of information indicating the ameliorative effect of hardness on metal toxicity, beginning as early as the 1930s (Erichsen-Jones, 1938). The progressive accumulation of this database led in the 1970s to the adoption by the US Environmental Protection Agency and other regulatory bodies in Canada and Europe of separate water quality criteria for metals such as Cd, Cu, Pb, Ni and Zn in hard or soft waters (Adams *et al.*, 2020); Section 12.3.1. Note that the protective effect of Ca^{2+} and Mg^{2+} is entirely consistent with the predictions of the BLM (Section 6.3.2), as these divalent cations would be expected to compete with other divalent metals such as Cd^{2+}, Cu^{2+}, Ni^{2+}, Pb^{2+} and Zn^{2+} for binding to the biotic ligand (Meyer *et al.*, 1999).

6.7.4 Salinity

In contrast to studies on many other water chemistry parameters, the study of the effect of salinity on metal uptake and toxicity is characterized by many field studies, often conducted in estuaries. As focal points of commerce and industry, estuaries may become contaminated by a variety of chemicals that interact with salinity to varying degrees with respect to their bioaccumulation and their toxicity to aquatic organisms.

Studies of the effect of salinity on trace metal accumulation by biota have indicated the influence of both biological and chemical factors. The physiology of an organism may affect trace metal uptake and toxicity in several ways. **Stenohaline** organisms can only survive within a narrow range of salinities whereas **euryhaline** and **anadromous** species clearly have a wide range of salinity tolerance, although even for these, optimal salinity may vary from species to species. Outside its optimal salinity range, an organism will devote a disproportionate amount of energy to ionic or cell volume regulation and may be more susceptible to toxic stress. Interaction between trace metal uptake and ionoregulation sometimes occurs. In such cases, pre-exposure to increased or decreased salinity may lead to acclimation of the organism to the changed conditions and to a change in its sensitivity to the test metal. As mentioned earlier for temperature, there is thus a possible link between the antecedent exposure of the test organism to salinity and its sensitivity to metals.

For several metals such as Cd and Hg, the effect of salinity on their bioavailability is related to chemical speciation. For example, in fresh waters the dominant inorganic form of Cd is the free ion (Cd^{2+}); as we have seen earlier (Section 6.3.2), the free metal ion is a consistently good predictor of metal bioavailability. In increasingly saline media, Cd forms chloro-complexes such as $CdCl^+$ and $CdCl_2^0$ (Section 6.2.1; Figure 6.4) and the overall bioavailability of Cd declines (Sunda *et al.*, 1978). In contrast, increasing soil salinity has been consistently linked to greater bioavailability of Cd to rooted terrestrial plants (e.g., potatoes, wheat, Swiss chard), probably because chloride complexation increases the mobility of Cd within the soil matrix (Smolders *et al.*, 1998).

In experiments with the marine diatom *Thalassiosira weisflogii*, Mason *et al.* (1996) demonstrated that the growth rate of the diatom when exposed to mercury was inversely related to salinity. However, in contrast to the changes described for Cd, salinity-related Hg toxicity was most closely related to the uncharged and relatively non-polar $HgCl_2^0$ form. The proportion of total dissolved mercury represented by $HgCl_2^0$ fell from 23% at a salinity of 10 parts per thousand (ppt) to 3% at 35 ppt. Higher salinities and the accompanying higher chloride concentrations led to an increased formation of anionic complexes such as $HgCl_3^-$; unlike $HgCl_2^0$, this anion could not simply diffuse across the diatom's cell membrane.

6.7.5 Dissolved Organic Matter

Natural dissolved organic matter (DOM) is found in a variety of forms and concentrations in the aquatic environment. In fresh water, recalcitrant forms such as humic and fulvic acids comprise 50–80% of dissolved organic matter. Dissolved organic molecules may range from amino acids having a molecular weight (MW) of less than 200 Da to much larger entities that have been described in the literature as either macromolecules or supramolecular associations of smaller molecules bound together by weak (non-covalent) bonds (Vialykh *et al.*, 2019). These dissolved organic compounds possess a wide variety of oxygen-containing functional groups, with phenolic and carboxylic groups dominating. Other functional groups associated with DOM include amine, ketone, quinone and sulphydryl moieties. This marked heterogeneity contrasts with the simple synthetic ligands (ethylenediamine tetraacetate, EDTA; nitrilotriacetate, NTA) that are normally used to control metal speciation in laboratory studies of metal bioavailability.

Most published studies demonstrate the attenuation of metal toxicity through the formation of metal–organic complexes at the expense of the free metal ion, yet a small but significant number of studies indicate that metal bioavailability may actually be increased through the formation of organic complexes (Campbell, 1995). In considering this variety of responses, it is important to remember that DOM cannot be viewed simply as an amorphous entity, even though it is often measured and reported as such (as dissolved organic carbon or 'DOC'). For example, as described in Section 6.3.2, some organic ligands form small lipophilic metal complexes that can enter cells by simple diffusion across the cell membrane, resulting in enhanced metal uptake. In other cases, where the organic ligand can be taken up in its own right, the metal may form a stable complex with the assimilable ligand and enter cells by transmembrane transport of the intact metal complex. Examples of such 'piggy-back' uptake have been documented for such assimilable ligands as citrate and several amino acids (Section 3.1.5).

Further complications may arise through interactions between metal–organic complexes and other water chemistry variables such as pH, Al and Fe(III). In other words, the same elements that compete with potentially toxic trace metals for receptor sites on an organism's epithelial surface may also compete for the organic ligands that would otherwise sequester metals in an unavailable form. Note, too, that natural DOM may be biologically active in its own right. It has been shown to provoke adverse effects in some cases, but positive responses have also been reported (Steinberg, 2014).

To make sense of the sometimes contradictory information concerning the effect of DOM on metal bioavailability, we need to better understand events happening at or close to the surface of the receptor membrane. It is known that some dissolved organic molecules, including humic acid, behave as surfactants, accumulate at both organic and inorganic surfaces, and may be adsorbed to the surface. For organic (living) surfaces, this process is thought to involve hydrophobic interactions or the formation of hydrogen bonds. The degree and strength of the surface interaction is determined by the nature of the available surface, in particular its electrical charge and chemical composition. Surface adsorption of natural organic matter to cell surfaces may also be affected by pH. Campbell *et al.* (1997) demonstrated that adsorption of fulvic acid to whole *Chlorella* algal cells was inversely related to pH. How such phenomena at the membrane surface affect its permeability remains relatively unexplored (Wood *et al.*, 2011), though there is some evidence that the adsorption of DOM increases membrane permeability towards lipophilic metal complexes (Boullemant *et al.*, 2011). Questions yet to be answered include: (i) does the presence of DOM alter the charge on the cell surface in a way that alters its receptivity to other chemicals? (ii) Does the presence of DOM change the nature or configuration of the surface ligands on the cell surface in a way that alters their chemical receptivity? (iii) Does the DOM–metal complex present the metal to the cell membrane in such a way as to enhance its passage across the membrane?

The balance of this chapter deals with the ecotoxicology of individual metals and metalloids, of which Hg is presented in the greatest detail. For each metal, we have considered its occurrence, sources and properties, its environmental transport and behaviour, and its ecotoxicity.

6.8 Metal-specific Sections

In this section, we consider eight metals or metalloids that were chosen because of their recognized importance as potential environmental stressors: four non-essential elements (Hg, Pb, Cd, As) and four essential elements (Cu, Ni, Zn, Se). For each element, we consider how it enters the

Table 6.3 Abundances, emissions ratios and toxicities of arsenic, cadmium, copper, lead, mercury, nickel, selenium and zinc[a].

Element	Crustal abundance (ppm or mg/kg)[b]	Annual mining production in 2019 (metric tonnes × 1,000)[c]	Emissions ratio (anthropogenic/ natural)[d]	LC$_{50}$ *Hyalella azteca* (µM)[e]	Water quality guidelines (freshwater) (µM)[f]	Covalent index[g] $\chi_m^2 r$
Arsenic	1.8	33	1.7	7.8	0.32(III) 0.16(V)	2.34 1.37
Cadmium	0.15	25	0.2	0.0013	0.0018	2.71
Copper	60	20,400	5.9	0.56	0.022	2.64
Lead	14	4,500	5.2	0.005	0.016	3.86
Mercury	0.085	4	18.5	NA	0.003	3.68
Nickel	84	2,700	2.3	1.3	0.19	2.52
Selenium	0.05	-	0.2	0.54	0.14	3.25
Zinc	70	13,000	0.9	0.86	0.12	2.01
Aluminium	8.23×10^4	64,000	0.4	3.3	2.0	1.40

[a] Estimates of relative toxicity are for a named freshwater invertebrate (column 5) and for freshwater biota in general (column 6).

[b] Crustal abundances taken from www.knowledgedoor.com/. Aluminium is included as a reference non-trace crustal constituent.

[c] Data from USGS (2020).

[d] Emissions ratios (estimated ratios of anthropogenic emissions/natural emissions to the Earth's surface environment) taken from Klee and Graedel (2004) – Figure 6.5.

[e] Data taken from Borgmann *et al.* (2005); lower value = more toxic.

[f] Data taken from ANZECC and ARMCANZ (2018), for level of species protection = 95%; lower value = more toxic.

[g] χ_m = cation electronegativity; r = ionic radius. See Nieboer and Richardson (1980) and Section 6.2.1.

environment (sources), how it behaves in the environment (biogeochemistry), how it behaves within living organisms (biochemistry) and how it exerts its toxicity (ecotoxicity). This group of eight was also selected because the individual elements include both cations and oxyanions, and they exhibit a wide range of behaviours.

Table 6.3 serves as an introduction to the eight elements and compares their abundance in the Earth's crust, the annual mining production of each element, the importance of anthropogenic inputs to the environment, some measures of their relative toxicity, and their covalent index.

We have positioned mercury at the beginning of Section 6.8, in keeping with its undisputed ranking as the most ecotoxicologically problematic metal, and also because of its incredibly complex behaviour in the environment – yet its crustal abundance and mining production (with the exception of Se) are the lowest of all eight chosen elements!

6.8.1 Mercury

Metal/metalloid Mercury, Hg
Atomic number 80

Atomic weight 200.59

Estimated ratio [anthropogenic emissions/natural emissions] 18.5

6.8.1.1 Occurrence, Sources and Uses

Mercury can be found in the Periodic Table in Group 12 (Figure 6.1), just below cadmium (Cd), with which it shares some properties. Note that the unusual chemical symbol for mercury (Hg) originates from the Greek *hydrargyrum*, which means 'liquid silver'; its common name is quicksilver.

Occurrence, Sources

The estimated crustal abundance of mercury is 0.085 mg/kg (ppm), a lower value than for the other true metals considered in this section (Table 6.3). Its mining and purification have a very long history, going back at least 2,000 years. Unlike the case for many other metals, there are relatively few areas of the globe where Hg is mined; nearly three-quarters of the world's supply comes from just five Hg belts (Rytuba, 2003). Cinnabar (HgS) is the major mineral in most Hg deposits. Mercury is extracted from cinnabar by roasting in air or by heating with quicklime, CaO, in the absence of air. Both processes produce elemental Hg, which then undergoes further purification. Because of the high temperatures involved, these processes lead to the loss of some elemental Hg to the atmosphere.

Release to the atmosphere may also occur during other high-temperature processes, such as the smelting of sulphide ores, which often contain Hg as an impurity, and also as a result of natural phenomena (Box 6.6). The chemical form of Hg that is released into the environment varies from elemental through inorganic to organic (Section 6.8.1.2), with subsequent physical, chemical and biochemical changes in the receiving environment, which further affect its behaviour and bioavailability (6.8.1.3).

The volatility of elemental mercury and in particular its long residence time in the atmosphere ($t_{1/2}$ = 0.8–1.7 yr), along with its re-entrainment from historical Hg deposits, have resulted in widespread contamination.

Box 6.6 Some sources of mercury to the environment

Natural

- Forest fires.
- Geological degassing.
- Weathering of granitic and other geological media.
- Discharge from under-sea hydrothermal vents.
- Volcanic activity.

Anthropogenic

- Mining and purification of mercury.
- Burning of coal, other fossil fuels and wood.
- Smelting and other metal processing.
- Releases from the chlor-alkali industry and pulp and paper mills using chlorine bleaching.
- Releases from chemical factories producing fertilizers, solvents and other industrial chemicals.
- Incineration of municipal and medical waste.
- Disposal of thermometers, manometers and barometers.
- Disposal of and spills from light bulbs including fluorescent tubes.
- Agricultural application of Hg-based fungicides.

Re-emission of legacy Hg

- Re-entrainment of Hg deposited in the past (natural or anthropogenic).

Many of the concerns for the hazards represented by this element relate to the fact that the distribution of Hg released into the environment has become widespread, such that release at one location can have effects outside of the jurisdictions from which the release occurred. Mercury is often detected in areas far from the locations of its original release, and global Hg cycling results in its distribution to the most remote regions of the planet. For example, although Hg levels in lake sediments in the Arctic are generally lower than those at more southern latitudes, they show similar increasing patterns since the 1850s, suggesting that significant anthropogenic Hg inputs are reaching the Arctic (Muir *et al.*, 2009).

Uses

Certain properties of mercury and its compounds make it very 'useful' for humans. These characteristics include its high density, good conductance of electricity, its ability to amalgamate with other metals and its biocidal activity (the very property that makes it rank high on the list of substances of concern in ecotoxicology). These properties have led to an impressively large number of applications, often with environmental and human health consequences. Box 6.7 summarizes some of the many uses to which Hg has been put.

The relatively recent improved understanding of the hazards and risks that mercury and its compounds represent for human and environmental health has resulted in many of the applications shown in Box 6.7 being discontinued. The mining, production and purification of Hg have also decreased (Rytuba, 2003). Nevertheless, the legacy of these uses often remains in the environment (Section 6.8.1.2; Capsule 13.1). Furthermore, some applications are current, albeit regulated. Jurisdictions vary in their regulation of the uses of Hg, and some uses are outside any regulatory framework.

One characteristic property of elemental mercury that has contributed to its usefulness is its capacity to amalgamate with other elements, including copper, gold and silver. This capacity to amalgamate has been applied for the extraction of precious metals from certain types of ores, e.g., the *patio* process for extracting silver (Capsule 11.1), as well as a technically distinct process for extracting gold. Amalgamation is still used in the purification of gold by the informal or artisanal industries in many parts of the world, and this unregulated application is now considered to be the largest current source of Hg pollution on Earth (Esdaile and Chalker, 2018).

Box 6.7 Some uses of mercury, both extant and discontinued

- Elemental mercury:
  - in measuring devices: thermometers, barometers, manometers;
  - in the electrochemical production of chlorine, hydrogen and sodium hydroxide in the 'chlor-alkali' industry;
  - in the amalgamation process to extract gold and silver from ores; still used in artisanal gold extraction;
  - as amalgams in filling dental caries.
- Mercuric sulphide, HgS:
  - used by Palaeolithic painters to decorate caves in Spain and France 30,000 years ago;
  - still used to colour paints as a pigment and for its antifouling properties;
  - one of the red colouring agents used in tattoo dyes.
- Mercurous chloride (calomel, Hg_2Cl_2):
  - primary means of treatment for syphilis until the early twentieth century.
- Mercuric salts (e.g., $HgCl_2$, $Hg(NO_3)_2$):
  - historical use in medicinal products, including laxatives, worming medications and teething powders;
  - historical use as a topical disinfectant (merbromin); some topical ointments for skin lightening still contain mercury;
  - historical use in the making of felt hats from animal pelts (process known as 'carroting' and the cause of 'mad hatters' disease');
  - historical use in batteries and in photography.
- Present use as a preservative in some vaccines as thimerosal (thiomersal) (CH_3CH_2-Hg-S-C_6H_4-COOH).
- Historical use as a slimicide (phenylmercuric acetate) in pulp and paper manufacture, and as a fungicide (alkylmercurials) in sprays and seed coatings.

6.8.1.2 Biogeochemistry

In common with all metals, mercury occurs naturally, and the absolute amount on the planet does not change. This element is unique in existing as solid, liquid and volatile forms under normal environmental conditions. In common with other metals, the behaviour of Hg in the environment and its effects depend on its chemical form (speciation) and the medium in which it occurs.

Forms of Mercury

Mercury exists in three oxidation states: Hg(0), Hg(I) and Hg(II). Conversions among these oxidation states require electron transfers, in what are referred to as oxidation and reduction reactions. These reactions may occur abiotically, often under the influence of sunlight, or microbially. Elemental or metallic mercury, Hg(0), is liquid at ambient temperatures, has appreciable vapour pressure and exhibits volatility; it is the major form of Hg in the atmosphere and is slightly soluble in water (~56 µg/l at 25 °C). Divalent inorganic mercury, symbolized as Hg(II), is the dominant form in oxic surface waters, where it forms strong complexes with the chloride and hydroxide anions.

$$Hg^{2+} + 2\,Cl^- \rightleftharpoons HgCl_2^0 \qquad (6.7)$$

$$Hg^{2+} + 2\,OH^- \rightleftharpoons Hg(OH)_2^0 \qquad (6.8)$$

$$Hg^{2+} + Cl^- + OH^- \rightleftharpoons HgOHCl^0 \qquad (6.9)$$

The equilibria shown in equations (6.7)–(6.9) all lie very strongly to the right, such that the concentration of the free Hg^{2+} ion in aqueous solutions is vanishingly small. For example, speciation calculations for an Hg-contaminated freshwater stream yielded [Hg^{2+}] estimates of 10^{-27} M to 10^{-28} M (Dong *et al.*, 2010). In this respect, Hg differs markedly from the other free cations considered in this section (Cd^{2+}, Cu^{2+}, Ni^{2+}, Pb^{2+}, Zn^{2+}), all of which are present in natural waters at biologically relevant concentrations. However, despite this very low concentration of the Hg^{2+} ion, inorganic Hg(II) is biologically available, at least in part because one of the complexes shown above ($HgCl_2^0$) is lipophilic and can move across biological membranes by simple diffusion (Section 6.3.2; Figure 6.6, mechanism 3). In oxic waters, inorganic Hg(II) also binds strongly to the thiol functional groups in natural DOM. Under anoxic conditions, which can occur in sediment pore waters and in the hypolimnion of stratified lakes, inorganic Hg(II) will tend to bind to sulphide and precipitate out of solution.

Divalent mercury also occurs in methylated forms (See Section 6.8.1.3 for a discussion of the very important methylation process). Monomethylmercury (CH_3HgX, with X = Cl^- or OH^-) is soluble in water and is rather stable because of the presence of a covalent carbon–mercury bond. A second methylated species, dimethylmercury (CH_3HgCH_3), can also be formed but it is less stable, less water-soluble and more volatile than the monomethyl form. From an ecotoxicological perspective, monomethylmercury is the most important chemical form of the element. Frequently the abbreviated term 'methylmercury' means this form, and that convention is used throughout the present section.

Mercury Behaviour in Air, Water, Sediments and Soil

Mercury enters the atmosphere from natural sources (e.g., volcanoes) and as a result of anthropogenic activities (e.g., pyrometallurgical processes). In the atmosphere, Hg is largely present as gaseous Hg(0), but small quantities of Hg(I) and Hg(II) may be present in aerosol droplets or adsorbed to dust particles, as a result of photo-oxidation processes (Ariya *et al.*, 2015). Particulate and reactive forms travel relatively short distances (tens of km) and have relatively short residence times in the atmosphere, compared with the much longer residence time of gaseous Hg(0). The return of atmospheric Hg to the earth may occur by wet deposition (rain, snow) and by dry deposition (e.g., onto terrestrial vegetation). As has been mentioned earlier, these deposition processes tend to be relatively ineffective, resulting in long residence times for Hg in the atmosphere. However, in both the Arctic and Antarctic regions, during the 'polar sunrise' period that occurs in the spring of each polar year, marked Hg deposition events have been recorded (now known by their own acronym as 'AMDEs', or Atmospheric Mercury Depletion Events). Bromine free radicals were identified as the principal oxidant responsible for the conversion of Hg(0) to Hg (II), a change in oxidation state that favours Hg dissolution in atmospheric water and subsequent deposition.

Freshwater concentrations of total dissolved mercury in areas not affected by contamination generally range from 1 to 10 ng/l (0.5–5 pM); divalent inorganic Hg(II) dominates, with only 1–10% as methylmercury, and elemental Hg(0) <0.1% (0.01 to 0.10 ng/l). In pelagic marine environments, total dissolved Hg is lower (~0.2 ng/l), and in subsurface seawater, methylmercury can be >50% of total Hg. Given these extremely low concentrations, there is a very real danger of inadvertent contamination of natural water samples, particularly during the sampling procedure (Box 6.1). In water bodies affected by mining or industrial pollution, total Hg concentrations range from 10 to 40 ng/l (50–200 pM), and in drainage from geologically enriched areas total Hg can exceed 100 ng/l. Natural DOM influences the geochemical mobility of Hg; surface waters with high concentrations of DOM tend to have higher concentrations of dissolved Hg than those with less DOM. In addition to the effects of complexation by inorganic and organic ligands, Hg speciation in surface waters is also affected by pervasive photochemical reactions. For example, production of dissolved gaseous Hg in freshwater lakes is induced by solar radiation in the 400–700 nm range and is also influenced by the nature and concentration of the DOM present in the lake.

As is the case for other metals, sediments are the main repository for mercury in aquatic systems and are known to be major sites of microbial methylation of Hg (Section 6.8.1.3). Fine-grained sediments tend to have higher Hg concentrations than do coarser sediments, a result that reflects both the high **specific surface area** of fine-grained sediments (expressed as surface area/mass of the sediment) and their high content of organic matter. Fine-grained and organic-rich sediments normally support an active microbial community in the surface sediments, underlain by an anoxic layer. In anoxic sediments, Hg(II) will be scavenged from solution by the sulphide present in the sediment pore water.

In soils, mercury occurs mainly as Hg(0) and Hg(II). Depending on the redox conditions, some mercurous Hg(I) may also be present. Under natural conditions, most of the Hg(II) in the soil is either bound in the soil minerals or adsorbed onto organic or inorganic solids, with only a very small portion present in the soil solution. In soils with high organic matter, complexation with fulvic acids occurs, increasing the porewater Hg concentration. In neutral and low-organic-matter soils, iron and clay minerals are important adsorption sites. In a similar way to that mentioned for surface waters, Hg may also be released from soils to the atmosphere. Volatilization increases with pH and temperature, but this process also depends on the physical characteristics of the soil (e.g., its pore space), its organic matter content, its moisture content and the local meteorological conditions (Schluter, 2000).

The overriding message to retain from this brief description of the environmental behaviour of mercury is that this element is far more mobile than other metals and that this mobility is bidirectional. Mercury deposition to snow, surface waters, soil or terrestrial vegetation can be followed by re-emission to the atmosphere from these same sites. One of the consequences of this bidirectional mobility is that Hg doesn't 'stay put'. Legacy sites that were contaminated in the past continue to release Hg to the surrounding environment. As a result, the effectiveness of measures to limit new anthropogenic Hg emissions may well be more modest than was originally anticipated.

6.8.1.3 Mercury Methylation

Biochemistry

In the natural environment, mercury undergoes various transformations, the most significant of which for human health and ecotoxicology is **methylation**. Methylation, a type of alkylation, involves the addition of one or two methyl groups to divalent Hg, resulting in two organo-metallic molecules, methylmercury (CH_3HgX), where X is usually Cl or OH, and dimethylmercury ($(CH_3)_2Hg$). Methylation of certain metals and metalloids in the environment is predominantly mediated by microorganisms. The process was first described for Hg in 1967 and published in English in 1969 (Jensen and Jernelov, 1969). Abiotic Hg methylation has also been demonstrated, with natural organic humic matter as an agent, but in natural (as compared with experimental) situations, abiotic methylation is considered to be quantitatively unimportant.

Although mercury is not unique among metals and metalloids in undergoing methylation, the process for this metal has profound ecotoxicological significance because methylmercury is the most toxic and bioavailable form of Hg. Not only is methylmercury the form of the metal that has the most harmful effects on living organisms, it is also the form that accumulates in living tissues and biomagnifies in aquatic and terrestrial ecosystems (Section 6.8.1.5). In the environment, microbial methylation of Hg is normally mediated by anaerobic prokaryotic organisms, mainly in sediments but also at the oxic/anoxic interface in stratified lakes with anoxic hypolimnia and in wetland soil. However, the discovery of methylmercury enrichment in subsurface seawater (e.g., at depths from 150 to 1,000 m) cannot be readily explained by anaerobic microbial methylation (Wang *et al.*, 2020).

Since microbial mercury methylation is an intracellular process, inorganic Hg(II) must enter the microbial cell before methylation can occur. Both passive and facilitated transport pathways have been proposed, but the precise uptake mechanism has not been elucidated. Metals other than Hg normally enter cells via cation transporters (Section 6.3.2); since Hg is a non-essential metal, it would have to 'fool' a transporter designed for the uptake of an essential metal.

Once inside the microbial cell, the mercury ion is presumed to be shunted towards a methyl group donor. Transmethylation with vitamin B12 (methylcobalamin) was the first mechanism proposed for methylating Hg. Although this route has been shown to produce methylmercury abiotically, at unrealistically low pH values, it is no longer considered to be a viable hypothesis for biologically mediated methylation. The identity of the methyl group donor and the details of the transmethylation reaction remain elusive, but recent evidence does still point to a cobalamin-like corrinoid as the methyl donor (see below). As a group of organic biomolecules, corrinoids include a cyclic system containing four pyrrole rings, similar to porphyrins.

Microbiology

Considerable effort and emphasis have been devoted to the study of the biological production of methylmercury at the microbial level. Early reports demonstrated that anaerobic sulphate-reducing bacteria and possibly iron-reducing bacteria were responsible for the methylation, and sulphate reducers are still considered to be the main actors in this process. Recent studies on bacteria and **archaea** that methylate Hg have provided new information about the process and the organisms that have the capacity to produce monomethylmercury.

In 2013, Parks *et al.* reported on the newly discovered two-gene cluster *hgcA* and *hgcB*, required for mercury methylation by two sulphate-reducing bacteria (SRB), *Desulfovibrio desulfuricans* ND132 and *Geobacter sulfurreducens* PCA. The authors concluded that the two genes encode for a putative corrinoid protein, HgcA, and a 2[4Fe-4S] ferredoxin, HgcB; these two products are plausible candidates for the methyl donor and the electron donor required for reduction of the corrinoid cofactor, respectively. In the overall process, Hg(II) is taken up by facilitated transport, methylation takes place in the cytosol by an enzyme-catalysed process and CH_3Hg-X is exported from the cell.

With the availability of the genetic marker, *hgcAB*, Gilmour *et al.* (2013) tested whether the presence of the gene cluster was an accurate predictor of the capacity for methylation in other microorganisms. They confirmed significant methylmercury production in all of 15 bacteria that carried the *hgcAB* homologue, but none in three strains from which it was lacking. A phylogenetic reconstruction of the distribution of the gene pair revealed that, as previously suggested but not demonstrated, the capacity for methylation is more widespread than just sulphate- and iron-reducers. The Hg-methylating microorganisms that were described in this work showed considerable diversity, including Fermicutes (a phylum of bacteria not previously known to methylate) and methanogenic archaea. Gilmour *et al.* (2013) also point out that identification of additional methylators "significantly expands our knowledge of methylmercury-producing habitats".

The overall importance of methanogenic microorganisms in methylmercury production is still not well understood. Gilmour *et al.* (2018) conducted a detailed study of nine laboratory-cultured methanogenic archaea, including four methanogens that contain *hgcAB* but had not previously been tested for methylation. Sulphide inhibited Hg methylation, as is the case for bacterial methylators. In laboratory culture, rates of Hg methylation by these methanogens were comparable to those of sulphate-reducing and iron-reducing bacteria.

It has been suggested that bacterial methylation functions as a primary detoxification mechanism for mercury (e.g., Trevors *et al.* (1986)). However, this suggestion has not been confirmed, and data from pure cultures of Hg-methylating sulphate-reducing bacteria suggest that the ability to methylate Hg is constitutive and does not confer resistance to Hg (Gilmour *et al.*, 2011). Some bacteria have the capacity to demethylate and volatilize Hg; this part of the bacterial cycle is discussed, along with other possible detoxification mechanisms, in Section 6.8.1.8.

6.8.1.4 Biogeochemical Cycle

Before considering the bioaccumulation of mercury and its environmental toxicity, we should step back and briefly consider the global Hg biogeochemical cycle, many components of which have been described in the previous sections. As indicated in Figure 6.17, the atmosphere plays a very important role in global Hg circulation. During its passage through the atmosphere, elemental Hg(0) may be oxidized to Hg(II), a reaction that favours its removal in wet or dry deposition. Except in cases of major point-source contamination, Hg delivery to terrestrial and freshwater systems is dominated by this deposition from the

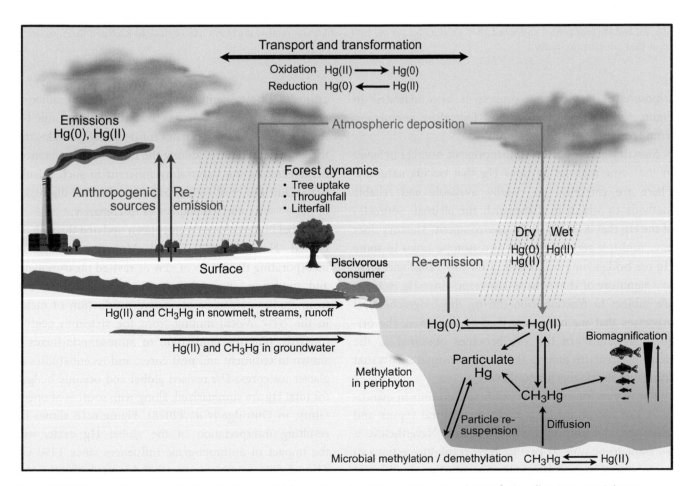

Figure 6.17 Mercury biogeochemical cycle. Composite figure based on Figures 1.1, 1.2 and 2.1 of *Canadian Mercury Science Assessment* (ECCC, 2016). Reproduced with permission from Environment and Climate Change Canada.

Table 6.4 Comparison of different estimates of mercury masses in air, soil and oceans.

Data taken from Outridge et al. *(2018)[a] and expressed in kilotonnes*

	Mason *et al.* (2012); AMAP/UNEP (2013)	Amos *et al.* (2013)	Zhang *et al.* (2014)	Lamborg *et al.* (2014)	Outridge *et al.* (2018)
Atmospheric Hg					
Total	5.1	5.3	4.4	n/a	4.4
Anthropogenic	3.4–4.1	4.6	3.6	n/a	3.6
Natural	1.0–1.7	0.7	0.8	n/a	0.8
Soil Hg (organic layers)					
Total	201	271	n/a	n/a	150
Anthropogenic	40	89	92	n/a	20
Natural	161	182	n/a	n/a	130
Oceanic Hg					
Total	358	343	257	316	313
Anthropogenic	53	222	66 (38–106)	58 ± 16	55
Natural	305	122	191	258	258

[a] The time point for designation of the 'natural' mercury state, and thus the quantification of 'natural' and 'anthropogenic' Hg masses, differed among studies: 2000 BCE in the 'pre-anthropogenic period' (Amos *et al.*, 2013); prior to 1450 CE (Zhang *et al.*, 2014), which preceded New World Ag, Au and Hg production; and about 1840 CE (Lamborg *et al.*, 2014), which was prior to the North American Gold Rush and the expansion of coal-fired combustion sources.

atmosphere. The reverse process is also indicated in Figure 6.17, with vertical arrows indicating re-emission from terrestrial and aquatic sources.

Mercury emitted from anthropogenic sources behaves in the same manner as does Hg that occurs naturally. There are currently no readily available and reliable methods by which to distinguish the ultimate source(s) of the Hg that is found in the environment. Mercury does have multiple stable isotopes, and isotope ratios in some Hg ore bodies may occur over a narrow range and serve as a signature of that source. However, since Hg isotopes are subject to fractionation during the biogeochemical processes that we have described to this point, the original 'signature' of the ore becomes obscured as the element circulates among the various compartments that constitute its biogeochemical cycle.

The two major areas of scientific uncertainty in quantifying the global mercury cycle are natural inputs and processes, and anthropogenic emissions. Nevertheless, if we want to be able to predict how regulatory actions to reduce emissions will affect Hg concentrations in different environmental compartments, biota and humans, quantification of the global Hg cycle is necessary. Recent

estimates of total, anthropogenic and 'natural' amounts of Hg in air, soils and oceans are provided in Table 6.4. A comparison of the estimates immediately indicates that there are discrepancies among the five selected references, demonstrating the uncertainty inherent in such calculations. Outridge *et al.* (2018) provide a detailed discussion of these values and the sources of discrepancy.

Global budgets have recently been updated through the United Nations 2018 Global Mercury Assessment, incorporating the results of new or revised measurements and modelling. Revised estimates have included data from studies of consumption and production of metals in the New World mining from the sixteenth century onwards, plus re-examination of atmospheric fluxes as shown in sediment and peat cores, and recent studies on glacier ice cores. The revised global and oceanic budgets for total Hg are summarized, along with sources of uncertainty, in Outridge *et al.* (2018). Figure 6.18 shows the resulting interpretation of the global Hg cycle, with the impact of anthropogenic influences since 1450 CE. This figure complements the biogeochemical cycle depicted in Figure 6.17 and clearly illustrates the quantitative importance of Hg re-emission.

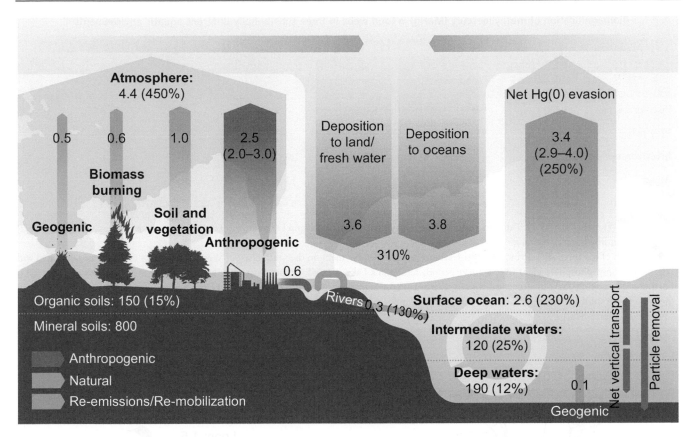

Figure 6.18 Updated global mercury budget showing the anthropogenic impact on the Hg cycle since the pre-anthropogenic period (prior to 1450 CE). Ranges (where available) are given in brackets after the best estimate values; percentages in brackets represent the estimated increase in mass or flux due to human activities since the pre-anthropogenic period. Mass units in kilotonnes, fluxes in kilotonnes per year. Figure reproduced with permission from Outridge *et al.* (2018). Copyright 2018 American Chemical Society.

Since mercury contamination does not respect geographical or political boundaries, and because of the resulting global distribution, regulating its emission is particularly problematic and remains an ongoing challenge. Paraphrasing Outridge *et al.* (2018), quantification of the global Hg cycle is a prerequisite for being able to predict how regulatory actions to reduce emissions will affect Hg concentrations in different environmental compartments, biota and humans. An initiative in response to this challenge is the **Minamata Convention**, a global treaty designed to control and reduce Hg emissions (UNEP, 2017).

6.8.1.5 Mercury Biomagnification

The behaviour of methylmercury in the ecosystem is the key to all major concerns about the risk posed by mercury to humans and the environment. Although there were some early examples of Hg reaching toxic levels in terrestrial food webs, the focus here is on aquatic environments. At one time there was a risk of exposure for

terrestrial birds through consumption of agricultural seeds that had been treated with methylmercury, but now the main Hg source for birds is through the aquatic food chain, for piscivorous birds and their predators. Similarly, terrestrial mammals that consume fish, such as mink and otter, accumulate much higher Hg concentrations than do herbivores or omnivores.

Methylmercury bioaccumulates in living organisms and can be transferred from primary producers, through primary and secondary consumers, to top-level consumers, as illustrated in Figure 6.19. More insidious and highly significant is the **biomagnification** of methylmercury by successive bioaccumulation and transfer steps in the aquatic food chain (bioconcentration, bioaccumulation and biomagnification are defined in Table 3.3). As a result, the high-level consumers can accumulate Hg concentrations that are many orders of magnitude greater than the concentrations in water. For example, if the concentration of methylmercury in lake water is considered to have an absolute value of 1, then approximate

Table 6.5 Biomagnification of methylmercury (MeHg) in food webs in three substantially different aquatic environments[a,b].

Food web component	Australian Marine Embayment		Wisconsin seepage lake		Tropical lake	
	MeHg (ng/g wet wt)	Total Hg present as MeHg (%)	MeHg (ng/g wet wt)	Total Hg present as MeHg (%)	MeHg (ng/g wet wt)	Total Hg present as MeHg (%)
Piscivorous fish	2,300	>95	650	>95	392	87
Prey fish	450	93	100	>90	26	55
Invertebrates	150	45	20	29	–	–
Algae	7	10	4	13	<0.3	<1
Water (μg/l)	nd[c]	nd	0.00005	5	0.000067	5

[a] Data from Wiener *et al.* (2003), Table 16.2. Australian embayment, Francesconi and Lenanton (1992). Wisconsin lake, Watras and Bloom (1992). Tropical lake, Bowles *et al.* (2001).

[b] Description of the three sites: Princess Royal Harbour, a marine embayment on the south coast of Western Australia that was contaminated with Hg from a super-phosphate plant over a 30-yr period; Little Rock Lake, a small, temperate seepage lake (no inflowing or outflowing streams) in northern Wisconsin (USA) that received Hg largely from atmospheric deposition directly onto the lake surface; and Lake Murray, a tropical lake in the remote Western Province of Papua New Guinea.

[c] nd = not determined.

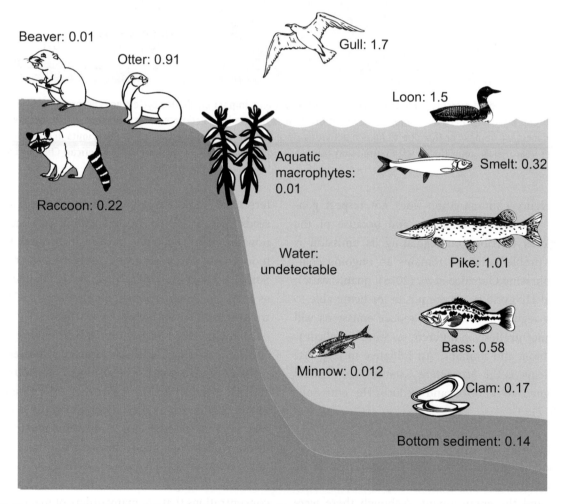

Figure 6.19 Mercury biomagnification in the aquatic ecosystem. The data were collected for total Hg in different compartments of a remote lake ecosystem and show increasing concentrations with trophic level (i.e., biomagnification). Mercury concentrations in biota are expressed as μg Hg/g fresh weight and as μg Hg/g dry weight for the sediment. Reproduced from Wright and Welbourn (2002) with permission from Cambridge University Press.

bioconcentration factors for microorganisms such as phytoplankton are 10^5; the bioaccumulation factors for zooplankton and planktivores are $\sim 10^6$, and those for piscivores like fish, birds and humans are $\sim 10^7$ (Watras *et al.*, 1998). These factors are rough estimates and will vary from one site to another, but the general trend is consistently reproducible. As an example, biomagnification values for methylmercury at three different sites (one contaminated by industrial Hg releases, one that received Hg from atmospheric deposition, and a third remote lake) are presented in Table 6.5.

Aquatic plants also bioaccumulate mercury, but herbivorous consumers such as beaver, and omnivores such as raccoons, generally have lower Hg concentrations than do piscivores, suggesting that Hg accumulation by aquatic plants may not contribute a great deal to biomagnification.

The accumulated mercury in high-level consumers has a very long biological residence time. For a piscivorous and high-level consumer fish such as pike (*Esox lucius*), the half-life of methylmercury is as long as 170 d. Several factors influence the actual concentrations of methylmercury in fish. These factors include the size (age) of the fish, the species, and lake water chemistry as well as the extent of Hg contamination. Figure 6.19 illustrates some fairly typical data for a single lake, one that had no known anthropogenic Hg sources. The data show the effects of species and size such that the predatory fish species have higher Hg concentrations than those lower in the trophic chain, and older or larger fish have higher concentrations than do younger smaller ones. It is significant that the form of Hg that accumulates in fish muscle is almost entirely the highly toxic methylmercury. However, for ease and economy of analytical determination, most fish Hg concentrations are expressed as total Hg, and for fish, this is presented on a wet weight basis. X-ray absorption spectroscopy has been used to probe how the methylmercury is bound in fish tissue, and unsurprisingly the Hg is bound to thiol groups (CH_3Hg-S-R) in a complex that spectroscopically resembles the methylmercury–cysteine complex.

Biomagnification often results in mercury concentrations in fish muscle that render the piscivorous fish unfit for human consumption, even though the fish may appear palatable and show no obvious adverse effects. Fish tend to be more tolerant of Hg than are the wildlife that are likely to consume them. It has been suggested that storage of methylmercury in fish muscle, a tissue that seems less sensitive to methylmercury than other tissues, has a protective effect. The question of human consumption of Hg-contaminated fish is addressed in Box 12.4, where we consider regulatory efforts to minimize the risk to human consumers, as well as in Sections 4.3.5 and 12.3.3.

Before leaving the topic of biomagnification, we should ask the question: "Why is mercury special? Why is it the only metal that enters the aquatic environment in inorganic form and for which trophic transfer consistently leads to biomagnification?" The answer to this question is 'hidden' in Chapter 3, where we address the biodynamics of metal uptake, storage and loss. For biomagnification to occur, the overall uptake of the metal must consistently exceed the rate of metal efflux plus any growth dilution that occurs. Because methylmercury is lipophilic, it can cross epithelial membranes with ease, and once it enters the intracellular environment, it binds strongly (with a covalent bond) to thiol groups. As a result, uptake rates are high and efflux rates are very low, and the Hg taken up by organisms in lower trophic levels tends to be conserved within the consumer organism. No other metal exists in the environment in a lipophilic, thiol-seeking form analogous to methylmercury.

6.8.1.6 Environmental Factors Affecting Mercury Bioaccumulation

Net methylation of mercury generally varies as a function of (i) the overall microbial activity at a particular site, which will depend on the amount of labile organic matter and the ambient temperature, and (ii) the concentration of 'bioavailable' Hg(II), which will be influenced by such factors as the pH and the availability of inorganic and organic ligands (i.e., the speciation of Hg(II)). Within this general framework, a number of reproducible trends have been identified.

Lakewater pH

In the course of research on lake acidification, studies in different countries showed that for soft-water lakes in Scandinavia and North America there is a statistical relationship between pH of the water and mercury in fish; all other things being equal, fish from acidic lakes have higher Hg concentrations than do fish in more alkaline pH lakes. This trend is seen over a pH range that supports normal fish populations (i.e., there is apparently no harmful effect of acidity *per se* on fish growth or reproduction). Interpreting this effect is not straightforward, since changes in pH can in principle affect both the biogeochemical behaviour of Hg and the physiology of

Box 6.8 Mechanisms that have been advanced to explain the increase in fish mercury concentrations in lakes with low pH

Abiotic

1. Low pH mobilizes existing inorganic mercury from the sediment and watershed, resulting in a net increase in the total amount of Hg available for methylation.
2. The pH-dependent partitioning of methylmercury between sediment and water column may favour the water column at low pH.

Biotic

3. Acid deposition alters the structure and function of food webs such that methylmercury bioaccumulation is enhanced.
4. Low pH has a direct effect on Hg bioaccumulation by aquatic biota.
5. Slower growth rates of fish in low-pH lakes result in increased Hg *concentrations* although there is no difference in the *amount* of fish Hg compared with larger faster-growing fish.

the native fish species. Note, too, that in field studies, it is often not possible to relate cause and effect because some factors co-vary. Box 6.8 summarizes some of the hypotheses that have been put forward to account for this relationship with pH and still remain viable, including geochemical effects (mainly through bioavailability), effects on food webs, and direct effects on the capacity of fish to bioaccumulate Hg. More than one of these mechanisms may be in operation simultaneously. Some of the more obvious explanations, such as the idea that low pH might stimulate microbial production of methylmercury, have not been borne out by experimental studies.

Flooding

Increased concentrations of mercury in fish have consistently been reported for recently flooded areas, particularly in those affected by hydroelectric dam construction (Bilodeau *et al.*, 2017; Cebalho *et al.*, 2017) where there is no known additional Hg source. This phenomenon has also been demonstrated in an experimentally flooded wetland area in the Canadian Experimental Lakes Area (ELA) (Heyes *et al.*, 2000).

The most widely accepted explanation for the 'reservoir problem' is that dissolved organic matter (DOM) from the vegetation in the original terrestrial ecosystem stimulates the microbial methylation of existing soil mercury to such an extent that, combined with biomagnification, predatory fish in recently impounded reservoirs accumulate unacceptably high Hg concentrations. Factors that promote methylation include increased microbial activity and increased prevalence of anoxic conditions. Note too that reservoirs are typically subject to an annual drawdown and refilling cycle. It has been suggested that this cycle of wetting and drying the littoral zone of the reservoir stimulates Hg methylation by regenerating key reactants such as sulphate, labile DOM and bioavailable Hg(II). There is no simple remedy because there is no controllable Hg source. Aside from a proposal to strip vegetation from the land prior to flooding, the only resort to date has been to impose restrictions on consumption (Box 12.4).

Mercury Interactions with Selenium

Studies in areas where there is a natural or anthropogenic gradient in selenium concentrations have revealed an inverse correlation between Se concentrations in the exposure medium and mercury concentrations bioaccumulated in the aquatic biota. For example, Belzile *et al.* (2006) collected zooplankton, mayflies, amphipods and various fish species in lakes located downwind from several major smelters that constituted an important point source of selenium. As distance from the smelters increased, aqueous Se concentrations in the lakes decreased but dissolved Hg concentrations showed no spatial trend. This observation is consistent with the release of Se during the smelting process being more important than the release of Hg (on a molar basis, the concentrations of Se in the nickel ores were more than 1,000 times higher than the Hg concentrations). In contrast, tissue concentrations of methylmercury and total dissolved Se showed clear negative correlations, indicating that higher Se concentrations were protecting the biota from Hg accumulation. Belzile *et al.* (2006) suggested that the Se might decrease the methylation of Hg or its accumulation at the lower levels of the food chain, but the exact mechanism(s) involved remain obscure. Laboratory experiments have shown that dietary Se (added as selenite, selenocysteine or selenomethionine) reduces the retention of methylmercury in freshwater and marine test species (Bjerregaard *et al.*, 2011; Bjerregaard and Christensen, 2012).

Other Factors

In addition to the factors mentioned to this point (ambient pH, selenium concentrations and immersion of terrestrial vegetation), several other environmental factors influence methylmercury bioaccumulation. For example, within the range of ambient conditions, an increase in temperature increases the rate of microbial Hg methylation. However, such a temperature increase would also be expected to increase the rate of bacterial demethylation, meaning that the overall effect of an increase in temperature is unclear. The influence of dissolved sulphate is similarly ambiguous. When added to an anoxic sediment, sulphate will act as an electron acceptor and promote the action of sulphate-reducing bacteria, which are known Hg methylators. However, the product of sulphate reduction is sulphide, which will tend to immobilize the Hg(II) as mercuric sulphide and render it less available for methylation. Recent work from Sweden suggests that increased sulphate loading both increases net methylmercury formation and favours redistribution of the methylmercury from the peat soils to their pore water (Åkerblom *et al.*, 2020).

The 'age' of the Hg(II) substrate also appears to influence the rate of mercury methylation, in that freshly deposited Hg is more readily methylated than Hg that has resided in the system for some time. This ageing effect was elegantly demonstrated in the METAALICUS experiment carried out in the Experimental Lakes Area in Ontario, Canada. Enriched stable Hg isotopes were added to a whole lake, and the isotopic signature of the methylmercury produced in the lake was shown to be enriched in the freshly added Hg (Harris *et al.*, 2007). Similar ageing effects are seen when other metals are spiked into soils or sediments. They are usually explained in terms of the slow conversion of initially adsorbed metals into more stable or occluded forms.

6.8.1.7 Mercury Bioaccumulation and Monitoring

Given the capacity of mercury to bioaccumulate and biomagnify, researchers have published a great deal of information on body burdens or other measures of Hg accumulation in biota. Uptake of Hg by biota is clearly a precursor to effects, although it does not automatically result in adverse effects (Section 6.5). A useful metric in this context is the tissue threshold-effect level (t-TEL), which has been developed to relate body Hg concentrations in fish or their prey to potentially harmful effects (the t-TEL is similar to the critical body 'burden', as defined in Section 2.3.4.3). In this way, risk can be evaluated from body concentration values such as those collected in field studies. For example, Beckvar *et al.* (2005) evaluated studies that reported tissue Hg residues and associated biological effects in freshwater fish.

Accumulation, aside from its potential to be 'translated' into harmful effects, is used in **biomonitoring** of exposure (Sections 4.3.3.1, 4.3.5). The shunting of mercury into non-living components such as fur and feathers (which represents a form of 'shedding' and thus detoxification) provides a non-destructive means of monitoring exposure. To interpret such data, it is necessary to establish a relationship between fur or feather concentrations and bioaccumulation and/or toxic effects in living tissues of the same species. For fish, the well-established process of Hg accumulation in muscle has been used extensively as a means of monitoring exposure. For the terrestrial environment, non-vascular plants, lichens and mosses provide a very effective means of monitoring metal deposition from the atmosphere. Although this approach does not measure soil concentrations (these plants have no roots) or biological effects, nevertheless it provides a type of monitoring that has proved effective in tracing trends in environmental concentrations and deposition patterns (Harmens *et al.*, 2004). Beyond their use in biomonitoring, measurements of Hg concentrations in tissues are also of value in terms of assessing the safety for consumers.

6.8.1.8 Ecotoxicity

In its metallic and inorganic forms, mercury has long been recognized as being a risk for occupational health, but the significance of methylmercury and its environmental toxicology came to the fore much later, with the recognition of food-chain transfer in the events related to Minamata disease (Section 1.2). In the late 1950s and the following decade, serious outbreaks of non-occupational Hg poisoning occurred in Minamata, Japan (Clarkson, 2002). Fish and shellfish from Minamata Bay and other areas around the Shiranui Sea, polluted by the discharge of methylmercury from an industrial operation, contained up to 15 µg/g (wet weight) methylmercury. For comparison, Hg concentrations in unpolluted marine fish and shellfish ranged from 0.01 to 1.5 µg/g.

Methylmercury poisoning through fish consumption in Minamata, as well as the 1971 Iraqi incident of consumption of grain that had been treated with methylmercury as a preservative (Chapter 1), led to illness and deaths in the human population. The focus of these early incidents of Hg poisoning was on human health, but the exposure was not occupational and the compound of concern was organometallic Hg. This marked a change in focus for our concern about Hg toxicology. By the late 1950s it was recognized that piscivorous fish, birds and mammals were accumulating methylmercury and showing neurological disorders (Jernelov *et al.*, 1975). Some of the early examples resulted from the consumption of Hg-treated seed by small mammals, from which the Hg was passed up the food chain and biomagnified, but eventually it was recognized that the problem was more widespread (Section 6.8.1.5).

Mechanisms of Toxicity

Mercury has a high affinity for the sulphydryl groups in endogenous biomolecules, notably peptides, proteins and enzymes, and its binding to these molecules may result in their dysfunction and subsequently produce multiple effects (Section 6.4). This interaction of Hg with sulphydryl groups is thought to be responsible for the well-established neurotoxic effects of Hg and its compounds. Mercury can also induce oxidative stress, indirectly, and organomercury compounds such as methylmercury and phenylmercury are among the strongest known inhibitors of cell division. Any or all of these interactions may translate into acute or chronic toxic effect at the organism level, although links between biochemical changes and whole organism effects are not consistently available (Box 6.9).

General Effects on Fish and Wildlife

Scheuhammer *et al.* (2012) reviewed in some detail a number of studies that have shown effects of mercury on behaviour, neurochemistry and endocrine function in fish and wildlife at realistic levels of environmental exposure. These authors show that current levels of environmental methylmercury exposure in some ecosystems are sufficient to cause significant biological impairment, both in individuals and in whole populations. They noted that toxicological studies on fish and wildlife need to focus on linking biomarkers of methylmercury exposure and associated oxidative stress to effects on reproduction and the resulting population changes. At the level of the individual animal, there is a lack of information about the genetic basis for Hg-related neurotoxic and other biological changes, including the genetic basis for species differences in sensitivity to methylmercury.

In the following sections on the toxicity of mercury and populations at risk, we frequently quote Hg concentrations in water, food and consumers. Unless otherwise stated, all these concentrations have been determined as total Hg rather than as methylmercury.

Fish

Much of our understanding of the harmful effects of mercury on fish comes from experimental work, the design of which can factor out the relative importance of dietary versus direct exposure from water. For fish, methylmercury is more toxic than inorganic Hg, and dietary exposure is responsible for most of the uptake and toxicity. In common with many other toxicants, Hg affects fish reproduction and early life stages more than it does adult fish.

Reduced spawning success and fecundity in fish have been demonstrated in laboratory studies with mercury. For example, Hammerschmidt *et al.* (2002) reported that dietary methylmercury concentrations over a range from 0.88 to 8.5 µg Hg/g dry weight all reduced gonad size, delayed spawning and reduced the reproductive success of female fathead minnows (*Pimephales promelas*). These effects occurred at dietary concentrations encountered by predatory fishes in actual aquatic systems with methylmercury-contaminated food webs, implying that exposed fish populations in the field could be adversely affected by this contaminant. More subtle effects of methylmercury on fish behaviour have also been reported. For example, experiments on a common forage species, the golden shiner (*Notemigonus crysoleucas*), demonstrated that diet-borne methylmercury altered the predator-avoidance behaviour of this species, making them potentially more vulnerable to predation (Webber and Haines, 2003).

Based on such results, and specifically on discussion by Hammerschmidt *et al.* (2002), methylmercury

Box 6.9 Signs of mercury toxicity

- Induction of oxidative stress (inhibition of antioxidant processes).
- Impairment of the immune system.
- Impaired kidney function (fish, birds, mammals).
- Neurological changes (fish, birds, mammals).
- Behavioural changes (e.g., attenuation of predator avoidance in fish).
- Reproductive failure (birds, fish, mammals).

concentrations in some natural fish populations would appear to be high enough to cause damage to reproductive processes, including effects on the production of sex hormones. However, there are rarely any assessments of the number of fish populations potentially at risk from methylmercury. In an exception to this statement, Sandheinrich *et al.* (2011) used a threshold effects concentration for whole-body methylmercury of 0.30 µg Hg/g (wet weight) to evaluate the ecotoxicological risk for fish in a field study of sexually mature female fish. They collected walleye (*Sander vitreus*), northern pike (*Esox lucius*), smallmouth bass (*Micropterus dolomieu*) and largemouth bass (*Micropterus salmoides*) in the Laurentian Great Lakes and in interior lakes, impoundments and rivers of the Great Lakes region. With more than 43,000 measurements of mercury in fish muscle from more than 2,000 locations, they found that sexually mature female fish at 3% to 18% of sites were at risk of injury. Most fish at increased risk were from interior lakes and impoundments. The authors concluded that "fish at a substantive number of locations within the Great Lakes region are potentially at risk from methylmercury contamination and would benefit from reduction in mercury concentrations". These data give credence to the idea that not only piscivorous birds and mammals, but fish as well may be at risk from current Hg levels in some freshwater ecosystems.

Wild Mammals

Piscivorous and predatory birds and mammals are at risk as a result of biomagnification of mercury in aquatic food chains. Although Hg accumulation in the liver may be diagnostic of exposure, it is not a reliable indicator of effects. Effects can be related to Hg concentrations in the brain, where methylmercury dominates. The most widely studied wild mammals are otter (*Lontra canadensis*) and mink (*Neovison vison*). Brain Hg concentrations in these animals with overt methylmercury poisoning (i.e., with symptoms resembling those noted in the Minamata case described earlier) are typically >5 µg/g wet weight (Wiener *et al.*, 2003). Tissue concentrations in experimentally exposed dietary-dosed otter and mink with overt symptoms of Hg poisoning are reported as 20–100 µg/g wet weight in the liver and >10 µg/g in the brain.

Effects of mercury on marine mammals include neurotoxicity, immunosuppression and endocrine disruption. In their review of the biological effects of contaminants on Arctic wildlife and fish, Dietz *et al.* (2019) concluded that Hg posed a risk for some marine top predators,

notably the polar bear (*Ursus maritimus*), beluga whale (*Delphinapterus leucas*), pilot whale (*Globicephala melas*) and hooded seal (*Cystophora cristata*). This type of study not only indicates potential risk for mammals and human consumers, but also illustrates the long-range transport of Hg far from its sources.

Birds

In birds, as in fish, methylmercury acts as an endocrine disrupter and affects reproduction. Experimental treatments for a limited range of bird species, along with field studies of wild birds, have produced some fairly consistent data on mercury concentrations in tissues and eggs that relate to toxic effects, with an emphasis on chronic effects, notably reproduction. Non-lethal chronic effects of lower Hg concentrations are less well-studied than those related to reproduction. In a number of bird species, egg-Hg concentrations >1 µg/g wet weight are associated with impaired hatchability and embryonic mortality, and brain-Hg concentrations >3 µg/g wet weight have been linked to mortality in developing bird embryos. Fuchsman *et al.* (2017) reported thresholds for adverse effects of methylmercury on avian reproductive success; these threshold values ranged from 0.2 µg/g to >1.4 µg/g in diet, 0.6 µg/g to 2.7 µg/g in eggs, and 2.1 µg/g to >6.7 µg/g in parental blood (all concentrations expressed on a wet weight basis).

For birds, surveys such as measurements of mercury in blood and feathers can be diagnostic of exposure and thus used for biomonitoring. Examples of such an approach come from studies of loons (*Gavia immer*). Loon chicks, because of their limited territories, restricted movement and relative ease of sampling, provide useful models for monitoring exposure and assessing the effects of Hg on piscivorous birds in their natural habitats. Sampling loon blood is non-destructive and has yielded valuable information. For example, Hg concentrations in blood and eggs are correlated; blood reflects dietary uptake and eggs are a well-established depuration route for methylmercury (Evers *et al.*, 2008). Blood and feather Hg levels in loon chicks are indicative of Hg obtained from prey items acquired almost entirely in their natal lakes. Blood Hg levels in adult loons also reflect recent dietary uptake, but concentrations in feathers integrate Hg that has been acquired over longer periods, including at migratory sites. Studies in northeastern North America have shown consistently that these piscivorous birds have blood-Hg concentrations high enough to reduce loon production below levels required to sustain stable population sizes

(Evers *et al.*, 2003). Numerous studies of other piscivorous, predatory or scavenger birds have reported similar results. Extensive reviews can be found in Scheuhammer *et al.* (2012, 2015).

Terrestrial Plants

Mercury concentrations in plants are generally low and, except when organomercury fungicides have been applied, methylmercury is not a major proportion of total Hg in plants. Most studies on Hg and plants have arisen from concern for contaminated soils, e.g., from mining and coal burning, artisanal metal extraction, the application of mercurial fungicides, or as experimental work. In these cases, most concern has been for human health through exposure to plants as food. Rice in particular has been identified in parts of China as a possible source of excess dietary Hg for humans.

Soil Microorganisms and Invertebrates

Relatively little information is available about the toxicity of mercury to terrestrial microorganisms and invertebrates. This contrast with the situation for aquatic environments presumably reflects the low bioavailability of Hg in the soil environment and the relative unimportance of Hg biomethylation in soils (with the exception of wetland and paddy soils). In their review of the relative sensitivity of different terrestrial organisms, Mahbub *et al.* (2017) suggested that soil microorganisms were the most sensitive to Hg contamination, based on microbial community measurements such as respiration and nitrification, and they speculated that the biogeochemical cycles of carbon and nitrogen might be affected. Although the design of experiments has made it difficult to derive any realistic effect concentrations for Hg in soils, such biogeochemical effects could indirectly affect terrestrial plants and soil microfauna.

6.8.1.9 Detoxification and Tolerance

The two main biochemical mechanisms of detoxification for mercury are binding to metallothionein (Sections 6.5.1, 6.5.4.4) and binding to selenium through the formation of metabolically inert forms (6.8.8.3). Inorganic Hg(II) and methylmercury are handled differently, and detoxification of methylmercury appears to be less effective than for inorganic Hg.

Metallothioneins

Inorganic mercury is a potent inducer of metallothionein (MT) synthesis, and as a soft metal it binds very strongly to the MT molecule, displacing all other divalent metals. Accordingly, one might anticipate that Hg and MT would remain in the cytosol and that their concentrations would be strongly correlated. This is sometimes the case, but more often the correlation is weak or non-existent. For example, Romero *et al.* (2016) measured concentrations of Hg and MTs in foetuses, calves, juveniles and adults of the Franciscana dolphin (*Pontoporia blainvillei*) from Argentina. Mercury concentrations were highest in liver for all specimens, but they found no significant relationship between Hg and MT concentrations in any tissues.

Similarly, in studies where the subcellular partitioning of mercury has been determined in a given organ, using the type of approach that is described in Section 6.5.2 (Figure 6.14), the HSP fraction containing MT-like peptides typically does contain some Hg, but most of the Hg is found elsewhere. Insoluble granules play a role in sequestering Hg in some cases (Bebianno *et al.*, 2007), and in others the Hg is associated with the heat-denatured protein (HDP) fraction, which is usually interpreted as an indication of incomplete detoxification (Barst *et al.*, 2016). It seems likely that MT is involved in handling Hg in the intracellular environment, but not as the ultimate detoxification sink. This behaviour contrasts markedly with what happens to silver or cadmium, two other soft metals that remain in the cytosol as complexes with MT. This difference in behaviour for Hg is probably attributable, once again, to the ubiquitous presence of methylmercury. If Hg were present in the intracellular environment in this form, it would bind strongly to thiol groups in proteins (i.e., the HDP fraction referred to above), whereas inorganic Hg(II) would remain with the MT.

Selenium

To this point, we have emphasized the propensity of mercury to seek out and bind to thiol groups. However, the affinity of Hg for selenium is about 10^6 times stronger than for sulphur. Numerous observations from laboratory and field studies, including studies on dietary Se, have shown a relationship between Se and Hg in biota, with indications of alleviation of Hg toxicity and, in some examples, a lowering of the proportion of organic to total Hg in the presence of Se. Studies of the mechanism by which such detoxification occurs are often confounded by the fact that Se is a required nutrient as well as a potential toxicant (Section 6.8.8). However, despite this potential complication, all the proposed mechanisms involve the formation of Hg–Se compounds with limited internal bioavailability and mobility. In most cases the precise identity of the Hg–Se compounds has not been established, but indications are that a range of different compounds may be involved (Khan and Wang, 2009). Candidate molecules include mercury selenide (HgSe), similar biominerals containing both Se and S ($HgSe_xS_{1-x}$) and methylmercury selenide (($CH_3Hg)_2Se$).

In agreement with the earlier suggestion that detoxification pathways for inorganic mercury and methylmercury differ, Kehrig *et al.* (2015) examined Hg, methylmercury, metallothionein and Se levels in the Magellanic penguin (*Spheniscus magellanicus*). They concluded that both Se and metallothionein are involved in the detoxification of inorganic Hg in the penguin liver, whereas only Se is involved for methylmercury.

Demethylation

In addition to the detoxification of Hg by binding to reduced sulphur (e.g., metallothionein) or selenium, the demethylation of methylmercury has also been proposed as a detoxification mechanism in some animals. Piscivorous birds demethylate methylmercury, mainly in their liver, and similar results have been found for otter and mink. There are inter-species differences in the apparent ability to demethylate methylmercury, which may affect their relative sensitivity to mercury toxicity. Otters (*Lontra canadensis)* generally accumulate higher concentrations of Hg than mink (*Neovison vison*), but in otter brains, organic Hg comprises only 74% of total Hg, compared with 90% for mink (Scheuhammer *et al.*, 2007). Haines *et al.* (2010) sampled otter and mink from eastern Canada and corroborated this observation, indicating that otters may be better able to metabolize organic Hg into an inorganic form. Their studies also demonstrated a significant correlation between Hg and Se in brains of otter but not in mink. Indeed, there have been suggestions in the literature that Se may be involved in the demethylation reaction, but no putative mechanism has been proposed.

6.8.1.10 Mercury Highlights

Originally known as an occupational hazard, mercury was identified as an environmental toxicant in the 1960s; despite decades of intensive research and various regulatory interventions, it remains at the top of the list of metals of environmental and ecotoxicological concern. Two major processes are responsible for current risks: microbial methylation forming methylmercury, the most toxic form of the element, and food chain biomagnification, resulting in toxic concentrations in the diet of consumers at the top of the food chain, including birds, terrestrial mammals and humans.

The environmental behaviour of mercury and the related ecotoxicological risks are unique among metals. The microbial methylation of Hg is largely responsible for this special status, in part because of the lipophilicity and intrinsic toxicity of methylmercury, but also because it forms strong covalent bonds with protein thiol groups. It follows that Hg taken up by organisms in lower trophic levels tends to be conserved within the consumer organisms, leading to progressive increases in internal Hg concentrations (i.e., biomagnification).

Except in cases of major point-source contamination, the major impacts of mercury on ecosystems and their components result from atmospheric deposition of Hg in various forms. Mercury moves in the atmosphere from primary sources and furthermore can be re-emitted from previously contaminated sites – in both cases, it can travel over long distances before redeposition. This complicates the calculation and identification of sources of environmental Hg, the identities of which are significant in terms of its control and regulation.

Arguably the most complex and hazardous of all metals, mercury remains incompletely understood in terms of its biochemistry and ecotoxicology. Along with its ease of atmospheric transport and resulting ubiquitous distribution, these knowledge limitations affect our ability to regulate the element effectively. The risk related to Hg and its compounds cannot be entirely eliminated, but scientific understanding of the physico-chemistry, transport and biology of Hg continues to advance. This can only lead to better regulation and management practices, and a potential for remediation or recovery from damage that has already been done.

6.8.2 Cadmium

Metal/metalloid Cadmium, Cd
Atomic number 48
Atomic weight 112.4
Estimated ratio [anthropogenic emissions/natural emissions] 0.2

6.8.2.1 Occurrence, Sources and Uses

Cadmium (Cd), a relatively rare but widely dispersed trace metal (crustal abundance 0.15 ppm; CCME (2014)), has assumed importance as an environmental contaminant only within the past 60 years or so. It is a common by-product of mining and smelting operations for zinc, lead and other non-ferrous metals. Other sources of release include coal combustion, refuse incineration, steel manufacture, and the application of some phosphate fertilizers in agriculture (Cd is present as an impurity in some phosphate ores). It is used in the electroplating of steel, as a component of various alloys, as a pigment in plastics manufacture, and in nickel–cadmium batteries. This latter use has diminished with the introduction of lithium–hydrogen batteries (Section 14.3), but for some applications Ni–Cd batteries are still favoured because of their greater reliability and stability. Worldwide production of Cd in 1935 totalled only 1,000 metric tonnes per year, but since the 1990s it has been reasonably stable at about 24,000 metric tonnes per year.

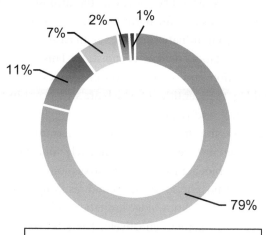

2%   1%
7%
11%
79%

Data from 2005, cited in CCME (2014)

- Ni–Cd batteries
- Pigments
- Coatings
- Stabilizers (plastics)
- Alloys

6.8.2.2 Biogeochemistry

Cadmium appears below zinc in the second row of transition elements in the Periodic Table (Figure 6.1) and shares some common properties with Zn. In uncontaminated fresh waters, Cd concentrations are normally very low (<1 μg/l; <9 nM), but much higher values have been reported in waters affected by the mining and smelting of sulphide-rich ores. Like Zn, Cd is not a redox-sensitive element, and only the Cd(II) oxidation state is environmentally relevant. In fresh waters, when present at low background concentrations (e.g., less than 10^{-11} M), most of the dissolved Cd is normally bound to natural dissolved organic matter (DOM), whereas in Cd-contaminated systems the proportion of Cd present as the free Cd^{2+} ion increases to values as high as or greater than 90%. As salinity increases, cadmium progressively forms a series of chloro-complexes ($CdCl^+$, $CdCl_2{}^0$, $CdCl_3{}^-$) until, in seawater, the proportion of free Cd^{2+} is less than 5% (Figure 6.4).

When introduced into the aquatic environment, cadmium tends to bind to suspended particles, both biotic (e.g., plankton) and abiotic (e.g., eroded soils), and then settle to the bottom sediments. For example, in a whole-lake Cd addition experiment, carried out in the Canadian Experimental Lakes Area from 1987 to 1992, 93% of the Cd added to the epilimnion of the lake during the ice-free season ended up in the bottom sediments (Lawrence *et al.*, 1996). In surface oxic sediments, Cd associates with iron and manganese oxyhydroxides and with detrital solid organic matter, and its concentrations in the sediment pore water are controlled by adsorption equilibria with these phases. Adsorption reactions for cations are inherently sensitive to pH. As the pH decreases, Cd^{2+} is displaced from the surface of the sediment particles and its concentration in the pore water increases (Section 11.1). In contrast, in anoxic sediments lying beneath the oxic surface layer, Cd concentrations in the pore water are controlled by precipitation reactions with sulphide (Section 6.3.3).

6.8.2.3 Biochemistry

Although cadmium is not generally regarded as having any physiological function, it has been shown to ameliorate zinc deficiency successfully in some marine phytoplankton species (Xu and Morel, 2013). This example, which involves a Cd carbonic anhydrase enzyme replacing the Zn anhydrase that is normally present, appears to be the only demonstration of an apparent nutritive function for Cd. As a non-essential metal, Cd^{2+} must mimic an essential metal in order to enter a living cell; both calcium (Ca) and Zn transporters have been implicated as routes for Cd uptake. These transporters are sensitive to changes in pH and water hardness, with the proton (H^+) and hardness cations (Ca^{2+}, Mg^{2+}) competing with Zn and Cd for binding to the transporter; as a result, Cd uptake tends to decrease as the ambient pH drops (i.e., increasing H^+) or as hardness increases (Section 6.3.2). In both freshwater and euryhaline crustaceans, i.e., irrespective of salinity, a major influence on Cd uptake is the activation of the ATP-driven epithelial Ca pump immediately following moult.

Cadmium is a soft metal, with a high affinity for thiol groups (R-SH), and once inside a cell it may displace Zn from various subcellular ligands (zinc finger proteins, metallothionein, glutathione, phytochelatins – Section 6.8.6.3). Despite its status as a redox-insensitive element, Cd is known to induce oxidative stress, probably by inhibiting one or more of the various enzymes that are involved in the cell's protection against reactive oxygen species (ROS) (Section 6.4).

Defence mechanisms against cadmium may involve shutting down the transporter that is responsible for allowing Cd to enter the cell, or more generally binding the Cd^{2+} ion internally and limiting its availability within the cell. This latter mechanism usually involves binding the Cd^{2+} to thiol groups in metallothioneins (ubiquitous in animal cells) or phytochelatins (present in plant cells and in some invertebrates) – Section 6.5. Cadmium may also be detoxified by sequestration into granules, a mechanism that is prevalent in unionid molluscs where the Cd replaces the calcium in the granules that normally act as a Ca reserve in the mollusc gills. There is also some evidence that selenium can protect against Cd toxicity in freshwater fish, but the exact mechanism of this protection has not been determined. Recall that Se also protects against Hg toxicity (Section 6.8.1.9).

Exceptions to the BLM for cadmium are rare, but can occur when Cd is bound to a bioavailable ligand, i.e., if the ligand can cross the biological membrane and 'piggy-back' the Cd into the cell (Figure 6.6, mechanism 2); citrate and thiosulphate can play this role. Another exception occurs when the ligand forms a neutral and lipophilic complex with Cd, in which case the complex can bypass membrane-bound transporters and simply diffuse across the cell membrane (Figure 6.6, mechanism 3); diethyl-dithiocarbamate is such a ligand.

Once inside the cell, cadmium can exert its toxicity by a variety of means, most of which involve substitution for calcium or zinc (Box 6.10). For example, Cd can perturb normal Ca homeostasis by affecting the activity of Ca-ATPase, a membrane-bound transporter responsible for Ca efflux. Alternatively, it may interfere with the functioning of catalase, glutathione reductase and superoxide dismutase, all of which normally act to limit the build-up of reactive oxygen species within the cell. Cadmium can also act as an endocrine disruptor, notably in freshwater fish, by inhibiting cortisol secretion. In laboratory experiments on aquatic organisms, these biochemical responses to chronic Cd exposure have been shown to result in adverse effects on growth, reproduction, development and behaviour. In an analysis of North American data for chronic toxicity of Cd to freshwater species, Mebane (2010) indicated that the four most sensitive genera to chronic exposures were *Hyalella* (amphipod), *Cottus* (sculpin), *Gammarus* (amphipod) and *Salvelinus* ('char' trout).

In the field, cadmium often co-occurs with other metals, such as copper and zinc. In such cases, it is often difficult to attribute observed negative effects on the

6.8.2.4 Ecotoxicity

As is the case for copper, nickel and zinc, the best predictor of toxicity for waterborne cadmium toxicity is the concentration of the free Cd^{2+} ion in the exposure solution (i.e., the biotic ligand model, BLM; Section 6.3.2). In natural waters containing chloride or natural organic matter, both of which can form complexes with Cd^{2+} and thus reduce free Cd^{2+} concentration, the toxicity of Cd is also reduced. Similarly, increased Ca^{2+} concentrations in hard waters afford some protection against Cd toxicity, as a result of competition between Ca^{2+} and Cd^{2+} at the epithelial surface.

Box 6.10 Signs of cadmium toxicity

- Induction of oxidative stress (e.g., inhibition of antioxidant processes).
- Perturbation of homeostasis of essential cations (Ca^{2+}, Zn^{2+}).
- Enzyme inhibition induced by binding of Cd^{2+} to protein thiol groups (R-SH).
- Endocrine disruption (e.g., impaired cortisol secretion).
- Inhibition of fish olfactory responses to chemical stimuli (Capsule 6.1).

resident biota to a single metal. Nevertheless, there are several field examples of aquatic and terrestrial systems where Cd has been identified as the 'bad actor'. For example, Case Study 6.1 showed that Cd detoxification was incomplete in yellow perch that had been collected from lakes representing a Cd concentration gradient. Consistent with this diagnosis of incomplete detoxification, progressive biochemical and physiological responses were detected in yellow perch collected along the metal concentration gradient and these responses could be related to the internalized Cd dose (Couture *et al.*, 2015). Among these responses were an attenuated cortisol stress response *in vivo* and a lower secretory capacity in response to adrenocorticotropic hormone *in vitro*. During the study period (2002–2008), the ambient dissolved Cd concentrations in the study lakes ranged from 0.3 to 6.6 nM; in seven of the eight lakes, the ambient Cd concentration exceeded the Canadian chronic water quality guideline for Cd in soft waters (0.04 µg/l; 0.4 nM) with hardness <17 mg/l $CaCO_3$.

A terrestrial field example was reported by Larison *et al.* (2000), who investigated a link between bone fragility and kidney failure in the snow quail (*Lagopus leucurus*) and cadmium present in their diet. The snow quail is a small member of the grouse family that is found in high-altitude habitats in western North America. The birds were sampled in an area of Colorado, USA, known as the 'ore belt', a zone characterized by high soil concentrations of various metals, and in a reference area. One of the genera favoured by the snow quail in its diet, the willow (*Salix* sp.), concentrated Cd in its leaf buds and newly grown shoots to levels that were an order of magnitude higher than those found in other possible food sources. These high Cd concentrations were accompanied by low background calcium concentrations, another factor favouring Cd uptake. The ingested Cd accumulated in the quail kidney over the birds' lifetimes, and levels high enough to be toxic were found in 44% of adult birds. Confirmatory evidence of toxicity was obtained by histopathological analysis, which indicated that 57% of the adult birds from the ore belt area showed signs of renal damage. Cadmium-induced renal failure has been linked to Ca balance and shown to affect skeletal integrity in other vertebrate organisms. Many other herbivores frequent this habitat, and they too may be at risk. Inversely, it is also possible that Cd plays a role as a deterrent to herbivore grazing, i.e., that Cd accumulation is an advantage to the plant itself (Stolpe *et al.*, 2017)!

6.8.2.5 Cadmium Highlights

Cadmium is a relatively rare trace element but is often present as an impurity in copper and zinc ores. The exploitation of such sulphidic ores constitutes the major anthropogenic source of Cd, but it is also present in some phosphate fertilizers.

As a non-essential metal, Cd must mimic an essential metal to enter a living cell; both calcium and zinc transporters have been implicated as routes for Cd uptake. It is a soft metal with high intrinsic toxicity (Table 6.3), but under most exposure scenarios detoxification of Cd is reasonably effective: metallothionein induction in animals, phytochelatin synthesis in plants. Nevertheless, there are examples of Cd toxicity at the ecosystem scale in the vicinity of mining and smelting activities.

6.8.3 Lead

Metal/metalloid Lead, Pb
Atomic number 82
Atomic weight 207
Estimated ratio [anthropogenic emissions/natural emissions] 5.2

6.8.3.1 Occurrence, Sources and Uses

Metallic lead is dense, soft, non-corrosive and malleable, with a relatively low melting point. The symbol Pb comes from the Latin *plumbum*. The most common naturally occurring form is lead sulphide, PbS, called galena. Other natural species include lead carbonate ($PbCO_3$), sulphate ($PbSO_4$) and chromate or crocosite ($PbCrO_4$). Anthropogenic Pb in the environment includes salts and oxides released from mining, primary and secondary metal processing, and fossil fuel combustion. An additional anthropogenic source in petrol (gasoline) is the additive tetraethyllead ($(CH_3CH_2)_4Pb$) and combustion products from this compound. Since many of these processes occur at high temperature, much of the Pb is emitted to the surrounding environment via the atmosphere in the form of fine particles, where it is subject to long-range transport, though less so than in the case of Hg.

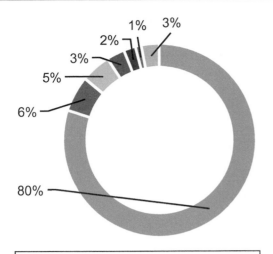

- Batteries
- Rolled - extruded products
- Pigments
- Ammunition
- Alloys
- Cable sheathing
- Other

The history of lead applications goes back at least 7,000 years in that we know the ancient Egyptians used Pb for weights and anchors, cooking utensils and piping, solder and pottery glaze. The Romans were probably the greatest users of Pb, with applications in plumbing, ceramics and pewter, up to the fifth century CE, until the Industrial Revolution in Europe. Beginning late in the eighteenth century in mainland Europe, Great Britain and the New World, Pb use increased in building materials (plumbing, roofing, paints) as well as in many industrial products. In the nineteenth century, Pb started to be used in large storage batteries, and at present this industrial sector accounts for about 80% of lead usage.

The large-scale industrial uses of and resulting demand for lead in the eighteenth and nineteenth centuries resulted in increases in mining and refining of the element, processes that exposed many people to its effects before there was any widespread recognition of the risks. Lead mining expanded in Great Britain and soon after this in Germany and the United States. In the twentieth century, Canada, Mexico and Australia also became major producers of Pb, by which time some of the European sources had been depleted. Mining and smelting have left a legacy of Pb contamination in the environment.

Modern uses of lead underwent another major increase beginning in the early twentieth century with the development and widespread use of tetraethyllead (TEL), an anti-knock agent and lubricant that was added to gasoline to raise its octane rating. The combustion of TEL in the internal combustion engine results in the presence of very small particles of Pb as a bromide or other halide in the emitted exhaust. These Pb forms originate from the addition of chlorides and bromides to remove lead from the combustion chamber of the engine. Large Pb-containing particles emitted from internal combustion engines tend to contaminate the local roadside environment, whereas small particles can be entrained in the atmosphere and transported over long distances. A similar reasoning applies to smelter emissions, where the Pb 'footprint' of a smelter can extend 50 km or more downwind (Telmer *et al.*, 2006). Much of the available information about the behaviour of Pb in the environment comes from such 'legacy' sites. Box 6.11 shows the main uses of Pb, indicating both historical and current applications. Changes in Pb use in Europe in the 2000–2015 period can be found in Table 6.6.

Recognition of the adverse human health effects of lead dates back several centuries (Pattee and Pain, 2003). Occupational exposure must have occurred through the mining and processing of Pb, and personal exposure resulted from the widespread use of Pb in plumbing and domestic articles. In biohistory, the solubilization of Pb from cooking pots and the use of 'sugar of lead' (lead acetate) to sweeten wine have been suggested as a major reason behind the fall of the Roman Empire (due to mental impairment of the rich who could afford the Pb salt). Exhumed bones of Roman aristocrats do show high Pb content, which supports this hypothesis, but it is by no means universally accepted. Nevertheless, the now well-known biochemical behaviour of ingested or inhaled Pb accumulating in bones certainly indicates that the Romans were exposed to it.

Even though many uses of lead have been controlled or banned, past actions have left a legacy of Pb contamination, particularly in soils and sediments. Research on its environmental chemistry and biochemistry continues to expand, aided by improved sampling and analytical methods and other technical advances.

Box 6.11 Uses of lead

Historical uses

- Lead plumbing for water distribution.
- As a sweetener: sugar of lead, lead acetate: $Pb(-O-C(O)CH_3)_2$.
- In traditional medicine – some still in use and spreading as naturopathic medicine becomes popular in the Occident.
- Early western medical uses from Roman times – lead plasters were used to treat sores, boils and other skin complaints (still in limited use today: lead oxide, PbO).
- Oral medications: lead acetate mixed with sulphur was prescribed for tuberculosis.
- Oral medications: lead acetate and opium were given to cure diarrhoea and as a sedative to treat hysteria and convulsive cough: $Pb(-O-C(O)CH_3)_2$.

Current uses

- Acid–lead batteries for automobiles, trucks, etc.
- Manufacture of corrosion and acid-resistant materials used in the building industry.
- Pigments in paint for artists and domestic and industrial use: lead(II) chromate, lead(II, IV) oxide, and lead(II) carbonate.
- Ceramic glazes: lead oxide PbO.
- Lead alloys including pewter and solder: elemental Pb
- Ammunition: elemental Pb.
- Fishing weights: elemental Pb.
- Radiation shields in medical equipment: elemental Pb.
- To dampen noise and vibration in industrial machinery: elemental Pb.
- Anti-knock additive in gasoline: tetraethyllead, $Pb(CH_2CH_3)_4$ (limited use in the 21st century).

Table 6.6 **Changes in lead use in Europe from 2000 to 2015.**

Use	2000	2005	2015
Storage batteries	64%	71%	84%
Rolled and extruded products	14%	14%	6%
Shot and ammunition	4%	3%	4%
Cable sheathing	1%	1%	1%
Lead compounds	13%	7%	4%
Alloys, solders, miscellaneous	4%	4%	1%
Total (kilotonnes)	1,515	1,457	1,517

Data from the International Lead Zinc Study Group, Lisbon, Portugal (ILZSG, 2017).

6.8.3.2 Biogeochemistry

Lead is a relatively unreactive post-transition metal. The metal and its oxides react with acids and bases, and it can form both coordinate and covalent bonds. It exists in inorganic compounds as Pb(II) and less frequently as Pb(IV), typically in organometallic species with covalent carbon–lead bonds. Three of its isotopes are endpoints of major nuclear decay chains of heavier elements, and they have been used to trace fallout from nuclear tests and accidents (Table 10.2). The sources of the metal accumulated in tissues can be characterized by the isotopic 'fingerprint' in environmental compartments (Komarek et al., 2008). In another type of application, the isotope ^{210}Pb provides reference dates for palaeo-ecological investigations of sediments and ice cores.

In freshwater systems, dissolved lead tends to be present at relatively low concentrations (<0.2–2 µg Pb/l; <1–10 nM) but with higher concentrations at industrially contaminated sites; dissolved Pb concentrations tend to increase at low pH and in the presence of organic ligands such as humic and fulvic acids. The divalent cation (Pb^{2+}) dominates under very acidic conditions, with lead carbonate and hydroxide ($Pb(OH)_2$) complexes becoming more important as the pH approaches neutrality and becomes slightly alkaline; complexation to natural dissolved organic matter also increases as the pH rises. Because of its tendency to form complexes with inorganic and organic ligands, the proportion of Pb present in fresh waters as the free Pb^{2+} ion is consistently low. For example, Vega and Weng (2013) determined that less than 0.5% of the total dissolved Pb in the Rhine River, Germany, was present as the free Pb^{2+} ion. Dissolved lead concentrations in the open ocean are typically $\leq$10 ng/l ($\leq$0.05 nM) and in the inorganic Pb fraction, lead chloride and carbonate complexes dominate.

Lead accumulates in soils and sediments, with sediments considered to be the ultimate sink. Sequential extraction studies (Section 6.3.3) suggest that most Pb in soils and sediments is bound to iron and manganese oxides, as well as to organic matter. Background Pb concentrations in sediments are typically less than 40 mg/kg, but in surficial sediments close to point sources of contamination, mean concentrations can be as high as 3,000 mg Pb/kg in freshwater sediments and even higher in marine harbours receiving various industrial and sewage inputs.

Not only does lead bind to suspended particles and settle out of water bodies, it also tends to remain immobilized in the settled sediments (Section 6.2.2). This means that sediments collected from deep lakes can be viewed as a temporal record of past Pb concentrations in the water column. Environmental scientists have used sediments in this manner to monitor how changes in environmental regulations have affected Pb contamination of the aquatic environment. An example of this use of sediment cores is shown in Figure 6.20, which illustrates how Pb concentrations increased as a result of the Industrial Revolution (e.g., use of Pb in paints, plumbing and as an anti-knock agent in gasoline) and then declined in response to the adoption of various environmental regulations, including the progressive removal of TEL from gasoline.

In the 1980s, following the discovery of mercury methylation (6.8.1.3), considerable research was devoted to investigating other metal methylation reactions,

including those for lead. Methylated Pb compounds have been detected in the environment, but since such products are used commercially, their detection in the environment cannot be taken as proof that they are derived from the biomethylation of Pb(II). Thayer (2002) noted that the methyl–lead linkage is very labile; unless the monomethyllead is stabilized in some manner, this lability undermines the idea that it could persist long enough to be used as the substrate for subsequent methylation steps. Thayer concluded that "the role of this reaction (i.e., biomethylation) in the biogeochemical cycling of lead is probably marginal". Researchers seem to have lost interest in this question, and the 2002 conclusion remains valid.

6.8.3.3 Biochemistry

As a non-essential metal, lead must enter living cells via divalent metal transporters designed to take up essential metals such as Ca^{2+}, Cu^{2+}, Mn^{2+} or Zn^{2+}. When they are present in the exposure medium, the hardness cations (Ca^{2+} and Mg^{2+}) compete with Pb^{2+} for binding to these transporters and protect against Pb accumulation and toxicity. Possible detoxification mechanisms for Pb include its complexation by phytochelatins

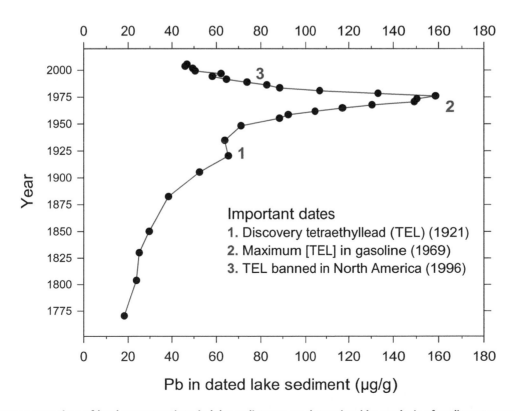

Figure 6.20 Changes over time of lead concentrations in lake sediments, as determined by analysis of sediment cores (Lake Tantaré, Quebec, Canada). Figure redrawn from data supplied by André Tessier and originally published in Gobeil *et al.* (2013).

or metallothioneins (Section 6.5.4) and its intracellular precipitation with (poly)phosphate. For example, on entering a plant cell Pb can induce the synthesis of phytochelatins, but it is unclear whether the binding of Pb(II) to phytochelatins is an effective detoxification mechanism. Scheidegger *et al.* (2011) determined that in long-term exposures of the unicellular alga *Chlamydomonas reinhardtii* to Pb, the accumulated Pb concentrations exceeded the measured phytochelatin concentrations. Subsequent work with the same alga, using synchrotron-based resonant X-ray emission spectroscopy, confirmed that intracellular Pb was not bound to reduced sulphur, but rather to oxygen, likely to phosphate groups (Stewart *et al.*, 2015). In yeasts, Pb^{2+} enters the cytoplasm, binds to glutathione and is shunted into vacuoles (Jarosławiecka and Piotrowska-Seget, 2014).

In most vertebrates, exposure to lead results in accumulation of the metal in liver, kidney and bone. Lead storage in bone is a form of detoxification, although bone Pb can be remobilized in immature animals. The metabolism of Pb in lower animals and terrestrial plants is less well understood than for vertebrates. For plants, there is some evidence that Pb may enter plant roots via Ca channels; extracellular Ca protected against Pb toxicity, and in the presence of a known Ca-channel blocker, Pb uptake decreased (Clemens, 2019).

Within the cell, lead can mimic and replace calcium or zinc, and because of this, Pb affects enzyme systems, notably δ-aminolevulinic acid dehydratase (ALAD). This Zn enzyme is essential in the biosynthesis of haem, and the inhibition of its activity (in blood) is diagnostic for Pb toxicity in birds, fish and mammals. Inhibition of ALAD activity is one of the very few biomarkers that is specific to a single metal. Lead affects other stages of haem synthesis, producing variations in some metabolites (e.g., zinc protoporphyrin) and resulting in anaemia. Lead may also affect other metalloenzymes, by displacing the essential metals that are needed as cofactors, such as Mg^{2+} and Mn^{2+}.

6.8.3.4 Ecotoxicity

In laboratory toxicity tests with a variety of aquatic organisms, lead conforms to what the biotic ligand model predicts for a divalent, non-essential metal. The hardness cations afford some protection against Pb toxicity, but the major factor limiting its toxicity is complexation by natural organic matter (DeForest *et al.*, 2017). Signs of Pb toxicity are summarized in Box 6.12.

Box 6.12 Signs of lead toxicity

- Inhibition of ALAD; perturbation of haem and chlorophyll synthesis.
- Perturbation of homeostasis of essential cations.
- Renal oedema and intranuclear inclusions.
- Interference with calcium in its intracellular signalling role.
- Interference with exocytosis of neurotransmitters in neuronal cells.
- Behavioural changes (locomotion, balance, depth perception, visual recognition).
- Neurotoxicity and a loss of control of vital functions, including muscle tonus.
- Cognitive impairment.

Plants

Terrestrial plants. Rooted plants accumulate lead from soil into roots, through the apoplast pathway or via transport across the plasma membrane into the symplast. For most reported examples, there is little translocation into the shoots; Pb is bound in the cell walls and as well its movement is blocked by **Casparian strips** within the endodermis of the root (Section 3.2.6). For above-ground plant parts, direct deposition of particulate Pb onto leaves has been reported and some of this is incorporated into the leaf cuticle. Although this does not appear to have toxic effects on the plant, this foliar Pb is nevertheless of possible significance for consumers, although this exposure route is mainly of concern for human consumers of vegetables.

Lead toxicity to terrestrial plants has been demonstrated under experimental conditions but at very high doses. Phytotoxicity results from changes to cell membrane permeability, reactions of Pb with active groups of different enzymes involved in plant metabolism or with the phosphate groups of ADP or ATP, and from the displacement of essential ions by Pb (Pourrut *et al.*, 2011). At contaminated sites such as mining areas, Pb-tolerant populations of plants have resulted from selection of tolerant individuals.

Aquatic plants. Similar to terrestrial plants, rooted emergent or submerged aquatic and wetland plants accumulate lead mostly in their rooting portions. Lead in sediment is often not bioavailable to plants, and toxic effects only occur at concentrations much higher than those found in most freshwater systems (Table 6.7). Experimental studies with selected aquatic plants, e.g.,

Table 6.7 **Dissolved lead in surface water compared with effects concentrations for aquatic organisms**[a].

	Typical lead concentrations in the environment ($\mu g/l$)	EC_{50} ($\mu g/l$)	EC_{10} or NOEC ($\mu g/l$)
Fresh water	0.18–1.00		
Algae, aquatic plants		27–388	6.1–348
Aquatic invertebrates		26.4–202,530	2.4–963
Fish		52–3,598	18–1,559
Salt water	0.02–0.05		
Algae, aquatic plants		73–1,232	12–1,234
Aquatic invertebrates		45–356,625	9.2–1,410
Fish		1,427–405,592	44–437

[a] Based on ILA (2015).

Eichhornia crassipes (Ali *et al.*, 2013), have shown the potential for phytoremediation by aquatic plants that accumulate Pb but do not show any toxic effects.

Wildlife

There is an extensive literature, including field and laboratory studies, concerning the toxicity of lead to waterfowl, other birds and their predators and scavengers. The environmental sources are often Pb in ammunition, and the pathways include feeding by birds in sediments and soils in shot-over areas. Lead fishing weights also poison loons and swans, and dabbling ducks that sieve sediments for food. If death is not immediate, birds exhibit signs of toxicity including emaciation, paralysis and limbs or necks that mimic the characteristic 'drop-wrist' condition of human Pb poisoning. These effects are probably due to nerve damage and the loss of neural control of muscle function. Langner *et al.* (2015) studied blood lead levels (BLL) in the free-flying golden eagle, *Aquila chrysaetos*, in western North America and found that 58% of the birds had BLL > 0.1 mg Pb/l. Ten per cent were clinically lead-poisoned with BLL > 0.6 mg Pb/l, and 4% were lethally exposed with BLL > 1.2 mg Pb/l. These and similar studies show that blood Pb concentrations can be linked to exposure for fish, birds and mammals, and that these BLLs can be related to overt toxicity (Pain *et al.*, 2005; Fallon *et al.*, 2017). Such studies have resulted in the banning of Pb gunshot in some places or for some uses (Pain *et al.*, 2019), but regulation and control are still patchy, and there are as yet no formal criteria or guidelines as to the use of BLLs to assess risk.

Fish and Aquatic Invertebrates

The toxicity of lead to fish and invertebrates in water is generally low compared with other trace metals. For example, Mebane *et al.* (2012) conducted controlled experiments with stream invertebrates and fish, and found that for Pb the lowest EC_{50} values were on the order of 100 times higher than ambient dissolved Pb concentrations, with the range of EC_{50} values for cutthroat trout (*Oncorhynchus clarkii lewisi*) and invertebrates overlapping. Table 6.7 shows the typical range of dissolved Pb concentrations in surface water and supports the contention that these are normally far lower than the effects concentrations for aquatic organisms. Possible exceptions might include aquatic sites affected by uncontrolled or legacy sources of Pb contamination.

Food Chain and Ecosystem Effects

Ecosystems affected by extremely high concentrations of lead may show decreased biodiversity through impoverishment of species or chronic effects such as reproductive failure. Decomposition of soil organic matter may be decreased by Pb contamination. Such ecosystem effects have been reported for individual and extreme situations, but Pb-specific generalizations of whole ecosystem effects are difficult to establish. What has been repeatedly shown is that with little accumulation of Pb in leaves, food chain transfer from plants is limited, and although there may be limited trophic transfer for other links in the food chain, there is no evidence that Pb biomagnifies (Pattee and Pain, 2003). Indeed, the term biominification or bio-dilution has been used to describe the behaviour of Pb (and many other metals) in food webs.

6.8.3.5 Lead Highlights

Lead is a relatively unreactive post-transition metal, occurring naturally in predominantly sulphide ores. Although Pb has been mined and used for many centuries, its applications have evolved continually since 7000 BCE to the present day. Until 1900, mining and processing of the metal were the major sources of Pb contamination of local soils and waterways. Early in the twentieth century, tetraethyllead (TEL) was introduced as an anti-knock agent in gasoline for automobile engines. This resulted in much more widespread distribution of Pb via long-range atmospheric transport of the small Pb-containing particles produced in the internal combustion engine.

Typical lead concentrations in surface water are normally much lower than the effects concentrations for aquatic organisms. The hardness cations afford some protection against the toxicity of dissolved Pb, but the major factor limiting its toxicity is complexation by natural dissolved organic matter (DOM). On the other hand, a major source for wildlife is Pb shot and Pb fishing weights, which accumulate and persist in sediments. Lead uptake and intoxication have been measured in sediment-feeding birds, and this issue is still of major concern.

Lead is a neurotoxin, and the potential effects of occupational Pb exposure to human health are well recognized. Many uses have been phased out or are strictly regulated. The banning of TEL by many jurisdictions has resulted in a dramatic decrease in the atmospheric deposition of Pb, both locally and from long-range transport. However, even with the banning of Pb in plumbing and from most paints, a Pb legacy remains. Drinking water in older plumbing systems as well as soils contaminated from flaking and dust from old painted structures remain as a source for Pb for humans and the environment.

6.8.4 Copper

Metal/metalloid Copper, Cu
Atomic number 29
Atomic weight 63.54
Estimated ratio [anthropogenic emissions/natural emissions] 5.9

6.8.4.1 Occurrence, Sources and Uses

The element occurs as metallic (native) copper, Cu(0), as well as sulphide, carbonate and silicate compounds. The principal ores are chalcopyrite ($CuFeS_2$), malachite and azurite ($CuCO_3(OH)_2$), cuprite (Cu_2O), copper glance (Cu_2S) and atacamite ($Cu_2Cl(OH)_3$). Pure Cu has excellent electrical conductance, and this has led to its use in electrical cables, transformers, electrical motors, electromagnets and as a thick coating for lightning conductors. The metal is also a good conductor of heat, and this property has favoured its use in cooking vessels (as a coating on pans), for plumbing and roofing and in heat sinks. Copper has long been used in pure metallic form or alloyed with other metals, as bronze (copper–tin) and brass (copper–zinc). In pure or alloyed form, Cu has been used to make tools, ornaments, statuary, jewellery, coins and vessels for food and beverages. Copper salts have been used and some continue to be used as fungicides, molluscicides (notably for antifouling paints on ship hulls and other submerged marine items) and algicides.

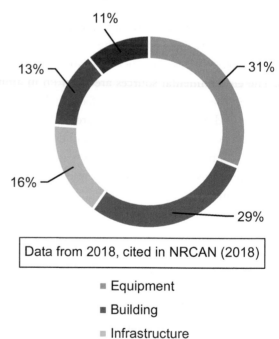

Data from 2018, cited in NRCAN (2018)

■ Equipment
■ Building
▪ Infrastructure
■ Transport
■ Industrial

The discovery of the fungicidal effect of copper is an example of a 'happy accident'. The grape vine mildew *Peronospora viticola* showed up as a major problem in French vineyards around 1878. The study of plant pathology was relatively young at that time, having been born 40 years earlier with the discovery of the fungal pathogen that caused potato blight. The vine mildew disease spread rapidly, with devastating effect on the vines. In 1882, Professor Millardet, who had studied the life history of the fungus, was walking in affected vineyards of St.-Julien in Médoc and noticed that the vines along the path were still in leaf, whereas elsewhere they were bare from the mildew disease. Upon enquiry, he learned that the growers used Cu salts and lime to mar the appearance of the grapes and deter passers-by from stealing them. Millardet experimented with solutions of copper sulphate and quicklime to control the mildew and in 1885 he communicated his successful findings to the Society of Agriculture for la Gironde (Large, 1940). This cocktail became known as the Bordeaux mixture and was the first fungicide to be used on a large scale, worldwide.

The major processes that result in copper mobilization are linked to its extraction from the parent ore (mining, milling, smelting and heap leaching). Weathering of slag material to oxides and silicates combined with **acid mine drainage** result in the mobilization of Cu that would otherwise be insoluble. Copper is also dispersed into the environment in agriculture (pest control products), through fossil fuel burning and by the use of waste from sewage treatment operations to fertilize soils. Natural sources include smoke plumes from forest fires and volcanic activity, with small particles carried by wind.

Copper mining and processing, the latter including thermal, electrochemical and hydro-metallurgical processes, have led to contamination of soils, water and sediments. Although the use of the metal goes back many centuries, demands for Cu were relatively small until the eighteenth century with the rise of the Industrial Revolution. At that time, the demand for copper (and for many other metals) increased and mining grew in response (Rainbow, 2018). For the most part, mining and processing have fairly local effects, but the legacy of old mining activities remains permanently in soil and sediments. Much of our information about the behaviour of Cu in the environment comes from such sites.

6.8.4.2 Biogeochemistry

Copper occurs in the first row of transition metals, lying between nickel and zinc (Figure 6.1). In this series, the stability of complexes formed between the divalent metal ion and various ligands generally increases from left to right, reaching a maximum with Cu and then declining: $Mn(II) < Fe(II) < Co(II) < Ni(II) < Cu(II) > Zn(II)$. This sequence, first recognized by Irving and Williams (1953), implies that at equivalent molar concentrations, Cu(II) will out-compete other divalent ions in binding to available ligands. In addition to its divalent state, Cu also exists as monovalent Cu(I). The reduction from Cu(II) to Cu(I) is known to occur photochemically and under the influence of the superoxide radical that is excreted by many microorganisms, but the cuprous ion, Cu^{1+}, is unstable in aqueous solution in the presence of oxygen and tends to revert to the Cu(II) state. However, as is discussed in the next section, Cu(I) can persist in the intracellular environment.

In uncontaminated oxic surface waters, copper concentrations generally range from 0.5 to 30 µg/l (8–470 nM) in fresh waters, with a median of about 10 µg/l (160 nM), and from 0.03 to 0.23 µg/l (0.5–3.6 nM) in seawater. Divalent Cu^{2+} ions tend to form relatively weak inorganic complexes with carbonate and hydroxide anions, but bind more strongly with organic ligands, both synthetic and naturally occurring (Table 6.2). In alkaline and circumneutral waters, virtually all Cu in the operationally defined 'dissolved' phase in rivers, lakes and the oceans is bound in such organic complexes, notably with natural fulvic acids; the concentrations of inorganically complexed Cu and the free Cu^{2+} ion are several orders of magnitude lower than that of total dissolved Cu. For example, Vega and Weng (2013) reported that only about 0.1% of dissolved Cu in the River Rhine was present as the free Cu^{2+} ion.

The affinity of copper for natural organic matter is also evident in oxic freshwater sediments, where sequential extraction techniques similar to those used for soils (Box 6.4) consistently show a large proportion of Cu bound to particulate organic carbon. In anoxic sediments, Cu will tend to bind to any available sulphide, including amorphous FeS, forming very insoluble CuS. Total Cu concentrations in sediments in polluted environments can be as high as 2,000 mg Cu/kg, in which case the capacity of the strong binding sites in the sediment (either particulate organic matter or amorphous iron sulphides) may well be exceeded, rendering the Cu more bioavailable.

Copper also binds to the natural organic matter in soils, but other solid phases, notably iron and manganese oxyhydroxides and clays, frequently play a role in retaining Cu. The average Cu concentrations in different types of soil in the world range generally fall between 10 and 30 mg/kg dry weight, but close to mines and smelters Cu in surface soils can exceed several thousand mg/kg, and on military training grounds in some regions of the world, soil Cu concentrations may be even higher. In contrast, there are also extensive areas in the world where the soils are Cu-deficient, notably in Europe, Australia and India. In these areas, animals grazing on or fed forage crops from such soils do not obtain sufficient Cu for their nutritional needs.

6.8.4.3 Biochemistry

Although required as a nutrient, copper can become toxic at concentrations exceeding that which is required metabolically (Figure 6.2). As an essential micronutrient, Cu functions as a cofactor for a number of biological processes including mitochondrial oxidative phosphorylation, free radical detoxification, neurotransmitter synthesis and maturation, and iron metabolism. The main biochemical functions of Cu relate to the cellular redox state (a leading Cu-containing protein is Cu/Zn superoxide dismutase or SOD), and in plants it is involved in electron transport in chloroplasts and mitochondria. In molluscs and crustaceans, Cu is a constituent of the blood pigment haemocyanin.

In its interactions with epithelial cells, copper behaves much like the other divalent metals considered in this chapter; its uptake varies as a function of the free Cu^{2+} ion in the exposure medium, and for a given Cu^{2+} concentration, uptake is lower in hard waters than in soft waters. However, Cu differs from these other metals in that it normally enters living cells in its reduced state, as Cu(I). Since Cu(I) concentrations in the ambient environment are vanishingly low, transport of Cu(I) requires the action of a Cu(II) → Cu(I) reductase at the cell surface. Examples of this sequence of extracellular reduction of Cu(II) followed by transmembrane transport of Cu(I) have been studied in great detail for unicellular algae (Blaby-Haas and Merchant, 2012) and the roots of terrestrial vascular plants (Printz *et al.*, 2016).

The Cu(I) entering a cell is captured by metallothioneins or copper-chaperones and enters an intricate intracellular trafficking network that is designed to maintain Cu homeostasis. The process of Cu transport into the cell and its insertion into metalloenzymes results from an elaborate interplay between the Cu(I) transport protein and associated chaperone proteins. This sequence of Cu transfers between proteins serves to protect the organism from the toxic effects of inappropriate Cu binding and to deliver the metal accurately to the correct enzyme. Both oxidation states may exist within a living cell, but they exhibit very different binding preferences. In its reduced state, Cu(I) is a soft metal and preferentially binds to reduced sulphur atoms, i.e., to thiols (R-SH) or thioethers (R-S-R'). In contrast, divalent Cu(II) is more 'promiscuous' and tends to seek out oxygen and nitrogen donor groups.

Given copper's role as an essential element and its potential toxicity, organisms have developed biochemical means for maintaining Cu at the required nutrient level, while protecting sensitive sites from toxic effects. These strategies include extracellular detoxification, intracellular detoxification, and efflux. Studies of Cu-tolerant organisms have contributed to our understanding of metal detoxification. For example, some Cu-tolerant unicellular algae respond to high Cu concentrations by excreting Cu-binding ligands. On the other hand, populations of *Silene vulgaris*, a common wildflower, are able to export Cu^{2+} ions across the plasma membrane of their root cells (Printz *et al.*, 2016). The sodium–proton antiporter (Na^+/H^+ exchanger) is a membrane protein that simultaneously mediates the influx of H^+ and efflux of Cu^{2+}; this Cu extrusion mechanism presumably contributes to the development of Cu tolerance.

Another intracellular detoxification strategy that is fairly widespread is the sequestering of copper in insoluble metal-rich granules. Transplant experiments, involving the transfer of aquatic animals from a clean site to a Cu-contaminated site, have shown that metallothionein synthesis is rapidly induced, but after this initial response, in some species, the excess Cu is shunted to insoluble granules. For example, Khan *et al.* (2012) exposed the freshwater crustacean *Gammarus pulex* to metal-contaminated and control sites. In animals caged at the contaminated site, Cu was associated with metallothionein-like proteins after 2 to 4 days, but on day 16 it was also found in spherical calcium-rich granules. However, granule formation was not related to Cu exposure – the granules serve as a calcium reserve for the moulting period, and they (passively) accumulate Cu during the pre-moult period. Similar results were reported for Cu in the gills of the freshwater bivalve, *Pyganodon grandis* (Bonneris *et al.*, 2005).

Lichens have been shown to detoxify copper by the production of norstictic acid by the fungal symbiont, which results in mostly insoluble granules (Hauck *et al.*,

2010). Some lichens also produce copper oxalates as a means of sequestering Cu in insoluble form (Rainbow, 2018). In summary, a relatively varied and large number of strategies have evolved that accomplish metal detoxification, some specifically for Cu and others for trace metals more generally.

6.8.4.4 Ecotoxicity

Across the range of phyla and species examined, copper affects a number of specific biochemical processes (Box 6.13). These effects reflect the direct involvement of excess Cu ions in free radical generation, i.e., in provoking generalized oxidative stress, and its interaction with specific targets such as photosystem II in plants and neurosensory cells in aquatic animals. Also across the entire range, the production of phytochelatins or metallothionein-like peptides is a generalized response to Cu exposure.

Laboratory-derived toxicological data for waterborne copper are particularly numerous, and the analysis of these data was instrumental in the early development of the biotic ligand model, BLM (Section 6.3.2) – at least in part because analytical chemists had devised a means to measure the free Cu^{2+} ion directly in the exposure media, with an ion-selective electrode that is remarkably sensitive to cupric ions. In this context, it is no surprise that the factors shown to modify the toxicity of waterborne Cu are those that were taken into account in the elaboration of the BLM, namely water hardness, pH and the concentration of natural organic matter. In addition to these water quality variables, increased sodium concentrations help to counteract Cu toxicity in a variety of animals. This physiological effect is consistent with the effect of Cu on Na^+/K^+ exchange, which leads to a drop in blood Na levels. Dietary Na also protects fish against Cu-induced olfactory impairment (Azizishirazi et al., 2015).

Box 6.13 Signs of copper toxicity

- Induction of oxidative stress.
- Perturbation of photosynthesis.
- Perturbation of homeostasis of essential cations.
- Disruption of sodium/potassium exchange.
- Inhibition of fish olfactory responses to chemical stimuli (Capsule 6.1).
- Chemosensory and behavioural impairment in some aquatic invertebrates.

Diet-borne copper can also contribute to toxicity. In their comprehensive review of the toxicity of dietary metals to aquatic animals, DeForest and Meyer (2015) concluded that small crustaceans (cladocerans, copepods, amphipods) and echinoderm larvae were relatively sensitive to Cu in their diet. Fish, on the other hand, proved to be relatively insensitive to diet-borne Cu. In fish, it has been shown that uptake of waterborne Cu via the gills is linked to the load of dietary Cu; in laboratory experiments, high levels of diet-borne Cu led to a depressed uptake of waterborne Cu, illustrating the interplay between the two uptake routes (Kamunde et al., 2002).

Freshwater and marine planktonic algae are exposed directly to copper in water, with the free Cu^{2+} ion being the best predictor of copper toxicity. Quigg (2016), reviewing the adverse effects of Cu on phytoplankton, identified reductions in growth, photosynthesis and respiration. Factors including pH, light, oxic status and the presence of other cations including iron, can influence the effect. Algal cells have been shown to control intracellular Cu, and research has identified Cu tolerance in some. For example, the use in the past of Cu as an algicide, including as a constituent of antifouling paints for use on marine structures, has resulted in the selection of Cu-tolerant ecotypes of both micro- and macro-algal species. Known mechanisms of tolerance include those mentioned earlier in the discussion of Cu detoxification, namely exclusion, accelerated efflux, and intracellular isolation or sequestration.

Among aquatic invertebrates, there is a wide range of toxic responses to copper. There are some consistent trends in that certain taxonomic groups do show some patterns, e.g., chironomids are less sensitive to Cu than are mayflies. The response of invertebrates to metals, including Cu, has been reviewed in some detail by Rainbow (2018). This author discusses the effects of metals among different functional groups of invertebrates, such as herbivores, detritivores, predators and parasites, but mainly in the context of metal bioaccumulation and detoxification.

Given that the gills have long been considered the main target for metal toxicity in fish, much research has focused on the effects of copper on osmoregulation, loss of sodium being a well-recognized and often fatal consequence of short-term exposures of fish to high Cu concentrations. However, there is now evidence that Cu can also inhibit fish olfaction, at moderate aqueous concentrations. Inhibition of olfactory receptors has implications for fish nutrition (predatory fish detecting their prey) and for social interactions among fish (mating; schooling) – see Capsule 6.1.

The toxicity of copper for mammals is rarely of concern; most mammals tested can tolerate high Cu concentrations in their diets. Indeed, in some places, domestic animals are provided with Cu supplements in their feed. Two reasons have been advanced to explain the rarity of examples of Cu toxicity in mammals. The primary explanation is that mammals are able to detoxify Cu effectively in their liver and kidney. The second factor relates to the primary route by which mammals are exposed to Cu, i.e., through their diet. The accessibility of diet-borne Cu is inherently low, largely because of its affinity for organic matter, and in addition dietary exposures are limited by the volume of material that can be ingested.

Terrestrial systems are somewhat susceptible to copper toxicity, although crop plants are more often Cu-deficient than Cu-inhibited. Soil microbial activity has been shown to be adversely affected by contamination by metals, particularly at pH values below neutrality, and Cu is one of the metals that have been implicated. With its known antifungal and algicidal properties, it is not surprising that other microorganisms are affected by Cu. Functional changes such as decreased carbon mineralization in soils, a rather general indicator of soil microbial activity, have been correlated with increased cupric ion activity.

Highly contaminated terrestrial areas show a paucity of vegetation or even desertification, but as pointed out for other metals, it is difficult to attribute specific changes to a named metal. Some of the most definitive results for Cu toxicity to plants have come from laboratory experiments on crop plants such as maize (*Zea mays*) and rice (*Oryza sativa*), often in the context of soil contamination from the application of sewage biosolids as a fertilizer, or from pest control products. Most concern about Cu has been for effects on yield and quality of a crop product, with little relevance to ecotoxicology.

Dietary ingestion is the major route of exposure for terrestrial consumers; the forms of ingested copper include the intracellularly detoxified forms described in the preceding section. However, there is no evidence for food-chain biomagnification of Cu, and indeed the concept of biominification may apply (Cardwell *et al.*, 2013). Łanocha-Arendarczyk and Kosik-Bogacka (2019) present data for renal and hepatic Cu in birds with different diets (Table 6.8). Herbivorous birds had the highest concentrations of Cu in the liver, and predators the lowest,

Table 6.8 Influence of avian diet on the levels of renal and hepatic copper[a].

Wild bird category	Kidney Cu (mg/kg dry wt)	Liver Cu (mg/kg dry wt)
Predator (medium-sized birds, mammals)	16	29
Predator (small-sized birds, mammals)	14	19
Piscivorous birds of prey	13	24
Piscivorous birds	22	54
Omnivores	32	120
Herbivores	49	3,160

[a] Data extracted from Łanocha-Arendarczyk and Kosik-Bogacka (2019).

with omnivores intermediate in this respect. Much of this renal and hepatic Cu is bound to metallothionein.

Field Observations

Van Genderen *et al.* (2008) applied the biotic ligand model (BLM) for copper to estimate bioavailable Cu concentrations in surface waters from sites in Europe, North America and Chile. The crustacean *Daphnia magna* and the bivalve *Mytilus edulis* were considered representative of the most Cu-sensitive aquatic species in fresh and estuarine waters, respectively. The sites chosen were ones for which the complementary water quality data needed to run the BLM were available (Section 6.3.2); 336 surface water samples were considered. Based on the BLM simulations, the authors calculated that in >90% of the water samples the Cu concentration was less than the regulatory limits in Europe and the United States, and thus too low to be of concern. This exercise provides an example of how the BLM can be used to identify potential trouble areas and guide management decisions.

At a smaller scale, Vinot and Pihan (2005) provide an interesting field example of the effects of copper contamination of an aquatic ecosystem. These researchers studied how Cu was distributed among various trophic levels in a small reservoir (100 ha) that had been subjected to Cu inputs for more than 10 years; the source of the Cu was water from the cooling system of an electricity generating plant in northeastern France (Mirgenbach Lake). During their study, the mean concentration of dissolved Cu (38 ± 12 µg/l; 0.6 µM) was well above the European

PNEC (predicted no-effect concentration) value for Cu in fresh water. They collected samples from both the autotrophic and detritivorous food chains (phyto- and zooplankton, aquatic plants, invertebrates and fish). Levels of bioaccumulated Cu varied greatly among different species, even among those at the same trophic level, but there was no evidence of biomagnification. Among the animals, highest Cu levels were observed in the viscera of crayfish, followed by mussels and aquatic snails, whereas the lowest Cu level was observed in fish muscle. By comparing the species distribution in the 1998–2001 sampling period with the distribution that had been documented before the cooling water was first introduced into the reservoir, Vinot and Pihan concluded that there had been long-term effects on the benthic and planktonic communities.

6.8.4.5 Copper Highlights

Copper, like nickel and zinc, is a transition metal, and both its mono- and divalent states are environmentally and biologically important. In nature, Cu occurs in sulphide and carbonate ores and, rather unusually, as native (metallic) Cu. As a biocide, Cu is one of the oldest pest control agents, used to control fungal disease and spoilage, limit the growth of nuisance algae and prevent fouling of ship hulls. Such uses inevitably lead to increased Cu loadings of the environment.

Virtually all the dissolved copper in rivers, lakes and oceans is bound to natural dissolved organic matter (DOM); the concentrations of inorganically complexed Cu and the free Cu^{2+} ion are several orders of magnitude lower than that of total dissolved Cu. Copper is an essential nutrient, functioning as a cofactor in many key biochemical processes, including iron metabolism. In contrast to the uptake of the divalent metals cadmium, nickel and zinc, Cu normally enters cells as Cu(I), a form that is relatively uncommon in the environment; uptake thus requires an initial reduction of Cu(II) at the cell surface.

Binding with peptides, such as phytochelatins and metallothioneins, as well as other means of compartmentalization mean that cellular copper is well regulated within living cells. However, when these cellular processes are overwhelmed, Cu toxicity results from oxidative stress, impairment of photosynthesis and disruption of sodium/potassium exchange, among other perturbations. In the ecosystem, dietary exposure to Cu can be significant.

6.8.5 Nickel

Metal/metalloid Nickel, Ni
Atomic number 28
Atomic weight 58.69
Estimated ratio [anthropogenic emissions/natural emissions] 2.3

6.8.5.1 Occurrence, Sources and Uses

Common mineral forms include those with nickel bound to sulphide (NiS, millerite) and to sulphide in the presence of iron (($NiFe)_9S_8$, pentlandite). Nickel is also associated with iron in nickeliferous limonite (($Fe,Ni)O(OH)$), a common mineral found in laterite deposits in tropical areas. The crustal abundance of Ni is estimated to be 84 ppm, a value similar to those for copper and zinc (Table 6.3), but in serpentine soils that result from the intense weathering of igneous ultramafic rocks Ni concentrations can be as high as 10,000 ppm. Such soils are relatively sterile and unproductive as farmland or timber stands, but they often support unusual plant life (Section 6.5.4.1). This is a good example of geobotanical prospecting – the serpentine flora is so distinctive that it identifies for prospectors the enrichment of useful elements.

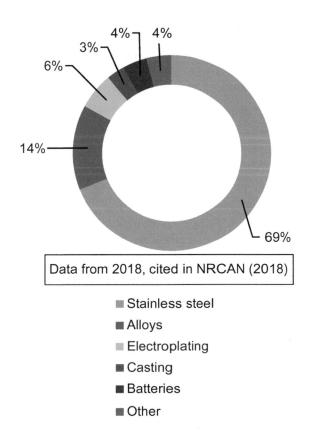

Data from 2018, cited in NRCAN (2018)

■ Stainless steel
■ Alloys
■ Electroplating
■ Casting
■ Batteries
■ Other

Industries linked to nickel production include mines, mills, smelters and refineries. Those involved in Ni use and processing include alloy and stainless steel production, electroplating, and the manufacture of catalysts, paint pigments and batteries (Buxton *et al.*, 2019); alloy and stainless steel production alone account for over 80% of global Ni use. These activities all contribute to Ni loadings of the aquatic and terrestrial environments. Emissions to the atmosphere accrue from smelting, refining, alloy processing, scrap metal reprocessing, fossil fuel combustion and waste incineration. Nickel is present in the atmosphere in aerosols or dust as Ni(II). Unlike mercury, its residence time in the atmosphere is quite short and thus it does not travel long distances before it returns to the Earth's surface as wet and dry deposition. Direct additions of Ni to the soil come principally from the disposal of smelting and refining wastes and from the application of sewage biosolids as a fertilizer. Weathering and erosion of Ni-bearing minerals contribute to background or natural environmental loadings, but these are estimated to be only about 30% of the total Ni flux (Klee and Graedel, 2004).

6.8.5.2 Biogeochemistry

Nickel belongs to the first row of transition metals and occurs in Group 10, flanked by cobalt and copper (Figure 6.1). The only oxidation state of environmental importance is divalent Ni(II). In pristine streams, rivers and lakes, total dissolved Ni concentrations typically fall in the 0.2–10 µg/l range (3–170 nM), and typical concentrations in pelagic seawater are 0.13–0.3 µg/l (0.2–5 nM). In contrast, concentrations in surface waters near nickel mines and smelters may be more than an order of magnitude higher; for example, dissolved Ni concentrations in lakes close to the Ni smelters near Sudbury, Ontario, Canada approached 100 µg/l (1.7 µM) in the early 2000s. In fresh waters, Ni speciation is dominated either by the free aquo ion (Ni^{2+}), when the total dissolved concentration is high, or by Ni complexes with natural dissolved organic matter (DOM) in lakes and rivers with low total dissolved Ni concentrations. In seawater, estimates of DOM complexation of Ni range from 10 to 60%, and the two major inorganic species are Ni^{2+} and the $NiCO_3$ carbonate complex.

In common with all other cationic trace metals, nickel that is released to the aquatic environment tends to accumulate in sediments. In the oxic surface layer of the sediments, Ni binds strongly to the iron oxyhydroxides that are invariably present in such environments. The Ni associated with this binding phase is in equilibrium with the sediment pore water; this equilibrium is pH-sensitive, with Ni^{2+} being released if the porewater pH decreases. Under anoxic conditions, e.g., in the sediment lying beneath the oxic surface layer, Ni is released by reductive dissolution of the iron oxyhydroxides, as is the case for As, Pb and Se. The Ni released in this manner is then captured by any available sulphide. In other words, Ni that is bound to particles and settles to the bottom of a water body tends to stay there.

The behaviour of nickel in soils resembles that in sediments, with iron and manganese oxyhydroxides again playing a key role in binding and retaining the metal cation, rendering it less bioavailable; this binding becomes stronger as the soil pH increases. As is the case for other cationic metals (cadmium, copper, lead, zinc), binding of Ni to the soil solid matrix is favoured by high values of the effective cation exchange capacity (CEC). In such soils, Ni will be less mobile and less bioavailable than in soils with low CEC values. The 'age' of the soil is also important, especially in cases where Ni has been spiked into a soil, either experimentally or by an inadvertent spill; in both cases, Ni availability decreases as the soil ages, reflecting the slow migration of the Ni^{2+} cations onto stronger and/or less accessible binding sites within the soil matrix.

6.8.5.3 Biochemistry

Nickel is an essential micronutrient for microorganisms, algae and indeed for all plant life. The range between the essential, beneficial and toxic concentrations to plants is rather narrow, and it varies with plant species. The role of Ni in vertebrates is not yet fully understood, but the observation of negative effects when Ni concentrations are very low has been interpreted to mean that Ni plays an essential role in the metabolism of at least some vertebrates (Muyssen *et al.*, 2004).

Factors affecting the uptake of dissolved nickel by freshwater and marine species include the usual BLM suite, i.e., water hardness, pH and the concentration of dissolved organic matter (DOM). However, unlike the case for most other cationic metals, for which the hardness correction is largely driven by the concentration of Ca^{2+}, for Ni both Ca^{2+} and Mg^{2+} can play a protective role. For example, experiments on a range of aquatic and terrestrial animals have shown that increasing the aqueous Ca^{2+} concentration reduces Ni uptake. In contrast, similar experiments on aquatic and terrestrial plants showed no protective effect attributable to Ca^{2+}, suggesting that a different uptake mechanism may be operative in plants. Increased Mg^{2+} concentrations led to reduced uptake and/or toxicity of Ni in experiments with various aquatic animals (e.g., daphnids, oligochaetes, rainbow trout) and with terrestrial plants, but not with aquatic plants (Buxton *et al.*, 2019).

Nickel is known to enter living cells by a variety of routes, including a Ni/Co transporter (NiCoT) in prokaryotes and a relatively unselective iron transporter (IRT1, a member of the **ZIP-transporters** – Box 3.4) that transports several divalent metals into plant cells. Known intracellular roles for Ni include functioning as a cofactor in the activity of at least eight enzymes, which are involved in such reactions as the **dismutation** of the superoxide radical, carbon dioxide and carbon monoxide metabolism, hydrogen uptake and production, and urea hydrolysis.

When it enters a cell, nickel must be delivered to the **apoenzymes** that require it as a cofactor. This is a rigorously orchestrated process, involving both Ni-chaperones and molecular chaperones. The latter ensure that the apoenzyme is in a conformation that is appropriate for insertion of the Ni ion from the chaperone. In addition to this intracellular trafficking, Ni is also subject to transport to different locations within the plant or animal. For plants, histidine is a common ligand for Ni chelation and transport to different plant parts, but citrate, malate and nicotinamide may also be involved. In vertebrates, Ni circulates in the blood bound to albumins, which control the tissue distribution or eventual elimination of the metal.

Nickel differs from most of the other metals considered in the present section (cadmium, copper, lead, mercury)

in its intracellular behaviour. These four metals are 'softer' than Ni and bind preferentially to ligands bearing thiol groups (R-SH). In contrast, the Ni^{2+} ion tends to associate with biomolecules that have oxygen and nitrogen as the donor groups, as is the case for Ca^{2+} and Mg^{2+}, and in so doing it may displace Ca and Mg. The Ni^{2+} ion normally forms octahedral complexes, as do Ca^{2+} and Mg^{2+}, and its ionic radius is similar to that of Mg^{2+}, which will favour such displacements. This example illustrates once again the overall link between the intrinsic toxicity of a metal and its ionic 'character', as discussed in Section 6.2.1 (QICARs).

As is the case for other essential metals such as copper and zinc, living organisms must be able to protect against excessive influxes of nickel. However, unlike Cu and Zn, Ni is not an effective inducer of metallothionein or phytochelatin synthesis. In studies on juvenile yellow perch (*Perca flavescens*) that had been chronically exposed to high concentrations of Ni in their natural environment, Caron *et al.* (2018) demonstrated that hepatic Ni was present in a low-molecular-weight (LMW) heat-stable form but not bound to metallothionein. In contrast, Ag, Cd and Cu were all bound to metallothionein in these same fish. This difference in metal partitioning is consistent with the known preference of soft metals such as Ag, Cd and Cu(I) for thiols and the contrasting tendency of Ni(II) to bind to ligands with oxygen and nitrogen as donor atoms. It is tempting to suggest that the LMW Ni-containing molecule found in the liver cytosol might correspond to one of the chaperones that are involved in Ni trafficking in the subcellular environment, but the possible role of chaperones in Ni detoxification remains speculative.

6.8.5.4 Ecotoxicity

At a fixed pH and constant hardness, the toxicity of nickel is best predicted by the concentration of the free Ni^{2+} ion in the exposure medium. Similar statements apply to the other BLM-compatible metals (cadmium, copper, lead, zinc). In the same manner, the toxicity of waterborne Ni decreases when either the water hardness or the concentration of dissolved organic carbon increases, since in both cases these changes would be expected to decrease the binding of Ni^{2+} to the 'biotic ligand' (Figure 6.8). Nickel toxicity also tends to decrease as the pH is lowered, a trend that is consistent with the general predictions of the biotic ligand model (Section 6.3.2). However, attempts to model this effect as a simple competition between the H^+-ion and Ni^{2+} for a single binding site have not been successful (Nys *et al.*, 2016), suggesting that the proton may be interfering non-competitively with Ni binding and uptake.

For marine species, the bioaccumulation and toxicity of Ni tend to decrease as salinity increases. The protection afforded by increased salinity is often greater than would have been predicted solely on the basis of the changes in Ni speciation that occur as salinity rises. In their review of these results, Blewett and Leonard (2017) note that changes in salinity also affect the physiology of euryhaline organisms (in particular those that are osmoregulators). They suggest that these physiological changes may contribute to the greater-than-predicted protection afforded by increases in salinity.

The intrinsic toxicity of nickel, which can be deduced by comparing EC_x values for different metals and the same test organisms (e.g., Borgmann *et al.* (2005)) or by comparing HC_5 values as derived from species sensitivity distributions, is generally lower than that of cadmium, copper and lead (Table 6.3). However, this inherent 'advantage' is somewhat illusory, because the complexation constants for Ni binding to the range of inorganic and organic ligands present in natural waters (Table 6.2) tend to be lower than those for Cd, Cu and Pb. It follows that at equivalent molar concentrations of different metals, the proportion of free Ni^{2+} will normally be higher than for the other metals.

Most of the mechanisms that have been proposed to explain nickel toxicity are generic in nature, i.e., they apply to most cationic metals and are not specific to this metal (Section 6.4). For example, Ni^{2+} could inhibit an enzyme by binding to the active site (with or without displacing another essential metal, such as iron), or it could bind outside the catalytic site, provoke a change in the enzyme's conformation and inhibit its action **allosterically**. Some of the signs of Ni toxicity, which could result from such molecular initiating events (MIEs), are listed in Box 6.14 (Brix *et al.*, 2017). In

Box 6.14 Signs of nickel toxicity

- Induction of oxidative stress.
- Perturbation of homeostasis of essential cations.
- Disruption of Fe^{2+}/Fe^{3+} homeostasis.
- Allergic-type response of respiratory epithelia.
- Impairment of gas exchange (respiration inhibition).

addition to these effects, Ni can cause allergic responses in epithelial cells, and it is also a recognized respirable carcinogen in humans. In animals, ingested Ni does not bioaccumulate to high levels and has a relatively short half-life in the body.

Laboratory testing suggests that the freshwater species most sensitive to waterborne nickel are unicellular algae, macrophytes and invertebrates (notably *Ceriodaphnia dubia*, a cladoceran, and *Lymnaea stagnalis*, a snail). These trends were confirmed in a mesocosm study in which a freshwater community of phytoplankton, periphyton, zooplankton and snails was exposed to a range of Ni concentrations for a 4-month period (Hommen *et al.*, 2016). The experiment was designed to test Ni under conditions of high bioavailability (e.g., a low dissolved organic carbon concentration of ~4 mg/l and an alkaline pH of 8.6). The snail *L. stagnalis* proved to be the most sensitive species, as would have been predicted from the single-species toxicity tests.

The sensitivity of marine species to nickel covers a very wide range. In data compiled by DeForest and Schlekat (2013), the most sensitive species was an early life stage of a tropical sea urchin from the Caribbean region, *Diadema antillarum*, for which the EC_{10} value was only 2.9 µg Ni/l. This value was almost 10^4 times lower than that for the least sensitive organism, *Cyprinodon variegatus*, the sheepshead minnow. This variability could not be attributed to differences in the experimental exposure media, which were based on seawater and accordingly were inherently much more constant than freshwater environments. The only variable water quality parameter was dissolved organic matter, which ranged from 1 to 5 mg C/l, and this variation could only explain about a 2-fold change in bioavailable Ni. In other words, the ~10^4 range represents true species-specific variability.

Field Observations

High nickel concentrations in the environment are often accompanied by elevated levels of other metals, such as copper and zinc. If negative effects are observed on the resident biota, it is often challenging to assign responsibility for these adverse effects to a specific metal. Despite this constraint, studies conducted around the Ni smelting complex in Sudbury, Ontario, Canada have provided good examples of the effects of Ni exposure in the aquatic environment (Capsule 12.1). For example, Borgmann *et al.* (2001) documented decreased abundances of a variety of benthic invertebrates in non-pH-stressed lakes in this region and with sediment toxicity tests they demonstrated severe toxicity to mayflies and amphipods. Using the amphipod *Hyalella azteca* as their test animal, they were able to show that among the four metals present in elevated concentrations in the lake sediments (i.e., cadmium, cobalt, copper and nickel), only Ni was bioaccumulated in sufficient amounts to cause toxicity. In experimental studies, the amphipod was caged above the sediment or allowed to contact the sediment directly; the observed toxicity was the same for both exposure scenarios, suggesting that the toxicity was caused by waterborne Ni only.

Since all plants require and retain nickel in their tissues, all herbivores are necessarily exposed to Ni in their diet; however, in undergoing this trophic transfer, Ni is not subject to biomagnification. As mentioned earlier, concentrations of Ni, magnesium and chromium in serpentine soils are very high, but these soils nevertheless support specialized plant life. These plants are often hyperaccumulators, i.e., plants that can accumulate very high levels of a metal, yet still grow and reproduce. In the case of Ni hyperaccumulators, the metal is often sequestered in vacuoles in the epidermal cells, where it can be isolated far from the physiologically active parts of the plant, allowing photosynthesis and respiration to continue with minimal interference. Concentrations as high as 24,000 mg/kg dry weight have been reported for such Ni hyperaccumulators. Such plants are of interest not only because of their novelty, but also because they may be of use in **phytomining** or **phytoremediation** (Section 13.9.3).

Although we referred earlier to serpentine soils being a 'harsh' environment (Section 6.8.5.1), it is also possible that plants found on such soils have evolved to use magnesium in the place of calcium, and that the high nickel concentrations in their tissues serve to discourage herbivore grazing. A similar anti-herbivore role has also been suggested for cadmium hyperaccumulators (Stolpe *et al.*, 2017).

In a comprehensive review of nickel hazards to fish, wildlife and invertebrates, Eisler (1998) noted that in wildlife collected from Ni-contaminated areas, the Ni concentrations in internal organs were usually similar to those measured in the same species collected from an uncontaminated environment. In contrast, tissues exposed to the external environment (skin, fur, feathers) tended to reflect contamination gradients. Such a result might simply be a result of contact between these external tissues and the Ni-contaminated environment, or it may reflect the elimination of ingested Ni. In mammalian wildlife, renal Ni concentrations are usually somewhat higher or similar to those in the liver, whereas in birds the highest internal concentrations are often found in their bones (Binkowski, 2019).

6.8.5.5 Nickel Highlights

Nickel, like the other metals in the first transition series (copper, zinc), is a moderately abundant trace element, the biogeochemical cycle of which has been perturbed by mining and smelting activities. It is an essential metal for microorganisms and all plant life, but its role in animal metabolism is still uncertain.

Geochemically and biochemically, nickel behaves as a **borderline** cation; unlike soft cations such as Cd, Cu(I), Pb and Hg, Ni does not bind strongly to the reduced sulphur (thiol) donor groups that characterize phytochelatins (plants) and metallothioneins (animals). Instead, Ni binds preferentially to ligands bearing nitrogen and oxygen donor groups, and in doing so it competes with Fe, Mn, Mg and Ca, all of which are also essential.

The intrinsic toxicity of Ni is low (Table 6.3), but several clear examples of Ni toxicity in field situations have been documented in the vicinity of Ni mining and smelting activities. High ambient Ni concentrations in terrestrial environments, notably in serpentine soils, may select for Ni-hyperaccumulator plants.

6.8.6 Zinc

Metal/metalloid Zinc, Zn
Atomic number 30
Atomic weight 65.38
Estimated ratio [anthropogenic emissions/natural emissions] 0.9

6.8.6.1 Occurrence, Sources and Uses

Zinc is among the more common elements in the Earth's crust, ranking 24th in crustal abundance (70 mg/kg). It occurs in metamorphic, igneous and sedimentary rocks in many parts of the world. Zinc sulphide (ZnS) and sphalerite (zinc iron sulphide) are its commonest ores, and it often co-occurs with copper (chalcopyrite) and lead (galena). Zinc may also be found in secondary minerals (e.g., carbonates, silicates), which result from the weathering of the original sulphide primary minerals. The mining and refining of these ores result in release of particulate and dissolved Zn. When exposed to oxygen, the sulphidic mine tailings will generate **acid mine drainage** (Section 11.2) and release dissolved Zn into the environment. Mine tailings are also a potential source of airborne particulate Zn. Beyond metal processing and electroplating, Zn contamination of soil and water also results from fossil fuel and waste combustion and the subsequent return of Zn from the atmosphere in the form of wet and dry deposition.

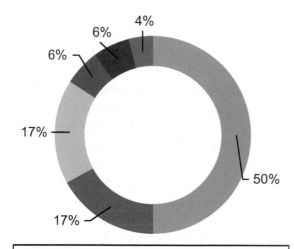

Data from 2018, cited in NRCAN (2018). 'Semi-manufacturers' refers to the use of Zn in intermediate products that are then incorporated into finished products.

■ Galvanizing

■ Alloys

■ Brass/bronze

■ Semi-manufacturers

■ Chemicals

■ Other

Properties of zinc that have favoured its use include its anticorrosive properties and its capacity to bond with other metals in alloys. Zinc galvanizing, a process that is used to coat iron and steel and prevent their corrosion, currently accounts for about half of worldwide Zn use. Zinc also has value alloyed with copper to form brass, and when alloyed with other metals it has uses in parts for automobiles, electrical components and household fixtures. A third significant use of Zn is in the production of zinc oxide (the most important zinc chemical by production volume), which is used in rubber manufacturing and as a protective skin ointment. Zinc chloride and sulphate were also used in older herbicides.

6.8.6.2 Biogeochemistry

Zinc occurs in the first series of transition metals, right after copper (Figure 6.1). In its reactions with ligands, it is classified as a 'borderline' metal, with a behaviour intermediate between those of 'soft' and 'hard' metals (Section 6.2.1); it can form coordinate bonds with a range of ligand types in aqueous systems. In contrast to other abundant transition metals such as Fe and Cu, Zn lacks redox activity, a reflection of its complete d-shell of electrons; the only oxidation state of environmental importance is Zn(II).

Natural background levels of zinc in fresh water vary quite widely as a function of the local geology, generally in the range from <1 to 15 µg/l (<15–230 nM). In fresh waters, the free Zn^{2+} ion is present at appreciable concentrations, with reported measured values ranging from about 10% (pristine Swiss lakes) to >80% (lakes on the Canadian Precambrian Shield) of the total dissolved metal. Most of the rest of the dissolved Zn, i.e., the part that is not free, is bound to the natural dissolved organic matter present in the water. Inorganic complexes $(Zn(OH)_n^{(2-n)+}$ and $ZnCO_3^0)$ make a modest contribution to the dissolved inorganic Zn pool as the pH rises above 7 (e.g., about 25% at pH 8). Dissolved Zn concentrations in the pelagic ocean are generally <65 ng/l (<1 nM) in surface waters and increase to about 650 ng/l (10 nM) at intermediate depth. Most of this Zn is bound to natural organic matter, but little is known about the identity of the ligands involved. Note that measurements of dissolved Zn, especially in ocean waters, are prone to inadvertent contamination issues due to the presence of Zn in lab materials (Box 6.1).

Consistent with what we have seen for the other cationic metals considered in Section 6.8, most of the zinc that finds its way into the aquatic environment will interact with suspended matter in the water column. This suspended matter includes both plankton and abiotic particles (e.g., eroded soil), and it follows the pull of gravity towards the bottom of the water body. In the bottom sediments, Zn behaves in a similar manner to cadmium and nickel. In oxic surface sediments, it binds to iron and manganese oxyhydroxides, and to a lesser extent to detrital organic matter, whereas in the underlying anoxic sediments it will be scavenged by sulphide.

Among the suite of metals that settle out of the water column, Zn and Ni are somewhat more mobile (i.e., less strongly bound) than copper, lead and mercury. For example, if the surface sediments are subject to acidification (Section 11.1), Zn will be desorbed before the other metals. Similarly, if oxygen is introduced into the anoxic sediments, Ni and Zn will be released from the sulphide pool before Cu, Pb or Hg.

The mobility of zinc in soils is similarly influenced by its tendency to sorb to inorganic surfaces (e.g., hydrous Fe and Mn oxides, clays) and to solid organic matter. The balance between sorption and desorption is affected by factors such as composition of the soil (clay mineralogy, organic matter content, pH) and the hydrological regime (rainfall, snowmelt, infiltration, drainage). Leaching of Zn from soils is favoured by an acidic pH (<5) and by a sandy texture that promotes rapid drainage. In a study of 66 contaminated soils from Europe and North America, Stephan *et al.* (2008) extracted the soils with a dilute solution of KNO_3 (0.01 M) and determined Zn speciation in the soil extracts. Most of the Zn found in the soil extract was bound to DOM (60–98%); the remaining Zn included the free Zn^{2+} ion and various inorganic complexes, in approximately equal proportions.

6.8.6.3 Biochemistry

Zinc is an essential element for all microbial, plant and animal life (Box 6.15). It was initially recognized for its

---

Box 6.15 Examples of the numerous metabolic roles of Zn

- Acts as a cofactor in >300 enzymes (the only metal to appear in all six enzyme classes).
- Binds protein subunits together and maintains proper protein folding.
- Binds to metal-responsive transcription factors and affects gene expression (e.g., induces synthesis of metallothionein).
- Acts to help control oxidative stress, by maintaining high intracellular levels of metallothionein.
- Acts a signalling ion for intracellular communication.
- Acts as a signalling ion for communication between cells (e.g., neurons).
- Is essential for DNA replication, cell proliferation and differentiation.

---

catalytic role (notably in erythrocyte carbonic anhydrase, in 1939), and there are now more than 300 known enzymes that require Zn as a cofactor. However, following this discovery it progressively became evident that Zn also plays a structural role, stabilizing enzymes by serving as a bridge between protein subunits. Zinc is well suited for this function because it is redox-inert and it binds to amino acids more strongly than do the other essential-metal ions (other than copper). This stabilizing function extends to various transcription factors (Zn finger proteins). These latter molecules are small peptides with repeating units of cysteine and histidine; Zn plays a role in maintaining these domains in a conformation that allows interaction between the zinc finger protein and nucleic acids, other proteins and in some cases with lipids. More recently, it has been demonstrated that pulses of the free Zn^{2+} ion play an important regulatory/signalling role for various cellular physiological functions (proliferation, differentiation, ion transport and secretion). Zinc is now recognized as a second messenger, i.e., a 'signalling ion', the regulatory functions of which are comparable in importance to those of calcium (Maret, 2013).

Not surprisingly, given the multiple roles played by zinc, the uptake of this element by living organisms is a highly regulated process. Within the cell, homeostatic mechanisms control concentrations of this crucially important nutrient, and this has a bearing on its potential toxicity. Two strategies are used to control cellular and subcellular concentrations of Zn. The first involves the membrane proteins or transporters that control uptake and efflux of Zn across the plasma membrane, whereas the second relies on the compartmentalization of Zn within the cell, i.e., the shunting of excess Zn into intracellular vesicles for storage. The transporters found in plasma membrane include the ZIP family of proteins (ZIP = Zrt- and Irt-like protein), which play an important role in bringing Zn into the cell, and a second family of ZnT proteins that moves Zn in the opposite direction, out of the cell. These transporters are not specific to Zn; other essential divalent metals such as Mn^{2+}, Ni^{2+} and Fe^{2+}, as well as the non-essential Cd^{2+} ion, are known to share the same routes of transmembrane transport. Similar influx and efflux transporters are involved in shunting excess Zn into intracellular vesicles and then recovering this Zn when levels in the cytosol drop below the required level. Indeed, the transporters themselves can move between the plasma membrane and the vesicle membrane in response to changes in the cell's zinc status. The Zn shunted into the vesicle may remain in solution, or it may be sequestered by precipitation, notably with phosphate.

Total cellular zinc concentrations are typically in the 100–200 µM range, well above the levels of other essential metals (other than iron), with most of this Zn being bound to proteins; concentrations of unbound Zn are in the 10–100 pM range, about 10^7 times lower than total cellular Zn. The majority of the bound Zn is held by metallothionein (MT). Each MT molecule can bind seven Zn ions, but the affinities of these binding sites for Zn are not identical, even though all the Zn(II) ions are in similar coordination environments, bound to four cysteinyl thiol groups. Because of these differences in binding site affinities for Zn, MT can play a dual role as a Zn acceptor and a Zn donor, i.e., it can act as Zn buffer. In addition, although Zn itself is redox-inert in living cells, MT is subject to oxidation and reduction of its thiol groups, meaning that the coordination environment of Zn is sensitive to redox changes within the cell. Note, too, that gene expression leading to the synthesis of MT is under the control of a Zn-responsive transcription factor, MTF-1. Zinc is thus a potent inducer of MT synthesis, meaning that if the unbound Zn concentration in the cell increases, additional MT will be synthesized and the concentration of unbound Zn will be brought back into the 'safe' range. The same Zn-responsive transcription factor also triggers the synthesis of Zn transporter ZnT1, which will accelerate Zn efflux from the cytosol through the plasma membrane.

In multicellular organisms, subcellular partitioning studies using protocols such as those described in Section 6.5.2 show that zinc is spread across all the operationally defined fractions (nuclei and debris, mitochondria, heat-stable proteins and heat-denatured proteins). This dispersed distribution differs markedly from what is found for non-essential metals such as silver, cadmium or lead, where the offending metals tend to be concentrated in a detoxified fraction (metallothionein-like proteins or metal-rich concretions). The broad distribution of Zn is, however, entirely consistent with its status as an essential metal with a very wide range of biochemical functions.

6.8.6.4 Ecotoxicity

As a general rule, there is more concern for zinc deficiency than for Zn toxicity in aquatic and terrestrial ecosystems (Box 6.16). For example, in the pelagic marine

---

Box 6.16 Signs of zinc deficiency or toxicity

Deficiency
- Hindered/stunted growth.
- Impairment of immune system.
- Neurosensory impairment.

Toxicity
- Perturbation of iron homeostasis and chlorophyll synthesis in plants (chlorosis).
- Perturbation of calcium homeostasis (interference with gill Ca uptake).
- Impairment of invertebrate reproduction.
- Avian pancreatitis.

---

environment Zn has been considered as one of the micronutrients for which extreme depletion may limit phytoplankton productivity in surface water; phytoplankton tend to deplete Zn in surface water, with a deepwater regeneration (e.g., Ellwood et al., 2020). Similarly, there are large regions of the world where soils are Zn-deficient (e.g., China, India, Turkey and Western Australia) and where the addition of Zn as a fertilizer has led to markedly improved crop yields (Broadley et al., 2007).

The intrinsic toxicity of zinc is low (Table 6.3) and, as described in the preceding section, cellular Zn is subject to very tight biochemical and physiological control. Acute and chronic toxicity thresholds for Zn are typically well above the concentrations measured in water, sediments or soil, both for freshwater and marine environments. Exceptions do occur for highly contaminated sites, which in the case of Zn are for the most part associated with metal mining and processing.

Aquatic Systems

Most of our knowledge about the toxicity of zinc in aquatic systems comes from laboratory studies. Careful studies with algae, crustaceans and fish demonstrated that Zn toxicity is influenced by the usual suite of water quality parameters (pH, hardness, dissolved organic matter) and that their influence fits the predictions of the biotic ligand model, BLM (De Schamphelaere et al., 2005). Increases in water hardness or in the DOM concentration attenuate Zn toxicity, whereas an increase in pH leads to an increase in toxicity (attributed to less competition between the proton (H^+ ion) and Zn^{2+} for binding at Zn transport sites).

As is the case for cadmium (Section 6.8.2), several exceptions to the BLM for Zn have been noted in laboratory studies, in cases where complexation of the Zn^{2+} ion leads not to a decrease in Zn uptake (as would be predicted by the BLM) but rather to an increase. For example, when Zn is bound to citrate, the Zn–citrate complex can be transported across the plasma membrane via a citrate transporter (Figure 6.6, mechanism 2). Similarly, when Zn binds to a ligand to form a neutral lipophilic complex, the intact complex can diffuse passively across the plasma membrane – diethyldithiocarbamate and ethyl xanthate are such ligands (Figure 6.6, mechanism 3). However, these exceptions are best seen as laboratory phenomena rather than major flaws in the BLM approach.

Zinc accumulates moderately in aquatic organisms, with bioconcentration factors (BCFs) higher in crustaceans and bivalve species than in fish. More importantly, the accumulated concentrations of Zn along a gradient of Zn exposure concentrations show very little variation. For example, in the set of lakes in northern Quebec and Ontario (Canada) that is discussed in Case Study 6.1, the ambient dissolved Zn concentrations increased 105-fold between the least and the most contaminated lake (0.023 → 2.4 µM), whereas in the native yellow perch, the hepatic liver Zn concentrations increased only 1.4-fold (1.4 → 1.9 µmol/g dry wt), an eloquent demonstration of homeostatic control (Giguère et al., 2004). It follows that BCFs for Zn tend to vary inversely as a function of the Zn concentration in the exposure medium, decreasing markedly as the exposure concentration increases. A comprehensive data analysis of field studies by DeForest et al. (2007) also found that BCFs (and bioaccumulation factors) for Zn were inversely related to combined waterborne and dietary Zn exposures.

Such a response is expected for an essential metal under tight homeostatic control, as we have seen to be the case for zinc. With reference to Section 6.1, where we compared inorganic and organic contaminants, and in anticipation of what is discussed in Section 7.3, the difference between Zn and a lipophilic and persistent organic pollutant (POP) should be highlighted. For POPs, the BCF remains more or less constant for a given organism, and thus if the exposure concentration doubles, so does the steady-state body concentration. In contrast, for Zn (and indeed for all metals, but a lesser degree), the BCF decreases progressively as the exposure concentration increases. Note, too, that tissue concentrations of Zn that do not harm sensitive freshwater organisms present no risk to consumers of such aquatic organisms.

Van Genderen et al. (2009) applied the biotic ligand model (BLM) for zinc to estimate bioavailable Zn concentrations in fresh surface waters from over 800 sites in different regions across the world (Europe, North America and South America). Using chronic no-observable-effect concentrations (NOEC; reproduction for Daphnia magna) or 10% effective concentrations (EC_{10}; growth rate for the unicellular alga Pseudokirchneriella subcapitata) as the toxicity thresholds, they calculated that at 83% of the sites considered, the ambient Zn concentrations were below these thresholds. However, at 17% of the sites the ambient Zn concentrations, in combination with the water chemistry prevailing at the site, led to predicted exceedances. This exercise illustrates how the BLM can be used to identify and target potential trouble areas. A similar exercise was carried out for Cu (Van Genderen et al., 2008) – Section 6.8.4.4.

Sediments

Sediment toxicity tests with benthic invertebrates and zinc-contaminated sediments indicate that Zn bioavailability is frequently low and that pore water is a major source of Zn for benthic organisms. Concern for Zn toxicity to sediment-dwelling organisms is mainly related to highly contaminated sites such as those affected by mining and metal processing, and in some instances, the dumping of sewage biosolids. Burton et al. (2005) showed no relationship between toxicity and total Zn in sediments, but found that the AVS-SEM model (Section 6.3.3) was able to predict effects on benthic macroinvertebrates.

Terrestrial Systems

In common with its behaviour in aquatic systems, zinc does not biomagnify through the terrestrial food chain. There have been indications that soil microorganisms are the terrestrial organisms that are the most sensitive to Zn, but there appears to be a great deal of variability in measured toxicities. Rout and Das (2003) reviewed Zn toxicity in plants and showed a number of adverse effects at high Zn concentrations, including root growth inhibition, chlorosis and 'interference' with iron translocation.

Zinc Hyperaccumulator Plants and Excluders

A number of vascular plant species are known to accumulate high zinc concentrations, e.g., pennycress (Thlaspi *caerulescens*). There has been considerable interest in the mechanisms for Zn uptake and root-to-shoot translocation in such species, and in their ability to detoxify excessive Zn. The results of this research may be relevant to our understanding of Zn metabolism and detoxification in more typical plants. Furthermore, researchers have suggested these hyperaccumulators have potential for phytoremediation of contaminated sites (Section 13.9.3). At the other end of the spectrum, some vascular plants maintain very low Zn concentrations in their aerial parts, even when growing on Zn-rich soils and despite the fact that their roots accumulate high concentrations (e.g., bladder campion or *Silene vulgaris*). These plants are (perhaps misleadingly) termed excluders. Selection for Zn tolerance has been demonstrated in grasses growing on mine waste sites, but there has been no consistent agreement about the mechanisms by which the tolerance is acquired.

6.8.6.5 Zinc Highlights

Zinc occurs fairly commonly in the Earth's crust, often co-occurring with copper and lead. One of its major uses is in the galvanizing of other metals, to limit their corrosion. However, a consequence of this use is that Zn is leached out of these protected structures into the surrounding environment. Zinc is mined as sulphide and other ores, the processing of which often results in contamination of soils and aquatic systems. Its chemical behaviour is intermediate between that of 'soft' and 'hard' metals. The metal forms coordinate bonds with a range of ligand types in aqueous systems, and it lacks redox activity.

As a well-recognized required micronutrient, zinc operates in living cells as a cofactor for at least 300 enzymes and plays a structural role as well. Total cellular Zn concentrations are typically in the 100–200 μM range, well above the levels of other essential metals (other than iron). The uptake and intracellular distribution of Zn are highly regulated, such that the intracellular concentration of free Zn^{2+} is normally extremely low. Another example of this tight homeostatic control is the observation that Zn normally accumulates only moderately in exposed organisms, with the accumulated Zn concentrations showing very little variation along broad Zn exposure gradients. Although toxicity can occur at highly contaminated sites, Zn toxicity is normally of relatively low concern; Zn deficiency may be more important.

6.8.7 Arsenic

Metal/metalloid Arsenic, As
Atomic number 33
Atomic weight 74.92
Estimated ratio [anthropogenic emissions/natural emissions] 1.7

6.8.7.1 Occurrence, Sources and Uses

Arsenic occurs along with phosphorus in Group 15 in the Periodic Table and is a relatively common element in the Earth's crust (crustal abundance 1.8 mg/kg). The element occurs free in nature (rarely) and most commonly as a constituent of different compounds; over 400 minerals containing As have been identified. Arsenic is typically found in geological formations containing sulphide deposits, notably those bearing gold but also in cobalt, copper, nickel, lead and zinc ores. Arsenic concentrations in such deposits may range from a few parts per million to percentage quantities. In sulphidic ores, As is generally found in association with pyrite, either as arsenopyrite (FeAsS) or as arsenic-rich pyrite with variable ratios of Fe and other metals bound to As and sulphide. Elevated As concentrations may also be found in some oxide minerals, as part of the mineral structure, or associated with (i.e., 'sorbed on') the surface of hydrous metal oxides.

Total arsenic concentrations in uncontaminated fresh waters are typically less than 1 μg/l ($<$13 nM), and open ocean concentrations are about 1.5 μg/l (20 nM). However, much higher concentrations can be found in surface waters that have been affected by mine drainage or by fallout from smelters, as well as in some groundwaters. Smedley and Kinniburgh (2002) provide a useful compilation of As concentrations that have been reported in various environmental matrices (atmospheric precipitation, surface water, groundwater, seawater, sediment pore water) around the world. They note that more than 70 countries worldwide including parts of Argentina, Chile, Mexico, China, United States and Hungary, as well as in India (West Bengal), Bangladesh and Vietnam are dealing with elevated dissolved As levels in groundwater. In many parts of the world, groundwater is a primary source of drinking water and its consumption has proved to be a key route of human exposure to As. The scale of the problem is difficult to grasp, with more than 40 million people in the Bengal Basin alone drinking water containing 'excessive' As (Smedley and Kinniburgh,

2002). The use of groundwater as a drinking water supply was initially promoted in the Bengal Basin in an attempt to limit the spread of diseases by microbially contaminated surface waters. The current WHO guideline for As in drinking water in 10 µg/l.

In addition to leaching into water, arsenic can also be released to the atmosphere. For example, the As present as an impurity in sulphidic ores is released during the ore-roasting or smelting process and escapes with the stack gases (Section 11.2). The burning of coal also introduces As into the atmosphere, as do volcanic emissions. However, in most industrialized nations, anthropogenic sources of airborne As are now limited as a result of air-pollution control measures.

Historically, arsenic has been used as a pesticide and herbicide, in the production of pigments, fireworks and chemical weapons (e.g., mustard gas), in semiconductors, in wood preservatives and as a growth promoter for livestock, especially poultry and pigs. Because of widespread human health concerns, present uses of As are primarily restricted to metallurgical applications.

6.8.7.2 Biogeochemistry

Forms of Arsenic

Arsenic geochemistry is considerably more complex than that of the common metal cations. Multiple oxidation states exist for arsenic (As($-$III), As($-$I), As(0), As(III), As(V)), corresponding to arsines and methylarsines ($-$III), arsenopyrite ($-$I), elemental arsenic (0), arsenite (III) and arsenate (V), with the latter two being the most relevant in ecotoxicology. Unlike complexation reactions (Section 6.2.1), which for most cations are rapid (allowing equilibrium to be approached in natural waters), changes among oxidation states, i.e. 'redox reactions', are often slow. Such couples as As(III) and As(V) are frequently not at equilibrium and individual chemical species can be found at concentrations that differ markedly from those predicted thermodynamically. However, countering the inherent slowness of some of these redox changes, many redox reactions can be microbiologically catalysed, in which case the reaction rate will be strongly influenced by microbial growth and metabolic activity. This is certainly the case for arsenic, where the rates of As(III) oxidation and As(V) reduction are effectively controlled by microorganisms (Smedley and Kinniburgh, 2002).

Dominant forms of arsenic in solution under oxic conditions are the oxyanions arsenite and arsenate.

Arsenite, $As(III) = As(OH)_3^0 \leftrightarrow As(OH)_2O^- \leftrightarrow As(OH)O_2^{2-} \leftrightarrow AsO_3^{3-}$

Arsenate, $As(V) = As(O)(OH)_3^0 \leftrightarrow As(O)(OH)_2O^- \leftrightarrow As(O)(OH)O_2^{2-} \leftrightarrow As(O)O_3^{3-}$

The chemical form shown in the unnumbered figure for arsenite corresponds to $As(OH)_3^0$, the shaded neutral form that predominates over a wide pH range (3–7). The formula for arsenate corresponds to the deprotonated free anion $As(O)O_3^{3-}$, but over the natural range of pH values, the mono and diprotic anions predominate (these are the shaded forms shown above for arsenate). Under reducing conditions and in the presence of dissolved sulphide, mono-, di- and tri-thioarsenates may also be formed, where sulphur replaces one or more of the oxygen atoms surrounding the arsenic atom. Marshes and bogs have been identified as 'hotspots' for the formation of these thioarsenicals, and once formed they tend to be more mobile than arsenate.

Arsenite   Arsenate
(protonated)   (unprotonated)

In addition to the inorganic forms described above, arsenic also forms several methylated species. These are not metal complexes, where a cation and a ligand are bound together by a weak coordinate bond. On the contrary, they are true biogenic molecules where the As atom and the methyl group are linked by a covalent bond (either as As-O-CH$_3$ bonds or as direct As-CH$_3$ bonds). Because of the presence of a covalent bond, these methylated forms are much more stable than typical metal cation complexes. Common methylated forms of arsenic include monomethylarsonic acid (MMA(V), CH_3-As(O)(OH)$_2$) and dimethylarsinic acid (DMA(V), $(CH_3)_2$-As(O)OH). Other than these methylated forms, arsenic can also be incorporated into natural biomolecules such as arsenosugars, where again there is a covalent carbon–arsenic bond.

Generic arsenosugar

In lake and river waters, arsenic is mainly present as inorganic As(V) and As(III), with As(V) the dominant species. Methylated forms are sometimes detected in surface waters (e.g., MMA(V) and DMA(V)), but normally such organic forms of As are of minor importance. In seawater, arsenate also dominates with ratios of As(V)/As (III) that typically range from 10:1 to 100:1 in open seawater (Smedley and Kinniburgh, 2002). These ratios decrease in zones of high primary productivity, presumably as a result of selective uptake of arsenate by phytoplankton (Section 6.8.7.3). The persistence of As(III) in oxic surface waters may reflect its replenishment by the biological reduction of As(V), particularly during summer months. In groundwaters, As(V)/As(III) ratios vary more widely than in surface water bodies, as a function notably of the abundance of redox-active substances such as labile organic carbon and the resulting microbial activity. In reducing aquifers, where reduction of Fe(III) and sulphate is occurring, As(III) dominates over As(V).

Mobilization and transport

The geochemical mobility of arsenic in the environment is markedly affected by pH and the presence or absence of oxygen. Conditions that can lead to the release of As from rocks, soils and sediments on a large scale include (i) oxidizing conditions and an elevated pH, which result in minimal sorption of the arsenic oxyanions, or (ii) reducing conditions, a circumneutral pH and the presence of arsenic bound to iron oxyhydroxides. These latter conditions favour the reductive dissolution of the iron oxyhydroxide solid phase and the release of the previously bound arsenic (Smedley and Kinniburgh, 2002) (Figure 6.21).

Consider the fate of arsenic that is sequestered in a sulphidic ore. When bound to sulphide in such an ore and protected from oxygen, As exhibits minimal mobility. Following exposure to oxygen and water, and under the influence of sulphide-oxidizing microorganisms, the As in such ores is converted to arsenite and arsenate, leading to a marked increase in its mobility. However, these oxidized forms have a high affinity for solid iron oxyhydroxides, with the arsenate binding more strongly than does arsenite, meaning that arsenite tends to be the more mobile oxyanion. In other words, As will initially be mobilized as the sulphidic ore oxidizes, but since this process also produces iron oxyhydroxides (an excellent

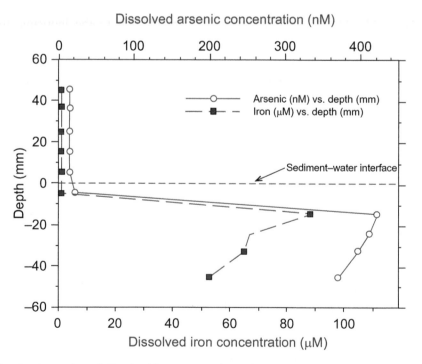

Figure 6.21 Profiles of dissolved arsenic and dissolved iron, across the sediment–water interface of a lake on the Canadian Precambrian Shield. The interface is shown by the horizontal dotted line. The water samples were collected by *in situ* diffusion, with a porewater peeper. Note the close association of the two elements. Reductive dissolution of Fe(III) oxyhydroxide at depths below 1 cm yields dissolved Fe(II) and also releases As into solution. Figure redrawn with data extracted from De Vitre et al. (1991), with permission from the authors and John Wiley & Sons Ltd.

Box 6.17 Common oxyanions that do not involve arsenic

Antimonite SbO_4^{4-}
Chromate CrO_4^{2-}
Molybdate MoO_4^{2-}
Selenite SeO_3^{2-}
Selenate SeO_4^{2-}
Vanadate VO_4^{3-}

An oxyanion, as its name implies, is an anion containing one or more oxygen atoms bonded to another element. Those of interest in the present context contain an element with properties intermediate between those of typical metals and non-metals. Such elements are sometimes referred to as **metalloids**, but there is no standard definition for this term.

sorbent), the subsequent mobility of As will depend on the prevailing pH and redox conditions. Note that the thioarsenicals mentioned in the preceding section also bind to iron oxyhydroxides, but somewhat less strongly than arsenate.

The degree of oxyanion adsorption on substrates like Fe(III) oxyhydroxides tends to increase as the ambient pH decreases, which is the *opposite* trend to that generally observed for cations (Section 6.2.2). Accordingly, arsenic (as well as other elements that form oxyanions such as selenium, molybdenum and antimony – Box 6.17) is more mobile at circumneutral or alkaline pH values than under acidic conditions, whereas the opposite behaviour is observed for cationic metals.

In addition to the pH effect, the mobility of the arsenic oxyanions will be affected by the redox potential. For example, As previously captured by the iron oxyhydroxides present in oxic surface sediments in a lake will be released back into the overlying water when conditions favouring the reductive dissolution of the Fe(III) phase prevail. Under such conditions, porewater concentrations of molybdenum and selenium oxyanions tend to increase in a similar manner. These reducing conditions also favour the conversion of sulphate to sulphide. The other oxyanions show limited mobility in the presence of dissolved sulphide, whereas As remains relatively mobile under such conditions (mg/l As concentrations can be observed, whereas the other oxyanions are generally present at µg/l values). In other words, As is more mobile than most other oxyanions under reducing conditions.

The development of reducing conditions in the upper strata of submerged sediments is favoured by the presence of microbiologically labile organic matter. Given the redox sensitivity of the arsenic oxyanions themselves and their affinity for the redox-sensitive sorbents (e.g., Mn(IV) and Fe(III) oxyhydroxides) present in aquatic sediments, it is clear that the nature and amount of sedimentary organic matter will have a strong influence on arsenic mobility. This is not just of geochemical interest – aquatic sediments also act as an important route of exposure to As for aquatic organisms.

6.8.7.3 Biochemistry
Uptake and Biotransformation

Let us now consider the interaction of arsenic with living organisms. Given its presence in natural waters as an oxyanion rather than as a cation, As clearly does not fit into the biotic ligand model (BLM) template, where the emphasis is on the role of the free metal cation in controlling metal uptake (Section 6.3.2). Intuitively, in an analogy with the free-ion model for cations, one might have expected a free-anion approach to work for the oxyanions, but this idea does not seem to have been tested.

One transmembrane transport mechanism that has been studied in some detail for arsenic involves **aquaporins**. Aquaporins are protein-based channels that allow the bidirectional passage of water molecules across biological membranes (Section 3.1.2.2). Since their discovery in erythrocytes and renal tissue (Denker *et al.*, 1988), aquaporins have been found in all animal taxa studied to date and are also present in the plasma and intracellular membranes of higher plants. In addition to water, some types of aquaporins may facilitate the transmembrane transport of non-polar solutes, such as urea or glycerol, the reactive oxygen species hydrogen peroxide, and gases such as ammonia, carbon dioxide and nitric oxide. The neutral polyhydroxy species $As(OH)_3^0$ behaves as an inorganic molecular mimic of glycerol, which allows it to enter and exit cells through aquaporin channels (Mukhopadhyay *et al.*, 2014); this pathway is also exploited by antimonite, $Sb(OH)_3^0$.

The distribution of a given oxyanion among various polyhydroxy species is determined by protonation and deprotonation reactions, which are very rapid and inherently pH sensitive.

$$As(OH)_3^0 \leftrightharpoons As(OH)_2O^- \leftrightharpoons As(OH)O_2^{2-} \leftrightharpoons AsO_3^{3-}$$

It follows that the bioavailability of the oxyanion may also be influenced by the ambient pH. For example, the proportion of the neutral $M(OH)_3^0$ species is close to 100% in the pH range from 3 to 7 for arsenic(III) but drops to 66% at pH 9.

Some oxyanions may cross biological membranes via anion transporters (Chapter 3). For example, arsenate (AsO_4^{3-}) has been shown to enter biological cells via transporters that normally supply the cells with phosphate (PO_4^{3-}); thioarsenates can also cross biological membranes in this manner. In the case of phytoplankton cells at the base of the aquatic food chain, the rate of uptake of arsenate will thus be affected by the ambient phosphate concentration. Indeed, in surface waters (lakes and oceans) depletions of arsenate and phosphate often co-occur, reflecting the uptake by phytoplankton of these two anions. Arsenate concentration minima in lakes often coincide with photosynthetic maxima (Cullen and Reimer, 1989).

Once it enters a living cell, arsenate is reduced to arsenite, which facilitates its subsequent elimination from the cells via membrane-bound aquaporins. Aquaporins operate as bidirectional channels, which allow the passage of solutes by simple diffusion down a concentration gradient; in other words, they may function both as toxicity and depuration pathways (Tamás, 2016). In the intracellular environment, As is subject not only to reduction but also to a variety of biochemical reactions, including biomethylation, a process known to occur in various microorganisms, algae, plants and animals (Bentley and Chasteen, 2002). Biomethylation involves the formation of arsenic-carbon bonds that attach one or more methyl groups to the central As atom. This process results in the synthesis of MMA(V) and DMA(V), the simple methylated forms of arsenic sometimes found in natural waters (Section 6.8.7.2), but it may also lead to the formation of more complex molecules, such as arsenobetaine. This molecule, distinguished by the presence of both a positive and a negative charge, was discovered in the Australian rock oyster in 1977 and proved to be the dominant form of arsenic in marine zooplankton and in higher invertebrates and vertebrates. The biochemical role of arsenobetaine is a matter of some debate, but it probably serves as an **osmolyte** in marine animals. It has been suggested that it acts as an analogue of glycine-betaine, a N-containing betaine that is a naturally abundant osmolyte in some marine animals (**betaine** = an internal salt, with both a positive and negative charge).

Arsenobetaine   Glycyl-betaine

6.8.7.4 Ecotoxicity
Trophic Transfer

The uptake of arsenic into phytoplankton and other primary producers opens the door to its trophic transfer, a phenomenon that has been studied quite intensively in marine environments, but less so in freshwater systems. It is clear that trophic transfer of As does occur, but in contrast to the case for Hg, this transfer does not lead to biomagnification; As concentrations decrease as one moves up freshwater, marine or terrestrial food chains.

In the marine environment, MMA(V), DMA(V) and arsenosugars are present in phytoplankton. The first traces of arsenobetaine only appear in herbivorous zooplankton grazers, and it becomes the dominant arsenic molecule in carnivorous zooplankton and in higher consumers. In their review of the origins and role of arsenobetaine in planktonic organisms, Caumette *et al.* (2012) suggest that arsenobetaine in zooplankton is derived from the degradation of ingested arsenosugars and is accumulated by the consumer organism as an osmolyte. Organisms that are subject to salinity changes, notably those living in estuarine environments, accumulate intracellular organic osmolytes to mitigate the potentially detrimental effects of salinity fluctuations on native protein structures within cells. Arsenobetaine remains intact once ingested by a consuming organism and, if not retained as an osmolyte, it can be excreted unchanged.

Far less research has been carried out in freshwater environments, but Caumette and co-workers conducted a study along an arsenic exposure gradient in a gold mining area in the Canadian Northwest Territories. They demonstrated that freshwater phytoplankton contained inorganic As (with As(V) >> As(III)), but unlike marine phytoplankton they contained negligible amounts of arsenosugars (Caumette *et al.*, 2011). These authors concluded that freshwater phytoplankton could not methylate or otherwise transform inorganic As into organoarsenic species. Total As concentrations were lower in the freshwater zooplankton than in their food, ruling out any biomagnification. In the contaminated lakes the As burden in the zooplankton was dominated by inorganic arsenic (As(V) > As(III)), as was the case in their food, whereas in the reference lakes arsenosugars were predominant in the zooplankton, with traces of arsenobetaine in one lake (9% of total As). One interpretation of these data would be that freshwater zooplankton possess the ability to transform inorganic arsenic into a variety of organic forms, but that this ability had been inhibited in the most contaminated lakes.

General Toxicity

The role of arsenic as a nutrient is still controversial, but it appears to be an essential ultratrace nutrient for red algae and perhaps some animals (chickens, rats and pigs), where a deficiency results in inhibited growth. However, the beneficial effect of arsenicals in farm animal diets may not be an indication of its essentiality. It has been suggested that growth promotion may result from an effect of As on the intestinal microbiome that increases the mobilization and absorption of nutrients.

The element is best known, however, for its adverse effects (Box 6.18). The toxicity of As has been attributed both to its similarity to phosphorus when it is present as the arsenate anion, and to its ability to form covalent bonds with thiols when it has been reduced intracellularly to arsenite. Because of its high affinity for thiol groups, arsenite can inactivate enzymes in the intracellular environment. Inhibition of enzymes involved in DNA repair may be responsible for the DNA damage caused by As exposure in laboratory experiments. On the other hand, if arsenate is not completely reduced but is at least partially conserved within the cell, it may substitute for phosphate and interfere with the cell's normal energy-transfer phosphorylation reactions. Arsenic can cause oxidative stress (Section 6.4.3), either directly through participation in Fenton-like redox reactions or indirectly by causing a depletion of subcellular antioxidants such as glutathione. It can also perturb gene expression by interfering with

Box 6.18 Signs of arsenic toxicity

- Induction of oxidative stress.
- Enzyme inhibition induced by binding of As(III) to protein thiol groups (R-SH).
- Substitution for phosphate in ATP.
- Inhibition of DNA repair.
- Disruption of gene expression.

signal transduction. Finally, inorganic As is recognized as a Class 1 human carcinogen and is cytotoxic and genotoxic. It is a weak mutagen and cannot induce gene mutations on its own, but it has been shown to enhance the mutagenicity of other agents.

Inorganic forms of arsenic are more toxic than organo-arsenic molecules such as MMA(V), DMA(V), arsenobetaine and arsenosugars. Among the inorganic forms, As(III) is normally more toxic than As(V), but due to the interconversion of the two oxidation states in the intracellular environment, the distinction between them is blurred once they have entered a living cell. Note the marked contrast here with Hg, where organo-Hg compounds are notoriously more toxic than inorganic Hg.

Aquatic

In controlled laboratory tests, phytoplankton and higher plants are more sensitive to arsenic than are aquatic invertebrates or fish, both for freshwater and marine species. Indeed, microscopic algae are some of the most sensitive organisms and show decreases in productivity and growth when exposed to As at very low concentrations (e.g., concentrations as low as 5 µg/l have been shown to reduce photosynthesis in a marine diatom). Competition between arsenate and phosphate is particularly evident for phytoplankton, where increased phosphate concentrations have consistently been shown to protect against arsenate uptake and toxicity. This competition may occur extracellularly, at the level of uptake across the **plasmalemma**, or internally where arsenate may interfere with photophosphorylation reactions. There is evidence from laboratory cultures, and from studies of phytoplankton communities in lakes representing a gradient in As contamination, that phytoplankton may develop tolerance to arsenate exposure upon prolonged exposure. This may be due to the development of tolerance within individual species or to shifts in the phytoplankton community in favour of arsenate-tolerant species (Knauer *et al.*, 1999). This change in community structure towards more tolerant phytoplankton species may have an indirect effect on herbivorous zooplankton (Sanders, 1986).

Except in cases of extreme contamination, dissolved arsenic concentrations in surface waters are generally too low to exert deleterious effects on aquatic invertebrates and fish. However, dietary uptake of As may lead to growth inhibition in fish. For example, in a careful study of the relative importance of waterborne and diet-borne As, Erickson *et al.* (2011) exposed juvenile rainbow trout

(*Oncorhynchus mykiss*) for 28 d to a range of As concentrations in water and in a live **oligochaete** diet, either separately or in combination. The oligochaetes had been pre-exposed to the same waterborne arsenate concentration as the fish; the worms accumulated As from the water but it was largely reduced from arsenate to arsenite within their tissue. When exposed in clean water, fish fed with worms that had previously been exposed to arsenate at 4 or 8 mg As/l showed pronounced reductions in growth. In parallel experiments, fish that were exposed to these same water concentrations but to a low arsenic diet experienced less or no effect. This striking result is a clear indication that assessments of the environmental risks of As contamination should consider exposure not only from the ambient water but also from the consumption of As-contaminated prey.

Terrestrial

In most terrestrial environments, arsenic does not pose a risk for the plants growing in the soil or for the invertebrates living there. However, in extreme environments (e.g., mine tailings or soils that have been irrigated with As-contaminated groundwater), As may be present at levels that are problematic. The *Ecological Soil Screening Level* for As, defined by the US EPA as "the concentration of a contaminant in soil that is protective of ecological receptors that commonly come in contact with or consume biota that live in or on the soil" (i.e., plants, soil invertebrates, birds and mammals), is 18 mg As/kg for plants, 43 mg As/kg for birds and 46 mg As/kg for wildlife, but no value could be derived for soil invertebrates due to insufficient data.

Excessive transfer of most metals and metalloids from plants to animals (including humans), via the food chain, is limited by a 'soil–plant barrier'; in other words, before metal concentrations in the plant reach levels that would be dangerous for animal consumption, they reach levels that are toxic to the plant itself (Chaney, 1980). However, one recently discovered exception is rice, which when grown in flooded paddy fields, can accumulate arsenic in its grains to levels that compromise consumers (including humans). This problem is particularly acute in Southeast Asia (Meharg, 2004).

Much of our current knowledge of the effects of arsenic on soil invertebrates has come from research on earthworms, which play an important role in aerating and mixing soil constituents and also serve as prey to many small terrestrial vertebrates (birds, moles, shrews). Some earthworms live and feed at or near the soil surface, whereas others burrow more deeply but return to the surface layer to feed. By engaging in this activity, they may accumulate As by dermal contact with soil water or by ingestion of solid matter. It has been suggested that earthworms may have the capacity to detect foreign chemicals in their environment and indeed they are known to exhibit avoidance behaviour when exposed to As-contaminated soils. Earthworms living in As-contaminated soil may acquire some tolerance to arsenic, but it is unclear whether this resistance represents physiological acclimation or genetic adaptation.

6.8.7.5 Arsenic Highlights

Arsenic is typically found in sulphidic geological formations containing such elements as gold, cadmium, copper, nickel, lead, antimony and zinc. However, it does not behave like these metals (i.e., as a cation) but falls instead in the metalloid category. Arsenite and arsenate are important forms of As in the aquatic and terrestrial environments; they do not enter living cells via cation transporters, but instead cross the cell membrane as an anion (e.g., arsenate masquerading as phosphate) or as a neutral polyhydroxo-species (arsenite moving through an aquaporin channel).

As anions, arsenate and arsenite tend to be more mobile at neutral pH values than under acidic conditions; cations, on the other hand, become more mobile as the pH decreases. Under reducing conditions and in the presence of dissolved sulphide (for example, in wetlands and bogs), thioarsenates may be formed, where sulphur replaces one or more of the oxygen atoms surrounding the arsenic atom. In addition to these inorganic forms, As also forms several methylated species, where the arsenic atom and the methyl group are linked by a covalent bond (both as $As-O-CH_3$ bonds and direct $As-CH_3$ bonds). Because of the presence of a covalent bond, these methylated forms are much more stable than typical metal cation complexes.

Despite the existence of these methylated forms, trophic transfer of arsenic does not lead to biomagnification; unlike the case for methylmercury, As concentrations decrease as one moves up freshwater, marine or terrestrial food chains. Also in contrast with Hg, the methylated forms of As are generally less toxic than the inorganic forms; among the inorganic forms, As(III) is normally more toxic than As(V).

Finally, despite the emphasis in the preceding summary on the differences between arsenic and Hg, they do share one unfortunate characteristic, namely that they are both environmental toxicants for humans (through drinking water and rice in the case of As).

6.8.8 Selenium

Metal/metalloid Selenium, Se.
Atomic number 34
Atomic weight 78.97
Estimated ratio [anthropogenic emissions/natural emissions] 0.2

6.8.8.1 Occurrence, Sources and Uses

Selenium is among the less common elements in the Earth's crust. Its global distribution is uneven and estimates of its crustal abundance are imprecise, varying between 0.05 and 0.30 ppm. In contrast to many metals and oxyanions, for which anthropogenic sources are dominant, total emissions of Se to the environment are dominated by natural sources (Table 6.3). Natural sources include weathering of selenium-rich rocks, volcanic activity, wildfires, volatilization from plants and water bodies, and sea-salt spray in coastal regions. As suggested by this enumeration of natural sources, the Earth's atmosphere is an important transient reservoir for Se, much as we have seen earlier for Hg. However, the residence-time of Se in the atmosphere is much shorter than that for Hg.

Selenium also enters the environment from many different industrial sources, which include coal mining and combustion, mining and smelting of Cu–Pb–Zn sulphide-rich ores, and the manufacture of glass, paint, petroleum, textiles and electrical components (Lemly, 2004). Selenium is used in the agricultural sector as a feeding supplement for livestock when its concentration in forage is inadequate for proper livestock nutrition. Among these anthropogenic activities, the major contributors to Se mobilization are the mining of coal and sulphide-seeking metals, and the irrigation of Se-rich soils. Historically, Se has been occasionally employed as a pesticide, but this use has been largely discontinued.

Selenium concentrations in uncontaminated fresh waters are typically <1 µg/l (<13 nM) but elevated concentrations may occur where the catchment soil is naturally rich in Se or in waterbodies downwind or downstream from metal mining or smelting activities. The Se content of soils is highly variable, with reported values varying from <0.01 mg/kg in Se-deficient areas up to 1,200 mg/kg in seleniferous areas.

6.8.8.2 Biogeochemistry

Selenium occurs in Group 16 in the Periodic Table, together with sulphur (S) and tellurium (Te); as we have already seen in the previous metal-specific sections, elements that co-occur in the same Periodic Table group tend to behave at least somewhat similarly in the environment and in living organisms.

Forms of Selenium

Selenium exists in three oxidation states in surface waters and sediment pore waters: selenide, Se($-$II); selenite, Se(IV); selenate, Se(VI). In addition, it may occur in sediments in the elemental form Se(0), which is insoluble in water.

Se(0), elemental selenium
Se($-$II), selenide: Se^{2-}
Se(IV), selenite: $SeO_3^{2-} \leftrightarrow SeO_2(OH)^- \leftrightarrow Se(O)(OH)_2^0$
Se(VI), selenate: $SeO_4^{2-} \leftrightarrow SeO_3OH^-$

We referred earlier to the similarities between the chemistry of Se and that of sulphur. However, it should be noted that selenate is a far stronger oxidizing agent than sulphate (it is more comparable to chromate) whereas selenide is a stronger reducing agent than sulphide.

In oxygenated lake and river waters, selenium is mainly present as selenate and selenite, with minor contributions from dissolved organic selenides. In common with arsenic, the oxyanion described in the previous section, abiotic conversions among these different oxidation states are slow, and redox disequilibrium may result. However, microorganisms are able to accelerate these reactions, especially in organic-rich sediments where selenate and selenite can be reduced to elemental Se, which can then become the dominant form of the element. Also in common with As (Section 6.8.7), Se can be methylated by soil and aquatic microorganisms, both heterotrophic and autotrophic. The methylation yields dimethyl selenide (CH_3-Se-CH_3), dimethyl diselenide (CH_3-Se-Se-CH_3) and methylselenol (CH_3-Se-H), all of which are volatile

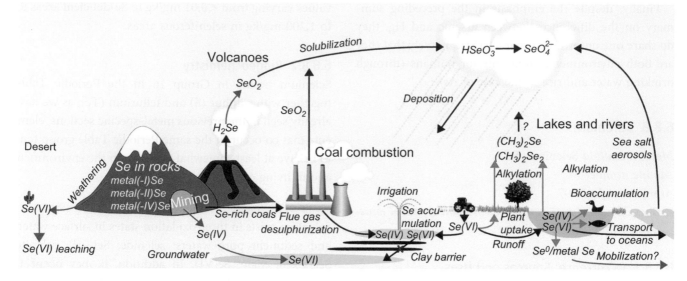

Figure 6.22 Schematic global cycle of selenium with main focus on the terrestrial environment. Blue arrows indicate processes that involve oxidation of Se species and green arrows indicate processes that involve reduction of Se species. Figure adapted with permission from Winkel *et al.* (2012). Copyright 2012 American Chemical Society.

and can find their way into the atmosphere (Figure 6.22). Note the close parallel here with the behaviour of sulphur, which also can be methylated, converted into dimethyl sulphide and transferred to the (marine) atmosphere (Luxem *et al.*, 2017).

Mobilization and Transport

Geochemically, selenium is typically associated with sedimentary rocks, and it is often found in sulphidic minerals such as pyrite. Weathering of these natural materials results in the mobilization of Se as selenate or selenite, both of which are highly water soluble and mobile. In the presence of particles, however, these oxyanions will tend to adsorb to the particle surface. As is the case for the analogous arsenic oxyanions, selenate and especially selenite also have a strong affinity for iron and manganese oxyhydroxides. The degree of Se oxyanion adsorption increases as the ambient pH decreases, a trend that is similar to the behaviour of As oxyanions (arsenite, arsenate). In contrast, the adsorption of cations decreases as the pH decreases. Selenium associated with particles will tend to settle out of the water column and accumulate in the bottom sediments. In the sedimentary environment and in the absence of oxygen, the Se initially bound to iron oxyhydroxides will be released by the reduction of the Fe (III) host phase. The Se oxyanions thus released are subject to abiotic and microbial reduction to elemental Se(0). Under such reducing conditions, the Se may also substitute for sulphide in pyrite and form FeSe, $FeSe_2$ and mixed

FeS_nSe_m species. If oxygen is reintroduced into sediment, the reverse reactions will be favoured, i.e., oxidation.

6.8.8.3 Biochemistry
Uptake and Biotransformation

Trace amounts of selenium are required for normal cellular function in almost all living organisms, with the exception of higher plants and yeasts. Selenium deficiency has been demonstrated in phytoplankton, invertebrates, fish and terrestrial animals (including livestock). It is an essential element but one with a very narrow range between sufficiency and toxicity (See Figure 6.2, plot B) for the typical profile of a biological response as a function of the concentration of an essential element). As a micronutrient, Se plays a key role in the protection of cells from reactive oxygen species. It is also known to afford protection against exposure to mercury, arsenic and cadmium, the most likely mechanism for Hg and Cd being the formation of metabolically inert Hg and Cd selenides. Indeed, mercury selenide (HgSe) has been identified in bacterial cultures, but for higher organisms the actual structure of the putative detoxified forms has not yet been conclusively demonstrated for Hg or Cd. Selenium has been shown to interact with As and accelerate its elimination via the bile into the intestinal tract. There are also reports that exposure to low levels of Se can increase **immunocompetence**, but at higher levels the opposite response is observed.

Selenate normally enters living cells by the same route as sulphate, setting up a competition between these two anions for the same membrane-bound anion transporters; uptake of selenate decreases when the sulphate concentration is increased. Less is known about selenite uptake, but it has been shown to compete with phosphate for transport into yeast cells. When present at low (micronutrient) levels in the exposure environment, Se will follow its normal biochemical pathways once it has entered the intracellular environment. These pathways include its incorporation into the amino acids selenocysteine and selenomethionine, and the synthesis of a variety of functional organo-Se metabolites. Examples of such molecules include glutathione peroxidase, an enzyme that is involved in the destruction of hydrogen peroxide and thus plays a key role in limiting the development of oxidative stress, and selenoprotein P, which is involved in Se distribution and homeostasis. When Se is present at high concentrations in the exposure environment, bacteria and primary producers shunt much of the excess Se into selenomethionine which can then be incorporated non-specifically into proteins in the place of methionine.

Selenocysteine   Selenomethionine

6.8.8.4 Ecotoxicity

Trophic Transfer

A discussion of selenium biogeochemistry and ecotoxicity would be incomplete if we did not address its trophic transfer. The ecotoxicological repercussions of Se exposure are inextricably linked both to the very narrow range between sufficiency and toxicity, and to its propensity for movement between trophic levels. Uptake of waterborne Se by primary producers (phytoplankton and periphyton) is a key step at the base of the aquatic food web, with bioconcentration factors (BCF = concentration in the algal cell/concentration in water) ranging from several hundred to tens of thousands. Consumption of these primary producers by herbivorous zooplankton, and to a lesser degree the ingestion of sediment-bound Se by benthic invertebrates, constitute the next step in the food chain. In this and in all subsequent steps, the uptake of waterborne Se is typically negligible in comparison with the uptake of diet-borne Se. Note that the presence of selenomethionine in benthic invertebrates, fish and piscivorous birds is a clear indication of the key role played by primary producers in initiating the dietary transfer of Se to higher trophic levels; plants (and bacteria) can synthesize selenomethionine, but animals are unable to do so.

The bioaccumulation of diet-borne selenium is influenced by the concentration and form of Se in the diet, the assimilation efficiency in the consumer's digestive tract and the internal fate of the Se after assimilation. Steady-state concentrations of Se in some bivalve taxa have been shown to be much higher than those in other invertebrates inhabiting the same environment, but these high internalized concentrations do not appear to be toxic to the bivalves. The ability of bivalves to accumulate Se to this extent has been attributed to their role as filter-feeders, which exposes them to both the Se in the suspended particles and to a constant stream of dissolved Se. In addition, bivalves tend to hold onto the Se once it has been assimilated, exhibiting a slow rate of elimination.

Consumption of selenium-contaminated primary producers can also be a problem in the terrestrial environment. For example, some birds may be at risk for Se toxicity, via trophic transfer from Se-accumulating plants. Horses, sheep and cattle (which cannot detoxify Se) show classic signs of Se poisoning after grazing on Se-rich vegetation. A fascinating example of this phenomenon is provided by Se-accumulating species of plants in the genus *Astragalus* (the Se-accumulating species of which are known as locoweed, where loco means crazy and refers to the effect of their consumption on horses). Several species of the genus are able to accumulate very high concentrations of Se and store it as selenomethionine, which protects the plant from Se toxicity; non-tolerant species of *Astragalus* do not produce this amino acid. The Se accumulators are tolerant of high concentrations of Se in the soil in which they grow, even to the extent of having a requirement for high Se. The accumulator plants, which may contain up to several thousand parts per million of Se, are toxic to herbivores that graze on them, causing signs of Se toxicity in horses and cattle. Hence, the consequence of this detoxification, although advantageous to the population of plants, can be seen as a potential disadvantage to the ecosystem as a whole.

Consumption of Se-rich vegetation can also pose a problem for some birds. Because of the low energy content of plant material, birds that rely on such material for sustenance will consume greater quantities of Se than would birds that consume a diet of energy-rich invertebrates.

General Toxicity

The toxicity of selenium has been attributed to its substitution for sulphur in two amino acids, selenocysteine and selenomethionine, with the hypothesis that such substitution in various proteins would alter their conformation and affect their functionality. However, recent research has cast some doubt on this long-standing hypothesis.

In normal proteins, disulphide bonds between two cysteine amino acids play a key role in maintaining the proper tertiary structure (3D form or folding) of the protein. Given the ability of selenium to substitute for sulphur in amino acids, it had been suggested that this substitution could lead to improper folding of the selenoprotein and result in a dysfunctional molecule. However, more recent research has cast doubt on this explanation since the excess Se is largely incorporated into selenomethionine, which does not have a free selenol group and thus cannot form a Se–S bond. The structure and function of proteins do not appear to be changed by the replacement of methionine by selenomethionine.

As mentioned earlier, selenium when present at low concentrations protects cells from reactive oxygen species. Ironically, when present at higher concentrations it can in fact induce oxidative stress, as indicated by lower ratios of reduced glutathione to oxidized glutathione (GSH/GSSG) and by increased indices of oxidative damage such as lipid peroxidation. These findings have resulted in a shift towards the belief that oxidative stress is the most likely mechanism that leads to Se toxicity (Janz et al., 2010) – Box 6.19. The current consensus is that the Se ecotoxicity results from dietary exposure to organo-Se compounds, predominantly selenomethionine,

and the resulting increase in the production of reactive oxygen species. Because of the importance of the dietary route of exposure, and the varying degree of Se accumulation in different food items, the feeding habits of target species inhabiting a Se-contaminated area will greatly influence their potential vulnerability. There is a certain parallel here with the case of the snow quail and its exposure to Cd (Section 6.8.2.4).

When selenium concentrations increase above required or protective levels, bacteria, algae, invertebrates and mammals are all reasonably tolerant. However, embryos of egg-laying vertebrates (e.g., birds; fish) prove to be particularly vulnerable to the increased exposure. They exhibit various signs of embryo-toxicity and teratogenesis at Se concentrations that are well tolerated by other organisms. The transition from Se levels that are biologically essential to those that are toxic occurs across a relatively narrow range of exposure concentrations with a very steep dose–response curve. For both birds and fish, Se is subject to maternal transfer, a process that serves as a detoxification pathway for fish but less so for birds. Selenium replaces sulphur in vitellogenin (a precursor protein of egg yolk), which is transported to the ovary and incorporated into the developing ovarian follicle. In fish, the hatched larvae experience exposure to organo-Se when they undergo yolk absorption. The most sensitive indicators of Se toxicity in fish appear in the developing larvae as teratogenic deformities that hinder normal development and as various forms of oedema that result in mortality. In birds, however, most of the Se in the egg does not find its way there by maternal transfer from the female's tissues but rather from the diet that she consumes during the ovulation period. The Se found in the bird egg is present not in the yolk but rather in the albumin and accordingly exposure to organo-Se occurs earlier, before hatching. In birds, as in fish, embryonic deformities resulting in impaired reproduction are commonly observed in response to high Se exposure.

Box 6.19 Signs of selenium toxicity

- Induction of oxidative stress.
- Teratogenesis (the process by which congenital malformations are produced in an embryo or foetus).
- Impairment of immune system.
- Inhibition of photosynthetic electron transport (chloroplasts).
- Behavioural effects, impaired social cognition.

Field Observations

Early evidence of selenium toxicity to native fish and water birds dates from the 1970s, when deformities in fish in Belews Lake, North Carolina, USA, and the loss of certain fish species were linked to the discharge into the lake of a Se-rich effluent from a coal-ash disposal basin. Sixteen of the fish species originally present in Lake Belews disappeared within a 3-year period after initiation of the effluent discharge, but four species remained

(Lemly, 1985). The disappearance of fish was linked to embryo mortality and to developmental abnormalities that compromised the recruitment of the malformed individuals into the adult population. Similar results were reported for Hyco Reservoir, also in North Carolina, and for Kesterson Reservoir, California, where the source of Se was recycled irrigation drainage water. There are also indications of elevated Se concentrations in the biota living in streams downstream from coal mining activities, e.g., on the eastern slopes of the Canadian Rocky mountains (Orr *et al.*, 2012). Interestingly, field investigations at these sites did not reveal major adverse effects on the resident macroinvertebrates, a result that is consistent with the earlier statement that fish and birds are inherently more sensitive to Se than other animal classes. Macroinvertebrates are essential prey items in aquatic food webs; their tolerance to Se and their persistence mean that they remain an important vector for the transfer of organic Se to higher trophic levels.

To quote from Janz *et al.* (2010), "These established linkages between the molecular/cellular mechanism of toxicity (oxidative stress), effects on individuals (early life stage mortality and deformities), and negative effects on populations and community structure provide one of the clearest examples in ecotoxicology of cause–effect relationships between exposure and altered population dynamics."

6.8.8.5 Selenium Highlights

Selenium is a relatively uncommon and unevenly distributed element in the Earth's crust, which results in both Se-deficient soils (<0.01 mg Se/kg) and seleniferous areas (up to 1,200 mg Se/kg). The behaviour of Se in the environment resembles in some respects that of sulphur, with selenite and selenate anions as the dominant forms in surface waters (analogous to sulphite and sulphate), and dimethyl selenide as a volatile methylated form (cf. dimethyl sulphide). Adsorption of the Se oxyanions increases as the ambient pH decreases, as is also observed for arsenite and arsenate; the opposite behaviour is generally observed for cations.

Selenium is an essential element but exhibits a very narrow range between sufficiency and toxicity. On entering living cells, Se is incorporated into the amino acids selenocysteine and selenomethionine, and is involved in the synthesis of a variety of essential organo-Se molecules such as glutathione peroxidase. As a micronutrient, Se protects against oxidative stress and also against the toxicity of cadmium and mercury.

Total emissions of selenium to the environment are dominated by natural sources, but in proximity to coal mines and coal-burning power plants, and downstream from areas where Se-rich soils have been irrigated, Se may reach toxic levels. This ecotoxicity is inextricably linked to the trophic transfer of Se from primary producers to herbivores to omnivores and eventually to piscivorous fish and birds. In such situations, most of the links in the food chain are reasonably tolerant of Se. However, egg-laying vertebrates (e.g., birds; fish) prove to be particularly vulnerable to the increased exposure. Maternal transfer of Se leads to embryo mortality and to developmental abnormalities that compromise the recruitment of the malformed individuals into the adult population.

Summary

In the early parts of this chapter, we summarized how inorganic contaminants differ from synthetic organic chemicals. This led to an overview of how metals and metalloids behave in the environment and how these naturally occurring substances interact with and affect living organisms. We distinguished between essential and non-essential elements, and explored how mimicry allows non-essential elements to gain entry to living cells and perturb normal cell functioning. If present at high concentrations in the exposure medium, essential elements can also disturb cellular homeostasis, leading to toxicity. The importance of metal speciation has been emphasized throughout the chapter, as it affects not only the fate of a metal or metalloid in the environment but also its bioavailability. Of particular importance here was the distinction between cationic metals and metalloid oxyanions.

The eight elements considered in detail in the second half of the chapter were selected to demonstrate the diversity of behaviours and toxicities that exist within the metals and metalloids group, but also to illustrate some of their commonalities. For each element we have considered its sources, both anthropogenic and natural, and discussed the relative importance of these two as an indication of the degree to which the biogeochemical cycling of the element has been affected by human activities. For each element we have explored how it crosses biological membranes and enters living cells. Our focus then shifted to the intracellular fate of the metal or metalloid, how it might be detoxified or provoke toxicity. We summarized some of the recognized signs of metal- or metalloid-induced toxicity, but most of the information presented for the individual elements was derived from laboratory exposures. In some cases we were able to refer the reader to field evidence of deleterious effects on individual species, but for the most part community- and ecosystem-level effects remain to be examined.

Review Questions and Exercises

1. How do inorganic contaminants differ from organic contaminants? In answering this question, consider such aspects as sources, environmental fates, half-lives, mechanisms of uptake, mechanisms of toxicity.

2. In Figure 6.5, we have compared anthropogenic loadings of various metals with what we have called 'natural loading'. Give examples of natural processes that lead to metal releases to the Earth's surface environment and include specific metals for which these natural sources are known to be important.

3. Once a metal has been introduced into the environment, its 'speciation' becomes a significant factor that influences its biological activity. Explain what is meant by speciation in this context. How can it be measured/calculated?

4. Define 'ligand' as it is used in this chapter. Give examples of naturally occurring and synthetic ligands that are known to bind with metals. Explain what is meant by 'metal complexes' and provide an equation for a general type of metal–ligand interaction. Illustrate what is meant by multi-dentate and chelation.

5. Using diagrams, illustrate the generalized dose–response relationships exhibited by living organisms to (a) a major nutrient, (b) a micronutrient and (c) a non-essential toxic element. Use the same diagram(s) to illustrate the concept of threshold and discuss the reasons why it may be difficult to determine the absolute threshold for these responses.

6. Within the class of inorganic contaminants, how do cationic metals differ from metalloids (environmental mobility; biological uptake; detoxification)?

7. Define what is meant by 'metal bioavailability'. In this chapter, we state that metal bioavailability is not an intrinsic property of a given metal. Elaborate on this topic, considering such factors as pH, hardness, salinity, natural dissolved organic matter, and uptake routes.

8. How would you explain the 'biotic ligand model' to someone working in a governmental agency concerned with environmental regulations?

9. How do metal cations and oxyanions (metalloids) enter living cells (refer to Chapter 3 and Figure 6.6)?

10. What is the difference between the dietary uptake and the trophic transfer of metals? Similarly, what is the difference between trophic transfer and biomagnification?

11. What are the properties of methylmercury that lead to its biomagnification?

12. Describe how a metal may induce toxicity in a living organism. Prepare a list of molecular mechanisms by which non-essential metals can cause toxicity in living organisms.

13. The chronic exposure of living organisms to sublethal metal concentrations can result in the development of metal tolerance in the exposed population. Provide examples of situations where metal tolerance has evolved and describe the mechanisms to which tolerance has been attributed. Are there ecological advantages or costs to the possession of tolerance?

14. What are the known functions of metallothioneins and phytochelatins? By what biochemical mechanisms do they function?

15. With reference to Figure 6.15, explain what is meant by 'metal spillover'. Why is this laboratory observation rarely observed in the field?

16. At your new position in regulatory toxicology, you are approached by your immediate superior and asked to identify metals that are currently understudied but that may become future targets of environmental regulations. How would you proceed?

17. Reports have been made of mass deaths of flocks of a species of predatory bird close to a water body that is known to be contaminated with multiple substances as a result of inadequate waste management. As a team charged with an investigation, including a wildlife ecologist, an environmental toxicologist, a representative of the local regulatory agency and a member of an ENGO that is concerned with wildlife protection, describe the approaches that your group is going to apply.

Element-specific questions

Hg: With reference to Table 6.3, the crustal abundance of mercury and its annual mining production are very low, yet it is recognized as the most ecotoxicologically problematic metal. How would you explain this to a sceptical politician?

Cd: In Figure 6.1, cadmium is indicated as an element that is 'possibly essential' – what is the justification for this unexpected classification?

Cd: Describe factors affecting the bioavailability of cadmium in the aquatic environment and ways employed by organisms to detoxify this metal.

Cu: Among the divalent cations, copper is unique in that it exists in the external environment as Cu(II) but largely as Cu(I) within living cells. Explain where and how this reduction from Cu(II) → Cu(I) occurs.

Ni: Explain how the internal handling and detoxification of nickel differ from that of cadmium, and link this to the intrinsic properties of the two divalent cations (e.g., the distinction between 'hard' and 'soft' metals).

Pb: What factors explain the changes in lead concentrations in the sediment core depicted in Figure 6.20? Under what conditions could this palaeolimnological approach be used to trace the environmental history of metals or metalloids other than Pb?

Pb: Summarize the ecotoxicological and health effects of lead. What is the significance of δ-aminolevulinic acid dehydratase (ALAD) in the context of the ecotoxicological effects of lead?

As, Se: Why doesn't the BLM apply to arsenic or selenium? How might it be adapted to predict the bioavailability of As or Se?

As: The methylation of mercury increases its toxicity and also leads to its biomagnification. However, the methylation of arsenic leads to a decrease in its toxicity. Explain why the effect of methylation is different in the two cases.

Se: For which metals or metalloids does the atmosphere play an important role in the global distribution of the element? What are the properties of these elements that favour atmospheric transport over long distances?

Zn: Zinc is a well-known required micronutrient. Describe the ways in which Zn is regulated in living cells. What are the implications of this regulation in terms of its potential ecotoxicity?

Abbreviations

ALAD	delta-Aminolaevulinic acid dehydratase
AOP	Adverse outcome pathway
ATP	Adenosine triphosphate
AVS	Acid-volatile sulphide
BCE	Before the Common Era
BLL	Blood lead levels
BLM	Biotic ligand model
CCME	Canadian Council of Ministers of the Environment
CE	Common Era
CEC	Cation exchange capacity
CuNP	Copper nanoparticle
DMA	Dimethylarsinic acid
DNA	Deoxyribonucleic acid
DOC	Dissolved organic carbon
DOM	Dissolved organic matter
DTPA	Diethylenetriaminepentaacetic acid
EDTA	Ethylenediaminetetraacetic acid
EOG	Electro-olfactogram
FIAM	Free-ion activity model
GSH	Glutathione
HDP	Heat-denatured proteins
HMW	High molecular weight
HSP	Heat-stable proteins
kDa	Kilodalton; a measure of molecular weight
LMW	Low molecular weight
M	Molar (moles per litre)
MIE	Molecular initiating event
ML	Metal–ligand complex
MMA	Monomethylarsonic acid
MT	Metallothionein
MTLP	Metallothionein-like peptide
NMDA	N-methyl-D-aspartate
NOM	Natural organic matter
NTA	Nitrilotriacetic acid
OSN	Olfactory sensory neuron
PC	Phytochelatin
PNEC	Predicted no-effect concentration
QICAR	Quantitative ion character–activity relationship
RNA	Ribonucleic acid
R-SH	Thiol (SH = sulphydryl group)
SEM	Simultaneously extracted metal
TEL	Threshold effect level
TRIS	Tris(hydroxymethyl)aminomethane, $(HOCH_2)_3CNH_2$
TTF	Trophic transfer factor
UNEP	United Nations Environment Programme
ZIP	Zrt- and Irt-like Protein

References

Adams, W., Blust, R., Dwyer, R., *et al.* (2020). Bioavailability assessment of metals in freshwater environments: a historical review. *Environmental Toxicology and Chemistry*, 39, 48–59.

Adams, W. J., Stewart, A. R., Kidd, K. A., Brix, K. V. & DeForest, D. K. (2005). Case histories of dietborne exposure to mercury and selenium in aquatic ecosystems. In J. S. Meyer, *et al.* (eds.) *Toxicity of Dietborne Metals to Aquatic Biota*. Pensacola, FL: Society of Environmental Toxicology and Chemistry, SETAC Press, 263–274.

Åkerblom, S., Nilsson, M. B., Skyllberg, U., *et al.* (2020). Formation and mobilization of methylmercury across natural and experimental sulfur deposition gradients. *Environmental Pollution*, 263, e114398.

Ali, H., Khan, E. & Sajad, M. A. (2013). Phytoremediation of heavy metals-Concepts and applications. *Chemosphere*, 91, 869-881.

AMAP/UNEP (2013). *Technical Background Report for the Global Mercury Assessment 2013*. AMAP-UNEP, Oslo, Norway.

Amos, H. M., Jacob, D. J., Streets, D. G. & Sunderland, E. M. (2013). Legacy impacts of all-time anthropogenic emissions on the global mercury cycle. *Global Biogeochemical Cycles*, 27, 410–421.

ANZECC & ARMCANZ. (2018). *Australian and New Zealand guidelines for fresh and marine water quality* [Online]. Available: www.waterquality.gov.au/anz-guidelines [Accessed 2022-02-21].

Ariya, P. A., Amyot, M., Dastoor, A., *et al.* (2015). Mercury physicochemical and biogeochemical transformation in the atmosphere and at atmospheric interfaces: a review and future directions. *Chemical Reviews*, 115, 3760–3802.

Armbruster, W. S. (2014). Multiple origins of serpentine-soil endemism explained by preexisting tolerance of open habitats. *Proceedings of the National Academy of Sciences of the United States of America*, 111, 14968–14969.

Azizishirazi, A., Dew, W. A., Bougas, B., Bernatchez, L. & Pyle, G. G. (2015). Dietary sodium protects fish against copper-induced olfactory impairment. *Aquatic Toxicology*, 161, 1–9.

Barst, B. D., Rosabal, M., Campbell, P. G. C., *et al.* (2016). Subcellular distribution of trace elements and liver histology of landlocked Arctic char (*Salvelinus alpinus*) sampled along a mercury contamination gradient. *Environmental Pollution*, 212, 574–583.

Bebianno, M. J., Santos, C., Canario, J., *et al.* (2007). Hg and metallothionein-like proteins in the black scabbardfish *Aphanopus carbo*. *Food and Chemical Toxicology*, 45, 1443–1452.

Beckvar, N., Dillon, T. M. & Read, L. B. (2005). Approaches for linking whole-body fish tissue residues of mercury or DDT to biological effects thresholds. *Environmental Toxicology and Chemistry*, 24, 2094–2105.

Belzile, N., Chen, Y. W., Gunn, J. M., *et al.* (2006). The effect of selenium on mercury assimilation by freshwater organisms. *Canadian Journal of Fisheries and Aquatic Sciences*, 63, 1–10.

Bentley, R. & Chasteen, T. G. (2002). Microbial methylation of metalloids: arsenic, antimony, and bismuth. *Microbiology and Molecular Biology Reviews*, 66, 250–271.

Bilodeau, F., Therrien, J. & Schetagne, R. (2017). Intensity and duration of effects of impoundment on mercury levels in fishes of hydroelectric reservoirs in northern Québec (Canada). *Inland Waters*, 7, 493–503.

Binkowski, Ł. J. (2019). Nickel, Ni. In E. Kalisińska (ed.) *Mammals and Birds as Bioindicators of Trace Element Contaminations in Terrestrial Environments: An Ecotoxicological Assessment of the Northern Hemisphere*. Cham: Springer International Publishing, 281–299.

Bjerregaard, P. & Christensen, A. (2012). Selenium reduces the retention of methyl mercury in the brown shrimp *Crangon crangon*. *Environmental Science & Technology*, 46, 6324–6329.

Bjerregaard, P., Fjordside, S., Hansen, M. G. & Petrova, M. B. (2011). Dietary selenium reduces retention of methyl mercury in freshwater fish. *Environmental Science & Technology*, 45, 9793–9798.

Blaby-Haas, C. E. & Merchant, S. S. (2012). The ins and outs of algal metal transport. *Biochimica et Biophysica Acta: Molecular Cell Research*, 1823, 1531–1552.

Blanck, H. (2002). A critical review of procedures and approaches used for assessing pollution-induced community tolerance (PICT) in biotic communities. *Human and Ecological Risk Assessment*, 8, 1003–1034.

Blewett, T. A. & Leonard, E. M. (2017). Mechanisms of nickel toxicity to fish and invertebrates in marine and estuarine waters. *Environmental Pollution*, 223, 311–322.

Bonneris, E., Giguère, A., Perceval, O. *et al.* (2005). Sub-cellular partitioning of metals (Cd, Cu, Zn) in the gills of a freshwater bivalve, *Pyganodon grandis*: role of calcium concretions in metal sequestration. *Aquatic Toxicology*, 71, 319–334.

Borgmann, U., Couillard, Y., Doyle, P. & Dixon, D. G. (2005). Toxicity of sixty-three metals and metalloids to *Hyalella azteca* at two levels of water hardness. *Environmental Toxicology and Chemistry*, 24, 641–652.

Borgmann, U., Norwood, W. P., Reynoldson, T. B. & Rosa, F. (2001). Identifying cause in sediment assessments: bioavailability and the sediment quality triad. *Canadian Journal of Fisheries and Aquatic Sciences*, 58, 950–960.

Boullemant, A., Le Faucheur, S., Fortin, C. & Campbell, P. G. C. (2011). Uptake of lipophilic cadmium complexes by three green algae: influence of humic acid and its pH dependence. *Journal of Phycology*, 47, 784–791.

Bowles, K. C., Apte, S. C., Maher, W. A., Kawei, M. & Smith, R. (2001). Bioaccumulation and biomagnification of mercury in Lake Murray, Papua New Guinea. *Canadian Journal of Fisheries and Aquatic Sciences*, 58, 888–897.

Bridges, C. C. & Zalups, R. K. (2005). Molecular and ionic mimicry and the transport of toxic metals. *Toxicology and Applied Pharmacology*, 204, 274–308.

Brix, K. V., Schlekat, C. E. & Garman, E. R. (2017). The mechanisms of nickel toxicity in aquatic environments: an Adverse Outcome Pathway (AOP) analysis. *Environmental Toxicology and Chemistry*, 36, 1128–1137.

Broadley, M. R., White, P. J., Hammond, J. P., Zelko, I. & Lux, A. (2007). Zinc in plants. *New Phytologist*, 173, 677–702.

Bundy, J. G., Kille, P., Liebeke, M. & Spurgeon, D. J. (2014). Metallothioneins may not be enough – the role of phytochelatins in invertebrate metal detoxification. *Environmental Science & Technology*, 48, 885–886.

Burgess, R. M., Berry, W. J., Mount, D. R. & Di Toro, D. M. (2013). Mechanistic sediment quality guidelines based on contaminant bioavailability: Equilibrium partitioning sediment benchmarks. *Environmental Toxicology and Chemistry*, 32, 102–114.

Burton, G. A., Nguyen, L. T. H., Janssen, C. *et al.* (2005). Field validation of sediment zinc toxicity. *Environmental Toxicology and Chemistry*, 24, 541-553.

Buxton, S., Garman, E., Heim, K. E. *et al.* (2019). Concise review of nickel human health toxicology and ecotoxicology. *Inorganics*, 7, e7070089.

Cain, D. J., Luoma, S. N. & Wallace, W. G. (2004). Linking metal bioaccumulation of aquatic insects to their distribution patterns in a mining-impacted river. *Environmental Toxicology and Chemistry*, 23, 1463–1473.

Campbell, P. G. C. (1995). Interactions between trace metals and organisms: a critique of the free-ion activity model. In A. Tessier & D. Turner (eds.) *Metal Speciation and Bioavailability in Aquatic Systems*. Chichester, UK: J. Wiley & Sons, 45–102.

Campbell, P. G. C., Clearwater, S. J., Brown, P. B., *et al.* (2005a). Digestive physiology, chemistry and nutrition. In J. S. Meyer, W. J. Adams, K. V. Brix, *et al.* (eds.) *Toxicity of Dietborne Metals to Aquatic Biota*. Pensacola, FL: Society of Environmental Toxicology and Chemistry, SETAC Press, 13–57.

Campbell, P. G. C. & Couillard, Y. (2004). Prise en charge et détoxication des métaux chez les organismes aquatiques. In E. Pelletier, P. G. C. Campbell & F. Denizeau (eds.) *Écotoxicologie moléculaire - Principes fondamentaux et perspectives de développement*. Ste-Foy, Quebec, Canada: Les Presses de l'Université du Québec, 9–61.

Campbell, P. G. C. & Fortin, C. (2013). Biotic ligand model. In J. F. Férard & C. Blaise (eds.) *Encyclopedia of Aquatic Ecotoxicology*. Heidelberg: Springer-Verlag, 237–245.

Campbell, P. G. C., Galloway, J. N. & Stokes, P. M. (1985). *Acid Deposition – Effects on Geochemical Cycling and Biological Availability of Trace Elements*, Washington, DC, USA: National Academy Press.

Campbell, P. G. C., Giguère, A., Bonneris, E. & Hare, L. (2005b). Cadmium-handling strategies in two chronically exposed indigenous freshwater organisms – the yellow perch (*Perca flavescens*) and the floater mollusc (*Pyganodon grandis*). *Aquatic Toxicology*, 72, 83–97.

Campbell, P. G. C. & Tessier, A. (1996). Ecotoxicology of metals in the aquatic environment – geochemical aspects. In M. C. Newman & C. H. Jagoe (eds.) *Quantitative Ecotoxicology: A Hierarchical Approach*. Boca Raton, FL: Lewis Publishers, 11–58.

Campbell, P. G. C., Twiss, M. R. & Wilkinson, K. J. (1997). Accumulation of natural organic matter on the surfaces of living cells: implications for the interaction of toxic solutes with aquatic biota. *Canadian Journal of Fisheries and Aquatic Sciences*, 54, 2543–2554.

Cardwell, R. D., DeForest, D. K., Brix, K. V. & Adams, W. J. (2013). Do Cd, Cu, Ni, Pb, and Zn biomagnify in aquatic ecosystems? In D. M. Whitacre (ed.) *Reviews of Environmental Contamination and Toxicology*, Vol. 226, 101–122.

Caron, A., Rosabal, M., Drevet, O., Couture, P. & Campbell, P. G. C. (2018). Binding of trace elements (Ag, Cd, Co, Cu, Ni and Tl) to cytosolic biomolecules in livers of juvenile yellow perch (*Perca flavescens*) collected from lakes representing metal-contamination gradients. *Environmental Toxicology and Chemistry*, 37, 576–586.

Caumette, G., Koch, I., Estrada, E. & Reimer, K. J. (2011). Arsenic speciation in plankton organisms from contaminated lakes: transformations at the base of the freshwater food chain. *Environmental Science & Technology*, 45, 9917–9923.

Caumette, G., Koch, I. & Reimer, K. J. (2012). Arsenobetaine formation in plankton: a review of studies at the base of the aquatic food chain. *Journal of Environmental Monitoring*, 14, 2841–2853.

CCME (2014). *Canadian Water Quality Guidelines for the Protection of Aquatic Life – Cadmium*, Report no. 1515.

Cebalho, E. C., Díez, S., dos Santos Filho, M., *et al.* (2017). Effects of small hydropower plants on mercury concentrations in fish. *Environmental Science and Pollution Research*, 24, 22709–22716.

Chaney, R. L. (1980). Health risks associated with toxic metals in municipal sludge. In G. Bitton, B. L. Damro, G. T. Davidson & J. M. Davidson (eds.) *Sludge – Health Risks of Land Application*. Ann Arbor, MI: Ann Arbor Sci. Publ, 59-83.

Chapman, P. M. (2016). Toxicity delayed in cold freshwaters? *Journal of Great Lakes Research*, 42, 286–289.

Chen, Q. Y., DesMarais, T. & Costa, M. (2019). Metals and mechanisms of carcinogenesis. *Annual Review of Pharmacology and Toxicology*, 59, 537–554.

Chiang, H. C., Lo, J. C. & Yeh, K. C. (2006). Genes associated with heavy metal tolerance and accumulation in Zn/Cd hyperaccumulator *Arabidopsis halleri*: a genomic survey with cDNA microarray. *Environmental Science & Technology*, 40, 6792–6798.

Clarkson, T. W. (2002). The three modern faces of mercury. *Environmental Health Perspectives*, 110, 11–23.

Clearwater, S. J., Handy, R. D. & Hogstrand, C. (2005). Interactions of dietborne metals with digestive processes in fish. In J. S. Meyer, W. J. Adams, K. V. Brix, *et al.* (eds.) *Toxicity of Dietborne Metals to Aquatic Biota*. Pensacola, FL: Society of Environmental Toxicology and Chemistry, SETAC Press, 205–225.

Clemens, S. (2019). Safer food through plant science: reducing toxic element accumulation in crops. *Journal of Experimental Botany*, 70, 5537–5557.

Courbot, M., Willems, G., Motte, P., *et al.* (2007). A major quantitative trait locus for cadmium tolerance in *Arabidopsis halleri* colocalizes with HMA4, a gene encoding a heavy metal ATPase. *Plant Physiology*, 144, 1052–1065.

Couture, P., Pyle, G. G., Campbell, P. G. C. & Hontela, A. (2015). Using *Perca* as biomonitors in ecotoxicological studies. In P. Couture & G. G. Pyle (eds.) *Biology of Perch*. Boca Raton, FL: CRC Press, 271–303.

Croteau, M. N., Hare, L. & Tessier, A. (2002). Increases in food web cadmium following reductions in atmospheric inputs to some lakes. *Environmental Science & Technology*, 36, 3079–3082.

Cullen, W. R. & Reimer, K. J. (1989). Arsenic speciation in the environment. *Chemical Reviews*, 89, 713–764.

De Schamphelaere, K. A. C., Lofts, S. & Janssen, C. R. (2005). Bioavailability models for predicting acute and chronic toxicity of zinc to algae, daphnids, and fish in natural surface waters. *Environmental Toxicology and Chemistry*, 24, 1190–1197.

De Vitre, R. R., Belzile, N. & Tessier, A. (1991). Speciation and adsorption of arsenic on diagenetic iron oxyhydroxides. *Limnology and Oceanography*, 36, 1480–1485.

DeForest, D. K., Brix, K. V. & Adams, W. J. (2007). Assessing metal bioaccumulation in aquatic environments: the inverse relationship between bioaccumulation factors, trophic transfer factors and exposure concentration. *Aquatic Toxicology*, 84, 236–246.

DeForest, D. K. & Meyer, J. S. (2015). Critical review: toxicity of dietborne metals to aquatic organisms. *Critical Reviews in Environmental Science & Technology*, 45, 1176–1241.

DeForest, D. K., Santore, R. C., Ryan, A. C., *et al.* (2017). Development of biotic ligand model-based freshwater aquatic life criteria for lead following US Environmental Protection Agency guidelines. *Environmental Toxicology and Chemistry*, 36, 2965–2973.

DeForest, D. K. & Schlekat, C. E. (2013). Species sensitivity distribution evaluation for chronic nickel toxicity to marine organisms. *Integrated Environmental Assessment and Management*, 9, 580–589.

Degryse, F., Buekers, J. & Smolders, E. (2004). Radio-labile cadmium and zinc in soils as affected by pH and source of contamination. *European Journal of Soil Science*, 55, 113–121.

Degryse, F. & Smolders, E. (2012). Cadmium and nickel uptake by tomato and spinach seedlings: plant or transport control? *Environmental Chemistry*, 9, 48–54.

Denker, B. M., Smith, B. L., Kuhajda, F. P. & Agre, P. (1988). Identification, purification, and partial characterization of a novel Mr 28,000 integral membrane protein from erythrocytes and renal tubules. *Journal of Biological Chemistry*, 263, 15634–15642.

Dietz, R., Letcher, R. J., Desforges, J.-P., *et al.* (2019). Current state of knowledge on biological effects from contaminants on Arctic wildlife and fish. *Science of The Total Environment*, 296, 133792.

Dong, W., Liang, L., Brooks, S., Southworth, G. & Gu, B. (2010). Roles of dissolved organic matter in the speciation of mercury and methylmercury in a contaminated ecosystem in Oak Ridge, Tennessee. *Environmental Chemistry*, 7, 94–102.

Ebbs, S., Lau, I., Ahner, B. & Kochian, L. (2002). Phytochelatin synthesis is not responsible for Cd tolerance in the Zn/Cd hyperaccumulator *Thlaspi caerulescenes* (J. and C. Presl). *Planta*, 214, 635–640.

ECCC (2016). *Canadian Mercury Science Assessment*. Ottawa, ON, Canada: Environment and Climate Change Canada.

Echevarria, G., Massoura, S. T., Sterckeman, T., *et al.* (2006). Assessment and control of the bioavailability of nickel in soils. *Environmental Toxicology and Chemistry*, 25, 643-651.

Eisler, R. (1998). *Nickel Hazards to Fish, Wildlife, And Invertebrates: A Synoptic Review*. Laurel, MD: US Department of the Interior, US Geological Survey.

El Mehdawi, A. F. & Pilon-Smits, E. A. H. (2012). Ecological aspects of plant selenium hyperaccumulation. *Plant Biology*, 14, 1–10.

Ellwood, M. J., Strzepek, R., Chen, X., Trull, T. W. & Boyd, P. W. (2020). Some observations on the biogeochemical cycling of zinc in the Australian sector of the Southern Ocean: a dedication to Keith Hunter. *Marine and Freshwater Research*, 71, 355–373.

Erichsen-Jones, J. R. (1938). The relative toxicity of salts of lead, zinc and copper to the stickleback (*Gasterosteus aculeatus* L.) and the effect of calcium on the toxicity of lead and zinc salts. *Journal of Experimental Biology*, 15, 394–407.

Erickson, R. J., Mount, D. R., Highland, T. L., Hockett, J. R. & Jenson, C. T. (2011). The relative importance of waterborne and dietborne arsenic exposure on survival and growth of juvenile rainbow trout. *Aquatic Toxicology*, 104, 108–115.

Ernst, W. H. O. (2006). Evolution of metal tolerance in higher plants. *Forest Snow Landscape Research*, 80, 251–274.

Esdaile, L. J. & Chalker, J. M. (2018). The mercury problem in artisanal and small-scale gold mining. *Chemistry – A European Journal*, 24, 6905–6916.

Evers, D. C., Savoy, L. J., DeSorbo, C. R *et al.* (2008). Adverse effects from environmental mercury loads on breeding common loons. *Ecotoxicology*, 17, 69–81.

Evers, D. C., Taylor, K. M., Major, A., *et al.* (2003). Common loon eggs as indicators of methylmercury availability in North America. *Ecotoxicology*, 12, 69–81.

Faburé, J., Dufour, M., Autret, A., Uher, E. & Fechner, L. C. (2015). Impact of an urban multi-metal contamination gradient: metal bioaccumulation and tolerance of river biofilms collected in different seasons. *Aquatic Toxicology*, 159, 276–289.

Fallon, J. A., Redig, P., Miller, T. A., Lanzone, M. & Katzner, T. (2017). Guidelines for evaluation and treatment of lead poisoning of wild raptors. *Wildlife Society Bulletin*, 41, 205–211.

Farina, M. & Aschner, M. (2019). Glutathione antioxidant system and methylmercury-induced neurotoxicity: an intriguing interplay. *Biochimica et Biophysica Acta – General Subjects*, 1863, e129285.

Finney, L. A. & O'Halloran, T. V. (2003). Transition metal speciation in the cell: insights from the chemistry of metal ion receptors. *Science*, 300, 931–936.

Fisher, N. S. & Hook, S. E. (2002). Toxicology tests with aquatic animals need to consider the trophic transfer of metals. *Toxicology*, 181, 531–536.

Fortin, C. & Campbell, P. G. C. (2001). Thiosulfate enhances silver uptake by a green alga: role of anion transporters in metal uptake. *Environmental Science & Technology*, 35, 2214–2218.

Fortin, C., Couillard, Y., Vigneault, B. & Campbell, P. G. C. (2010). Determination of free Cd, Cu and Zn concentrations in lake waters by in situ diffusion followed by column equilibration ion-exchange. *Aquatic Geochemistry*, 16, 151–172.

Francesconi, K. A. & Lenanton, R. C. J. (1992). Mercury contamination in a semienclosed marine embayment – organic and inorganic mercury content of biota, and factors influencing mercury levels in fish. *Marine Environmental Research*, 33, 189–212.

Frausto da Silva, J. J. R. & Williams, R. J. P. (2001). *The Biological Chemistry of the Elements: The Inorganic Chemistry of Life*, Oxford, UK: Oxford University Press.

Fuchsman, P. C., Brown, L. E., Henning, M. H., Bock, M. J. & Magar, V. S. (2017). Toxicity reference values for methylmercury effects on avian reproduction: critical review and analysis. *Environmental Toxicology and Chemistry*, 36, 294–319.

Giguère, A., Campbell, P. G. C., Hare, L., McDonald, D. G. & Rasmussen, J. B. (2004). Influence of lake chemistry and fish age on cadmium, copper, and zinc concentrations in various organs of indigenous yellow perch (*Perca flavescens*). *Canadian Journal of Fisheries and Aquatic Sciences*, 61, 1702–1716.

Gilmour, C. C., Bullock, A. L., McBurney, A., Podar, M. & Elias, D. A. (2018). Robust mercury methylation across diverse methanogenic *Archaea*. *mBio*, 9, e02403–17.

Gilmour, C. C., Elias, D. A., Kucken, A. M., *et al.* (2011). Sulfate-reducing bacterium *Desulfovibrio desulfuricans* ND132 as a model for understanding bacterial mercury methylation. *Applied and Environmental Microbiology*, 77, 3938–3951.

Gilmour, C. C., Podar, M., Bullock, A. L., *et al.* (2013). Mercury methylation by novel microorganisms from new environments. *Environmental Science & Technology*, 47, 11810–11820.

Gobeil, C., Tessier, A. & Couture, R. M. (2013). Upper Mississippi Pb as a mid-1800s chronostratigraphic marker in sediments from seasonally anoxic lakes in Eastern Canada. *Geochimica et Cosmochimica Acta*, 113, 125–135.

Gustafsson, J. P. (2019). *Visual MINTEQ*. 3.1 ed. Stockholm, Sweden: KTH Royal Institute of Technology, Department of Land and Water Resources Engineering.

Haines, K. J. R., Evans, R. D., O'Brien, M. & Evans, H. E. (2010). Accumulation of mercury and selenium in the brain of river otters (*Lontra canadensis*) and wild mink (*Mustela vison*) from Nova Scotia, Canada. *Science of the Total Environment*, 408, 537–542.

Hammerschmidt, C. R., Sandheinrich, M. B., Wiener, J. G. & Rada, R. G. (2002). Effects of dietary methylmercury on reproduction of fathead minnows. *Environmental Science & Technology*, 36, 877–883.

Hansen, D. J., Berry, W. J., Boothman, W. S., *et al.* (1996). Predicting the toxicity of metal-contaminated field sediments using interstitial concentration of metals and acid-volatile sulfide normalizations. *Environmental Toxicology and Chemistry*, 15, 2080–2094.

Hanson, B., Garifullina, G. F., Lindblom, S. D., *et al.* (2003). Selenium accumulation protects *Brassica juncea* from invertebrate herbivory and fungal infection. *New Phytologist*, 159, 461–469.

Hanson, B., Lindblom, S. D., Loeffler, M. L. & Pilon-Smits, E. A. H. (2004). Selenium protects plants from phloem-feeding aphids due to both deterrence and toxicity. *New Phytologist*, 162, 655–662.

Harmens, H., Buse, A., Büker, P., *et al.* (2004). Heavy metal concentrations in European mosses: 2000/2001 survey. *Journal of Atmospheric Chemistry*, 49, 425–436.

Harris, R. C., Rudd, J. W. M., Amyot, M., *et al.* (2007). Whole-ecosystem study shows rapid fish-mercury response to changes in mercury deposition. *Proceedings of the National Academy of Sciences of the United States of America*, 104, 16586–16591.

Hauck, M., Jurgens, S. R. & Leuschner, C. (2010). Norstictic acid: correlations between its physico-chemical characteristics and ecological preferences of lichens producing this depsidone. *Environmental and Experimental Botany*, 68, 309–313.

Henry, R. P., Lucu, C., Onken, H. & Weihrauch, D. (2012). Multiple functions of the crustacean gill: osmotic/ionic regulation, acid–base balance, ammonia excretion, and bioaccumulation of toxic metals. *Frontiers in Physiology*, 3, e00431.

Heyes, A., Moore, T. R., Rudd, J. W. M. & Dugoua, J. J. (2000). Methyl mercury in pristine and impounded boreal peatlands, experimental Lakes Area, Ontario. *Canadian Journal of Fisheries and Aquatic Sciences*, 57, 2211–2222.

Hodson, P. V. & Sprague, J. B. (1975). Temperature-induced changes in acute toxicity of zinc to Atlantic salmon (*Salmo salar*). *Journal of the Fisheries Research Board of Canada*, 32, 1–10.

Hommen, U., Knopf, B., Rüdel, H., *et al.* (2016). A microcosm study to support aquatic risk assessment of nickel: community-level effects and comparison with bioavailability-normalized species sensitivity distributions. *Environmental Toxicology and Chemistry*, 35, 1172–1182.

Hoppe, S., Gustafsson, J. P., Borg, H. & Breitholtz, M. (2015). Can natural levels of Al influence Cu speciation and toxicity to *Daphnia magna* in a Swedish soft water lake? *Chemosphere*, 138, 205–210.

Horowitz, A. J., Lemieux, C., Lum, K. R., *et al.* (1996). Problems associated with using filtration to define dissolved trace element concentrations in natural water samples. *Environmental Science & Technology*, 30, 954–963.

ILA (2015). *Lead in Aquatic Environments – Understanding the Science*. London, UK.

ILZSG (2017). *Main First Uses of Lead and Zinc in Europe 2017*. International Lead Zinc Study Group, Lisbon, Portugal.

Irving, H. & Williams, R. J. P. (1953). The stability of transition-metal complexes. *Journal of the Chemical Society (Resumed)*, 3192–3210.

Janz, D. M., DeForest, D. K., Brooks, M. L., *et al.* (2010). Selenium toxicity to aquatic organisms. In P. M. Chapman, W. J. Adams, M. L. Brooks, *et al.* (eds.) *Ecotoxicology of Selenium in the Aquatic Environment*. Boca Raton, FL: CRC Press, 141–231.

Jarosławiecka, A. & Piotrowska-Seget, Z. (2014). Lead resistance in micro-organisms. *Microbiology*, 160, 12–25.

Jenkins, K. D. & Mason, A. Z. (1988). Relationships between subcellular distributions of cadmium and perturbations in reproduction in the polychaete *Neanthes arenaceodentata*. *Aquatic Toxicology*, 12, 229–244.

Jensen, S. & Jernelov, A. (1969). Biological methylation of mercury in aquatic organisms. *Nature*, 753–754.

Jernelov, A., Landner, L. & Larsson, T. (1975). Swedish perspectives on mercury pollution. *Journal Water Pollution Control Federation*, 47, 810–822.

Kamunde, C., Grosell, M., Higgs, D. & Wood, C. M. (2002). Copper metabolism in actively growing rainbow trout (*Oncorhynchus mykiss*): interactions between dietary and waterborne copper uptake. *Journal of Experimental Biology*, 205, 279–290.

Kamunde, C. & MacPhail, R. (2008). Bioaccumulation and hepatic speciation of copper in rainbow trout (*Oncorhynchus mykiss*) during chronic waterborne copper exposure. *Archives of Environmental Contamination and Toxicology*, 54, 493–503.

Kehrig, H. A., Hauser-Davis, R. A., Seixas, T. G. & Fillmann, G. (2015). Trace-elements, methylmercury and metallothionein levels in Magellanic penguin (*Spheniscus magellanicus*) found stranded on the Southern Brazilian coast. *Marine Pollution Bulletin*, 96, 450–455.

Khan, F. R., Bury, N. R. & Hogstrand, C. (2012). Copper and zinc detoxification in *Gammarus pulex* (L.). *Journal of Experimental Biology*, 215, 822–832.

Khan, M. A. K. & Wang, F. Y. (2009). Mercury-selenium compounds and their toxicological significance: Toward a molecular understanding of the mercury-selenium antagonism. *Environmental Toxicology and Chemistry*, 28, 1567–1577.

Klee, R. J. & Graedel, T. E. (2004). Elemental cycles: a status report on human or natural dominance. *Annual Reviews of Environment and Resources*, 29, 69–107.

Knauer, K., Behra, R. & Hemond, H. (1999). Toxicity of inorganic and methylated arsenic to algal communities from lakes along an arsenic contamination gradient. *Aquatic Toxicology*, 46, 221–230.

Komarek, M., Ettler, V., Chrastny, V. & Mihaljevic, M. (2008). Lead isotopes in environmental sciences: a review. *Environment International*, 34, 562–577.

Lamborg, C. H., Hammerschmidt, C. R., Bowman, K. L., *et al.* (2014). A global ocean inventory of anthropogenic mercury based on water column measurements. *Nature*, 512, 65–68.

Langner, H. W., Domenech, R., Slabe, V. A. & Sullivan, S. P. (2015). Lead and mercury in fall migrant golden eagles from western North America. *Archives of Environmental Contamination and Toxicology*, 69, 54–61.

Łanocha-Arendarczyk, N. & Kosik-Bogacka, D. I. (2019). Copper, Cu. In E. Kalisińska (ed.) *Mammals and Birds as Bioindicators of Trace Element Contaminations in Terrestrial Environments: An Ecotoxicological Assessment of the Northern Hemisphere*. Cham: Springer International Publishing, 125–161.

Large, E. C. (1940). *The Advance of the Fungi*, London, UK: Jonathan Cape.

Larison, J. R., Likens, G. E., Fitzpatrick, J. W. & Crock, J. G. (2000). Cadmium toxicity among wildlife in the Colorado Rocky Mountains. *Nature*, 406, 181–183.

Lawrence, S. G., Holoka, M. H., Hunt, R. V. & Hesslein, R. H. (1996). Multi-year experimental additions of cadmium to a lake epilimnion and resulting water column cadmium concentrations. *Canadian Journal of Fisheries and Aquatic Sciences*, 53, 1876–1887.

Lemly, A. D. (1985). Toxicology of selenium in a freshwater reservoir: Implications for environmental hazard evaluation and safety. *Ecotoxicology and Environmental Safety*, 10, 314–338.

Lemly, A. D. (2004). Aquatic selenium pollution is a global environmental safety issue. *Ecotoxicology and Environmental Safety*, 59, 44–56.

Liu, Y., Vijver, M. G., Pan, B. & Peijnenburg, W. J. G. M. (2017). Toxicity models of metal mixtures established on the basis of "additivity" and "interactions". *Frontiers of Environmental Science & Engineering*, 11, 10.

Luxem, K. E., Vriens, B., Behra, R. & Winkel, L. H. E. (2017). Studying selenium and sulfur volatilisation by marine algae *Emiliania huxleyi* and *Thalassiosira oceanica* in culture. *Environmental Chemistry*, 14, 199–206.

Mahbub, K. R., Krishnan, K., Naidu, R., Andrews, S. & Megharaj, M. (2017). Mercury toxicity to terrestrial biota. *Ecological Indicators*, 74, 451–462.

Maret, W. (2013). Zinc biochemistry: from a single zinc enzyme to a key element of life. *Advances in Nutrition (Bethesda, Md.)*, 4, 82–91.

Marrugo-Madrid, S., Turull, M., Zhang, H. & Díez, S. (2021). Diffusive gradients in thin films for the measurement of labile metal species in water and soils: a review. *Environmental Chemistry Letters*, 19, 3761–3788.

Mason, A. Z. & Jenkins, K. D. (1995). Metal detoxification in aquatic organisms. In A. Tessier & D. Turner (eds.) *Metal Speciation and Bioavailability in Aquatic Systems*. Chichester, UK: J. Wiley & Sons, 479–608.

Mason, R. P., Choi, A. L., Fitzgerald, W. F., *et al.* (2012). Mercury biogeochemical cycling in the ocean and policy implications. *Environmental Research*, 119, 101–117.

Mason, R. P., Reinfelder, J. R. & Morel, F. M. M. (1996). Uptake, toxicity, and trophic transfer of mercury in a coastal diatom. *Environmental Science & Technology*, 30, 1835–1845.

Mebane, C. A. (2010). *Cadmium Risks to Freshwater Life: Derivation and Validation of Low-Effect Criteria Values Using Laboratory and Field Studies*. Reston, VA: US Department of the Interior, US Geological Survey.

Mebane, C. A., Dillon, F. S. & Hennessy, D. P. (2012). Acute toxicity of cadmium, lead, zinc, and their mixtures to stream-resident fish and invertebrates. *Environmental Toxicology and Chemistry*, 31, 1334–1348.

Meharg, A. A. (2004). Arsenic in rice – understanding a new disaster for South-East Asia. *Trends in Plant Science*, 9, 415–417.

Meyer, J. S., Adams, W. J., Brix, K. V *et al.* (2005). *Toxicity of Dietborne Metals to Aquatic Biota*. Pensacola, FL: Society of Environmental Toxicology and Chemistry, SETAC Press.

Meyer, J. S., Santore, R. C., Bobbitt, J. P., *et al.* (1999). Binding of nickel and copper to fish gills predicts toxicity when water hardness varies, but free-ion activity does not. *Environmental Science & Technology*, 33, 913–916.

Milne, C. J., Kinniburgh, D. G., Van Riemsdijk, W. H. & Tipping, E. (2003). Generic NICA-Donnan model parameters for metal-ion binding by humic substances. *Environmental Science & Technology*, 37, 958–971.

Morel, F. M. M. (1983). Trace metals and microorganisms. *Principles and Applications of Aquatic Chemistry*. New York, NY: J. Wiley & Sons Ltd, pp. 300–308.

Mounicou, S., Szpunar, J. & Lobinski, R. (2009). Metallomics: the concept and methodology. *Chemical Society Reviews*, 38, 1119–1138.

Muir, D. C. G., Wang, X., Yang, F., *et al.* (2009). Spatial trends and historical deposition of mercury in eastern and northern Canada inferred from lake sediment cores. *Environmental Science & Technology*, 43, 4802–4809.

Mukhopadhyay, R., Bhattacharjee, H. & Rosen, B. P. (2014). Aquaglyceroporins: generalized metalloid channels. *Biochimica et Biophysica Acta*, 1840, 1583–1591.

Muyssen, B. T. A., Brix, K. V., DeForest, D. K. & Janssen, C. R. (2004). Nickel essentiality and homeostasis in aquatic organisms. *Environmental Reviews*, 12, 113–131.

Neuhierl, B. & Böck, A. (1996). On the mechanism of selenium tolerance in selenium-accumulating plants – purification and characterization of a specific selenocysteine methyltransferase from cultured cells of *Astragalus bisculatus*. *European Journal of Biochemistry*, 239, 235–238.

Nieboer, E. & Richardson, D. H. S. (1980). The replacement of the nondescript term 'heavy metals' by a biologically and chemically significant classification of metal ions. *Environmental Pollution (Series B)*, 1, 3–26.

Niyogi, S., Couture, P., Pyle, G., McDonald, D. G. & Wood, C. M. (2004). Acute cadmium biotic ligand model characteristics of laboratory-reared and wild yellow perch (*Perca flavescens*) relative to rainbow trout (*Oncorhynchus mykiss*). *Canadian Journal of Fisheries and Aquatic Sciences*, 61, 942–953.

NRCAN. (2018). *Natural Resources Canada, Mineral and Metals Facts* [Online]. Available: www.nrcan.gc.ca/our-natural-resources/minerals-mining/minerals-metals-facts/20507 [Accessed 2020].

Nriagu, J. O. (1979). Global inventory of natural and anthropogenic emissions of trace metals to the atmosphere. *Nature*, 279, 409–411.

Nys, C., Janssen, C. R., Van Sprang, P. & De Schamphelaere, K. A. C. (2016). The effect of pH on chronic aquatic nickel toxicity is dependent on the pH itself: extending the chronic nickel bioavailability models. *Environmental Toxicology and Chemistry*, 35, 1097–1106.

Oguma, A. Y. & Klerks, P. L. (2017). Pollution-induced community tolerance in benthic macroinvertebrates of a mildly lead-contaminated lake. *Environmental Science and Pollution Research*, 24, 19076–19085.

Orr, P. L., Wiramanaden, C. I. E., Paine, M. D., Franklin, W. & Fraser, C. (2012). Food chain model based on field data to predict westslope cutthroat trout (*Oncorhynchus clarkii lewisi*) ovary selenium concentrations from water selenium concentrations in the Elk Valley, British Columbia. *Environmental Toxicology and Chemistry*, 31, 672–680.

Outridge, P. M., Mason, R. P., Wang, F., Guerrero, S. & Heimbürger-Boavida, L. E. (2018). Updated global and oceanic mercury budgets for the United Nations global mercury assessment 2018. *Environmental Science & Technology*, 52, 11466–11477.

Pain, D. J., Mateo, R. & Green, R. E. (2019). Effects of lead from ammunition on birds and other wildlife: a review and update. *Ambio*, 48, 935–953.

Pain, D. J., Meharg, A. A., Ferrer, M., Taggart, M. & Penteriani, V. (2005). Lead concentrations in bones and feathers of the globally threatened Spanish imperial eagle. *Biological Conservation*, 121, 603–610.

Paquin, P. R., Gorsuch, J. W., Apte, S., *et al.* (2002). The biotic ligand model: a historical overview. *Comparative Biochemistry and Physiology C: Pharmacology Toxicology and Endocrinology*, 133, 3–35.

Parks, J. M., Johs, A., Podar, M., *et al.* (2013). The genetic basis for bacterial mercury methylation. *Science*, 339, 1332–1335.

Pattee, O. H. & Pain, D. J. (2003). Lead in the environment. In D. J. Hoffman, B. A. Rattner, G. A. Burton, Jr & J. Cairns, Jr (eds.) *Handbook of Ecotoxicology*. Boca Raton, FL: CRC Press, 373–399.

Perceval, O., Couillard, Y., Pinel-Alloul, B., Giguère, A. & Campbell, P. G. C. (2004). Metal-induced stress in bivalves living along a gradient of Cd contamination: relating sub-cellular metal distribution to population-level responses. *Aquatic Toxicology*, 69, 327–345.

Phinney, J. T. & Bruland, K. W. (1994). Uptake of lipophilic organic Cu, Cd and Pb complexes in the coastal diatom, *Thalassiosira weissflogii*. *Environmental Science & Technology*, 28, 1781–1790.

Pourrut, B., Shahid, M., Dumat, C., Winterton, P. & Pinelli, E. (2011). Lead uptake, toxicity, and detoxification in plants. In D. M. Whitacre (ed.) *Reviews of Environmental Contamination and Toxicology*. New York, NY: Springer New York, 113–136.

Prévéral, S., Gayet, L., Moldes, C., *et al.* (2009). A common highly conserved cadmium detoxification mechanism from bacteria to humans. Heavy metal tolerance conferred by the ATP-binding cassette (ABC) transporter SPHMT1 requires glutathione but not metal-chelating phytochelatin peptides. *Journal of Biological Chemistry*, 284, 4936–4943.

Printz, B., Lutts, S., Hausman, J. F. & Sergeants, K. (2016). Copper trafficking in plants and its implication on cell wall dynamics. *Frontiers in Plant Science*, 7, e00601.

Quigg, A. (2016). Micronutrients. In A. M. Borowitzka, J. Beardall & A. J. Raven (eds.) *The Physiology of Microalgae*. Cham: Springer International Publishing, 211–231.

Rainbow, P. S. (2018). *Trace Metals in the Environment and Living Organisms – The British Isles as a Case Study*, Cambridge, UK: Cambridge University Press, sections 2.5.5.1, 4.4.1, 4.7–4.9.

Romero, M. B., Polizzi, P., Chiodi, L., Das, K. & Gerpe, M. (2016). The role of metallothioneins, selenium and transfer to offspring in mercury detoxification in Franciscana dolphins (*Pontoporia blainvillei*). *Marine Pollution Bulletin*, 109, 650–654.

Rosabal, M., Hare, L. & Campbell, P. G. C. (2012). Subcellular metal partitioning in larvae of the insect *Chaoborus* collected along an environmental metal exposure gradient (Cd, Cu, Ni and Zn). *Aquatic Toxicology*, 120–121, 67–78.

Rout, G. R. & Das, P. (2003). Effect of metal toxicity on plant growth and metabolism: I. Zinc. *Agronomie*, 23, 3–11.

Rytuba, J. J. (2003). Mercury from mineral deposits and potential environmental impact. *Environmental Geology*, 43, 326–338.

Sanders, B. M. & Jenkins, K. D. (1984). Relationship between free cupric ion concentrations in sea water and copper metabolism and growth in crab larvae. *Biology Bulletin*, 167, 704–711.

Sanders, J. G. (1986). Direct and indirect effects of arsenic on the survival and fecundity of estuarine zooplankton. *Canadian Journal of Fisheries and Aquatic Sciences*, 43, 694–699.

Sandheinrich, M. B., Bhavsar, S. P., Bodaly, R. A., Drevnick, P. E. & Paul, E. A. (2011). Ecological risk of methylmercury to piscivorous fish of the Great Lakes region. *Ecotoxicology*, 20, 1577–1587.

Scheidegger, C., Behra, R. & Sigg, L. (2011). Phytochelatin formation kinetics and toxic effects in the freshwater alga *Chlamydomonas reinhardtii* upon short- and long-term exposure to lead(II). *Aquatic Toxicology*, 101, 423–429.

Scheuhammer, A., Braune, B., Chan, H. M., *et al.* (2015). Recent progress on our understanding of the biological effects of mercury in fish and wildlife in the Canadian Arctic. *Science of the Total Environment*, 509, 91–103.

Scheuhammer, A. M., Basu, N., Evers, D. C., *et al.* (2012). Ecotoxicology of mercury in fish and wildlife – recent advances. In M. S. Bank (ed.) *Mercury in the Environment, Pattern and Process*. Berkeley, CA: University of California Press, 223–238.

Scheuhammer, A. M., Meyer, M. W., Sandheinrich, M. B. & Murray, M. W. (2007). Effects of environmental methylmercury on the health of wild birds, mammals, and fish. *AMBIO: A Journal of the Human Environment*, 36, 12–19.

Schluter, K. (2000). Review: evaporation of mercury from soils. An integration and synthesis of current knowledge. *Environmental Geology*, 39, 249–271.

Schreck, E., Foucault, Y., Sarret, G., *et al.* (2012). Metal and metalloid foliar uptake by various plant species exposed to atmospheric industrial fallout: mechanisms involved for lead. *Science of the Total Environment*, 427–428, 253–262.

Sigel, H. & Sigel, A. (2019). The bio-relevant metals of the Periodic Table of the elements. *Zeitschrift für Naturforschung B – A Journal of Chemical Sciences*, 74, 461–471.

Smedley, P. L. & Kinniburgh, D. G. (2002). A review of the source, behaviour and distribution of arsenic in natural waters. *Applied Geochemistry*, 17, 517–568.

Smol, J. P. (2002). *Pollution of Lakes and Rivers. A Paleoenvironmental Perspective*, New York, NY: Oxford University Press Inc.

Smolders, E., Lambregts, R. M., McLaughlin, M. J. & Tiller, K. G. (1998). Effect of soil chloride on cadmium availability to Swiss chard. *Journal of Environmental Quality*, 27, 426–431.

Steinberg, C. E. W. (2014). NOM as natural xenobiotics. In F. L. Rosario-Ortiz (ed.) *Advances in the Physicochemical Characterization of Dissolved Organic Matter: Impact on Natural and Engineered Systems*. Washington, DC: American Chemical Society, 115–144.

Stephan, C. H., Courchesne, F., Hendershot, W. H., *et al.* (2008). Speciation of zinc in contaminated soils. *Environmental Pollution*, 155, 208–216.

Stewart, T. J., Szlachetko, J., Sigg, L., Behra, R. & Nachtegaal, M. (2015). Tracking the temporal dynamics of intracellular lead speciation in a green alga. *Environmental Science & Technology*, 49, 11176–11181.

Stolpe, C., Krämer, U. & Muller, C. (2017). Heavy metal (hyper)accumulation in leaves of *Arabidopsis halleri* is accompanied by a reduced performance of herbivores and shifts in leaf glucosinolate and element concentrations. *Environmental and Experimental Botany*, 133, 78–86.

Sunda, W. G., Engel, D. W. & Thuotte, R. M. (1978). Effect of chemical speciation on toxicity of cadmium to grass shrimp, *Palaemonetes pugio*: importance of free cadmium ion. *Environmental Science & Technology*, 12, 409–413.

Sunda, W. G. & Guillard, R. R. L. (1976). The relationship between cupric ion activity and the toxicity of copper to phytoplankton. *Journal of Marine Research*, 34, 511–529.

Szivak, I., Behra, R. & Sigg, L. (2009). Metal-induced reactive oxygen species production in *Chlamydomonas reinhardtii* (Chlorophyceae). *Journal of Phycology*, 45, 427–435.

Tamás, M. J. (2016). Cellular and molecular mechanisms of antimony transport, toxicity and resistance. *Environmental Chemistry*, 13, 955–962.

Telmer, K. H., Daneshfar, B., Sanborn, M. S., Kliza-Petelle, D. & Rancourt, D. G. (2006). The role of smelter emissions and element remobilization in the sediment chemistry of 99 lakes around the Horne smelter, Quebec. *Geochemistry - Exploration Environment Analysis*, 6, 187–202.

Templeton, D. M., Ariese, F., Cornelis, R., *et al.* (2000). Guidelines for terms related to chemical speciation and fractionation of elements. Definitions, structural aspects, and methodological approaches. *Pure and Applied Chemistry*, 72, 1453–1470.

Thayer, J. S. (2002). Biological methylation of less-studied elements. *Applied Organometallic Chemistry*, 16, 677–691.

Tipping, E., Lofts, S. & Sonke, J. E. (2011). Humic Ion-Binding Model VII: a revised parameterisation of cation-binding by humic substances. *Environmental Chemistry*, 8, 225–235.

Trevors, J. T., Stratton, G. W. & Gadd, G. M. (1986). Cadmium transport, resistance, and toxicity in bacteria, algae, and fungi. *Canadian Journal of Microbiology*, 32, 447–464.

UNEP (2013). *Environmental Risks and Challenges of Anthropogenic Metals Flows and Cycles*. Nairobi, Kenya.

UNEP (2017). *Minamata Convention on Mercury – Fact Sheet*.

USGS (2020). *Mineral Commodity Summaries 2020*. Reston, VA: US Department of the Interior, US Geological Survey.

Van Genderen, E., Adams, W., Cardwell, R., *et al.* (2008). An evaluation of the bioavailability and aquatic toxicity attributed to ambient copper concentrations in surface waters from several parts of the world. *Integrated Environmental Assessment and Management*, 4, 416–424.

Van Genderen, E., Adams, W., Cardwell, R., *et al.* (2009). An evaluation of the bioavailability and aquatic toxicity attributed to ambient zinc concentrations in fresh surface waters from several parts of the world. *Integrated Environmental Assessment and Management*, 5, 426–434.

Van Genderen, E., Adams, W., Dwyer, R., Garman, E. & Gorsuch, J. (2015). Modeling and interpreting biological effects of mixtures in the environment: introduction to the metal mixture modeling evaluation project. *Environmental Toxicology and Chemistry*, 34, 721–725.

Vega, F. A. & Weng, L. (2013). Speciation of heavy metals in River Rhine. *Water Research*, 47, 363–372.

Vialykh, E. A., Salahub, D. R., Achari, G., Cook, R. L. & Langford, C. H. (2019). Emergent functional behaviour of humic substances perceived as complex labile aggregates of small organic molecules and oligomers. *Environmental Chemistry*, 16, 505–516.

Vinot, I. & Pihan, J. C. (2005). Circulation of copper in the biotic compartments of a freshwater dammed reservoir. *Environmental Pollution*, 133, 169–182.

Waisberg, M., Joseph, P., Hale, B. & Beyersmann, D. (2003). Molecular and cellular mechanisms of cadmium carcinogenesis. *Toxicology*, 192, 95–117.

Wang, K., Munson, K. M., Armstrong, D. A., Macdonald, R. W. & Wang, F. (2020). Determining seawater mercury methylation and demethylation rates by the seawater incubation approach: a critique. *Marine Chemistry*, 219, e103753.

Wang, Z., Meador, J. P. & Leung, K. M. Y. (2016). Metal toxicity to freshwater organisms as a function of pH: a meta-analysis. *Chemosphere*, 144, 1544–1552.

Wang, Z. S. & Yang, C. F. (2019). Metal carcinogen exposure induces cancer stem cell-like property through epigenetic reprograming: a novel mechanism of metal carcinogenesis. *Seminars in Cancer Biology*, 57, 95–104.

Watras, C. J., Back, R. C., Halvorsen, S., *et al.* (1998). Bioaccumulation of mercury in pelagic freshwater food webs. *Science of the Total Environment*, 219, 183–208.

Watras, C. J. & Bloom, N. S. (1992). Mercury and methylmercury in individual zooplankton – implications for bioaccumulation. *Limnology and Oceanography*, 37, 1313–1318.

Webber, H. M. & Haines, T. A. (2003). Mercury effects on predator avoidance behavior of a forage fish, golden shiner (*Notemigonus crysoleucas*). *Environmental Toxicology and Chemistry*, 22, 1556–1561.

Weng, L. P., Temminghoff, E. J. M. & van Riemsdijk, W. H. (2001). Contribution of individual sorbents to the control of heavy metal activity in sandy soil. *Environmental Science & Technology*, 35, 4436–4443.

Weston, D. P., Judd, J. R. & Mayer, L. M. (2004). Effect of extraction conditions on trace element solubilization in deposit feeder digestive fluid. *Environmental Toxicology and Chemistry*, 23, 1834–1841.

Wiener, J. G., Krabbenhoft, D. P., Heinz, G. H. & Scheuhammer, A. M. (2003). Ecotoxicology of mercury. In D. J. Hoffman, B. A. Rattner, G. A. Burton, Jr & J. Cairns, Jr (eds.) *Handbook of Ecotoxicology*. Boca Raton, FL: CRC Press, 409–463.

Wilkinson, K. J., Campbell, P. G. C. & Couture, P. (1990). Effect of fluoride complexation on aluminum toxicity towards juvenile Atlantic salmon (*Salmo salar*). *Canadian Journal of Fisheries and Aquatic Sciences*, 47, 1446–1452.

Williams, R. J. P. & Rickaby, R. (2012). Outline of the main chemical factors in evolution. *Evolution's Destiny: Co-evolving Chemistry of the Environment and Life*. Cambridge, UK: Royal Society of Chemistry, pp. 1–31.

Winkel, L. H. E., Johnson, C. A., Lenz, M., *et al.* (2012). Environmental selenium research: from microscopic processes to global understanding. *Environmental Science & Technology*, 46, 571–579.

Wood, C. M., Al-Reasi, H. A. & Smith, D. S. (2011). The two faces of DOC. *Aquatic Toxicology*, 105, 3–8.

Wright, D. A. & Welbourn, P. M. (2002). *Environmental Toxicology*, Cambridge, UK: Cambridge University Press.

Wu, X., Cobbina, S. J., Mao, G., *et al.* (2016). A review of toxicity and mechanisms of individual and mixtures of heavy metals in the environment. *Environmental Science and Pollution Research*, 23, 8244–8259.

Wysocki, R. & Tamas, M. J. (2010). How *Saccharomyces cerevisiae* copes with toxic metals and metalloids. *FEMS Microbiology Reviews*, 34, 925–951.

Xu, Y. & Morel, F. M. M. (2013). Cadmium in marine phytoplankton. In A. Sigel, H. Sigel & R. K. O. Sigel (eds.) *Cadmium: From Toxicity to Essentiality*. Dordrecht, The Netherlands: Springer Science, 509–528.

Zhang, Y. X., Jaegle, L., Thompson, L. & Streets, D. G. (2014). Six centuries of changing oceanic mercury. *Global Biogeochemical Cycles*, 28, 1251–1261.

7 Organic Compounds

Christopher D. Metcalfe, David A. Wright and Peter V. Hodson

Learning Objectives

Upon completion of this chapter, the student should be able to:

- Describe the global distillation phenomenon in terms of the long-range transport of persistent contaminants to colder regions.
- Describe Phase I, Phase II and Phase III metabolic pathways and their role in the detoxification and elimination of chemicals.
- Identify the various classes of legacy contaminants and describe why they are still present in the environment.
- Compare and contrast the chemical characteristics that affect the environmental distribution and toxic effects of major classes of polycyclic aromatic compounds, current use pesticides, flame retardants, perfluorinated alkyl compounds and plasticizers.
- Describe how regulations to control the production of hazardous chemicals resulted in the marketing of other compounds that may be equally hazardous.
- Identify the ways that regulatory agencies screen new chemicals to determine whether they may be a potential hazard to the environment before they enter the marketplace.

In this chapter, we consider the ecological threat posed by carbon-containing chemicals, known as organic compounds. The ability of carbon to form stable bonds with hydrogen, oxygen and nitrogen results in the formation of a vast number of chemicals that are released into the environment from both natural sources and an array of human activities. This chapter is primarily concerned with synthetic organic compounds or anthropogenic 'xenobiotics', although some naturally occurring organics are processed and refined for human use. In Chapter 9, several naturally occurring organic compounds that are toxic are discussed. Here we describe how the physicochemical properties of organic contaminants influence their distribution and fate in the environment, their bioavailability and their toxicity. Part of the chapter is devoted to describing how different compounds interact with biological receptors at the molecular level, which affects their **mode of action (MoA)** and how they are metabolized. We discuss the residual impact of 'legacy contaminants' produced several decades ago but still persisting in the environment as well as several classes of more recently developed chemicals, including pesticides and industrial chemicals.

7.1 Classes of Organic Compounds

Natural organic compounds are formed through the photosynthetic activity of plants, protists and some bacteria whereby light energy from the Sun is converted into chemical energy through the reaction of carbon dioxide and water, with the release of oxygen. Synthetic chemicals are produced by industry for various purposes that are meant to benefit society. However, as history has shown, some of these synthetic chemicals can be harmful to humans or to the environment. Throughout evolutionary history, prokaryotes and eukaryotes have been exposed to a variety of natural organic chemicals from the external environment (**exogenous**) or produced by the organism itself (**endogenous**). However, exogenous compounds that are synthesized for medical, industrial or domestic use and to which organisms have not been exposed over evolutionary history are termed xenobiotics (i.e., foreign to biological life). It can be considered a tribute to the versatility and efficacy of the enzymatic and physiological systems of organisms that the great majority of these new compounds can be metabolized, detoxified and excreted.

The simplest organic compounds contain only carbon and hydrogen and are called hydrocarbons. Aliphatic compounds are hydrocarbons in which the carbons are joined by single, double or triple covalent bonds. The alkane class of aliphatics includes hydrocarbons in which carbon and hydrogen are linked by single covalent bonds consisting of two shared electrons. They may exist as straight chains, as illustrated by pentane in Figure 7.1, or branched chains, as illustrated by 2-methylbutane in Figure 7.1. They may also form ring-like structures called cycloalkanes. Alkenes, also called olefins, are aliphatics with double bonds consisting of four shared electrons (e.g., ethylene, $H_2C=CH_2$). Alkynes have triple carbon bonds consisting of six shared electrons (e.g., acetylene, $HC\equiv CH$). Substitutions for hydrogen in aliphatic compounds include single oxygen (e.g., aldehydes, ketones, ethers), hydroxyl groups (i.e., alcohols), nitroso groups (N=O), nitro groups (NO_2), azo groups (N=N) and amine groups (NH_2). Some of these substitution patterns are illustrated in Figure 7.1.

Aliphatic compounds with oxygen-containing functional groups, including alcohols, aldehydes, ketones, carboxylic acids, epoxides and ethers, have a wide range of properties. Many are readily metabolized by biota, although, in some cases, the transformation products formed by metabolism may be more toxic than the parent compound. Exposure to these compounds at high concentrations may affect cellular functions through the general mechanism of disruption to the integrity of membranes and proteins, called narcosis (Section 2.3.4.3).

Aromatic compounds are a class of hydrocarbons based on a carbon ring structure, which is usually the six-carbon benzene ring. Aromatics are characterized by strong bonds between adjacent carbon atoms and a low (1:1) hydrogen-to-carbon ratio relative to aliphatics. The bonding between carbon and adjacent hydrogen atoms may be conceptualized as an equilibrium state

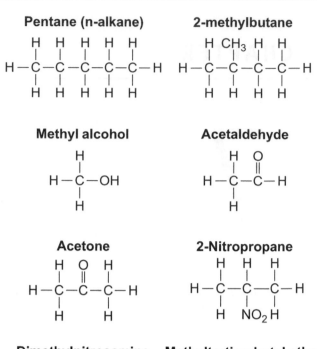

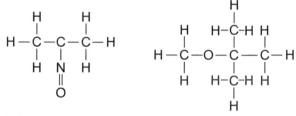

Figure 7.1 Structures of alkane hydrocarbons and derivatives showing important functional groups: methyl -CH_3; hydroxyl -OH; carbonyl -C=O; nitro -NO_2; Oxo -O-. Reproduced from Wright & Welbourn (2002), with permission from Cambridge University Press.

between two resonant forms and is conventionally portrayed as a circle within a hexagon (Figure 7.2).

Single-ring aromatic compounds such as benzene, toluene (i.e., methylbenzene), ethylbenzene and xylene, or the so-called **BTEX** compounds, are constituents of crude oil and are widely used as industrial solvents. Substitution of a hydrogen atom in benzene by a hydroxyl (OH) group forms an aromatic alcohol, or phenol (Figure 7.3). Phenolic compounds are used in a variety of chemical synthesis processes and in the manufacture of polymers. Substitution with a carboxylate group yields benzoic acid, which is also widely used for chemical syntheses (Figure 7.3). Polycyclic aromatic compounds (PACs) comprise a diverse range of natural compounds with two or more fused aromatic rings. They include compounds containing only C and H (i.e., polycyclic aromatic hydrocarbons; PAH) and heterocyclic compounds with

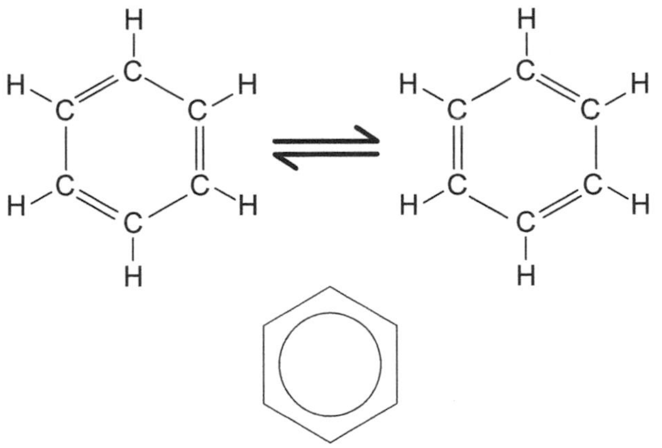

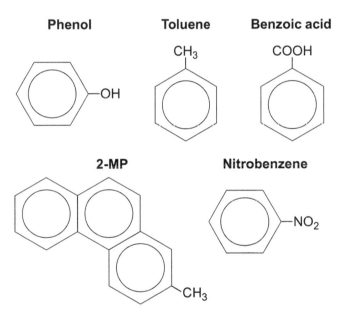

Figure 7.2 Structure of the benzene ring. Reproduced from Wright & Welbourn (2002), with permission from Cambridge University Press.

Phenol   **Toluene**   **Benzoic acid**

2-MP   **Nitrobenzene**

Figure 7.3 Structures of different classes of aromatic compounds and their side groups. Shown here are single-ring compounds phenol (OH), benzoic acid (COOH), nitrobenzene (NO_2) and toluene (CH_3), and the three-ringed derivative 2-methylphenanthrene (CH_3). Classes may contain hundreds of different compounds depending on the number, size, composition and location of side groups. Adapted from Wright & Welbourn (2002), with permission from Cambridge University Press.

additional elements (i.e., C and H plus O, N, or S) incorporated in the ring structures. Aromatic compounds may also include substituents of aryl, alkyl, nitro (NO_2), amino (NH_2) or thiol (S) groups, oxygen (hydroxyl, carboxylic acid) groups, or halogens (Br, Cl, F). An aryl group is any functional group or substituent that includes an aromatic ring (e.g., phenyl group). The **halogenated aromatic hydrocarbons** (HAHs) include many legacy contaminants (e.g., PCBs, chlorinated dioxins) that are discussed in

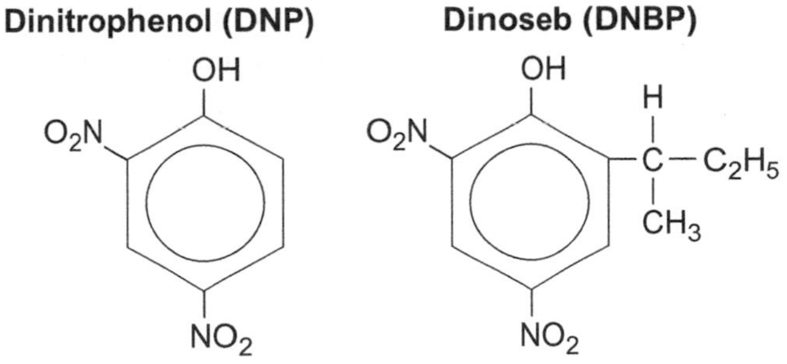

Dinitrophenol (DNP)   **Dinoseb (DNBP)**

Figure 7.4 Structure of two nitro-substituted phenolic compounds developed as herbicides.

Section 7.7 and some of the brominated flame retardants that are discussed in Section 7.9. The PAHs range in size from two-ringed compounds (e.g., naphthalene) to multi-ringed compounds, such as benzo(a)pyrene or BaP (see also Section 7.6.1.1). Three-ringed compounds such as phenanthrenes are used in the manufacture of a wide range of synthetic products.

Carboxylate substitution of benzene at two adjacent carbons yields the aromatic compound phthalic acid. As discussed later in this chapter (see Section 7.11), phthalates are esters of phthalic acid that are widely used as plasticizers. Phthalates are used to increase the flexibility of polyvinyl chloride (PVC) plastics, contributing to a global annual market of 8.5 million tonnes for this class of compounds. They are also used as additives in a variety of other products, such as cosmetics. There is regulatory interest in this common class of compounds because of the ability of phthalates to disrupt sexual development (Chapter 8).

There are several aromatic compounds with nitro (NO_2) substitution that are of environmental concern. These include the herbicides dinitrophenol (2,4-DNP) and dinoseb (4,6-dinitro-2-sec-butylphenol) illustrated in Figure 7.4. These nitro-substituted phenolic compounds are now banned in the United States and the European Union because of evidence that they induce developmental abnormalities. Anilines are also called phenylamines because their basic structure consists of a phenyl group attached to an amino (NH_2) group. These aromatic amine compounds are used in a wide variety of industrial products, including as dyes for colouring clothing and other textiles.

7.2 Fate in the Environment

An understanding of how organic compounds are distributed among environmental media is critically important to our understanding of the risk of toxicity to ecosystems posed by these chemicals (Section 4.5). Figure 7.5 illustrates a simplified concept of the compartments available for partitioning of organic contaminants in the

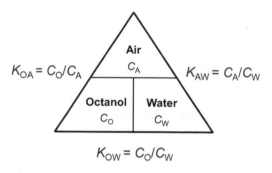

Figure 7.5 Simplified conceptual framework for the distribution of organic chemicals among the compartments of air, water and octanol in the environment and the partition coefficients for air/water (K_{OA}), octanol/air (K_{OA}) and octanol/water (K_{OW}).

environment: that is, air, water and octanol. Octanol is a synthetic surrogate for any medium that contains organic carbon, including the lipids in the tissues of organisms and the organic matter associated with soils and sediments. The ratios of the concentrations of organic compounds observed at equilibrium in these environmental compartments, referred to as **partition coefficients** (Section 2.3.3), are also shown in Figure 7.5.

Relatively small organic molecules with high vapour pressure may be transported long distances in the atmosphere if they are persistent (Hermanson *et al.*, 2020). **Global distillation** is a transport phenomenon whereby there is volatilization of **persistent organic pollutants** (POPs) in warmer latitudes, followed by their transport in the atmosphere and deposition in colder regions, which favours the partitioning of these compounds into rain or snow (Figure 7.6). This sequence of events often

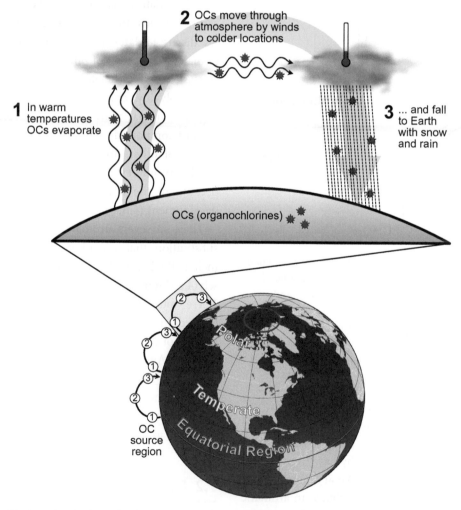

Figure 7.6 Global distillation of volatile and persistent chemicals caused by volatilization in warm latitudes (i.e., equatorial regions) and deposition in cold latitudes (i.e., polar regions).

takes place in multiple cycles of volatilization and deposition as POPs move from temperate mid-latitudes to the poles, so it is sometimes called the 'grasshopper effect'. The net result is the transport of several classes of POPs from warmer to colder regions, a process that may be exacerbated by climate change whereby increased volatilization leads to more widespread dispersal of POPs (Ma, 2010).

Similar profiles of volatilization and deposition have been observed in alpine regions, where there is deposition of POPs at higher elevations. Profiles of POPs in ice cores in alpine or polar regions have provided a historical record of deposition in the post-industrial era. Many of these POPs have been subject to partial or complete bans, yet a history of their use and application can be tracked through ice core analysis. In general, these profiles indicate declines in usage, although continuing use of DDT (1,1'-(2,2,2-trichloroethane-1,1-diyl)bis(4-chlorobenzene); see **dichlorodiphenyltrichloroethane**; Section 7.7.1) for malaria control in some countries can be detected through traces in some shallower (i.e., more recent) samples. These profiles may also alert us to re-mobilization of POPs resulting from climate change (Miner *et al.*, 2018).

The mobility of organic compounds from water into air is affected by molecular size, volatility and water solubility. Even relatively volatile compounds will not partition into air if they are highly water soluble and therefore have a low K_{AW}. The capacity of a compound to partition from water to air is defined by a physical parameter called a **Henry's Law** constant (H), which is calculated as the ratio of the vapour pressure to the solubility of the compound in water. Solubility in water is dependent on the polarity of the compound. Water is a polar compound, which has a net **dipole moment** because of the presence of unpaired electrons in s and p orbitals. Therefore, the water matrix is stabilized by dipole–dipole interactions between adjacent water molecules (Figure 7.7). For a compound to be dissolved in the water matrix, it must be sufficiently polar to overcome these interactions. Water-soluble polar compounds are said to be **hydrophilic**. Of course, organic chemicals that ionize are more highly water soluble as they form aqua ion complexes with several water molecules.

Non-polar organic compounds have an affinity for non-aqueous phases and are said to be **hydrophobic**. These compounds tend to partition from water into the organic fraction of soils or sediments, or partition into the lipids in biological tissues. Hydrophobic chemicals are

Water

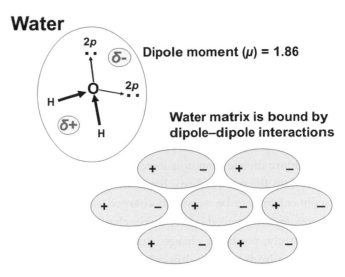

Dipole moment (μ) = 1.86

Water matrix is bound by dipole–dipole interactions

Figure 7.7 The polarity of the water molecule and an illustration of dipole–dipole interactions in a liquid water matrix. The δ defines a partial negative or positive charge within the water molecule.

sometimes referred to as being **lipophilic** because of their affinity for lipids, but the reverse is not always true (Box 2.9). A commonly used measure of the hydrophobicity of organic compounds is the **octanol–water partition coefficient**, which is usually presented in log units (i.e., log K_{OW}). This is an empirical measure of the ratio of the concentrations of a chemical between water and n-octanol at equilibrium. Compounds with log K_{OW} values above 3 have the potential to **bioaccumulate** but this is dependent upon whether these compounds are readily metabolized and eliminated by organisms, as discussed in Section 3.6.2.

Particulate matter in the aquatic or terrestrial environments comprises a wide range of both organic and inorganic materials, which have a variety of sorptive sites. Hydrophobic compounds preferentially adsorb to the organic fraction of sediments or soils. Monitoring data are often normalized to sediment or soil organic carbon concentrations to avoid bias created by different proportions of organic material in the substrates, such as the low organic content of sandy sediments. In the aquatic environment, particulates may take the form of suspended solids or bulk sediment. The sorption capacity of a sediment or soil for hydrophobic organic chemicals is determined by its organic carbon content. For sediments or soils with an organic carbon content greater than 0.2% by weight, the organic carbon fraction is the predominant phase for chemical sorption. The K_{OC} value is the **sediment (soil)–water partition coefficient** that is normalized to the content of organic carbon. If K_{OC} data are not available, this property can often be estimated from log K_{OW} values.

Most moderately buffered surface waters have a pH range between 7.5 and 8.5, but the pH of less well-buffered surface waters can be as low as 5, especially in areas affected by acid precipitation (Section 11.1). Organic chemicals present in ionic form over this pH range tend to be more mobile than non-ionic forms but are often less toxic than the neutral form. Ionic compounds in soil and sediment tend to be negatively charged (i.e., anions) and, therefore, are repelled by negatively charged clay particles. Conversely, positively charged compounds (i.e., cations) tend to adsorb to clay, although this may be blocked by protons (H^+) under acidic conditions. Therefore, low pH (that is, high concentrations of H^+) may increase the mobility of organic cations in soils and sediments. As an example of how ionization can affect mobility, the antibacterial compound triclosan, which is added to many soaps and surface cleaners, has a pK_a of 7.8. It is present primarily in ionized form at a pH greater than 7.8 (Figure 7.8). Therefore, in more basic surface waters and in marine environments, it is highly soluble in its ionic form and subject to transport. In more acidic surface waters, triclosan is present in neutral form and tends to partition into sediments.

The degradation of a compound is dependent not only on the properties of the compound itself, but also on the ambient conditions, such as the organic content of sediment or soil, and the redox potential, temperature and pH of environmental media. These conditions will determine whether degradation can take place through physicochemical processes (e.g., hydrolysis, photolysis) or through microbial activity. In the case of microbial degradation, the density of microorganisms and the presence of nutrients and other metabolic substrates will influence the rates of degradation. Chemical persistence in environmental media is often recorded in terms of chemical half-lives ($t_{1/2}$). For instance, compounds with a half-life of <5 d in sediment or soil are regarded as non-persistent, whereas those with a $t_{1/2}$ > 60 d are regarded as persistent. In the United Kingdom, environmental regulators use the **groundwater ubiquity score** (GUS) to assess the potential for a compound to move from soil into groundwater. This index predicts the capacity of a chemical to be transported into groundwater from its half-life in soil ($t_{1/2}$) and its soil/water partition coefficient on an organic carbon basis (K_{OC}):

$$GUS = \log_{10}\left(\text{soil } t_{1/2}\right) \times \left(4 - \log_{10}K_{OC}\right) \quad (7.1)$$

7.3 Uptake into Organisms and Bioaccumulation

The outer epithelium of most plants and animals is a relatively impermeable barrier because of the need to regulate the balance of water, ions and nutrients between the outside environment and the organism. Therefore, in plants the primary routes for uptake of chemicals are through the roots and respiratory stomata in the leaves (Section 3.2.6). In animals, uptake is primarily through the respiratory epithelium (e.g., lungs, gills) or through the gut endothelium. **Bioaccumulation** of organic chemicals is strongly influenced by their relative distribution among atmospheric, terrestrial and aquatic media, as well as their affinity for suspended particulates, soils and sediments. Compounds present in environmental media in forms that are readily taken up by biota are said to be **bioavailable**. Low-molecular-weight organic compounds are generally more volatile than larger molecules and have a greater capacity for uptake via the lungs in terrestrial organisms. Water-soluble molecules will be transported from the terrestrial into the aquatic environment, exposing aquatic organisms to these chemicals. In the aquatic environment, including in the pore water within sediments, larger and more hydrophobic molecules tend

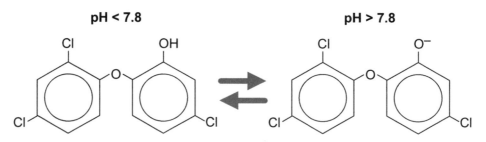

Figure 7.8 Triclosan is largely in neutral form at a pH below 7.8 and in ionized form at a pH above 7.8.

to adsorb to particles. As discussed previously, these molecules will also bioaccumulate by partitioning into lipids. For these lipophilic compounds, the diet is the primary route of uptake, which is an important consideration for accumulation through food chains, as illustrated in Case Study 4.6 on PCB accumulation in killer whales (*Orcinus orca*).

For very hydrophobic chemicals, dietary uptake plays a major role in their accumulation in organisms. As mentioned in Section 3.6.2, **bioaccumulation factors** (BAFs) refer to the ratio between the concentration of a chemical in an organism and the concentration of that chemical in the medium from which the chemical is transferred, which could be water, air or the diet. **Bioconcentration factors** (BCFs) specifically refer to the ratio of the concentrations of a compound in an aquatic organism and in water. Some very persistent and hydrophobic compounds with log K_{OW} values above 5 have a tendency to increase in concentration with each trophic level through a process called **biomagnification**. Biomagnification is demonstrated when lipid-normalised concentrations increase with each step in the food chain (Walters *et al.* (2016) as discussed also in Section 3.6.2. The physiological basis for biomagnification is still a matter of scientific inquiry. Bioaccumulation of many hydrophobic chemicals is related to the lipid content of the organism. For example, a fatty fish like a mackerel will have higher concentrations of a hydrophobic chemical on a wet weight basis than a leaner fish on which it feeds, such as a herring. Therefore, to evaluate whether biomagnification is truly occurring through the food web, the data on the biotic concentrations of these compounds are often 'lipid normalized'; that is, the concentrations in tissues are calculated on a lipid weight basis (e.g., microgram of the contaminant per gram of lipid) rather than a wet weight basis. Furthermore, it is now clear that some organic compounds show potential for bioaccumulation and biomagnification because of their affinity for proteins rather than lipids. These **proteinophilic** compounds include some **perfluorinated alkyl compounds** (Kelly *et al.*, 2009), as discussed later in this chapter and in Section 3.6.2.3. The role of dynamic mass balance models in predicting chemical uptake and transfer among biota is discussed further in Section 4.5.2.

7.4 Cellular Receptors

Cellular receptors are molecular constituents of a cell that recognize and bind to specific types of exogenous or endogenous compounds. Usually these receptors are proteins, although they can also be nucleic acids or lipids. The outcome of receptor binding may be toxicity, but binding can also initiate a signal that is important for physiological functions. Intracellular receptors that regulate gene expression belong to a family of receptor proteins that follow a similar evolutionary lineage. These receptors include those that bind with hormones that regulate endocrine function, such as thyroid hormones, sex steroids (e.g., oestrogens, androgens) and corticosteroids. These receptors are discussed in detail in Chapter 8. Receptors may be associated with the cell membrane, but many are found intracellularly because their substrates (i.e., 'ligands') are sufficiently lipid-soluble to cross the plasma membrane. Ligands can take a variety of forms, including endogenous chemicals (e.g., hormones), natural exogenous compounds (e.g., phyto-oestrogens) and xenobiotics. Following complexation with a ligand, the receptor goes through a change in its molecular configuration. After binding intracellularly with a ligand, some receptors can translocate into the nucleus itself. As such, they are commonly referred to as nuclear receptors. They consist of a carboxyl-terminal ligand-binding domain (LBD), an amino-terminal transactivation domain (NTD) and a central DNA binding domain (DBD). The NTD provides the transcriptional activation function. Examples of receptors that control a wide variety of cellular functions are summarized in Table 7.1.

One of the earliest and most intensively studied examples of receptor-based toxicity is the binding of **2,3,7,8-tetrachlorodibenzo-p-dioxin** (TCDD) or structurally related ligands (i.e., 'dioxin-like' chemicals) to the **aryl hydrocarbon receptor** (AhR), as illustrated in Figure 7.9. After TCDD diffuses through the cell membrane, this xenobiotic ligand binds to the AhR. Following the release of heat-shock proteins, the ligand–AhR complex binds with an **aryl hydrocarbon receptor nuclear translocating** protein (ARNT protein) that promotes transport through the nuclear membrane. In the nucleus, this complex binds with a specific DNA sequence within the genome, the **xenobiotic regulatory element**, which triggers transcription of a variety of genes, including the gene expressing the mRNA for **cytochrome P450** from the 1A class (i.e., CYP4501A), followed by translation to the CYP1A protein (Figure 7.9). As discussed below in the section on metabolism (Section 7.5), this is an important metabolic enzyme that is upregulated in organisms exposed to PAHs, TCDD or other compounds

Table 7.1 Cellular receptors mediating responses to endogenous biological molecules and xenobiotics.

Homologues exist in vertebrate and invertebrate animals unless otherwise stated.

Receptor	Characteristics
Aryl hydrocarbon receptor (AhR)	The ligand bound to the AhR forms a complex with a nuclear translator protein, prior to DNA binding. The complex activates the transcription of genes responsible for production of various proteins, including metabolic enzymes. Xenobiotic ligands include the chlorinated dioxins and related compounds.
Oestrogen receptor (ER)	There are at least two major receptor classes in humans (ERα and ERβ). Although their primary activator is the female sex hormone 17β-oestradiol, oestrogen receptors can interact with a broad range of chemicals (i.e., **xeno-oestrogens**), as discussed in Chapter 8.
Androgen receptor (AR)	This is a cytoplasm-based receptor responsible for mediating responses to androgens, including testosterone, dihydrotestosterone and 11-ketotestosterone, depending on the species. As discussed in Chapter 8, most xenobiotics are not **agonists** but can be **antagonists**.
Thyroid receptor (TR)	Thyroid hormone receptors are associated with the plasma membranes, mitochondrial membranes and nuclear membranes in cells of hormone-responsive tissues. The principal endogenous activators are **thyroxine (T4)** and **triiodothyronine (T3)**. Xenobiotic ligands may include some brominated flame retardants, as discussed in Chapter 8.
Retinoic acid receptor (RAR); retinoid X receptor (RXR)	Retinoids are important for reproduction and development. They inhibit β-hydroxysteroid dehydrogenase, which converts oestradiol to oestrone. The peroxisome proliferator receptor, the RAR and the thyroid receptor are capable of association with the RXR, resulting in the formation of a heterodimer.
Peroxisome proliferator-activated receptor (PPAR)	**Peroxisomes** are subcellular organelles in most plant and animal cells, performing various metabolic functions, including cholesterol and steroid metabolism and oxidation of fatty acids. Endogenous activators of the **peroxisome proliferator-activated receptors** include fatty acids and prostaglandins. Some pharmaceuticals that are prescribed to reduce cholesterol and xenobiotics such as phthalates have an affinity for isoforms of the receptor.
Constitutive androstane receptor (CAR); pregnane X receptor (PXR)	Primarily expressed in the liver and intestinal tract, these receptors act to eliminate cholesterol, lipids, sugars, steroid hormones and bile acids, but CAR and PXR also are involved in regulating the absorption, metabolism and excretion of xenobiotics.
Invertebrate ecdysteroid receptor, farnesoid receptor	Invertebrates possess unique hormones, such as the moulting hormone (ecdysone) and the juvenile hormone (methyl farnesoate). These hormones show a significant degree of homology with the RAR and can form functional heterodimers with RXR.

of similar size and shape. Hundreds of closely related CYP proteins with a wide array of functions are involved in the normal physiology of different species and, like CYP1A, their synthesis is induced by exposure to specific xenobiotics.

7.5 Metabolism

The metabolism of xenobiotics is defined as the process of chemical transformation, detoxification and excretion of these compounds (see also Section 3.5.2). The first step in metabolism (i.e., Phase I) is carried out by a diverse family

of enzymes collectively known as monooxygenases, of which the cytochrome P450 (CYP450) system is the most important. This system is found in all phylogenetic groups of organisms, including plants, and is probably at least 450 million years old in evolutionary history. The system has evolved to deal with the biotransformation and biosynthesis of endogenous compounds produced by the organisms themselves (e.g., fatty acid metabolism, steroid biosynthesis). CYP450 enzymes also transform and detoxify natural exogenous compounds to which organisms are exposed (e.g., plant alkaloids, mycotoxins). As such, this enzyme system long pre-dated the advent of man-made

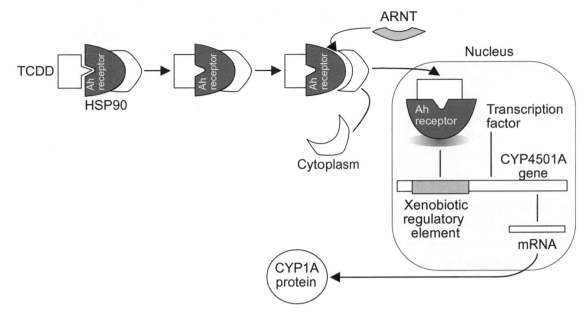

Figure 7.9 Binding of TCDD with AhR and the process for upregulation of the expression of the gene for CYP1A1. ARNT = aryl hydrocarbon (Ah) receptor nuclear translator; HSP90 = heat shock protein 90 released during complexation which stabilizes the unbound AhR. Adapted from Wright & Welbourn (2002), with permission from Cambridge University Press.

organic compounds. In vertebrates, the CYP450 enzymes are found in many different tissues, including the lungs or gills, kidneys and skin, but are primarily located in cells of the liver and adrenal glands where they are associated with the **endoplasmic reticulum** and mitochondria. In mammals, about 70% of CYP450 activity is in the liver, although significant levels are found in many other tissues. Most compounds are metabolized, detoxified and excreted through an enzyme-mediated process. The presence of xenobiotic compounds at a variety of receptor sites can result in increased synthesis and activity of these metabolizing enzymes, which is termed enzyme induction. In some cases, this metabolic process may result in the formation of intermediary compounds that are more toxic than the original compounds, some of which may cause **oxidative stress** or are **carcinogenic**.

Three phases of energy-requiring reactions are recognized in the metabolism of xenobiotics by living organisms:

(1) **Phase I** reactions that are catabolic in nature and involve the addition of a functional group to the parent compound, thereby creating a more water-soluble intermediate that is sometimes also more reactive with biological molecules;

(2) **Phase II** reactions that involve the conjugation of the intermediate with molecules that further increase the water solubility of the xenobiotic to facilitate excretion in bile or urine;

(3) **Phase III** reactions that are mediated by transporter molecules that facilitate the transport of the hydrophilic products of Phase I and Phase II reactions across lipid cell membranes. In mammals, these transporters are found in liver, kidney, intestine and neural cells and play critical roles in the excretion of xenobiotics by transferring the water-soluble conjugates to the bile or to the bloodstream.

Homologous metabolic pathways are responsible for xenobiotic metabolism in plants, culminating in the sequestration of Phase II metabolites in cellular vacuoles, which may then be transferred to the intercellular space via **exocytosis**.

As described in Section 7.4, the interactions of xenobiotics with AhR receptors within cells can enhance or suppress expression of CYP450 enzymes. About 50 membrane-bound and cellular receptors from this class have been recognized in vertebrates. Similar receptors have been detected in insects, cnidarians, molluscs, nematodes and annelids.

7.5.1 Phase I Reactions

In Phase I reactions, reactive functional groups are introduced into organic contaminants, making them more hydrophilic and favouring their further transformation by Phase II enzymes. Examples of such reactive functional groups include the following:

Monoatomic oxygen	–O
Superoxide	$-O_2^-$
Hydroxyl	–OH
Carboxylic acid	–C(O)OH
Amine	$-NH_2$
Sulphydryl or thiol	–SH

Some products of Phase I metabolism may be excreted directly, but in many cases, they must undergo further transformation before being eliminated. A typical substrate for a Phase I reaction would be a lipophilic xenobiotic. The introduction of a functional group into such a molecule is often achieved by enzymatic oxygenation following the generalized reaction:

$$RH + NADPH + O_2 + H^+ \rightarrow ROH + NADP + H_2O$$

where RH is the substrate, ROH is its hydroxylated form and NADP is **nicotinamide adenine dinucleotide phosphate** (i.e., NADPH is the reduced form).

The Phase I enzymes responsible for this reaction are called **monooxygenases**, which catalyse oxidative reactions involving the insertion of monoatomic oxygen into substrates. Two major monooxygenase groups have been identified: the flavoprotein monooxygenases (FMOs), also known as the amine oxygenase system, and the CYP450 family comprising approximately 400 proteins. Both groups are found intracellularly in prokaryotes and are membrane-bound in eukaryotes. The FMOs, which have flavin adenine dinucleotide as a prosthetic group, have been identified from mammals, fish and molluscs. Substrates include thiols and a wide range of aliphatic amines and secondary and tertiary aromatic amines. Other Phase I enzymes include esterases, epoxide hydrolases, amidases, and aldehyde and alcohol dehydrogenases. However, the most important and phylogenetically diverse group of enzymes involved in Phase I metabolism is the CYP450 family. Preparations from homogenized cells that have high activities of these enzymes strongly adsorb light at a wavelength of 450 nm; hence the name CYP450. These enzymes are ubiquitous in many invertebrate phyla, including Arthropoda, Cnidaria, Mollusca, Annelida, Insecta and Echinodermata, although they display varying degrees of evolutionary divergence, even within the same phylum. For example, CYP450 enzymes in many marine invertebrates appear to share little homology with respective insect enzymes, although some homology can be demonstrated among some insects and crustaceans (Rewitz *et al.*, 2006). Groups of CYP450 enzyme groups have also been described in plants (Morant *et al.*, 2003), with the capacity for metabolizing endogenous and exogenous compounds, although their homology with animal CYP450 enzymes remains uncertain.

In eukaryotes, the CYP450 enzyme system consists of a complex of molecules embedded in the lipid membrane of the endoplasmic reticulum. Because these enzymes are associated with the fraction of cell homogenates that is derived from preparations in the laboratory that contains ruptured endoplasmic reticulum membranes (i.e., 'microsomes'), they are sometimes referred to as 'microsomal' enzymes. The CYP450 enzymes can accept electrons that originate from NADPH or NADH as a result of the activity of either one of two proteins. One is cytochrome P450 NADPH reductase, and the other is cytochrome b5, which, in turn, receives electrons from cytochrome b_5NADH reductase (Figure 7.10). The figure also shows the relative position of the Phase II enzyme UGT, which utilizes products from the CYP450 system as substrates. Although each component is shown singly, in fact the CYP450 NADPH reductase molecule is associated with 10 to 20 CYP450 enzyme molecules and 5 to 10 molecules of cytochrome b5. Non-polar substrates are associated with the lipid bilayer of the membrane and consequently have access to the active site of the CYP450 enzyme. This interaction causes a change in the absorption spectrum of the haem moiety of the enzyme to a maximum at 450 nm and triggers a flow of electrons from NADPH via NADPH cytochrome-P450 reductase to the iron core of the haem molecule, reducing it from Fe(III) to Fe(II). The strong affinity of oxygen for Fe(II) results in the binding of molecular oxygen to CYP450. A second electron is then transferred from NADH via NADH reductase.

The reduced electrons are added to molecular oxygen to produce the superoxide anion (O_2^-), which is an extremely powerful oxidizing agent. The substrate is then spontaneously oxidized. One oxygen atom (i.e., monoatomic oxygen) is inserted into the substrate, while the other oxygen combines with hydrogen ion to produce water. Additions of monoatomic oxygen to carbon atoms substituted with halogens are difficult, with the strength of these bonds increasing in the order Cl < Br < F. Resistance to the addition of an oxygen atom is due to the electronegative repulsion between the halogen and the entering oxygen atom. Halogens also withdraw electrons from carbon atoms in the same ring, reducing the ease with which carbon atoms

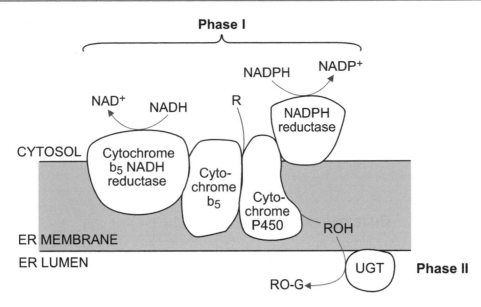

Figure 7.10 The components of the CYP450 system. ER = endoplasmic reticulum; NAD = nicotinamide adenine dinucleotide; NADH = nicotinamide adenine dinucleotide hydride ('reduced NAD'); NADP = nicotinamide adenine dinucleotide phosphate; NADPH = reduced NADP; UGT = UDP glucuronosyltransferase; R = substrate; ROH = hydroxylated substrate; RO-G = oxygenated substrate conjugated to glucuronide. Adapted from Wright & Welbourn (2002), with permission from Cambridge University Press.

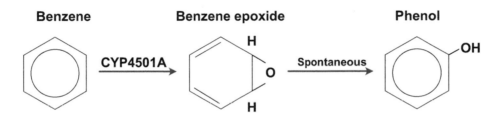

Figure 7.11 Hydroxylation of benzene by CYP450 monooxygenase system, followed by spontaneous hydrolysis to phenol. Adapted from Wright & Welbourn (2002), with permission from Cambridge University Press.

can be oxidized. This explains the refractory nature of halogenated aromatic hydrocarbons (HAHs) such as PCBs. Dechlorination reactions are involved in the metabolism of many chlorinated compounds. However, dechlorination is a very slow process, measured in years, which accounts for the environmental persistence of these compounds. An important consequence of the induction of high activities of CYP450 enzymes by HAHs is the presence of increased (but transient) concentrations of reactive oxygen, which initiates oxidative stress in cells.

Many important metabolic transformations are carried out by CYP450-mediated monooxygenase reactions, including dehalogenations, dealkylations, deaminations and oxidations. There are several variants of these reactions. In hydrolytic dehalogenation, for example, the halogen atom is replaced by –OH, not H. In alkyl halides, this involves the addition of water, but in aryl halides,

O_2 is added. Under low-oxygen conditions, electrons produced by the CYP450 system are also capable of carrying out reductions. These include azo- and nitro-reductions. Nitro-reductases are present in mammalian gut microflora and are active under the reducing conditions in the gastrointestinal system.

CYP450-mediated metabolism may result in the production of reactive intermediates that are more toxic than the parent compound. For instance, the oxidation of the thiophosphate moiety of the pesticide parathion is required for activation to its toxic analogue, paraoxon. The metabolic hydroxylation of aromatic rings requires an intermediate epoxidation step (Figure 7.11). Most **epoxides** are only transient intermediates in reactions that are followed by spontaneous hydrolysis to phenolic compounds.

However, epoxides may be more persistent in cases where the molecular configuration of the metabolite

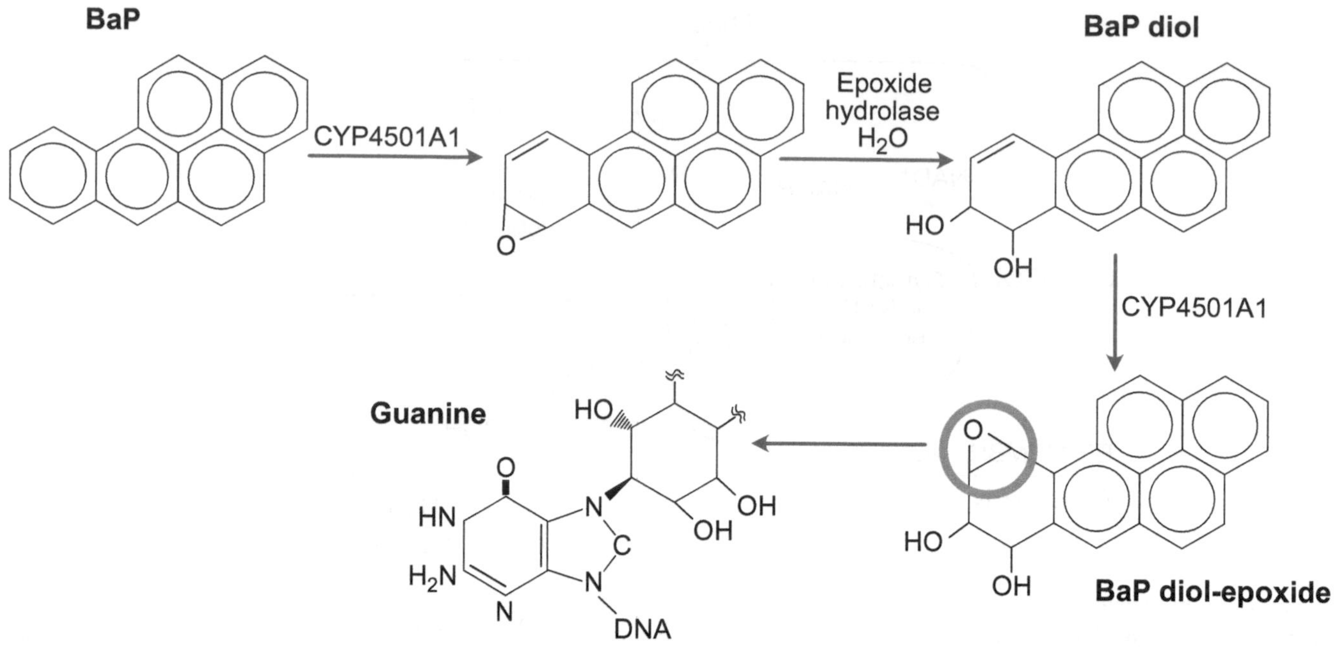

Figure 7.12 Transformation of benzo(a)pyrene (BaP) to a diol-epoxide that can bind to the guanine nitrogenous base of DNA. Adapted from Wright & Welbourn (2002), with permission from Cambridge University Press.

retards further transformation. A typical case is illustrated by the activation of benzo[a]pyrene (BaP) to a carcinogenic derivative where the formation of a diol is an intermediate step in the production of the compound 7,8-dihydrodiol-9,10-epoxide, which shows high affinity for nitrogenous bases in DNA. The 7,8-epoxide is hydrolysed to a diol through activity of an epoxide hydrolase enzyme, but the hydrolysis of the 9,10-epoxide is retarded by the proximity of the adjacent benzene ring in the 'bay area' of the molecule. The BaP-7,8-diol-9,10-epoxide has a high affinity for guanine (Figure 7.12). The formation of adducts with nitrogenous bases results in errors in the replication of DNA, causing mutations that can lead to malignancy. Other examples of Phase I epoxidations that result in transformations to carcinogenic metabolites include epoxidation of the natural mycotoxin, aflatoxin B1 (Section 9.3.3.3). The carcinogenicity of vinyl chloride, discovered because of the high cancer rates among workers in the polyvinyl chloride (PVC) plastics industry, is due to epoxidation of the monomer.

For vertebrate species that express the CYP1A gene, induction or inhibition of CYP4501A enzyme activity can increase or reduce the toxicity of some organic contaminants, depending on the compound. For example, exposure of mummichog (*Fundulus heteroclitus*) embryos to the PCB congener 126 caused embryotoxicity, but co-exposure of embryos to PCB 126 and to α-naphthoflavone,

an inhibitor of CYP4501A activity, decreased embryotoxicity relative to PCB 126 alone (Wassenberg and Di Giulio, 2004). Conversely, partial inhibition of CYP450 increased the toxicity of an alkyl phenanthrene compound (retene) to rainbow trout (*Oncorhynchus mykiss*) embryos (Hodson *et al.*, 2007). Thus, the CYP450 activity of vertebrates exposed to mixtures of organic compounds will reflect the interactions among compounds that induce CYP450 synthesis and those that inhibit this enzyme's activity.

The CYP450 enzyme system may be inhibited by carbon monoxide and cyanide, which bind to the Fe(II) of the haem moiety of CYP450, blocking access to oxygen. Other substances act as inhibitors by occupying the active site of CYP450. This includes piperonyl butoxide, which is often added to insecticide formulations to increase the potency of the product. Chlorinated alkanes (e.g., carbon tetrachloride) and alkenes (e.g., vinyl chloride) may be transformed by the CYP450 enzyme system into reactive metabolites, which in turn bind covalently to the haem moiety of cytochrome P450, thereby inactivating the enzyme.

Changes in CYP450 enzyme activity have proven to be useful biomarkers of exposure to specific classes of chemicals, and this biomarker approach for monitoring exposure is discussed in Section 4.3.3. In addition, there is a growing acceptance of an association between the activity of CYP450 enzymes and endogenous steroid synthesis, with implications for endocrine disruption (Chapter 8).

7.5.2 Phase II Reactions

Phase II reactions are biosynthetic in nature, leading to the transformation of hydrophobic organic compounds into more water-soluble conjugates that are non-toxic and readily excreted once they enter the bile or urine. In such reactions, the conjugating agent is an endogenous compound, and the substrate is a xenobiotic chemical having the requisite functional groups for metabolism by conjugating (Phase II) enzymes. The required functional groups include hydroxyl, epoxide, carboxyl and amino groups. Some xenobiotics may already possess such functional groups, while others require prior transformation by Phase I enzymes to add or unmask the reactive functional groups, thereby facilitating the elimination of the conjugate from the cell. The major conjugating agents for nucleophilic (electron-rich) compounds are **glucuronides**, whereas the principal conjugating agent for electrophilic (electron-deficient) compounds is **glutathione** (GSH). Oxygenated metabolites may also be conjugated to sulphate ions.

Glucuronides are the most common endogenous conjugating agents. The acidic sugar, glucuronic acid, is conjugated to O, N and S heteroatoms after initial activation to uridine diphosphate glucuronic acid (UDPGA), as illustrated in Figure 7.13. The conjugation of xenobiotics with UDPGA is catalysed by UDP glucuronyl transferase, which is found in the microsomal membrane.

Carboxylic acids and alcoholic oxygen may also be conjugated as glucuronides. The resulting conjugates are much more polar and hydrophilic and therefore more readily excreted than the original xenobiotic, but less likely to diffuse across cell membranes because they are bulkier and no longer lipophilic. Specific transport proteins (Phase III) facilitate the rapid excretion of glucuronides across cell membranes and into urine and bile.

Glutathione (GSH) is a sulphur-containing tripeptide of glutamate, cysteine and glycine that is a critically important conjugating agent for reactive electrophiles, which would otherwise attack vital macromolecules. It forms conjugates with a wide variety of xenobiotic compounds, including aromatic hydrocarbons, aryl halides, alkyl halides, aromatic nitro-compounds, alkenes, alkyl epoxides and aryl epoxides. Glutathione transferases are a family of enzymes that catalyse these conjugations. Following initial binding of the substrate to GSH, the glutamyl and glycinal peptides are lost (Figure 7.14). The remaining cysteine conjugate is transported to the bile by specific transport proteins. However, this conjugate is rarely excreted directly and usually undergoes further enzymatic acetylation to a mercapturic acid, which is a readily excreted polar compound. GSH-mediated conjugation of organic compounds has also been demonstrated in plants where conjugates may become isolated in cellular vacuoles or are moved systemically for exudation through the root tip (Dixon et al., 2010).

Sulphate conjugation requires substantial energy input from ATP but provides an effective means of eliminating a variety of compounds through the urine, since the sulphur

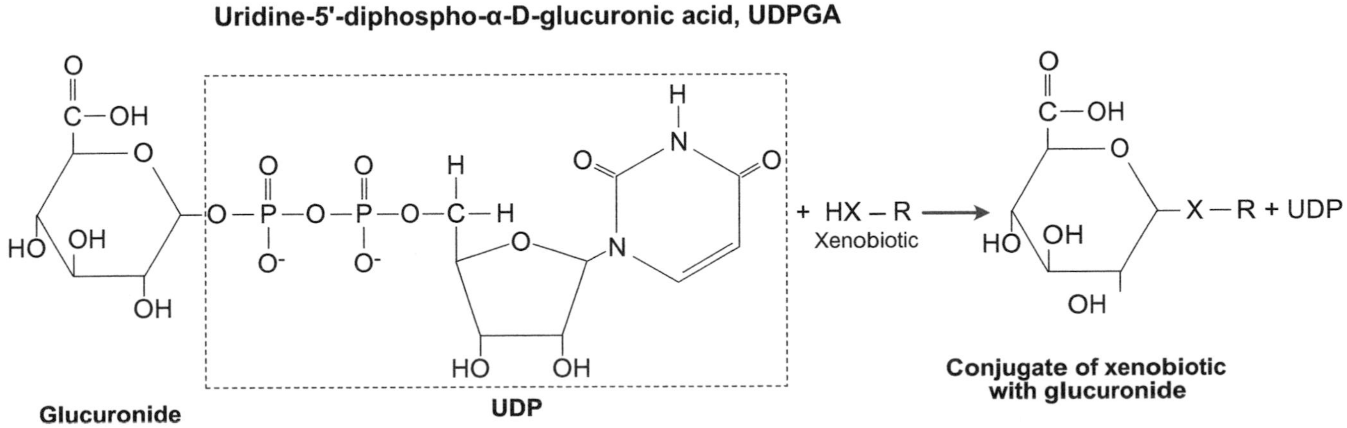

Figure 7.13 Activation of glucuronic acid by binding with UDP and subsequent conjugation of the UDPGA complex with a xenobiotic (HX-R).

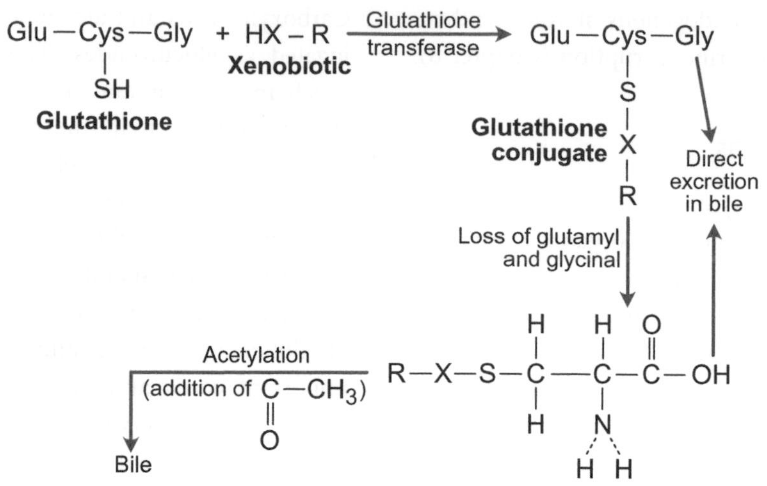

Figure 7.14 Binding of a xenobiotic substrate (HX-R) with glutathione and subsequent formation of a cysteine conjugate.

conjugates are highly water-soluble. The types of compounds conjugated by sulphur are alcohols, phenols and aromatic amines. Enzymes that catalyse these reactions are called sulphotransferases. Other pathways play minor roles in conjugation reactions and involve only a few compounds. For example, amino acid conjugation of carboxylic acids is an enzymic acetylation reaction resulting in the formation of an amide, a process known as amidation.

7.5.3 Phase III Reactions

Whereas non-polar compounds can easily diffuse into cells across lipid membranes, the polar, water-soluble nature of oxygenated products of Phase I and Phase II metabolism means that they require active transport to facilitate their passage across cell membranes. This role is performed by specific transporter proteins known as **multidrug resistance proteins** (MRPs), which facilitate the passage of polar organic compounds into cells prior to Phase I and Phase II metabolism processes and the elimination of conjugated products from Phase II reactions into the bile via the liver or into the lumen of the intestine. The excretion of Phase II conjugates by specific ATP-dependent MRPs represents a distinct third detoxification stage referred to as Phase III metabolism. MRPs have been identified in both prokaryotes and eukaryotes as a family of P-glycoproteins. They have moderate substrate specificity and are encoded by a small family of genes that differ in number from species to species. In humans, two genes have been identified, and in other mammalian species there are up to five genes. These P-glycoproteins are induced by exposure to a variety of

endogenous and xenobiotic compounds and by X-ray irradiation and heat shock. Several different types of transporter proteins have been recognized. In mammals, MRP2 is responsible for transporting glucuronide and glutathione conjugates into bile, whereas MRP3 and MRP4 control the export of these conjugates into the bloodstream for excretion via the kidneys.

7.5.4 Induction of Metabolism

Phase I, Phase II and Phase III metabolic processes are all inducible, and in some cases the induction of enzymatic and transporter functions may be traced to a common factor. The **nuclear factor erythroid 2-related factor 2** (Nrf2) is a transcription factor found in the cytoplasm as a complex with a cofactor called **Kelch-like ECH-associated protein 1** (Keap-1). The Nrf2/Keap-1 complex responds to changes in oxidative state to promote a variety of antioxidant functions. As part of this protective function, Nrf2 influences DNA transcription to regulate Phase I, Phase II and Phase III enzyme activity. Figure 7.15 summarizes the regulation by Nrf2 of Phase I and II detoxifying enzymes and Phase III MRPs. These elements of the biotransformation process are regulated through the interaction of Nrf2 with the antioxidant response elements (AREs) within the genome.

Much of the research over the past 20 years relating to the role of the Nrf2/Keap-1 complex in detoxification has focused on human health implications as a result of its antioxidant activity. These transcription factors and other cofactors found in the cells of most animals have been the subject of great scrutiny due to their involvement with

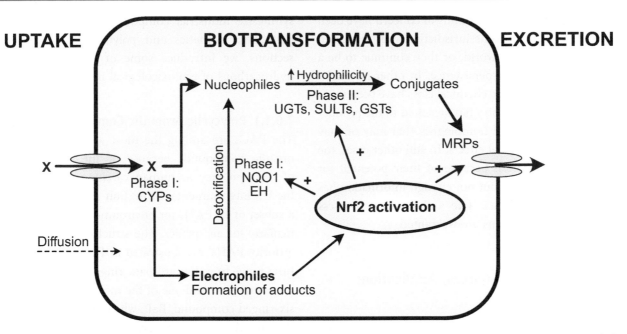

UPTAKE **BIOTRANSFORMATION** **EXCRETION**

Figure 7.15 Illustration of the central role of activation of nuclear erythroid 2-related factor 2 (Nrf2) in regulating induction of Phase I cytochrome P450 (CYP) enzymes, epoxide hydralase and NQO1 (NADPH:Quinone acceptor oxidoreductase 1), Phase II conjugation enzyme systems (i.e., GSTs – glutathione S-transferases; SULTs – sulphotransferases; UGTs – uridine 5'-diphospho-glucuronosyltransferases) and Phase III multidrug resistance-associated proteins (MRPs). Adapted from Maher *et al.* (2007) with permission from John Wiley and Sons, Publishers.

inflammation, immune response, apoptosis (i.e., programmed cell death) and cellular proliferation. Of critical importance is the induction of enzymes with antioxidant and anti-inflammatory properties, such as the Phase I enzyme, NADPH:quinone acceptor oxidoreductase 1 (NQO1), which is a ubiquitous flavoprotein that promotes a variety of reduction reactions. The DNA transcription responsible for the induction of these stress response enzymes is, in turn, dictated by the ability of the Nrf2/Keap-1 complex to initiate 'signalling' activity relating to the degree of oxidative stress. Much of this research is related to human toxicology and oncology and is beyond the scope of this book. However, it is worth noting that natural phytochemicals with antioxidant properties, including terpenoids, phenolics and flavonoids, can cause upregulation of NQO1 and other detoxifying enzymes. This serves as a reminder that toxicology remains a normal part of biological interactions within the natural world.

7.6 Compounds of Particular Concern

The CAS Registry® maintained by the American Chemical Society currently contains information on more than 161 million unique chemical substances, with many

being synthetic organic compounds. In addition, there are many thousands of organic substances in the natural environment that are not included in the registry but may cause biological effects. It would be impossible to review the ecotoxicological characteristics of all these substances. Therefore, the following review will focus on specific classes of organic compounds that cause concern because they:

- are mass-produced for commercial use, with the potential for significant contamination of the environment;
- are present in a form or in a matrix that is mobile in the environment and therefore subject to transport into air, water or soil;
- are persistent in the environment. This includes compounds that are 'pseudo-persistent', i.e., they are continuously released into the environment at a rate that exceeds the rate of degradation (examples include pharmaceuticals discharged from wastewater treatment plants);
- demonstrate a potential for bioaccumulation in organisms; and
- cause toxicity or biological effects in organisms exposed to concentrations or doses that are environmentally relevant.

Some of these classes of compounds of concern have been regulated or banned in some jurisdictions but remain in use in other parts of the world, or they continue to be a problem because of their persistence. In other instances, some compounds from the chemical class have been regulated or banned, but industry has switched to the production and use of structural homologues that may or may not cause similar biological effects. In still other cases, the chemicals are of concern because of their potential for ecotoxicological impacts, but not enough information has yet been collected on their fate and toxicity to assess whether regulations or bans are warranted.

7.6.1 Hydrocarbons: Sources, Applications and Concerns

Many aliphatic and aromatic hydrocarbons derived from petroleum are the feedstocks for production of synthetic chemical products (i.e., petrochemicals). Petroleum originated from decayed compounds of plant origin. Aromatic hydrocarbons are also produced naturally in plants, or through processes of combustion. The aliphatic and aromatic hydrocarbons are used to synthesize industrial compounds and pharmaceuticals, pesticides, cosmetics and polymers. In the following sections, we introduce some of the major classes of hydrocarbons of ecotoxicological interest.

7.6.1.1 Polycyclic Aromatic Compounds

The PACs are among the most widely studied class of organic compounds because of the early recognition of their potency as human carcinogens. In the 1970s, the US Environmental Protection Agency recommended a subset of 16 PAHs for environmental monitoring, particularly for air quality. The structures of some of these 'priority PAHs' are illustrated in Figure 7.16. The structures range from two aromatic rings (e.g., naphthalene) to more than ten, and one of the most widely studied is the six-ringed compound, BaP, which is a potent carcinogen (Figure 7.16). The PACs are classified as **pyrogenic, petrogenic** or **biogenic**, depending on their source.

Pyrogenic PACs are derived from the incomplete combustion of organic matter, which transforms linear, branched or cyclic hydrocarbons into aromatic ring structures under conditions of low oxygen. Most are unsubstituted two- to six-ringed compounds, but they

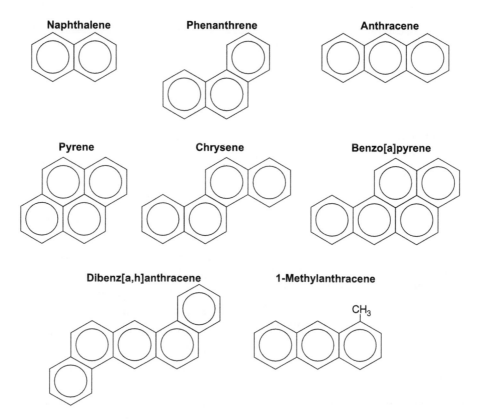

Figure 7.16 Structures of 7 of the 16 US EPA priority polycyclic aromatic hydrocarbons (PAHs) and an alkyl-PAH recommended for monitoring by the US EPA (Andersson and Achten, 2015).

include significant proportions of alkyl PACs, such as retene (7-isopropyl-1-methyl phenanthrene), a chemical marker of burning vegetation (Harner *et al.*, 2018). Natural sources of pyrogenic PACs include forest and brush fires, and anthropogenic sources include fossil fuel combustion and emissions from industries. For instance, **coke** is used in the steel industry because it burns at a higher temperature than coal and can melt iron. Coke is formed industrially by heating coal under conditions of low oxygen; the pyrolysis process generates aromatic compounds from the primarily aliphatic hydrocarbons in coal. The PACs released from coking plants are significant sources of contamination in aquatic environments. Volatile hydrocarbons emitted from coke ovens also condense to form coal tar, which is used in a variety of industrial products. All of these products are enriched with PACs, and their production and use provide a direct route for PAC emissions to air, water and soil.

Petrogenic PACs are released by the production, refining and spillage of crude, synthetic crude and refined oils, the production of creosote and asphalt from the heavy residues of oil refining, and the use of these products. The ratios of the concentrations of selected PACs can be used to identify whether PAHs in environmental matrices (e.g., sediments, soils) are of pyrogenic or petrogenic origin (Yunker *et al.*, 2002). For instance, a ratio of anthracene to anthracene-plus-phenanthrene (i.e., An/An+Ph) >0.1 indicates a combustion source, while a ratio <0.1 indicates a petroleum source. A characteristic of petrogenic PACs is the predominance of alkyl PAHs (80–95% of total PAHs), which is far greater than the proportion in pyrogenic PACs.

Biogenic PACs are those produced by organisms as a normal part of their metabolism. An interesting example is the conversion of pulp-mill by-products abietic acid (AA) and dehydroabietic acid (DHAA) to retene, an alkyl PAC that is toxic to embryonic fish in receiving waters (Section 11.4.1). Alkylated PACs are of growing concern because they are often more embryotoxic than non-alkylated PACs, and toxicity increases with the degree of alkylation (Hodson, 2017). One of the major challenges in the study of alkyl-PAHs is their structural diversity. A good example is phenanthrene, which has 10 ring carbons available for alkyl substitution. The alkyl substitutions can include methyl (C1), ethyl (C2), propyl (C3), isopropyl (branched, C3), butyl (C4) and isobutyl (branched, C4) forms. The permutations and combinations of their distribution on the 10 available carbons

of phenanthrene translate to 575 possible C1 to C4 congeners. Thus, the universe of PACs probably approaches 100,000, of which fewer than 100 have been thoroughly assessed toxicologically, and most are not considered in analyses of environmental samples.

The fate of PACs is governed by environmental conditions and the chemical properties of each congener. In air, the partitioning of PACs between particulate and gaseous phases depends on the number, type and size of particles present, the ambient air temperature and the vapour pressure of individual compounds. Most PACs are removed from the atmosphere by dry particle deposition, by vapour exchange between the atmosphere and terrestrial surfaces and water bodies, and by scavenging by rain and snowfall. PACs can be photodegraded by ultraviolet light or converted to nitro-PACs in urban air by ultraviolet photochemical reactions with oxides of nitrogen (NO_X). For example, nitropyrene, a potent carcinogen, is formed in urban air and persists longer than higher-molecular-weight PACs that bind to particulates and are removed from the atmosphere by settling or impaction. Because of their high K_{OC} values, most PACs partition rapidly from water onto surfaces, particularly on particles that settle out of the water column to aquatic sediments. Most PACs can be biodegraded by aerobic microbes, but they persist in anoxic sediments. They also persist in terrestrial ecosystems, but the extent of persistence depends on weathering by the combined effects of volatilization, photodegradation and biodegradation. Higher-molecular-weight PACs usually persist longer in the environment than their smaller homologues.

All organisms can accumulate PACs, but bioaccumulation factors are highest for species that have low activities of CYP450 enzymes, such as crustaceans and bivalve molluscs. Because most food webs include vertebrate species with active CYP450 systems, PACs do not biomagnify. Their primary routes of accumulation are via the diet, inhalation (gills or lungs), dermal absorption from contaminated surfaces, intake from grooming of contaminated fur, feathers or skin, and maternal transfer to offspring during egg formation, embryo development or lactation.

High concentrations of PACs in the environment are usually observed in areas of intense industrialization and little regulatory control of discharges into air, soil or water. For instance, in China, the summed concentrations of PACs in surface waters of seven river basins were as high as 38 μg/l, with significantly higher concentrations in water in industrialized areas (Guo *et al.*, 2012).

From these data, the highest ecological risk of PAC emissions was for aquatic species in the Liaohe and Huaihe River Basins. In North America and Europe, areas historically contaminated with PACs are often subject to remediation (e.g., Case Study 13.1). Sediments of some 'Areas of Concern' in the Laurentian Great Lakes, including the Black and the Ashtabula Rivers in Ohio, along the south shore of Lake Erie, and Hamilton Harbour, Ontario, an embayment of western Lake Ontario, are highly contaminated with PACs from historical inputs from coking plants and steel industries. Past remedial actions to reduce the concentrations of sediment PACs met with mixed success, despite heavy investments (Meier *et al.*, 2015; Milani and Grapentine, 2016).

A high prevalence of malignant cholangiomas and hepatocarcinomas in the livers of brown bullhead catfish (*Ictalurus nebulosus*) collected in the 1980s and 1990s from the Black and Ashtabula Rivers provided compelling evidence of the carcinogenicity of the sediment PACs (Rafferty *et al.* (2009) and Case Study 4.2). The extent of PAC contamination and the severity and prevalence of liver lesions decreased in both rivers following remedial dredging. However, a low prevalence of lesions seemed to persist in bullheads from the Ashtabula River (Meier *et al.*, 2015), perhaps because of co-contamination by other carcinogenic compounds. We know from research focused on cancer in humans that there are many factors involved in cancer aetiology, including hereditary and epigenetic factors, viral oncogenesis, mitogenic stimulation of cell proliferation, genotoxicity, DNA repair and immune surveillance of cancer cells. However, in the case of liver tumours in benthic fish exposed to PAC-contaminated sediments, the aetiology of these tumours is consistent with the 'classical' model of carcinogenesis:

- metabolism of PACs to reactive intermediates (e.g., diol-epoxides) by CYP450 enzymes;
- formation of DNA adducts that result in point mutations;
- deletions or substitutions within the genome (i.e., genotoxicity) during DNA replication;
- mutations that occur within multiple oncogenes and tumour suppressor genes; and
- progression of affected cells towards malignancy.

7.6.1.2 Petroleum Hydrocarbons

Crude oil is a complex mixture of thousands of hydrocarbons formed from the burial and transformation of

marine organisms over hundreds of millions of years (Lee *et al.*, 2015). The transformation involves heat and pressure, and may be catalysed by mineral constituents of the Earth's crust, such as aluminosilicates. As it forms, oil migrates to the Earth's surface where it becomes trapped in sedimentary rock formations beneath impervious layers, or it seeps through cracks to the surface. Early uses of highly weathered oil (i.e., tar and bitumen) included waterproofing ships and road surfacing. The world's first commercial oil well was completed in Petrolia, Ontario, in 1858, followed shortly thereafter in 1861 by Canada's first large oil spill, an estimated 1 million barrels, much of which drained into Lake St. Clair in the Laurentian Great Lakes basin. As methods developed for finding and refining crude oil for an ever-expanding array of fuels and petrochemicals, new oil fields and refineries were developed on every continent except Antarctica, where international treaties prohibit development. Today, petroleum accounts for about 40% of the world energy needs, half of which is consumed for transportation.

Of all crude oil constituents, 98% are organic, classified as **saturates, aromatics, resins** and **asphaltenes** (SARA). Saturates are aliphatic hydrocarbons (i.e., linear, branched and cyclic alkanes) and are the most highly valued components of oil. The aromatics group refers to cyclic compounds with double bonds, including monoaromatics such as BTEX and the polycyclic PACs (Sections 7.1, 7.6.1.1), which are dominated by two- to four-ringed PAHs. They also include smaller amounts of five- to six-ringed congeners, plus the sulphur heterocyclics abundant in high sulphur oils (e.g., dibenzothiophenes). However, the sum of all PACs comprises only 0.5–6.5% by weight of crude and refined oils. Resins and asphaltenes are very large molecules that include polymers of aliphatic, aromatic and heterocyclic compounds. They are the primary components of the black tar-like residue that is left after refining.

Crude oils vary widely in their chemical composition and properties. The chemical composition of crude oils determines their characteristics, such as density, viscosity and dispersibility in water, all of which affect the fate of spilled oil. Oil spilled on land will spread and pool and may cause soil and groundwater contamination if not recovered within a few days. In colder climates, winter oil spills are relatively easy to clean up because frozen soils prevent oil penetration and heavier oils become sufficiently viscous to scoop up. Viscosity decreases as temperatures rise, enabling a more rapid spread of oil and penetration into soil and groundwater, and a more rapid evaporation of low-molecular-weight hydrocarbons that are a hazard for fire and explosions. Less viscous oil is also more difficult to clean up, requiring absorbent materials and pumps, although viscosity increases with weathering and evaporation, allowing the residual oil to be recovered more easily.

In marine or freshwater systems, ultralight crude and fuel oils spread rapidly on the water's surface, making them difficult to recover but increasing the rate of evaporation of volatile components. The rapid mixing of ultralight products in turbulent water facilitates dilution and biodegradation, whereas medium and heavy oils take longer to spread, are slower to evaporate and are less easily dispersed in water. If weather permits, these oils can be corralled with booms and removed with mechanical skimmers. However, wind and wave conditions are rarely ideal, and mechanical recovery is known as the '10% solution' because, at most, 10% of a spill is recovered. Surface slicks can also be reduced by applying agents that 'herd' oil (prevent oil from spreading) and igniting agents that promote burning.

For light to medium oils, chemical dispersants containing surfactants are often applied to oil slicks to reduce the surface tension of floating oil and to promote rapid dilution. The dispersed oil breaks up into micrometre-sized droplets that are neutrally buoyant and are mixed into subsurface water by wind, waves and currents. In contrast, in 2010 dispersant was successfully injected into a jet of oil escaping from the Deepwater Horizon oil wellhead at more than 1,500 metres below the surface of the Gulf of Mexico, reducing the formation of surface oil slicks. Removing surface oil protects marine birds, mammals and turtles that must access the surface to breathe and feed, and dispersion increases the rate of microbial degradation of oil and its transfer to sediments (Capsule 13.2). The trade-offs include a period of hours to days when the oil dispersed in water exceeds concentrations toxic to different aquatic species. Dispersion also creates a high potential for contaminating shallow sediments and the potential for toxicity from any surfactant that has not mixed with the oil.

The environmental impacts of spilled oil include physical effects, such as oil coatings on birds, marine mammals, shoreline and wetland habitats, and shoreline animal and plant species. For freshwater spills, or spills into coastal wetlands, a major hazard is habitat destruction caused by clean-up activities. These include steam-cleaning of rocks and gravel beaches, detergent washing of shorelines, physical removal of sand and gravel, damage to riverbanks and riparian vegetarian, increased erosion and sedimentation, and destruction of benthic habitats by dredging. Spilled oil can also be highly toxic. Low-molecular-weight aliphatic and aromatic compounds (e.g., BTEX, two-ringed PAC, C6–C10 aliphatics) may cause acute lethality before evaporating, and the more persistent three- to six-ringed PACs are chronically toxic, particularly to the embryos of fish and invertebrates. Birds and mammals suffer mortality when feathers or fur are oiled sufficiently to destroy flight characteristics, buoyancy and thermal insulation. Oil contamination can also cause starvation in animals such as sea otters (*Enhydra lutra*) that spend too little time feeding because they groom their fur obsessively to maintain its buoyancy. The oil removed by grooming may also be ingested at toxic doses. For nesting birds, small amounts of oil on feathers can be transferred to incubating eggs. As little as 1.0 µl (one drop) on the surface of an egg can be acutely lethal to chicks or cause deformities that prevent normal development. Aquatic plants are somewhat resistant to spilled oil. However, high concentrations of oil and clean-up procedures can damage marine macrophytes such as brown algae (e.g., kelp). Kelp forests provide critical spawning substrates for marine fish species and habitat for invertebrates such as sea urchins and molluscs that form the basis of the food webs for marine birds and mammals. Exposure of aquatic organisms to crude or fuel oils may also impair chemoreception, with subsequent effects on behaviour or avoidance of polluted areas (Brown *et al.*, 2017; Schlenker *et al.*, 2019), or even reduced reproductive success (Seuront, 2011).

The extraction of petroleum products from bitumen introduces unique environmental problems, as these products require considerable processing to produce synthetic oils of sufficient quality. At bitumen mines in northern Alberta, Canada, oil is extracted from a sand matrix by hot water in surface refineries, or by steam-assisted gravity drainage in deep subsurface deposits *in situ*. Although the extracted bitumen is immiscible in water, water-soluble compounds such as naphthenic acids partition into the water. These saturated aliphatic and cyclic carboxylates are acutely toxic to aquatic organisms, and some are oestrogenic or androgenic at sublethal concentrations.

7.7 Legacy Contaminants

Several classes of organic contaminants have been tightly regulated for many decades, but they are still present in the environment because of their persistence, and so have been called 'legacy contaminants'. The first global initiative to regulate the manufacture and use of persistent organic pollutants (POPs) was the Stockholm Convention on POPs, which was adopted in May 2004 and by 2022 had been signed by 185 nations (see Table 1.2). The original version of the convention was designed to protect humans and other organisms from the toxic effects of a group of 12 chlorinated compounds of concern ('The Dirty Dozen'). These compounds were targeted because of their persistence, their tendency to bioaccumulate in organisms or to biomagnify through food chains, and their acute and chronic toxicity. The long-range transport of POPs is another concern, and the convention sought to eliminate or severely restrict the production and use of POPs that showed evidence of transport to remote regions, including polar ecosystems.

The initial list of POPs selected by the convention was a group of 12 organochlorine pesticides and several compounds used in industry, including hexachlorobenzene (HCB), polychlorinated biphenyls (PCBs), polychlorinated dibenzodioxins (PCDDs) and polychlorinated dibenzofurans (PCDFs). These compounds are all chlorinated, as illustrated in Figure 7.17. The convention made provision for adding new substances at a later date, and so the pesticides methoxychlor, chlordecone and **hexachlorocyclohexane (HCH)** were later added as annexes to the convention. As discussed later in this chapter, more recent additions to the convention include some classes of brominated flame retardants, perfluorinated substances and chlorinated paraffins. The histories of these compounds serve as examples of the time-lag involved with gathering evidence of toxic effects and appropriate regulation. However, as an illustration of the successful regulatory control of a specific POP through the Stockholm Convention, more than 90% of HCH was lost from Lake Superior between 1985 and 2015. First-order dissipation rates ($t_{1/2}$s) for isomers α-HCH and γ-HCH were 5.7 years and 8.5 years respectively (Bidleman *et al.*, 2021).

7.7.1 Organochlorine Insecticides

Of all the organochlorine pesticides produced over the past 80 years, none has received as much notoriety as DDT. Its use in agriculture and forestry was associated with widespread records of eggshell thinning and reproductive failure among several bird species, including raptors and pelicans, and effects on fish populations (Chapter 1). The structure of DDT is illustrated in Figure 7.17. Many of these effects were associated with DDE (**dichlorodiphenydichloroethylene**), which is a stable metabolite of DDT. The bans occurring in most countries that are signatories to the Stockholm Convention greatly restricted the use of DDT, and as a result, most populations of raptors and pelicans have recovered. However, because this insecticide is still the most effective compound available for killing the larvae of mosquitoes, it continues to be used for malaria control in several countries, including South Africa, India and Mexico. The World Health Organisation (WHO) has endorsed this practice because of the potential for saving human lives, while recognizing the potential negative effects on the environment. Global manufacture of DDT averaged over 3,500 metric tonnes per year between 2008 and 2014, with India accounting for 86% of its usage.

Another class of organochlorine insecticides that was banned under the Stockholm Convention is the cyclodiene group, which includes aldrin, dieldrin, endrin, heptachlor and different isomers of chlordane (Figure 7.17). Like DDT, the use of these insecticides resulted in adult bird mortalities, and, to a lesser degree, they were implicated in eggshell thinning. Heptachlor, which was widely used as a seed treatment, was responsible for mortalities among non-target mammalian species such as foxes through food chain accumulation. The insecticides within the **HCH** class were banned in the 2009 annex of the Stockholm Convention. These HCHs had been developed to replace heptachlor as a seed treatment but this class of compounds was subsequently widely used as a broad-spectrum insecticide. It consists of several stereoisomers of which the compound with the greatest insecticidal activity is the gamma isomer, lindane. However, this insecticide is usually sold as the cheaper 'technical HCH' that contains all the isomers. The alpha isomer of HCH is the most prevalent form transported to remote locations because of its persistence in the atmosphere.

Although now banned in many countries through the Stockholm Convention, some of these compounds are still used in countries with developing economies, such as the BRIC (i.e., Brazil, Russia, India, China) group of countries. They are still major contaminants in the cold

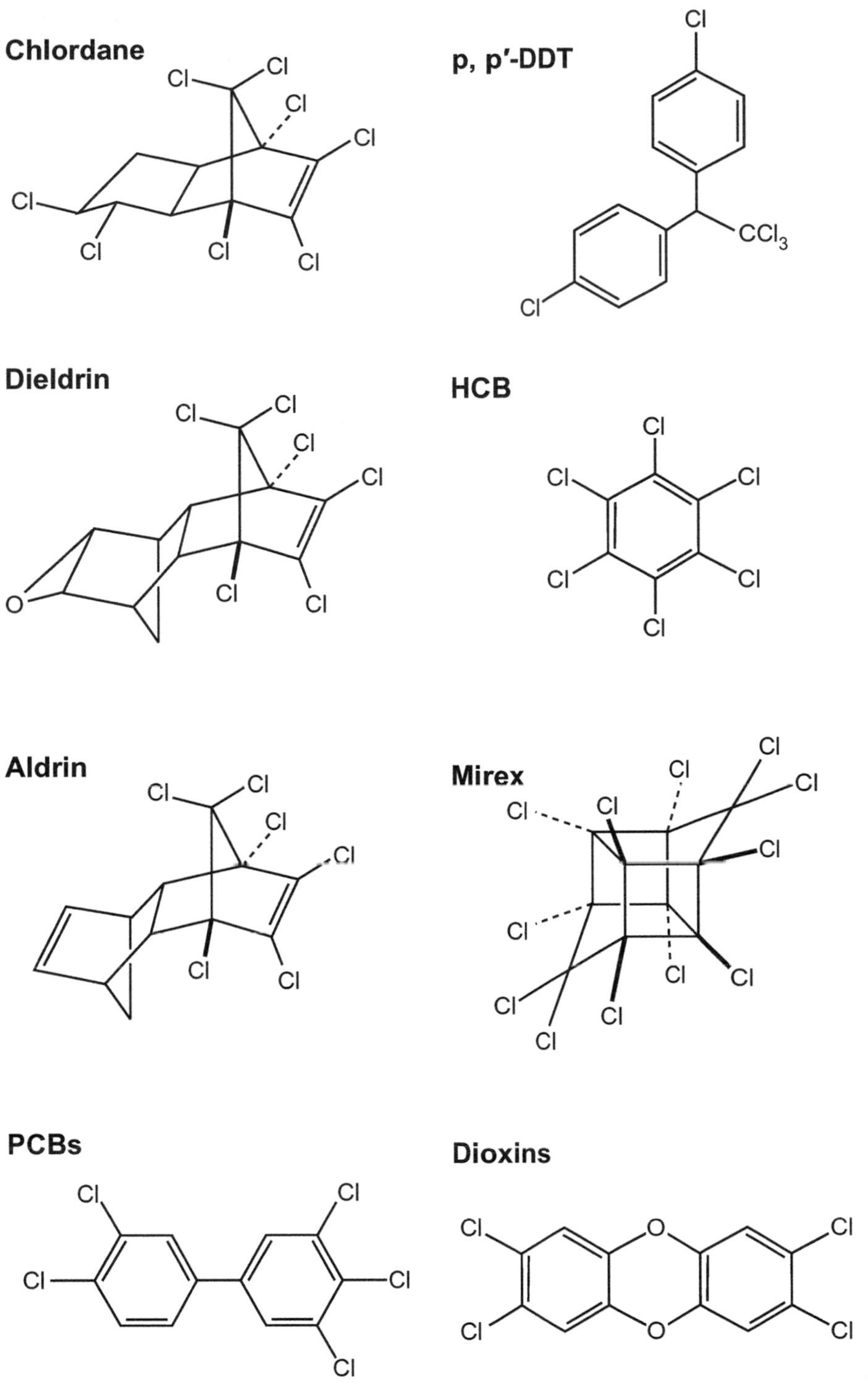

Figure 7.17 Chemical structures of eight of the 12 compounds ('The Dirty Dozen') initially banned in 2001 under the original Stockholm Convention, which was fully adopted in 2004. Additional compounds were added to the list in 2009 and 2013. The compounds illustrated include five organochlorine insecticides (i.e., aldrin, dieldrin, chlordane, mirex, p,p'-DDT), HCB, a PCB congener (PCB 126; 3,3,4,4,5-pentachlorobiphenyl) and a chlorinated dioxin congener (2,3,7,8-tetrachlorodibenzo-p-dioxin).

Arctic region, which acts as a condensation trap for compounds having intermediate volatility and high persistence (Section 7.2). Mirex and chlordecone (kepone) are two heavily chlorinated compounds with a boxlike structure that were marketed for eradication of fire ants in the southern United States. Both were marketed as insecticides, although the nonflammable property of mirex also led to its use as a fire retardant. Although fire ants are not a pest in colder climates, widespread contamination by mirex and its transformation product photomirex in Lake Ontario resulted from its production at the Hooker Chemical Plant on the Niagara River until the early 1970s. Mirex and photomirex residues in fish from Lake Ontario have declined greatly and are expected to soon be below detection limits (Gandhi *et al.*, 2015). Kepone was widely used as an insecticide in banana plantations in the French Caribbean islands of Martinique and Guadeloupe until the early 1990s, and contamination in the terrestrial and marine environments persists to this day.

7.7.2 Polychlorinated Dibenzodioxins and Dibenzofurans

Polychlorinated dibenzodioxins (PCDDs) and polychlorinated dibenzofurans (PCDFs) are a group of toxic halogenated aryl hydrocarbons (HAHs) formed as by-products in a variety of manufacturing processes and during incineration of plastics. Although they may be released through forest fires and volcanic activity, they mainly result from industrial activity, which goes back over 200 years.

The number of possible chlorine substitutions on the dioxin molecule (Figure 7.17) may vary from one to eight, giving a possible 72 positional isomers. Substitution at the lateral (i.e., 2, 3, 7 and 8) positions is associated with a particularly high degree of toxicity. Most of the available toxicological information is derived from studies of this congener. Historically, PCDDs and PCDFs were notable as by-products in the manufacture of PCBs and chlorophenoxy herbicides. In 1970, as much as one-half of the environmental release of these compounds came from the use of contaminated chlorophenoxy herbicides. The pulp and paper industry has also been a source of PCDDs and PCDFs, which are formed through the bleaching of 'kraft' wood pulp with elemental chlorine (Cl_2). The global recognition of this source of dioxin contamination and bans by various nations on the importation of paper products from kraft bleaching has led to the widespread application of other bleaching processes in pulp and paper mills (Section 11.4). Currently, the principal sources of PCDD and PCDF emissions are medical and municipal waste incinerators. Other sources include the burning of wood treated with chlorophenols.

Following numerous accidental human and wildlife exposures to PCDDs and similar compounds, the US EPA issued its first health assessment of 2,3,7,8-tetra-chlorodibenzo-p-dioxin (TCDD) in 1985. In 1991, the National Institute of Occupational Health and Safety confirmed a direct relationship between cancer mortality and occupational exposure to TCDD. Effects on wildlife have been less easily confirmed. For example, in areas of the Great Lakes that have been polluted with dioxin and dioxin-like compounds, some species of birds, reptiles, fish and mammals have exhibited developmental toxicity, reproductive impairment, compromised immunologic function and other adverse effects that correlated with dioxin exposures. However, the unequivocal attribution of these effects to dioxin toxicity is complicated by the simultaneous presence of other contaminants such as PCBs (see below).

7.7.3 Polychlorinated Biphenyls

Polychlorinated biphenyls (PCBs) are molecules consisting of two benzene rings linked by a carbon–carbon bond (i.e., biphenyls) and having chlorine atoms substituted at one or more of the ten available carbons in the molecule (Figure 7.17). There are 209 possible PCB compounds that vary in the number and position of chlorination of the biphenyl structure, but only about 90 of these congeners are typically detected in environmental matrices. Initially employed as plasticizers, the use of PCBs increased dramatically in the 1960s and particularly in the electrical industry. Because of their high boiling point and resistance to heat, they were used in many closed applications, such as in capacitors, transformers and hydraulic fluids. Later, these compounds were used in a variety of open applications, including as additives to paper products, cosmetics and plastics. Various commercial formulations were manufactured in several countries, varying in the degree of chlorination. The products manufactured by Monsanto in the United States included Aroclor 1242, Aroclor 1254 and Aroclor 1260, with the first two numbers referring to the 12-carbon structure and the last two numbers reflecting the per cent contribution of chlorine to the product (e.g., 60 = 60% by weight of chlorine).

Beginning in the 1960s, there have been several, well-documented instances of PCB pollution at specific sites, resulting in a variety of toxic effects in exposed humans, livestock and fish and wildlife (Case Study 13.2). Pathologies included carcinogenicity, teratogenicity and inhibition of reproduction. In the USA, uses of PCBs were restricted in 1974, and their production by Monsanto was halted altogether in 1977, 27 years prior to their eventual ban under the Stockholm Convention. Recent declines in PCB levels have been reported in fish from North America (Section 13.7.1) and Europe, although their persistence in the environment still poses problems in several areas (Carlson *et al.*, 2010). Proposed links between PCB exposures and effects have been complicated by variable mixtures of congeners having different properties and the co-occurrence of trace amounts of PCDDs and PCDFs formed during their manufacture and use.

Improvements in analytical methods permitted the detailed study of a small number of PCB congeners that are typically minor constituents of the commercial PCB mixtures but have a high degree of toxicity. These **coplanar PCBs** may be responsible for the majority of the toxicity of PCB mixtures. Coplanar PCB congeners lack chlorine substitutions at the ortho $(2, 2', 6, 6')$ positions, which allows the biphenyl molecule to adopt a planar configuration that is similar to the molecular configuration of TCDD (Figure 7.17). As a result, some coplanar PCBs, including $3,3',4,4'$- tetrachlorobiphenyl (congener 77), $3,3',4,4',5,5'$-pentachlorobiphenyl (congener 126) and $3,3',4,4',5,5'$-hexachlorobiphenyl (congener 169) share similar toxic properties with PCDDs and PCDFs by binding strongly to the AhR.

It is now clear that a broad range of HAHs are capable of binding with the AhR, causing a variety of adverse effects. These compounds include the coplanar congeners of PCBs as well as polychlorinated naphthalenes, polyhalogenated diphenyl ethers, polychlorinated anisole, polychlorinated fluorenes and polychlorinated dibenzothiophenes. This observation has led to a structure–activity approach to assigning toxicity of complex mixtures of HAHs. Where such mixtures occur in biota or other environmental matrixes (e.g., soil, sediment), the expression of the overall toxicity of the mixture in terms of **toxic equivalents** using PCDD as the 'reference toxicant' is a recognized approach for assessing overall risk (Section 2.3.4.2). Thus, the total **TCDD toxic equivalent quantity** (TEQ) of a mixture is estimated by summing the products of the relative potencies (**toxic equivalent factors**, TEFs) of each compound multiplied by their concentrations in the mixture. The TEF values were originally developed several decades ago (Van den Berg *et al.*, 1998), but continue to be refined (Lee *et al.*, 2013). This approach provides an overall toxicity index of the various active compounds detected at different concentrations in the environmental matrix (e.g., biota, sediments, soils). This model has been demonstrated to predict the degree of enzyme induction, embryo mortality and teratogenicity in exposed organisms.

7.8 Current Use Pesticides

The first synthetic insecticides were manufactured in the 1940s. Much of this early work was initiated in wartime Germany because the Allied embargo restricted access to tobacco from which the natural insecticide nicotine was extracted. Worldwide, it has been estimated that, as of 2020, 3.5 million tonnes of pesticides will be applied annually (Zhang, 2018). The majority of current use pesticides (CUPs) are used in agriculture, but bactericidal and fungicidal agents are also employed in industry where they are added as biocides to cooling fluids or to lubricants, or they are added to building materials as biocides. Forestry applications include the control of insect pests such as the Colorado pine beetle, spruce budworm, Asian longhorn beetle and emerald ash borer. Herbicides and fungicides are commonly used on recreational areas such as golf courses, and herbicides are employed to defoliate the margins of roads, railroad tracks and electricity power line corridors. Antimicrobial and fungicidal compounds are also used in personal care products and pharmaceuticals.

Agricultural pesticides may be broadly categorized as herbicides (40% of total usage by weight), insecticides (17%), and fungicides (10%). Ecological risks from pesticide use will always be balanced against their impact on ecological services, some of which may be difficult to quantify. The annual cost of pesticide application in the United States in 2016 was approximately $9 billion, resulting in an estimated saving of $60 billion in crops that would otherwise have been destroyed, a cost/benefit ratio of 6.6. Since 1960, wheat production in India has undergone approximately a 10-fold increase, whereas arable land only increased 2- to 3-fold over the same period. Much of this increased efficiency is attributed to the use of chemical pesticides, and similarly impressive figures are available from many other parts of the world. The domestic (household) market for pesticides can also be very large, particularly in developed countries. Very little is known about the contribution of non-agricultural uses of pesticides and fertilizers to surface water run-off, although as much as one-third of all pesticide use may be for non-agricultural purposes.

Most insect and related arthropod species are beneficial and play essential roles in natural ecosystems. Only 1% of the about 900,000 species are agricultural pests or are irritants or disease vectors for humans. However, insect pests are responsible for the worldwide destruction of about 13% of food and fibre crops. Negative aspects of pesticide use have focused on their persistence in the biosphere and their chronic toxicity to non-target species. Subtle but highly important effects include the loss of species that would otherwise exercise biological control over pests and the negative impacts on insect pollinators. In the USA, crops worth approximately $1 billion are ruined each year through the over-application or inappropriate use of pesticides, and many countries are currently adopting aggressive initiatives designed to reduce pesticide usage while maintaining crop yields.

The results of pesticide management programmes have been mixed. Pesticide use in the United States has remained stable between 0.4 million and 0.5 million tonnes since 1990. However, despite initiatives to rationalize the application of pesticides in China, their use in that country increased from 0.76 million tonnes to 1.8 million tonnes over the past 30 years. Pesticide usage in Brazil increased from 0.04 to 0.37 million tonnes over the same period. A similar trend is noted in Argentina, where pesticide application has risen from 0.07 to 0.26 million tonnes over the past decade. However, pesticide usage in India has fallen from 0.075 to 0.04 million tonnes since 1990. Consumer interest in purchasing 'organic' fruits, vegetables and cereals that have been produced without the use of synthetic chemical pesticides may further reduce pesticide use. However, this is only likely to happen in more affluent countries where consumers are able to pay the higher prices for organic food products.

Bearing in mind the history of organochlorine insecticides as persistent in the environment and with high potential for bioaccumulation in many organisms, most CUPs have been developed to only persist in the environment and in biota for a matter of days, or at most, weeks. They are also less lipophilic than organochlorines and therefore less susceptible to bioaccumulation. In addition, pyrethroid and neonicotinoid insecticides were developed to be extremely toxic to invertebrates, including the 'target' insect pests, but much less toxic to vertebrates, including humans, fish and wildlife. However, these CUPs can still have ecological effects, including the eradication of 'non-target' arthropods that carry out important ecological functions. For instance, the application of neonicotinoid insecticides can affect important pollinator insects, as discussed below (Section 7.8.5). The insecticides that are currently used include compounds from the organophosphate, carbamate, phenylpyrazole, pyrethroid and neonicotinoid classes, and these are described below in the approximate chronological order in which they were developed. A wide variety of herbicides and fungicides is currently used to control plant and fungal pests; the primary herbicides and fungicides in use today are also described below.

7.8.1 Organophosphate Insecticides

The first organophosphate pesticide, tetraethyl-pyrophosphate (TEPP) was developed in Germany during the early 1940s. Following the discovery that this compound is also extremely toxic to humans, research on organophosphates diverged from the development of insecticides to the development of 'nerve gas' for chemical warfare. For instance, Sarin (O-isopropyl methylphospho-nofluoridate) was developed in Germany and never used in World War II, but it was used in 2013 by the Syrian military to gas civilian populations within rebel-held territories in that country. The first insecticide developed in 1944 was parathion (Figure 7.18). This compound is a phosphorothioate ($R_3P=S$) ester, which, after ingestion, is

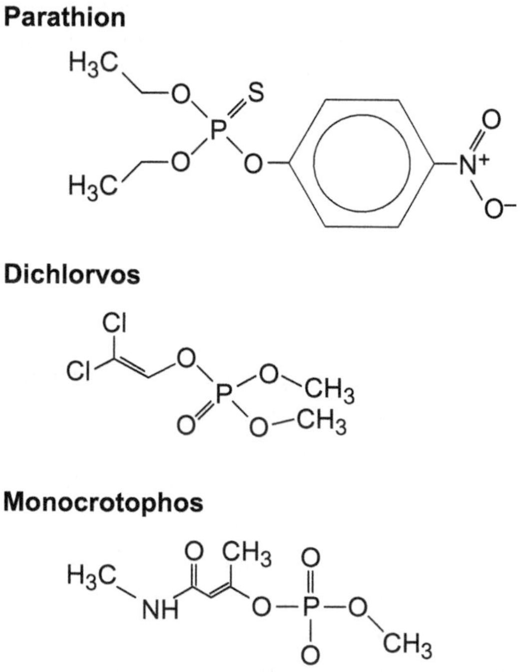

Figure 7.18 Structures of the organophosphate insecticides parathion, dichlorvos and monocrotophos.

metabolically converted (activated) to paraoxon by the substitution of oxygen for sulphur (i.e., R$_3$P=O). This is an advantage for control of pests, because insects may consume large amounts of plant material containing the pesticide and then subsequently be killed by the metabolite. Paraoxon binds to the active site of the enzyme acetylcholinesterase (AChE). Unlike the natural substrate, acetylcholine, which binds to the enzyme in a reversible manner, paraoxon forms a stable covalent complex that inhibits the activity of this enzyme. Chlorpyrifos is another organophosphate pesticide requiring activation to the chlorpyrifos-oxon form, whereas malathion and dichlorvos are organophosphate compounds (R$_3$P=O) that do not require metabolic activation (Figure 7.18).

Case Study 7.1 Toxicity of Insecticide, Monocrotophos, to Swainson's Hawks

It is generally assumed that organophosphate pesticides degrade quickly in the environment and so pose little threat to ecosystems. However, the mass mortalities among Swainson's hawks (*Buteo swainsoni*) in the Pampas region of Argentina during the austral summer of 1995–96 illustrate how even transient exposure to organophosphates can be a hazard to wildlife. During the breeding season in North America, pairs of Swainson's hawks spread out over large territories, but when these birds of prey migrate to South America during the austral summer, they congregate in large groups. Flocks of the hawks forage on grasshoppers and caterpillars throughout the Pampas region of Argentina. In 1995–96, large numbers of these hawks were exposed to an organophosphate insecticide, monocrotophos (Figure 7.18), either by direct contact with spray or by ingesting grasshoppers recently sprayed with this insecticide by farmers protecting

their crops of sunflowers, alfalfa and soy. A total of 19 mortality incidents and an estimated 5,095 dead hawks were recorded, representing about 1% of the global population (Goldstein *et al.*, 1999). Monocrotophos residues ranging from 0.05 to 1.08 µg/g were found in the contents of the gastrointestinal tract.

As discussed below for the MoA of both organophosphate and carbamate insecticides, reducing the activity of AChE means that the transmission of stimuli across nerves to muscle fibres by the neurotransmitter substance acetylcholine cannot be controlled, resulting in convulsions, inability to maintain respiratory movements and death by asphyxiation. Determination of brain AChE activity has been widely used to diagnose and evaluate the effects of organophosphate in birds, and to a lesser extent in amphibians, reptiles, small mammals and fish. In birds, a reduction of brain AChE of 50% is generally regarded as life-threatening. In the case of the dead Swainson's hawks, brain AChE activity was reduced by as much as 99% (Goldstein *et al.*, 1999). As a result of this incident and bans on the use of monocrotophos in the United States and European Union because of concerns over extreme toxicity to vertebrates, the original manufacturer, Novartis, ceased production in 1998. However, this insecticide is still available from other commercial sources, particularly in India.

7.8.2 Carbamate Insecticides

Carbamate pesticides (Figure 7.19) are derivatives of carbamic acid and, like organophosphate pesticides, they act as AChE inhibitors. However, they have a much lower toxicity than most organophosphates and are more easily metabolized. Two carbamate insecticides in common usage to control insect pests on cereal grains and other plants are carbaryl and carbofuran. In Canada and the

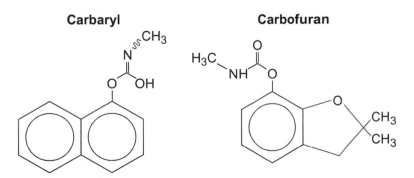

Figure 7.19 Structures of the carbamate insecticides carbaryl and carbofuran.

United States, both are commonly employed to control grasshopper populations in the western plains. As with organophosphates, instances of accidental bird poisonings have been frequently reported.

In many passerine bird species, carbofuran has an LD_{50} close to 1 mg/kg body mass. Preening, ingestion of poisoned insects and inhalation are all significant sources of exposure. Young birds may be up to 100 times more sensitive than adults. Although some field deaths have been reported for gulls, owls and pigeons, the lethal dose for most species generally exceeds that which might be expected to occur through the ingestion of contaminated food.

As illustrated in Figure 7.20, the cholinergic neurotransmitter acetylcholine is secreted from synaptic vesicles and diffuses across the pre-synaptic membrane of the axon and then the short distance (the synaptic cleft) to the post-synaptic membrane of the muscle fibre, where it binds to a receptor protein. As we will discuss later with respect to neonicotinoid insecticides, this receptor protein is called the nicotinic acetylcholine receptor (nAChR). The binding of acetylcholine to the receptor protein opens up the cation channels in the post-synaptic membrane, and the influx of Na^+ ions across the membrane causes contraction of the myofibrils within the muscle fibre. Acetylcholinesterase rapidly catalyses the degradation of acetylcholine to acetate and choline, and these constituents are re-absorbed into the pre-synaptic nerve axon for re-synthesis of acetylcholine and packaging in vesicles. In this way, the stimuli across neuromuscular junctions can

Neuromuscular junction

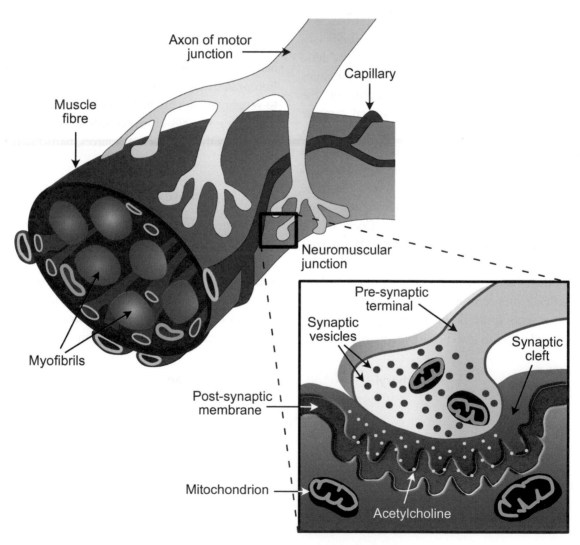

Figure 7.20 The neuromuscular junction between nerve axons and striated muscle fibres.

be controlled and coordinated to facilitate muscle movement. However, if organophosphate or carbamate pesticides bind to the enzyme and reduce its activity, the action of acetylcholine is prolonged and coordination of muscle contractions is lost. This can be lethal if there is loss of the control of the muscles controlling respiratory movements.

7.8.3 Phenylpyrazole Insecticides

The phenylpyrazole pesticide fipronil is the most frequently applied member of a group of broad-spectrum insecticides that are used to treat a variety of crops and for the control of veterinary pests. Phenylpyrazole pesticides act as potent inhibitors of **gamma-amino butyric acid (GABA)**-gated chloride ion channels, causing hyperexcitation and death in insects. Fipronil was first registered for use in the United States in 1985. Concerns about its toxicity to non-target organisms have led to a greater than 95% reduction in its use in agriculture. However, domestic use for control of termites and cockroaches, and for flea and tick control on pets, remains steady. Thus, fipronil has been widely detected in urban run-off as a result of its domestic use.

7.8.4 Pyrethroid Insecticides

The pyrethrum extract from dried chrysanthemum flowers contains the natural plant esters pyrethrins, jasmolin and cinerin, and has been used for centuries as an insecticide. The insecticidal components of the extract, pyrethrin I and pyrethrin II, were isolated and characterized in 1924. Use of the rapidly degraded natural pyrethrins as alternatives to the persistent organochlorine pesticides was superseded in the 1970s by the development of synthetic pyrethroids such as fenvalerate. Ironically, they were designed to be more persistent and less volatile than the natural compounds to increase their pesticidal efficiency (Figure 7.21).

Pyrethroids are extremely potent neurotoxins in arthropods, causing a variety of neurophysiological effects, including repetitive neuronal discharges, similar to the action of DDT, and disturbances in synaptic transmission. Pyrethroid esters such as permethrin have also been shown to inhibit Ca–Mg ATPase, resulting in an increase in intracellular calcium in neurons and enhanced neurotransmitter release. These properties combine to elicit rapid 'knock-down' and mortality in insects, with the result that these compounds are commonly used in agriculture as insecticides on crops such as cotton. Because of their low toxicity to vertebrates, they are extremely popular as domestic and horticultural insecticides.

Pyrethroids currently account for about 25% of worldwide insecticide usage. Their use in many developing countries is often limited by the fact that they are expensive relative to many other more easily synthesized insecticides. Their popularity as household insecticides is due to their low acute toxicity to mammals. Pyrethroids are rapidly metabolized by the CYP450 monooxygenase system in insects, and certain formulations contain CYP450 monooxygenase inhibitors such as piperonyl butoxide to potentiate the action of the insecticide. These compounds have effects on ecological functions by killing non-target arthropods that are sources of forage for higher

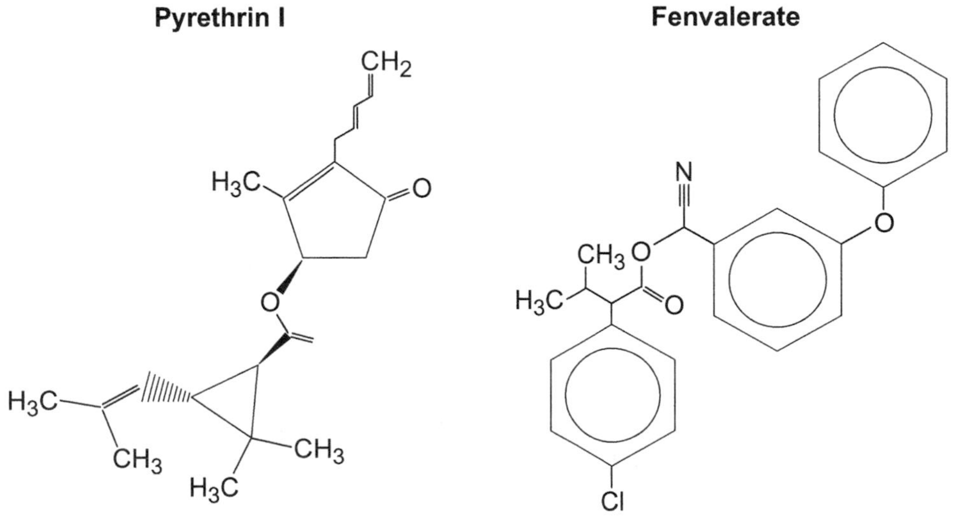

Figure 7.21 Structure of the natural insecticidal compound, pyrethrin I, and the synthetic pyrethroid insecticide, fenvalerate.

trophic levels (e.g., fish, birds), or are ecological competitors for the target insects (Antwi and Reddy, 2015).

7.8.5 Neonicotinoid Insecticides

Neonicotinoids (Figure 7.22) are a family of insecticides similar in structure to the plant alkaloid nicotine. Developed in the 1990s to replace the organophosphate and carbamate pesticides, they are now the most widely used insecticides in the world. They particularly target insect pests in the Homoptera, Coleoptera and Lepidoptera families, where they act as potent acetylcholine mimics that bind to the post-synaptic nicotinic acetylcholine receptor (nAChR). However, they have a high affinity for the arthropod form of this receptor relative to the vertebrate receptor, making them relatively non-toxic to humans, fish and wildlife. Nevertheless, concerns have been raised over their toxicity to pollinating insects such as honeybees, which appear to be deficient in detoxification enzymes, including CYP450 monoxygenases, glutathione-S-transferases and carboxylesterases. Consequently, in 2018, the European Union prohibited the outdoor application of the three most commonly used neonicotinoids: Clothianidin, Imidacloprid and Thiamethoxam. These compounds continue to be widely used on crops (e.g., corn, soy) and as seed coatings in the United States and Canada. It has been predicted that there will be a 'neonicotinoid tsunami' in terms of global use of this class of insecticides for rice and other crops in Asia transitions from small family farms to mechanized production.

7.8.6 Chlorophenoxy Herbicides

Compounds from this class of herbicides are used in several formulations as herbicides that accelerate plant growth by mimicking plant **auxins** (Figure 7.23). The two best-known compounds are 2,4-dichlorophenoxy acetic acid (2,4-D) and 2,4,5-trichlorophenoxyacetic acid (2,4,5-T). These compounds are moderately water-soluble, at least when present as anions under circum-neutral and basic conditions.

Their most notorious usage was in Agent Orange, an approximately 1:1 mixture of 2,4-D and 2,4,5-T, which was employed in large quantities as a defoliant by the US military in Vietnam during the late 1960s and early 1970s. However, toxic effects were linked primarily to contamination of the formulation by TCDD and other dioxins. The presence of TCDD as a contaminant in 2,4,5-T formulations led to the banning or strict control of this compound in most developed countries, whereas 2,4-D continues in widespread use as a domestic and agricultural herbicide, where it is particularly effective at killing broadleaf plant pests (e.g., dandelions). Structural analogues that are also in use include MCPP or

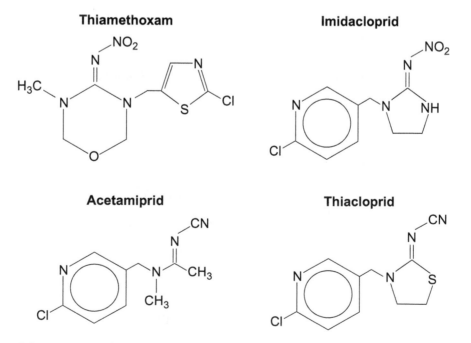

Figure 7.22 Structure of the neonicotinoid insecticides, Thiamethoxam, Imidacloprid, Acetamiprid and Thiacloprid.

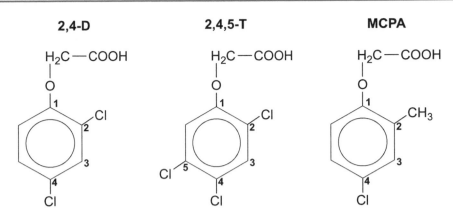

Figure 7.23 Structures of the chlorophenoxy herbicides, 2,4-D, 2,4,5-T and MCPA.

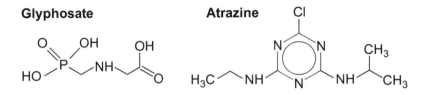

Figure 7.24 Structures of the herbicides glyphosate and atrazine.

Mecoprop (2-methyl-4-chlorophenoxy propionic acid), Dicamba (3,6-dichloro-2-methoxybenzoic acid) and MCPA (2-methyl-4-chlorophenoxy acetic acid).

Reports in the 1980s of a link between contact with 2,4-D by farmers and the incidence of non-Hodgkin's lymphoma have led to many municipalities in North America and Europe banning the use of these herbicides for domestic use and for weed control in recreational areas.

7.8.7 Bipyridilium Herbicides

The two most important members of this group of compounds are diquat and paraquat. Paraquat, which was registered for use in 1965, is still the most widely used post-emergent herbicide in Latin America, the Caribbean, and many other tropical areas. Bipyridilium compounds quickly destroy plant cells and have a high binding affinity for inorganic minerals such as clay, which rapidly reduces their bioavailability. This property enables rapid replanting following their application. These compounds are activated by exposure to sunlight.

Although accidental human exposure to paraquat has resulted in severe illness and some fatalities, there are few field data available on the potential toxic effects of these compounds on terrestrial animals as a result of normal application. Feeding experiments using paraquat-dosed

syrup fed to honeybees (*Apis melifera*) have indicated adverse effects on the development of larvae at concentrations as low as 1 ng/kg food (Cousin *et al.*, 2013). High water solubility of paraquat has also led to concerns over adverse effects on larval amphibians and fish resulting from run-off from treated fields and from application for treatment of aquatic plant species, where such treatment is allowed (US EPA, 2009; Salazar-Lugo *et al.*, 2011). However, dissipation of these compounds in the aquatic environment is considered to be rapid ($t_{1/2} < 15$ h).

7.8.8 Glyphosate Herbicide

Glyphosate has been commercially manufactured as a general herbicide since 1974 and is marketed by Monsanto under the name RoundUp® (Figure 7.24). The active ingredient in this formulation is N-(phosphonomethyl) glycine, which interferes with the production of essential aromatic amino acids in plants, such as phenylalanine, tyrosine and tryptophan through its specific inhibition of 5-enolpyruvylshikimic acid 3-phosphate synthetase. This biosynthetic pathway is not shared by mammals. Hence, it has low acute toxicity to humans and to rodent species commonly employed as laboratory test animals. Its attraction as a non-specific herbicide stems from its effectiveness in no-till crop farming and its absorption through the leaves of plants rather than roots, thereby facilitating its targeted use between rows of orchard trees and other crops, including maize, soy, canola and sugar beets.

The glyphosate usage pattern changed from approximately 60% agricultural and 40% domestic in 1974 to 92% agricultural and 8% domestic in 2014. The expansion of glyphosate into the agricultural market coincided with the development by Monsanto of genetically engineered crops that are tolerant to this herbicide (e.g., 'RoundUp Ready' corn), beginning in the mid-1990s. This has led to a rapid expansion of its use worldwide. Between 1995 and 2014, global applications of glyphosate increased more than 12-fold, from 67 million kg to 826 million kg. This expansion has brought about an increase in the emergence of herbicide-resistant plants, which in turn has resulted in higher application rates of the herbicide. This sharply accelerated application has prompted recent concern over increased exposure to the herbicide and its primary breakdown product, aminomethylphosphonic acid (AMPA). Particular attention has been paid to possible chronic effects such as carcinogenicity, teratogenicity and endocrine disruption (Case Study 5.1). Although the US EPA has classified glyphosate as a non-carcinogen, the toxicity of this compound to humans and wildlife at realistic environmental concentrations remains a topic of vigorous debate. Austria, Sri Lanka and Brazil became the first countries to issue outright bans on glyphosate in 2019. Partial and inconsistent bans elsewhere reflect uncertainties regarding the toxicity of this herbicide, and so it remains widely used.

7.8.9 Triazine Herbicides

The herbicidal properties of the s-triazines were discovered in 1952 by the J. R. Geigy Chemical Company of Basel, Switzerland. Since then, several major products, including atrazine, simazine, cyanazine and prometryn, have been developed for the control of broad-leaved annual and perennial weeds in corn, sorghum and sugar cane crops. These compounds kill plants by blocking photosystem II, which prevents the electron transport chain from functioning in plant cells, leading to the destruction of chlorophyll and an inhibition of carbohydrate synthesis. The most widely used triazine herbicide, atrazine (Figure 7.24), has been used since the 1960s and accounts for more than 75% of triazine usage. In 2014, approximately 34 million kg of atrazine were applied in North America, principally in the mid-western United States, second only to glyphosate in

rate of application. Because of its highly specific mode of action, atrazine and other triazines are generally not regarded as toxic threats to animals, as reflected in the relatively high acute LC_{50} values seen in most animal species tested. Much of the controversy over the potential toxic effects of atrazine stems from studies of Hayes and co-workers (Hayes *et al.*, 2011), who documented feminization of male frogs exposed in the laboratory to atrazine. However, two studies funded by the US EPA were unable to replicate these results. Other independent laboratories have shown developmental effects in amphibians (Langlois *et al.*, 2010), with evidence of its endocrine disrupting activity (Chapter 8). Regardless, there is ample evidence that atrazine is one of the most widely detected chemical contaminants of surface waters and groundwater in North America, South America and Asia. Atrazine has been banned since 2004 in the European Union because of concerns over widespread contamination of sources of drinking water.

7.8.10 Fungicides

Until recently, there was little focus on the potential environmental impacts of chemicals used to control fungal pathogens of fruit, vegetables and turf grass. Fungicides are also used as biocides in paints, paper production, building materials, and in pharmaceuticals and personal care products. The structures of fungicides are diverse (Figure 7.25), but the fungicides most commonly used in agriculture include propiconazole, tebuconazole and other members of the triazole class of compounds, as well as myclobutanil, azoxystrobin, iprodione, thiophanate-methyl and carbendazim. In several of these applications, organic fungicides were introduced to replace the widespread use of metal salts and organometals such as phenylmercuric acetate.

Several classes of fungicides have come under regulatory scrutiny because of concerns about their toxicity. The use of quintozene (pentachloronitrobenzene) is now highly regulated in the United States, Canada and the European Union because of concerns about the teratogenicity of this compound. Vinclozolin, which has anti-androgenic activity, is now regulated by the US EPA and has been banned in several European countries. Carbendazim is a systemic fungicide widely used to control fungal pathogens in agriculture and forestry. It is also used as a biocide in building materials. Because of concerns that carbendazim causes embryotoxicity,

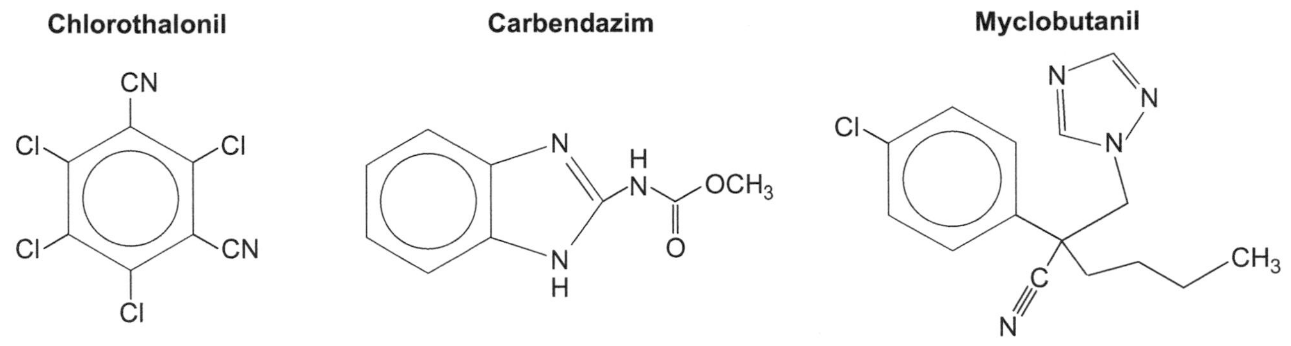

Chlorothalonil   **Carbendazim**   **Myclobutanil**

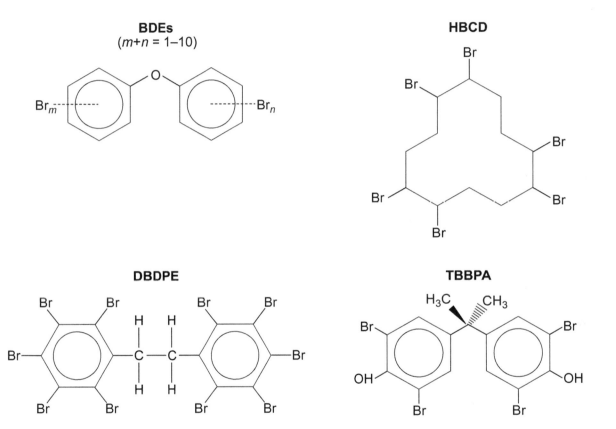

Figure 7.25 Structures of several fungicides that are widely used for agricultural applications and for the care of turf grass.

BDEs
(*m+n* = 1–10)

HBCD

DBDPE

TBBPA

Figure 7.26 Structures of the four main classes of brominated flame retardants in current production.

teratogenicity and endocrine disrupting effects, this compound has been regulated in the European Union and in some jurisdictions in North America, but continues to be widely used in South America, Asia and Africa.

7.9 Flame Retardants

Brominated flame retardants (BFRs) and organophosphate flame retardants are the two primary classes of flame retardants that are produced in volumes of hundreds of thousands of metric tonnes annually. Of these, the BFRs are the most widely used class of compounds. Whereas

North America and Europe contribute a significant amount to the global consumption of BFRs, now Asia and especially China consume over 60% of BFRs for incorporation into a variety of products that are marketed domestically and globally (Yu *et al.*, 2016). The BFRs include 'novel brominated flame retardants' (NBFRs) that are beginning to enter the market, such as bis(2-ethylhexyl) tetrabromophthalate and 2-ethylhexyl 2,3,4,5-tetrabromobenzoate. Examples of the four major classes of BFRs are shown in Figure 7.26.

Brominated diphenyl ethers (BDEs) were the predominant BFR product manufactured throughout the 1970s to the mid-1990s. Uses included incorporation in upholstery, carpets, automobile trim and electronics. Given that BDEs are very lipophilic, with log K_{OW} values from 5.0 for di-BDE to 8.6 for octa-BDE, they bioaccumulate in lipid-rich tissues and show evidence of biomagnification through food chains. They are also subject to global transport and are present at relatively high concentrations in marine mammals and seabirds in the Arctic. There is also evidence that the metabolites of these compounds can disrupt thyroid function, as discussed in Chapter 8.

Following the discovery that the concentrations of BDEs were increasing at an alarming rate in the tissues of wildlife, fish and humans with doubling times every 3 to 8 years, depending on the species, there has been a global initiative to reduce the use of this class of BFRs. The European Union started restricting the use of penta-BDEs and octa-BDEs in electronic and electrical equipment in July 2006. The North American manufacturers of the penta-BDE voluntarily removed this product from the market in 2007. In 2009 BDEs were added to Annex A of the Stockholm Convention, along with several other classes of POPs (the 'Nasty Nine'). However, BDEs are still being manufactured and used in products sourced from China. In addition, China accepts large amounts of e-waste from around the world that is a source of BFRs, including BDEs. For instance, the cities of Guiyu and Taizhou in China have developed industrial-scale facilities for dismantling and recycling e-waste (Yu *et al.*, 2016).

Although there is some evidence that the levels of BDEs are declining globally, the temporal trends are not uniform across all BDE congeners. For example, the concentrations of total BDEs in adipose samples from adult female polar bears in East Greenland declined after 2000, but the concentrations of congener BDE153 continued to increase (Dietz *et al.*, 2013). In addition, there was no evidence that total BDEs were declining in adult male bears over the period between 1990 and 2010. Addison *et al.* (2020) observed a similar trend for total BDEs in the blubber of ringed seals from the Canadian Arctic, with some declines in female seals after BDE bans were imposed and no evidence of declines in male seals (Figure 7.27). These observations probably illustrate the capacity of female mammals to eliminate lipophilic contaminants through transplacental transfer to the developing foetus and through lactational transfer to the young (see Case Study 4.6 on PCBs in killer whales). Male mammals, including humans, do not have this capacity.

As a result of the regulatory changes to reduce the use of BDEs, industry has moved towards using other BFRs such as hexabromocyclododecane (HBCD), tetrabromobisphenol A (TBBPA) and decabromodiphenylethane (DBDPE). However, there is evidence that the concentrations of some of these compounds are also increasing in wildlife (Dietz *et al.*, 2013; Miller *et al.*, 2014). As a consequence of these observations and other lines of evidence, HBCD was added to an annex of the Stockholm Convention in 2011. Now, yet another

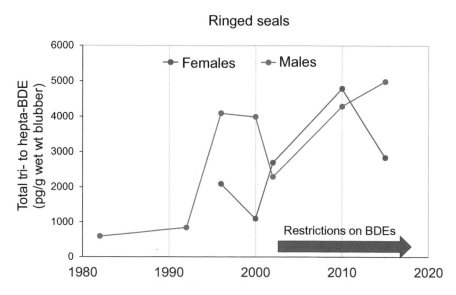

Figure 7.27 Temporal trends in the concentrations of total BDEs (pg/g lipid weight) in the blubber of adult female and adult male ringed seals (*Pusa hispida*) from Ulukhaktok, Northwest Territories, Canada, collected between 1981 and 2015. Figure drawn from data in Addison *et al.* (2020), with permission from the authors and Springer Nature.

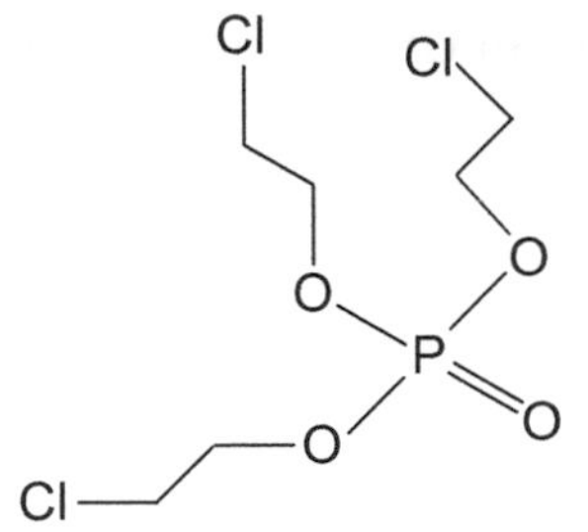

Figure 7.28 Trichloroethylphosphate, which is one of the organophosphate flame retardants commonly detected in the environment.

compound, DBDPE, has been widely detected in wildlife and humans in China, where most of the global production of this compound occurs (Covaci *et al.*, 2011).

Owing to the phase-out of several BFRs, the use of organophosphate flame retardants (OPFRs) has increased exponentially. Compounds from this class of flame retardants include trichloroethylphosphate, tris(2-butoxyethyl) phosphate and tris(2-chloroisopropyl) phosphate (Figure 7.28). These OPFRs have been widely detected in abiotic environmental compartments, such as water, air and dust. Even though this group of compounds appear to be rapidly metabolized in vertebrates, some OPFR compounds have been detected at relatively low concentrations in biota. Temporal data from the analysis of archived samples of herring gull eggs collected from colonies around the Laurentian Great Lakes indicate that these compounds have been present in the environment for several decades (Greaves *et al.*, 2016). There is evidence from laboratory exposure experiments that OPFRs can cause a variety of biological responses, including thyroid disruption, so more research is needed to evaluate whether this class of flame retardants has the potential to be an environmental hazard.

7.10 Perfluoroalkyl Compounds

Polyfluorinated hydrocarbons have multiple sites where hydrogen has been substituted by fluorine (e.g., telomer alcohols). Perfluorinated species have had all of the hydrogens substituted with fluorine (e.g., perfluorooctanoic acid (PFOA); perfluorooctane sulphonic acid (PFOS)). **Perfluoroalkyl compounds** (PFCs; also called PFACs) first came to the attention of the public when it was discovered that DuPont, the manufacturer of two products, Teflon® and the additive PFOA (known commercially as 'C8'), had dispersed almost 1 million kilograms of

waste containing PFOA generated from its manufacturing plant into the Ohio River Valley in the United States. Disposal included burying the waste in drums and depositing contaminated sludge in landfills. These practices came to light when a West Virginia cattle rancher named Wilbur Tennant sued DuPont in 1998 over the loss of hundreds of head of his cattle because of pollution from a landfill next to his farm. However, the contamination was not isolated to the Tennant property, and it was later discovered that PFOA had seeped into the water supply of at least six public water systems in the states of West Virginia and Ohio, USA. The subsequent class action lawsuits on behalf of about 80,000 people in the six water districts have resulted in awards of millions of dollars paid out by DuPont. A lawsuit by the US EPA was only able to secure from DuPont a voluntary phase-out of PFOA by 2015. The 3M company, which manufactured the structurally related compound PFOS under the trade name Scotchguard®, voluntarily phased out that product in 2002, responding to mounting evidence that it was bioaccumulating in humans and in wildlife.

PFOS was added to an annex of the Stockholm Convention in 2009, although this product, as well as PFOA, is still being manufactured in China and elsewhere in Asia. As described above, the basic structural element of PFCs is a partially or fully fluorinated alkyl chain, which is hydrophobic, and a terminal functional group (carboxylate, sulphonate, sulphonamide, phosphonate), which is hydrophilic (Figure 7.29). Because of the presence of both hydrophobic and hydrophilic parts of the molecule, PFCs exhibit surfactant properties. The carbon–fluorine bonds are the strongest bonds in organic chemistry and are highly resistant to degradation. PFCs are also non-flammable and resistant towards acids and bases. Thus, PFC compounds are used for numerous applications, including in firefighting foams, and as coatings for fabrics, leather and the paper products used for food packing. The most commonly encountered PFCs in the environment are perfluoroalkyl carboxylic acids, perfluoroalkyl sulphonates and sulphonamides, and fluorotelomer alcohols, acids, olefins and sulphonates. Until recently, research attention has focused on PFOA and PFOS because these classes of compounds are distributed globally (Giesy and Kannan, 2001; Houde *et al.*, 2011). However, a range of PFCs has been detected in the environment. Because of their structural heterogeneity, these compounds are now usually described as per- and polyfluoroalkyl substances (PFAS). A more recent

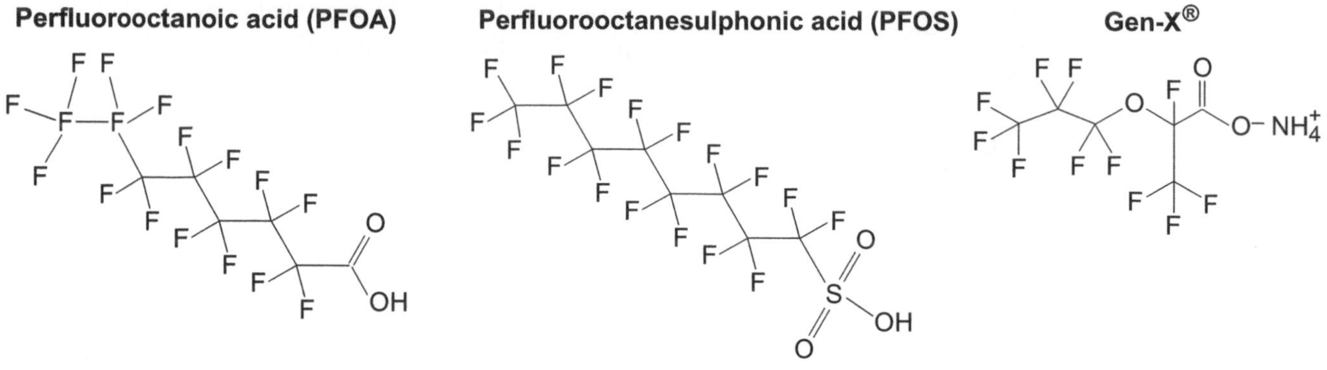

Perfluorooctanoic acid (PFOA)   **Perfluorooctanesulphonic acid (PFOS)**   **Gen-X®**

Figure 7.29 Structures of perfluorooctanoic acid (PFOA) and perfluorooctanesulphonic acid (PFOS), and Gen-X®, the ammonium salt of perfluoro-2-propoxypropionate.

addition to the PFAS class is the ammonium salt of perfluoro-2-propoxypropionate (Figure 7.29). **GenX®** is the trade name of this compound, which was developed by DuPont (now Chemours) in 2008 as a substitute for the PFOA that was previously added to Teflon products.

There are numerous direct and indirect sources of PFAS in the natural environment. These include losses from the synthesis of various fluorinated products by electrochemical fluorination, which has been used since 1950s to produce fluoropolymers. Past emissions of PFAS included contamination from the use of firefighting foams, as well as losses from a variety of consumer and industrial products. Fluorotelomer alcohols (FTOHs) are precursor compounds for the synthesis of various fluorinated products. They are present in these products at low concentrations as unreacted residues but are subject to release into the atmosphere because of their volatility. Degradation of FTOHs in the atmosphere yields a homologous series of fluorinated polymers that vary in chain length. PFOS is still being released into the environment because of its use in specialized industrial products (e.g., semiconductors, medical devices). However, the primary current source of PFOS is from the manufacture of perfluorooctane sulphonyl fluoride (POSF), which is used as a surface coating for paper products and food packaging. These semi-volatile compounds are subject to long-range atmospheric transport, and when they break down in the atmosphere, they are degraded to PFOS.

The bioaccumulation of PFAS compounds may be limited to long-chain products that have perfluoroalkyl chains with eight or more carbons and two fluorine atoms associated with each carbon, such as 8:2 FTOH (i.e., PFOA). Recently, some FTOH manufacturers have committed to replace long-chain products with short-chain alternatives (e.g., 6:2 FTOHs) that may have lower potential

for bioaccumulation. The mechanism for bioaccumulation of PFAS compounds cannot be attributed to partitioning into lipids, since these compounds are not lipophilic. Rather, these 'proteinophilic' compounds bioaccumulate in organisms by binding to cellular proteins (Kelly *et al.*, 2009; Ng and Hungerbuhler, 2013; Section 3.6.2.3).

The case history of PFOA contamination in the Ohio River Valley demonstrated that these compounds were acutely toxic to beef cattle exposed to high concentrations in drinking water. However, the evidence for sublethal toxicity from PFAS compounds at the lower concentrations that are typically observed in biota is less definitive. Some rodent studies suggest that these compounds can act as 'obesogens', by binding with PPAR receptors (Table 7.1) in the developing foetus as a result of prenatal exposures. There is also some evidence that PFAS compounds disrupt the endocrine system, as discussed in Chapter 8 (Table 8.2).

The fact that PFAS compounds are detected in a wide range of organisms, including humans, indicates that they are bioavailable and can be taken up by organisms through ingestion of contaminated water, inhalation of contaminated air or through food-chain pathways. As noted in Section 3.6.2.3, these compounds are proteinophilic (i.e., bind to proteins) and therefore favour tissues of high protein content. Mechanisms of transport of PFASs across cell membranes have only been investigated for a few compounds (principally PFOS and PFOA). Results indicate that these compounds may be transported across cell membranes by anion transporters in animals or aquaporins in plants. Short-chain PFAS appear to be more easily taken up by plants from soil and more readily translocated to the shoots or storage organs than long-chain compounds, probably because of their weaker binding to soil and to cell walls in the roots.

Capsule 7.1 Mobility, Bioavailability and Remediation of PFAS Compounds in Soils

Michael J. McLaughlin

Soil Science, School of Agriculture, Food and Wine, University of Adelaide, South Australia, Australia

As noted in Section 7.10, there are many different compounds included under the umbrella term of PFAS, with now more than 3,000 different compounds from this class that have been detected in the environment (Wang *et al.*, 2017). While some of these compounds are cationic (possessing a positive charge at environmentally relevant pH values) or zwitterionic (possessing both negative and positively charged groups in the same molecule), many are relatively strong acids. With low acid dissociation constants (pK_a), they are present in the environment as negatively charged anions. Hence, many PFAS compounds have high mobility through the landscape, as most soils are predominantly negatively charged, repelling the majority of anionic PFASs. This high mobility, coupled with environmental persistence, is one of the major reasons that contamination of surface and groundwaters by soil-applied PFAS is common (Crone *et al.*, 2019). Contamination of soil can be from atmospheric emissions, irrigated wastewaters, sewage biosolids applied to land, or application of other industrial or urban wastes to land (Oliaei *et al.*, 2013; Zhang *et al.*, 2019; Schroeder *et al.*, 2021).

Mobility in Soils

Mobility in soils is governed by both hydrophilic and hydrophobic interactions with soil surfaces (Figure 7C.1). As noted above, most PFAS compounds carry either a negative (most common) or a positive charge on the molecule, usually associated with the head group of the compound. This charge interacts with negatively charged surfaces on aluminosilicate clay minerals (e.g., kaolinite, illite, montmorillonite) and

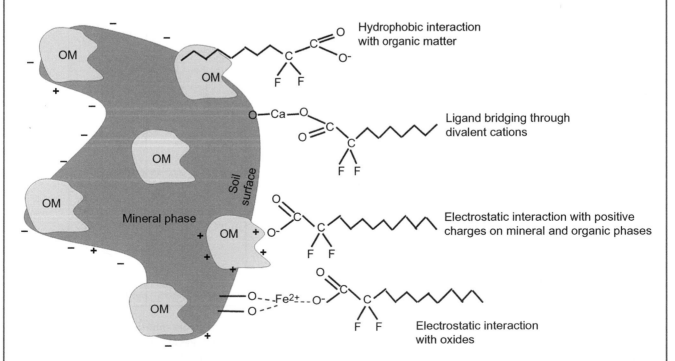

Figure 7C.1 Possible sorptive interactions of PFAS with soil surfaces. OM = organic matter. Reprinted from Li *et al*. (2018), with permission from Elsevier.

organic matter in soil, or with positive charges on the edge faces of aluminosilicate minerals or surfaces of oxidic minerals (aluminium, iron, manganese oxyhydroxides). Hydrophobic interactions occur via the lipophilic C–F tail of the compound interacting with uncharged C-rich domains on soil organic matter.

Effect of PFAS Chemical Structure on Sorption

Longer-chain PFAS compounds with six or more C atoms in the molecule tend to bind to soils more strongly than short-chain compounds (Figure 7C.2), probably as a result of stronger hydrophobic interaction of the C-F tail with soil organic matter (Nguyen *et al.*, 2020). Overall, however, sorption of anionic PFAS to most soils appears to be very weak, especially for the short C-chain compounds, compared with other common organic contaminants in soils, such as polycyclic aromatic hydrocarbons (PAHs), polychlorinated biphenyls (PCBs), organochlorine pesticides, etc. Compounds developed more recently to replace PFOA or PFOS in domestic materials and fire-fighting foams, such as GenX ($C_6F_{11}O_3^-$), ADONA ($C_7HF_{12}O_4^-$) and 9Cl-PF3ONS ($C_8ClF_{16}O_4S-$), also appear to be quite weakly retained by soil and thus will be highly mobile in the environment (Nguyen *et al.*, 2020).

A further property that is key to migration through soils is the propensity of these compounds to accumulate at phase interfaces, such as the air/water interface in unsaturated soils, or the interface between water/non-aqueous phase liquid (NAPL), such as oils in contaminated soils. This is due to the dual hydrophobic/hydrophilic properties of these compounds. This partitioning tends to retard transport through soil and lead to retention of PFAS within the soil matrix (Brusseau, 2018). Retention in soil is therefore a mix of sorption to soil surfaces combined with retention at air/water interfaces.

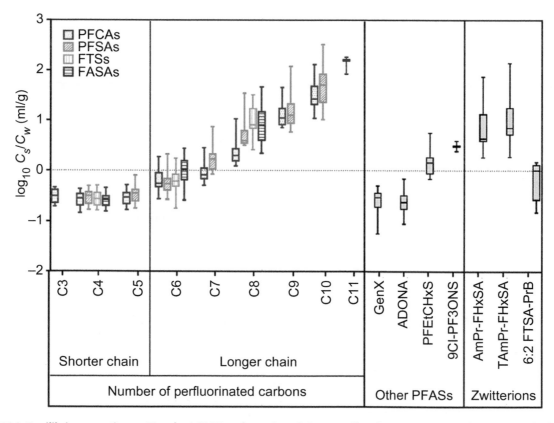

Figure 7C.2 Equilibrium sorption or K_d values (C_s/C_w, where C_s and C_w are soil and water concentrations, respectively) for various PFAS compounds in ten soils. PFCAs = perfluoroalkyl carboxylates ($C_nF_{2n+1}COO^-$), PFSAs = perfluoroalkyl sulphonates ($C_nF_{2n+1}SO_3^-$), FTSs = fluorotelomer sulphonates ($C_nF_{2n+1}CH_2CH_2 SO_3^-$), FASAs = perfluoroalkane sulphonamides ($C_nF_{2n+1}-SO_2NH^-$). Reproduced from Nguyen *et al.* (2020), with permission of the American Society of Chemistry (ACS), DOI: 10.1021/acs.est.0c05705. Further permissions related to excerpted material should be directed to the ACS.

Effect of Soil Properties on PFAS Sorption

Unlike other persistent organic contaminants that are generally highly hydrophobic and therefore bind strongly to organic matter in soils, PFAS compounds exhibit both hydrophobic and electrostatic (charge) interactions with soil surfaces. Hence, partitioning between the aqueous phase and solid surfaces in soils, as defined by the K_d value, is not just dependent on the organic carbon (OC) content of the soil, but is also influenced by other soil characteristics. These parameters include soil pH, which influences soil surface charge and electrostatic interactions which lead to greater retention in acid soils; and soil texture, which leads to greater retention in clay-rich soils (Knight *et al.*, 2019; Nguyen *et al.*, 2020). For cationic and zwitterionic PFAS compounds with positively charged moieties, the cation-exchange capacity (CEC) of soils governed by the interactions with negatively charged surfaces is very important in controlling mobility.

Effects of Transformation on Mobility

All C atoms in the alkyl chain of perfluorinated compounds are bonded with F. Because the C–F bond is very strong, PFAS compounds are highly resistant to degradation and very persistent in the environment. However, as stated in Section 7.10, polyfluorinated compounds are not fully fluorinated and are therefore susceptible to degradation and transformation in the environment. Carbon atoms bonded to H, O or other elements are locations in the molecular structure of polyfluorinated compounds that can be attacked by microorganisms or by abiotic processes (i.e., hydrolysis, photolysis, oxidation). Thus, many polyfluorinated compounds are termed 'precursors' and over time will transform in the environment to perfluorinated compounds (e.g., PFOA, PFOS). Little is known about the mobility of many precursor compounds in soils, although it is known that they do co-migrate with the perfluorinated compounds. Furthermore, precursors can comprise the dominant pool of PFAS in a contaminated soil and may be cationic or zwitterionic (Adamson *et al.*, 2020). The presence of precursor compounds in samples of soil pore water is often determined using the total oxidizable precursor (TOP) assay. This is achieved by measuring the concentrations of perfluorinated compounds before and after a chemical oxidation procedure that converts the polyfluorinated compounds to perfluorinated carboxylates (Houtz and Sedlak, 2012). If we want to understand the potential for mobility of PFAS compounds in the environment, it is clear that we need to advance our knowledge of the behaviour of these precursor compounds.

Toxicity

PFOS and PFOA are toxic to many organisms at high exposure concentrations, but in general, these compounds are not highly toxic to terrestrial organisms at the levels generally encountered in the environment, as shown in Figure 7C.3 (Hekster *et al.*, 2003; Kwak *et al.*, 2020). Now that the use of PFOS and PFOA is restricted, exposures and body burdens of these compounds appear to be declining for many organisms (Sunderland *et al.*, 2019). However, we know little regarding the toxicity of the many other PFAS compounds in the environment, including compounds that have been introduced as substitutes for PFOS and PFOA, as well as shorter C-chain compounds.

Remediation

Whereas polyfluorinated compounds can be degraded and transformed by biotic or abiotic processes to perfluorinated compounds, these transformation products cannot be readily degraded in soils via bioremediation processes, unlike other alkyl or aryl contaminants, such as petroleum hydrocarbons. The C–F bond is very resistant to degradation, even by aggressive chemical oxidation procedures (e.g., persulphate oxidation). Thermal decomposition is possible, but it only occurs at very high temperatures, thus increasing the cost of remediation considerably. Hence, most commercial soil remediation approaches used so far have focused either on desorption (thermal or solute) with collection of the concentrated PFAS compounds in another medium for final disposal/destruction, or on *in situ* stabilization using sorbents.

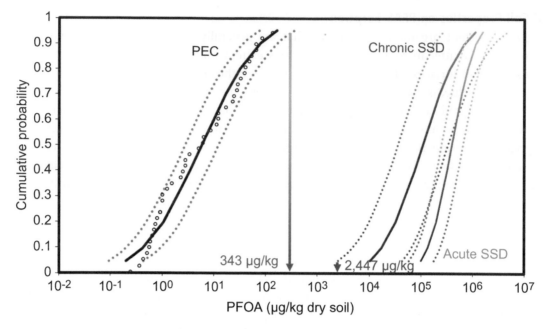

Figure 7C.3 Cumulative probability distribution of predicted (measured) environmental concentrations (PEC) of PFOA in soil and acute/chronic predicted no-effect concentrations (PNEC) derived using a species sensitivity distribution (SSD) approach. All effect concentrations are much higher than measured soil concentrations. Dotted lines are the 95% confidence intervals. These data show that the maximum PEC of 343 µg/kg in soil is an order of magnitude below the lowest PNEC value of 2,447 µg/kg for plants, algae and invertebrates in soil. Reprinted from Kwak *et al.* (2020), with permission from Elsevier.

Desorption methods include the use of high temperatures (i.e., thermal desorption), electrical current (i.e., electrokinetic extraction) or solutes (i.e., soil washing) to remove PFAS compounds from soil. A key feature of all these approaches is that the released PFAS compounds must be extracted, concentrated and trapped. For thermal desorption this is usually by scrubbing of exhaust gases, for electrokinetic desorption this is usually concentration on the anode (for anionic compounds) and for soil washing this is usually by sorption from the washing solution to a solid medium (e.g., activated carbon or resins).

Stabilization of PFAS-contaminated soil is perhaps the most cost-effective remediation option, where the solubility of the compounds is reduced by adding sorbents that bind the contaminants strongly. There has been research on a wide range of sorbents, including activated carbon, resins, modified clays, biochar, graphene and other nanomaterials. However, remediation on a commercial scale usually adopts sorbents that are low cost and readily available, such as modified clays, mixed minerals or activated carbon. A key issue with many sorbents is that they usually target the anionic PFAS compounds, while in reality a multimode remediation material is needed which binds anions, cations, zwitterions and highly hydrophobic compounds that are usually present in PFAS-contaminated soils (Lath *et al.*, 2018). A further issue for regulatory authorities is the longevity of this sorptive remediation approach, as assurance is needed that the immobilization will be robust to changes in environmental conditions over time (Kabiri *et al.*, 2021); this is a subject of ongoing research and regulatory evaluation.

In summary, the group of compounds identified as PFAS comprise perhaps the most diverse range of anthropogenic compounds that contaminate soils, with a wide range of fates and behaviours due to the range of their chemical structures. The key points illustrated by this Capsule are that the C–F bond is extremely resistant to degradation in the soil and these chemicals will therefore persist for decades or centuries. Short C-chain/anionic PFAS compounds are extremely mobile in soils and can move long distances both vertically to groundwaters and laterally to surface waters. While the concentrations of the most common PFAS compounds, PFOA and PFOS, in most soils are below thresholds for acute and chronic toxicity for soil organisms, plants and animals can accumulate these chemicals, potentially posing a risk to wildlife and humans through the food chain. Little is known regarding the mobility, bioavailability and toxicity of most precursor polyfluorinated compounds. Presently there are many options for remediating contaminated soils, but most techniques focus on extraction and/or containment, owing to the high cost and technical complexity of destructive remediation methods.

Capsule References

Adamson, D. T., Nickerson, A., Kulkarni, P. R., *et al.* (2020). Mass-based, field-scale demonstration of PFAS retention within AFFF-associated source areas. *Environmental Science & Technology*, 54, 15768–15777.

Brusseau, M. L. (2018). Assessing the potential contributions of additional retention processes to PFAS retardation in the subsurface. *Science of the Total Environment*, 613, 176–185.

Crone, B. C., Speth, T. F., Wahman, D. G., *et al.* (2019). Occurrence of per- and polyfluoroalkyl substances (PFAS) in source water and their treatment in drinking water. *Critical Reviews in Environmental Science & Technology*, 49, 2359–2396.

Hekster, F. M., Laane, R. W. P. M. & de Voogt, P. (2003). Environmental and toxicity effects of perfluoroalkylated substances. *Reviews of Environmental Contamination and Toxicology*. New York, NY: Springer, 99–121.

Houtz, E. F. & Sedlak, D. L. (2012). Oxidative conversion as a means of detecting precursors to perfluoroalkyl acids in urban runoff. *Environmental Science & Technology*, 46, 9342–9349.

Kabiri, S., Centner, M. & McLaughlin, M. J. (2021). Durability of sorption of per- and polyfluorinated alkyl substances in soils immobilised using common adsorbents: 1. Effects of perturbations in pH. *Science of the Total Environment*, e144857.

Knight, E. R., Janik, L. J., Navarro, D. A., Kookana, R. S. & McLaughlin, M. J. (2019). Predicting partitioning of radiolabelled C-14-PFOA in a range of soils using diffuse reflectance infrared spectroscopy. *Science of the Total Environment*, 686, 505–513.

Kwak, J. I., Lee, T. Y., Seo, H., *et al.* (2020). Ecological risk assessment for perfluorooctanoic acid in soil using a species sensitivity approach. *Journal of Hazardous Materials*, 382, e121150.

Lath, S., Navarro, D. A., Losic, D., Kumar, A. & McLaughlin, M. J. (2018). Sorptive remediation of perfluorooctanoic acid (PFOA) using mixed mineral and graphene/carbon-based materials. *Environmental Chemistry*, 15, 472–480.

Li, Y. S., Oliver, D. P. & Kookana, R. S. (2018). A critical analysis of published data to discern the role of soil and sediment properties in determining sorption of per and polyfluoroalkyl substances (PFASs). *Science of the Total Environment*, 628–629, 110–120.

Nguyen, T. M. H., Bräunig, J., Thompson, K., *et al.* (2020). Influences of chemical properties, soil properties, and solution pH on soil–water partitioning coefficients of per- and polyfluoroalkyl substances (PFASs). *Environmental Science & Technology*, 54, 15883–15892.

Oliaei, F., Kriens, D., Weber, R. & Watson, A. (2013). PFOS and PFC releases and associated pollution from a PFC production plant in Minnesota (USA). *Environmental Science and Pollution Research*, 20, 1977–1992.

Schroeder, T., Bond, D. & Foley, J. (2021). PFAS soil and groundwater contamination via industrial airborne emission and land deposition in SW Vermont and Eastern New York State, USA. *Environmental Science: Processes and Impacts*, 23, 291–301.

Sunderland, E. M., Hu, X. D. C., Dassuncao, C., *et al.* (2019). A review of the pathways of human exposure to poly- and perfluoroalkyl substances (PFASs) and present understanding of health effects. *Journal of Exposure Science and Environmental Epidemiology*, 29, 131–147.

Wang, Z., DeWitt, J. C., Higgins, C. P. & Cousins, I. T. (2017). A never-ending story of per- and polyfluoroalkyl substances (PFASs)? *Environmental Science & Technology*, 51, 2508–2518.

Zhang, G., Pan, Z., Wu, Y., *et al.* (2019). Distribution of perfluorinated compounds in surface water and soil in partial areas of Shandong Province, China. *Soil and Sediment Contamination*, 28, 502–512.

7.11 Plasticizers

Plasticizers are defined broadly as chemicals that are added to manufactured plastic polymers to reduce 'wear and tear' or to alter their physical properties, such as making them more flexible or reducing viscosity. Only approximately 50 plasticizers are in commercial use, but the global market is over 8 million tonnes. Plasticizers function to increase the longevity of plastics, protect against ultraviolet radiation, and improve flexibility and tensile strength. They are widely added to polyvinyl chloride (PVC) pipes to make them more malleable and flexible. Other products that incorporate plasticizers include synthetic rubber, plastic toys, food wrap and storage containers, electrical cables, adhesives and product coatings. Plasticizers can be incorporated directly into the polymers (internal) or they can be added as a coating (external). Although external applications facilitate broader use, they also result in a weaker physiochemical interaction with the product. Plasticizers that are not bonded covalently with polymers can readily leach into aquatic and terrestrial environments.

Phthalates are the most widely used plasticizers on a global scale. These esters of phthalic acid were first introduced into the marketplace in the 1930s. They are substituted with esters having different structural moieties (e.g., methyl, benzyl, pentyl alcohols, etc.). Some commonly used phthalates include benzyl butyl phthalate, dibutyl phthalate, diethyl phthalate, iso-octyl phthalate and di-n-octyl phthalate. **Di(2-ethylhexyl)phthalate** or DEHP is the dominant phthalate used in PVC piping (Figure 7.30). There was some initial debate within chemical regulatory agencies as to whether phthalates were carcinogenic, but in the 1990s, these compounds were designated as chemicals with low carcinogenic potential. More recently, it has become clear that these chemicals induce other biological responses, including endocrine disruption (see Chapter 8). Phthalates are metabolized relatively rapidly, but the mono-esters that are the primary metabolites appear to persist in organisms for longer periods. Figure 7.30 illustrates MEHP, which is the mono-ester metabolite of DEHP.

Chlorinated paraffins, also known as polychlorinated n-alkanes, consist of alkyl chains of 10–30 carbon atoms that are chlorinated in proportions of between 30% and 70% by mass. These compounds are classified into three categories based on the length of the carbon chain: short-chain chlorinated paraffins (SCCPs; C_{10}–C_{13}), medium-chain chlorinated paraffins (C_{14}–C_{17}), and long-chain chlorinated paraffins (C_{18}–C_{30}). Owing to their thermal and chemical stability, SCCPs have been widely used as additives to plastic polymers, but they are also used in coatings, sealants, adhesives and rubber. Because of their persistence, potential for bioaccumulation and toxicity, SCCPs were added in 2017 as an annex to the list of POPs to be regulated under the Stockholm Convention, and these compounds are prohibited chemicals in the European Union, Canada and Japan. The US EPA has restricted the importation of SCCPs and banned production in the United States. However, China and some other countries produce and market large quantities of SCCPs. The global production of these compounds has increased yearly to over 165,000 tonnes per year.

As China is the largest producer of SCCPs, these compounds are widely detected in soils, sediments, food, water, biota and human breast milk in that country, as reviewed by Zheng *et al.* (2020). Emission of SCCPs into

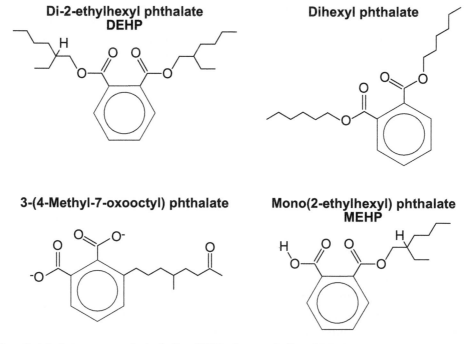

Figure 7.30 Examples of phthalate compounds, including MEHP, the metabolite of DEHP.

the atmosphere, primarily from the industrialized eastern parts of China, contribute to the global transport of these chemicals. The concentrations of SCCPs detected in the Yangtze River and some other surface waters in China are in the µg/l range, which corresponds with the concentrations that are toxic to aquatic organisms (Zheng *et al.*, 2020). As indicated by the recent addition of SCCPs to the Stockholm Convention, there is an urgent need to reduce the global production and use of these compounds.

7.12 Pharmaceutically Active Compounds

Pharmaceuticals and personal care products that enter the environment include drugs and their metabolites, known as **pharmaceutically active compounds** (PhACs), which are excreted by humans, discarded as unused medicines, or excreted by domestic animals treated with veterinary medicines (Figure 7.31). Pharmacokinetic data indicate that human excretion of unchanged drugs usually exceeds 50% of the daily dose. PhACs of human origin that enter domestic wastewater treatment systems may or may not be removed effectively, depending on the treatment system. Another source of PhACs is waste from

drug manufacturing operations, and this source is especially problematic in countries where environmental regulations or enforcement are lax (Arnold *et al.*, 2013).

In wastewater treatment plants, some drugs are removed effectively by sewage treatment technologies, but some compounds, such as the anti-epileptic drug carbamazepine, pass through the wastewater treatment process with essentially no removal. Some drugs partition into sewage sludge and can enter the environment when the treated sludge (i.e., biosolids) is deposited in landfills or applied onto agricultural land for soil amendment. Contamination of agricultural soils is also possible through the application of manure from animals (e.g., chickens, pigs, cattle) treated with veterinary medicines. Many of the animals that are raised under crowded industrial farming conditions are fed medicated feeds containing antibiotics or antiparasitic drugs for prophylactic treatment of bacterial diseases or parasitic infections. In marine coastal waters, fish farms are sources of veterinary medicines that are used as antibiotics and antiparasitics, especially for compounds used to combat sea lice in salmon aquaculture pens.

Once PhACs make their way into the aquatic environment, they may affect aquatic organisms. Howard and

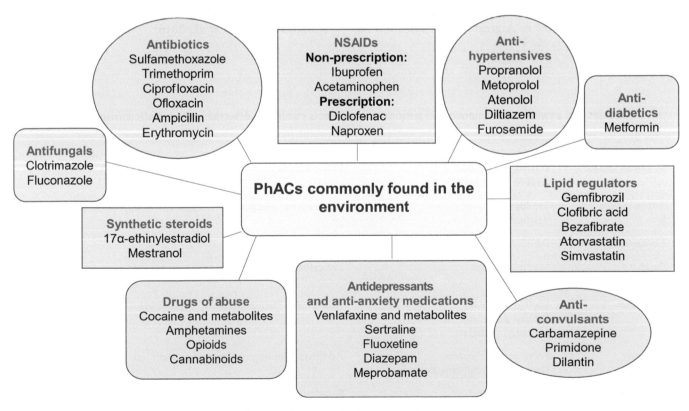

Figure 7.31 Pharmaceutically active compounds (PhACs) commonly detected in the environment.

Muir (2011) listed 275 pharmaceutical products that have been detected in aquatic organisms. However, the majority of PhACs are hydrophilic and non-persistent relative to legacy contaminants such as PCBs and organochlorine pesticides, so they do not bioaccumulate in aquatic organisms. Because of the sheer volume of wastewater that is discharged daily from municipal wastewater treatment plants, these compounds are often referred to as 'pseudo-persistent' because they are ubiquitous in the aquatic environment at ng/l concentrations in locations contaminated by discharges of domestic wastewater (Cizmas *et al.*, 2015).

Several PhACs act on the central nervous system and have been implicated in altered behaviour in exposed fish and wildlife. A focus of recent attention has been the effects of antidepressants, including **selective serotonin reuptake inhibitors** (SSRIs), which modulate the neurotransmitters serotonin, dopamine and norepinephrine. These molecular targets are evolutionarily conserved in many aquatic species and modulate biological processes such as reproduction, growth, metabolism, immunity, feeding, locomotion, colour physiology and behaviour. Effects at environmentally relevant concentrations of antidepressants have been observed in fish, molluscs and other aquatic invertebrates (Mezzelani *et al.*, 2018). PhACs also have the potential for endocrine disruption, as described in Chapter 8. A great deal of attention has been focused on the potential endocrine effects in the aquatic environment of 17α-ethinyloestradiol (EE2), one of the active ingredients in the birth control pill.

In addition to PhACs, wastewater treatment plants are also sources of contamination from chemicals that are included in various personal care products, including fragrances, antibacterial compounds, sunscreens and insect repellents (Table 7.2). Several of these compounds may be endocrine disrupting chemicals, as discussed in Chapter 8. Some of these compounds are relatively hydrophobic (i.e., log $K_{OW} > 4$) and have the potential to bioaccumulate in aquatic organisms. Examples include the synthetic musk compounds HHCB (1,3,4,6,7,8-hexahydro-4,6,6,7,8-hexamethylcyclopenta-γ-2-benzopyran, 7-acetyl-1,1,3,4,4,6-hexamethyl-1,2,3,4-tetrahydronaphthalene) and the antibacterial compound triclosan (Section 7.2; Figure 7.8).

A prime example of effects of PhACs on wildlife in terrestrial ecosystems is the collapse of vulture populations in several South Asian countries, caused by the ingestion of the **non-steroidal anti-inflammatory drug** (NSAID) diclofenac (see Case Study 7.2). Government action to limit the use of diclofenac has resulted in the recovery of many of these vulture populations. However, there is still concern about exposure of vultures to diclofenac obtained illegally or about exposure to other NSAIDs (Cuthbert *et al.*, 2016).

Table 7.2 **Classes and examples of compounds in personal care products commonly detected in the aquatic environment.**

Class	Compounds	Example products
Synthetic musks	HHCB (Galaxolide®), AHTN (Tonalide®)	Fragrances in perfumes, deodorant, etc.
Antibacterials	Triclosan Triclocarban	Antibacterial compounds in soaps, toothpaste, surface cleaners, etc.
Toluamides	DEET *N,N*-diethyl-meta-toluamide	Insect repellent
Parabens	Methyl paraben Propyl paraben	Preservatives in cosmetics, etc.
Phthalates	Diethyl phthalate Dibutyl phthalate Dimethyl phthalate	Additives to shampoo, hair spray, cosmetics, etc.
Siloxanes	Polydimethylsiloxane Decamethylcyclopentasiloxane	Additives to skin emollients, sunscreens, etc.
Benzophenones	Benzophenone-3 Oxybenzone	Sunscreens

Case Study 7.2 Decline of Populations of *Gyps* Vultures in South Asia

Beginning in the mid-1990s, there was a catastrophic decline in the numbers of vultures of the *Gyps* genera in South Asian countries, including India, Pakistan and Nepal. Populations of the white-rumped vulture (*Gyps bengalensis*), Indian vulture (*G. indicus*) and slender-billed vulture (*G. teniurostris*) had declined to such low numbers that they were all added to the list of Critically Endangered Species by the International Union for the Conservation of Nature (IUCN). Necropsies showed that the birds all had 'visceral gout', characterized by accumulation of uric acid in the viscera, and that all birds had died from kidney failure caused by the build-up of uric acid. A multinational investigation to determine the cause of this pathology eventually linked the condition to high concentrations in the bird tissues of the NSAID diclofenac. Birds were ingesting lethal doses of the drug by feeding on the carcasses of cattle that had been given the drug as a veterinary therapy. Treatment of cattle with anti-inflammatories is a common practice in South Asian countries where cattle are used as beasts of burden, to maintain the working life of the animals. When the animals die, the carcasses are typically left in the field or are deposited in bone yards, where vultures feed on them, thereby ingesting the drug. Following the discovery that diclofenac was causing these mortalities, the governments of India, Pakistan and Nepal banned the sale of this drug for veterinary use in 2006. Figure 7.32 shows that this resulted in at least stabilization of the populations of some of these vulture species by 2015.

However, other NSAIDs may also induce visceral gout in vultures, including Nimesulide (Cuthbert *et al.*, 2016). In addition, other bird species that feed on carrion but are not from the *Gyps* genus may also be under threat from exposure to NSAID residues in cattle. There appears to be a novel metabolic pathway in sensitive species of birds involving the Phase I pathway catalysed by cytochrome B that results in uric acid build-up. The important message from this case study is that we cannot assume that drugs that are safe for use in humans or in domestic animals are also safe for other species that co-exist in our environment, particularly because there is no control over exposure or dose.

Another line of investigation of PhACs concerns the potential effects of antibiotics released into the environment on the global increase in **antimicrobial resistance**

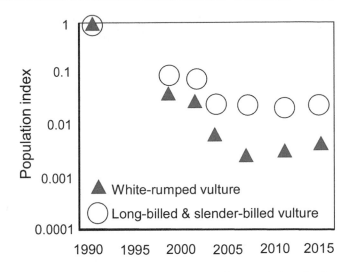

Figure 7.32 Population index relative to the population size in 1992 for the white-rumped vulture (filled triangles) and combined long-billed and slender-billed vultures (open circles) in India. Adapted from data presented by Prakash *et al.* (2019) with the permission of Cambridge University Press.

(AMR). One of the environmental factors contributing to the increase in AMR is the widespread use of antibiotics for human therapy or as veterinary medicines. Exposure to these antibiotics provides selection pressure for the AMR genes in bacteria (Singer *et al.*, 2016). Worldwide, approximately 73% of total antibiotic production is for agricultural use, and it has been estimated that 30% to 90% of antibiotics are excreted in urine from domestic animals as biologically active substances. The discharge of antibiotics in municipal wastewater may contribute to antibiotic resistance in bacteria entering drinking water treatment plants (Bergeron *et al.*, 2015). There is now evidence for the spread of AMR among wildlife populations living in proximity to human habitation, but also migratory bird species (Arnold *et al.*, 2016). Because of growing concern over the development of AMR, the use of antibiotics for prophylactic use in medicated feeds has been banned in European countries.

One class of compounds that has attracted public attention is sunscreens, because of the potential for these compounds to negatively impact corals and other marine species. **Benzophenone** compounds, including benzophenone-3 (BP-3) and oxybenzone, are often used as active ingredients in sunscreen lotions, as well as in many personal care products. Because these compounds wash off when sunbathers enter the water, benzophenones have been detected in coastal waters near tourist beaches. In addition, benzophenones are present in discharges from wastewater treatment plants. These compounds have been associated with a variety of negative effects in marine organisms, including effects on corals. Danovaro *et al.* (2008) first reported that direct exposure of a stony coral species (*Acropora* spp.) to sunscreens, including a BP-3 based product, increased viral infections and resulted in coral bleaching and loss of diversity of all forms of marine life associated with corals. Several subsequent studies have shown that benzophenone compounds are toxic to the polyps and the planulae (i.e., larval stages) of corals, as well as other marine species (Sanchez-Quiles and Tovar-Sanchez, 2015). Concerns regarding the adverse impacts of exposure on coral reefs and other marine species have led to the banning of sunscreen products in some managed marine areas and, most recently, the banning of these products in Hawaii and Florida.

7.13 Toxicovigilance

New organic chemicals are being developed and patented every day, so vigilance is needed to ensure that these products do not cause harm to the environment. Table 7.3 illustrates the basic dataset that is usually provided by industry to government agencies, so that regulators can conduct a preliminary (i.e., Tier I) ecological risk assessment. Until recently, toxicity data for aquatic and terrestrial organisms were typically derived from 24-h to 96-h acute lethality tests. Information on sublethal

Table 7.3 **Example of a base dataset provided by industry to regulatory agencies for Tier I ecological risk assessments of new chemicals before they are approved to enter the marketplace (see Section 12.3.2.2).**

Physical-chemical studies	Environmental fate studies
• Water solubility	• Soil adsorption coefficient (K_{OC})
• Dissociation constant (pK_a)	• Degradation in soil ($t_{1/2}$)
• Ultraviolet–visible adsorption spectrum	• Degradation in water ($t_{1/2}$)
• Melting temperature	• Photolysis (optional)
• Vapour pressure	• Hydrolysis (optional)
• Octanol–water partition coefficient (K_{OW})	

Aquatic effects studies	Terrestrial effects studies
Acute toxicity	Acute toxicity
• Algae	• Microorganisms
• Aquatic invertebrate (e.g. *Daphnia*)	• Terrestrial plants
• Fish	• Soil invertebrates (e.g. earthworm)

responses that may occur after chronic exposures over days or weeks (e.g., reduced reproduction) is not typically provided, unless these tests are specifically requested by the regulatory agency. The fate in the environment and potential for bioaccumulation must be predicted from the data generated by physical-chemical studies and environmental fate studies. This process is supported by further refinements to analytical methods and instrumentation to better detect ambient chemicals in environmental media. 'Non-targeted' analytical approaches with high-resolution mass spectrometry are now being used to screen for chemicals that we previously may not have suspected were present in the environment (Section 14.4). Advances may also include adjustments to the paradigms that have governed our screening of chemicals for potential hazards. For instance, up until recently, bioaccumulation was thought to only occur for highly lipophilic compounds, such as PCBs and DDT. However, the discovery that proteinophilic compounds such as long-chain PFCs can bioaccumulate has challenged that paradigm, and we now need models to predict whether there are other chemicals that can accumulate in biota through this mechanism (Section 3.6.2.3). In other cases, such as with BDEs, it was determined that metabolites cause sublethal effects that are not detected by acute screening studies of the parent compound. Where adverse effects of specific chemicals have been demonstrated, it is also important to pay attention to their substitutes. As can be seen from earlier discussion points, a frequent response by industry to chemical regulations is to switch to chemicals that are structural analogues or from the same chemical class that may or may not have similar ecotoxicological effects. As also pointed out in Section 12.2.2, countries around the world are not uniform in their response to the bans on the manufacture and use of chemicals. In the meantime, new classes of chemicals are making their way into commerce without adequate regulatory attention. For instance, there has been a rapid increase in the production and use of 'ionic liquids' as candidate compounds to replace volatile organic solvents as 'green solvents' for use in chemical synthesis, biotechnology, chemical engineering and other applications. These compounds have high thermal and chemical stability and are highly soluble in both water and lipids (i.e., amphiphilic); they resemble PFACs in their potential for widespread contamination of the environment (Oskarsson and Wright, 2019). Such concerns represent a major challenge for regulatory agencies.

Summary

In this chapter, we discuss how the physicochemical characteristics of organic compounds affect their mobility and persistence in the environment, and the climatological and geological processes that influence their partitioning in the geosphere, hydrosphere and atmosphere, with important implications for their bioavailability and bioaccumulation. The chapter also includes an account of the MoA and metabolism of organic compounds, with reference to specific molecular receptors and excretory pathways. The chapter describes the chemical nature and uses of a variety of industrial and agricultural compounds, including 'legacy contaminants' that are still a hazard today. We also review some of the new classes of industrial chemicals that are recognized as hazardous to the environment, including flame retardants and plasticizers. The endocrine disrupting properties of some of these compounds are explored in more detail in Chapter 8. We also consider by-products that are inadvertently released into the environment as a result of the manufacturing process, product usage and end-of-use disposal. In some cases, we note that product development involves the refinement of natural compounds, such as the distillation of petroleum hydrocarbons to produce fuels and the production of 'coke' from coal for use in the steel industry. These processes can result in contamination of the environment. Chapter 9 extends this topic to include an overview of natural toxins and their adaptation for human use, as well as a consideration of the many naturally occurring toxic chemicals that impact ecosystems with and without the influence of humans.

Review Questions and Exercises

1. What is the difference between polycyclic aromatic hydrocarbons and polycyclic aromatic compounds?

2. Give an account of the principal factors affecting the mobility of organic chemicals in the environment.

3. What distinguishes pyrogenic from petrogenic hydrocarbons?

4. Describe the most important factors affecting the bioaccumulation and toxicity of halogenated aromatic hydrocarbon (HAH) compounds, giving specific examples.

5. The exposure of vertebrate species to some PAHs induces CYP1A enzyme activity. Under what circumstances is this a benefit to an organism or a problem?

6. Describe the history, usage, ecotoxicology and regulation of the following compounds: (a) DDT, (b) PCBs, (c) perfluoroalkyl compounds.

7. Give an account of the chemistry and usage of brominated flame retardants, with an assessment of their environmental toxicity and regulation.

8. Write an essay on the structure, mode of action and environmental toxicology of the following herbicides: (a) glyphosate, (b) chlorophenoxyacids, (c) atrazine.

9. Classroom discussion: What are the characteristics of an 'ideal' pesticide, considering its intended use and the need to protect the environment?

10. Describe the cellular components and catabolic activity of the CYP450 system. Give examples of some common metabolic reactions carried out by this enzyme system.

11. When a vertebrate is exposed to a mixture of PAHs or PACs and CYP1A enzymes are induced, is it only the inducing compound that is subsequently oxygenated? Are endogenous hydrocarbons (e.g., steroid hormones) susceptible to oxygenation? If so, what are the potential consequences?

12. Discuss how our understanding of the environmental threat of organic chemicals has evolved in recent decades. Explain how these changes have affected the regulation of these compounds.

Abbreviations

2,4-D	2,4-dichlorophenoxy acetic acid	HSP90	Heat shock protein 90
2,4,5-T	2,4,5-trichlorophenoxyacetic acid	IUCN	International Union for the Conservation of Nature
AhR	Aryl hydrocarbon receptor		
AHTN	7-acetyl-1,1,3,4,4,6-hexamethyl-1,2,3,4-tetrahydronaphthalene	Keap-1	Kelch-like ECH-associated protein 1
AMPA	Aminomethylphosphonic acid	K_{OC}	Sediment (soil)–water partition coefficient
AMR	Antimicrobial resistance		
AR	Androgen receptor	K_{OW}	Octanol–water partition coefficient
ARNT	Aryl hydrocarbon receptor nuclear translocating protein	LBD	Ligand-binding domain (of receptor)
BAF	Bioaccumulation factor	MoA	Mode of action
BaP	Benzo(a)pyrene	MRPs	Multidrug resistance proteins (e.g., MRP1, MRP2, MRP3)
BCF	Bioconcentration factor		
BDEs	Brominated diphenyl ethers	nAChR	Nicotinic acetylcholine receptor
BFRs	Brominated flame retardants	NAD	Nicotinamide adenine dinucleotide
BRIC	Acronym for Brazil, Russia, India, China group of countries	NAD(H)	Nicotinamide adenine dinucleotide (reduced form)
BTEX	Compounds benzene, toluene, ethylbenzene and xylene	NADP(H)	Nicotinamide adenine dinucleotide phosphate (reduced form)
CAR	Constitutive androstane receptor	NAPL	Non-aqueous phase liquid
CAS Registry®	Chemical Abstracts Service Registry (American Chemical Society)	NQO1	NADPH:quinone acceptor oxidoreductase 1
CUPs	Current use pesticides	Nrf2	Nuclear factor erythroid 2-related factor 2
CYP450	Cytochrome P450		
DBD	DNA binding domain (of receptor)	NSAID	Non-steroidal anti-inflammatory drug
DBDPE	Decabromodiphenylethane		
DDE	Dichlorodiphenyldichloroethylene (IUPAC ID: 1,1′-(2,2,2-dichloroethylene-1,1-diyl) bis(4-chlorobenzene)	NTD	Amino (NH2)-terminal transactivation domain (of receptor)
		OPFRs	Organophosphate flame retardants
DDT	Dichlorodiphenyltrichloroethane (IUPAC ID: 1,1′-(2,2,2-trichloroethane-1,1-diyl) bis(4-chlorobenzene)	PACs	Polycyclic aromatic compounds
		PAHs	Polycyclic aromatic hydrocarbons
		PCBs	polychlorinated biphenyls
		PCDDs	Polychlorinated dibenzodioxins
		PCDFs	Polychlorinated dibenzofurans
ER	[o]estrogen receptor	PFCs (PFACs)	Perfluoroalkyl compounds
FMOs	Flavoprotein monooxygenases	PFOA	Perfluorooctanoic acid
FTOHs	Fluorotelomer alcohols	PFOS	Perfluorooctane sulphonic acid
GSH	Glutathione	pK_a	Acid dissociation constant
GST	Glutathione transferase	POPs	Persistent organic pollutants
GUS	Groundwater ubiquity score	POSF	Perfluorooctane sulphonyl fluoride
HBCD	Hexabromocyclododecane	PPAR	Peroxisome proliferator-activated receptor
HCB	Hexachlorobenzene		
HCH	Hexachlorocyclohexane	PVC	Polyvinyl chloride
HHCB	1,3,4,6,7,8-hexahydro-4,6,6,7,8-hexamethylcyclopenta-γ-2-benzopyran	PXR	Pregnane X receptor
		RAR	Retinoic acid receptor
		RXR	Retinoid X receptor

SARA	Acronym describing crude oil constituents, saturates, aromatics, resins and asphaltenes	TEPP	Tetraethyl-pyrophosphate
		TR	Thyroid receptor
		UDP	Uridine diphosphate
SSRIs	Selective serotonin reuptake inhibitors	UDPGA	Uridine diphosphate glucuronic acid
SULT	Sulfotransferase	UGT	UDP glucuronosyltransferase
$t_{1/2}$	Chemical half-life in the environment	UGT	Uridine 5′-diphospho-glucuronosyltransferase (UDP-glucuronosyltransferase)
TBBPA	Tetrabromobisphenol A		
TCDD	2,3,7,8-tetrachlorodibenzo-p-dioxin		
TEF	Toxic equivalent factor	XRE	Xenobiotic regulatory element

References

Addison, R. F., Muir, D. C. G., Ikonomou, M. G., et al. (2020). Temporal trends in polybrominated diphenylethers (PBDEs) in blubber of ringed seals (*Pusa hispida*) from Ulukhaktok, NT, Canada between 1981 and 2015. *Archives of Environmental Contamination and Toxicology*, 79, 167–176.

Andersson, J. T. & Achten, C. (2015). Time to say goodbye to the 16 EPA PAHs? Toward an up-to-date use of PACs for environmental purposes. *Polycyclic Aromatic Compounds*, 35, 330–354.

Antwi, F. B. & Reddy, G. V. P. (2015). Toxicological effects of pyrethroids on non-target aquatic insects. *Environmental Toxicology and Pharmacology*, 40, 915–923.

Arnold, K. E., Boxall, A. B. A., Brown, A. R., et al. (2013). Assessing the exposure risk and impacts of pharmaceuticals in the environment on individuals and ecosystems. *Biology Letters*, 9, e20130492.

Arnold, K. E., Williams, N. J. & Bennett, M. (2016). 'Disperse abroad in the land': the role of wildlife in the dissemination of antimicrobial resistance. *Biology Letters*, 12, e20160137.

Bergeron, S., Boopathy, R., Nathaniel, R., Corbin, A. & LaFleur, G. (2015). Presence of antibiotic resistant bacteria and antibiotic resistance genes in raw source water and treated drinking water. *International Biodeterioration and Biodegradation*, 102, 370–374.

Bidleman, T. F., Backus, S., Dove, A., et al. (2021). Lake Superior has lost over 90% of its pesticide HCH load since 1986. *Environmental Science & Technology*, 55, 9518–9526.

Brown, K. E., King, C. K. & Harrison, P. L. (2017). Lethal and behavioral impacts of diesel and fuel oil on the Antarctic amphipod *Paramoera walkeri*. *Environmental Toxicology and Chemistry*, 36, 2444–2455.

Carlson, D. L., De Vault, D. S. & Swackhamer, D. L. (2010). On the rate of decline of persistent organic contaminants in lake trout (*Salvelinus namaycush*) from the Great Lakes, 1970–2003. *Environmental Science & Technology*, 44, 2004–2010.

Cizmas, L., Sharma, V. K., Gray, C. M. & McDonald, T. J. (2015). Pharmaceuticals and personal care products in waters: occurrence, toxicity, and risk. *Environmental Chemistry Letters*, 13, 381–394.

Cousin, M., Silva-Zacarin, E., Kretzschmar, A., et al. (2013). Size changes in honey bee larvae oenocytes induced by exposure to paraquat at very low concentrations. *PLoS ONE*, 8, e65683.

Covaci, A., Harrad, S., Abdallah, M. A. E., et al. (2011). Novel brominated flame retardants: a review of their analysis, environmental fate and behaviour. *Environment International*, 37, 532–556.

Cuthbert, R. J., Taggart, M. A., Saini, M., et al. (2016). Continuing mortality of vultures in India associated with illegal veterinary use of diclofenac and a potential threat from nimesulide. *Oryx*, 50, 104–112.

Danovaro, R., Bongiorni, L., Corinaldesi, C., et al. (2008). Sunscreens cause coral bleaching by promoting viral infections. *Environmental Health Perspectives*, 116, 441–447.

Dietz, R., Riget, F. F., Sonne, C., et al. (2013). Three decades (1983–2010) of contaminant trends in East Greenland polar bears (*Ursus maritimus*). Part 2: Brominated flame retardants. *Environment International*, 59, 494–500.

Dixon, D. P., Skipsey, M. & Edwards, R. (2010). Roles for glutathione transferases in plant secondary metabolism. *Phytochemistry*, 71, 338–350.

Gandhi, N., Tang, R. W. K., Bhavsar, S. P., *et al.* (2015). Is Mirex still a contaminant of concern for the North American Great Lakes? *Journal of Great Lakes Research*, 41, 1114–1122.

Giesy, J. P. & Kannan, K. (2001). Global distribution of perfluorooctane sulfonate in wildlife. *Environmental Science & Technology*, 35, 1339–1342.

Goldstein, M. I., Lacher, T. E., Woodbridge, B., *et al.* (1999). Monocrotophos-induced mass mortality of Swainson's hawks in Argentina, 1995–96. *Ecotoxicology*, 8, 201–214.

Greaves, A. K., Letcher, R. J., Chen, D., *et al.* (2016). Retrospective analysis of organophosphate flame retardants in herring gull eggs and relation to the aquatic food web in the Laurentian Great Lakes of North America. *Environmental Research*, 150, 255–263.

Guo, G. H., Wu, F. C., He, H. P., *et al.* (2012). Distribution characteristics and ecological risk assessment of PAHs in surface waters of China. *Science China-Earth Sciences*, 55, 914–925.

Harner, T., Rauert, C., Muir, D., *et al.* (2018). Air synthesis review: polycyclic aromatic compounds in the oil sands region. *Environmental Reviews*, 26, 430–468.

Hayes, T. B., Anderson, L. L., Beasley, V. R., *et al.* (2011). Demasculinization and feminization of male gonads by atrazine: Consistent effects across vertebrate classes. *Journal of Steroid Biochemistry and Molecular Biology*, 127, 64–73.

Hermanson, M. H., Isaksson, E., Hann, R., Teixeira, C. & Muir, D. C. G. (2020). Deposition of polychlorinated biphenyls to firn and ice cores at opposite polar sites: Site M, Dronning Maud Land, Antarctica, and Holtedahlfonna, Svalbard. *ACS Earth and Space Chemistry*, 4, 2096–2104.

Hodson, P. V. (2017). The toxicity to fish embryos of PAH in crude and refined oils. *Archives of Environmental Contamination and Toxicology*, 73, 12–18.

Hodson, P. V., Qureshi, K., Noble, C. A. J., Akhtar, P. & Brown, R. S. (2007). Inhibition of CYP1A enzymes by alpha-naphthoflavone causes both synergism and antagonism of retene toxicity to rainbow trout (*Oncorhynchus mykiss*). *Aquatic Toxicology*, 81, 275–285.

Houde, M., De Silva, A. O., Muir, D. C. G. & Letcher, R. J. (2011). Monitoring of perfluorinated compounds in aquatic biota: an updated review. PFCs in aquatic biota. *Environmental Science & Technology*, 45, 7962–7973.

Howard, P. H. & Muir, D. C. G. (2011). Identifying new persistent and bioaccumulative organics among chemicals in commerce II: Pharmaceuticals. *Environmental Science & Technology*, 45, 6938–6946.

Kelly, B. C., Ikonomou, M. G., Blair, J. D., *et al.* (2009). Perfluoroalkyl contaminants in an Arctic marine food web: trophic magnification and wildlife exposure. *Environmental Science & Technology*, 43, 4037–4043.

Langlois, V. S., Carew, A. C., Pauli, B. D., *et al.* (2010). Low levels of the herbicide Atrazine alter sex ratios and reduce metamorphic success in *Rana pipiens* tadpoles raised in outdoor mesocosms. *Environmental Health Perspectives*, 118, 552–557.

Lee, K., Boufadel, M., Chen, B., *et al.* (2015). *Expert Panel Report on the Behaviour and Environmental Impacts of Crude Oil Released into Aqueous Environments*. Ottawa, ON: Royal Society of Canada.

Lee, K. T., Hong, S., Lee, J. S., *et al.* (2013). Revised relative potency values for PCDDs, PCDFs, and non-ortho-substituted PCBs for the optimized H4IIE-luc in vitro bioassay. *Environmental Science and Pollution Research*, 20, 8590–8599.

Ma, J. (2010). Atmospheric transport of persistent semi-volatile organic chemicals to the Arctic and cold condensation in the mid-troposphere – Part 1: 2-D modeling in mean atmosphere. *Atmospheric Chemistry and Physics*, 10, 7303–7314.

Maher, J. M., Dieter, M. Z., Aleksunes, L. M., *et al.* (2007). Oxidative and electrophilic stress induces multidrug resistance-associated protein transporters via the nuclear factor-E2-related factor-2 transcriptional pathway. *Hepatology*, 46, 1597–1610.

Meier, J. R., Lazorchak, J. M., Mills, M., Wernsing, P. & Baumann, P. C. (2015). Monitoring exposure of brown bullheads and benthic macroinvertebrates to sediment contaminants in the Ashtabula River before, during, and after remediation. *Environmental Toxicology and Chemistry*, 34, 1267–1276.

Mezzelani, M., Gorbi, S. & Regoli, F. (2018). Pharmaceuticals in the aquatic environments: evidence of emerged threat and future challenges for marine organisms. *Marine Environmental Research*, 140, 41–60.

Milani, D. & Grapentine, L. C. (2016). Prioritization of sites for sediment remedial action at Randle Reef, Hamilton Harbour. *Aquatic Ecosystem Health and Management*, 19, 150–160.

Miller, A., Elliott, J. E., Elliott, K. H., *et al.* (2014). Spatial and temporal trends in brominated flame retardants in seabirds from the Pacific coast of Canada. *Environmental Pollution*, 195, 48–55.

Miner, K. R., Campbell, S., Gerbi, C., *et al.* (2018). Organochlorine pollutants within a polythermal glacier in the interior eastern Alaska range. *Water*, 10, e1157.

Morant, M., Bak, S., Moller, B. L. & Werck-Reichhart, D. (2003). Plant cytochromes P450: tools for pharmacology, plant protection and phytoremediation. *Current Opinion in Biotechnology*, 14, 151–162.

Ng, C. A. & Hungerbuhler, K. (2013). Bioconcentration of perfluorinated alkyl acids: how important is specific binding? *Environmental Science & Technology*, 47, 7214–7223.

Oskarsson, A. & Wright, M. C. (2019). Ionic liquids: New emerging pollutants, similarities with perfluorinated alkyl substances (PFASs). *Environmental Science & Technology*, 53, 10539–10541.

Prakash, V., Galligan, T. H., Chakraborty, S. S., *et al.* (2019). Recent changes in populations of critically endangered *Gyps* vultures in India. *Bird Conservation International*, 29, 55–70.

Rafferty, S. D., Blazer, V. S., Pinkney, A. E., *et al.* (2009). A historical perspective on the "fish tumors or other deformities" Beneficial Use Impairment at Great Lakes Areas of Concern. *Journal of Great Lakes Research*, 35, 496–506.

Rewitz, K. F., Styrishave, B., Lobner-Olesen, A. & Andersen, O. (2006). Marine invertebrate cytochrome P450: emerging insights from vertebrate and insect analogies. *Comparative Biochemistry and Physiology C – Toxicology and Pharmacology*, 143, 363–381.

Salazar-Lugo, R., Mata, C., Oliveros, A., *et al.* (2011). Histopathological changes in gill, liver and kidney of neotropical fish *Colossoma macropomum* exposed to paraquat at different temperatures. *Environmental Toxicology and Pharmacology*, 31, 490–495.

Sanchez-Quiles, D. & Tovar-Sanchez, A. (2015). Are sunscreens a new environmental risk associated with coastal tourism? *Environment International*, 83, 158–170.

Schlenker, L. S., Welch, M. J., Meredith, T. L., *et al.* (2019). Damsels in distress: oil exposure modifies behavior and olfaction in bicolor damselfish (*Stegastes partitus*). *Environmental Science & Technology*, 53, 10993–11001.

Seuront, L. (2011). Hydrocarbon contamination decreases mating success in a marine planktonic copepod. *Plos ONE*, 6, e26283.

Singer, A. C., Shaw, H., Rhodes, V. & Hart, A. (2016). Review of antimicrobial resistance in the environment and its relevance to environmental regulators. *Frontiers in Microbiology*, 7, e1728.

US EPA (2009). *Risks of Paraquat Use to Federally Threatened California Red-legged Frog (*Rana aurora draytonii*).* Washington, DC: Office of Pesticide Programs, Environmental Fate and Effects Division, US Environmental Protection Agency.

Van den Berg, M., Birnbaum, L., Bosveld, A. T. C., *et al.* (1998). Toxic equivalency factors (TEFs) for PCBs, PCDDs, PCDFs for humans and wildlife. *Environmental Health Perspectives*, 106, 775–792.

Walters, D. M., Jardine, T. D., Cade, B. S., *et al.* (2016). Trophic magnification of organic chemicals: a global synthesis. *Environmental Science & Technology*, 50, 4650–4658.

Wassenberg, D. M. & Di Giulio, R. T. (2004). Synergistic embryotoxicity of polycyclic aromatic hydrocarbon aryl hydrocarbon receptor agonists with cytochrome P4501A inhibitors in *Fundulus heteroclitus*. *Environmental Health Perspectives*, 112, 1658–1664.

Wright, D. A. & Welbourn, P. M. (2002). *Environmental Toxicology*. Cambridge: Cambridge University Press.

Yu, G., Bu, Q. W., Cao, Z. G., *et al.* (2016). Brominated flame retardants (BFRs): a review on environmental contamination in China. *Chemosphere*, 150, 479–490.

Yunker, M. B., Macdonald, R. W., Vingarzan, R., *et al.* (2002). PAHs in the Fraser River basin: a critical appraisal of PAH ratios as indicators of PAH source and composition. *Organic Geochemistry*, 33, 489–515.

Zhang, W. (2018). Global pesticide use: profile, trend, cost/benefit and more. *Proceedings of the International Academy of Ecology and Environmental Sciences*, 8, 1–27.

Zheng, X., Sun, Q. H., Wang, S. P., *et al.* (2020). Advances in studies on toxic effects of short-chain chlorinated paraffins (SCCPs) and characterization of environmental pollution in China. *Archives of Environmental Contamination and Toxicology*, 78, 501–512.

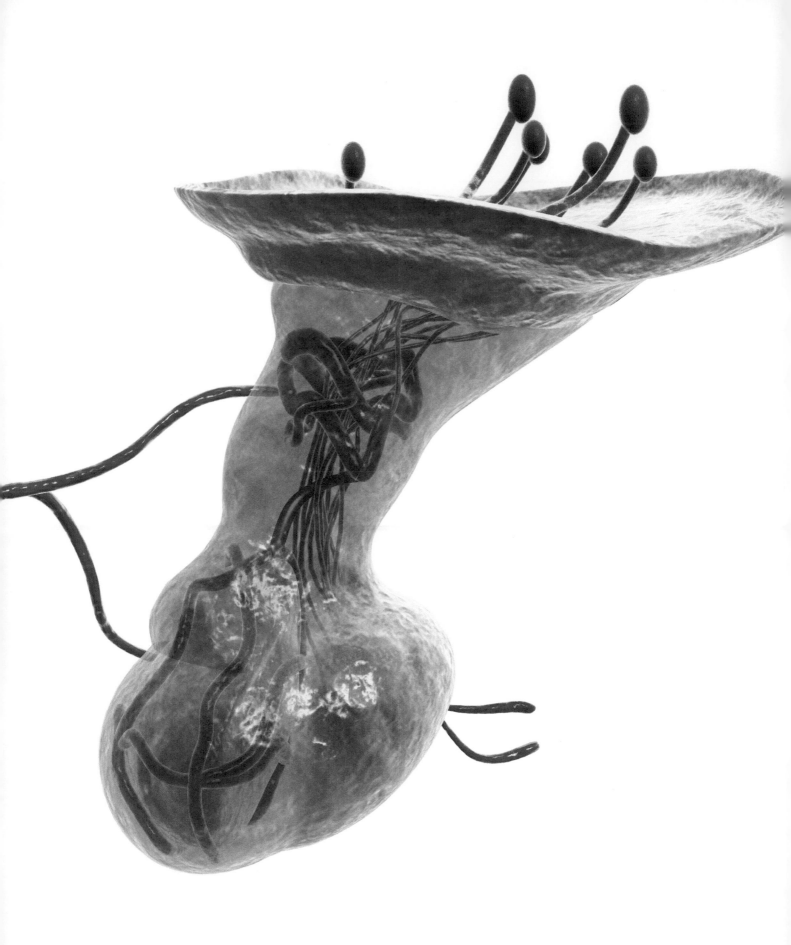

CHAPTER

8 Endocrine Disrupting Chemicals

Christopher D. Metcalfe, Christopher J. Martyniuk, Valérie S. Langlois and David A. Wright

Learning Objectives

Upon completion of this chapter, the student should be able to:

- Describe the primary endocrine axes known to be sensitive to EDCs.
- Understand the fundamentals of hormone synthesis and how hormones are transported within the body.
- Describe the mechanisms of action of hormone agonists and antagonists.
- Understand the neuroendocrine control of hormone biosynthesis.
- Provide examples of EDCs and the mechanisms by which they disrupt the endocrine system.
- Conceptualize and discuss the ecological impacts of exposure to EDCs.

The focus of this chapter is exogenous compounds that cause negative effects in vertebrates and invertebrates by interfering with chemical communication mediated by natural hormones and cellular receptors. Such compounds are called **endocrine disrupting chemicals** (EDCs). At the receptor level, disruption may result from EDCs mimicking the action of the natural hormone (agonism), thereby initiating inappropriate signalling. Alternatively, EDCs may block the action of the natural hormone (antagonism). However, recent research has shown that EDCs may affect organisms through a range of other mechanisms. They can affect many different tissues, as hormones are responsible for a broad range of actions, including physiological homeostasis and regulation of reproduction, development and behaviour. EDCs are also very heterogeneous in structure, varying from organometals (e.g., tributyltin) to synthetic oestrogens (e.g., **17α-ethinyloestradiol**, EE2). Initially identified as the cause of reproductive disorders in fish and wildlife, often in organisms exposed to very low concentrations or doses, EDCs have attracted increased research and regulatory interest. This has been prompted by concern over the subtle effects of EDCs on populations and natural ecosystems. This chapter introduces the breadth of this active area of research.

8.1 Endocrine Disruption

By the 1990s, it was clear that the reproductive performance of many species of fish and wildlife was being compromised by exposure to chlorinated compounds and other xenobiotics. The book *Our Stolen Future*, written by Theo Colborn, Dianne Dumanoski and John Peterson Myers and published in 1996, was a seminal contribution to environmental science, and increased awareness among the public and researchers of a worsening environmental issue with global significance (Section 1.2.6). Much of the early research on this topic focused on xenobiotics that bind to the oestrogen receptor and mimic the natural hormone, 17β-oestradiol (E2). These compounds are called oestrogen **agonists**, or **xeno-oestrogens**. Subsequent research identified xenobiotics that can block the binding of E2, called oestrogen **antagonists** or **anti-oestrogens**. However, research has advanced beyond this early work to identify compounds that disrupt the endocrine system through a range of mechanisms and hormone targets. There is solid evidence that xenobiotics can also alter gene expression through epigenetic mechanisms, and this altered genetic material may be passed on to offspring, causing decreased survival or fitness of subsequent generations. Changes in the genotype of succeeding generations may also lead to subtle shifts in the capacity of organisms to adapt to environmental stressors that are not necessarily related to exposures to chemical contaminants (e.g., climate change).

Concern over the adverse effects of EDCs was initially focused on human cases. However, instances of adverse effects of EDCs on fish and wildlife have an equally long history. Not surprisingly, EDC research continues to be dominated by the study of developmental and reproductive anomalies related to human exposures, and much of what we know about the mechanisms of endocrine disruption stems from research with humans and mammalian model species (e.g., rodents). However, there is a rich literature on laboratory and field studies with fish, amphibians and, to a lesser extent, reptiles and birds. There is evidence that some xenobiotics can disrupt the signalling of plant hormones, and indeed some herbicides including atrazine and 2,4-D have been designed to affect specific plant hormone systems (Couée *et al.*, 2013). The primary focus on the effects of EDCs on vertebrates in this chapter reflects the large and expanding body of evidence for the developmental and reproductive effects of EDCs on vertebrate fauna. Nevertheless, the disruption of plant signalling systems by xenobiotics represents an expanding, yet still largely unexplored line of research. A major challenge for EDC research focused on both humans and wildlife is to establish the links between biological responses observed in the laboratory and the potential for effects under environmental exposure scenarios. Concentrations of EDCs found in the environment are often lower than the levels of exposure in the laboratory. This uncertainty is the basis for much of the controversy over whether EDCs should be regulated or even banned outright (Kassotis *et al.*, 2020).

EDCs are heterogeneous in structure and disrupt the endocrine system through several different **modes of action** (MoAs). Most EDCs are organic contaminants, but some metals, metalloids and organometals can also disrupt the endocrine system. In some cases, a single chemical may have multiple effects on the endocrine system through different MoAs. In addition, some chemicals can exert effects through non-endocrine MoAs (e.g., oxidative stress, mitochondrial dysfunction), as well as endocrine-mediated effects. Over 800 chemicals are now known or suspected to exert disrupting effects on the endocrine system. However, only a small fraction of these chemicals has been investigated in experiments with organisms under environmentally realistic conditions. Evidence for the influence of EDCs from an ecotoxicological perspective generally falls into two categories:

- A high incidence and/or increasing trend of endocrine-related disorders observed in natural populations of invertebrates, fish, amphibians, reptiles, birds or mammals;
- Detection of chemicals in environmental matrices (e.g., water, air, soil, diet) that have been linked to endocrine disrupting effects observed *in vivo* or *in vitro* in laboratory studies.

Prior to discussing the different chemicals that exert hormone disruption in animals, it is important to briefly review the elements of the endocrine system. In the section below, we highlight only the major hormones and briefly discuss their synthesis and physiological actions within the individual. When a chemical disrupts a hormone system, downstream physiological events are altered, which can compromise both the short-term and long-term health of the organism. Therefore, it is important for those interested in studying endocrine disruption to be versed in both ecotoxicology and comparative endocrinology if they are to understand the significance of chemical effects and the long-term consequences to organisms.

8.2 The Endocrine System and its Disruption

The endocrine system is a chemical communications network that functions in animals by transporting hormones through the circulatory system. This system controls a range of biological responses in both vertebrates and invertebrates. There are also hormones in plants, such as auxins, that control growth. In some cases, hormones are highly conserved throughout animal and plant phyla. For instance, **serotonin**, best known for its neuroendocrine role in vertebrate animals, is present in several species of plants, as well as in invertebrates, where it controls a range of other physiological functions. In vertebrates, the endocrine system controls cell differentiation and organ formation in early life stages, whereas in adults this system controls a range of physiological processes. Exposures of embryos to hormones or to EDCs that alter hormone levels or function can 'programme' tissues for life, whereas exposures in adults usually have transient effects that return to normal once exposures cease.

Autocrine hormones are produced by and act on the same cell. For example, epidermal growth factor is produced by epidermal cells and stimulates proliferation of the same cells. **Paracrine** hormones act on adjacent cells. For example, the insulin produced by the Islets of Langerhans affects glucose metabolism in adjacent pancreatic cells. **Endocrine** hormones act over longer distances, relative to signals that are transmitted by the nervous system. For example, **vasopressin**, which is released from the pituitary gland near the brain, acts to regulate water resorption in the kidney. Hormones can have powerful effects that regulate growth, development and physiological functions, and there are positive and negative feedback systems within organisms that tightly regulate the production of hormones. In addition, hormones are often transported from the site of synthesis to the target organ in association with transport proteins, so that the cells in non-target organs are not affected.

One of the most widely studied endocrine systems in vertebrates is the **hypothalamic–pituitary-adrenal axis (HPA)**. The hypothalamus is a region of the vertebrate brain that integrates sensory information (e.g., photoperiod, temperature, smell) from other brain centres and controls visceral functions through the production of **hypophysiotropic** hormones that travel through the circulatory system to act on tropic secreting cells in the pituitary. This axis controls several endocrine systems, as illustrated in Figure 8.1. **Corticotropin-releasing hormone** (CRH) is secreted by the hypothalamus and controls the release of **adrenocorticotropic hormone** (ACTH) by the pituitary, which in turn initiates the release of **corticosteroids** (glucocorticoids and mineralocorticoids) by the adrenal glands. These hormones include cortisol, which increases blood glucose levels, and aldosterone, which regulates water reuptake in the kidney and blood pressure. The **HPA axis** refers to the stress axis and the feedback loops that also involve cortisol. **Gonadotropin-releasing hormone** (GnRH) is synthesized in the preoptic area of the hypothalamus and is responsible for stimulating the release of gonadotropins from the anterior pituitary. The **hypothalamic-pituitary-gonadal axis** (HPG) regulates all aspects of reproduction, from mating behaviours to the maturation of the gonads and development of secondary sex characteristics. The synthesis of **thyroid-releasing hormone** (TRH) in the paraventricular nuclei of the hypothalamus stimulates secretion of **thyroid-stimulating hormone** (TSH) from the pituitary, which in turn acts to regulate thyroid hormone release from the thyroid gland. The **hypothalamic–pituitary–thyroid axis** (HPT) regulates

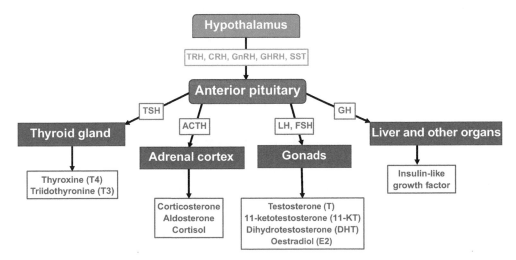

Figure 8.1 The endocrine axes controlling the synthesis and release of hormones by the thyroid gland, adrenal cortex, gonads (i.e., testis and ovary), liver and other organs. GHRH, growth hormone-releasing hormone; TRH, thyrotropin-releasing hormone; CRH, corticotropin-releasing hormone; GnRH, gonadotropin-releasing hormone; SST, somatostatin (growth hormone inhibitory hormone); TSH, thyroid-stimulating hormone; ACTH, adrenocorticotropic hormone; LH, luteinizing hormone; FSH, follicle-stimulating hormone; GH, growth hormone.

body temperature, metabolism, and bioenergetics in adult vertebrates. The HPT and stress axes act in tandem with the HPG axis to control sexual differentiation in the early life stages of vertebrates (Castañeda Cortés *et al.*, 2014). Another important hormone system is the growth axis mediated by pituitary **growth hormone** (GH) that regulates all aspects of growth in vertebrates (Figure 8.1).

The presence of high circulating levels of thyroid, steroid, corticosteroid and GH hormones has a negative feedback effect on the secretion of hypophysiotropic hormones in the hypothalamus and the secretion of **tropic hormones** in the pituitary, thereby regulating the balance of hormones within the organism. Control of growth is particularly important, so **somatostatin** (SST) secreted by the hypothalamus has an inhibitory effect on GH release in the pituitary. Any disruption to this delicately balanced system can upset the secretion of hormones which can subsequently affect numerous endocrine organs. For instance, Qiu *et al.* (2019) observed that exposure of zebrafish (*Danio rerio*) to bisphenol F, used as a replacement plasticizer for bisphenol A (BPA) in polycarbonate plastics, caused alterations to the production of ACTH, but also reduced levels of **luteinizing hormone** (LH) and **follicle-stimulating hormone** (FSH) in fish. These effects were observed at environmentally relevant bisphenol-F concentrations.

8.2.1 Neuroendocrine Control

Various regions of the brain provide the initial signals that trigger endocrine functions through the secretion of hormones or direct innervation of endocrine organs. Sensory stimuli such as temperature, photoperiod and even certain sights and smells can act as stimuli within these brain centres.

8.2.1.1 The Hypothalamic–Pituitary Axis

The hypothalamus region of the brain is the major control centre for the release of hormones by the pituitary. In general (although this varies across animal taxa), neurosecretory cells located in the brain synthesize hormone releasing factors that are transported through the **portal blood system** to act on tropic secreting cells ('-trophs') in the pituitary, as illustrated in Figure 8.2.

Other pituitary hormones include **prolactin** (PRL), which stimulates milk production in mammals but has important roles in metamorphosis in amphibians and ion regulation in fish. The catecholamine neurotransmitter **dopamine** is produced in several areas of the brain, including the hypothalamus. In mammals, dopamine synthesized in the hypothalamus inhibits the release of PRL by the pituitary, but in lower vertebrates, it can have other functions. For instance, dopamine inhibits the release of

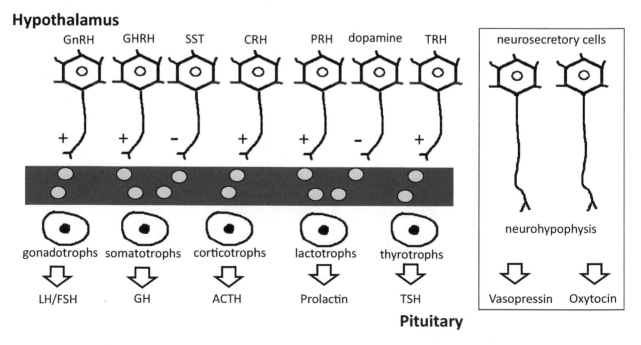

Figure 8.2 Illustration of the nuclei of the hypothalamus that synthesize different vertebrate neurohormones and releasing factors. A plus sign (+) indicates a stimulatory effect on hormone secretion and a minus sign (−) indicates an inhibitory effect on hormone secretion.

gonadotropins by the pituitary in fish, thereby regulating reproduction. Two other major hormones are produced by the action of neurosecretory cells that directly connect the hypothalamus with the pituitary gland (Figure 8.2). These hormones are **vasopressin** (also called antidiuretic hormone) which regulates ion and water balance to modify blood pressure and hydration, and **oxytocin** which regulates parturition and reproduction in higher vertebrates. Like many of the natural compounds mentioned in the previous section, these hormones can have a range of physiological effects in different vertebrate taxa. For instance, isotocin (i.e., the 'oxytocin' in fish) is involved in ion regulation in the fish gills. Some of these releasing hormones travel from the median eminence of the hypothalamus to the pituitary to stimulate different -trophs, whereas direct connection with the pituitary results in secretion of vasopressin and oxytocin.

There are several important points to make regarding neuroendocrine axes in vertebrates. A detailed discussion of this topic is beyond the scope of this chapter, but questions surrounding hormone systems fall within the discipline of **comparative endocrinology**, which is a branch of endocrinology devoted to identifying and characterizing neuroendocrine and endocrine systems in a diverse array of animal taxa. Readers interested in learning more about the similarities and differences in neuroendocrine systems in invertebrates and vertebrates are encouraged to seek out these resources and additional readings on this topic. Chemicals released into the environment can affect neuroendocrine control of the pituitary. We are only beginning to scratch the surface in our understanding regarding environmental contaminants and their effects on the release of pituitary hormones. This has led to a new field of research in ecotoxicology called **neuroendocrine disruption**.

The neuroendocrine system does not act in isolation. The central nervous system (CNS) receives feedback from the peripheral system through the transport of steroid and peptide hormones. These chemical signals circulating in the blood can have either a positive or a negative feedback on the neuroendocrine system (i.e., positive and negative feedback loops). For example, E2 can bind to oestrogen receptors in the CNS to modulate the release of GnRH, which subsequently affects the release of gonadotropins from the pituitary. This interaction controls the synthesis of either more or less E2 in the gonad. This finely tuned control is maintained throughout the reproductive cycle. Thus, the CNS and specifically the

hypothalamus integrate both internal and external information to control the timing and number of hormones that circulate within the body.

The neuroendocrine system contains hundreds of neuro-hormonally active molecules. The function of many neurons can be modulated by different hormonal input. There are numerous neurotransmitters and neuro-hormones that interact with the major secretory neurons illustrated in Figure 8.2. For example, a reproductive neuropeptide called **kisspeptin** stimulates neurons to activate gonadotropins from the pituitary. Other hormones such as **neuropeptide Y** and **leptin** can modify growth and stress responses. Once again, the effects of these hormones can vary across taxa. For instance, the neurotransmitter serotonin can influence sleep patterns, appetite, locomotion, cognition, arousal and a variety of behaviours in both vertebrates and invertebrates, depending upon the taxon (Bacqué-Cazenave *et al.*, 2020). Evolutionary history has yielded a diverse array of receptor systems and enzymes that are involved in hormone signalling. It is important to remember that hormone synthesis, release and target action can differ across taxa, from insects to mammals.

8.2.1.2 Neuroendocrine Disruption

Neuroendocrine disruption is a relatively new term that describes the mechanisms and outcomes related to chemical interactions with **neurohormones** and **neuropeptides**. Within the context of ecotoxicology, chemicals that cause neuroendocrine disruption have been defined as "...*pollutants in the environment that are capable of acting as agonists/antagonists or altering the synthesis and/or metabolism of neuropeptides, neurotransmitters, or neurohormones, which subsequently alter diverse physiological, behavioural, or hormonal processes to affect an animal's capacity to reproduce, develop and grow, or deal with stress and other challenges*" (Waye and Trudeau, 2011). Thus, environmental contaminants can alter the synthesis, metabolism and release of neurohormones in the CNS, affecting hormone concentrations and receptor abundance. Such disruption can have profound effects on the entire endocrine system since the hypothalamus is the control centre for pituitary hormone release. The concept of neuroendocrine disruption has been further reviewed in the literature (León-Olea *et al.*, 2014; Rosenfeld *et al.*, 2017).

Examples from a range of animal taxa demonstrate that environmental contaminants can interact with the neuroendocrine system, leading to adverse physiological effects. These negative outcomes can occur through different mechanisms, such as gene expression modulation in the brain and altered neurotransmitter or neurohormone synthesis. For instance, an additive to polycarbonate plastics, BPA, modulated the expression of genes involved in the synthesis of steroid hormones in the freshwater cypriniform fish *Catla catla* exposed for 14 d to concentrations of 10 µg/l (Faheem *et al.*, 2019). In another study, injection of methylmercury into largemouth bass (*Micropterus salmoides*) at a dose of 2.5 µg/g suppressed expression of a subtype of GnRH in the hypothalamus. These same alterations in transcription were also observed in bass captured from a site with elevated levels of methylmercury (Richter *et al.*, 2014). Martyniuk *et al.* (2013) showed that dietary exposures of fish to the organochlorine pesticide dieldrin for 2 months perturbed the expression of gene networks involving **dopamine** and **gamma-amino butyric acid** (GABA). These two neurotransmitters play essential roles in regulating the release of pituitary gonadotropins in teleost fishes and may explain the capacity of persistent chemicals such as organochlorines to impair reproduction in fish and other poikilothermic ('cold blooded') vertebrates.

Other examples of disruption of neurotransmitters by chemicals have been observed in both invertebrates and vertebrates. Gesto *et al.* (2006) reported disruption of brain monoaminergic neurotransmitters (e.g., **serotonin, dopamine**) in rainbow trout (*Oncorhynchus mykiss*) exposed by intraperitoneal injection to the polycyclic aromatic hydrocarbon (PAH) naphthalene at doses of 10 and 50 mg/kg. As an example of effects on neurotransmitters in invertebrates, Gagné *et al.* (2007) reported that clams exposed *in situ* to municipal wastewater in the estuary of the St. Lawrence River in eastern Canada showed changes to serotonin and dopamine metabolism and monoamine adenylate cyclase activity in synaptosomes, affecting the overall physiology of these bivalves. Disruptions in dopamine and **glutamate** have also been associated with oxidative stress and cellular damage in freshwater mussels exposed to municipal wastewater (Gagné *et al.*, 2011).

Neuroendocrine disruption can also have a significant impact on the social behaviours of animals, such as predator avoidance, shoaling, feeding behaviour and courtship rituals. Many behaviours are related to neurotransmitters, such as dopamine and serotonin, as well as neurohormones that include prolactin, vasopressin and oxytocin. Environmental contaminants, specifically organophosphate pesticides, affect such behaviours in mammals (Venerosi *et al.*, 2012). As discussed in Section 7.12, fish can be exposed to wastewater treatment discharges containing pharmaceuticals that are formulated to specifically target neurotransmitters. Exposures to **fluoxetine** (i.e., Prozac®), a serotonin reuptake inhibitor that is prescribed as an antidepressant in humans, altered burrowing behaviour in freshwater mussels (Hazelton *et al.*, 2014). Brooks (2014) summarized the research on the effects of antidepressants on fish, including effects on predator–prey behaviour, and concluded that concentrations in water as low as 0.1 µg/l are a hazard to fish.

Clearly, much remains to be learned about how environmental contaminants can modify neuroendocrine systems, and an integrated approach that considers the entire hormonal axis is warranted. Within an organism, multiple levels of biological organization can be altered by a chemical. For example, a chemical can dysregulate GnRH signalling in the hypothalamus, block release of gonadotropins from the pituitary, or antagonize steroidogenic enzymes in the gonad. Each of these mechanisms can lead to reduced sex steroid production. Chemicals have the potential to interact with receptors that are expressed in all tissues, and the direct targets of EDCs can remain elusive.

8.2.2 Gonadotropins

Gonadotropins regulate sexual development in early life stages and reproductive function in adult vertebrates. In mammals, gonadotropins include **luteinizing hormone** (LH) and **follicle-stimulating hormone** (FSH) secreted from the pituitary, as well as the **chorionic gonadotropin** secreted from the placenta of mammals. Both LH and FSH are glycoproteins made of an alpha and a beta subunit. These subunits come from the expression of a different gene; that is, *LHβ*, *FSHβ*, and a shared subunit used by both hormones called *gonadotropin subunit α*. Fish and other lower vertebrates have their own forms of gonadotropins that are closely related in structure to the mammalian LH and FSH. The role of LH is to signal the ovaries and the testes that it is time to produce oestrogens and androgens, respectively. A rise in oestrogen levels will trigger ovulation in female mammals and will start egg production (i.e., **oogenesis**) in oviparous lower vertebrates. In male vertebrates, an increase in testosterone levels triggers the development of mature sperm cells (i.e., **spermatogenesis**). FSH and LH work together to modulate both oogenesis and spermatogenesis.

Some EDCs are known to directly alter the expression of the *LH* and *FSH* genes. For example, Lorenz *et al.* (2011) exposed tadpoles of the African clawed frog (*Xenopus laevis*) to various concentrations of synthetic **gestagen**, levonorgestrel (LNG), which is used in human birth control formulations. These authors observed that exposures at concentrations of approximately 300 and 3,000 ng/l repressed the expression of the *LHβ* gene, and this response persisted past the metamorphosis of tadpoles to adult frogs. They also showed that *FSHβ* mRNA levels were altered by the treatment. This and similar studies provide insight into the possible MoA of artificial oestrogens; that is, disruption of the regulation of any of the gonadotropins could lead to altered reproduction in vertebrates. However, these experimental concentrations are well above the highest recorded LNG concentration in a domestic wastewater discharge of 30 ng/l, as reported by Viglino *et al.* (2008). Nevertheless, masculinization of female fish has been demonstrated at LNG concentrations as low as 1 ng/l (Zeilinger *et al.*, 2009).

Hormones are a diverse group of compounds that are synthesized from amino acids, fatty acids, sterols and aromatic amines. For instance, the basic building block for the synthesis of thyroid hormones is the amino acid tyrosine, whereas steroid hormones are synthesized from the sterol compound cholesterol. All steps in the biosynthetic pathways are catalysed by a range of enzymes located in endocrine organs. Note that in addition to the specific endocrine glands that have been mentioned to this point, the brain too is an important endocrine organ. Two important targets for endocrine disruption by xenobiotics are the biosynthetic pathways for steroid hormones (i.e., testosterone, oestradiol) and for thyroid hormones (i.e., **triiodothyronine**, T3, and **thyroxine**, T4).

8.2.3 Steroid Hormones

Steroids regulate a vast array of cell signalling pathways in all tissues during development and maturation of organisms. Steroid biosynthesis is a complex process, involving several enzymes that coordinate hormonal homeostasis. Figure 8.3 illustrates the biosynthetic pathway for the production of oestrogens and androgens, which takes place in the mitochondria of cells in vertebrate tissues. These include theca and luteal cells in the ovary and Leydig cells in the testis, but there are also centres for steroid synthesis in the brain. Many of the steps in this process are catalysed by **cytochrome P450 enzymes** from the CYP11 and CYP17 classes.

Also, the rate-limiting step in steroid hormone synthesis is the movement of cholesterol across the mitochondrial membrane and this is mediated by the **steroidogenic acute regulatory** (StAR) protein. Chemicals that disrupt the gene expression or enzymatic activity of StAR and other CYP450 enzymes can disrupt the endocrine system.

Particular attention has been given to the role of CYP19 in the final step of synthesis of oestradiol (Figure 8.3). This metabolic cytochrome is often referred to as **aromatase** because it catalyses the aromatization of the terminal cyclic ring of the **androgens**, testosterone and androstenedione. There are several xenobiotics that inhibit or induce the activity of aromatase, and therefore, affect the conversion of androgens to oestrogens. Among them are the aromatase inhibitor drugs developed for therapy of oestrogen-responsive breast cancers, but also many other xenobiotics, such as plasticizers and pesticides. For example, exposures of zebrafish for 60 d to the widely used fungicide Flutolanil, at waterborne concentrations as low as 0.25 µg/l, resulted in the upregulation of aromatase expression and an increase in the levels of E2 and a decrease in testosterone levels (Teng *et al.*, 2020).

Although not as widely studied as aromatase, but equally important, the **5α-reductases** catalyse the synthesis from testosterone of **5α-dihydrotestosterone** (5α-DHT), which is a more potent androgen than testosterone. In mammals, 5α-DHT is produced in the genitalia, prostate gland, seminal vesicles, skin and hair follicles. In lower vertebrates, 5α-DHT is the dominant androgen produced in all tissues, except in fish, where 11-ketotestosterone is a more potent **agonist** for the androgen receptor. Several subtypes of steroid 5α-reductases exist (types 1, 2 and 3) and an increasing number of studies are showing that these 5α-reductases are targets for inhibition by xenobiotics. As an example, embryos of the Western clawed frog (*Silurana tropicalis*), exposed to low (µg/l) concentrations of the plasticizer **di(2-ethylhexyl)phthalate** (DEHP), expressed significantly elevated levels of mRNA for all three 5α-reductases (Bissegger *et al.*, 2018).

8.2.4 Thyroid Hormones

These unusual biomolecules incorporate a halogen, iodine, as a substituent of their diphenyl ether structure.

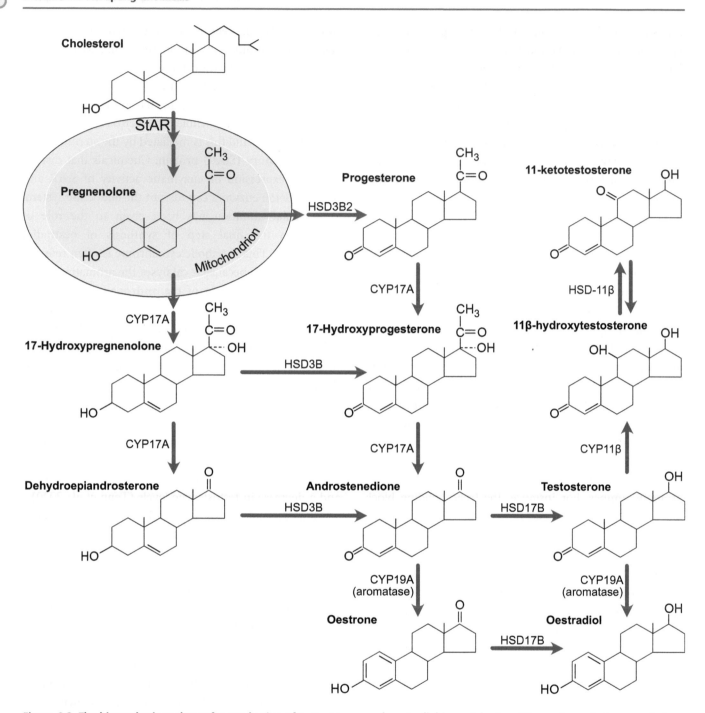

Figure 8.3 The biosynthetic pathway for production of testosterone and oestradiol in vertebrates. (Figure drawn to be generic for vertebrates, and to point out the phenolic group involved in binding with oestrogen receptor.)

Figure 8.4 illustrates the basic elements of the biosynthetic pathway for thyroid hormones and their transport to target cells. The outer membranes of the epithelial cells that surround the thyroid follicles (i.e., **thyrocytes**) have iodide pumps that mediate the flow of iodine ions into the colloidal space (i.e., **lumen**) of the follicles. The activity of these ion pumps and indeed almost all steps involved in the synthesis and release of thyroid hormones are controlled by TSH. **Goitre**, characterized by an enlarged thyroid, is a pathological condition in humans that was observed historically in people with insufficient dietary intake of iodine. For this reason, most table salt is now iodized to ensure sufficient iodine intake. Iodine is also conserved within the body through the activity of deiodinase enzymes in the target organs that catalyse the release of iodine ion from thyroid hormones so that the iodine can be recycled back to the thyroid gland. Another important influence on T3 and T4 production relates to the iodide transporter responsible for iodine uptake by the thyroid gland. This transmembrane Na^+/I^- symporter is subject to competitive inhibition by thiocyanate (SCN^-), nitrate (NO_3^-) and perchlorate (ClO_4^-). Excessive exposures to any of these ions inhibit iodine uptake by the thyroid gland, and all have been implicated as a cause of hypothyroidism in humans. Their contribution to decreased thyroid function has been difficult to determine since mixtures of these ions originate from natural geogenic and anthropogenic sources. For example, thiocyanates have a range of industrial uses, including as herbicides and fungicides, and perchlorates are commonly used as oxidizers in explosive and in solid propellants, including in fireworks. Dietary iodine content is another contributory variable. Perchlorate contamination in surface and groundwater is now recognized as a global problem (Cao *et al.*, 2019), and its effect on aquatic organisms is a subject of vigorous research, including studies of effects on amphibian metamorphosis (Bulaeva *et al.*, 2015).

Through the activity of **iodinases** and **peroxidases** associated with the internal membrane of the thyrocytes, there is synthesis of a mixture of compounds known as **I-thyroglobulin** that contains tyrosine residues with between one (i.e., mono-iodotyrosine) and four (i.e., tetra-iodotyrosine)

iodine substituents. The thyroglobulin is stored in the colloidal space of the follicles until the thyroid hormones are needed. **Triiodothyronine** (i.e., T3) and **tetra-iodotyrosine** (i.e., T4 or **thyroxine**) are the active thyroid hormones that are transported out of the follicles. The **thyroglobulin** is first taken up from the colloidal space by the thyrocytes and conveyed to lysosomal compartments for proteolytic cleavage that releases T4 and T3 from their peptide linkages. These hormones are then released from the thyrocytes by exocytosis into extracellular spaces.

Because T3 and T4 are relatively non-polar (i.e., hydrophobic), they are bound to thyroid serum transport proteins for transport via the circulatory system to the target organs. In humans, three serum transport proteins, **thyroxine-binding globulin** (TBG), **transthyretin** (TTR) and human serum albumin, are involved, and T4 is the predominant thyroid hormone that is transported (Figure 8.4). The types of serum transport proteins vary across vertebrate taxa and even among mammalian species. Once the thyroid hormones reach the target cell, they may diffuse across the cell membrane, but it is now known that transport is primarily mediated by binding to transporter proteins that are specific for T3 or T4. Once in the cell, T4 is deiodinated intracellularly to T3, which is the more potent thyroid receptor agonist.

Thyroid hormones are some of the main targets of EDCs. Over 2,000 scientific papers published up to 2020

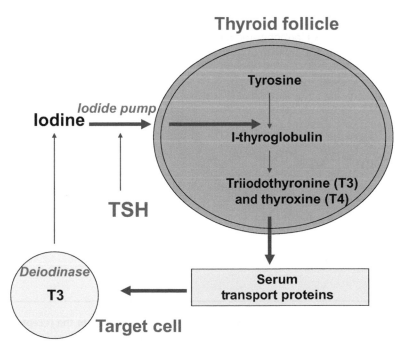

Figure 8.4 The basic elements for the biosynthesis and transport of thyroid hormones. TSH = thyroid-stimulating hormone.

were identified from PubMed using the keywords 'thyroid hormone disruption'. Pesticides are among the many compounds that can disrupt the activity of thyroid hormones. As reviewed by Leemans *et al.* (2019), a long list of pesticides, including compounds from the **organophosphate**, **carbamate**, **pyrethroid**, **phenylpyrazole** and **neonicotinoid** classes of insecticides (Section 7.8), interfere with thyroid hormone activity in vertebrate species. Leemans *et al.* (2019) concluded that a complete toxicological screening is needed for all pesticides prior to placing them on the market.

8.3 Hormone Receptors

Hormonal activity is regulated precisely with the involvement of highly specific receptors, which are discussed in general in Section 7.4. The receptors for steroid hormones are nuclear receptors, and once the **hormone–receptor complex** passes into the nucleus, it binds to specific regions of DNA to regulate the process of gene transcription. However, it is important to bear in mind that these hormones regulate different genes in different cell types, or at different times during development. Furthermore, the same receptor can have unique effects in different cell types. For example, an oestrogen may regulate different proteins in liver cells compared with brain cells. Figure 8.5 illustrates the process of binding of steroids to the receptor, transport to steroid responsive elements in the genome and regulation of gene expression. Xenobiotics can bind to the steroid receptor and either act as agonists by inducing responses, similar to those induced by the endogenous ligand, or act as antagonists by blocking the binding of the endogenous ligand.

Screening assays are needed, both in the field and in the laboratory, to detect whether organisms have been exposed to EDCs. These assays are often called **biomarkers** of EDC effects. One widely used assay system is the production of **vitellogenin** (VTG) by oviparous vertebrates. In females, the release of E2 from the ovary causes the liver to produce large amounts of this high-density lipoprotein, which is a precursor for egg yolk. The VTG produced by the liver is transported in the bloodstream and incorporated into developing oocytes. Chemicals that can mimic the endogenous ligand (i.e., 'xeno-oestrogens')

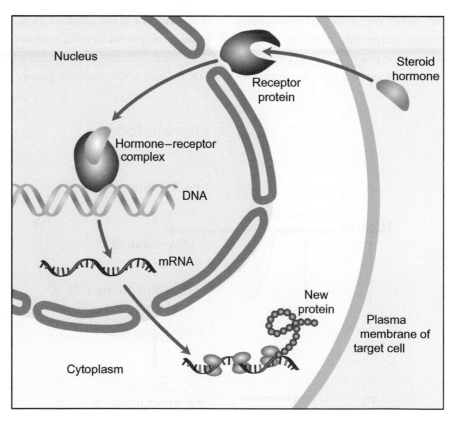

Figure 8.5 Process of diffusion of a steroid hormone across the cell membrane, binding to the receptor protein, translocation of the steroid–receptor complex into the nucleus, and binding to specific steroid responsive elements in the genome that regulate gene expression and translation of proteins.

can also stimulate VTG production. As illustrated in Case Study 8.1 for exposures of fish downstream from wastewater treatment plants, detection of high levels of VTG in male or immature fish indicates that these organisms have been exposed to xeno-oestrogens. A number of convenient assays, including real-time PCR (Section 5.2) and **enzyme-linked immunosorbent assays** (ELISA), have been developed for VTG analysis. Transgenic fish and amphibians have also been developed with gene editing to indicate the presence of xeno-oestrogens in water. In the era of ecotoxicogenomics, there is a new toolbox of methodologies (transcriptomics, proteomics, metabolomics, lipidomics, epigenetics, etc.) available to researchers to enable the discovery of the molecular mechanisms by which EDCs affect organisms. These methods are described in Chapter 5.

8.4 Modes of Action of EDCs

The large structural diversity of EDCs indicates that endocrine disruption may operate through a variety of mechanisms, and the MoAs include:

- mimicking the effects of endogenous hormones (agonists);
- blocking the effects of endogenous hormones (antagonists);
- altering the synthesis and metabolism of endogenous hormones;
- binding to hormone transport proteins and preventing their function;
- modifying hormone receptor levels and transcriptional processes.

8.4.1 Agonists and Antagonists

Many EDCs or their oxidative metabolites possess a common structural relationship to the phenolic A ring in E2 (Figure 8.3). Initially, reports of this property led to a range of EDCs being identified as oestrogen agonists or antagonists. Subsequently, binding of chemicals to the androgen receptor has been reported, although this receptor appears to have greater fidelity for the endogenous ligand, in comparison to the oestrogen receptor that binds with a range of xenobiotic ligands (i.e., 'promiscuous' binding). Therefore, it is more common for xenobiotics that bind to the androgen receptor to be antagonists than agonists. For instance, there is evidence

that perfluorooctanoic acid (PFOA) has anti-androgenic activity. In this way, EDCs may mimic or block the action of natural steroid hormones by activating or antagonizing nuclear hormone receptors. Several compounds have also been identified as agonists for the thyroid receptors. For instance, the brominated BPA compound, **tetrabromobisphenol A** (TBBPA), which is now widely used as a replacement flame retardant for the now-banned **brominated diphenyl ethers** (BDEs), has been shown to be a thyroid agonist, as discussed later in this chapter.

8.4.2 Altered Biosynthesis of Hormones

As mentioned previously, any alterations to the steroid biosynthesis pathway can affect the balance of steroid hormones. Arukwe (2005) reported that exposure of juvenile Atlantic salmon (*Salmo salar*) for 7 d to the surfactant **nonylphenol** at concentrations as low as 15 µg/l caused transcriptional changes in the expression of StAR protein. At the same time, nonylphenol suppressed cholesterol side-chain cleavage by CYP450 enzymes, and these changes altered steroidogenesis in the brain. Lee *et al.* (2010) showed that exposures of female oriental fire-bellied toads (*Bombina orientalis*), by intraperitoneal injection of 10 mg/kg of either nonylphenol or BPA, reduced gonadal aromatase activity by more than 50%. Disruption of the 5α-reductases during embryogenesis could lead to dysregulation of gonadal development. Duarte-Guterman *et al.* (2009) demonstrated that exposures of early life stages of western clawed frogs to micromolar concentrations of a clinical 5α-reductase inhibitor caused the appearance of oocytes in the testes of mature males (i.e., gonadal intersex).

Disruption of thyroid hormone synthesis could also disrupt thyroid function. Exposures of early life stages of zebrafish to waterborne Thiram, a dithiocarbamate fungicide, at concentrations as low as 24 µg/l induced notochord curvatures (Chen *et al.*, 2019). The curvatures were correlated with increased activity of a deiodinase enzyme, and a decrease in the activity of thyroid peroxidase, an important enzyme for thyroglobulin synthesis. Dong *et al.* (2013) observed reduced deiodinase activity in the brain and the pre-nephric duct of zebrafish embryos exposed from 4 to 22 h post-fertilization to the hydroxy-metabolite of the brominated flame-retardant compound, BDE-47. These responses were observed in embryos exposed at waterborne concentrations as low as 0.5 µg/l.

8.4.3 Binding to Hormone Transport Proteins

Many hormones are relatively non-polar and poorly soluble in blood serum. Although this characteristic facilitates their transport across cell membranes, it means that these compounds are not transported readily in blood to the target organs. In addition, these potent compounds could have effects on organs other than the target organ. Therefore, many hormones bind to transport proteins to be carried from the site of synthesis to the target organ. Binding of xenobiotics to transporter proteins can alter the levels of circulating hormones by competitive inhibition of binding with the endogenous ligand. For example, several classes of xenobiotics are neither thyroid agonists nor antagonists, but they are known to affect thyroid function. Weiss *et al.* (2009) concluded that some perfluorinated compounds disrupt thyroid homeostasis by binding to thyroid serum transport proteins. There is a consensus that hydroxy-metabolites of the BDE class of flame retardants interfere with thyroid homeostasis by binding to the proteins that transport T4 and, to a lesser extent, T3 (Legler and Brouwer, 2003). In amphibians, the metamorphosis of tadpoles to adults is triggered by a spike in levels of thyroid hormones. Balch *et al.* (2006) showed that dietary exposure of *Xenopus laevis* tadpoles to a commercial BDE mixture, DE-71, at doses of 1,000 and 5,000 μg/g in the food completely inhibited metamorphosis, as illustrated by the photograph in Figure 8.6.

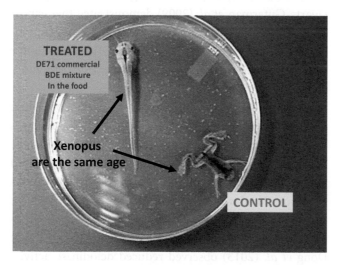

Figure 8.6 Photograph of sibling early life stages of *Xenopus laevis* from a control group and from a treatment group. Tadpoles were exposed via the diet to a commercial BDE mixture, DE71. Note that metamorphosis was completely inhibited in the tadpole exposed to the BDE mixture. Photograph provided by Gordon Balch of Fleming College, Peterborough, ON, Canada.

These authors speculated that the mechanism for inhibiting metamorphosis was binding of hydroxy-BDEs to thyroid serum transport proteins, competitively inhibiting the transport of T4 and T3 to the tissues that are reorganized during metamorphosis (e.g., tail and limb muscles).

8.4.4 Altered Hormone Receptor Levels and Gene Expression

The proposed MoAs of EDCs have now been expanded to include effects on a range of other nuclear receptors, and on the resultant transcription processes and signalling pathways. The **steroid and xenobiotic receptor/pregnane X receptor** (SXR/PXR) and the **constitutive androstane receptor** (CAR) regulate xenobiotic and steroid hormone metabolism through the mediation of cytochrome P450 (**Phase I**) enzymes, **Phase II** enzymes (e.g., UDP glucuronyl transferase) and **Phase III** multidrug resistance proteins (MRPs) that are described in Section 7.5.

Endocrine disrupting compounds affect transcriptional activity through the alteration of nuclear receptors and their co-regulators. Xenobiotics may also act to enhance receptor activity by inhibiting histone deacetylase activity and stimulating protein kinase activity, thereby enhancing gene expression.

Organisms are exposed to a 'cocktail' of many different EDCs in the environment. The different EDCs have the potential to interact to produce a range of effects across multiple endocrine systems. Table 8.1 shows the potential competing, additive and/or synergistic effects of just three compounds, BPA and two phthalate esters on a range of hormones, proteins and enzymes that influence developmental, reproductive and stress functions. The presence of these compounds at concentrations in the low μg/l range in some sewage treatment and industrial effluents has been implicated in causing adverse effects in both vertebrates and invertebrates.

8.5 Examples of EDCs

Many chemicals, including those listed in Table 8.2, have been identified as potential EDCs. Some have already been mentioned in this chapter, including methylmercury, dieldrin, serotonin reuptake inhibitor drugs, BPA and analogues, brominated flame retardants, nonylphenol and phthalate plasticizers. Specific examples and case histories for EDC effects on fish and wildlife are highlighted in Section 8.5.1.

Table 8.1 **Effects of bisphenol A and two phthalate esters on hormones, proteins and enzymes associated with the HPT, HPG and stress axes.**

Suppression and activation of transcription, translation and activity are represented by downward and upward pointing arrows, respectively. A dash indicates that no data were available. The table was adapted from Mathieu-Denoncourt et al. (2015).

Biological targets	Bisphenol A	Dibutyl phthalate	Di(2-ethylhexyl)phthalate
Development			
Thyroid stimulating hormone	–	↑	–
Thyroid hormones	↓	–	↑
Thyroid receptors	↓	↓	↓
Retinoid X receptor γ	↓	↓	–
Insulin-like growth factor	↓	↑	–
Reproduction			
Cytochrome P450 C17	↓	↓	–
Cytochrome P450 19 (aromatase)	↓	–	↓
Insulin-like growth hormone 3	–	↓	↓
17β-oestradiol	–	↑	↓
Follicle stimulating hormone	–	↑	–
Luteinizing hormone	↓	↑	–
Testosterone	↓	↓	↓
Dihydrotestosterone	↓	↓	–
Vitellogenin	↑	↑	↑
Oestrogen receptors	↑	↓	↓
Reproduction and stress			
Scavenger receptor class B-1	–	↓	↓
Steroidogenic acute regulatory protein	–	↓	↓
Peripheral benzodiazepine receptor	–	↑	↓
Cytochrome P450 11 (side-chain cleavage)	–	↓	↓
3β-hydroxysteroid dehydrogenase	–	↓	–
Cortisol	↓	–	↓
Peroxisome proliferator-related receptor α	–	–	↑
Peroxisome proliferator-related receptor γ	–	–	↑
Heat shock protein 70	–	–	↑
Glutathione peroxidase	–	–	↑
Superoxide dismutase	–	–	↓
Catalase	–	–	↓
Glutathione	–	–	↓

Table 8.2 **Examples of endocrine disrupting chemicals (EDCs).**

See also Chapter 7.

Chemical	History and mode of action
p,p′-DDT: dichlorodiphenyltrichloroethane [IUPAC name 1,1′-(2,2,2-trichloroethane-1,1-diyl)bis(4-chlorobenzene)] p,p′-DDE: dichlorodiphenyldichloroethylene [IUPAC name 1,1′-(2,2-dichloroethylene-1,1-diyl)bis(4-chlorobenzene)]	DDT was widely used as an insecticide in the 1940s–1960s. It was found to cause eggshell thinning in birds through inhibition of prostaglandin by its metabolite DDE, which has also been linked to sex reversal in alligators.
Organochlorine pesticides	Many organochlorine pesticides (e.g., chlordane, dieldrin) are now banned under the Stockholm Convention because they are linked to feminization in wildlife.
Polychlorinated biphenyls (PCBs) – 209 possible congeners Polychlorinated dibenzodioxins (PCDDs) –75 possible congeners Polychlorinated dibenzofurans (PCDFs) – 135 possible congeners	PCBs were banned globally following the Stockholm Convention in 2004; PCDDs, PCDFs restricted under the Stockholm Convention for intentional release. These halogenated aromatic hydrocarbons (HAHs) and their hydroxy-metabolites are implicated in reproductive failures in birds, reptiles, mustelids and cetaceans through oestrogen agonist and antagonist activity. They disrupt thyroid homeostasis in polar bears.
Bisphenol A (BPA): 4,4-dihydroxy-2,2-diphenylpropane	BPA is a constituent of polycarbonate plastics and epoxy resins. It has oestrogenic activity affecting sexual development, as demonstrated in laboratory studies.
Nonylphenol	Nonylphenol is a degradation product of alkylphenol ethoxylate surfactants used in detergents. Laboratory studies indicate that these chemicals are oestrogenic and may also affect thyroid function.
Polybrominated diphenyl ethers (PBDEs) – 209 possible congeners	Widely used as flame retardants, although now restricted under the Stockholm Convention. Adverse effects include neurotoxicity in birds and mammals, and delayed amphibian metamorphosis.
Perfluorooctanesulphonic acid (PFOS) Perfluorooctanoic acid (PFOA)	Used to treat textiles, to coat surfaces and in fire-fighting foams. They are persistent compounds found in tissues of fish, reptiles, birds and mammals. Oestrogenic and anti-androgenic activity and thyroid disruption have been observed, although the exact modes of action are unknown. Both compounds are now restricted under the Stockholm Convention, although several nations are out of compliance.
Tributyltin (TBT) Triphenyltin (TPT)	Tributyltin and triphenyltin are used as biocidal antifouling agents. Use is now restricted to large ocean-going vessels. Linked to masculinization ('imposex') in gastropods and declining populations of bivalves and crustaceans. Both organometal compounds are agonists of the retinoid X receptor (RXR) and peroxisome proliferator-activated receptors (PPAR), both of which are mediators of aromatase gene expression. Where restrictions have been enforced, subsequent recoveries of sensitive species have been reported.
Phthalates	Implicated in 'phthalate syndrome' in exposed male rodents, with reduced production of testosterone in Leydig cells in the testis and altered expression of genes regulating pregnancy. These plasticizers alter fatty acid metabolism through binding to the PPAR.
Pharmaceuticals. Examples include: Ibuprofen (non-prescription analgesic) 17α-ethinyloestradiol (synthetic oestrogen) Carbamazepine (anti-epileptic) Fluoxetine (antidepressant) Sulfathiazole (antibiotic)	Ibuprofen induced VTG and reduced reproduction in fish and affected tadpole metamorphosis in amphibians. 17α-ethinyloestradiol (EE2) is a potent oestrogen agonist. Carbamazepine reduced 11-ketotestosterone levels in fish. Fluoxetine induced VTG and affected behaviour in male fish. Sulfathiazole increased levels of oestradiol and increased aromatase activity in fish.
Personal care products. Examples include: Triclosan (antibacterial) Parabens (preservatives) Benzophenones (sunscreens)	Triclosan induced VTG in fish and decreased T3 thyroid hormones in amphibians. Parabens and benzophenones induced VTG in fish.

8.5.1 Xenobiotics in Wastewater as Sex Steroid Mimics

Many organic environmental contaminants act as steroid agonists or antagonists, leading to the inappropriate transduction of steroid signalling and subsequent gene expression. Perhaps there is no better example of an endocrine disruptor that alters oestrogenic signalling than the active ingredient of the birth control pill, 17α-ethinyloestradiol (EE2). This pharmaceutical is an environmental contaminant discharged in domestic wastewater that affects all aspects of animal reproductive health, including VTG synthesis, steroid biosynthesis, secondary sex characteristics, gonadal development and reproductive behaviour. Researchers in the United Kingdom were alarmed when it was observed that male fish swimming in rivers in which low concentrations of EE2 were detected exhibited female characteristics (i.e., **feminization**). The fish exhibited a condition known as intersex which involves the co-occurrence of both female and male gametes in the gonad, as well as changes to secondary (i.e., external) sex characteristics. Since then, **gonadal intersex** has been observed in several other fish populations. Figure 8.7 is a photograph of gonadal intersex observed in a darter species collected from the Grand River in Ontario, Canada at a location downstream from a municipal wastewater treatment plant

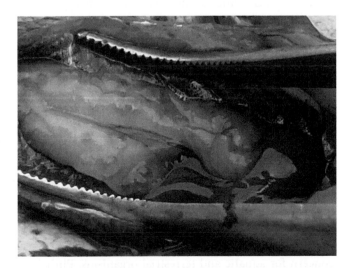

Figure 8.7 Gonadal intersex in a rainbow darter (*Etheostoma caeruleum*) collected from the Grand River in Ontario, Canada. Note the presence of mature oocytes (eggs) in the male testis, as indicated by the yellow blotches. The milky white area is the testis. Photograph provided by Mark Servos of the University of Waterloo, ON, Canada.

(WWTP). Although the effects of intersex on reproductive performance, particularly behavioural effects, are difficult to quantify, decreased egg production has been shown to be a consequence of exposure to oestrogenic effluent (Thorpe *et al.*, 2009). Some of the more eye-opening data regarding the effects of EE2 in aquatic environments were derived from a whole-lake addition study conducted at the Experimental Lakes Area in Canada (Kidd *et al.*, 2007). These investigations revealed significant and rapid declines in the abundance of native populations of fathead minnows (*Pimephales promelas*) following EE2 additions to Lake 260 over two field seasons at a very low concentration of about 5 ng/l. This was a dramatic example of how endocrine disruptors can jeopardize wild populations.

A significant number of xenobiotics that are agonists or antagonists of sex steroids are present in treated domestic wastewater and treated sewage sludge (i.e., **biosolids**) and are released into surface waters or applied onto agricultural soils. These include numerous pharmaceuticals and personal care products, biocides and chemicals produced by industry. It has often been said that aquatic organisms downstream from wastewater treatment plants are swimming in a 'sea of oestrogens' that can activate or block oestrogen receptor signalling. Examples of oestrogen receptor activators discharged in wastewater include EE2, the natural steroid hormone, E2 and its metabolites (i.e., oestrone, oestriol), BPA and structural analogues that are now being used as replacement compounds in polycarbonate plastics (e.g., bisphenol K and bisphenol S), paraben compounds, alkyphenols and benzophenones, to name a few. Other compounds like triclosan (bactericide), linuron (herbicide) and vinclozolin (fungicide) can act as anti-androgens and produce similar responses.

The good news is that treatment methods at municipal wastewater facilities are improving over time. However, it is important to stay diligent, as these technologies are not 100% effective at removing EDCs. In addition, most of these technological advances are taking place in highly developed regions of the world, whereas countries with developing economies have been left behind in controlling the release of EDCs into the environment.

Case Study 8.1 Gonadal Intersex in Fish

Recent studies in the Grand River watershed in Ontario, Canada have provided a test case for evaluating the effects on fish exposed to EDCs of wastewater origin. The most striking response was the high prevalence and severity of intersex in male rainbow darters (*Etheostoma caeruleum*) (Figure 8.7), as first described by Tetreault *et al.* (2011). These small bottom-dwelling fish spend their entire life history within a territory of a few square metres (i.e., **philopatric**), and thus they are a reliable indicator species for local exposure to oestrogenic chemicals and other EDCs. The WWTPs serving the cities of Kitchener and Waterloo were identified as the primary sources of the oestrogens. A key **biomarker** response noted in male rainbow darters was the induction of the egg-yolk protein, VTG, which is an indicator of exposure to oestrogenic chemicals (see Section 8.4). Figure 8.8 illustrates the pathway for the synthesis of this protein in oviparous female fish. Several EDCs have been detected in the Grand River, including the synthetic oestrogen EE2 and the anti-androgen triclosan. Upgrades in sewage treatment at the Kitchener and Waterloo WWTPs have resulted in dramatic changes in various biological endpoints in male rainbow darters, such as lowered levels of VTG and reduced prevalence of intersex (Hicks *et al.*, 2017; Marjan *et al.*, 2017). These studies show

that investments to improve sewage treatment will have benefits in terms of reducing the impacts of EDCs on aquatic organisms.

8.5.2 Phthalates as EDCs

Studies with rats have shown that prenatal exposures to phthalates produces a range of developmental responses in male pups termed '**phthalate syndrome**', which includes a high incidence of hypospadias (congenital deformities of the urethral opening), and cryptorchidism (non-descent of the testicles at birth), shortened anogenital distance and reduced penis size. This syndrome also includes reduced steroidogenesis in the Leydig cells that are present in the developing male testis, and this can have long-term effects, such as reduced sperm counts when the rodents become sexually mature (Svechnikov *et al.*, 2010). Although the majority of the toxicological studies with phthalates have focused on male development and reproduction in mammals, phthalates may have a broad range of endocrine disrupting effects in aquatic organisms exposed to environmentally relevant (i.e., low µg/l) concentrations (Mathieu-Denoncourt *et al.*, 2015). In a review of toxicological research with fish species exposed to the phthalate DEHP, Golshan and Alavi (2019) documented that exposures of some fish species to this compound at concentrations as low as 1 µg/l can modulate levels of androgens and oestrogens and increase levels of VTG. Other studies with the Murray rainbow fish (*Melanotaenia fluviatilis*) indicate that exposures for 90 d to phthalates at concentrations as low as 5 µg/l resulted in changes in E2/11-ketotestosterone (11-KT) ratios, feminization of the gonad and reduced reproduction (Bhatia *et al.*, 2014).

Recent bans for some phthalates (e.g., DEHP) in the European Union have resulted in the production of new phthalate compounds in the plastics industry that differ in the ester groups attached to the same phthalic acid backbone (Park *et al.*, 2019). With the emerging issue of micro- and nanoplastics in our oceans (Section 14.2), the global environmental impact of plasticizers and the plastics that contain them is becoming more of an ecological concern for aquatic and terrestrial organisms. Phthalates have been classified as 'obesogens', based upon the interactions of these compounds with various **peroxisome proliferator-activated receptors** (PPARs), which alter lipid regulation and adiposity in rodents (Wassenaar and Legler, 2017).

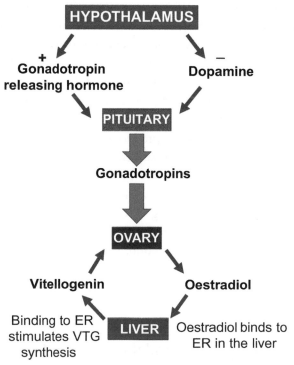

Figure 8.8 Pathway for production of vitellogenin (VTG) in female oviparous fish under the control of the HPG axis.

8.5.3 Atrazine as an EDC

Endocrine disruption by atrazine has been shown to be complex, involving the alteration of many essential physiological and molecular activities. Researchers from eight countries joined efforts to compile these data and concluded that atrazine acts as an endocrine disruptor by demasculinizing and feminizing males across all vertebrate taxa (Hayes *et al.*, 2011). Several MoAs for atrazine were suggested, including the alteration of mRNA and/or the hormone levels of several key targets, such as the oestrogen and androgen receptors. Exposures to atrazine at environmentally relevant concentrations have also been shown to elicit transgenerational reproductive dysfunction (Cleary *et al.*, 2019), behavioural changes and brain abnormalities in fish (Horzmann *et al.*, 2018), disruption to metamorphosis in amphibians (Langlois *et al.*, 2010), and changes in immune responses (Oliveira *et al.*, 2018). The list of physiological and molecular effects of atrazine across vertebrate taxa continues to grow yearly.

Based on these and other studies indicating that atrazine alters vertebrate reproduction and development, and because it is ubiquitous in the environment, this herbicide was banned from use in the European Union in 2003. However, most other countries around the world continue to use this herbicide in large volumes for weed control on agricultural lands (Section 7.8.9).

8.5.4 Flame Retardants as EDCs

Brominated flame retardants (BFRs) are widely used as additives to slow ignition in a variety of manufactured materials, such as plastics and textiles. Under higher temperatures, brominated compounds release hydrogen bromide, which reacts with the H⁻ and OH⁻ radicals in flames. This results in more stable halogen radicals that are less reactive than the precursor radicals, thus limiting the process of combustion. The BDE class of BFRs is now restricted under the Stockholm Convention (see Chapters 1, 7 and 12), but because these chemicals are persistent, they are still being detected in environmental matrices, such as animal tissues, and in sediment, soil, water and dust in the air. In Brazil and elsewhere, landfills have been identified as active sources of flame retardants that are released into the surrounding environment and into groundwater (Cristale *et al.*, 2019). Wildlife are attracted to landfills for foraging and thus may accumulate high concentrations of flame retardants.

Brominated flame retardants can cause a wide range of endocrine disrupting activities. In a study with the American kestrel (*Falco sparverius*), exposures in the diet for 82 d to the β-isomer of 1,2-dibromo-4-(1,2-dibromoethyl) cyclohexane at a daily dose of 0.239 ng/g of the bird (which was judged to be environmentally relevant), enhanced androgen-dependent behaviour such as copulation and aggression. This observation indicated that this compound is an androgen agonist in male birds (Marteinson *et al.*, 2015). This compound also increased testosterone levels in male kestrels and yielded higher free T4 levels in this same species (Marteinson *et al.*, 2017). Tetrabromobisphenol A (TBBPA) is now the brominated flame retardant with the highest worldwide sales volume, at over 210,000 tonnes per year. TBBPA was first employed to replace BDEs because its half-life in mammals is relatively short (<24 h). However, TBBPA has been detected repeatedly in the tissues of wildlife and humans. As illustrated in Figure 8.9, TBBPA shares structural similarities with thyroid hormones (i.e., T4, T3), immediately raising the question of its potential to

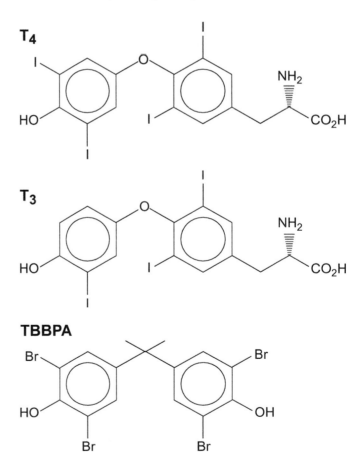

Figure 8.9 Structures of T4 and T3 thyroid hormones (iodinated) and TBBPA (brominated).

disrupt thyroid homeostasis. Fini *et al.* (2012) observed that TBBPA inhibited metamorphosis in *Xenopus laevis* tadpoles exposed to low µM concentrations. In associated *in vitro* tests, TBBPA displaced T3 from the thyroid receptor (i.e., TRα) in human, *Xenopus* and zebrafish cell lines. These *in vitro* tests provide an effective screening tool for thyroid hormone disruption in vertebrates.

8.5.5 Legacy Contaminants as EDCs

Over time, we have phased out many synthetic organic compounds because they persist in the environment, have a high potential for bioaccumulation and show sublethal toxicity, including evidence of endocrine disruption. These organic contaminants include BDEs and organochlorine pesticides, as well as PCBs (Chapter 7). However, owing to their environmental persistence, these contaminants continue to be distributed on a global scale. Many of these compounds are highly lipophilic and are readily transferred to higher trophic levels in food webs by the process of biomagnification (Section 4.5.2). This results in significantly high **body burdens** in the fat of top predators, which leads to impaired reproduction, immune system dysfunction and altered hormone signalling.

There are two dramatic and recent examples of how these persistent organic pollutants (POPs) and endocrine disruptors can negatively affect large marine and terrestrial mammals. The first example comes from our world's oceans. A study published in the journal *Science* warns of a global decline in killer whales (orcas) due to PCB contamination (Desforges *et al.*, 2018). The authors collected data on PCBs in blubber and related these residues to population-level responses using computer modelling. The PCBs were suspected of causing reproductive dysfunction and compromising immunity in these cetaceans. Figure 8.10 illustrates the routes of exposure of orcas to PCBs. In this study, researchers from around the globe reported the concentrations of PCBs in subpopulations of killer whales, including resident orcas in Alaska, the eastern Pacific (Case Study 4.6), the Canadian north, Greenland and Norway, and other transient populations of orcas off the coasts of Iceland, Hawaii, Japan and the United Kingdom. Some populations have higher concentrations of PCBs due to their diet and proximity to contaminated environments, whereas other populations inhabit sites with less pollution. Strikingly, model simulations revealed that the higher the concentrations of PCBs in blubber, the lower the predicted annual growth of the subpopulation (Desforges *et al.*, 2018). Moreover, the study concluded that there is a high risk of global collapse of killer whale populations due to PCB accumulation, and the long-term prognosis for this cetacean species is a 50% reduction in numbers within 30–50 years. This illustrates how an organic contaminant from a class that was banned more than 40 years ago can still impact wildlife populations today.

The second example of how POPs can place large mammals at risk is the case history of polar bears

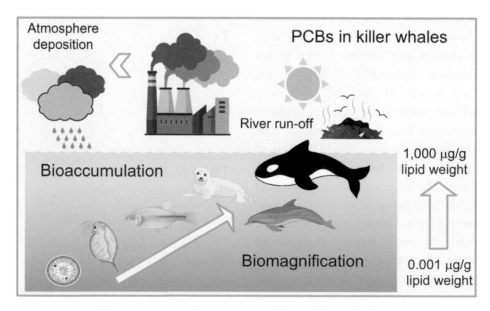

Figure 8.10 Routes for exposure and bioaccumulation of PCBs in populations of killer whales (*Orcinus orca*). Adapted from Desforges *et al.* (2018). Generated with BioRender.

(*Ursus maritimus*) in the Arctic. Many different types of organic chemicals have made their way into the Arctic through long-range transport via wind and water currents. Interestingly, a significant source of pollution in the Arctic can be attributed to migratory fish and birds that end their life cycle in the Arctic, thereby releasing the organic chemicals that they carry into the local Arctic environment. Nations with polar regions, including Canada, the United States (Alaska), Russia and Scandinavia, have been monitoring organic pollutants in polar bears for some time. Although there is little doubt that climate change and habitat loss are partly responsible for declining numbers of polar bears, environmental pollution also plays a significant role. A recent literature review on polar bears concluded that legacy POPs remain the primary contaminant found in tissues of polar bears today (Routti *et al.*, 2019). These contaminants include PCBs, organochlorine insecticides, perfluorinated compounds and BFRs, which are all readily detected in individuals from 19 polar bear subpopulations. Similar to the situation with killer whales, polar bears are at the top of the food chain in the Arctic ecosystem, feeding predominantly on seals that are rich in the fats that sequester many of these compounds.

Studies summarized by Routti *et al.* (2019) show that testosterone levels in male polar bears are negatively correlated with plasma concentrations of PCBs and organochlorine pesticides, and there are other data showing that these legacy contaminants can affect steroid homeostasis in the bears. The impacts of POPs in the Arctic may also apply to other animals in the region, including the Arctic fox, walrus, ring seal and beluga whale. Many of the contaminated animals in the Arctic are a source of food for Indigenous peoples, and accordingly there are concerns about endocrine and other health effects in human populations living in the region, and in particular, the impacts of exposure in women of childbearing age (Binnington *et al.*, 2016).

8.5.6 Organotins as EDCs

Organotins are organometallic compounds in which an atom of tin (Sn) is covalently bonded to one or more organic chains. Butyltins and phenyltins are used as insecticides and fungicides in the PVC plastics industry and as industrial catalysts. However, their widest use is as additives to antifouling paints that are applied to the hulls of boats and ships to inhibit encrustation by marine organisms, such as barnacles and mussels. As a result of this usage, high concentrations of **organotins**, and in particular tributyltin (TBT), have been detected in water and sediments around harbours and marinas. Use of TBT in marine antifouling paints was restricted by an international convention in 2008, but these compounds and their degradation products (i.e., dibutyltin, monobutyltin) are still present in contaminated sediments around the world.

The most obvious effect of endocrine disruption associated with exposure to TBT in polluted marine areas is the development of a syndrome called '**imposex**' in female gastropod molluscs. This masculinization condition is characterized by the development in female gastropods of male sexual characteristics, such as the presence of a penis and/or vas deferens. By the early 1980s, several reports linking high TBT concentrations with unnatural shell thickening and poor reproductive performance in oysters from the French coast began to appear. Laboratory experiments showed that TBT levels as low as 2 ng/l had detrimental effects on oysters. Further reports demonstrated that TBT caused imposex in the dog whelk (*Nucella lapillus*) characterized by the masculinization of females leading to reproductive failure. By 2009, this condition had been reported in 268 species of marine gastropods (Fernandez, 2019). The exact mechanism for the development of imposex is currently unclear, but it could result from the increase in levels of testosterone that have been observed in gastropods exposed to TBT.

Crustaceans are also affected by TBT, but unlike gastropods, females exposed to TBT do not develop complete male sex organs. Nevertheless, female crustaceans exposed to TBT show altered ovarian morphology, as well as **atresia** of oocytes. However, the primary cause for concern for crustaceans exposed to TBT is its effect on moulting (Vogt *et al.*, 2018). Crustacean growth occurs through repetitive moulting of the exoskeleton that is regulated by ecdysteroids. Ecdysone and 25-deoxyecdysone are inactive **ecdysteroids** that are secreted by the Y-organ in crustaceans and are converted in peripheral tissues to 20-hydroxyecdysone (20-HE) and ponasterone A, the active forms. Ecdysteroids bind to the arthropod ecdysteroid receptor (EcR) that complexes with RXR. The heterodimer EcR:RXR binds to the ecdysteroid response element within the genome to regulate the transcription of genes involved in development, growth and reproduction. Incomplete ecdysis leading to death occurs in crustaceans exposed to TBT, and this is thought to occur by the interference of TBT with the expression of RXR (Vogt *et al.*, 2018).

As reports of imposex in gastropods mounted in the literature, the problem of contamination by TBT received global attention and this was also extended to concerns about triphenyltin (TPT), an alternative to TBT as a biocide. Bans on the use of TBT on boats less than 25 m long first started in the 1980s. However, regulations on complete bans did not occur until 2008 when organotin biocides in antifouling paints were banned under the Rotterdam Convention. Enforcement of these bans among countries that are signatories to the convention has been inconsistent, notably in some Caribbean countries. Unregulated use of these compounds has been documented in other regions, such as maritime countries in Latin America and North Africa. In several European countries that have banned TBT-based antifouling paints, recoveries of populations of gastropods and crustaceans have been documented (Fernandez, 2019; see Table 8.2).

8.6 EDCs as a Human Health Concern

An important final point is that humans are just as susceptible to EDCs as the organisms we study. Although examples above come from the field of ecotoxicology, we must not forget that humans are integrated with our environment and the wildlife that co-inhabit it (i.e., the **One Health** concept). In addition, much of the work to determine the MoAs of EDCs is rooted in research conducted with rodent model species (i.e., rats, mice), human clinical trials and epidemiological studies. The research focused on endocrine disruption in humans can point ecotoxicologists in the right direction for similar studies with other vertebrates and even invertebrates. For instance, reports of infertility in humans addicted to opioid drugs recently led to a study that showed that female fish exposed to the opioid drug codeine produced fewer eggs than the unexposed controls and had low levels of oestradiol (Fischer *et al.*, 2021).

The accumulated health data, as well as laboratory studies that show similar responses in rodents, are beginning to show that exposures to environmental contaminants, particularly during foetal development, may have long-lasting effects on the health of humans. These observations highlight the need for parallel studies of similar MoAs in mammalian and non-mammalian species (including invertebrates) exposed to xenobiotics in the environment.

8.7 Conclusions

Endocrine disrupting compounds are ubiquitous in the environment and range from legacy POPs to contaminants of emerging concern, such as perfluorinated compounds, brominated flame retardants, plasticizers, pharmaceuticals and personal care products. These EDCs can affect reproduction, immune systems, growth, lipid metabolism and other processes essential for healthy populations of both animals and humans. Ecotoxicologists rigorously evaluate the potential for toxicity and endocrine disrupting properties of new chemicals. Recent studies with humans have also provided better information on the identity of EDCs and their sources. For instance, Di Nisio *et al.* (2019) studied men from areas in northern Italy in which there are high concentrations of perfluorinated chemicals (PFCs) in drinking water. They observed that the levels of these compounds in plasma were correlated with reduced semen quality and anomalies in male sex organs. While there has been an initial emphasis on human health effects for these compounds, concern has expanded to include potential detrimental effects caused by contamination of terrestrial and aquatic ecosystems. For example, exposure of fish to PFCs has been associated with disruption of the HPG and HPT axes, with adverse effects on the sexual development of fish larvae and fish reproduction, including multi-generational effects (Lee *et al.*, 2017). As observed in mammalian species, several PFCs including PFOA and PFOS have been shown to activate nuclear receptors involved in lipid metabolism, such as PPARs in fish (Yang *et al.*, 2014).

It will be essential to diligently evaluate new products that enter the marketplace for potential effects on endocrine systems. We need safe chemicals for use in industry, for pharmaceuticals that contribute to our health and wellbeing, and for products that make our life easier and more productive, but we must always recognize that these chemicals have the potential for negative impacts when released into the environment.

While the focus of this book is primarily on non-human organisms, it is not always possible to draw a sharp distinction between effects in humans and wildlife. Much of our knowledge of the MoAs of EDCs has been derived from mammalian test species used as human surrogates (e.g., rats, mice). At the molecular level, *in vitro* studies using mammalian cell lines have provided valuable information on endocrine function and molecular signalling pathways.

Information from mammalian studies may be extrapolated to mammalian wildlife and to livestock, and vice versa. Growing evidence points to some common responses across vertebrate taxa in the effects of EDCs, as indicated by the successful use of lower vertebrates as model organisms for EDC action, including zebrafish (Segner, 2009) and *Xenopus* frogs (Fini *et al.*, 2012). Amphibian species have been recognized as sensitive laboratory models for extrapolating the effects of EDCs from wildlife to humans (Holbech *et al.*, 2020). Nevertheless, it is important to recognize that the dosing regimens adopted in laboratory-based investigations of EDCs are not necessarily comparable with the levels and the routes of exposure of these chemicals in the environment.

In this chapter, we have highlighted clear instances of adverse effects in populations of fish and wildlife that can be linked to exposures to xenobiotics. In many cases, it is difficult to identify specific active compounds from the mix of environmental contaminants that could be causing these effects. On the other hand, laboratory research and human epidemiological studies have provided unique insights into the MoAs of specific chemicals, including multigenerational effects. Chemical risk assessors will need to acknowledge that EDCs do not behave like classic contaminants; they do not necessarily exhibit classic dose–response behaviour, and therefore EDCs should be assessed differently from contaminants that do not disrupt the endocrine system. Also, LC_{50} determinations are no longer needed for EDCs; evidence of the sublethal effects of EDCs at environmentally realistic concentrations is what risk assessment models should be incorporating.

Summary

In this chapter, we discussed the evidence from laboratory and field studies for endocrine disruption in wildlife as caused by exposure to xenobiotics. We evaluated the wide range of EDCs from persistent 'legacy' compounds to more recently manufactured chemicals. We provided an inventory of the main classes of EDCs and their sites of action at the cellular and tissue level. We described the various MoAs of EDCs, including their roles as agonists or antagonists, and the cellular receptors and signalling pathways involved in the functioning of endocrine systems. We also described how the endocrine system is involved in cellular differentiation, growth and development in early life stages of organisms, and regulates a range of physiological functions in adults, including reproduction. Finally, we gained an appreciation for the difficulties involved in extrapolating from effects observed in laboratory studies with EDCs to effects in fish and wildlife that are exposed in the natural environment to xenobiotics, including mixtures of these compounds.

Review Questions and Exercises

1. Describe the different endpoints that have been used to identify the effects of EDCs on ecosystems. Give examples of how studies of EDCs with rodents and humans have elucidated and expanded our knowledge of their effects on wildlife.

2. Describe the role played by the hypothalamic–pituitary–adrenal (HPA) axis and the hypothalamic–pituitary–thyroid (HPT) axis in controlling the physiology of adult vertebrate organisms.

3. Give an example of how exposure to EDCs during early life stages (e.g., during foetal development) can irreversibly alter development or the physiology of organisms over their entire lifetime.

4. Describe the various functions of the pituitary hormone prolactin in mammals, amphibians and fish.

5. What is 'aromatase' and why is the activity of this enzyme so important in regulating the development of vertebrates?

6. What are the principal receptors regulating the function of sex steroids? Give examples of EDCs that are agonists and antagonists of these receptors.

7. Describe the synthesis and transport of thyroid hormones. Give three examples of EDCs that disrupt thyroid hormone function and their mode of action.

8. Describe why induction of the egg-yolk protein vitellogenin in male oviparous vertebrates can be used as a biomarker of exposure to EDCs.

9. Give two examples of classes of xenobiotics that are recognized as EDCs in vertebrate organisms, including humans, and describe their mode of action.

10. Give examples of wildlife populations that are at risk due to chronic exposure to EDCs.

Abbreviations

11-KT	11-ketotestosterone	LH	Luteinizing hormone
2,4-D	2,4-dichlorophenoxyacetic acid	LNG	Levonorgestrel
5α-DHT	5α-dihydrotestosterone	MoA	Mode of action
ACTH	Adrenocorticotropic hormone	MRPs	Multidrug resistance proteins
BDE	Brominated diphenyl ether	PAH	Polycyclic aromatic hydrocarbon
BFRs	Brominated flame retardants	PCBs	Polychlorinated biphenyls
BPA	Bisphenol A	PCDDs	Polychlorinated dibenzodioxins
CAR	Constitutive androstane receptor	PCDFs	Polychlorinated dibenzofurans
CNS	Central nervous system	PFOA	Perfluorooctanoic acid
CRH	Corticotropin-releasing hormone	PFOS	Perfluorooctanesulphonic acid
CYP19	Aromatase enzyme	POPs	Persistent organic pollutants
CYP450	Cytochrome P450 enzyme	PPAR	Peroxisome proliferator-activated receptor
DDE	1,1-dichloro-2,2-bis(4-chlorophenyl)ethylene	PRH	Prolactin-releasing hormone
DDT	1-chloro-4-[2,2,2-trichloro-1-(4-chlorophenyl)ethyl]benzene	PRL	Prolactin
		PXR	Pregnane X receptor
DEHP	Di(2-ethylhexyl)phthalate	RXR	Retinoid X receptor
DHT	Dihydrotestosterone	SST	Somatostatin
E2	17β-oestradiol	StAR	Steroidogenic acute regulatory (protein)
EDCs	Endocrine disrupting chemicals	SXR	Steroid and xenobiotic receptor
EE2	17α-ethinyloestradiol	T3	Triiodothyronine
ELISA	Enzyme-linked immunoabsorbent assay	T4	Thyroxine
FSH	Follicle-stimulating hormone	TBBPA	Tetrabromobisphenol A
GABA	Gamma-aminobutyric acid	TBG	Thyroxine-binding globulin
GH	Growth hormone	TBT	Tributyltin
GHRH	Growth-hormone-releasing hormone	TPT	Triphenyltin
GnRH	Gonadotropin-releasing hormone	TRH	Thyroid-releasing hormone
HAHs	Halogenated aromatic hydrocarbons	TSH	Thyroid-stimulating hormone
HPA	Hypothalamic–pituitary–adrenal axis	TTR	Transthyretin
HPG	Hypothalamic–pituitary–gonadal axis	VTG	Vitellogenin
HPT	Hypothalamic–pituitary–thyroid axis	WWTP	Wastewater treatment plant

References

Arukwe, A. (2005). Modulation of brain steroidogenesis by affecting transcriptional changes of steroidogenic acute regulatory (StAR) protein and cholesterol side chain cleavage (P450 scc) in juvenile Atlantic Salmon (*Salmo salar*) is a novel aspect of nonylphenol toxicity. *Environmental Science & Technology*, 39, 9791–9798.

Bacqué-Cazenave, J., Bharatiya, R., Barrière, G., *et al.* (2020). Serotonin in animal cognition and behavior. *International Journal of Molecular Sciences*, 21, e1649.

Balch, G. C., Vélez-Espino, L. A., Sweet, C., Alaee, M. & Metcalfe, C. D. (2006). Inhibition of metamorphosis in tadpoles of *Xenopus laevis* exposed to polybrominated diphenyl ethers (PBDEs). *Chemosphere*, 64, 328–338.

Bhatia, H., Kumar, A., Ogino, Y., *et al.* (2014). Di-n-butyl phthalate causes estrogenic effects in adult male Murray rainbowfish (*Melanotaenia fluviatilis*). *Aquatic Toxicology*, 149, 103–115.

Binnington, M. J., Curren, M. S., Chan, H. M. & Wania, F. (2016). Balancing the benefits and costs of traditional food substitution by indigenous Arctic women of childbearing age: Impacts on persistent organic pollutant, mercury, and nutrient intakes. *Environment International*, 94, 554–566.

Bissegger, S., Castro, M. A. P., Yargeau, V. & Langlois, V. S. (2018). Phthalates modulate steroid 5-reductase transcripts in the Western clawed frog embryo. *Comparative Biochemistry and Physiology Part C: Toxicology and Pharmacology*, 213, 39–46.

Brooks, B. W. (2014). Fish on Prozac (and Zoloft): ten years later. *Aquatic Toxicology*, 151, 61–67.

Bulaeva, E., Lanctot, C., Reynolds, L., Trudeau, V. L. & Navarro-Martin, L. (2015). Sodium perchlorate disrupts development and affects metamorphosis- and growth-related gene expression in tadpoles of the wood frog (*Lithobates sylvaticus*). *General and Comparative Endocrinology*, 222, 33–43.

Cao, F., Li, H., Zhao, F., *et al.* (2019). Parental exposure to azoxystrobin causes developmental effects and disrupts gene expression in F1 embryonic zebrafish (*Danio rerio*). *Science of the Total Environment*, 646, 595–605.

Castañeda Cortés, D. C., Langlois, V. S. & Fernandino, J. I. (2014). Crossover of the hypothalamic pituitary–adrenal/interrenal,–thyroid, and–gonadal axes in testicular development. *Frontiers in Endocrinology*, 5, e139.

Chen, X., Fang, M. L., Chernick, M., *et al.* (2019). The case for thyroid disruption in early life stage exposures to thiram in zebrafish (*Danio rerio*). *General and Comparative Endocrinology*, 271, 73-81.

Cleary, J. A., Tillitt, D. E., Vom Saal, F. S., *et al.* (2019). Atrazine induced transgenerational reproductive effects in medaka (*Oryzias latipes*). *Environmental Pollution*, 251, 639–650.

Colborn, T., Dumanoski, D. & Myers, J. P. (1996). *Our Stolen Future – Are We Threatening Our Fertility, Intelligence, and Survival? A Scientific Detective Story*, New York, NY: Dutton.

Couée, I., Serra, A.-A., Ramel, F., Gouesbet, G. & Sulmon, C. (2013). Physiology and toxicology of hormone-disrupting chemicals in higher plants. *Plant Cell Reports*, 32, 933–941.

Cristale, J., Belé, T. G. A., Lacorte, S. & de Marchi, M. R. R. (2019). Occurrence of flame retardants in landfills: a case study in Brazil. *Environmental Research*, 168, 420–427.

Desforges, J.-P., Hall, A., McConnell, B., *et al.* (2018). Predicting global killer whale population collapse from PCB pollution. *Science*, 361, 1373–1376.

Di Nisio, A., Sabovic, I., Valente, U., *et al.* (2019). Endocrine disruption of androgenic activity by perfluoroalkyl substances: Clinical and experimental evidence. *Journal of Clinical Endocrinology and Metabolism*, 104, 1259–1271.

Dong, W., Macaulay, L. J., Kwok, K. W., Hinton, D. E. & Stapleton, H. M. (2013). Using whole mount in situ hybridization to examine thyroid hormone deiodinase expression in embryonic and larval zebrafish: a tool for examining OH-BDE toxicity to early life stages. *Aquatic Toxicology*, 132, 190–199.

Duarte-Guterman, P., Langlois, V. S., Hodgkinson, K., *et al.* (2009). The aromatase inhibitor fadrozole and the 5-reductase inhibitor finasteride affect gonadal differentiation and gene expression in the frog *Silurana tropicalis*. *Sexual Development*, 3, 333–341.

Faheem, M., Jahan, N., Khaliq, S. & Lone, K. P. (2019). Modulation of brain kisspeptin expression after bisphenol-A exposure in a teleost fish, *Catla catla*. *Fish Physiology and Biochemistry*, 45, 33–42.

Fernandez, M. A. (2019). Populations collapses in marine invertebrates due to endocrine disruption: a cause for concern? *Frontiers in Endocrinology*, 10, e721.

Fini, J.-B., Riu, A., Debrauwer, L., *et al*. (2012). Parallel biotransformation of tetrabromobisphenol A in *Xenopus laevis* and mammals: *Xenopus* as a model for endocrine perturbation studies. *Toxicological Sciences*, 125, 359–367.

Fischer, A. J., Kerr, L., Sultana, T. & Metcalfe, C. D. (2021). Effects of opioids on reproduction in Japanese medaka, *Oryzias latipes*. *Aquatic Toxicology*, 236, e105873.

Gagné, F., André, C., Cejka, P., Hausler, R. & Fournier, M. (2011). Evidence of neuroendocrine disruption in freshwater mussels exposed to municipal wastewaters. *Science of the Total Environment*, 409, 3711–3718.

Gagné, F., Blaise, C., Pellerin, J. & André, C. (2007). Neuroendocrine disruption in *Mya arenaria* clams during gametogenesis at sites under pollution stress. *Marine Environmental Research*, 64, 87–107.

Gesto, M., Tintos, A., Soengas, J. L. & Míguez, J. M. (2006). Effects of acute and prolonged naphthalene exposure on brain monoaminergic neurotransmitters in rainbow trout (*Oncorhynchus mykiss*). *Comparative Biochemistry and Physiology Part C: Toxicology & Pharmacology*, 144, 173–183.

Golshan, M. & Alavi, S. M. H. (2019). Androgen signaling in male fishes: examples of anti-androgenic chemicals that cause reproductive disorders. *Theriogenology*, 139, 58–71.

Hayes, T. B., Anderson, L. L., Beasley, V. R., *et al*. (2011). Demasculinization and feminization of male gonads by atrazine: consistent effects across vertebrate classes. *Journal of Steroid Biochemistry and Molecular Biology*, 127, 64–73.

Hazelton, P. D., Du, B., Haddad, S. P., *et al*. (2014). Chronic fluoxetine exposure alters movement and burrowing in adult freshwater mussels. *Aquatic Toxicology*, 151, 27-35.

Hicks, K. A., Fuzzen, M. L., McCann, E. K., *et al*. (2017). Reduction of intersex in a wild fish population in response to major municipal wastewater treatment plant upgrades. *Environmental Science & Technology*, 51, 1811–1819.

Holbech, H., Matthiessen, P., Hansen, M., *et al*. (2020). ERGO: Breaking down the wall between human health and environmental testing of endocrine disrupters. *International Journal of Molecular Sciences*, 21, e2954.

Horzmann, K. A., Reidenbach, L. S., Thanki, D. H., *et al*. (2018). Embryonic atrazine exposure elicits proteomic, behavioral, and brain abnormalities with developmental time specific gene expression signatures. *Journal of Proteomics*, 186, 71–82.

Kassotis, C. D., Vandenberg, L. N., Demeneix, B. A., *et al*. (2020). Endocrine-disrupting chemicals: economic, regulatory, and policy implications. *The Lancet Diabetes & Endocrinology*, 8, 719–730.

Kidd, K. A., Blanchfield, P. J., Mills, K. H., *et al*. (2007). Collapse of a fish population after exposure to a synthetic estrogen. *Proceedings of the National Academy of Sciences USA*, 104, 8897–8901.

Langlois, V. S., Carew, A. C., Pauli, B. D., *et al*. (2010). Low levels of the herbicide atrazine alter sex ratios and reduce metamorphic success in *Rana pipiens* tadpoles raised in outdoor mesocosms. *Environmental Health Perspectives*, 118, 552–557.

Lee, J. W., Lee, J. W., Kim, K., *et al*. (2017). PFOA-induced metabolism disturbance and multi-generational reproductive toxicity in *Oryzias latipes*. *Journal of Hazardous Materials*, 340, 231–240.

Lee, K.-M., Yang, W., Rhee, J.-S., *et al*. (2010). Effects of endocrine disruptors on *Bombina orientalis* P450 aromatase activity. *Zoological Science*, 27, 338–343.

Leemans, M., Couderq, S. C.-H., Demeneix, B. A. & Fini, J.-B. (2019). Pesticides with potential thyroid hormone-disrupting effects: a review of recent data. *Frontiers in Endocrinology*, 10, e743.

Legler, J. & Brouwer, A. (2003). Are brominated flame retardants endocrine disruptors? *Environment International*, 29, 879–885.

León-Olea, M., Martyniuk, C. J., Orlando, E. F., *et al*. (2014). Current concepts in neuroendocrine disruption. *General and Comparative Endocrinology*, 203, 158–173.

Lorenz, C., Contardo-Jara, V., Trubiroha, A., *et al*. (2011). The synthetic gestagen Levonorgestrel disrupts sexual development in *Xenopus laevis* by affecting gene expression of pituitary gonadotropins and gonadal steroidogenic enzymes. *Toxicological Sciences*, 124, 311–319.

Marjan, P., Martyniuk, C. J., Fuzzen, M. L., *et al*. (2017). Returning to normal? Assessing transcriptome recovery over time in male rainbow darter (*Etheostoma caeruleum*) liver in response to wastewater-treatment plant upgrades. *Environmental Toxicology and Chemistry*, 36, 2108–2122.

Marteinson, S. C., Letcher, R. J. & Fernie, K. J. (2015). Exposure to the androgenic brominated flame retardant 1,2-dibromo-4-(1,2-dibromoethyl)-cyclohexane alters reproductive and aggressive behaviors in birds. *Environmental Toxicology and Chemistry*, 34, 2395–2402.

Marteinson, S. C., Palace, V., Letcher, R. J. & Ferni, K. J. (2017). Disruption of thyroxine and sex hormones by 1,2-dibromo-4-(1,2-dibromoethyl)cyclohexane (DBE-DBCH) in American kestrels (*Falco sparverius*) and associations with reproductive and behavioral changes. *Environmental Research*, 154, 389–397.

Martyniuk, C. J., Doperalski, N. J., Kroll, K. J., Barber, D. S. & Denslow, N. D. (2013). Sexually dimorphic transcriptomic responses in the teleostean hypothalamus: a case study with the organochlorine pesticide dieldrin. *Neurotoxicology*, 34, 105–117.

Mathieu-Denoncourt, J., Wallace, S. J., de Solla, S. R. & Langlois, V. S. (2015). Plasticizer endocrine disruption: highlighting developmental and reproductive effects in mammals and non-mammalian aquatic species. *General and Comparative Endocrinology*, 219, 74–88.

Oliveira, S. E., Costa, P. M., Nascimento, S. B., *et al*. (2018). Atrazine promotes immunomodulation by melanomacrophage centre alterations in spleen and vascular disorders in gills from *Oreochromis niloticus*. *Aquatic Toxicology*, 202, 57–64.

Park, J., Park, C., Gye, M. C. & Lee, Y. (2019). Assessment of endocrine-disrupting activities of alternative chemicals for bis (2-ethylhexyl) phthalate. *Environmental Research*, 172, 10–17.

Qiu, W., Fang, M., Liu, J., *et al*. (2019). In vivo actions of bisphenol F on the reproductive neuroendocrine system after long-term exposure in zebrafish. *Science of the Total Environment*, 665, 995–1002.

Richter, C. A., Martyniuk, C. J., Annis, M. L., *et al*. (2014). Methylmercury-induced changes in gene transcription associated with neuroendocrine disruption in largemouth bass (*Micropterus salmoides*). *General and Comparative Endocrinology*, 203, 215–224.

Rosenfeld, C. S., Denslow, N. D., Orlando, E. F., Gutierrez-Villagomez, J. M. & Trudeau, V. L. (2017). Neuroendocrine disruption of organizational and activational hormone programming in poikilothermic vertebrates. *Journal of Toxicology and Environmental Health, Part B*, 20, 276–304.

Routti, H., Atwood, T. C., Bechshoft, T., *et al*. (2019). State of knowledge on current exposure, fate and potential health effects of contaminants in polar bears from the circumpolar Arctic. *Science of the Total Environment*, 664, 1063–1083.

Segner, H. (2009). Zebrafish (*Danio rerio*) as a model organism for investigating endocrine disruption. *Comparative Biochemistry and Physiology Part C: Toxicology and Pharmacology*, 149, 187–195.

Svechnikov, K., Izzo, G., Landreh, L., Weisser, J. & Söder, O. (2010). Endocrine disruptors and Leydig cell function. *Journal of Biomedicine and Biotechnology*, 2010, e684504.

Teng, M., Wang, C., Song, M., *et al*. (2020). Chronic exposure of zebrafish (*Danio rerio*) to flutolanil leads to endocrine disruption and reproductive disorders. *Environmental Research*, 184, e109310.

Tetreault, G. R., Bennett, C. J., Shires, K., *et al*. (2011). Intersex and reproductive impairment of wild fish exposed to multiple municipal wastewater discharges. *Aquatic Toxicology*, 104, 278–290.

Thorpe, K. L., Maack, G., Benstead, R. & Tyler, C. R. (2009). Estrogenic wastewater treatment works effluents reduce egg production in fish. *Environmental Science & Technology*, 43, 2976–2982.

Venerosi, A., Ricceri, L., Tait, S. & Calamandrei, G. (2012). Sex dimorphic behaviors as markers of neuroendocrine disruption by environmental chemicals: the case of chlorpyrifos. *Neurotoxicology*, 33, 1420–1426.

Viglino, L., Aboulfadl, K., Prevost, M. & Sauve, S. (2008). Analysis of natural and synthetic estrogenic endocrine disruptors in environmental waters using online preconcentration coupled with LC-APPI-MS/MS. *Talanta*, 76, 1088–1096.

Vogt, E. L., Model, J. F. & Vinagre, A. S. (2018). Effects of organotins on crustaceans: update and perspectives. *Frontiers in Endocrinology*, 9, e65.

Wassenaar, P. N. H. & Legler, J. (2017). Systematic review and meta-analysis of early life exposure to di(2-ethylhexyl) phthalate and obesity related outcomes in rodents. *Chemosphere*, 188, 174–181.

Waye, A. & Trudeau, V. L. (2011). Neuroendocrine disruption: more than hormones are upset. *Journal of Toxicology and Environmental Health, Part B*, 14, 270–291.

Weiss, J. M., Andersson, P. L., Lamoree, M. H., *et al.* (2009). Competitive binding of poly- and perfluorinated compounds to the thyroid hormone transport protein transthyretin. *Toxicological Sciences*, 109, 206–216.

Yang, S. L., Liu, S. C., Ren, Z. M., Jiao, X. D. & Qin, S. (2014). Induction of oxidative stress and related transcriptional effects of perfluorononanoic acid using an *in vivo* assessment. *Comparative Biochemistry and Physiology C: Toxicology and Pharmacology*, 160, 60–65.

Zeilinger, J., Steger-Hartmann, T., Maser, E., *et al.* (2009). Effects of synthetic gestagens on fish reproduction. *Environmental Toxicology and Chemistry*, 28, 2663–2670.

9 Natural Toxins

David A. Wright and Pamela M. Welbourn

Learning Objectives

Upon completion of this chapter, the student should be able to:
- Define natural toxins and to distinguish them, in terms of probable origin, from toxicants that result from human activity.
- Illustrate, with examples, the roles played by natural toxins in the natural environment, and to include, where known, the advantage(s) conferred on the organisms that produce these compounds.
- Understand the biological sources of the major toxins that are of concern for ecological or human health.
- Appreciate sources and modes of action of natural toxins that are used in pest management or biological warfare.

A large number of known organic compounds do not result directly from human activity, i.e., are natural. Of these, natural toxins are an important subset, defined as 'poisonous substances produced within living cells or organisms'. Some of these are among the most toxic substances known, for example amanitin, produced by the fungus *Amanita ocreata*, known as the 'Angel of Death'. In this chapter we provide examples from the range of organisms that produce toxins, their ecotoxicological effects and mode of action. We discuss the ecological significance of toxin production by a variety of microorganisms, animals and plants, including some explanations that are still speculative. Finally, we review some applications of natural toxins, including those that are still at the experimental stage.

9.1 What Is a Toxin?

Of the more than 20 million or so organic compounds that have been identified so far, only perhaps 200,000 xenobiotics and other organic substances have been or still are manufactured for a wide variety of uses. These may be deliberately or inadvertently released into the environment, where they can pose a risk to biological systems. Risks from some manufactured organic compounds were demonstrated dramatically by Rachel Carson in her book *Silent Spring* (Section 1.2.1), and the science of organic compounds as related to ecotoxicology is discussed in Chapter 7.

Some popular articles estimate that five of the seven most toxic substances known to us are natural toxins. Although this proportion is difficult to substantiate, and in any case depends on definitions, nevertheless some of the most toxic substances are those produced by living organisms. In some cases, no antidote is yet known, and exposure often leads to fatal results. While their synthesis generally does not result directly from human activity, we have long recognized their biochemical characteristics and have sought to adapt some toxins for human use in the form of pesticides, medicines and other products. 'Borrowing' these properties can involve either extracting the raw product or *de novo* synthesis.

In devoting a chapter to these compounds, we extend the definition of 'ecotoxicology' beyond simply 'the effect of anthropogenic pollutants on the environment' and recognize that natural toxins form the basis of many natural products. The distribution and effects of these toxins in their natural form can also be significantly influenced by phenomena associated with human activity such as eutrophication and climate change. Whether compounds are natural or manufactured, similar mechanisms of action and metabolism apply.

In general terms, the ecological roles of natural toxins can be defined as passive, typically as a defence against predators/consumers, or active, where prey immobilization/consumption is involved. In this chapter we explore some of the roles played by natural toxins in the natural environment, the hazards they pose and how some of them have been adopted for human use, usually as pharmaceuticals or pesticides. The chapter provides examples of some natural toxins, arranged according to the taxonomic source, namely harmful algal blooms (Section 9.3.1), vascular plants (9.3.2), fungi and bacteria (9.3.3) and animals (9.3.4). We also discuss the ecological advantages (Section 9.4) and some applications of natural toxins (Section 9.5). Appendix 9.1 summarizes the sources, chemical nature and properties of a wider range of natural toxins.

9.2 Evolutionary Perspective and Role of Natural Toxins

In studying the MoA of natural toxins and the adaptation of a small number for human use, we recognize that their biochemical activity and metabolism long pre-date human activity. The influence of humans on the geosphere and biosphere can be detected over a period of only a few thousand years. In contrast, the timescale over which toxins have evolved is several orders of magnitude longer than the era of human influence on the natural environment; the evolutionary processes whereby organisms have adapted to their physical, chemical and biological environment may be counted in billions of years. Such adaptive mechanisms have evolved in response to enormous prehistoric changes in the geosphere, including some natural phenomena still in evidence today. Tectonic plate disruptions, historic climate change and resulting events such as earthquakes, volcanoes and forest fires have all served to create shifts and transformations in the chemosphere that long pre-date human existence. In addition to adapting to their chemical environment, microorganisms, plants and animals have had to accommodate to other organisms, which may be predators, prey (food), competitors and pathological agents. So-called 'natural toxins' may, therefore, be endogenous or exogenous chemicals that are fundamental to this evolutionary process of adaptation and accommodation. To make an analogy, we might consider that there is 'chemical warfare' among many individual species and communities, or even an 'evolutionary arms race' (Walton, 2018b), the result of long-term adaptation.

Secondary metabolites or excretory products from one organism may be detrimental to the health, survival or reproduction of a second organism. This may result in a selective **ecological advantage** for the source of the toxicant (which we define here as the '**protagonist**'), particularly if the second organism is a potential competitor or consumer. **Allelopathy**, a term coined by Hans Molisch in 1937, is a specific type of competitive ecology. Interjit and Duke (2003) review various definitions of the term, which was originally defined as the effect of a plant on the growth of another plant through the release of chemical compounds into the environment. Whereas both positive and negative effects can be considered under the original definition, more commonly allelopathy refers only to negative effects. Interjit and Duke stress the importance of allelopathy as a motive force behind several metabolic and detoxification pathways underpinning the toxicology of current day xenobiotics.

It is generally assumed, and can sometimes be demonstrated, that natural toxins perform a protective function for protagonists, and in predatory organisms their function in prey capture and digestion is well known. Nevertheless, in some cases it is difficult to understand the exact 'role' played by the highly toxic chemical compounds produced by some organisms. The situation is complicated by collateral effects on non-target organisms.

9.3 Toxins and Their Mode of Action

Natural toxins are characterized by their enormous variety of structure and the huge taxonomic diversity of organisms that produce them, including bacteria, plants, insects, fish, frogs, reptiles and mammals. Also notable are examples of toxins with very similar chemical structure(s) but which are known to be produced by organisms that are not at all closely related. Bearing in mind their long evolutionary history, this is perhaps not surprising, and the parallel evolution of similar natural toxins reinforces the notion that their production confers an ecological advantage on the respective protagonists.

9.3.1 Toxins Produced by Harmful Algal Blooms

Natural toxins have been described and studied in several phytoplankton species which, under certain nutrient conditions, can reach extraordinarily high densities in the form of algal blooms. Shellfish, as filter feeders, are capable of ingesting very high cell numbers and consequently high toxin concentrations. As a result, shellfish may suffer a variety of pathological effects, including reproductive failure. Of even more ecological significance, consumption of contaminated shellfish may lead to the deaths of fish, seabirds and even humans. Such phytoplankton blooms are often given the generic name **harmful algal blooms (HABs)**. Box 9.1 provides some factors that have been suggested as contributing to the development of HABs. Increasing attention is now being paid to water quality changes related to human activity that may influence the exposure of organisms sensitive to HAB toxins. For example, blooms of harmful algal species resulting in major finfish and shellfish kills in several countries have been linked to coastal run-off containing elevated levels of major anthropogenic nutrients, particularly nitrogen. Experimental evidence suggests a positive correlation between water temperature and the densities of toxic algae (Peperzak, 2003).

Toxic shellfish poisons originate from algae or bacteria that are consumed by the shellfish, and subsequently ingested by other organisms through the food chain. They have often been categorized according to symptoms shown by human consumers of contaminated organisms, and such categories are commonly listed together with their primary causative agents. For example, domoic acid (DA) is often described as a major cause of amnesic shellfish poisoning (ASP) and saxatonin is commonly associated with paralytic shellfish poisoning (PSP). Other categories of shellfish poisoning together with their primary causative toxins include neurotoxic shellfish poisoning (NSP) (brevetoxin) and diarrhoeal shellfish poisoning (DSP) (okadaic acid). This categorization is somewhat narrow, however. Several toxic compounds are produced by a wide variety of taxonomic groups and exert their toxic effects in a variety of ways and to a range of organisms, not simply through shellfish ingestion. For example, the potent neurotoxin tetrodotoxin is best known as a toxin associated with vertebrate species and is not confined to the marine (or aquatic) environment. Karlotoxins are primarily known as ichthyotoxins rather than human toxicants. A brief outline of some of these compounds and their effects is given below, together with other toxins that may, or may not, fall into the aforementioned categories.

Box 9.1 Factors contributing to the development of harmful algal blooms (see also Section 14.1.2)

- *Nutrients.* HABs have been associated with coastal discharge of urea, used as an agricultural fertilizer (Glibert *et al.*, 2006). Toxic species are capable of reaching extraordinarily high densities under eutrophic conditions resulting from human activity (Burkholder *et al.*, 2008).

- *Climate change (temperature; salinity).* Multi-year trends in the abundance of *Prorocentrum* and *Noctiluca* have been correlated with the sea surface temperature (SST) in parts of the North Sea (Edwards *et al.*, 2006). Historical records of fossilized phytoplankton are sparse yet offer evidence of changes in prevalence of warmer-water species due to climatic warming (Dale and Nordberg, 1993; Mudie *et al.*, 2002). Boivin-Rioux *et al.* (2021) cite both temperature and salinity effects related to climate change as contributory factors to the projected twenty-first-century expansion of *Alexandrium catanella* blooms on the east coast of Canada.

9.3.1.1 Domoic Acid

A severe outbreak of amnesic shellfish poisoning that occurred in Prince Edward Island, Canada, in 1987 resulted from the consumption of mussels contaminated with the alga *Pseudo-nitzschia australis*, which had occurred as algal blooms (Ramsdell and Zabka, 2008). More than 100 people became acutely ill, and three died. Domoic acid is a neurotoxin responsible for this and other outbreaks of ASP in humans and for the deaths of a variety of seabirds and mammals. *P. australis* is known for its high toxicity and has caused the deaths of seabirds and sea lions, following their consumption of anchovies (*Engraulis mordax*) contaminated with DA. Figure 9.1 describes events leading to DA toxicity for marine mammals.

Domoic acid is one of a group of neuroexcitatory kainoid amino acids. It is an amino acid analogue of L-glutamate, which is the ionic form of L-glutamic acid (see illustrations of the respective chemical structures), with strong structural similarity, and acts as an antagonist capable of binding to the glutamate receptor in neuronal cells. Activation of this receptor leads to the uncontrolled influx of sodium and calcium into neuronal cells, resulting in massive neuroexcitatory activity and cell death. In

Domoic acid

L-glutamic acid

9.3.1.2 Saxitoxin

Saxitoxin is an alkaloid which is a potent neurotoxin occurring in a variety of marine **dinoflagellates**, including *Alexandrium*, *Gymnodinium* (=*Karenia*) and freshwater cyanobacteria such as *Anabaena* and *Aphanizomenon*. Together with tetrodotoxin (Section 9.3.1.6), it is also found in several species of puffer fish from southeast Asia and South America. Paralytic shellfish poisoning is caused by saxitoxin ingestion through the consumption of contaminated filter-feeding bivalves, with occasional fatal outcomes in humans (Wiese *et al.*, 2010).

Saxitoxin

humans, toxic effects include vomiting and diarrhoea, sometimes accompanied by headache, disorientation and memory loss. In rare cases symptoms include seizures, coma and death. The spread of DA contamination in coastal regions worldwide is related to warm ocean conditions (McKibben *et al.*, 2017). Figure 9.2 shows a model for the adverse outcome of the neurological effects of domoic acid on a population via behavioural effects.

This has led to frequent bans on commercial and recreational shellfish harvesting in temperate coastal

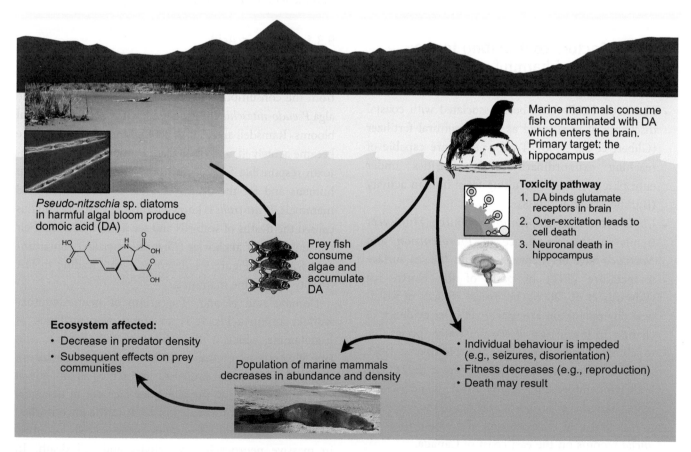

Figure 9.1 Food chain pathways for domoic acid affecting marine mammals and the accompanying ecosystem. Figure adapted from Villeneuve and Garcia-Reyero (2010), with permission from Wiley and Sons Ltd.

waters around the world including North America, western Europe, southeast Asia, Australia, New Zealand and South Africa. The first recorded case of PSP was in 1927 near San Francisco, USA, and was caused by a dinoflagellate, *Alexandrium catenella*. The incident involved the food-chain transfer of the toxin from the dinoflagellate to filter-feeding bivalves and then to human consumers. It resulted in 102 people displaying gastrointestinal and neuronal symptoms and six deaths. Whereas human illness and deaths understandably cause most concern, toxic effects on primary consumers are often also evident. For example, a large bloom of *A. catanella* on the coast of Chile in 2016 caused massive mortalities of surf clams, *Mesodesma donacium*, probably due to paralysis of burrowing and feeding activity (Alvarez *et al.*, 2019). The toxicity of saxitoxin results from its ability to bind to neuronal voltage-sensitive sodium channels along

the axon of a neuron, thereby blocking the passage of sodium ions responsible for the resting potential and action potentials. Blockage of the channel prevents normal neuronal function, resulting in paralysis.

9.3.1.3 Brevetoxin

Brevetoxins are a family of potent neurotoxins produced by the dinoflagellate *Karenia brevis* (formerly *Gymnodinium breve*). They have been responsible for isolated cases of human NSP through the consumption of contaminated bivalves and whelks and have been responsible for massive fish kills along the Florida coast in the United States and in the Gulf of Mexico. They are structurally similar to another group of neurotoxins, ciguatoxins. Both groups have a chain-like polyether structure, and their MoA is through the occupation of voltage-sensitive Na^+ channel receptor sites on neurons.

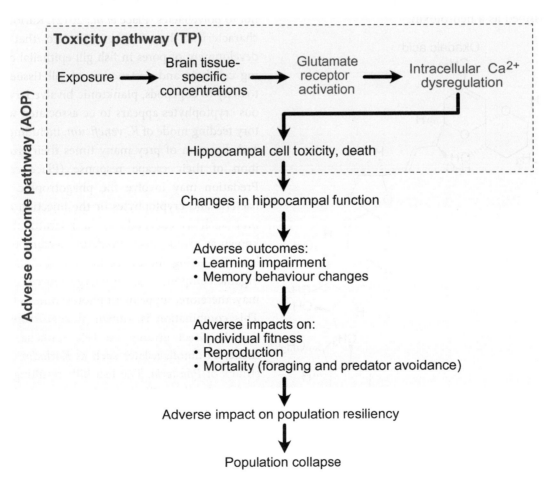

Figure 9.2 Flow chart for a domoic acid toxicity pathway model. Figure adapted from Watanabe *et al.* (2011), with permission from Wiley and Sons Ltd.

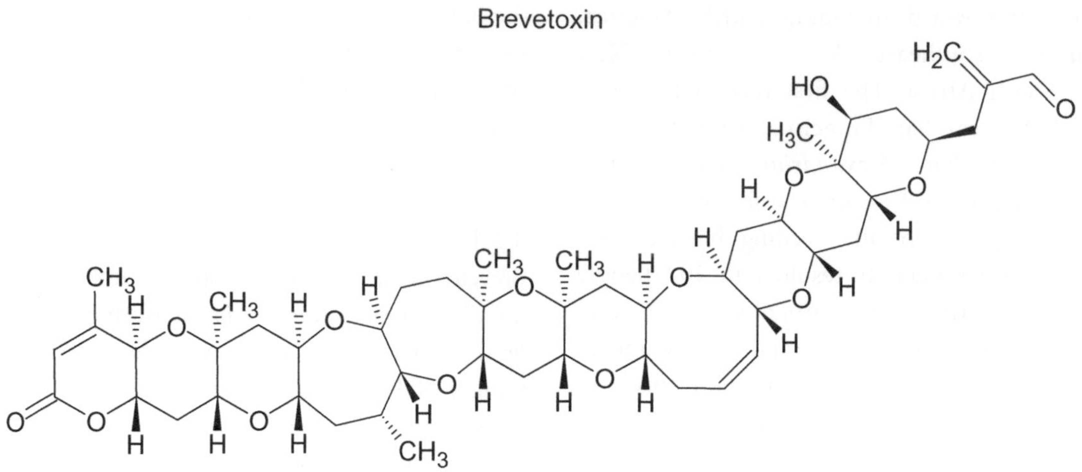

Brevetoxin

9.3.1.4 Okadaic Acid

Although recognized primarily as a cause of gastrointestinal illness in humans (DSP), okadaic acid (OA) has been shown to cause a variety of adverse effects including the induction of **apoptosis** in some cell types. OA has been shown to inhibit protein phosphatases and to cause oxidative stress in various cells, including neurons, although it is not classified as a neurotoxin.

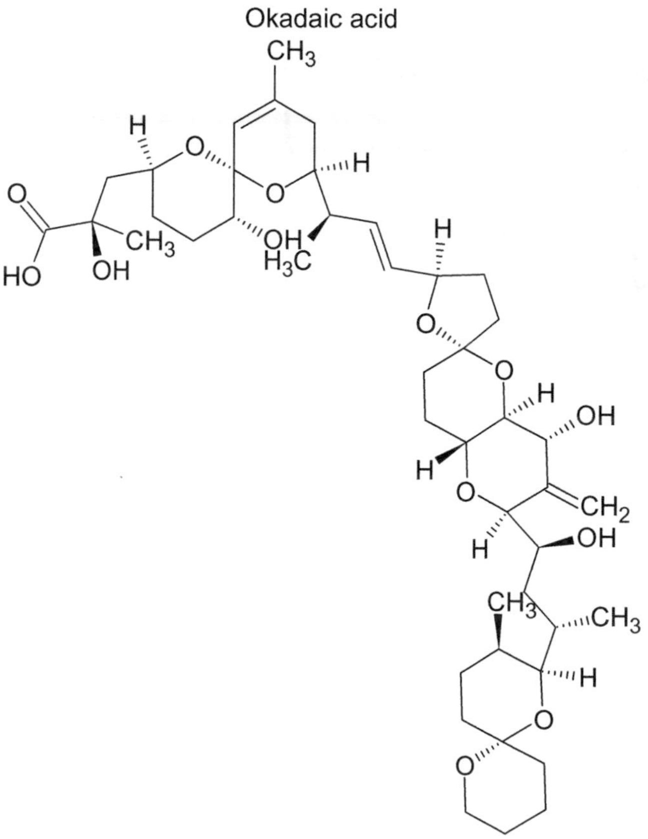

Okadaic acid

9.3.1.5 Karlotoxin

The toxic effects of this family of compounds were seen long before it was formally identified. In the 1950s a

powerful toxin was discovered, associated with the dinoflagellate *Gymnodinium veneficum*. Following a period of taxonomic uncertainty, this family of toxins is now recognized as a product of the species now identified as *Karlodinium veneficum*. The members of this genus are a source of powerful ichthyotoxins, as well as being selectively toxic to mussels, scallops and cuttlefish, but not to polychaetes (Place *et al.*, 2012). Karlotoxins have a characteristic hairpin-like structure that causes the development of pores in fish gill epithelial cells promoting cell lysis and destruction of gill tissue. Karlotoxin toxicity to copepods, planktonic bivalve larvae and various cryptophytes appears to be associated with a predatory feeding mode of *K. veneficum*, including the capture and digestion of prey many times their size and inhibition of their escape response (Place *et al.*, 2012). Predation may involve the phagotrophic ingestion of bacteria and cryptophytes or the injection of toxin into prey such as copepods via a microtubule, a process known as **myzocytosis**. Predatory feeding by dinoflagellates, involving the extracellular digestion of dissolved organic compounds as an energy source (**osmotrophy**), may, therefore, supplement photosynthesis (autotrophy). This combination is known as **mixotrophy**, and can include direct grazing on fish epithelia, notably gill tissue, by dinoflagellates such as *Karlodinium* and toxic strains of *Pfiesteria*. The fish kills resulting from karlotoxin poisoning will obviously reduce predation by fish and thus might be considered to have defensive effects. However, the role of powerful toxins such as karlotoxin and the factors responsible for massive HABs remain subjects of intense research interest.

Interestingly, earlier studies identified two species of *Karlodinium*, *K. veneficum* and *K. vitiligo*, collected from the same location but at different times (near Plymouth,

Karlotoxin

UK, in the 1950s). Although both species were very similar morphologically, the primary difference was that *K. veneficum* produced a highly potent toxin but *K. vitiligo* did not (Place *et al.*, 2012). This highlights the fact that closely related taxa can differ dramatically in their ability to produce toxins, the production of which is also dependent on environmental conditions.

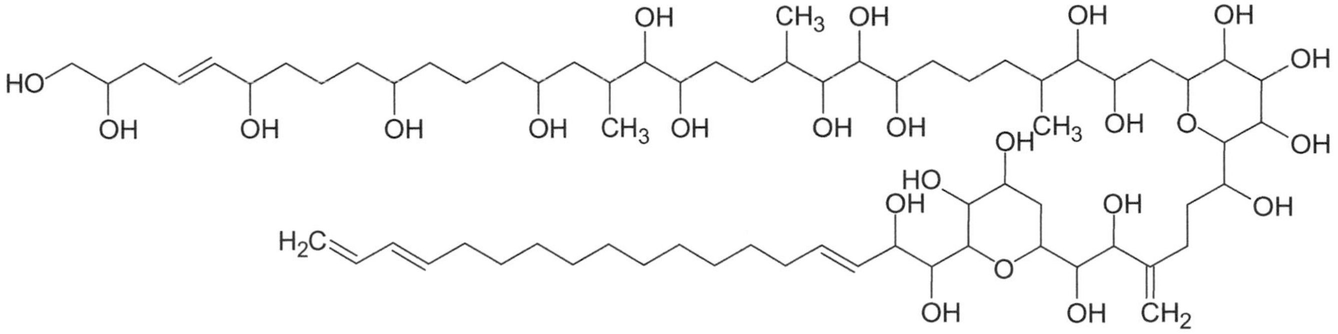

9.3.1.6 Tetrodotoxin

Tetrodotoxin is a potent neurotoxin known for thousands of years. It was originally thought to exist only in puffer fish but is now recognized in a wide range of organisms including both marine and terrestrial phyla. As in the case of saxitoxin, production of tetrodotoxin is presumed to be defensive in nature, i.e., toxic to potential predators. However, it seems to be produced by symbiotic bacteria rather than the host organisms themselves. Like saxitoxin, tetrodotoxin is an alkaloid, but it is structurally different from saxitoxin (Section 9.3.1.2) with fewer amine groups. Tetrodotoxin exhibits properties very similar to those of PSP and, like saxitoxin, it acts as a neuronal sodium blocker causing neuronal, gastrointestinal and cardiovascular symptoms. Although illness and deaths have historically been associated with puffer fish exposure, more recent cases of tetrodotoxin poisoning have involved ingestion of contaminated bivalve molluscs.

Tetrodotoxin

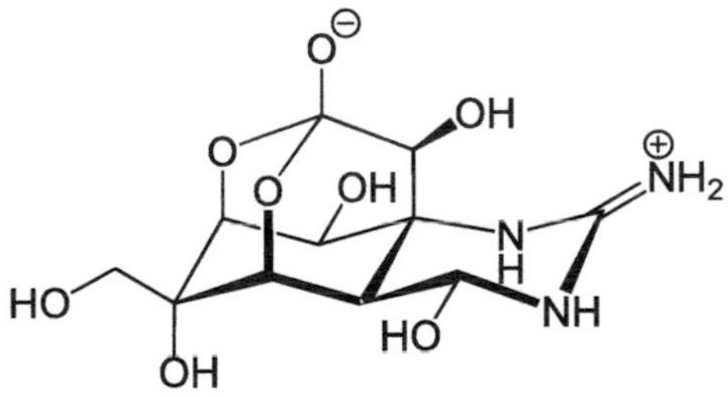

9.3.1.7 Microcystins

Microcystins are a family of over 50 variants of cyclic peptides produced by freshwater blooms of the **cyanobacterium** *Microcystis*, primarily *M. aeruginosa*, as well as the cyanobacteria *Oscillatoria* and *Anabaena*. Incidents of chronic liver necrosis occasionally resulting in fatalities have been demonstrated in wild animals and livestock exposed to water contaminated with cyanobacteria. Lower doses cause gastroenteritis, and cyanobacteria blooms have been recognized as a human health threat in contaminated water bodies, particularly where clean water is in short supply.

9.3.1.8 Anatoxins

Anatoxins are neurotoxins produced by cyanobacteria such as *Anabaena*, which have been linked to the deaths of both wildlife and domestic animals. Bird kills in Danish lakes in 1993–4 followed massive blooms of *Anabaena lemmermannii* that probably resulted from eutrophication. Tests on mice indicated anticholinesterase activity associated with this toxin, which has also been implicated in the deaths of dogs and pigs (Henriksen *et al.*, 1997).

9.3.2 Toxins Produced by Vascular Plants

Phytochemicals can be divided into a number of groups including terpenes, polyphenols, phytosterols, alkaloids, organosulphur compounds and quinones.

Terpenes and their oxygenated form, terpenoids, form a diverse group of over 10,000 organic compounds. Examples include carotenoids, of which about 600 are known, and neurotoxic pyrethrins (Section 7.8.4). Saponins are a large group of compounds with a triterpene or steroid core structure. They have been shown to act as scavengers of reactive oxygen intermediates, and some have shown antitumour activity through the induction of apoptosis and prevention of mutagenesis caused by environmental toxins. Antioxidant activity contributes to their effectiveness as antimicrobial agents, and antiviral activity against HIV (human immunodeficiency virus) and SARS (severe acute respiratory syndrome) viruses has been demonstrated. Plant steroids are produced from terpenoid precursors. Perhaps the best-known clinical use of this class of compounds relates to the development of oral contraceptives, notably the progesterone mimic norethisterone. The precursor for this product is the saponin glycoside diosgenin, extracted from the wild yam *Discorea mexicana*.

Polyphenols are a class of plant secondary metabolites found in a wide variety of plants, and include flavonoids, tannins and phenolic acids. Many act as protective agents against pathogens and their resultant pathologies. In this role they may act as antioxidants and, depending on dose, may perform an anticancer function through the induction of apoptosis and suppression of cell proliferation. The insecticide rotenone is a flavonoid derivative that acts as an inhibitor of mitochondrial respiration.

Plant alkaloids comprise a group of more than 3,000 nitrogen-containing compounds derived from amino acids. They include nicotine, quinine, morphine and cocaine. They are protective in nature as a result of their toxicity to herbivorous animals. The toxic action of many of these compounds involves the inhibition of protein synthesis. Natural and semi-synthetic opioids are made by extracting and processing the crude opium produced by poppies, and include morphine and heroin, as well as the pharmaceutical, codeine. These opiates, as well as cocaine prepared from extracts of the coca plant, interact with the receptors for brain neurotransmitters, such as dopamine. The resultant neurological responses can result in addictions that lead to the abuse of these compounds.

9.3.2.1 Naphthoquinones

Naphthoquinones are a group of plant-based derivatives of aromatic hydrocarbons identified by a double benzene ring structure, which distinguishes them from single-ring benzoquinones and three-ring anthroquinones. They are commonly characterized by hydroxyl, methyl and more complex substitutions at the C1 and C4 positions (1,4-naphthoquinone) or C1 and C2 positions (1,2-naphthoquinone). Probably the best-known examples of naphthoquinones with toxic properties include the simplest forms: juglone (5-hydroxy-1,4-naphthoquinone),

plumbagin (2-methyl-5-hydroxy-1,4-naphthoquinone), menadione (2-methyl-1,4-naphthoquinone) and naphthazarin (5,8-hydroxy-1,4-naphthoquinone). Many of these compounds display a variety of toxic properties primarily related to their strong oxidizing potential and which are assumed to be largely protective in nature. A well-known example is the allelopathic action of juglone (5-hydroxy-1,4-naphthoquinone) which is exuded from the roots and leaves of the black walnut (*Juglans nigra*) tree, thereby killing or inhibiting the growth of competing trees or shrubs. Plants in the family Solanaceae, such as tomato, peppers and potatoes, are particularly sensitive to juglone. There are also several accounts of the toxicity of juglone and other 1,4-naphthoquinones to herbivorous insects. A highly specialized property of naphthoquinone toxicity is demonstrated by the family Droseraceae, which comprises over 100 species of carnivorous perennial or annual herbs with active trapping mechanisms for insect prey. The primary naphthoquinones detected in almost all species of this family are plumbagin and 7-methyl juglone.

Fungal naphthoquinones are toxic to bacteria, fungi and yeast as potent inhibitors of electron transport, as uncouplers of oxidative phosphorylation, as intercalating agents in the DNA molecule and as producers of reactive oxygen radicals by redox cycling under aerobic conditions. Whereas cell damage results primarily from the formation of reactive oxygen species (ROS) caused by redox cycling between the quinone and the semiquinone form of the molecule, toxicity may also result from adduct formation between the semiquinone and macromolecules, and through glutathione depletion. In an extensive series of assays on 1,4-naphthoquinone derivatives on the parasitic protozoan *Trypanosoma*, Salmon-Chemin *et al.* (2001) concluded that trypanocidal activity was a combination of redox cycling activity and inhibition of the specific reductase system used by these organisms for the reduction of oxidative stress. Some of the applications of naphthoquinones are summarized later in Section 9.5.1.2.

9.3.2.2 Lectins

A number of plants produce carbohydrate-binding proteins called lectins. Although some lectins appear harmless, other (taxonomically unrelated) plants are capable of producing highly toxic lectins the toxicity of which is linked to ribosome-inactivating proteins (RIP II proteins). One of the best known is ricin from the castor oil plant *Ricinus communis. Sambucus ebulus* (dwarf elder) and *Viscum album* (mistletoe), among others, have also been shown to produce compounds similar to ricin (Olsnes and Kozlov, 2001). The mechanism of action is consistently described as the inactivation of ribosomes.

Human exposure to ricin can occur by ingestion, inhalation or injection. Symptoms include weakness and muscle pain, vomiting, fever and low blood pressure, and typically result in multisystem organ failure and death. These symptoms are caused by the failure to synthesize proteins and can take several hours or even days to appear, but the ingestion route results in the fastest development of symptoms. Ricin, which is usually extracted from the castor bean and prepared as a powder, a mist or a pellet, has been identified as an agent for bioterrorism and **biological warfare.**

The structure of ricin has been known for some time, enabling the development of antitoxins and other therapies. These toxic lectins belong to a chemical group that includes several bacterial toxins, e.g., diphtheria, cholera, *Shigella*, anthrax toxins and certain enteropathogenic strains of *Escherichia coli*, illustrating the point made earlier that organisms that are taxonomically widely separated may produce chemically similar toxins.

9.3.3 Toxins Produced by Microorganisms: Fungi and Bacteria

A number of bacterial toxins contribute to specific disease syndromes, e.g., *Staphylococcus aureus* (boils, impetigo, food poisoning), *Listeria monocytogenes* (listeriosis, a serious food-borne disease), *Bordetella pertussis* (whooping cough) and *Corynebacterium diphtheria* (diphtheria). Others have harmful effects in ecosystems or have the potential to cause such effects.

9.3.3.1 Anthrax Toxin

Bacillus anthracis, a soil-borne spore-producing bacterium, an aerobe and facultative anaerobe, produces spores that are heat-resistant and can survive for long periods in soil. Spore production is the main vector for transmission. The longevity of the spores provides long-term survival for the organism, whereas the role of the vertebrate host is as a dispersal mechanism, via passage through the gut. The pathogen multiplies in the host, and spores are shed and subsequently eliminated into soil or other external media. The spores contain a toxin that causes disease in humans, cattle and wildlife. Exposure pathways can be cutaneous, through inhalation and, most commonly, gastrointestinal.

Mock and Mignot (2003) describe three secreted proteins, acting in binary combinations. None of these proteins is toxic individually, indicating that they probably act together within the host. The toxins, produced when the spores germinate under experimental conditions or in living tissue, are the main weapon of *B. anthracis* used to weaken the host. A lethal infection is rapidly established, providing favourable conditions for the massive multiplication of the pathogen. The pathology and genetics of the toxin production are relatively well understood, but the role of the toxins in terms of advantage to the bacterium is less clear.

The main target organisms other than humans are cattle and other hoofed animals, with the main transmission path being soil intake through grazing. Domestic animals can be protected by veterinary vaccines, but sporadic cases of anthrax occur in wild animals worldwide, and many more cases probably go undetected. There are occasional outbreaks in Africa and central Asia, and if these are in national parks or other similar protected areas, they are more likely to be reported. A dieback of hippopotamuses and Cape buffalo in Bwabwata National

Park, Namibia, in 2017 was confirmed to be caused by *B. anthracis* (Cossaboom *et al.*, 2019). The World Health Organisation (WHO) is concerned because of the risk of transmission to humans from infected wildlife, exacerbated by the longevity and resistant nature of the spores.

The resistant nature, longevity and tolerance of anthrax spores to desiccation provide the potential for deliberate release and application in terrorism and other forms of biological warfare (Section 9.5.2). This was demonstrated in 2001, when bioterrorist attacks in the United States used *B. anthracis* as a biological weapon. This type of application highlights the need for a detailed understanding of the mode of action and for related research on therapeutic actions for anthrax infection. The current treatment of anthrax, however contracted, mainly relies on antibiotics.

9.3.3.2 Microbial Methylation of Mercury

In addition to their direct production of toxins, bacteria and fungi chemically transform elements and compounds into toxic forms. Perhaps the most dramatic example of this phenomenon is the ability of certain bacteria to methylate inorganic mercury to form monomethylmercury (CH_3Hg), the most toxic form of mercury. Bacterial methylation of mercury has been ascribed to sulphate-reducing bacteria. However, with the identification of the *hgcA* genes, it is now known to be mediated by a wider range of microorganisms. The ecotoxicology of mercury and specifically monomethylmercury is described in more detail in Section 6.8.1 and Capsule 13.1.

Bacterial methylation of other elements (e.g., arsenic, antimony and bismuth) leads to the possibility that microbial conversion can modify the toxic effects of these elements, either increasing or, in some examples, alleviating toxicity.

9.3.3.3 Fungal Toxins

Historically, macro fungi and moulds have been associated with poisoning in a variety of consumers. Secondary metabolites of several species of fungus are highly toxic to consumers and have been given the name **mycotoxins**. Among the most dramatic examples are cases of fatal mushroom poisoning in humans and domestic animals, notably dogs. Other mammals (e.g., squirrels and rabbits) consume the toxic mushrooms with no apparent ill effects, and appear to be far less susceptible (Walton, 2018b). Experimental data are, however, very limited. The most potent known fungal toxins are produced by

the Basidiomycete *Amanita phalloides*, the so-called 'death cap'. Amatoxins are bicyclic octapeptides that inhibit DNA-dependent RNA polymerase II (Wong and Ng, 2006; Walton, 2018a). As a potent inhibitor of RNA polymerase II, α-amanitin is toxic to eukaryotic cells (Matinkhoo *et al.*, 2018). The liver is the main target organ of toxicity, but other organs are also affected, especially the kidneys (Garcia *et al.*, 2015). Consumption is invariably fatal for susceptible species. The distribution of toxin production among unrelated or related species is discontinuous (patchy) (Walton, 2018b). Other Basiodiomycetes including *A. virosa*, *A. bisporigera* and *A. verna*, as well as species of *Cortinarius*, *Entoloma*, *Inocybe* and *Clitocybe*, are also toxic, producing gastrointestinal symptoms, but are typically not lethal.

Moulds and other filamentous (lower) fungi also produce toxins. Of these, the best known and most widely investigated are antibiotics. The lethal action of penicillin, cephalosporin and other antibiotics has been known for some time, beginning in the mid-twentieth century. Bacterial cell walls contain peptidoglycan. Penicillin inhibits the transpeptidase that catalyses the final step in bacterial cell wall biosynthesis, namely the cross-linking of peptidoglycan. Penicillin, by virtue of its highly reactive beta-lactam structure, irreversibly acylates the active site of the cell wall transpeptidase (Fernandes *et al.*, 2013).

Two species of the filamentous fungi, *Aspergillus flavus* and *A. parasiticus*, can infect a variety of crops, before and after harvest. The fungi produce aflatoxins, powerful mycotoxins. Infestations by these moulds have resulted in degradation of the cotton crop and poisoning of livestock feeding on infected peanuts and corn. Aflatoxin was identified as the cause of so-called 'turkey X disease' in England in 1960 and has since been linked with several cases of sickness and death in poultry fed on contaminated seed. Human illness and death have resulted from consumption of a variety of food crops contaminated with aflatoxins. Of a total of 14 in this group of mycotoxins, aflatoxins B_1, B_2, G_1 and G_2 have the highest degree of potency. Human pathologies associated with aflatoxin exposure include immunosuppression resulting in decreased resistance to infectious agents, mutagenicity leading to developmental defects and carcinogenicity, particularly liver cancer. Aflatoxin B_2 is among the most potent natural carcinogens known.

In the context of ecological advantage, there is speculation but little experimental proof concerning the ecological roles of fungal toxins. Walton (2018b) discusses the possible ecological function of amanitin and other Basidiomycete toxins. He considers that involvement in defence is the most plausible function, but questions, "Defence against what?" Some gastropods are **mycophagous**, and defence against their grazing is proposed as one 'plausible ecological function'. Some insects are also mycophagous. The mycophagous fruit fly *Drosophila* spp. is resistant to amatoxins (Mitchell *et al.*, 2014). It has been suggested that resistance to these toxins might be advantageous to the fruit flies, either by blocking the larval growth of competing insects such as gnats and craneflies, or by killing nematodes that parasitize fruit flies. The strongest argument for there being a selective advantage is the persistence of the amatoxin production through evolutionary time, in spite of an obvious metabolic cost. But this does not provide any further understanding of the ecological advantage. A slightly different interpretation might apply to aflatoxins, where plant tissue damage caused by insects may stimulate the growth of *Aspergillus*. *Aspergillus* has insecticidal properties despite the fact that insects themselves may be responsible for dispersal of fungal spores. In this case the mould may offer some protection for the host plant against insect attack. In the final analysis, some of these toxic properties could be relics or by-products from conditions in the past when they were advantageous. Even if the toxin has little obvious function today, if it is not costly or disadvantageous, it might persist.

The medical application of some mycotoxins can provide excellent examples of 'selective toxicity' since the toxins are effective against various pathogens but, within limits, are far less toxic to the recipient. The ecological advantage that these toxins confer to the proponent merits consideration. The phenomenon of fungal **antibiosis** can be considered as a 'classic' example of chemical warfare in ecology. Microorganisms obtain nutrients and discharge waste products and protective substances directly by exchange with their environment, in contrast to higher organisms that have more complex systems of ingestion, transport and excretion. The release of antibacterial chemicals into the surrounding environment provides an immediate and direct method of killing or at least immobilizing the potential pathogen (or competitor). Whereas demonstration of this phenomenon in field studies is problematic, laboratory studies have repeatedly shown the antibiotic effect of fungal toxins on bacteria. The specificity of antibiotics for control of known bacterial pathogens is well established and continues to be tested in the laboratory.

9.3.4 Toxins Produced by Animals

Section 9.3.1 includes descriptions of toxic effects on animals (typically fish and shellfish) through ingestion, with the protagonist in most examples being toxic algae or cyanophytes. In these examples, the toxin takes its effect through the food chain, often to deter consumers or competitors. In contrast, a number of animals produce toxins that are delivered directly to the target organisms. These are usually referred to as venoms. Protagonists include jellyfish, annelids, spiders, scorpions, insects, fish, amphibians, reptiles (notably snakes) and a few mammals. Usually, this application of toxins involves predation, and even the digestion of prey, but it may also be defensive.

9.3.4.1 Venoms

Venoms are produced by a wide range of vertebrate and invertebrate taxonomic groups and are injected, typically by a bite or a sting with variously adapted organs (e.g., fangs, bristles, nematocysts). Nelsen *et al.* (2014) suggested that the delivery method was related to the function of the toxins and the evolution of the organism, i.e., delivery by penetration is particularly suited for predation, but those delivered by contact or ingestion are more suited to defence.

Applications and chemical composition. Products derived from snake venoms have long been known to have therapeutic applications such as antimicrobial and anticancer activity (Samy *et al.*, 2017). The long-established uses of venoms by ancient cultures for hunting/warfare on the one hand, and as medicines on the other, provide interesting early examples of what may be termed 'applied toxicology'. The chemical composition of venoms has been studied primarily as a result of concern for human and animal poisoning (57,000 recorded human deaths in 2013). Potential applications of their pharmacological potential have also stimulated and focused research on venoms.

In contrast to the other toxins described in this chapter, for which one or a small number of chemical compounds has been identified as responsible for the toxic effect, animal venom is one of the most complex biochemical secretions known. Venom is typically a 'cocktail' containing hundreds of proteins and peptides as well as non-proteinaceous compounds. Some components are neurotoxic proteins that bind with varying degrees of specificity to members of the voltage-gated Na^+, K^+ and TRP (transient receptor potential) channels, affecting the function of the channels in the nervous and cardiac organ systems. Others are cytotoxins that cause cell death and liquify the food items, aiding in digestion. Although the major physiological effects of these compounds on prey are well known, their structural complexity often makes it difficult to ascribe cause–effect to individual chemicals.

Evolution and genetics Recent application of omics technology has revealed some ancient, fairly conservative forms and some more recent clades. Several protein groups have been convergently recruited in a number of distinct animal lineages for use as venom toxins, resulting in similarities among members of different taxonomic groups (Fry *et al.*, 2009). Genetic studies have been carried out to trace evolutionary relationships among venom-producing species. Venom producers and their prey also provide a convenient model for genetic studies of the inter-relationship. For example, Holding *et al.* (2016) used molecular markers to study the coevolution of rattlesnake venom and squirrel resistance.

Ecological function Arguably the primary ecological advantage of these venoms is related to the capture and immobilization of prey. However, venoms may serve other biological and ecological functions as well. Predation, defence or competition have all been attributed to venoms in a range of taxa. Furthermore, some venoms contain protease enzymes that may aid in digestion, and some venoms may maintain oral hygiene through antimicrobial effects. Arbuckle (2009) reviewed the ecological functions of venom and suggested that for the lizard genus *Varanus* an enhancement of digestive function may be an important element of the venom, possibly even the primary function, at least in some species. For viperids, the digestive function of their venoms is still speculative (Bottrall *et al.*, 2010). Table 9.1 summarizes the mode of delivery, effects and ecological functions of venoms from a range of animals.

9.4 Defining the Ecological Advantage of Toxin Production

A review of literature concerning natural toxins raises some questions that remain at least partially unanswered, certainly concerning their ecological advantage (EA). Attention in research and the literature is understandably focused on the detrimental effects of toxin exposure to organisms at the top of the food chain, including humans, our crops and livestock. Yet it is sometimes difficult to understand the 'role' of some toxins within the ecosystem as a whole. Clearly, many toxins are 'defensive' in nature,

deterring consumers (predators) and reducing grazing pressure. Others are clearly employed for prey capture (and digestion) among predatory organisms. Also, allelopathy is a phenomenon well known to scientists interested in competition among species, and usually applies to the establishment of a competitive advantage to the protagonist.

Although it seems reasonable to assume that toxin production confers some advantage in spite of the possible metabolic cost to the toxin producer, in some cases it is not easy to discern a clear relationship among protagonists and the organisms adversely affected by toxins. For example, the removal or depletion of a secondary consumer resulting from bioaccumulation of a toxin high up in the food chain could conceivably lead to an *increase* in numbers of potential (primary) consumers of the protagonist. There are several recorded examples where consumption of apparently healthy filter-feeders contaminated by high concentrations of toxic algae and dinoflagellates has resulted in massive kills of finfish and other vertebrates. Questions arising from this are "How do the primary consumers survive such high concentrations?" and "How do the protagonists themselves produce (and survive) such high levels of toxins?" Of course, there are many instances where the primary consumers, including filter-feeding shellfish, have been adversely affected. However, where deaths of non-consumers (of the protagonist) are concerned, the non-scientific term 'collateral damage' may be appropriate. In such cases, part of the explanation may relate to bioaccumulation at the top of the food chain, and the evolution of detoxification and excretory processes among primary consumers.

A related question arises from consideration of the methylation of mercury by bacteria. There has been no definitive demonstration of any ecological advantage of methylation to the bacteria responsible for this chemical transformation, given that this is ostensibly a change from a less toxic (inorganic) form of the metal to a more toxic form. Furthermore, the aquatic food-chain transfer and biomagnification of methylmercury results in high concentrations of the compound, particularly in predatory fish (Section 6.8.1), yet the obvious toxic effects are usually manifested only in top-level predators such as piscivorous birds and mammals, and humans.

Our conclusion is that the concept of ecological advantage, which we have sought to apply to toxin production, may be, at best, elusive in some cases or even non-existent; perhaps the toxin was derived from a 'neutral' chemical mutation conveying little or no advantage to its producer.

Table 9.1 **Effects and mode of delivery for venoms from different taxonomic groups.**

Taxonomic group/ subgroup	Effect on prey/target; mechanism when known	Mode of delivery	Ecological/health significance
Arachnids			
Spiders	Insecticidal and paralysing neurotoxins and cytotoxins.	Fangs- or pincer-like chelicerae.	Predation and liquification of prey tissues.
Scorpions	Neurotoxins.	*Stinger* mounted on the *telson* (posterior division of the body).	Toxicity to arthropods (for prey capture); also predator deterrence.
Insects			
Wasps and ants	Neurotoxins and antimicrobial secretions.	Lance-like stinger with attached venom sac.	Immobilized insects are used as live food for the wasp larvae. Defensive – protects from microorganisms.
Lepidopterans	Neuropeptide venom is active only in larval (caterpillar) stages.	Delivery through hairs. Venom is transferred from associated glands or specialized secretory cells.	Defence from predators (most venomous species are herbivorous, so the function of the venom is not for predation).
Molluscs			
Cone snails (family Conidae)	Neurotoxin (conopeptide).	Harpoon-like *radular* tooth.	Predation on small fish and marine invertebrates.
Octopus	Neurotoxic *tetrodotoxin*.	Beak and drill-like tongue breaks through the shell of its prey and injects venomous saliva.	Paralysis or death of prey. The octopus venom is used after prey capture.
Cnidarians (the most ancient venomous animals)			
	Lipolytic and proteolytic enzymes; pore-forming toxins cause cell death via osmotic lysis; neurotoxins.	Nematocysts with a microscopic lance.	The primary function of the venom is prey capture and immobilization.
Fish			
	The composition of fish venom is poorly known for most species but often contains potent neurotoxins.	Varies: stingrays have serrated spines and stonefish have needle-like spines.	Fish venoms appear to be mainly for defence.
Reptiles			
Snakes	Neurotoxins and hydrolytic enzymes.	Fangs that transmit secretions from salivary glands. Some snakes deliver venom by spitting.	Immobilization and digestion of *prey*, and defence against predators.
Lizards (very few species are venomous)	Haemorrhagins, gilatoxin, vasomotor-active peptides and cell-specific ion channel toxins.	Venom produced in oral glands, delivered by bite from serrated teeth.	Predation, defence, competitor deterrence, and in some, a role in aiding digestion.
Amphibians			
Salamanders – Pleurodeles, *Echinotriton*	The mechanism of toxicity is unknown.	Ribs are sharp-tipped and protrude through the skin, piercing poison glands and coating the ribs in toxins before puncturing the skin of a victim.	Injection of the toxins is lethal to many potential predators including mammals and other amphibians.

Table 9.1 (*cont.*)

Taxonomic group/ subgroup	Effect on prey/target; mechanism when known	Mode of delivery	Ecological/health significance
Frogs and toads	800 different alkaloid compounds have been identified in the skin of various species.	In two frog species, bony spines on the head are used to pierce the skin of prey and inject venom.	Defence against predators. Warning colouration combines with chemical defence.
Mammals			
Vampire bat, slow loris, shrew	Varied among taxonomic groups; generally not well-known.	Bite or sting, delivering venom produced by a gland.	Predation, defence from predators and conspecifics or in agonistic social encounters.

A counterargument might point to the parallel evolution of several closely similar or identical toxins among protagonists from disparate taxonomic groups. Why would the evolutionary process result in the same toxic compounds unless an ecological advantage was gained? On the other hand, we have noted the collection and identification of two morphologically similar species *of Karlodinium*, only one of which is capable of producing the highly potent toxin karlotoxin (Section 9.3.1.5).

9.5 Applications of Natural Toxins

Microbial, plant and animal products have been adapted to serve a number of uses, including the development of pest control products and medicines. The potential for these toxins in terrorism and other types of biological warfare has also been recognized.

The demand for increased agricultural production along with public concern for the risk of synthetic pesticides has stimulated this area of research and commercial production. Natural-product-based pesticides, as well as other natural toxins that show promise as pesticides, may prove to be the next generation of pest-control products (Section 1.5). However, despite claims from commercial producers, no pest-control products, including those based on natural toxins, are free from the potential to harm non-target organisms. In common with most synthetic pesticides, they have the potential to have adverse effects on ecosystems as well as human health. Furthermore, the development of resistance to a toxin in pest populations remains a problem, as it has been for most pest-control products in the past.

The previous sections provide illustrations of small but important differences that may exist between concentrations of chemical products that are beneficial and those that have negative consequences. Toxic plant products are considered to be components of defence mechanisms against competing organisms. Through a greater understanding of their mode of action we have been able to harness some of their chemical properties to create useful medicines and biocides, either *in vivo* or *in vitro*. Several synthetic therapeutic and biocidal products have been developed from natural toxins, and work continues on such applications. While many high-profile pollution problems are seen to arise from the mass production and distribution of these derivatives and *de novo* artificial products, it is also important to bear in mind that damaging effects to ecosystems by natural toxins may arise indirectly from human activity. Examples include contamination by toxic algal species and cyanobacteria of public water supplies and drinking water for livestock as a result of nutrient release and run-off, and the resulting eutrophication.

9.5.1 Pest-control Products

With the knowledge that microorganisms, plants and animals produce toxins as their own defence mechanisms against competitors or predators, it is not surprising that the potential for using natural toxins in pest management has been explored. So-called natural pesticides may be phyto- or microbial products *per se* or may be synthesized based on the chemical structures of natural products. Claims have been made that pesticides based on natural products are environmentally and toxicologically safer than many of the synthetic pesticides on the market and that natural toxins often have molecular target sites that are not exploited by currently marketed pesticides. Table 9.2 shows some examples of natural toxins in use, or with potential use as pest-control products.

Table 9.2 **Examples of natural toxins used or with potential for use as pest-control products.**

Natural compound	Source	Action	Product	Application
Bt toxin, called Cry toxins. Various strains with narrow insect specificity	*Bacillus thuringiensis* (bacterium)	Endotoxin causes osmotic imbalance in response to the formation of pores in a cell membrane of the insect gut, OR it causes an opening of ion channels that activate the process of cell death.	Various strains of Bt, as well as genetically engineered plants that contain the gene for Bt production.	Insecticide – one of the earliest commercially available biological controls for insect pests including Lepidoptera, Culicidae, Coleoptera, Simuliidae, Hymenoptera, Homoptera, Mallophaga and others.
Pyrethrin/ pyrethroids	Pyrethrins are produced by *Chrysanthemum cinerariaefolium*, *Tanacetum coccineum* and other Asteraceae (vascular plants)	Neurotoxins. Bind to and disrupt voltage-gated sodium channels of insect nervous system.	Thousands of household products contain synthetic pyrethroids, e.g., Cypermethrin®, Raid®, Permethrin®.	Insecticides, in agriculture, domestic use and mosquito control, including malaria. But there is concern for non-target organisms, particularly fish and aquatic invertebrates.
Phosphinothricin	*Streptomyces* spp. (actinomycete)	Inhibits glutamine synthase, causing build-up of ammonia.	Glufosinate®.	Herbicide.
Spinosad	*Saccharopolyspora spinosa* (actinomycete)	Neurotoxin for insects.	Spinosad®.	Insecticide controlling a wide range of insects including weevils, beetles, thrips, spider mites.
Strobilurin	*Strobiluris tenecellus* (fungus)	Inhibits fungal respiration.	Named commercial products, e.g., Quadris® (azoxystrobin), Flint® (trifloxystrobin), Headline SC® (pyraclostrobin).	Fungicide.

9.5.1.1 Bt Insecticide

Bacillus thuringiensis toxin (Bt) has been in use as a biogenic insecticide since the 1930s, and Bt toxins have been considered by some as the most successful bioinsecticides of the past century. Commercial use began in France in 1938, and by 1958 its use had spread to the United States (Fernández-Chapa *et al.*, 2019). The bacterium can be used directly as a pesticide, and also, more recently, as a source of genes for the construction of transgenic plants resistant to insects (Schünemann *et al.*, 2014). *Bacillus thuringiensis* produces endospores and a parasporal crystal, from which the term 'Cry toxins' has been derived. Cry is in general use for Bt toxins, with individual toxins identified as, e.g., Cry1Ac. Ingestion by the insect is an essential pathway for the toxin to be

effective. In the insect gut, the alkaline conditions result in dissolution of the crystal, releasing one or more crystalline proteins, known as delta endotoxins. The activated proteins then destroy the membrane of the midgut epithelial cells, with lethal results (Xiao and Wu, 2019).

Of additional interest in the context of pest control, and the related concern that frequently accompanies the use of toxic substances, is the use of Bt in genetic engineering. The Bt gene has been introduced into crop plants including potato, corn, soybean and rice, and also into cotton. In these and other applications, Bt appears to be effective as a control of specific insect pests. Transgenic crops whose cultivation would result in human exposure to the toxin have promoted research on the safety of the crops. Toxicity against lepidopterans, coleopterans, hemipterans, dipterans and also nematodes has been demonstrated for Bt toxins. They may also be effective as molluscicides, acaricides and even against human cancer cells (Fernández-Chapa *et al.*, 2019). Toxicity to non-target organisms such as larval amphibians has also been reported (Lajmanovich *et al.*, 2015).

Examples of some other commercial pesticides that are currently on the market, based on natural toxins, are shown in Table 9.2.

9.5.1.2 Quinones

Naphthoquinones (see Section 9.3.2.1) have been widely studied for allelopathic activity, and plants with naphthoquinone content are used worldwide in the traditional medicines of countries where they grow. Their cytotoxic action against what might be termed 'rival organisms', including harmful bacteria and parasitic protozoa such as *Plasmodium* and *Trypanosoma*, has long been recognized as having potentially therapeutic and other benefits.

Toxicity to a broad spectrum of aquatic organisms including larval fish, crustaceans, zooplankton, algae, dinoflagellates and bacteria has prompted consideration of naphthoquinones as natural aquatic biocides for control of invasive species in the marine environment (Wright *et al.*, 2007). More recent work has investigated the toxic effect of juglone on potentially harmful cyanobacteria (Park *et al.*, 2020). Recent research on naphthoquinones has focused on the biochemical mechanism(s) underlying their potent cytotoxicity. These include the induction of cell death (apoptosis) through the mediation of reactive oxygen species, depletion of glutathione and disruption of mitochondrial transmembrane potential. This line of enquiry is often founded on the use of herbal remedies for a variety of medical conditions. In several tropical countries, such plant-based medicines for anticancer, anti-inflammatory and antiparasitic treatments may go back hundreds of years. Prominent sources of therapeutic naphthoquinones have been fungi, lichens and *in vitro* tissue products from leaves, bark and roots from several genera of the family Bignoniaceae and Plumbaginaceae. Plumbagin and derivatives of lapachol (2-hydroxy-3-(3-methyl-2-butenyl)-1,4-naphthoquinone) and shikonin [1-(5,8-dihydroxy-1,4-dioxonaphthalen-2-yl)-4-methylpent-3-enyl] in particular have been investigated for their anticancer activity. The cytotoxic effects of naphthoquinones such as 2-plumbagin and lapachol have led to their promotion as anticancer agents through their induction of apoptosis.

9.5.2 Biological Warfare and Bioterrorism

Biological warfare or bioterrorism is defined as the deliberate release of viruses, bacteria or other agents used to cause illness or death in people, but also in animals or plants. Typically, humans are the target, but such releases can also adversely affect animals or plants. They are aimed at creating casualties, terror, societal disruption or economic loss, inspired by ideological, religious or political beliefs (Jansen *et al.*, 2014). In addition to ricin and anthrax (Sections 9.3.2.2, 9.3.3.1), other biological agents that have potential for warfare or terrorism include staphylococcal enterotoxin B, botulinum toxin and T-2 mycotoxin. In addition to the actual toxins, infective live agents, e.g., those responsible for brucellosis, cholera, pneumonic plague, smallpox, Venezuelan equine encephalitis and viral haemorrhagic fever, have been considered as biological 'weapons' (Kortepeter and Parker, 1999).

In contrast to more conventional weapons, large volumes of microorganisms can be cultured in relatively small spaces and with few overt signs of such an operation. These properties, along with the insidious nature of biological warfare/biological terrorism, are well recognized and have resulted in attempts to control, regulate or make unlawful the development and use of agents of biological warfare. However, delivery of the biological agents to the intended target population is technically difficult, bearing in mind the sensitivity of many microorganisms, particularly pathogens, to temperature and other environmental conditions. This is why anthrax and other spore-forming organisms represent favourable prospects in this context. Post-infection therapy and protective measures are also being developed. The Biological and Toxin Weapons Convention, signed in 1972 and brought into force in 1975, prohibits the use of biological agents in warfare.

9.6 Conclusions

It is perhaps ironic that we speak of 'chemical warfare' among organisms in the introductory section of this chapter and end with a reference to human use of biological agents as weapons. Does this illustrate that the antagonism that humans share with other organisms extends to competition among members of our own species? Scientists are understandably careful to confine their comments and judgement to scientific matters. Yet now, more than ever before, we are being called on to place our observations within the context of how humans interact with the

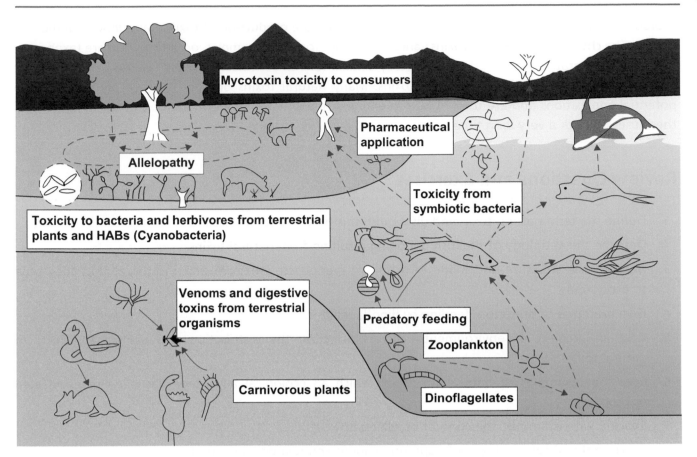

Figure 9.3 Pathways showing the origin and trophic transfer of natural toxins in ecosystems. Arrows with dashed lines make no distinction among the roles played by respective toxins, which may include deterrence of consumers and suppression of competition. Trophic transfer may involve bioaccumulation of toxins with differential toxicity at different trophic levels. Active transmission of toxins during predation (e.g., venoms) is indicated by arrows with solid lines.

natural environment, available resources and the ethical dilemmas governing how we manage those resources.

Figure 9.3 provides an overview of the propagation of natural toxins and their trophic transfer throughout ecosystems. The decision to include a chapter on natural toxins in this text was made in acknowledgement of the fact that ecotoxicology needs to be viewed as a broader subject than simply 'human influence on the ecosystem'. While human influence on the environment is of critical concern, it is important to place ecotoxicology in a wider context, and within a time frame that long pre-dates human existence. In so doing, we remind ourselves that human manufacturing endeavours often borrow chemical and toxicological characteristics that are part of a larger biosphere. That biosphere still contains untapped resources that can be beneficial to many aspects of human existence, yet remains vulnerable to critical imbalance related to human activity.

Summary

Toxins are produced by organisms at all levels of biological organization, from bacteria to vertebrates. The ecotoxicological effects of some natural toxins are clear and demonstrable, e.g., microcystins from cyanobacteria enter the food chain, as does domoic acid, produced by dinoflagellates. Toxins have also been shown to confer ecological advantages to the protagonist. These include allelopathy, protection from herbivory, pathogens and predation. For other toxins, no function has been demonstrated, and their biological and ecological roles remain speculative. There is evidence that the production of toxins has a long evolutionary history. The chemical structure of toxins varies widely, but for some examples, organisms that are taxonomically widely separated may produce chemically similar toxins. Conversely, closely related taxa can differ dramatically in their ability to produce toxins. Many natural toxins or their chemical derivatives have potential applications and are used in human and veterinary medicine, pest control and warfare. Some of these uses go back a very long time.

Review Questions and Exercises

1. Define the term natural toxin and distinguish this from the term toxicant.

2. Describe the **aetiology** of shellfish poisoning resulting from natural toxins.

3. Describe the structure and mode of action of cyanobacterial toxins and the threat that they pose to ecosystems.

4. Give examples of predatory feeding by dinoflagellates.

5. Give an account of environmental factors affecting the distribution and ecological effects of natural toxins.

6. Give four examples of natural toxins for which an ecological advantage has been demonstrated or proposed.

7. Explain, with examples, the concept of allelopathy.

8. Describe the history and uses of naphthoquinones and related compounds.

9. List four examples of phytochemicals and briefly explain the structure and mode of action for each of your examples.

10. Describe the source and mode of action of Anthrax toxin. How and why can this toxin be used in terrorism or warfare?

11. What are aflatoxins and how are they formed? What risk is associated with aflatoxins?

12. List species of Basidiomycetes that produce toxins and describe their mode of action and effect(s) on vertebrate animals.

13. What distinguishes venoms from most other natural toxins? List the properties and modes of delivery of venoms, and name three types of venomous organisms with their characteristic modes of attack and their ecological function(s).

14. Discuss the potential and current applications of natural toxins in pest management and human or veterinary medicine.

Abbreviations

ASP	Amnesic shellfish poisoning	OA	Okadaic acid
Bt	*Bacillus thuringiensis*	PSP	Paralytic shellfish poisoning
DA	Domoic acid	SST	Sea surface temperature
DSP	Diarrhoeal shellfish poisoning	TP	Toxicity pathway
HAB	Harmful algal bloom	TRP	Transient receptor potential
NSP	Neurotoxic shellfish poisoning		

References

Alvarez, G., Diaz, P. A., Godoy, M., *et al.* (2019). Paralytic shellfish toxins in surf clams *Mesodesma donacium* during a large bloom of *Alexandrium catenella* dinoflagellates associated to an intense shellfish mass mortality. *Toxins*, 11, e11040188.

Arbuckle, K. (2009). Ecological function of venom in *Varanus*, with a compilation of dietary records from the literature. *Biawak*, 3, 46–56.

Boivin-Rioux, A., Starr, M., Chassé, J., *et al.* (2021). Predicting the effects of climate change on the occurrence of the toxic dinoflagellate *Alexandrium catenella* along Canada's East Coast. *Frontiers in Marine Science*, 7, e608021.

Bottrall, J. L., Madaras, F., Biven, C. D., Venning, M. G. & Mirtschin, P. J. (2010). Proteolytic activity of Elapid and Viperid snake venoms and its implication to digestion. *Journal of Venom Research*, 1, 18.

Burkholder, J. M., Glibert, P. M. & Skelton, H. M. (2008). Mixotrophy, a major mode of nutrition for harmful algal species in eutrophic waters. *Harmful Algae*, 8, 77–93.

Cossaboom, C. M., Khaiseb, S., Haufiku, B., *et al.* (2019). Anthrax epizootic in wildlife, Bwabwata National Park, Namibia, 2017. *Emerging Infectious Diseases*, 25, 947.

Dale, B. & Nordberg, K. (1993). Possible environmental-factors regulating prehistoric and historic blooms of the toxic dinoflagellate *Gymnodinium catenatum* in the Kattegat-Skagerrak Region of Scandinavia. *Toxic Phytoplankton Blooms in the Sea*, 3, 53–57.

Edwards, M., Johns, D., Leterme, S., Svendsen, E. & Richardson, A. (2006). Regional climate change and harmful algal blooms in the northeast Atlantic. *Limnology and Oceanography*, 51, 820–829.

Fernandes, R., Amador, P. & Prudêncio, C. (2013). β-Lactams: chemical structure, mode of action and mechanisms of resistance. *Reviews in Medical Microbiology*, 24, 7–17.

Fernández-Chapa, D., Ramírez-Villalobos, J. & Galán-Wong, L. (2019). Toxic potential of *Bacillus thuringiensis*: an overview. In *Protecting Rice Grains in the Post-Genomic Era*. IntechOpen, 1–22.

Fry, B. G., Roelants, K., Champagne, D. E., *et al.* (2009). The toxicogenomic multiverse: convergent recruitment of proteins into animal venoms. *Annual Review of Genomics and Human Genetics*, 10, 483–511.

Garcia, J., Costa, V. M., Carvalho, A., *et al.* (2015). *Amanita phalloides* poisoning: mechanisms of toxicity and treatment. *Food and Chemical Toxicology*, 86, 41–55.

Glibert, P. M., Harrison, J., Heil, C. & Seitzinger, S. (2006). Escalating worldwide use of urea – a global change contributing to coastal eutrophication. *Biogeochemistry*, 77, 441–463.

Henriksen, P., Carmichael, W. W., An, J. & Moestrup, Ø. (1997). Detection of an anatoxin-a (s)-like anticholinesterase in natural blooms and cultures of cyanobacteria/blue-green algae from Danish lakes and in the stomach contents of poisoned birds. *Toxicon*, 35, 901–913.

Holding, M. L., Biardi, J. E. & Gibbs, H. L. (2016). Coevolution of venom function and venom resistance in a rattlesnake predator and its squirrel prey. *Proceedings of the Royal Society B: Biological Sciences*, 283, e20152841.

Interjit, K. & Duke, S. O. (2003). Ecophysiological aspects of allelopathy. *Planta*, 217, 529–539.

Jansen, H.-J., Breeveld, F. J., Stijnis, C. & Grobusch, M. P. (2014). Biological warfare, bioterrorism, and biocrime. *Clinical Microbiology and Infection*, 20, 488–496.

Kortepeter, M. G. & Parker, G. W. (1999). Potential biological weapons threats. *Emerging Infectious Diseases*, 5, 523–527.

Lajmanovich, R. C., Junges, C. M., Cabagna-Zenklusen, M. C., *et al.* (2015). Toxicity of *Bacillus thuringiensis* var. *israelensis* in aqueous suspension on the South American common frog *Leptodactylus latrans* (Anura: Leptodactylidae) tadpoles. *Environmental Research*, 136, 205–212.

Matinkhoo, K., Pryyma, A., Todorovic, M., Patrick, B. O. & Perrin, D. M. (2018). Synthesis of the death-cap mushroom toxin α-amanitin. *Journal of the American Chemical Society*, 140, 6513–6517.

McKibben, S. M., Peterson, W., Wood, A. M., *et al.* (2017). Climatic regulation of the neurotoxin domoic acid. *Proceedings of the National Academy of Sciences USA*, 114, 239–244.

Mitchell, C. L., Saul, M. C., Lei, L., Wei, H. & Werner, T. (2014). The mechanisms underlying α-amanitin resistance in *Drosophila melanogaster*: a microarray analysis. *PLoS ONE*, 9, e93489.

Mock, M. & Mignot, T. (2003). Anthrax toxins and the host: a story of intimacy. *Cellular Microbiology*, 5, 15–23.

Mudie, P. J., Rochon, A. & Levac, E. (2002). Palynological records of red tide-producing species in Canada: past trends and implications for the future. *Palaeogeography, Palaeoclimatology, Palaeoecology*, 180, 159–186.

Nelsen, D. R., Nisani, Z., Cooper, A. M., *et al.* (2014). Poisons, toxungens, and venoms: redefining and classifying toxic biological secretions and the organisms that employ them. *Biological Reviews*, 89, 450–465.

Olsnes, S. & Kozlov, J. V. (2001). Ricin. *Toxicon*, 39, 1723–1728.

Park, M.-H., Kim, K. & Hwang, S.-J. (2020). Differential effects of the allelochemical juglone on growth of harmful and non-target freshwater algae. *Applied Sciences*, 10, e2873.

Peperzak, L. (2003). Climate change and harmful algal blooms in the North Sea. *Acta Oecologica*, 24, S139–S144.

Place, A. R., Bowers, H. A., Bachvaroff, T. R., *et al.* (2012). *Karlodinium veneficum* – the little dinoflagellate with a big bite. *Harmful Algae*, 14, 179–195.

Ramsdell, J. S. & Zabka, T. S. (2008). In utero domoic acid toxicity: a fetal basis to adult disease in the California sea lion (*Zalophus californianus*). *Marine Drugs*, 6, 262–290.

Salmon-Chemin, L., Buisine, E., Yardley, V., *et al.* (2001). 2- and 3-substituted 1,4-naphthoquinone derivatives as subversive substrates of trypanothione reductase and lipoamide dehydrogenase from *Trypanosoma cruzi*: synthesis and correlation between redox cycling activities and in vitro cytotoxicity. *Journal of Medicinal Chemistry*, 44, 548–565.

Samy, R. P., Stiles, B. G., Franco, O. L., Sethi, G. & Lim, L. H. K. (2017). Animal venoms as antimicrobial agents. *Biochemical Pharmacology*, 134, 127–138.

Schünemann, R., Knaak, N. & Fiuza, L. M. (2014). Mode of action and specificity of *Bacillus thuringiensis* toxins in the control of caterpillars and stink bugs in soybean culture. *ISRN Microbiology*, Article ID 135675.

Villeneuve, D. L. & Garcia-Reyero, N. (2010). Vision & strategy: predictive ecotoxicology in the 21st century. *Environmental Toxicology and Chemistry*, 30, 1–8.

Walton, J. (2018a). Biosynthesis of the Amanita cyclic peptide toxins. *The Cyclic Peptide Toxins of Amanita and Other Poisonous Mushrooms*. Cham, Switzerland: Springer, 93–130.

Walton, J. (2018b). Ecology and evolution of the amanita cyclic peptide toxins. *The Cyclic Peptide Toxins of Amanita and Other Poisonous Mushrooms*. Cham, Switzerland: Springer, 167–204.

Watanabe, K. H., Andersen, M. E., Basu, N., *et al.* (2011). Defining and modeling known adverse outcome pathways: domoic acid and neuronal signaling as a case study. *Environmental Toxicology and Chemistry*, 30, 9–21.

Wiese, M., D'agostino, P. M., Mihali, T. K., Moffitt, M. C. & Neilan, B. A. (2010). Neurotoxic alkaloids: saxitoxin and its analogs. *Marine Drugs*, 8, 2185–2211.

Wong, J. H. & Ng, T. B. (2006). Toxins from basidiomycete fungi (mushroom): amatoxins, phallotoxins, and virotoxins. In A. Kastin (ed.), *Handbook of Biologically Active Peptides*. San Diego, CA: Elsevier, 131–135.

Wright, D., Dawson, R., Cutler, S., Cutler, H. & Orano-Dawson, C. (2007). Screening of natural product biocides for control of non-indigenous species. *Environmental Technology*, 28, 309–319.

Xiao, Y. & Wu, K. (2019). Recent progress on the interaction between insects and *Bacillus thuringiensis* crops. *Philosophical Transactions of the Royal Society B*, 374, e20180316.

Appendix 9.1 Summary of Some Toxins, Their Sources and Effects

Starred toxins are discussed in Section 9.3.

Toxin	Source	Effect(s)	Comments
*Domoic acid (DA)	Shellfish contaminated with the dinoflagellate *Pseudo-nitzschia*.	Amnesic shellfish poisoning (ASP) in humans, and the deaths of seabirds and mammals following algal blooms.	Presumed EA – protection from consumers. *Pseudo-nitzschia australis* is known for its high toxicity and responsibility for the deaths of seabirds and sea lions, following consumption of DA-contaminated anchovies. Spread of DA in coastal regions worldwide is related to warm ocean conditions.
*Saxitoxin	Shellfish contaminated with the dinoflagellate *Alexandrium*, *Gymnodinium* (= *Karenia*, *Karlodinium*) and freshwater cyanobacteria e.g. *Anabaena*, *Aphanizomenon*.	Paralytic shellfish poisoning (PSP).	Presumed EA – protection from consumers. PSP caused by consumption of contaminated shellfish.
*Karlotoxin	Family of toxins produced by dinoflagellate *Gymnodinium* (= *Karenia*, *Karlodinium*), notably *K. veneficum*.	Fish haemolytic activity following ingestion of *Karlodinium* results from the destruction of red blood cells and osmoregulatory chloride cells through changes in epithelial permeability.	EA probably primarily predatory, including inhibition of movement of prey, prey capture and digestion in some dinoflagellates (see Section 9.3 and discussion of **mixotrophy**). EA probably includes protection from consumers such as planktonic and **benthic** grazers. Predation on cryptophytes and copepods has also been demonstrated.
*Tetrodotoxin (TTX)	Found in at least seven different phyla, including vertebrates (puffer fish, amphibians), octopuses, crabs, echinoderms.	Potent neurotoxin, causing nausea, vomiting, paralysis, loss of consciousness, and respiratory failure. Can lead to death.	Presumed EA – protection from consumers. In most cases, symbiotic bacteria associated with metazoans are suspected as the source of toxin, but difficult to prove.
*Brevetoxin – lipid-soluble polycyclic ether compounds	Shellfish contaminated with dinoflagellate *Karenia* (= *Gymnodinium*) *brevis*.	Neurotoxic shellfish poisoning (NSP).	Presumed EA – protection from consumers.
*Okadaic acid (OA), heat-stable lipophilic polyether	Shellfish contaminated with dinoflagellates *Dinophysis* spp., *Prorocentrum* spp. and *Phalocroma rotundatum*.	Diarrhoeal shellfish poisoning (DSP). OA is a potent inhibitor of protein phosphorylase phosphatase 1 which is involved in the regulation of a wide range of biological functions including intermediary metabolism and apoptosis. OA has also been shown to act as a tumour promotor.	Presumed EA – protection from consumers.

(cont.)

Toxin	Source	Effect(s)	Comments
Palytoxin	Fish, crabs, sea urchins, crabs via ingestion of dinoflagellates.	Binds to Na^+, K^+ channels causing depolarization and loss of neural function. Opening of voltage gate Ca^{++} channels causing inhibition of neuromuscular function.	EA uncertain. While highly toxic to humans, it is largely non-toxic to primary consumers such as polychaetes, crustaceans and fish.
Azaspiracid (AZP)	Shellfish contaminated with marine dinoflagellate *Protoperidinium crassipes*.	AZP causes nausea, vomiting and severe diarrhoea. Effects on membrane-bound kinases and proteins responsible for cellular integrity and trans-epithelial Ca^{++} movement.	First outbreak of AZP toxicity reported in the Netherlands in 1995. Concerns exist over synergistic effects with OA and other DSP-related toxins.
*Microcystin: over 50 variants of cyclic peptides	Cyanobacteria *Microcystis*, primarily *M. aeruginosa*; *Oscillatoria*, *Anabaena*.	Chronic liver necrosis, occasional fatalities in wild animals and livestock exposed to water contaminated with cyanobacteria. Lower doses cause gastroenteritis.	EA – possible protection from herbivory.
Yessotoxin Lipophilic polyether	Scallops and other filter-feeding bivalves contaminated with dinoflagellates such as *Gonyaulax*.	Suspected binding to Ca^{++} channels causing inhibition of neuromuscular function. Evidence of apoptosis promotion.	First isolated in Japan in 1986. Regulated as DSP agent (1 mg/kg in seafood) by European Commission, but mode of action considered different from OA.
Histamine	Certain fish that are high in histidine: *Coryphaena hippurus* (mahi-mahi), *Thunnus* spp. (tuna), *Salmo* spp. (salmon).	Rapid onset of gastrointestinal upset termed Scombroid poisoning.	Bacterial conversion of histidine to histamine.
*Naphthoquinones, e.g. juglone (5 hydroxy-1,4 naphthoquinone), 7-methyl juglone, plumbagin	Vascular plants: black walnut (*Juglans nigrax*). *Diospyros* (persimmon), insectivorous plants (Droseraceae), fungi, actinomycetes and bacteria.	Root exudate inhibits competing shrubs. Broad-spectrum biocidal activity probably a defence against consumers and specialist insectivores. Cytotoxicity of naphthoquinones probably relates to high redox potential.	EA – competition, including allelopathy and protection against consumers. Several naphthoquinones have applications as biocides and have pharmacological (antibacterial, antifungal, antimalarial, anticancer, trypanocidal) properties.
*Lectin	Vascular plants: e.g. *Ricinus* (castor oil plant).	Extremely toxic when inhaled; less so when ingested.	EA – presumed to protect seeds from consumers. Different plant parts have different toxicity e.g., castor bean seeds are toxic, but castor oil is used in medicine.
Digoxin – cardiac glycoside	Vascular plant. *Digitalis purpurea* (foxglove).	Nausea, periventricular contractions, arrhythmia.	EA – presumed to protect plant against herbivory. Used in medical treatment of cardiac conditions. Inadvertent toxicity can result from overdosing the medical use or consumption of plants.

(cont.)

Toxin	Source	Effect(s)	Comments
*Anthrax toxin	Bacterium: *Bacillus anthracis.*	The mechanism involves three-protein exotoxin composed of the protective antigen, the lethal factor LF, and the oedema factor EF. Anthrax toxin causes internal bleeding and destroys the body's immune system, eventually causing septic shock and killing the host.	Spores of the bacterium are extremely resistant to heat and are long-lived in soil. Potential as an agent of biological warfare.
Botulinum	Bacterium *Clostridium botulinum.*	A protein that prevents the release of the neurotransmitter acetylcholine from axon endings at the neuromuscular junction, causing paralysis.	Organism is spore-forming, resistant to heat, and can grow anaerobically. Has some recent applications in human medicine.
Bt Cry toxins	Bacterium *Bacillus thuringiensis.*	Cytotoxic action - pores form in the plasma membrane of the receptor, which results in sequential binding and damages the osmotic balance.	EA – insectivorous properties have resulted in use as a natural pesticide, e.g., treatment of Dutch elm disease which is caused by a fungus with an invertebrate vector.
*Methylmercury	Sulphate-reducing bacteria, including *Desulfovibrio desulfuricans* and *Geobacter sulfurreducens.* Other bacterial species also recently implicated in methylation.	Methylmercury biomagnifies through the aquatic food chain and reaches high concentrations in carnivorous fish, resulting in toxicity to consumers – humans, fish-eating wildlife and domestic animals.	An example of microbial conversion of an inorganic substance that for mercury results in a compound that is toxic to high-level consumers, but, insidiously, often results in no overtly observable symptoms during transport up the food chain.
Cyclopeptides: amatoxins phallotoxins, virotoxins and many others	Basidiomycete fungi: *Amanita* spp. especially *A. phalloides* and several other genera.	Inhibition of RNA polymerase II, causing protein deficit and ultimately cell death. The liver is the main target organ of toxicity.	EA – presumed to protect fungus against herbivory. Alpha amanitin is invariably lethal in humans, domestic pets; less information for wildlife, but squirrels appear to have protective glycoproteins lining the gut and are not susceptible.
Ergotamine	Fungus: *Claviceps purpurea* is a disease of cereal plants.	Alkaloid produced by the fungus. Consumption of infected grains (usually inadvertent) results in constriction of vascular system; can lead to loss of limbs, abortion and cardiac arrest.	The same properties of causing vasoconstriction have been applied as medical treatment for migraine.
Aflatoxin	Fungus: *Aspergillus* spp. as a contaminant of peanuts, corn and any seed crop stored under damp conditions.	Carcinogenic to vertebrates.	Aflatoxin in peanut food products is regulated in some jurisdictions.
*Venoms	Gila monster, snakes, scorpions, spiders, ants, bees, hornets, wasps, some marine animals.	Neurotoxic, block peripheral nerve transmission. Variety of amino acids in the neurotoxic proteins. More than one toxic component of venom, but generally one major protein is known. Venoms also contain hydrolytic enzymes.	EA – procuring food and its digestion. Some venoms are used in folk medicine and some venoms or synthetic mimics have promise as pharmaceuticals.

10 Ionizing Radiation

Louise Winn

Learning Objectives

Upon completion of this chapter, the student should be able to:

- Define the meaning of ionizing radiation, including the units of measurement.
- Describe common sources of ionizing radiation including both natural and anthropogenic sources.
- Understand the effects of ionizing radiation at the molecular and cellular levels.
- Understand the risks and benefits of ionizing radiation.
- Discuss examples of radiotoxicity that affect ecosystems or components of ecosystems.

The aim of this chapter is to introduce students to the basic **toxicology** associated with **ionizing radiation**, including definitions, units of measurement, sources of exposure, molecular and cellular effects, and assessment of risk. Concern over this source of toxicity stems from several well-publicized major accidents at nuclear facilities, beginning as far back as the late 1950s, and it includes ongoing examples of accidental release of toxic radiation. These remind us of the need for extreme vigilance at all stages of what has come to be known as the nuclear cycle. In addition, extensive studies, including retrospective ones involving miners and mill workers, have demonstrated the need to understand the toxicity associated with ionizing radiation. It is also important to note that almost all the available information concerning the toxicology of ionizing radiation is derived from a number of inadvertent 'experiments' on living systems, stemming from accidental or even deliberate ionizing radiation exposure. As a result, the subject of ionizing radiation is dominated by the consequences of exposure to humans and other mammals, with much less known about the effects of ionizing radiation on other organisms and on ecosystems in general. The chapter concludes with a capsule discussing approaches for the development of environmental assessments with respect to ionizing radiation (Capsule 10.1).

10.1 Non-ionizing Versus Ionizing Radiation

To understand the concepts surrounding this field of toxicology, it is essential to understand the difference between **non-ionizing** and ionizing radiation. To begin, when energy is released from a source, that energy can be referred to as radiation. Depending on the amount of energy released, it can be either non-ionizing or ionizing (Figure 10.1).

The difference is that non-ionizing radiation has enough energy to cause atoms to move or vibrate, but it does not have enough energy to detach tightly bound electrons from their orbits around the atom, causing it to become **ionized**. Ionizing radiation, on the other hand, has enough energy to remove electrons, resulting in ionization of the atom. Although the main concern in this chapter is with the potential toxicity associated with

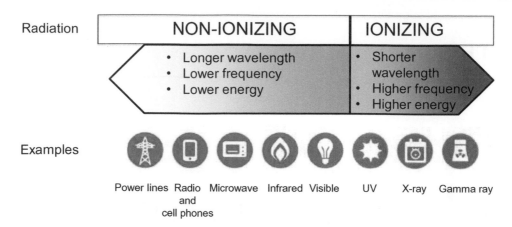

Figure 10.1 The difference between non-ionizing and ionizing radiation. Non-ionizing radiation has enough energy to cause atoms to move or vibrate, but the energy is not enough to detach tightly bound electrons from their orbit around an atom. In contrast, ionizing radiation has enough energy to remove electrons, resulting in ionization of the atom. Ionizing energy has a shorter wavelength, higher frequency and higher energy than non-ionizing energy.

ionizing radiation, understanding the use of this type of radiation for beneficial purposes is also discussed. There is also some concern surrounding the potential adverse effects caused by non-ionizing radiation such as from mobile phones (cell phones), wireless networks and exposure to power lines. Currently, however, there is no clear scientific evidence linking exposure to these sources of non-ionizing radiation with any adverse outcome. Although there is ongoing research in this area, it is beyond the scope of this chapter.

It is important to note that before the adverse health effects of ionizing radiation exposure were fully recognized, deliberate exposure to this type of radiation often formed part of the treatment for a variety of conditions. It was not until after intense nuclear weapons testing in the 1950s and 1960s, together with accumulated information from victims of the two wartime nuclear detonations (Hiroshima and Nagasaki), that the full extent of harmful effects began to be fully understood. Even then, the full realization of harmful impacts took many years to develop. As recently as the 1970s and 1980s, some aspects of what might be termed the 'nuclear industry', notably waste disposal, were very inadequately regulated. This oversight was particularly apparent in some military establishments where the deployment of weapons was given a higher priority than certain regulatory details. For example, in the former Soviet Union, low- and medium-level radioactive wastes were routinely dumped on the ground with few appropriate safeguards. Even in countries that paid some attention to containment of waste radiation materials, it is clear that earlier methods

were often critically inadequate, and leaks were common. As a society, we are still recipients of a legacy of past deficiencies of containment and disposal which have great impacts on all levels of the ecosystem, including non-human targets.

10.2 Definitions

To begin to understand the toxicity associated with ionizing radiation, it is imperative to start by defining ionizing energy and identifying the key components responsible for generating this powerful energy.

10.2.1 What Is Ionizing Radiation?

In nature, molecules can be subdivided into atoms, with each atom being characteristic of an element. The nucleus of an atom consists of positively charged protons and electrically neutral **neutrons**, which together make up the majority of the mass of the atom. Negatively charged electrons orbit the atom and normally balance the positive charge of the protons. Atoms of the same element have the same number of protons in the nucleus but can have a varying number of neutrons, thus having slightly different masses. These different forms of the same element are referred to as **isotopes**. The element carbon (C) is useful to illustrate this point. The dominant form (99%) of natural carbon contains six protons and six neutrons (^{12}C) whereas 1% contains an extra neutron in the nucleus (^{13}C), and therefore these two isotopes have slightly different atomic masses. In this example, both

Table 10.1 Essential differences between alpha and beta particles.

Particle	Symbol	Mass	Decay product	Penetrating power	Ionizing power	Shielding
Alpha	α	4 amu (atomic mass unit)	Two fewer protons and neutrons	Very low	Very high	Paper or skin
Beta	β	1/1,200 amu	One more proton and one fewer neutron	Intermediate	Intermediate	Aluminium

isotopes are stable, but many naturally occurring isotopes are unstable and as a result of this nuclear instability may emit energy in the form of ionizing radiation. These unstable isotopes are known as **radioisotopes**. All elements in the Periodic Table (see Figure 6.1) with greater atomic mass than bismuth (^{83}Bi) emit ionizing radiation and are therefore radioactive. There are also several lighter elements that have one or more radioactive forms, for example carbon-14 (^{14}C). Also of importance, as discussed later in this chapter, some radioisotopes are formed as a result of human activity.

Since radioisotopes are unstable, they naturally revert to a less energetic state. This process results in the release of ionizing radiation and is referred to as **radioactive decay**. Each radioisotope has a highly specific decay pattern, both in terms of the speed of decay and the type of energy released. Natural ionizing radiation can be characterized by three principal forms: **alpha (α) particles**; **beta (β) particles**; and **electromagnetic photons** consisting of gamma (γ) rays and X-rays. Neutron radiation can be produced synthetically from splitting atoms in **nuclear reactors**.

Alpha particles: Alpha particles consist of two neutrons and two protons and therefore hold a +2 charge, equivalent to a helium nucleus. As a result of alpha decay, the resulting atom or daughter element has an atomic number two units lower and a relative atomic mass four units lower than that of the parent **radionuclide**. In other words, it drops two places in the Periodic Table. Although alpha particles have very high ionizing capability, their energy is quickly dissipated when they collide with matter since they have mass and therefore are heavy particles. Alpha particles travel through only a few centimetres of air and can be stopped by a piece of paper or the outer epithelium of an organism. Because of this lack of penetrating ability, alpha emitters generally do not cause damage to organisms when encountered externally, but they are of significant concern when taken into the body by inhalation or ingestion.

Beta particles: Beta particles are negatively charged electrons (no mass) formed when an excess of neutrons in the nucleon causes a neutron to be changed into a proton. If nuclear imbalance results in an excess of protons in the nucleus, the emitted particle is the positively charged equivalent of an electron, known as a positron. This causes a proton to be changed to a neutron. Alternatively, the proton may be converted to a neutron with the capture of a satellite electron, a process known as electron capture. Beta emitters vary widely in their energy output but can generally be stopped by a few millimetres of plastic or aluminium, or about a centimetre of tissue. Essential differences between alpha and beta particles are noted in Table 10.1.

Neutrons: Neutrons are the particles in atoms that have a neutral charge and no mass. Aside from cosmic radiation, neutrons are only generated from splitting uranium or plutonium atoms in nuclear reactors to produce nuclear power. Neutrons are very powerful and can penetrate tissues and the human body.

Electromagnetic photons: Electromagnetic photons consist of gamma rays and X-rays, which are high-energy waves emitted from an unstable nucleus (gamma rays), or orbital electrons (X-rays). These rays are similar to photons of visible light and travel at the same speed as light rays. Since there is no mass or charge associated with these rays, this type of radiation can travel much further through air than either alpha or beta particles. Both gamma and X-rays are strongly ionizing, and a thick protective layer of dense material such as lead or concrete is required to absorb all their energy.

Each radionuclide has a characteristic decay rate, which is independent of temperature, physical form and chemical combination. The rate of decay of any radionuclide is usually expressed as the **radioactive half-life**, which is the time it takes for 50% of a particular radioisotope to decay to its daughter. Highly radioactive substances have short half-lives, whereas those with long half-lives have low radioactivity. The decay of one radioisotope may result

Table 10.2 Uranium radioactive decay series.*

Element	Half-life
Uranium-238 (^{238}U) $\quad \alpha \downarrow$	4.5×10^9 years
Thorium-234 (^{234}Th) $\quad \beta \downarrow$	24.1 days
Protactinium-234 (^{234}Pa) $\quad \beta \downarrow$	1.14 minutes
Uranium-234 (^{234}U) $\quad \alpha \downarrow$	2.4×10^5 years
Thorium-230 (^{230}Th) $\quad \alpha \downarrow$	7.7×10^4 years
Radium-226 (^{226}Ra) $\quad \alpha \downarrow$	1,600 years
Radon-222 (^{222}Rn) $\quad \alpha \downarrow$	3.82 days
Polonium-218 (^{218}Po) $\quad \alpha \downarrow$	3.05 minutes
Lead-214 (^{214}Pb) $\quad \beta \downarrow$	26.8 minutes
Bismuth-214 (^{214}Bi) $\quad \beta \downarrow$	19.9 minutes
Polonium-214 (^{214}Po) $\quad \alpha \downarrow$	1.64×10^{-4} seconds
Lead-210 (^{210}Pb) $\quad \beta \downarrow$	22.3 years
Bismuth-210 (^{210}Bi) $\quad \beta \downarrow$	5 days
Polonium-210 (^{210}Po) $\quad \alpha \downarrow$	138.4 days
Lead-206 (^{206}Pb)	Stable

* Gamma emitters are not indicated.

in the formation of a daughter element, which may itself be unstable. This, in turn, will decay to a further element and so on through a decay chain until a stable element is formed. An important and the most-studied decay chain is that of uranium-238 (^{238}U), shown in Table 10.2.

In addition to the radiation listed above, some ultraviolet (UV) radiation can also be considered ionizing, depending upon the strength of the energy released. This type of radiation primarily comes from the Sun and is divided into three types of rays, UVA, UVB and UVC. UVA rays have the lowest amount of energy and UVC rays have the highest. The ozone layer does protect the Earth from exposure to UVC rays, but exposure to both UVA and UVB rays has been linked to skin cancer.

10.2.2 Units of Measurement

As with the measurement of many things, units for defining the measurement of radioactivity are complicated because traditional units are being phased out around the world in favour of standard international (SI) units, whereas the older units remain in active use in some countries, including the United States. A comparison of these units is summarized in Table 10.3. Importantly, there are two ways to consider the measure of radioactivity. First, one can define the amount of radioactivity emitted by a substance. Second, the measure of radioactivity absorbed by a living organism, which accounts for the destructive potential of the different kinds of radiation, can be calculated.

In contrast to damage from non-radioactive chemicals, the harmful effects of radioisotopes are usually related to their ionization potential and the amount of radiation

Table 10.3 Units of measurement of radioactivity.

Term	Definition	Old unit	International (SI) unit
Activity	The amount of ionizing radiation released by a material as alpha or beta particles, gamma rays, X-rays or neutrons. It represents how many atoms in the material decay in a given time period.	Curie (Ci): 2.2×10^{12} nuclear disintegrations per minute (dpm)	Becquerel (Bq): 1 Bq = 27.03 pCi
Exposure	The amount of radiation travelling through the air.	Roentgen (R)	Coulomb/ kilogram (C/kg)
Absorbed	The amount of radiation absorbed by an object or person.	Radiation absorbed **dose** (rad)	Gray (Gy)
Effective dose	Combines the amount of radiation absorbed and the medical effects of that type of radiation.	Roentgen equivalent man (rem)	Sievert (Sv)

For practical purposes, 1 R (exposure) = 1 rad (absorbed dose) = 1 rem or 1,000 mrem (dose equivalent).

Sources of radiation exposures

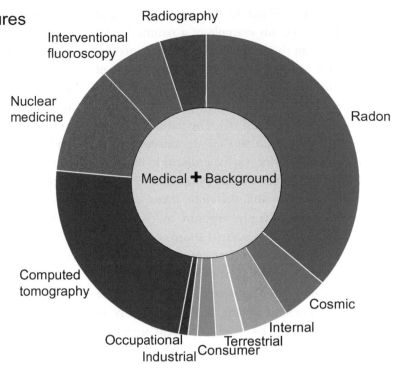

- Radon (37%)
- Cosmic (5%)
- Internal (5%)
- Terrestrial (3%)
- Consumer (2%)
- Industrial (<0.1%)
- Occupational (<0.1%)
- Computed tomography (24%)
- Nuclear medicine (12%)
- Interventional fluoroscopy (7%)
- Radiography (5%)

Figure 10.2 Sources of ionizing radiation exposure.

actually reaching the tissues. Therefore, for radioactive substances it is common to use a unit of dose (see Section 2.1.1.4) that represents the amount of energy deposited in tissues following exposure. The currently used SI unit for this is the gray (Gy) (Table 10.3), which is equivalent to 1 joule (J) of energy per kilogram of tissue. Enough is known of the ionization potential of different types of radiation to make further refinements to this concept of dose. When assessing radiation risk, it is important to consider that an alpha particle is capable of transferring 20 times more of its energy per unit distance travelled to a tissue than a gamma ray of similar energy. In other words, it has 20 times more linear transfer energy (LTE). Absorbed radiation doses are therefore normalized to dose equivalents by arbitrary factors known as quality factors or weighting factors (W_T), which take into account the LTE. The international measure of dose equivalent is the sievert (Sv) (= 100 rem). Thus,

$$H \text{ (equivalent dose in sieverts)} = \text{Dose (Gy)} \times W_T$$

$$(10.1)$$

10.3 Sources of Ionizing Radiation

When discussing the potential toxicity associated with ionizing radiation and how we manage the risk

Box 10.1 How do we manage risk?

Very little can be done to reduce levels of ionizing radiation that come from natural sources like the Sun, soil or rocks. However, some **remediation** and risk management through avoidance is possible for natural sources of radioactivity, such as radon gas which can enter residential homes.

Levels of ionizing radiation that come from manufactured sources and activities can be controlled much more carefully. In these cases, the balance between the societal benefits and the risks that ionizing radiation poses to people and the environment is considered carefully, and dose limits are set to restrict radiation exposures.

(Box 10.1), it is of utmost importance to discuss all potential sources of exposure including background and manufactured sources (Figure 10.2).

10.3.1 Background Ionizing Radiation

In addition to ultraviolet rays being released from the Sun, radioisotopes are present in all segments of the ecosystem, and radioactivity is distributed fairly homogeneously throughout the environment. Chemically, radioisotopes behave in the same way as their stable counterparts and are subject to the same **uptake** and physiological pathways. An example of a natural source of radiation is tritium (hydrogen-3, ^{3}H), which is present in the atmosphere where it is produced by the reaction of cosmic ray protons and neutrons with atmospheric gases such as oxygen (O), nitrogen (N) and argon (Ar). Another example is potassium (K). The naturally occurring radioisotope potassium-40 (^{40}K) is concentrated in certain plants such as coffee (*Coffea arabica*) in the same way as the normal, stable isotope potassium-39 (^{39}K). High coffee consumption will, therefore, increase potassium intake and, concomitantly, exposure to ^{40}K.

An important well-known natural source of radiation is uranium (principally occurring as uranium-238, ^{238}U), which is a ubiquitous component of rocks, particularly granite. In this example, the degree of human exposure will be heavily influenced by the relative proximity to different types of rock. The most important source of exposure to natural radiation for air-breathing animals is through inhalation of gaseous radioisotopes, with the largest dose coming from exposure to radon-222 (^{222}Rn). Radon is produced as part of the ^{238}U decay series (Table 10.2 and Figure 10.3) and is a naturally occurring gas, for which levels vary globally (Figure 10.3) and regionally (Gaskin *et al.*, 2018).

Even though the biological effects of radon are not fully understood, exposure of humans to radon is widely thought to play a large role in certain cases of lung cancer occurring in non-smoking patients. The United States Environmental Protection Agency (US EPA) states that radon is the second leading cause of lung cancer in the United States, and the leading cause among those who never smoked. Radon was first identified as a health concern when it was noted that underground uranium miners who were exposed to radon had a higher incidence of lung cancer than non-exposed workers. Radon can enter residential dwellings from the soil and be inhaled into the occupants' lungs. Since radon emits

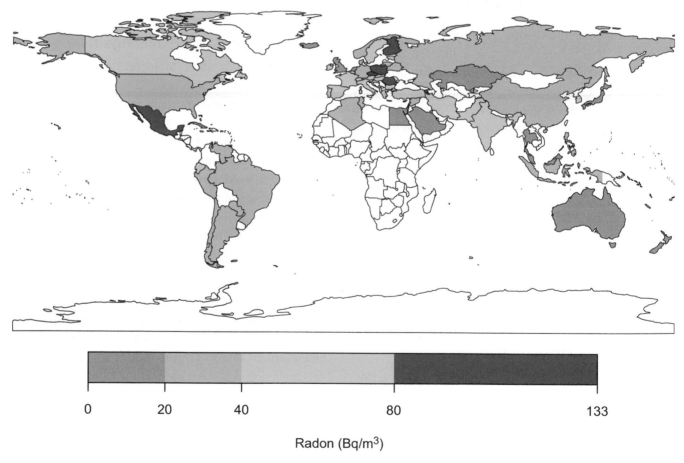

Figure 10.3 Global indoor residential levels (geometric mean) of radon-222 (^{222}Rn), a colourless, odourless radioactive gas produced naturally as part of the decay series of uranium-238 (^{238}U) found in soil and rocks. Adapted from Gaskin *et al.* (2018) with permission from the authors.

alpha particles, it can cause lung damage, potentially leading to lung cancer. In fact, the International Agency for Research on Cancer (IARC) classifies radon as a human **carcinogen** (IARC, 2012).

10.3.2 Manufactured Ionizing Radiation for Medical Use

X-rays can be artificially produced and lead to the release of energy in waves that can have useful medical purposes, for example diagnosing broken bones, and also damaging effects (Box 10.2).

Importantly, the levels of radiation emitted for medical purposes are highly regulated, and doses are kept as low as possible. Examples of medicinal uses of X-rays include dental and chest X-rays and mammograms to aid in diagnosis of disease conditions. In nuclear medicine, radioactive isotopes can also be used to treat and diagnose diseases such as cancer. Typically, small doses of radioactive tracers can be used to image the body for diagnostic purposes, whereas higher doses can be used to treat a diagnosed disease. For example, radioactive iodine-131 (^{131}I) can be used in small doses to image the thyroid and determine whether thyroid cancer is present. Certain forms of thyroid cancer can then be treated with higher doses of iodine-131 (^{131}I), which will accumulate in the thyroid and kill cancerous cells.

10.3.3 Nuclear Weapons

Historically, the release of radionuclides from nuclear weapons production and use (including testing) has substantially outstripped release from electricity generation.

Box 10.2 Historical example

It is important to note that historically, prior to the known damaging effects of X-rays, they were used for the treatment of ringworm of the scalp in children (mean age 7), during the period from 1905 to 1960. It is estimated that close to 200,000 children worldwide were irradiated. Follow-up studies on 2,224 children irradiated at the New York University Hospital during the period of 1940–1959 demonstrated that irradiated children were at higher risk for developing leukaemia, brain, thyroid, salivary gland and skin cancers than non-exposed controls.

For several elements, production of radioisotopes related to anthropogenic activity now greatly exceeds natural sources. For tritium (^{3}H), for example, production related to either electricity generation or the manufacture of weapons exceeds the global inventory from natural sources. Similarly, plutonium-239 (^{239}Pu) is a naturally occurring element, but the largest global source is through nuclear weapons production.

To date, nuclear weapons have only been used twice in warfare, both times by the United States. Towards the end of World War II, on 6 August 1945, the Japanese city of Hiroshima was hit with a uranium bomb deployed by the US Army Air Forces, and on 9 August 1945 a plutonium bomb was detonated by the same Forces over the Japanese city of Nagasaki. In both cases, massive destruction occurred in the immediate 1 km radius of the bombings, and approximately 200,000 civilian and military personnel were killed. In addition, there was an increased incidence of leukaemia rates in the areas in the 5- to 7-year period following the bombings. Since then, nuclear weapons have been detonated many times for testing purposes.

During the period from 1945 to 2006, 2,053 nuclear tests were conducted in the atmosphere, underground and underwater (Fedchenko and Hellgren, 2007), with 85% of these tests being conducted by the United States. The Bikini Atoll, a coral reef located in the Marshall Islands, was a site of 23 nuclear tests conducted by the United States between 1946 and 1958. Prior to these tests, a thorough documentation of the coral ecosystem was conducted, which provided the opportunity to investigate the long-term ecosystem impacts of nuclear testing, including coral biodiversity (Richards *et al.*, 2008). Immediate environmental impacts of the tests included increased surface water temperatures, blast waves (speeds up to 8 m/s), and shock and surface waves up to 30 m in height. In addition, the distribution of the natural sediment was altered, and both seawater and lagoonal water were contaminated with radioactivity, which was subsequently detected in marine species that colonize substrates, algae, clams, fish and coral skeletons (Richards *et al.*, 2008). The largest impact of these tests, however, resulted from the millions of tonnes of sediment that were pulverized and ultimately deposited in the surrounding ecosystem. Fifty years after the testing, Richards and colleagues compared the coral biodiversity pre- and post-testing and found that approximately 70% of the coral make-up demonstrated resilience to the effects of testing.

10.3.4 Nuclear Power

Electricity is produced from nuclear energy principally through a chain reaction originated by the bombardment of the uranium isotope uranium-235 (^{235}U) with slow neutrons. Fission (or splitting) of the uranium atom results in a number of products of lower atomic weight, and the release of alpha and beta particles and gamma radiation, together with high-energy neutrons and heat. These collide with other ^{235}U nuclei to produce a chain reaction. The 'cradle-to-grave' nuclear fuel cycle comprises the basic steps discussed below.

10.3.4.1 Mining and Extraction

The method of extracting uranium ore-bearing rocks is determined by their depth and distribution. Underground mines typically harvest only about 50% of the total ore present, whereas surface mining can remove as much as 90–95% of the ore in the deposit. Because the mean ^{235}U content of ore is 0.7% and the final product for commercial power generation must be 3–5% ^{235}U (much higher for military propulsion), considerable enrichment is required. This step requires massive quantities of ore and involves the production of large amounts of mining **overburden** and **mine tailings** from the extraction process. These mining activities lead to significant impacts on the surrounding ecosystem, including erosion, water contamination and effects on biodiversity. For example, habitat destruction, habitat modification (pH and temperature), and direct and indirect toxicity due to contaminated food and water will all affect animals, including aquatic species, vegetation and microorganisms.

Mine tailings (see Section 11.2), which are the materials remaining after the separation of the desired components from the ore, remain a source of uranium isotopes and decay products. Although **legislation** governing uranium mine and mill reclamation exists in the United States in the form of the Uranium Mill Tailings Act of 1978, contamination from tailings runoff and radioactive dust remain significant environmental issues. This is particularly relevant where **statutes** are not rigorously enforced or in countries where no such laws exist. Even where masks and high-efficiency particle-arresting (HEPA) filters are used to trap fine particulates, miners and mill workers are not protected from gaseous radioisotopes such as radon. As indicated earlier, several case studies have correlated the incidence of lung cancer with the inhalation of radon and its daughter isotopes by uranium miners and mill workers. The actual carcinogen is not the inert radon gas itself but several of its particulate and short-lived strong alpha-emitting daughter decay products such as polonium-218 (^{218}Po). When such isotopes are inhaled and deposited in the lungs, they can transfer a significant amount of their energy through the thin bronchial epithelium to the underlying target cells.

10.3.4.2 Enrichment, Conversion and Fuel Fabrication

Uranium has only two significantly abundant natural isotopes, ^{235}U and ^{238}U, with natural abundances of 0.71% and 99.28%, respectively. With the exception of the CANDU (Canadian deuterium uranium) system and reactors using Magnox fuel, nuclear reactors typically depend on enriched uranium which contains ^{235}U at levels of about 3.5–4.5%. Currently, there are three known commercial isotopic enrichment processes: gaseous diffusion, gas centrifugation and atomic vapour laser separation (AVLS). Enrichment by diffusion is based on diffusion kinetics at very high temperatures and uses gaseous uranium hexafluoride (UF$_6$) under pressure diffusing across a membrane. This method requires considerable energy and was considered an obsolete technology in 2013; the last commercial ^{235}U gaseous diffusion plant in the world (the Paducah facility in the United States) ceased operations in 2013. Gas centrifugation requires much less energy than gaseous diffusion and involves a large number of rotating cylinders in series and parallel formations; it currently accounts for close to 100% of the world's enriched uranium. Although AVLS is a promising method developed in the late 1990s using wavelength-tunable lasers to separate isotopes by differential photo-ionization, this process is not currently used by any country.

10.3.4.3 In-core Fuel Management

The two basic components of a nuclear fuel assembly are the nuclear fuel itself (either uranium, plutonium or a mixture) and a protective (stainless steel) pressure vessel to contain the **waste fissile** material that is produced. The heat produced in the fission process is exchanged with water under high pressure, which is circulated at high velocity through the reactor. The rate of the chain reaction can be controlled by inserting boron control rods (to absorb neutrons) among the fuel rods, all of which are immersed in water. The control rods, or moderators, can be raised or lowered to increase or decrease the reaction rate. When 90% or more of the rods need to be removed to maintain the efficiency of the fission reaction, the nuclear fuel is considered to be 'spent'.

An operating reactor will produce about 80 radionuclides, with even more produced by decay chains, yielding half-lives from a few milliseconds to billions of years. Of these, a subset of synthetic radionuclides has been the focus of particular concern due to their incorporation into biological systems and their significant longevity with respect to the total radiation emitted. These radionuclides include uranium-238 (^{238}U), thorium-232 (^{232}Th), radium-226 (^{226}Ra), caesium-137 (^{137}Cs), iodine-131 (^{131}I), strontium-90 (^{90}Sr), zinc-65 (^{65}Zn), potassium-40 (^{40}K), phosphorus-32 (^{32}P), carbon-14 (^{14}C) and tritium (^{3}H).

The release of some radioactivity during normal reactor operation is unavoidable. Some radioactive fission products leak through pinholes in the steel cladding surrounding the core into the cooling water. In addition to contamination by radioactive materials from the nuclear fuel, other radionuclides are formed by neutron activation of gases in the atmosphere, graphite (used as a moderator in some nuclear power plants) and impurities in water in the vicinity of the reactor. Such neutron activation products include caesium-137 (^{137}Cs), caesium-134 (^{134}Cs), krypton-85 (^{85}Kr), arsenic-76 (^{76}As), zinc-65 (^{65}Zn), copper-64 (^{64}Cu), cobalt-60 (^{60}Co), cobalt-58 (^{58}Co), manganese-56 (^{56}Mn), chromium-51 (^{51}Cr), phosphorus-32 (^{32}P), silicon-31 (^{31}Si) and tritium (^{3}H), some of which are often detected in measurable quantities in liquid effluents from nuclear power plants. It is important to note that radionuclide emissions from nuclear power plants that occur during normal operations are very small and are generally regarded as negligible in terms of their environmental effect.

10.3.4.4 Fuel Reprocessing

As indicated in the preceding section, nuclear fuel is considered to be spent when about 90% of the moderator/control rods must be removed to achieve a critical state. **Spent fuel** is then transferred from the reactor core to a holding tank where it is allowed to dissipate both heat and radioactivity before being shipped to a reprocessing facility. At this stage, it is either disassembled and discarded or reprocessed. Although reprocessing in connection with plutonium production for nuclear weapons has taken place at the United States Department of Energy weapons production sites, few

reprocessing plants exist worldwide for the regeneration of fuel for electric power production. The concept of nuclear reprocessing is politically challenging, given the potential for nuclear proliferation and vulnerability to nuclear terrorism. Despite this, nuclear fuel reprocessing does occur in Europe, Russia and Japan and potentially in Iran.

In the course of reprocessing, the outer layer of the fuel rod, known as the fuel cladding, is removed either by dissolution or stripping, followed by fuel dissolution in acid. The ultimate goal is to recover usable plutonium from spent nuclear fuel. The PUREX (plutonium and uranium recovery by extraction) process is almost universally used in the removal of unusable fission products from solution and solvent extraction of the desirable plutonium-239 (^{239}Pu) and uranium-235 (^{235}U). Very large amounts of acid are used in the dissolution of cladding and fuel pellets. Equally large amounts of solvents such as diethyl ether, isobutyl ketone, dibutyl carbitol and tributyl phosphate are used in extraction and repurification operations.

Reprocessing of uranium from irradiated fuel is different from processing natural uranium, in that the additional isotopes ^{232}U and ^{236}U are present, both of which must be removed: ^{232}U produces a strong gamma emitter, and ^{236}U is a neutron flux poison – that is, a substance that absorbs neutrons. The latter creates an economic penalty for re-enrichment of spent fuel. When compared with direct spent fuel disposal, recycling, if performed correctly, appears to have many obvious environmental benefits, including reduced waste volumes, reduced radioactive content of waste and reduced toxicity of wastes. However, in fact, the economic benefits of fuel reprocessing are highly debated, with arguments indicating a higher cost associated with the reprocessing–recycling option compared with that of direct geological disposal of spent fuel.

10.3.5 Nuclear Waste Management

There are several types of nuclear waste: high level, intermediate level, low level and very low level. High-level wastes encompass spent fuel and transuranic wastes and weapons processing effluents. Intermediate waste consists of resins, chemical sludges and contaminated material from reactor decommissioning. Low-level wastes comprise the high-volume/low-radioactivity waste rock and milling from the uranium extraction process but also include protective clothing, tools, laboratory research compounds, and other equipment associated with the operation and maintenance of nuclear facilities. Low-level waste is also generated from hospitals (Section 10.3.2) and other industries. Very low-level waste emits radiation at such a low level that it is not considered to be harmful to mammals, including humans, or the surrounding ecosystem, and therefore disposal practices for this waste are similar to those for domestic refuse, such as domestic smoke detectors. The time required for radioactive waste storage depends on the type of waste and the radioactive isotopes of concern.

It is important to note that plutonium is a major factor forcing the disposal rather than recycling of spent nuclear fuel. Although the plutonium contained in spent fuel is of significant political, economic and strategic military value (primarily as the critical trigger in constructing nuclear bombs that are easily human-portable), the spent fuel itself is, in effect, 'self-protecting' with respect to the high degree of radiation and shielding necessary for its transport: the radioactive nature of the spent fuel makes it almost impossible for it to be stolen by individuals seeking the plutonium for use in nuclear weapon development. There has been a certain amount of debate as to whether the plutonium metal itself is inherently toxic, or whether its toxicity derives from its strong alpha emission. However, this question will probably never be answered, since it is currently impossible to separate the chemical toxicity of the element from its radioactive effects.

10.3.5.1 Short-lived Intermediate and Low-level Waste

Short-lived intermediate and low-level wastes comprise 90% by volume of all radioactive wastes, yet they contain less than 1% of the total radioactivity that requires disposal. Different countries have adopted very diverse standards for the disposal of low-level waste. Until 1982, most countries dumped intermediate and low-level radioactive waste at deepwater oceanic sites beyond the continental shelf. Since then, land burial has been favoured. Usually, the waste is packed in drums that are placed in shallow trenches covered with earth. However, some countries (e.g., Germany and Sweden) prefer deep geologic disposal, similar to how they handle high-level waste. The United States relies on a system of landfills of varying degrees of packing and security depending on the type of waste. With groundwater and soil contamination being the most important concerns, the most sophisticated of these landfills are thick concrete bunkers with remote **monitoring** devices. The Waste Isolation Pilot Plant (WIPP) designed to permanently store low-level and transuranic military waste at an underground site in an area known as the southeastern New Mexico nuclear corridor is in operation in the United States. The explosion of a nuclear waste container at this site in 2014 led to a 3-year shutdown of the plant, which was formally reopened in January 2017.

A full investigation into the accident revealed a moderate release of radioactivity into the underground air, with a small portion of contaminated air escaping to the surface and being detected approximately 1 km away. Total radioactivity levels detected were localized and very low. Additional sampling of the surrounding vegetation, soil, water and animals indicated no enhanced radioactivity above background levels. Two weeks following the incident, radiation levels were at background levels, suggesting no long-term ecological effects of the accident (Thakur *et al.*, 2016).

10.3.5.2 Long-lived Intermediate and High-level Waste

High-level waste is generated by nuclear reactors and accounts for approximately 95% of the total radioactivity produced during the generation of electricity through nuclear reaction. Irradiated fuel rods and assemblies are even more radioactive than the new fuel because of the accumulation of fission products, and thus this waste is highly radioactive. Fuel rods remain dangerous for many thousands of years owing to emissions from radioisotopes with very long half-lives. In 2010, the worldwide inventory of high-level radioactive waste from nuclear power plant operation was about 250,000 metric tonnes with approximately 12,000 metric tonnes being added annually. The overall global agreement is that high-level waste should be buried in deep geological repositories, although no currently accepted repository exists. In 1987, amendments were made to the United States Nuclear Waste **Policy** Act which led to the start of construction of a single high-level waste site at Yucca Mountain in Nevada, to be known as the Yucca Mountain Nuclear Waste Repository. Excavation of approximately 160 km of tunnels in volcanic rock 300 m below the surface of the mountain yet well above the water table was started. It was anticipated that high-level waste would be placed in the repository over a 50-year period, after which time it would be sealed and marked. Federal United States funding for the site ended in 2011, after billions of dollars had been invested in the project; funding was cancelled for a variety of reasons, including strong state and regional opposition. Currently, high-level and long-lived intermediate wastes are kept on the sites where they are produced and stored using dry cask storage within steel and cemented casks.

10.4 Case Studies

Two case studies describing two different nuclear accidents are described in the next sections. The aim here is to highlight the complex toxicological issues associated with the unintended release of ionizing radiation.

Case Study 10.1 The Chernobyl Accident

On 26 April 1986, a reactor explosion occurred at unit four of the light water graphite-moderated reactor at Chernobyl, 80 km north of Kiev in Ukraine. Ukraine was then part of the former Union of Soviet Socialist Republics (USSR). At the time of the accident, this type of reactor was one of 14 such reactors in the USSR, comprising more than half of the nuclear-powered electricity generation capacity of that country. The accident followed a dramatic and cumulative series of errors incurred while conducting a set of tests. It included a blatant disregard of safety procedures, which was exacerbated by the design faults of this type of reactor itself. Several safety systems were deliberately disabled, resulting in an uncontrolled chain reaction. The single most important design fault was that steam generation in the fuel channels causes an increased number of neutrons to collide with the graphite, leading to an increased rate of fission. Under such circumstances, a last desperate attempt to stop the reaction by inserting the control rods had the opposite effect, as the control rods caused a power surge, owing to their unique design, and within 4 seconds the reactor power reached more than 100 times its capacity. The resultant explosion sheared all 1,661 water pipes, blew the 1,000-tonne cap off the top of the core and ruptured the concrete walls of the reactor hall. Radioactive dust, including pulverized fuel, was thrown at

least 7.5 km into the air and the intense heat started at least 30 fires. Radioactive release lasted about 10 days, and despite attempts to quench the reaction through air drops of boron carbide, sand, clay and lead, there was a complete meltdown of the reactor core. Nine days following the initial explosion, the daily release rate of radioactive material was nearly as high as it was at the time of the initial release. As an immediate result of the accident, 31 people died, and many cases of **acute** radiation sickness were recorded. In addition, 65 of the original staff of the reactor had died by 1991. As a group, they experienced a death rate more than 100 times that of a comparable, unexposed population. In the months following the explosion, more than half a million people were involved with the construction of a sarcophagus around the melted core. In some areas, the radiation fields were very high, and despite the stated intent to limit workers' exposure to radiation, it can be appreciated that, in such an environment, many were probably exposed to considerably high levels of radiation, and many probably received doses comparable to survivors of the Hiroshima and Nagasaki explosions.

Released radionuclides included essentially all the noble gases, volatile elements such as iodine-131 (^{131}I), caesium-134 and caesium-137 (^{134,137}Cs), and some refractory materials, for example strontium-89, strontium-90 (^{89,90}Sr), cerium-141, cerium-144 (^{141,144}Ce) and radionuclides of plutonium (^{238,239,240}Pu) (Anspaugh *et al.*, 1988). Through integration of environmental data, it is estimated that some 100 petabecquerels of ^{137}Cs (1 PBq = 10^{15} Bq) were released during and subsequent to the accident. The total release of radioactivity exceeded 3×10^{18} Bq. The highest radioactivity depositions occurred in Belarus, Russia and Ukraine. However, owing to the prevailing weather patterns at the time of the accident, the spread of the fission product plume from Chernobyl was sustained for 11 days and affected many countries including Finland, Sweden, Denmark, Germany, France, Italy, Austria, Poland, the United Kingdom and Czechoslovakia (Megaw, 1987). In Sweden, the fallout was identified as containing radioactive krypton-85 (^{85}Kr), xenon-133 (^{133}Xe), iodine-131 (^{131}I), caesium-134 and caesium-137 (^{134,137}Cs), and cobalt-60 (^{60}Co) (Megaw, 1987). Japan, the United States and Canada were also slightly affected. In Poland, the closest country to Ukraine, the government banned the sale of milk from cows on pasture, and children were treated with potassium iodide to reduce their uptake of iodine-131 (^{131}I). In parts of Sweden where rain or snow had fallen since the accident, people were advised not to drink tap water, which could be 100 times more radioactive than normal. In Finland, a national radioisotope monitoring programme established in response to post-war nuclear bomb testing was rejuvenated. Because of the decline in the atmospheric deposition of isotopes following the shift to underground bomb testing, the programme had been scheduled for elimination in the year following Chernobyl. Although no acute effects occurred outside the USSR, the risk for lifetime expectation of fatal radiogenic cancer increased from 0 to 0.02% in Europe and 0 to 0.003% in the northern hemisphere (Anspaugh *et al.*, 1988).

Human Health Effects

Numerous clinical studies of human populations exposed to radiation have been made during the years following the Chernobyl accident; some of these continue today. Several reveal **symptoms** both directly and indirectly related to the radiation. Distinct changes in the clinical picture of acute pneumonia were noted in patients subjected to the constant, prolonged (1986–1990) effect of small doses of ionizing radiation as a result of residing in the contaminated territory after the accident (Kolpakov *et al.*, 1992). These changes included increased duration of the disease and the frequency of protracted forms as well as suppression of the immune system. The small radiation doses in Kiev, 80 km away from the accident, had a significant impact on the humoral immunity of the population (Bidnenko *et al.*, 1992). Recent reviews have reported dose-related increases in papillary thyroid cancer and follicular adenoma of the thyroid among children and adolescents exposed to radioiodines in Ukraine and Belarus (Cardis and Hatch, 2011; Hatch and Cardis, 2017). Children and adolescents were particularly sensitive since on average they were exposed to higher relative radiation doses due to their smaller thyroid mass and greater consumption of ^{131}I-contaminated milk compared to adults. In addition, an elevated risk of cataracts, leukaemia and possibly cardiovascular disease among clean-up workers exposed to external radiation has been reported (Cardis and Hatch, 2011; Hatch and Cardis, 2017).

It should be noted that psycho-social effects are considered to be one of the important public health consequences following the Chernobyl accident. Post-traumatic **stress** and extreme anxiety about radiation exposure developed in many individuals despite many having had negligible exposure to radiation. Many of the evacuees faced difficulties and uncertainties following evacuation orders which for many led to depression, alcoholism and for some, suicide (Bromet, 2014).

Ecological Effects

Following the accident, many European species of mycorrhizal fungi, including several edible ones, were found to contain unacceptably high levels of ^{137}Cs (>1,000 Bq/kg dry wt). Concentrations of ^{137}Cs in lichens and mosses in some areas were significantly elevated 5 years after the accident, and levels still remain high today. These lower organisms do not have roots and get most of their nutrients (and contaminants) from atmospheric depositions. A study of earthworm populations in a 30-km zone around the Chernobyl nuclear power plant following the accident indicated a significant, but temporary, depression in recruitment relative to control plots; the populations had recovered by the summer of 1988 (Krivolutzkii *et al.*, 1992). Thirty years after the accident, there still remain high levels of ^{137}Cs in fish and marine mammals, particularly in shallow estuaries with slow water turnover such as the Baltic Sea (Saremi *et al.*, 2018).

Despite the high levels of radioactivity remaining, following the forced evacuations and decades of limited human presence, the area's ecosystem has flourished, reflecting at least in part the fact that it has been virtually undisturbed by humans during this period. Wildlife, including birds, wolves and moose, is now thriving in the area, and vegetation has taken over abandoned buildings and is blooming. The site of the accident has now also become part of the tourism industry. Pripyat, which was the closest city to Chernobyl and once home to 50,000 people, became a ghost town following the accident. Tours of the city of Pripyat have taken place since 2000. In 2019, the American television network Home Office Box (HBO) produced a mini-series about the tragedy, which caused a spike in the demand for tours to the contaminated site.

Case Study 10.2 Fukushima Daiichi Nuclear Power Plant

On 11 March 2011, another major nuclear accident occurred, this time at the Fukushima Daiichi Nuclear Power Plant in Okuma, Japan, which was operated by the Tokyo Electric Power Company (TEPCO). This began with an earthquake followed by a tsunami, resulting in a complete lack of coolant which led to three nuclear meltdowns, three hydrogen explosions and the release of radioactive contamination. Approximately 155,000 residents from the surrounding communities were evacuated and, because of the proximity of the plant to the Pacific Ocean, large amounts of radioactive isotopes entered the ocean. Fifteen days following the accident, trace radioactive particles were measured across the northern hemisphere, and within 1 month were detected by measuring stations in the southern hemisphere. Radioactive contamination in food was detected in items such as spinach, milk, tea leaves and beef. Although there are several discrepancies in the data, TEPCO published results of the amount of radioactive released into the air, which were a combined estimate of 538 PBq of iodine-131 (^{131}I), caesium-134 and caesium-137 (^{134,137}Cs) with an overall total release of 900 PBq when all radioactive isotopes were included (TEPCO, 2012). By comparison, the release of radioactivity was about 10–30% of that from the Chernobyl accident, depending upon the isotope. With respect to the amounts of caesium-137 (^{137}Cs) released into the ocean, this is estimated at over 27 PBq. Owing to the strong ocean currents along the Fukushima coast, this radioactivity was transported and ultimately dispersed through much of the ocean.

Several studies have analysed the ecological effects of caesium-134 and caesium-137 (^{134,137}Cs) contamination and found that radioactivity fallout has become strongly adsorbed within soil and has not migrated significantly, even with heavy rainfall. Furthermore, the transfer of this radioactivity from the soil into crop plants is small, especially when potassium fertilizer is used on farmlands to minimize uptake of radioactive caesium by plants (Nakanishi, 2018). These studies also indicate that since deciduous trees did not have leaves at the time of the accident, radioactive incorporation into these trees was lower than that for evergreens, which absorbed higher amounts of radioactive fallout because of their needle-like leaves.

Because of the known association of increased thyroid cancers following the Chernobyl accident, thyroid screening for children and adolescents living in Fukushima at the time of the accident was carried out, beginning in October 2011. As of 2018, over 324,000 individuals 18 years of age or younger had their thyroid glands screened (Ohtsuru et al., 2019). Thyroid cancer was diagnosed in 187 of these individuals with an estimated incidence of 29–64 per 100,000 depending on the age group. The majority of these cancers were papillary thyroid cancers, which has a baseline incidence of 4–5 per 100,000 in children and adolescents (Bernier et al., 2019). In addition, a recent study reports that the mental health status of adolescents following the accident was significantly affected, with an estimate of close to 9% of individuals experiencing psychological distress including extreme anxiety about the impacts of radiation exposure on their future health (Hayashi et al., 2020).

In the aftermath of this disaster, the decommissioning process is still ongoing, and some progress has been made. Robots are being deployed to assess the highly radioactive areas and to provide engineers and other scientists with a better picture of the location of radioactive material, which will ultimately help with the removal process. Even though tours of the disaster area began in 2018, and, despite all the efforts deployed, it is still anticipated that the full clean-up of the site will take another 30 years. Regardless of this, Japan is committed to its nuclear power industry and plans to generate 20% of its energy from other reactors by 2030.

10.5 Effects of Ionizing Radiation at the Molecular and Cellular Levels

Energy absorbed by tissues exposed to radiation causes a variety of chemical changes in cells as described in the next few sections, resulting in a range of damaging effects (Mu et al., 2018) (Figure 10.4). Although the effects illustrated in Figure 10.4 are for humans, the same types of physiological damage are likely to apply to other mammals as well.

Some changes may be sufficiently severe that they cause tissue death or **necrosis**, which in extreme cases will cause the death of the organism. Alternatively, exposure can lead to permanent changes in cellular metabolism, resulting in carcinogenesis or genetic disorders. These damaging effects can be the consequence of indirect or direct effects of ionizing radiation (Figure 10.5).

Direct damage occurs when ionizing radiation interacts with atoms of cellular components such as DNA; the resulting molecular damage, if not repaired, can cause a variety of adverse effects, including **teratogenesis** (birth defects) and cancer. Indirect effects result from the ionization of water molecules, the major component of cells, leading to the formation of high-energy radicals that,

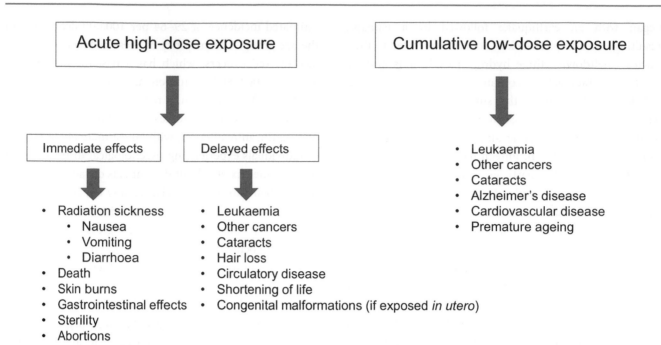

Figure 10.4 Potential human health effects following exposure to acute high-dose and cumulative low-dose ionizing radiation.

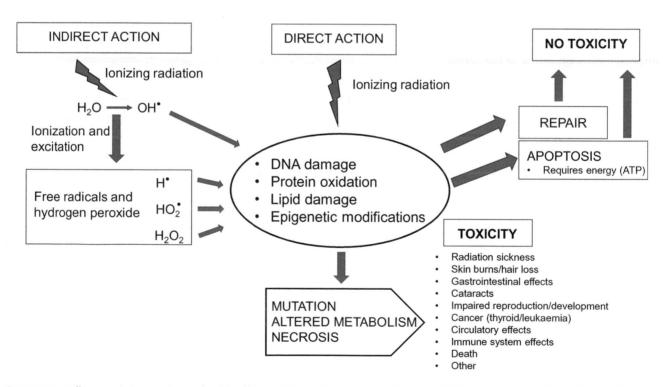

Figure 10.5 Indirect and direct adverse health effects at the molecular and cellular level following exposure to ionizing radiation.

although short-lived, are extremely reactive. The sequence of their formation follows:

Primary ionization:

$$H_2O \rightarrow H_2O^+ + e^- \qquad (10.2)$$

Electron capture by neutral water molecule:

$$H_2O + e^- \rightarrow H_2O^- \qquad (10.3)$$

Breakdown of ionized water molecules:

$$H_2O^+ \rightarrow H^+ + {}^{\bullet}OH \qquad (10.4)$$

$$H_2O^- \rightarrow {}^{\bullet}H + OH^- \qquad (10.5)$$

H^+ and OH^- combine to produce neutral water, resulting in the net reaction

$$H_2O \rightarrow {}^{\bullet}H + {}^{\bullet}OH \qquad (10.6)$$

Although these free radicals have a very short lifetime in water, they are extremely reactive. They may react with each other to form water, or two •OH radicals may combine to form hydrogen peroxide (H_2O_2), which is a stable molecule but a powerful oxidizing agent:

$$^{\bullet}OH + {}^{\bullet}OH \rightarrow H_2O_2 \qquad (10.7)$$

These radicals, together with H_2O_2, may interact with macromolecules such as nucleic acids, proteins, lipids and large polymers. Large molecules may be broken up, forming additional organic radicals that may initiate the formation of cross-linkages within the molecules or between molecules. In the latter case, molecular pairs, or dimers, may be formed, a process known as dimerization. The proximity of oxygen to the initial site of chemical change increases the probability of damage. This may be due to increased H_2O_2 production, or through direct interaction with the organic radical.

10.5.1 Cell Death

Radiation may cause cell death, either through mitotic death or interphase (**apoptotic**) death. As its name suggests, interphase death occurs when the cell is not dividing. Not to be confused with necrosis, it is called apoptosis, or programmed cell death, which is a deliberate, energy-dependent cell death. In the case of radiation exposure, cell death via apoptosis is thought to be a protective mechanism for the **elimination** of highly damaged cells. During the apoptotic process, cells shrink and cellular components are converted to fragments by caspases found in the cell, followed by **phagocytosis** of the fragments. This process is aimed at protecting any neighbouring cells and localizing the affected ones. Since apoptosis is an active process, it requires sufficient cellular energy in the form of adenosine triphosphate (**ATP**), which is produced in the mitochondria, to be successful. Depending on the amount of damage and radiation, ATP production in the cell may be compromised, and apoptosis may not be an option for the cell. In this case, the cell may undergo pathological cell death, otherwise known as necrosis. In the case of necrosis, a damaged cell will swell and release cellular contents triggering secondary processes such as inflammation, which can result in a chain reaction in which neighbouring cells also undergo necrosis. In addition, radiation-induced **reactive oxygen species** can lead to damage in non-targeted bystander cells through mechanisms involving intracellular communication. The persistence of this type of stress has been linked to secondary cancers that can occur following radiotherapy treatment.

Radiation can also damage cells that are in the process of dividing (mitosis). These cells are particularly sensitive to radiation and include bone marrow cells and stem cells of the gastrointestinal tract and skin, which are continually in a high state of replication to repair tissue damage. Young children and the developing foetus are particularly sensitive to radiation because of the high proportion of actively growing tissue. Mature, fully differentiated cells such as nerve and muscle cells are less sensitive to radiation. Even though blood cells such as erythrocytes and granulocytes are themselves **resistant** to radiation, their stem cells are not. This resistance eventually leads to a reduction in both red and white cell counts with attendant anaemia, reduction in oxygen-carrying capacity and loss of immune function.

The **lethal** effect of radiation on replicating cells has been utilized in the treatment of malignant cancer cells, which have a high proliferation rate. Precisely directed doses of high-energy gamma rays have been shown to selectively destroy malignant tumours. Damage to surrounding tissue is minimized by varying the angle of irradiation in controlled serial doses.

10.5.2 DNA Damage

It is well known that ionizing radiation can cause DNA damage. This can occur via direct interaction resulting in molecular distortion leading to double-strand breaks in the phosphodiester chains of the DNA molecule.

Alternatively, DNA damage can also occur indirectly by the production of reactive oxygen species through the **hydrolysis** of water as described above. Reactive oxygen species can directly modify DNA bases and also cause single-strand or double-strand DNA breaks. Accumulation of this damage, if left unrepaired, can lead to genomic instability, mutations and **carcinogenesis**. Whether or not a mutation occurs will depend on the type of damage caused and the response of the cell. There are two types of mutations that can occur, chromosomal aberrations and gene mutations. Chromosomal aberrations result in either a change in the number of chromosomes or structural changes in specific chromosomes that include deletions, duplications, inversions and translocations. Deletions and duplications result in changes in the amount of gene expression that occurs, either increasing or decreasing the amount. Inversions and translocations alter the position of gene(s), either within a chromosome (inversion) or to a different chromosome (translocation). In either case, since the location of a gene is important in its regulation and activity, these aberrations can be harmful to the cell. Gene mutations occur when there is a change in the nucleotide sequence of a gene, which can result in loss of the gene or a change in the gene product.

The cell has developed an array of DNA repair mechanisms to repair this damage. However, it is known that ionizing radiation can also lead to defects in DNA repair mechanisms themselves, including key proteins and enzymes that are required for successful repair of damaged DNA.

10.5.3 Protein Damage

As alluded to above, proteins represent another potential molecular target that can be damaged by ionizing radiation. Once a protein has been damaged, the structure and function of the protein may be compromised. Given the large number of proteins and their vital role in numerous cellular functions, protein damage can be very detrimental. Additionally, once a protein has been oxidized (by reactive oxygen species), this increases the likelihood that it will be denatured by proteolytic enzymes. The types of protein damage that ionizing radiation can cause include changes in bond length, breakage of disulphide bonds, decarboxylation and changes in the binding sites of specific metal ions.

A major concern about protein damage following exposure to ionizing radiation is the possibility of resulting effects on functionality and efficiency of enzymes

essential for cellular repair mechanisms; this will impair the cell's ability to recover from DNA damage caused by ionizing radiation (Case Study 10.2).

10.5.4 Lipid Damage

Radiation can also damage **cell membranes**, which are made up of phospholipids, proteins and carbohydrates (Section 3.1.1). The phospholipids contain saturated and unsaturated fatty acid chains that vary in length. Unsaturated fatty acids contain at least one double bond in the fatty acid chain which makes them susceptible to radiation-induced peroxidation. The process of lipid peroxidation starts with the attack of a free radical on the unsaturated lipid to produce a lipid radical, causing a chain reaction in which further radicals are formed. This cascade is only terminated when two lipid radicals join together or radicals are handled by the cell's detoxification pathways (e.g., **glutathione**, vitamin E). Lipid peroxidation leads to the destruction of the lipid, which can affect the integrity of the cell membrane, and furthermore gives rise to breakdown products that can also be toxic to the cell. In addition, lipid radicals that are formed can interact with other cellular molecules, such as proteins, further damaging cellular components including the inhibition of essential enzymes. Lipid damage is a significant issue for **poikilotherms** that rely on polyunsaturated fatty acids to maintain membrane fluidity at low temperatures, which may explain why some fish species are rich in antioxidants such as vitamin E.

10.5.5 Epigenetic Effects

Epigenetics refers to the regulation of genomic function that leads to heritable changes in gene expression that are outside the deoxyribonucleic acid (DNA) sequence. Epigenetic modifications are thought to be one of the links between environmental factors and changes in gene expression (see Section 5.5). Epigenetic modifications include DNA **methylation**, histone modifications (acetylation, methylation, phosphorylation) and ATP-dependent chromatin remodelling. The misregulation of these components, regardless of the DNA sequence, can lead to genomic instability. Inappropriate epigenetic control has also been shown to lead to an increased incidence in several diseases including Type II diabetes, cancer and Alzheimer's.

Recent evidence indicates that exposure to ionizing radiation can lead to global hypomethylation of DNA (Burgio *et al.*, 2018). Interestingly, further studies demonstrated that more differences in epigenetic modifications are observed following chronic exposure to low doses of radiation compared with acute exposures, with chronic exposures appearing to be more powerful inducers of changes to the epigenome. It is now thought that ionizing radiation-induced epigenetic effects may be related to the phenomenon known as the '**bystander effect**'. This is the observation that cells exposed to ionizing radiation can transmit cellular consequences of radiation exposure to non-exposed cells (Burgio *et al.*, 2018).

Examples of the bystander effect at the whole-animal scale include reports that exposure to ionizing radiation in one species of fish that is subsequently placed in the same location as another species of fish that has not been exposed to any ionizing radiation can lead to changes in protein expression in the non-exposed bystander fish (Mothersill *et al.*, 2018). Similarly, studies in rainbow trout have reported bystander effects, as measured by changes in cellular survival and growth, two generations following ionizing radiation exposure (Smith *et al.*, 2016). These results raise significant questions about the legacy of ionizing radiation and the long-term impacts following ecological exposures, especially on aquatic populations.

10.5.6 Effects on the Immune System

The principal cellular components of the mammalian immune system are natural killer (NK) cells (a type of lymphocyte), macrophages and a series of immunological mediators, formed from macrophages, which regulate the activity of the lymphocytes. Macrophages are responsible for non-specific immune reactions such as phagocytosis. Lymphocytes are highly specific in their immune response and are responsible for the recognition of foreign (non-self) antigens and for the initiation and degree of the response. Small lymphocytes are functionally divided into two groups: B-cells produced by the bone marrow and T-cells produced by the thymus gland. B-cells are primarily responsible for humoral immunity and the production of circulating antibodies. B-cells can differentiate into plasma cells. T-cells are mainly responsible for cellular immunity and hypersensitivity responses such as transplantation immunity and immunity to agents such as bacterial and viral antigens. Some subsets of T-cells known as helper or suppressor cells can modulate the effect of both T- and B-cells. A characteristic feature of T-cells is their high incidence of apoptosis prior to full differentiation.

For reasons that are not fully understood, some subpopulations of lymphocytes are extremely sensitive to radiation (CD4 helper-T lymphocytes), but macrophages and plasma cells are very resistant. Generally speaking, radiation has a dose-dependent immunosuppressive effect. However, under certain experimental conditions, exposure to radiation may be associated with an increased immune response. Such a phenomenon appears to be related to **hormesis**, where exposure to low levels of a **toxicant** can stimulate a beneficial response (see Box 2.4). In the case of the immune system, exposure to low levels of ionizing radiation can enhance the immune function of the body.

10.6 Risk Assessment of Ionizing Radiation

There are many different consequences of exposure to ionizing radiation; responses will depend upon a number of factors, including the dose, type of radiation, duration and target tissue. Exposure to very high levels of ionizing radiation over a short period of time (acute), which occur due to extreme events such as a nuclear explosion, can lead to death and symptoms such as nausea and vomiting referred to as radiation sickness. On the other hand, exposure to lower doses can increase the risk of cancer over a lifetime (Figure 10.4). As indicated earlier, exposure to very low doses can potentially have a hormetic effect, resulting in the opposite outcome of delaying cancer progression.

It is well known that not all cells are equally sensitive to radiation. For example, rapidly dividing cells, such as blood-forming cells and embryonic cells, are highly sensitive, whereas nerve and muscle cells are the least sensitive. In 1977, the International Commission on Radiological Protection (ICRP) introduced the term *effective dose equivalent* to compare human carcinogenic and genetic risks from different tissue or whole-body radiation doses (ICRP, 1977). The measure incorporates tissue weighting factors (W_T) in recognition of the fact that certain tissues are more susceptible than others to the effects of radiation (Table 10.4). The term effective dose (E), which is measured in the SI units sieverts (Table 10.3), replaced the *effective dose equivalent* in 1991 and is the main quantity by which dose limitation **guidelines** are set by international regulatory agencies and by the ICRP. Effective doses from different sources are assumed to be additive to calculate overall carcinogenic and genetic risk.

Table 10.4 **Values for tissue weighting factors, which are multipliers used to express relative biological damage (ICRP, 2007).**

Tissue/organ	Tissue weighting factor (W_T)
Gonads	0.08
Colon	0.12
Lung	0.12
Red bone marrow	0.12
Stomach	0.12
Breast	0.12
Bladder	0.04
Liver	0.04
Oesophagus	0.04
Thyroid	0.04
Bone surface	0.01
Skin	0.01
Brain	0.01
Salivary glands	0.01
Other tissue	0.12
Total	**1.00**

The use of effective dose assumes that the risk from ionizing radiation is a linear, non-**threshold** dose–response relationship. Estimates by the ICRP (2007) of human risk from radiation were based on data from Japanese atomic bomb survivors who had received doses between 0 and 4 Sv. In the earlier part of the study, observed effects were dominated by subjects who had received particularly high doses of radiation (>1 Sv). However, by 1990 attempts were being made to use the low-dose portion of the data to assess risks for a variety of different cancers (US NAS, 1990). In this regard, toxicologists are faced with problems similar to those encountered with chemical toxicology at very low doses. Is there a toxicity threshold, or is the dose–response linear? In the case of low-dose radiation, the possibility of protective or **hormetic** effects also had to be considered. For all cancers apart from leukaemia, the data corresponded to a linear, non-threshold dose response with no evidence to suggest that the slope in the lower dose portion of the curve was different from that based on the entire dose range. For leukaemia, there was some indication that the risk-versus-dose relationship might be

better described by a linear quadratic curve where the slope in the low-dose range is shallower than in the high-dose range. However, the wide confidence limits associated with these data mean that there is a high degree of uncertainty associated with low-dose risk.

To strengthen the assessment of risk from low-dose radiation, information has been gathered and analysed from a variety of industrial operations where workers face chronic exposure from external radiation. One of these studies was conducted at Hanford in southeast Washington State in the United States, which was established in the 1940s for plutonium production, but which has since been expanded as a power generation and research facility. At the Hanford facility, over 32,000 workers with more than 6 months' work experience at the site have been monitored. Most of these workers received only very small occupational doses (<10 mSv) and only 10% received more than 50 mSv. This large dataset gives no clear indication of cancer or genetic risks at these low exposures.

For the present, the relationship between radiation dose and harmful effects in humans is assumed by the ICRP to be linear with no threshold. This assumption remains despite the fact that, in addition to epidemiological evidence that a threshold exists, several experimental studies with animals point to a threshold. Indeed, some studies suggest that organisms are capable of responding positively to very low doses of radiation, which may afford them some degree of protection against higher doses. In simple terms, the mechanism(s) by which such a response might be beneficial would be the **induction** of proteins involved in the protection from and/or repair of DNA damage caused by radiation. Hill and Godfrey (1994) provide a review of some of this evidence, including data that suggest that the responsive process might be triggered by stresses other than radiation (e.g., heat-induced radiation resistance). In light of such experimental evidence, the current ICRP assumption of a linear relationship between low radiation dose and deleterious effect remains controversial. Advocates of a linear dose response maintain that it errs on the conservative side (i.e., it overestimates risk). It has for its mechanistic basis the **one-hit model** for carcinogenesis and embodies the assumption that a molecular 'hit' by a unit of radiation will result in a lesion. As contradictory evidence mounts, and more is learnt about repair processes related to radiation, it may be necessary to modify this approach. The problem is illustrated in Figure 10.6, where we might make the

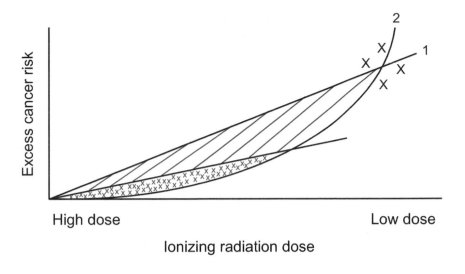

1 = Linear extrapolation of risk from high-dose data
2 = Linear quadratic dose–response relationship
▨ = Overestimate of risk using linear relationship versus linear quadratic relationship
⊠ = Overestimate of risk (relative to linear quadratic curve) if linear dose response is adjusted using a factor of 2

Figure 10.6 Comparison of a linear-non-threshold dose–response relationship to a linear + quadratic curve dose–response relationship for ionizing radiation-induced toxicity. Reproduced from Wright & Welbourn (2002), with permission from Cambridge University Press.

assumption that the 'true' dose–response is a linear quadratic represented by the curved line.

The 'currently accepted' linear relationship is based on non-controversial high-dose data. Overestimated risk is, therefore, represented by the shaded area. Given all the uncertainties, future remedies may include the application of an across-the-board correction factor. For example, it has been suggested by the ICRP that risk estimates for low dose exposures might be halved compared to estimates based on extrapolation from information from high doses at high dose rates. As more information is gathered on persons exposed to low-radiation doses, it is likely that risk estimates will be modified and may become more specific for different types of cancer and other radiation-induced disorders.

A special case arises where radionuclides are ingested and remain in the body as an irreversible radiation source subject to the decay half-life and the kinetics of elimination. The latter will depend on the particular tissues that accumulate the radionuclides. For example, those deposited in the bone will have a much longer residence time than those entering soft tissues. For this reason, the **committed equivalent dose** is used as the fundamental unit of risk for humans and takes into account the potential for a radiation dose to be delivered over a long period of time following ingestion. Radioisotopes of particular concern in this context often include those produced by nuclear fission. For example, radium-226 (^{226}Ra), a member of the uranium-238 (^{238}U) decay series (see Table 10.2) with a half-life of 1,600 years, is an alpha and beta emitter having similar characteristics to calcium. As such, it is readily incorporated into bone. Strontium-90 (^{90}Sr), a beta emitter with a half-life of 29 days, is also largely deposited in bone. Caesium-137 (^{137}Cs), a beta emitter (half-life 30.2 years), replaces potassium in soft tissues such as muscle. Iodine-131 (^{131}I) is an alpha and beta emitter with a half-life of 8.1 days and accumulates in the thyroid gland.

The ICRP has recommended an occupational exposure limit of 20 mSv per year averaged over a defined period of 5 years for workers in industries involving radionuclides but with stipulations relating to specific tissue exposures (ICRP, 2007). Detailed recommendations have been published on 240 different radionuclides that could be ingested or inhaled. For persons not exposed to radioactivity as a result of their profession, the recommended limit for annual whole-body dose is 1 mSv per year (ICRP, 1977). Risk factors published by the ICRP

Box 10.3 What is considered for cancer risk?

The probability of fatal cancer is calculated based on information obtained from scientific bodies to determine:

- The probability of fatal cancer
- The probability of non-fatal cancer
- The probability of severe hereditary effects
- The length of life lost if the harm occurs.

Effect	Risk factor (mSv)
Fatal cancer	5.0×10^{-5}
Non-fatal cancer	1.0×10^{-5}
Severe genetic effects	1.3×10^{-5}
Total deleterious effect	7.3×10^{-5}

estimate the probability of adverse effect per mSv of dose received. Four categories are recognized (Box 10.3).

Sensitivity to radiation also varies widely among species, with mammals appearing to be the most radiosensitive group of animals yet studied. Of mammals studied so far, humans appear to be the most sensitive. Evidence suggests that resistance to the damaging effects of ionizing radiation occurs in phylogenetically distant taxa including bacteria, archaea, fungi and plants (Shuryak, 2019). The organism most resistant to ionizing radiation is a bacterium known as *Deinococcus radiodurans*, which can survive very high radiation levels (Daly *et al.*, 1994). Adult fruit flies, the freshwater invertebrate rotifer *Philodina roseola* and the roundworm *Caenorhabditis elegans* are also known to be highly resistant to radiation (ICRP, 1977). In the case of fungi, it appears that these organisms may be highly resistant to ionizing radiation because of the unique properties of fungal melanin. Other proposed mechanisms of ionizing resistance include physiological adjustments such as increasing manganese complexes, reducing cellular iron concentrations and upregulation of DNA repair mechanisms. In general, invertebrates and plants are more resistant than vertebrates, and although all of the factors involved in resistance remain to be identified, combinations involving efficient DNA repair, cellular protection and highly condensed nucleoid structure are predicted

to be important (ICRP, 1977). Similarly, within taxonomic groups, it is generally found that younger developing organisms are more sensitive than older organisms, owing to the immaturity of protective mechanisms.

10.7 Ecological Effects of Radiation

Ecological effects of radiation are difficult to identify, and there is no evidence of adverse effects resulting from exposure at or near background levels. In 1957, the worst nuclear accident in Great Britain's history occurred at the Windscale Piles Facility on the northwest coast of England and resulted in the release of iodine-131 (^{131}I) and other radionuclides. It is estimated that the event resulted in 240 additional cases of thyroid cancer in humans, but there is no evidence of significant ecological effects from the fallout, despite years of studies (Wakeford, 2007). Likewise, studies of invertebrates, macrophytes and algae in ponds and streams within and beyond the perimeter of the Hanford site (Section 10.6) in Washington State in the northwest United States reveal some intersite population differences, but none that can be directly attributed to radiation exposure (Emery and McShane, 1980). This site, established in 1943, was where the first full-scale reactor for plutonium production in the world was commissioned. As part of the Manhattan Project, it produced the plutonium used in the bomb that was detonated over Nagasaki, Japan. The site is now decommissioned, with operation of the last functioning reactor ceasing in 1987, and it is almost certainly the most contaminated nuclear site in the United States (Burger *et al.*, 2019). Since this site is not open to the public, the ecosystem has not been exposed to human disruption and is considered to be unique, with extensive ecological resources. A recent review discusses the risk to the ecosystem associated with active remediation (Burger *et al.*, 2019). For example, endangered species and birds would be negatively impacted by habitat disruption and noise, whereas soil removal and compaction would harm other species.

Although it has been possible to detect small localized changes in biota very close to some sites of heavy radioisotope usage and disposal where leakages have been reported, the overwhelming volume of information relating to ecosystem effects of radiation exposure comes from very large releases such as the Chernobyl accident and aboveground detonation of nuclear weapons. Where such investigations extend into risk assessment, the database is expanded to include occupational radiation exposure. Such an approach is not unlike that used to characterize the environmental toxicology of non-radioactive substances where unintentional releases form the basis of field 'experiments' (sometimes called "spills of opportunity"). In view of the risks involved, field experiments using radioisotopes are limited but some have been conducted (Henner *et al.*, 2010). An interesting example is the long-term experiment conducted at the Brookhaven National Laboratory in New York, USA, where a natural oak-pine forest was irradiated for 15 years (1961–1976) by a gamma emitter, caesium-137 (^{137}Cs) source mounted on top of a 5-m tower. The principal aim of the investigators was to obtain estimates of ecosystem change in response to chronic exposures under controlled conditions. The authors of the study described a two-phase effect: initially a thinning out of sensitive, larger-bodied species such as trees, in favour of smaller, more resistant ground-hugging species, followed by a 'second sorting' of species wherein more uncommon or exotic species created new communities in conjunction with the survivors of the first sorting (Woodwell and Houghton, 1990). Investigators used a number of different parameters to compare the exposed site with reference sites. Depending on the endpoint used, they were able to discern differences due to radiation up to 50–100 m from the source. Interestingly, they found simple **diversity indices** (irrespective of actual species

found) to be less useful than parameters that included a quantitative comparative component such as community indices (Section 4.3.3). The coefficient of community is the number of species shared by two communities expressed as a percentage of the total number of species in both. The coefficient I is expressed as:

$$I = C/(A + B + C) \qquad (10.8)$$

where C equals the number of species shared by the two communities, one of which has A species and the other B. This coefficient is essentially the same as that used to compare animal communities upstream and downstream from a point source of pollution (see Section 4.3.3).

Other findings from experimental ecosystem studies have demonstrated that ecological responses to ionizing radiation include decreased species diversity, reduced productivity, altered species dominance and overall changes in community structure (Geras'kin, 2016). With respect to plant communities, coniferous forests appear to be the most sensitive to ionizing radiation since they have a high retention capacity and low turnover rate for contaminants in general, whereas herbaceous invaders seem to be the most resistant (Henner *et al.*, 2010). The need for and difficulties associated with providing ecosystem protection against ionizing radiation are described in Capsule 10.1.

Capsule 10.1 Radiological Protection of the Environment

Nicholas A. Beresford
UK Centre for Ecology & Hydrology, Lancaster Environment Centre, UK

David Copplestone
University of Stirling, Stirling, UK

The Need for Radiological Environmental Protection

The system for ensuring the protection of humans from ionizing radiation is internationally well established (ICRP, 2007), having begun development early in the twentieth century. However, there was an assumption that the control (or regulation) required to protect humans would also ensure the safety of other species. The 1990 Recommendations of the ICRP (International Commission on Radiological Protection) stated:

> The Commission believes that the standard of environmental control needed to protect man to the degree currently thought desirable will ensure that other species are not put at risk. Occasionally, individual members of non-human species might be harmed, but not to the extent of endangering whole species or creating imbalance between species. At

the present time, the Commission concerns itself with mankind's environment only with regard to the transfer of radionuclides through the environment, since this directly affects the radiological protection of man. (ICRP, 1991)

This assumption was generally accepted and adopted by those national authorities responsible for radiation protection, and the impact of authorized releases of radioactivity on the environment (or wildlife) was not routinely assessed. Consequently, there were no commonly used models for conducting radiological environmental impact assessments.

From about 1990, the statement of the ICRP with regard to protection of the environment was increasingly questioned. Criticisms included: a lack of supporting data for the statement; potential scenarios where wildlife may be more exposed than humans; and the need to demonstrate protection of the environment from any human activity. At about the same time, some countries began to establish requirements and guidelines for the protection of wildlife (often termed 'non-human species'). To support these requirements, approaches and models were developed to assess the potential impact of authorized releases of radionuclides on the environment.

Subsequently, the ICRP amended its recommendations to include the consideration of radiological protection of the environment with the objective '*to maintain biological diversity, conservation of species, protection of the health and status of natural habitats, communities and ecosystems, with targets related to populations or higher organisational levels rather than individual organisms*' (ICRP, 2007). The revised ICRP recommendations acknowledged the need to develop a framework to assess non-human species, which the ICRP has subsequently begun. The International Atomic Energy Agency (IAEA) has a role in transforming the ICRP's recommendations into practical guidance for application in regulatory frameworks, and the latest IAEA 'Safety Fundamentals' acknowledges the need to consider the environment within radiological assessment: '*People and the environment, present and in the future, must be protected against radiation risks*' (IAEA, 2006). In response to changes in international recommendations, radiological environmental impact assessments are being conducted in many countries for a wide range of facility types (e.g., see Brown *et al.*, 2016).

Environmental Radiological Assessment Approaches

A number of assessment approaches have been developed over the past 20 years. In common with approaches considering other non-radioactive environmental **stressors**, the more comprehensive methodologies use a tiered assessment approach beginning with a simplistic screening tier that requires comparatively little input and is highly conservative. An aim of this initial tier is to identify sites of negligible concern and remove them from the need for more refined higher-tier assessments that require increasingly more data and resources. Software implementing two of these tiered assessment approaches has been made freely available: the ERICA Tool (https://erica-tool.com/ Brown *et al.*, 2016) and RESRAD-BIOTA (https://resrad .evs.anl.gov/codes/resrad-biota/, US DOE, 2004). The goal of radiological assessment, in common with other areas of environmental regulation, is usually to protect populations.

The developed approaches are primarily for the assessment of 'planned releases' (i.e., from operating or planned sites) and existing contamination scenarios (e.g., sites contaminated by historical activities). The approaches are not designed to assess the dynamic nature of accidental releases, although the Fukushima accident demonstrated the desire to predict the potential exposure of wildlife following large-scale accidents (Strand *et al.*, 2014). Below, we give an overview of how the approaches estimate the exposure and risk of wildlife from ionizing radiation.

Simplifying the Ecosystem

It would be impossible to have screening tier assessment approaches that considered every potential species in any potential ecosystem. Therefore, the approaches make simplifications; ecosystems are usually simplified as generic marine, freshwater or terrestrial. Similarly, simplifications are made with regard to how organisms are represented within the approaches. The US Department of Energy RESRAD-BIOTA approach simplifies 'organism' as a generic plant or generic animal (US DOE, 2004). The ICRP (2008) proposed a set of Reference Animals and Plants which they defined as: '*A hypothetical entity, with the assumed basic biological characteristics of a particular type of animal or plant, as described to the generality of the taxonomic level of family, with defined anatomical, physiological, and life history properties, that can be used for the purposes of relating exposure to dose, and dose to effects, for that type of living organism*'. The ERICA Tool (Brown *et al.*, 2016) uses an approach similar to that of the ICRP, with 13 'Reference Organisms' for each of the three generic ecosystems. Reference Organisms are not specific species but are representative of an organism type (e.g., 'amphibian', 'reptile', 'macroalgae', 'mammal', 'tree', etc.). The ERICA Tool Reference Organisms encompass organism types that are: likely to be highly exposed; radiosensitive organisms; representative of different ecological niches; and representative of (European) protected species.

Dosimetry

To estimate the exposure (or dose rate) of organisms, dose coefficients (DCs), which relate dose rate (e.g., μGy/h) to the activity concentration (e.g., Bq/kg) in environmental media (water, soil, air, sediment), or within an organism, are used to calculate external or internal dose rate respectively (Vives i Batlle *et al.*, 2011). To calculate the DCs, organisms are typically assumed to have a homogeneous geometry (usually an ellipsoid). The ERICA Tool and ICRP approaches select geometries (dimensions and masses) representative of a representative species for the reference organism type. RESRAD-BIOTA takes a more conservative approach in its initial screening tier by assuming a small geometry for the external DC and a large geometry for the internal DC; these assumptions maximize both external and internal dose rate estimates, respectively. Weighted dose rates are estimated by applying a radiation-weighting factor to account for the relative biological effectiveness of different types of radiation. An overview of factors influencing DC values can be found in Vives i Batlle *et al.* (2011).

More complex and more realistic models (i.e., including individual organs) have been generated for a number of wildlife types (e.g., see Ruedig *et al.* (2015)). These are not proposed for regulatory assessments, but they have been useful in demonstrating whether simple homogeneous geometry assumptions are generally fit for purpose in the available regulatory assessment models. **Voxel** models could also have a useful role in interpreting wildlife dose–effect studies.

Estimating Organism Activity Concentrations

If organism activity concentrations need to be predicted, **equilibrium** concentration ratios ($CR_{wo-media}$) relating whole-organism radionuclide activity concentration to those in media (typically soil, water or air) are commonly used (Beresford *et al.*, 2008). This approach is pragmatic being simple to apply and some data are available (IAEA, 2014). However, $CR_{wo-media}$ values can be highly variable, ranging over four orders of magnitude for a given radionuclide–organism combination, leading to considerable uncertainty in predictions.

Benchmarks

For assessments, estimated dose rates need to be put into context with some form of risk criteria (i.e., we need to be able to judge if the estimated dose rate will potentially cause harm or not). Prior to the development of radiological environmental protection approaches, a number of publications had compiled data on the effects of radiation on wildlife from the available literature, considering population relevant **endpoints** such as mortality, fertility and fecundity (NCRP, 1991; IAEA, 1992; UNSCEAR, 1996). Through 'expert judgement', these reviews reached broadly similar conclusions (Box 10C.1; see original references for exact wording). Although these values were not originally proposed as benchmarks for environmental assessment, they are now sometimes being used as such.

Box 10C.1 Early estimates of dose rates below which population-level effects would not be expected in wildlife

IAEA (1992):
Terrestrial plants: 10 mGy/d
Terrestrial animals: 1 mGy/d
Aquatic organisms: 10 mGy/d

NCRP (1991):
Aquatic organisms: 10 mGy/d

UNSCEAR (1996):
Terrestrial plants: <10 mGy/d
Terrestrial animals: 400 µGy/h (mortality)
Terrestrial animals: 40–100 µGy/h (reproduction)
Aquatic organisms: 400 µGy/h

The ICRP have proposed 'derived consideration reference levels' (DCRLs) for their suite of Reference Animals and Plants (ICRP, 2008). These are defined as *one order of magnitude broad bands of dose rates covering the level where the dose rates warrant a more considered level of evaluation of the situation*. The DCRLs range from 0.1 to 1 mGy/d (for Reference Deer, Rat, Duck and Pine tree) to 10–100 mGy/d (for Reference Seaweed, Bee, Crab and Earthworm). As for the UNSCEAR, IAEA and ICRP reviews, the DCRLs were based on expert judgement, although the decision process was better documented.

To be consistent with approaches used for chemical regulation, Garnier-Laplace *et al.* (2010) applied the **species sensitivity distribution** approach as described in Chapters 2 (Section 2.3.3.2) and 12 (Section 12.3.1) to derive a screening dose rate (equating to a predicted no-effect concentration as used for risk assessment of chemical stressors). This approach provided a framework for a more transparent and objective derivation of the screening dose rate than the previous derivation of benchmarks using expert judgement. The resultant estimated screening dose rate was 10 µGy/h, and this value is used as the default in the ERICA Tool. The screening dose rate derived was generic across all ecosystem and organism types. It would be beneficial to be able to derive organism-specific (e.g., at the level of terrestrial vertebrates, plants, fish, etc.) screening dose rates as the application of a single screening dose rate identifies the most exposed and not necessarily the most at-risk organism. However, data availability precluded Garnier-Laplace *et al.* (2010) from being able to derive organism-specific values. The screening dose rate is for use in screening assessments to help screen sites out from the requirement for further assessment and to identify those that need more detailed consideration; it is not a regulatory 'limit'. The screening dose rate is applicable to the additional dose rate arising from the source(s) under assessment and not the total dose rate including natural background exposure; this is consistent with the radiological protection of humans. For comparison, weighted dose rates to terrestrial and aquatic wildlife due to naturally occurring radionuclides of the ^{238}U and ^{232}Th series and ^{40}K are typically in the region of 1 µGy/h or less; this does not include the exposure of burrowing animals to ^{222}Rn and daughter products which may be of the order of tens of µGy/h (Beresford *et al.*, 2012).

The Scientific Controversy

Three accidents, Chernobyl (Ukraine, 1986), Fukushima (Japan, 2011) and Kyshtym (Russian Urals, 1957), have resulted in releases of radioactivity sufficient to result in radiation-induced effects in local wildlife. Such sites provide an ideal opportunity to obtain data under realistic conditions of exposure with the potential to investigate population- to ecosystem-level impacts, and to improve and test our environmental assessment approaches. However, although it is accepted that radiation-induced effects have occurred in these areas, there are a number of reports of significant impacts on wildlife at extremely low dose rates, for example below the proposed screening dose rate or DCRLs discussed above, and in the range of typical background exposure rates (Beresford *et al.*, 2020a). Many factors might contribute to the reported observations at low dose rates, including: poor estimates of exposure; lack of consideration of confounding factors; residual influence of acute/high exposures soon after the accident; or interpretation of statistical results. Furthermore, some studies directly conflict in the findings, e.g., for mammals and leaf litter decomposition rates

(Beresford *et al.*, 2020b). These scientific disagreements on the impacts of radiation at contaminated field sites have a relatively high media profile and the potential to influence public opinion. This controversy needs to be resolved to maintain confidence in the environmental radiation protection approaches that have been developed over the past 20 years and are now being used for the regulation of radioactive releases into the environment from sources ranging from hospitals to nuclear power facilities. Recent studies have started to address these uncertainties, with priorities for future research being identified (Beresford *et al.*, 2020a).

Capsule References

Beresford, N. A., Barnett, C. L., Brown, J. E., *et al.* (2008). Inter-comparison of models to estimate radionuclide activity concentrations in non-human biota. *Radiation and Environmental Biophysics*, 47, 491–514.

Beresford, N. A., Barnett, C. L., Vives i Batlle, J., *et al.* (2012). Exposure of burrowing mammals to ^{222}Rn. *Science of The Total Environment*, 431, 252–261.

Beresford, N. A., Horemans, N., Copplestone, D., *et al.* (2020a). Towards solving a scientific controversy – the effects of ionising radiation on the environment. *Journal of Environmental Radioactivity*, 211, e106033.

Beresford, N. A., Scott, E. M. & Copplestone, D. (2020b). Field effects studies in the Chernobyl Exclusion Zone: lessons to be learnt. *Journal of Environmental Radioactivity*, 211, e105893.

Brown, J. E., Alfonso, B., Avila, R., *et al.* (2016). A new version of the ERICA tool to facilitate impact assessments of radioactivity on wild plants and animals. *Journal of Environmental Radioactivity*, 153, 141–148.

Garnier-Laplace, J., Della-Vedova, C., Andersson, P., *et al.* (2010). A multi-criteria weight of evidence approach for deriving ecological benchmarks for radioactive substances. *Journal of Radiological Protection*, 30, 215–233.

IAEA (1992). *Effects of Ionizing Radiation on Plants and Animals at Levels Implied by Current Radiation Protection Standards.* Technical Reports Series No. 332. Vienna, Austria: International Atomic Energy Agency.

IAEA (2006). *Fundamental Safety Principles – Safety Fundamentals.* Safety Standards No. SF-1. Vienna, Austria: International Atomic Energy Agency.

IAEA (2014). *Handbook of Parameter Values for the Prediction of Radionuclide Transfer to Wildlife.* Technical Reports Series No. 479. Vienna, Austria: International Atomic Energy Agency.

ICRP (1991). *1990 Recommendations of the International Commission on Radiological Protection.* ICRP Publication 60. Oxford, UK: International Commission on Radiological Protection.

ICRP (2007). *The 2007 Recommendations of the International Commission on Radiological Protection.* ICRP Publication 103. Oxford, UK: International Commission on Radiological Protection.

ICRP (2008). *Environmental Protection: The Concept and Use of Reference Animals and Plants.* ICRP Publication 108. Oxford, UK: International Commission on Radiological Protection.

NCRP (1991). *Effects of Ionizing Radiation on Aquatic Organisms.* NCRP Report 109. Bethesda, MD: US National Commission on Radiological Protection.

Ruedig, E., Beresford, N. A., Gomez Fernandez, M. E. & Higley, K. (2015). A comparison of the ellipsoidal and voxelized dosimetric methodologies for internal, heterogeneous radionuclide sources. *Journal of Environmental Radioactivity*, 140, 70–77.

Strand, P., Aono, T., Brown, J. E., *et al.* (2014). Assessment of Fukushima-derived radiation doses and effects on wildlife in Japan. *Environmental Science & Technology Letters*, 1, 198–203.

UNSCEAR (1996). *Sources and Effects of Ionizing Radiation.* Report to the General Assembly with scientific annex. New York, NY: United Nations Scientific Committee on the Effects of Atomic Radiation.

US DOE (2004). *RESRAD-BIOTA: A Tool for Implementing a Graded Approach to Biota Dose Evaluation.* User's guide, version 1. ISCORS technical report 2004-02. Washington, DC: United States Department of Energy, Office of Air, Water and Radiation Protection Policy and Guidance.

Vives i Batlle, J., Beaugelin-Seiller, K., Beresford, N. A., *et al.* (2011). The estimation of absorbed dose rates for non-human biota: an extended intercomparison. *Radiation and Environmental Biophysics*, 50, 231–251.

10.8 Conclusions

It is clear that our current understanding of both the benefits and harmful effects of ionizing radiation has increased over time owing to knowledge gained through accidents and scientific research. However, it is important to recognize that the knowledge about harmful effects on human health is far more advanced than knowledge about ecosystem-level effects. This stems, in part, from restrictions with respect to experiments purposely exposing ecosystems to ionizing radiation, which is in contrast to most non-radioactive contaminants.

Summary

This chapter introduces the reader to important concepts associated with the potential toxicity following exposure to ionizing radiation. These concepts include how we measure exposure, sources of exposure, mechanistic effects at the cellular level and how we balance the beneficial versus harmful effects. Understanding these risks is paramount given that exposure to ionizing radiation at some level is inevitable, as worldwide natural sources are ubiquitous. Furthermore, as humans we benefit greatly from manufactured sources of ionizing radiation, but it is of utmost importance to be cognizant and manage any potential harmful effects.

Review Questions and Exercises

1. Explain the difference between non-ionizing and ionizing radiation.

2. What are the principal forms of ionizing radiation?

3. What is the definition of an isotope?

4. Explain the term radioactive half-life.

5. What is the currently used unit of measurement for the amount of energy absorbed into tissue following exposure to ionizing radiation?

6. Compare and contrast the three sources of natural radiation.

7. Describe the different types of nuclear waste levels and explain how we manage them.

8. Describe the effects of ionizing radiation at the molecular level.

9. Identify two ways in which ionizing radiation may cause cell death.

10. Describe how ionizing radiation may affect the immune system.

11. What does the term 'bystander effect' mean?

12. Give an account of the chronic effects of ionizing radiation exposure.

13. What are the difficulties in assessing the risk from low levels of ionizing radiation?

14. Describe the immediate and long-term effects of the Chernobyl reactor accident.

15. What were the lessons from the Chernobyl incident for the nuclear power industry?

Abbreviations

AVLS	Atomic vapour laser separation	IAEA	International Atomic Energy Agency
Bq	Becquerel	IARC	International Agency for Research on Cancer
CANDU	Canadian Deuterium Uranium (reactor)		
Ci	Curie	ICRP	International Commission on Radiological Protection
C/kg	Coulomb/kilogram		
ERICA tool	Software provided by Environmental Risk from Ionising Contaminants: Assessment and Management (ERICA)	LTE	Linear transfer energy
		PUREX	Plutonium and Uranium Recovery by Extraction
Gy	Gray	R	Roentgen

rad	Radiation absorbed dose	UV	Ultraviolet radiation
rem	Roentgen equivalent man	UVA	Ultraviolet radiation, 315–400 nm
SI units	Standard international units		(nanometres)
Sy	Sievert	UVB	Ultraviolet radiation, 280–315 nm
UNSCEAR	United Nations Scientific Committee on	UVC	Ultraviolet radiation, 100–280 nm
	the Effects of Atomic Radiation	WIPP	Waste Isolation Pilot Plant

References

Anspaugh, L. R., Catlin, R. J. & Goldman, M. (1988). The global impact of the Chernobyl reactor accident. *Science*, 242, 1513–1519.

Bernier, M. O., Withrow, D. R., Berrington de Gonzalez, A., *et al.* (2019). Trends in pediatric thyroid cancer incidence in the United States, 1998–2013. *Cancer*, 125, 2497–2505.

Bidnenko, S. I., Nazarchuk, L. V., Fedorovskaia, E. A., Liutko, O. B. & Open'ko, L. B. (1992). [The antibacterial immunity of people under dynamic observation in an altered radiation situation]. *Zhurnal mikrobiologii, épidemiologii i immunobiologii*, 33–36.

Bromet, E. J. (2014). Emotional consequences of nuclear power plant disasters. *Health Physics*, 106, 206–210.

Burger, J., Gochfeld, M. & Jeitner, C. (2019). Importance of buffer lands to determining risk to ecological resources at legacy contaminated sites: A case study for the Department of Energy's Hanford Site, Washington, USA. *Journal of Toxicology and Environmental Health, Part A*, 82, 1151–1163.

Burgio, E., Piscitelli, P. & Migliore, L. (2018). Ionizing radiation and human health: reviewing models of exposure and mechanisms of cellular damage. an epigenetic perspective. *International Journal of Environmental Research and Public Health*, 15, e15091971.

Cardis, E. & Hatch, M. (2011). The Chernobyl accident – an epidemiological perspective. *Clinical Oncology* 23, 251–260.

Daly, M. J., Ouyang, L., Fuchs, P. & Minton, K. W. (1994). In vivo damage and recA-dependent repair of plasmid and chromosomal DNA in the radiation-resistant bacterium *Deinococcus radiodurans*. *Journal of Bacteriology*, 176, 3508–3517.

Emery, R. M. & McShane, M. C. (1980). Nuclear waste ponds and streams on the Hanford Site: an ecological search for radiation effects. *Health Physics*, 38, 787–809.

Fedchenko, V. & Hellgren, R. F. (2007). Appendix 12B: Nuclear explosions, 1945–2006. *SIPRI Yearbook 2007 Armaments, Disarmament and International Security*. Solna, Sweden: Stockholm International Peace Research Institute.

Gaskin, J., Coyle, D., Whyte, J. & Krewksi, D. (2018). Global estimate of lung cancer mortality attributable to residential radon. *Environmental Health Perspectives*, 126, e057009.

Geras'kin, S. A. (2016). Ecological effects of exposure to enhanced levels of ionizing radiation. *Journal of Environmental Radioactivity*, 162–163, 347–357.

Hatch, M. & Cardis, E. (2017). Somatic health effects of Chernobyl: 30 years on. *European Journal of Epidemiology*, 32, 1047–1054.

Hayashi, F., Sanpei, M., Ohira, T., *et al.* (2020). Changes in the mental health status of adolescents following the Fukushima Daiichi nuclear accident and related factors: Fukushima Health Management Survey. *Journal of Affective Disorders*, 260, 432–439.

Henner, P., Geras'kin, S. A. & Hinton, T. (2010). *State-of-knowledge Regarding the Effects of Ionizing Radiation on Plant Communities: Pre-requisite for the Definition of Experimental Studies on Chernobyl Exclusion Zone*. Final report of TRASSE 2009-1B project. Cadarache, France: IRSN/DEI-SECRE.

Hill, C. K. & Godfrey, T. (1994). Biological effects of low level exposures: dose–response relationships. In E. J. Calabrese (ed.), *Biological Effects of Low Level Exposures: Dose-Response Relationships*. Boca Raton, FL: Lewis Publishers.

IARC (2012). *GLOBOCAN 2012: Estimated Cancer Incidence, Mortality and Prevalence Worldwide in 2012*. Lyon, France: International Agency for Research on Cancer.

ICRP (1977). *Recommendations of the International Commission on Radiological Protection*, Report no. 26. New York, NY: Pergamon.

ICRP (2007). *The 2007 Recommendations of the International Commission on Radiological Protection*. ICRP publication 103. *Annals ICRP*, 37, 1–332.

Kolpakov, M., Mal'tsev, V. I., Iakobchuk, V. A., Shatilo, V. I. & Kolpakova, N. N. (1992). [The characteristics of the course of acute pneumonia in patients subjected to prolonged exposure to low doses of ionizing radiation as a result of the accident at the Chernobyl Atomic Electric Power Station]. *Lik Sprava*, 11–15.

Krivolutzkii, D. A., Pokarzhevskii, A. D. & Viktorov, A. G. (1992). Earthworm populations in soils contaminated by the Chernobyl atomic power station accident, 1986–1988. *Soil Biology and Biochemistry*, 24, 1729–1731.

Megaw, W. J. (1987). *How Safe? Three Mile Island, Chernobyl and Beyond*. Toronto, ON: Stoddart.

Mothersill, C., Smith, R., Wang, J., *et al*. (2018). Biological entanglement-like effect after communication of fish prior to X-ray exposure. *Dose Response*, 16, e1559325817750067.

Mu, H., Sun, J., Li, L., *et al*. (2018). Ionizing radiation exposure: hazards, prevention, and biomarker screening. *Environmental Science and Pollution Research*, 25, 15294–15306.

Nakanishi, T. M. (2018). Agricultural aspects of radiocontamination induced by the Fukushima nuclear accident – a survey of studies by the Univ. of Tokyo Agricultural Dept. (2011–2016). *Proceedings of the Japan Academy Series B-Physical and Biological Sciences*, 94, 20–34.

Ohtsuru, A., Midorikawa, S., Ohira, T., *et al*. (2019). Incidence of thyroid cancer among children and young adults in Fukushima, Japan, screened with 2 rounds of ultrasonography within 5 years of the 2011 Fukushima Daiichi nuclear power station accident. *Journal of the American Medical Association, Otolaryngology – Head and Neck Surgery*, 145, 4–11.

Richards, Z. T., Beger, M., Pinca, S. & Wallace, C. C. (2008). Bikini Atoll coral biodiversity resilience five decades after nuclear testing. *Marine Pollution Bulletin*, 56, 503–515.

Saremi, S., Isaksson, M. & Harding, K. C. (2018). Bio accumulation of radioactive caesium in marine mammals in the Baltic Sea – reconstruction of a historical time series. *Science of the Total Environment*, 631–632, 7–12.

Shuryak, I. (2019). Review of microbial resistance to chronic ionizing radiation exposure under environmental conditions. *Journal of Environmental Radioactivity*, 196, 50–63.

Smith, R. W., Seymour, C. B., Moccia, R. D. & Mothersill, C. E. (2016). Irradiation of rainbow trout at early life stages results in trans-generational effects including the induction of a bystander effect in non-irradiated fish. *Environmental Research*, 145, 26–38.

TEPCO (2012). *Estimation of Radioactive Material Released to the Atmosphere during the Fukushima Daiichi NPS Accident*. Tokyo, Japan: Tokyo Electric Power Company.

Thakur, P., Lemons, B. G. & White, C. R. (2016). The magnitude and relevance of the February 2014 radiation release from the Waste Isolation Pilot Plant repository in New Mexico, USA. *Science of the Total Environment*, 565, 1124–1137.

US NAS (1990). *Health Effects of Exposure to Low Levels of Ionizing Radiation: BEIR V*, Washington, DC: National Academies Press.

Wakeford, R. (2007). The Windscale reactor accident – 50 years on. *Journal of Radiological Protection*, 27, 211–215.

Woodwell, G. M. & Houghton, R. M. (1990). The experimental impoverishment of natural communities: effects of ionizing radiation on plant communities, 1961–1976. *The Earth in Transition: Patterns and Processes of Biotic Impoverishment*. Cambridge, UK: Cambridge University Press.

Wright, D. A. & Welbourn, P. (2002). *Environmental Toxicology*, Cambridge, UK: Cambridge University Press.

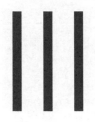

Complex Issues

11

Complex Issues, Multiple Stressors and Lessons Learned

Pamela M. Welbourn, Peter G. C. Campbell, Peter V. Hodson and Christopher D. Metcalfe

Learning Objectives

Upon completion of this chapter, the student should be able to:

- Understand what is meant by 'multiple stressors' and 'complex issues', and how cradle-to-grave (life-cycle) exposures to contaminants can occur.
- Communicate the concept of 'legacy effects'.
- Identify the sources of acidification in freshwater and marine environments, explore the resulting chemical and biological changes, including those at the ecosystem level, and understand the mechanisms involved.
- Appreciate the delicate balance between the economic benefits of natural resource utilization (e.g., minerals and forests) and the potential detrimental effects on the surrounding environment.
- Appreciate the marked changes that occur in the chemical and physical properties of materials on the nanoscale.
- Recognize why manufactured nanomaterials are inherently difficult to evaluate, in terms of both their environmental fate and their intrinsic toxicity.

Ecotoxicology is by definition a very complex subject. In attempting to understand the adverse effects of anthropogenic activities on our environment, we are confronted by the need to consider phenomena that occur across a wide range of scales, from the subcellular

to whole ecosystems. In considering this daunting task, Frost *et al.* (1999) observed that "singly operating stresses may actually be quite rare". In other words, adverse effects in the environment rarely involve a single chemical and rarely result from a single mode of action.

In the current chapter, we consider four global issues: acidification; metal mining and processing; nanomaterials; and pulp and paper processing. These issues qualify as 'complex' in that they involve multiple stressors that act independently or jointly, and which may change in nature and importance over time. A 'cradle-to-grave' or life-cycle approach is often used to follow exposures to contaminants that occur during the sequence from extraction of a raw material to processing, manufacturing and often disposal. Drawing on the basic concepts presented in earlier chapters, we show how these concepts can be applied to understand the 'complex issue' and identify possible remedial actions.

11.1 Acidification of Freshwater, Terrestrial and Marine Systems

In Chapter 1, we introduced the recognition of acidic deposition, with its effects on freshwater and terrestrial systems in Europe and North America, as a significant landmark in the history of ecotoxicology; although it qualifies as an 'old issue', it is still of current importance. The release of sulphur dioxide and oxides of nitrogen

(NO$_X$, e.g., nitric oxide, NO$^\bullet$, or nitrogen dioxide, NO$_2$) as atmospheric pollutants can have direct effects on ecosystems. These gases can also be oxidized to sulphuric and nitric acids, which are ultimately deposited on land or water as mist, rain or snow. In south and southwest China, there are reports of "the highest acidity of precipitation in the world" (Wang and Xu, 2009). As noted by Burns *et al.* (2016), "The inadvertent acid rain experiment continues to provide opportunities to study the long-term effects of air pollutants on terrestrial and aquatic ecosystems." The chemical changes that accompany acidification result in multiple stressors, and the direct and indirect responses to these multiple drivers, at all levels of the ecosystem, provide a seminal example of a complex issue.

Since acidic deposition came into high profile as a scientific and political topic in the 1980s, a mass of information has accumulated concerning its sources, chemistry and effects on ecosystems. We now have examples showing how regulation of sources can reduce acidic inputs and potentially lead to recovery from the stresses related to the problem (Section 13.7.1). In contrast, ocean acidification has only emerged as an environmental issue since the turn of the century. Increasing concentrations of carbon dioxide in the atmosphere from fossil fuel combustion can increase the partial pressure of CO$_2$, with resulting changes in pH and carbonate chemistry. The increase in atmospheric CO$_2$ also results in warming of the surface of the Earth and oceans. In comparison with freshwater acidification, much less is known from direct observations about ocean acidification, but predictions are well supported by theoretical considerations and models. Warming and acidification of oceans are two different phenomena, but are closely coupled, making ocean acidification another very complex issue.

Interestingly, although they did not discuss ocean acidification, as early as 1995 Wright and Schindler anticipated the interaction between the effects of gases that contributed to air pollution and acidic deposition, and anticipated further complexity with what was then called greenhouse warming. They also pointed out that while emissions of sulphur-containing gases were already being controlled, emissions of nitrogen-containing compounds and CO$_2$ were still increasing. They concluded, "Predicting the interactions of regional and global environmental factors in the coming decades thus poses new challenges to scientists, managers and policy-makers."

11.1.1 Freshwater Acidification

Uncontaminated rainwater should have a pH of 5.4–5.6 but the pH of acidic rain can be as low as 4. Acidic deposition affects freshwater systems that are poorly buffered, and therefore susceptible to titration by H$^+$ ions, resulting in lowering of the overall pH of the water. This is particularly significant during snowmelt in temperate regions. The acids accumulate in snow, and a flush enters the surface water during the spring thaw. Such episodic pulses of acid are particularly damaging if they coincide with sensitive life stages of aquatic biota (Section 11.1.1.3). Lowering the pH can have direct effects on ecosystems and their components, as well as causing other chemical changes, including mobilization of metals.

11.1.1.1 Chemical Effects

The dissolved concentrations of many metals and their chemical speciation/bioavailability are affected by pH (Sections 6.2 and 6.3). Chemical changes that accompany the acidification of watersheds include the mobilization of metals, notably those of natural geological origin such as aluminium. In response to acidification, the concentration of monomeric Al increases in waters draining from watersheds into surface waters, and there is a shift in speciation towards the free Al^{3+} species (Figure 11.1). Other metals that are present in the terrestrial and aquatic systems, either naturally or as contaminants, may also be mobilized. These include cadmium, manganese and zinc. Additional inputs of metals (e.g., mercury) may result from their atmospheric deposition.

A decline of calcium in lake water is part of a long-term natural process because of decreases in exchangeable Ca in watershed soils, but it is accelerated by acidic deposition. The explanation is complex, because acidic deposition initially increases surface-water Ca concentrations, but in watersheds with thin soils over granitic or other igneous bedrock, the leaching rate typically exceeds the replenishment rate from weathering and atmospheric inputs. Eventually there is a depletion of soil base saturation, and lake water Ca decreases. For example, between 1985 and 2005, there was a 13% decline of Ca in 36 lakes from the Muskoka region of the southern Canadian Shield in Ontario, Canada, concurrent with an increase in their average pH from 5.9 to 6.2 (Jeziorski *et al.*, 2008). Similar trends in Ca decline have been shown for other parts of North America, Europe and China, although patterns and the extent of loss vary among lakes and

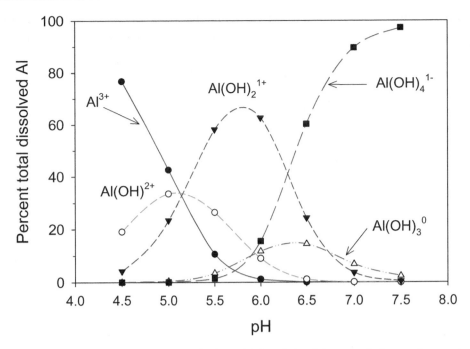

Figure 11.1 Equilibrium modelling results showing the marked sensitivity of aluminium speciation to changes in pH.

regions. Long-term Ca declines may also be affected by climate change and forest management in the watershed.

11.1.1.2 Physical Changes

While most studies have emphasized chemical changes, some have noted an increase in water clarity because of acidification. Water clarity is mainly controlled by the amount of (coloured) dissolved organic matter (DOM). Acidification can decrease the amount of DOM in lake water, largely by favouring the retention of fulvic and humic acids within the watershed. These natural organic acids adsorb strongly to soil particles at acidic pH, resulting in less input of DOM into surface waters.

Changes in water clarity can affect thermal stratification as well as the distribution of primary producers (Section 11.1.1.3) because light is able to penetrate into the cooler waters of the hypolimnion. Data from the experimental acidification of lakes illustrate the effects of acidification on water clarity, as shown in a whole-lake study in the Experimental Lakes Area in Ontario, Canada, and a comparable study on Little Rock Lake in Wisconsin. A more striking illustration is described by Arnott and Yan (2002), who measured changes following a very rapid drought-induced re-acidification of Swan Lake in the Sudbury, Ontario area. During a 2-year drought, reduced sulphur compounds present in the lake catchment and in the littoral sediments were exposed to atmospheric O_2 and oxidized. With the return of normal precipitation events in 1987 and 1988, the products of this oxidation were flushed into the lake, and the pH fell from a high of 5.8 to 4.5. In response, DOM decreased from 190 mmol C/l to 63 mmol C/l, and mean **Secchi depth** increased from 4.3 m to 7.6 m. In association with increased light penetration, mean bottom water temperature and oxygen concentration increased in 1988. Climate change and ozone depletion can also increase water clarity (Gunn *et al.*, 2001), but for the examples cited, the main driver was indisputably acidification.

11.1.1.3 Biological Effects and Risks for Sensitive Aquatic Systems

Early (pre-1980s) field observations showed the adverse effects of acidification on fish and other aquatic biota. Palaeolimnological studies have provided descriptive data on changes in community composition for algae and invertebrates, which can be used to infer changes in pH and reconstruct the chemical history of lakes (Sections 4.3.4, 11.2.4.2). Laboratory studies have addressed the mechanisms of these effects, and whole-lake acidification experiments have elucidated ecosystem effects that short-term field studies and laboratory experiments could not show (Schindler, 1988; Brezonik *et al.*, 1993).

One of the consistent chemical changes accompanying freshwater (and soil) acidification is the increased concentration of inorganic monomeric Al; in the field, H^+ and Al act in combination at pH values below neutrality. Studies with fish have shown that H^+ acts primarily on the permeability of the gill membrane, disrupting iono-regulation, whereas Al affects the gills by disrupting iono-regulation and impairing respiration. It may be academic, and irrelevant when considering recovery, to question the relative importance of H^+ and Al or their interaction for toxicity to aquatic biota in acid-stressed lakes, because outside the laboratory, they will always be acting together. In effect, H^+ and Al act in concert at intermediate pH levels (Kroglund *et al.*, 2008). Nevertheless, considerable emphasis has been put on the study of Al effects in acidified systems.

The biological effects of Ca deficiency on aquatic biota are still incompletely studied. Jeziorski *et al.* (2008) reported that as Ca concentrations decline, the Ca-rich *Daphnia* species, key components of many food webs, decline concurrently. In their study, the decline in Ca and daphnids continued after pH had stabilized, with the threshold for *Daphnia* being 1.5 mg/l Ca. They suggest that calcium decline will influence other aquatic biota as well. There are examples of molluscs and crustaceans with calcium requirements of >1 mg/l. Raddum and Fjellheim (2002) reported on field studies of invertebrates in three Norwegian lakes and reviewed some of the relevant literature on the distribution of invertebrates. Certain insect larvae were absent from lakes with low calcium content, but little is known about the critical limits of Ca for the most sensitive insect larvae. Some snails can tolerate a pH below 6 if the calcium concentration is favourable. The interaction with pH is difficult to uncouple in field studies unless low Ca persists after pH has started to recover. Another possible effect of the paucity of Ca in lakes is related to its protective effect against the toxicity of other metals to biota (Section 6.7.3), with implications for recovery when acidification has abated.

Following from the evidence that Ca deficiency adversely affects *Daphnia*, Hessen and Rukke (2000) carried out laboratory experiments and showed that "low Ca and UV radiation may act as strong mutual stressors for *Daphnia*, meaning that low levels of Ca increase the likelihood for UV damage." Although this link between Ca decline and ultraviolet (UV) damage has not been demonstrated in the 'real world', other stresses occurring at the same time as acidification reinforce the fact that this is a multi-stressor example. Acidification, less light-absorbing DOM in the water column, less Ca in the water and greater penetration of UV radiation are likely to be acting in concert. Table 11.1 provides examples of some of the general effects of acidification on freshwater biota.

11.1.2 The Effects of Acidification on Terrestrial Systems

Soil type determines the sensitivity of terrestrial systems to acidification, so there are similarities with the situation for aquatic systems. Chemical changes in poorly buffered soils include a decrease in pH, depletion of nutrient cations (calcium, magnesium) and mobilization of aluminium, all of which are in common with changes in surface waters. The deposition of sulphuric and nitric acids results in accumulation of sulphur and nitrogen in soils. The sulphur in soil has the potential to be leached into surface waters, adding to acidification or delaying recovery; nitrogen accumulates in soils beyond what forests can use and retain (Driscoll *et al.*, 2003). For grassland soils in northern China, from the central-southern Tibetan Plateau to the eastern Inner Mongolian Plateau, the pH in the surface layer has declined significantly since the 1980s, with an overall decrease of 0.63 units (Yang *et al.*, 2012). Regions with low soil carbonate were the most affected, but changes of soil pH showed no significant associations with soil cation exchange capacity. In China, nitrogen deposition appears to be one of the key factors influencing soil acidification (Huang *et al.*, 2015a, 2015b).

Much of the information on effects of acid deposition on terrestrial systems comes from studies of forest decline. Forest decline is complex and multi-causal, with nutrient deficiencies, nutrient imbalances and metal/nutrient antagonism all potentially involved. Beyond effects on soil nutrients, acid deposition can leach nutrients from the leaves of trees, which adds a layer of complexity to the study of nutrient deficiency. Land use and forest management have also been suggested as contributing to soil acidification, so acid deposition may be superimposed on or interacting with some practices of land use. Drought, which may be related to climate change, has also been implicated for some changes in terrestrial systems observed during the period for which acidic deposition has been known to occur. This provides another example of multiple stressors.

Severe forest decline was noted in Germany and other areas of Europe during the early studies on acidic

Table 11.1 **Some examples of effects of acidification on freshwater biota.**

Taxonomic group(s)	Sign/effect of lowering pH	Outcome/observation	Approach used
Freshwater fish	Increased mercury in muscle.	pH-related changes in the mercury cycle in lakes.	Surveys
Freshwater fish, Scandinavian lakes	Loss of fish populations, or reduction of population size, leaving only older fish, between the 1950s and 1970s.	Intermittent or chronic reproductive failure.	Historical records, surveys, data on success and failure of stocking experiments and *in situ* exposures of fish in field and laboratory situations
Juvenile fish and invertebrate food items	Losses of populations; possible lack of food for higher trophic levels.	Early life stages less tolerant to pH below 5.5 than mature fish.	Surveys
Fathead minnows (*Pimephales promelas*)	Larval fish did not survive at pH 6. Reproductive failure at pH 5.	Early life stages less tolerant to low pH than adult fish. Possible physiological effects on oocyte development.	Various laboratory experiments
Natural fish populations, Experimental Lakes Area, Ontario, Canada	Decline in abundance of many species during early years of acidification. Loss of prey items for lake trout (*Salvelinus namaycush*) resulted in emaciation. By 1982 recruitment had ceased for all Lake 223 fishes.	Primary cause of decline for many fish was recruitment failure.	Experimental whole-lake acidification starting at pH 6.49 in 1976 and reaching pH 5.02–5.13 by 1981 (Section 4.4.3; Figure 4.8)
Benthic fauna	Loss of mayflies and stoneflies.	Elimination of acid-sensitive species.	Surveys
Invertebrates	Decreased insect (mainly Chironomid) emergence to 13% at pH 5; macroinvertebrate taxa vary in response to low pH from the most sensitive to very tolerant.	Elimination of acid-sensitive species.	Outdoor experimental channels, pH ambient, 6 and 5; experiment ran for 17 weeks
Phytoplankton	Lower species diversity.	Elimination of acid-sensitive species.	Surveys
Periphyton/ benthic algae	Increased abundance at acidic pH.	Absence of grazers; light penetration to greater depths as a result of acidification.	Surveys; experimental lake acidification; laboratory culture experiments
Zooplankton	Lower species diversity; daphnids notably sensitive to low pH.	Elimination of acid-sensitive species.	Surveys; experimental lake acidification

deposition. At the time there was emphasis on aluminium toxicity as a major cause of toxicity to trees. Certainly, experiments with tree roots showed that Al was indeed toxic, but comparisons with the field situation, where the dose is difficult to determine, complicated the interpretation. More recently it has been shown that nutrient cation imbalance (particularly calcium depletion) is at least partly responsible for tree decline. The question is complex, given the demonstrated existence of other stressors on forests, beyond acidification. In fact, the German forest decline was extreme in comparison to the decline of forests in Canada and Eastern North America, and Al

may have played a more significant role in the former. Table 11.2 summarizes reported causes of tree decline.

11.1.3 Regulation of Acidic Emissions and Recovery of Aquatic and Terrestrial Systems

Although treatment after the fact has been applied to many acid-stressed systems, ideally recovery begins with control of sulphur emissions at the source. An understanding of cause–effect connections as well as dose–response relationships is needed for regulation of atmospheric pollutants, which is particularly challenging

Table 11.2 **Reported causes of tree decline.**

Tree species	Type and location of study	Effect(s)	Process(es) underlying decline
Red spruce (*Picea rubens*)	Forest surveys, New York State, Vermont, New Hampshire	Severe decline with death of many trees at higher elevations.	Acidification: leaching of calcium from needles of trees, making them more susceptible to freezing; low Ca and elevated Al in soil may cause reduction in biomass and limit root uptake of water and nutrients.
Sugar maple (*Acer saccharum*)	Forest surveys, eastern USA	Extensive mortality.	Deficiencies of nutrient cations, coupled with insect defoliation or drought.
Scots pine (*Pinus sylvestris*)	Scots pine as a model species for the study of drought-induced decline	Forest die-off events or partial mortality.	Drought: hydraulic failure, defoliation and stomatal closure; the latter two processes limit CO_2 assimilation.
Pine, spruce, and maple	Aluminium in tree rings, states in the USA bordering the Laurentian Great Lakes	Aluminium concentration increase in xylem after 1980.	Increase in soil Al concurrent with soil acidification, but not necessarily causing effects.
Soils and tree foliage	Nutrient status (from studies of soils and leaf samples)	Calcium concentration in leaves of deciduous trees correlated with soil Ca.	Soil deficiency, possible in combination with leaching of Ca from leaves by acidic deposition.

when considering long-range transboundary air pollution (LRTAP). Most regulation has addressed sources of SO_2 with attempts to quantify the relationship between emissions and effects – a challenging task when the receptors are often at a considerable distance from the source and in different jurisdictions. Hydrogen ions in precipitation are correlated much more closely with sulphate than with nitrate. As stated by Gorham *et al.* (1984), "The relation of H^+ concentration and precipitation pH to wet SO_4^{2-} deposition is of great interest in connection with the atmospheric loading of SO_4^{2-} that is likely to be the target of any emission control strategy."

11.1.3.1 Abatement

From the 1970/80s, national and international legislation began to take effect, resulting in significant reductions of emissions of acid precursors. Once acid deposition has been reduced, the surface water or soil generally shows a chemical response, recovering with increasing pH. A key question is whether biological and ecosystem recovery is also occurring. Biological recovery is less predictable than the geochemical response, and furthermore, relief from some of the multiple stressors may not be occurring at the same rates. For example, calcium deficiency that persists in previously acidified lakes is expected to affect biological recovery, for zooplankton in particular, which will affect the rest of the food chain. In their review of the current state of previously acidified lakes in the Sudbury, ON, Canada region, Keller *et al.* (2007) noted recovery

among fish, zooplankton, phytoplankton and zoobenthos. Importantly, they also pointed out that biological recovery was still at an early stage. Similar signs of biological recovery following abatement have been observed for Scandinavian lakes.

During the experimental acidification of Little Rock Lake, mercury concentrations in fish were observed to increase. This trend was retested during de-acidification (Hrabik and Watras, 2002). Fish Hg decreased between 1994 and 2000 in one basin by 5% and in the other basin by 30% (the basins de-acidified at very different rates). De-acidification could account for the decreases in fish Hg, even after normalizing for decreases in atmospheric Hg loading. The authors considered that the results were consistent with the hypothesis that depositional inputs of SO_2 and Hg(II) co-mediate the biosynthesis of methylmercury and thereby co-limit bioaccumulation.

In comparison with aquatic systems, responses of formerly impacted soils to decreasing acidic deposition are less well documented. Lawrence *et al.* (2015) sampled soils in eastern Canada and the northeastern United States at 27 sites where wet SO_4^{2-} deposition had decreased. They recorded decreases of exchangeable aluminium in the soil 'O horizon' (upper organic layer) and increases in pH in the O and B horizons (B being the subsoil) at most sites. They concluded that "the effects of acidic deposition on North American soils have begun to reverse". There is less information about the recovery from forest decline following abatement of acid deposition. This is perhaps not too surprising, bearing in mind the multiple stressors that are causing forest decline as well as the longer generation time of trees, compared with most aquatic biota.

11.1.3.2 Treatment

In addition to controlling the sources of acidification, a great deal of remedial work has been done from the 1970s onwards, as part of the overall scientific investigation of acidic deposition. Liming has been a part of acid rain science since the 1970s and has been of particular concern for fisheries. Adding lime to a watershed increases calcium availability and reduces or prevents mobilization of aluminium, an outcome that is beneficial to both terrestrial and aquatic ecosystems. Direct additions of lime to surface water are a relatively inexpensive approach to lake or stream de-acidification.

Beyond chemical recovery from liming, biotic recovery is likely to be more complex and slower. Increasing calcium availability for tree species with relatively high Ca demand is one potential benefit of liming, and some experimental work has been done to address biotic changes. Sugar maple (*Acer saccharum*), in particular, has been a focus of terrestrial liming studies, in part because its widespread decline has been linked to soil-Ca depletion from acidic deposition. Liming of hardwood forests led to an increase in the basal diameters of trees and reduced mortality of sugar maple (Lawrence *et al.*, 2016).

Continued declines in acidic deposition have led to a partial recovery of surface water chemistry, and the start of soil recovery. For such examples, liming might no longer be required to prevent further damage from acidic deposition, although it may accelerate recovery in calcium-depleted landscapes (Lawrence *et al.*, 2016). Litter decomposition is an integral link in nutrient cycling in forest ecosystems. Experimental and field studies have shown that acidification results in litter accumulation, i.e., there is limited decomposition, along with changes in the soil microflora. Liming is expected to reverse this trend, but results from experimental liming are inconclusive. Indeed, an experimental study of soils (Allen *et al.*, 2020) showed that liming forest soils increased Ca and Mg concentrations in litter, but did not restore its microbial decomposition.

More details about the processes involved in the recovery of impaired aquatic and terrestrial systems are presented in Chapter 13.

11.1.4 Acidification of Marine Systems: 'The Other CO_2 Problem'

Greenhouse gases (GHGs), carbon dioxide (CO_2), methane (CH_4) and nitrous oxide are contributing to climate change (Box 14.1). Carbon dioxide is playing an additional role, because the ocean has absorbed about 30% of the emitted anthropogenic CO_2, resulting in acidification (Pachauri *et al.*, 2014). Indeed, the current anthropogenic rate of carbon input appears to be greater than during any of the ocean acidification events identified so far from the geological record, including the Palaeo-Eocene Thermal Maximum (Zeebe, 2012). Recent reports from the Intergovernmental Panel on Climate Change (IPCC) indicate that about half of the anthropogenic CO_2 emissions between 1750 and 2011 have occurred in the past 40 years. Since the beginning of the industrial era, the pH of ocean surface water has decreased by 0.1 pH units, corresponding to a 26% increase in acidity, measured as hydrogen ion concentration (Pachauri *et al.*, 2014). During the period from 1971 to 2010, also as a result of increased atmospheric CO_2 concentrations, average ocean temperatures in the upper 75 m have increased at a rate of over 0.1 °C per decade (Section 14.1).

Carbon dioxide dissolves in seawater, with chemical reactions that reduce seawater pH, carbonate ion concentration and the saturation status of biologically important calcium carbonate minerals. Changes occur simultaneously for many inorganic carbon parameters, including total dissolved inorganic carbon (DIC), CO_2, HCO_3^-, CO_3^{2-}, H^+ and $CaCO_3$ saturation, i.e., multiple potential drivers/stressors (Hurd *et al.*, 2020). This inherent complexity is compounded by the co-occurrence of temperature increase, with the concomitant lowering of dissolved oxygen concentrations. Effects of ocean acidification may be further complicated by interacting effects with UVB radiation, as mentioned earlier for freshwater lakes (Section 11.1.1.3).

Monitoring and research on ocean acidification (OA) are mostly in their early stages; considerable information comes from relatively short-term laboratory studies and some from more realistic microcosm experiments (e.g., Riebesell *et al.*, 2008), but most studies have applied only one or two stressors (Gao *et al.*, 2019). Additional information comes from models and other predictions based on physical and chemical theory. The response of individual organisms, populations and communities to realistic

Table 11.3 Observed and potential effect(s) of ocean acidification.

Organism or system	Effect	Reference	Demonstrated (D) or predicted (P)
Surface water	pH decrease, carbonate ion decrease, decrease in buffering capacity.	Doney *et al.* (2009)	D: Well documented in field data
Arctic water	Changes in seawater chemistry exacerbated in Arctic waters because CO_2 solubility increases with lower temperature.	Deppeler and Davidson (2017)	P
Ocean water	Changes in inputs and cycling of iron and other biologically relevant trace metals.	Hoffmann *et al.* (2012)	P: Implications for inorganic trace metal chemistry from both acidification and warming
Ocean water	Increased Fe(III) solubility, decreased stability of some Fe–ligand complexes, and increasing Fe(II) stability.	Breitbarth *et al.* (2010)	P: Predictions
Ocean water	Decrease in dissolved oxygen with concurrent changes in stratification (related to variations in water temperatures and salinity) with increased visible and UV radiation for organisms.	Gao *et al.* (2019)	P: Predicted, related to warming as well as ocean eutrophication
Test amphipod	Increase in toxicity of metals in sediment under conditions that mimicked current and projected OA; proposal of important physiological effects of OA, more significant than effects on metal speciation.	Roberts *et al.* (2013)	D & P: Experiments and predictions
Highly calcified molluscs, echinoderms and reef-building corals	Depletion, with effects on food webs.	Doney *et al.* (2009)	P: Predictions based on laboratory experiments
Photosynthetic organisms (micro- and macro-algae)	Elevated CO_2 may stimulate photosynthesis but changes in trace metal availability may reverse such effects; effects of OA on photosynthesis are relatively small, and variable among taxa that have been examined.	Mackey *et al.* (2015); Gao *et al.* (2019)	D & P: Laboratory studies and predictions for field
Marine biota	Alteration of biological trace metal uptake rates and metal binding to organic ligands will affect trace metal uptake rates.	Hoffmann *et al.* (2012)	P: Predictions
Nitrogen cycle	Large differences between sources and sinks of N, with increase in N supply (bottleneck in nitrogen cycle).	Wannicke *et al.* (2018)	P: Meta-analysis

combinations of stressors and gradual changes is as yet largely unknown. An overview of the possible stressors and effects is provided in Table 11.3.

11.1.5 Lessons Learned

It is now well established that gases emitted into the atmosphere are not only potential air pollutants locally, but also through LRTAP are vectors for profound effects on aquatic and terrestrial systems, often at long distances from the source. Acid deposition sets off a series of chemical, biochemical and biological changes in sensitive systems, resulting in complex effects on ecosystems: fresh and salt water, sediments and soils.

In the course of studying the ecotoxicological effects of acidic deposition, our understanding of the relationship between watershed processes and surface water ecology has been advanced, and studies on ocean acidification have highlighted the key role of carbonate chemistry in marine ecosystems.

An understanding of the quantitative relationship between emissions of gaseous precursors of acid precipitation (particularly oxides of sulphur) and the intensity of their effects on target catchments can guide emission control measures, for strong acids and also for CO_2. With the cause–effect relationship well established, control measures for S emissions can lead to chemical recovery, and potentially to ecological recovery.

11.2 Metal Mining and Smelting

Mining, defined as the extraction of useful minerals from the surface of the Earth, has been practised since prehistoric times. The oldest known underground mine, dating from 40,000 BCE, is located in the Kingdom of Eswatini (formerly known as Swaziland). The mine served as a source of ochre, a mixture of ferric oxide with clay and sand that was used as a pigment and skin colouring. Jumping ahead to about 2,000 years ago, the Iberian Peninsula in Spain and the Cornwall area in southwestern England were the sites of relatively early major copper discoveries.

Metals and metallic products are fundamental to our contemporary way of life; indeed, progress in earlier and more primitive societies than ours was driven to a considerable degree by the discovery and usage of metals such as iron and copper, and alloys such as bronze. Precious metals have historically and in contemporary life provided aesthetic as well as practical and commercial enrichment to humans. Like other resource industries, the metals industries have acted as drivers of regional development, and they shaped and continue to shape the social fabric of human societies. At the regional level, the arrival of a metal mine can influence the development of an entire town, with direct employment, secondary industries and capital growth. These aspects are, for the most part, positive attributes of the industry. The negative aspects of metal mining and smelting include their potential adverse effects on the health of humans and of ecosystems. However, as discussed in the following section, these negative impacts are not inevitable. They are much better understood today than was the case as recently as 50 years ago, and with this improved understanding of the underlying mechanisms, better mining and smelting practices can be applied to minimize the negative impacts.

11.2.1 The Issue

In the past, the negative impacts of mining and the metallurgical industries appear to have been viewed as the necessary consequences of human development. Local damage, normally related to mining, was familiar to and, for the most part, accepted as normal by those who lived locally and usually depended upon the industry, whereas for the average member of society living elsewhere, the damage was scarcely visible. This was especially true in large, sparsely populated countries such as Australia, Canada and the western United States, where mines were developed, exploited and then abandoned. Metal processing, especially smelting, had more widespread effects, but even these were initially seen as relatively local and acceptable. This tacit acceptance of environmental damage in the past is discussed in Chapters 1 and 13.

However, as mining and smelting operations increased greatly in scope during the twentieth century, their impacts began to be noticed downstream and downwind. It became increasingly evident that the cost of remediating environmental damage far exceeded the cost of preventing the damage in the first place. As with many environmental issues, public awareness of the environmental impacts of metal extraction and purification began to rise in the late 1960s. Attitudes changed quite rapidly, especially concerning smelting. One factor that proved influential in bringing about changes in industrial practice, especially for smelters and their emissions, was the emergence of acid rain as a sociopolitical issue (Sections 1.2.3 and 11.1). This issue came to the fore in the 1970s, first in Europe, especially in the Nordic countries, and soon thereafter in North America. Scrubbing smelter stack gases to remove sulphur dioxide (SO_2, a precursor of the sulphuric acid in acid precipitation) was also linked to the removal of fine particles and resulted in a marked reduction in the emission of particulate metals.

Forty years later, in the twenty-first century, cradle-to-grave design is now required in codes of mining practice in most developed countries, the mantra being to design for closure and to operate for closure, leaving the site in an acceptable condition, even if not identical to pre-operation conditions. For example, the International Council on Mining and Minerals (ICMM) includes 27 corporations responsible for large-scale mining operations on the global scale. It has recently adopted a code of practice to which its members must adhere, and many countries have also adopted their own codes of practice. However, it was not always so, as illustrated in Capsule 11.1.

Capsule 11.1 Mercury and Silver: A History of Unexpected Environmental Consequences

Saul Guerrero

Visiting Fellow, School of Culture, History and Language, Australian National University, Canberra, Australia

How It Started...

For thousands of years, until the end of the fifteenth century, the only known refining process to reduce a metal to its elemental state from its chemical compound in metallic ores was due to the fortuitous dual role played by charcoal in a fire, as the source of energy (high temperatures) and of the reducing agent carbon monoxide. This process came to be known as smelting. To produce silver from the most readily available Ag ores found in Europe (argentiferous lead or copper), a more complex refining process was required. In the case of argentiferous galena (lead sulphide ores that contain Ag), as early artisans around 4,000 BCE found by trial and error, two distinct chemical reactions are required. The lead sulphide in the ore can either be reduced to Pb in the presence of litharge (lead oxide, PbO) or it can be roasted to produce PbO and subsequently smelted in a reducing atmosphere to produce liquid Pb. In both cases, Pb extracts Ag present in the galena ore. The Ag-rich Pb from the smelting stage is cooled to a solid, and then heated in a reverberatory furnace. In this furnace, Pb is separated by directing a stream of air onto its melting surface, producing litharge, which is skimmed off or absorbed by the cupel or vessel that holds the Ag-rich lead. Once all the Pb has been oxidized and removed in this manner, the Ag, which is impervious to this oxidizing treatment (together with any gold present), remains at the bottom of the cupel. The second stage of the process is called cupellation (Figure 11C.1).

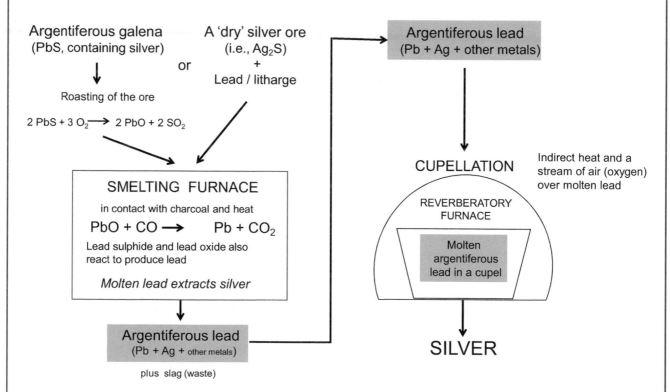

Figure 11C.1 A simplified outline of a historical two-stage refining process for silver ores by smelting with lead followed by cupellation.

For thousands of years, the combination of lead and charcoal was the only key able to unlock silver from its ores. Artisan-refiners were exposed to toxic Pb and its compounds (lead sulphide, lead sulphate, lead oxide, lead carbonate) in fumes released to the air from the Ag-refining furnaces. Livestock in the surroundings of Ag smelters were also deeply affected, with farmers demanding compensation. Hand in hand with the toxic emission of Pb fumes came the depletion of woodlands, since approximately one metric tonne (t) of charcoal was required to refine one kilogram of Ag (Guerrero, 2017).

By the beginning of the sixteenth century, two events combined to change the traditional metallurgy of ores containing silver. In 1492, the voyage by Columbus would lead to the Spanish conquest of the world's largest and unique reserves of primary Ag, strung along the mountainous spine close to the western seaboard of the American continent. In 1502, two Venetian gold prospectors would apply for a patent in Venice for a process that claimed to extract Ag from its ores *sine igne*, without fire, but instead with liquid mercury (Hg) (Vergani, 1994). The market for mercury would now expand to unexpected levels across the New World, a new metallurgical process at an industrial scale would be born, and the marriage between Ag and Hg would lead to anthropogenic releases of Hg and its compounds, the consequences of which are still felt in the present day.

How It Ended…

By the end of the Spanish colonial era in Hispanic America (1820s), some 58,000 t of Hg had been consumed in New Spain, and approximately 64,000 t in Peru (TePaske, 2010). From 1576 to 1650, the consumption of Hg averaged around 300 t per year at a single location, the city of Potosí in colonial Upper Peru (Guerrero, 2012). For over 70 continuous years, the refining *ingenios* that lay right next to a city of over 100,000 inhabitants consumed a yearly amount of Hg that would not be seen again until modern times, and then only as a countrywide total. For example, France in its entirety is estimated to have consumed just under 300 t of Hg in 1970 (Hylander and Meili, 2005). From 1550 to 1900, approximately 371,000 t of Hg were produced around the world (Hylander and Meili, 2005), of which approximately 250,000 t (67%) were destined for the refining of Ag in the Americas (Guerrero, unpublished). What had happened to convert Hg from an alchemical curiosity into the most heavily traded heavy metal across continents? What was the environmental impact of this unique anthropogenic event?

The answer to the first question is not straightforward. The enormous scale of deposits of silver sulphide ores in the American continent triggered a proportional demand for mercury up to the end of the nineteenth century, when refining via cyanidation displaced it. However, the parallel history of Japanese production of Ag between the sixteenth and seventeenth centuries is an excellent example of an equally possible alternative historical path in the New World without Hg. The Americas and Japan held major deposits of silver sulphides of equivalent composition, in contrast to Europe where argentiferous lead or copper deposits predominated. The Japanese successfully applied smelting with Pb to extract Ag in quantities that at times equalled the annual production from Potosí (Guerrero, 2019). Spanish refiners, however, chose to experiment on an industrial scale. They used a locally modified Hg recipe that added iron and copper sulphate to the traditional mix of sodium chloride, water and Hg used in Europe to extract gold from its mineral matrix. This came to be known as the *patio* process, and it involved two critical chemical reactions for the recipe to work with the Ag_2S ores (Johnson and Whittle, 1999).

First, Ag_2S had to be converted to silver chloride, in the presence of cupric and chloride ions in aqueous solution:

$$2Cu^{2+}(aq) + Ag_2S(s) + 8Cl^-(aq) \rightarrow 2AgCl(s) + 2[CuCl_3]^{2-}(aq) + S(s)$$

The AgCl was then reduced by mercury to Ag, with calomel (Hg_2Cl_2) as a solid by-product. Silver was finally extracted from the reaction slurry by amalgamation with excess Hg present:

$$(n + 2)\, Hg(l) + 2AgCl(s) \rightarrow Ag_2Hg_n(l) + Hg_2Cl_2(s)$$

The flowchart of the historic industrial process is summarized in Figure 11C.2. The chemical recipe could include iron, which also acts as reducing agent for AgCl. The two-fold role of charcoal in smelting now had competition from the two-fold role of Hg as a reducing and amalgamating agent, with one important caveat. While smelting could be applied to any Ag ore, the use of Hg was restricted to ores that did not have lead, since Pb also amalgamates with Hg. This new metallurgical method has been widely characterized in the English literature as the amalgamation process, although in Spanish it was known historically as silver by mercury, *plata de azogue*. The latter term is less confusing, since Hg is unable to directly amalgamate the Ag compounds present in an ore. It has been more correctly identified as a very early example of hydrometallurgy, in recognition of its chemical fundamentals (Johnson and Whittle, 1999).

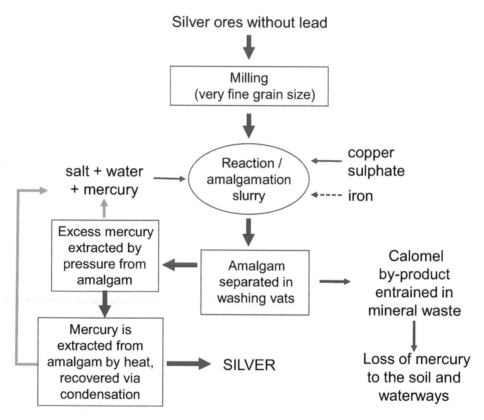

Figure 11C.2 The main physical and chemical stages of the *patio* process as practised in New Spain/Mexico up to the end of the nineteenth century. In the United States, iron vessels and heating would be employed to hasten the reactions and reduce the consumption of Hg, but the recipe would remain identical in its chemical components to the one created in the Andes in the late sixteenth century.

This complex chemical process was packaged as a deceptively primitive operation that at times resembled a farmyard after a heavy downpour (see Figure 11C.3). Workers and animals were constantly exposed to liquid mercury as they trod the reaction mixture over a period of weeks within what was essentially an array of open-air batch reactors of very flexible dimensions. It was a brilliant solution in the technical context of the colonial period and continued to be used in republican Mexico in the nineteenth century when the photograph was taken. Why did written sources of the time not comment on a widespread human toll from such an intensive use of Hg, from the Spanish colonial texts to the writings of Mark Twain from Virginia City, Nevada, in the nineteenth century?

The reason was that up to 85% of this mercury would have been transformed into insoluble Hg_2Cl_2, buried alongside tonnes of mineral waste either in landfills or entrained along waterways. Most of the

Figure 11C.3 A very wet process: the *patio* (courtyard) acted as an array of open-air batch reactors for the refining of silver ores with mercury. Unidentified refining *hacienda* in Guanajuato, Mexico. Reproduced from Rickard (1907).

remaining Hg would have been lost as percolations to the soil underneath the refining facilities, and less than 1% would have been discharged directly into the air (Guerrero, 2012, 2016). Workers and local communities were shielded in the short-term by the formation of calomel from an even greater exposure to elemental Hg. What happened over the longer term is still unclear. Did calomel ultimately disproportionate into Hg and mercuric chloride (a soluble and very potent poison), or mercury(II) sulphide? How soon was Hg released to the air from its initial entombment as Hg_2Cl_2? The amount of legacy Hg from historic silver refining that still contributes to current background levels is one important area of research within the context of the Minamata Convention (AMAP/UNEP, 2019).

Spain supplied most of the mercury consumed in New Spain from the world's largest cinnabar (HgS) deposit at Almadén. Its workers and nearby communities have been exposed to the effects of HgS smelting for centuries, making its soil "one of the most polluted sites in the world" (Higueras *et al.*, 2013). Over seven times more Hg is projected to have been released directly to the air at Almadén from 1560 to 1850 than from the silver refining sites in New Spain and Mexico during the same period. This is one of the rare historical cases in which Europe suffered a greater environmental impact than its overseas conquered territories, as the direct consequence of producing a colonial commodity (Guerrero, 2018).

The use of mercury instead of smelting in the New World had another unintended consequence: the forests of the Americas were spared the depletion that would have accompanied the widespread use of smelting with lead. Its use is calculated to have saved around 105 million hectares of woodland in New Spain alone during colonial times, if natural generation or forest husbandry is not taken into account. Had smelting with Pb been used to refine the whole colonial silver production in New Spain, an additional 140,000 to 280,000 t of Pb (as lead compounds) in fumes would have been released to the environment (Guerrero, 2017). The ecotoxicological impact of having to switch from Hg to Pb would have been even greater around Potosí, since in the Andes Hg played a major role in Ag refining.

Mercury has lost its alchemical allure, becoming a metal of ill repute. Its terrible impact on human health was first brought to worldwide attention in Minamata, Japan, in the 1960s, and soon after, its ecotoxicological effects were revealed. It seems therefore counterintuitive to argue that historic silver refining with Hg presented clear advantages, but it did: a lower level of immediate impact on workers and communities compared with lead and lead fumes, a deferred total ecotoxicological impact until such time that Hg_2Cl_2 disproportionates, and very low concomitant woodland depletion rates. The relentless European demand for bullion and coinage presented refiners with a devil's choice between Pb and Hg. Their efforts and metallurgical innovations flooded the markets with Ag, changing the course of global history at a lower environmental cost to the New World and its people.

Capsule References

AMAP/UNEP (2019). *Technical Background Report for the Global Mercury Assessment 2018*. Oslo, Norway and Geneva, Switzerland: AMAP-UNEP.

Guerrero, S. (2012). Chemistry as a tool for historical research: identifying paths of historical mercury pollution in the Hispanic New World. *Bulletin of the History of Chemistry*, 27, 61–72.

Guerrero, S. (2016). The history of silver refining in New Spain, 16c to 18c: back to the basics. *History and Technology*, 32, 2–32.

Guerrero, S. (2017). *Silver by Fire, Silver by Mercury. The Chemical History of Silver Refining in New Spain and Mexico, 16c to 19c*, Boston, MA: Brill, pp. 89, 93–96, 345, 358–365.

Guerrero, S. (2018). *The Environmental History of Silver Production, and Its Impact on the United Nations Minamata Convention on Mercury*. Boston, MA: World Economic History Congress.

Guerrero, S. (2019). Lead or mercury, *haifuki-ho* or *plata de azogue*: The environmental dilemma in the history of silver refining. *Asian Review of World Histories*, 7, 107–125.

Higueras, P., Esbri, J. M., Oyarzun, R., *et al.* (2013). Industrial and natural sources of gaseous elemental mercury in the Almaden district (Spain): an updated report on this issue after the ceasing of mining and metallurgical activities in 2003 and major land reclamation works. *Environmental Research*, 125, 197–208.

Hylander, L. D. & Meili, M. (2005). The rise and fall of mercury: converting a resource to refuse after 500 years of mining and pollution. *Critical Reviews in Environmental Science & Technology*, 35, 1–36.

Johnson, D. A. & Whittle, K. (1999). The chemistry of the Hispanic-American amalgamation process. *Journal of the Chemical Society, Dalton Transactions*, 4239–4243.

Rickard, T. A. (1907). *Journeys of Observation*, San Francisco, CA: Dewey Publishing Company, p. 146.

TePaske, J. J. (2010). *A New World of Gold and Silver*, Boston, MA: Brill, pp. 139, 211.

Vergani, R. (1994). La métallurgie des non-ferreux dans la république du Venise (XVe–XVIIIe siècles). In: P. Benoit (ed.) *Mines et Métallurgie*. Villurbanne, France: Programme Rhone-Alpes recherches en sciences humaines, 207–215.

11.2.2 Processes Involved in the Extraction and Purification of Metals

A typical mine cycle involves the following steps: exploration, discovery, economic evaluation; development of a mine plan, environmental impact assessment, construction; operation; closure. The initial steps involving such actions as deforestation and road construction may result in habitat fragmentation and barriers to migration. Nevertheless, from an ecotoxicological perspective, the impacts of the early investigative stage are minor compared with those that follow. Consequently, in the present section we shall focus on the construction and operation steps. The construction phase normally involves clearing the site, removing the vegetation and **overburden** and building the infrastructure needed to extract the desired mineral from the host rock. In the development of the overall mine plan, a choice will have been made between a surface mine, an underground mine or perhaps some combination of the two. Underground operations generally have a smaller ecological footprint than surface operations, and reactive sulphidic mine wastes can be handled by backfilling galleries that have been mined and emptied of ore. Surface operations such as strip or open-pit mines have a much larger environmental footprint than underground mines, and they generate huge quantities of waste rock that must be removed to access the mineral deposits. The choice between the two types of operation will typically have been determined mainly by the depth and location of the ore body. Deep deposits with localized and rich ore bodies lend themselves to an underground mining approach, whereas shallow and widely dispersed deposits will normally require a surface operation. Surface mining is often the more economically attractive procedure, even though it disturbs more surface than does underground mining. In some cases, non-entry mining may be a viable option (Box 11.1).

Box 11.1 Non-entry mining

Non-entry mining is carried out *in situ*, typically by leaching the desired metal from its parent material by means of microbial or chemical treatment. It may involve initial excavation to access the ore body and then injection of liquid into the underground cavity to solubilize the metal. In other cases, the ore may be brought to the surface and the leaching liquid may be dispersed through the collected ore heap. In both cases, the metal is recovered from solution.

Non-entry mining is used mainly above ground, and typically for low-grade ores containing uranium and copper, and sometimes for the extraction of 'new' metals from old tailings. *In situ* leaching avoids many of the environmental problems associated with excavating an ore and processing it by other methods, but it still runs the risk of contaminating groundwater if the leaching solution is not properly contained.

Operation of the mine will normally involve blasting to access and recover the ore. In most ores, metal concentrations are too low to allow the ore to be smelted directly, meaning that a **beneficiation** step is required. Beneficiation typically involves grinding, concentration and finally dewatering of the concentrate (Figure 11.2, steps 3, 4 and 5). The concentrate will normally undergo a smelting and refining step, designed to free the metal from its host mineral. This may be a pyrometallurgical process, in which high temperatures are used to heat the concentrate to temperatures above the metal's melting point. The heating will be done in the presence of an oxidizing agent (e.g., sulphide-based ores will be smelted in the presence of air) or a reducing agent (e.g., oxide-based ores will be heated in the presence of coke). As an alternative to the use of extreme heat, hydrometallurgical processes can also be used in the refining step; in such cases, aqueous reagents are used to leach metal from its ore. Electrometallurgical processes, involving the application of electrical energy, can then be used to collect metal from the aqueous solution, a process known as electrowinning.

Whether underground or at the surface, a mine's operation produces two categories of solid waste: mine spoil (soil and rock removed when the mine is excavated) and mill tailings (the finely ground rock resulting from the beneficiation process). These materials vary greatly in chemical composition, but most have in common a paucity of plant nutrients and concentrations of metals that may range from non-toxic to highly toxic. The waste

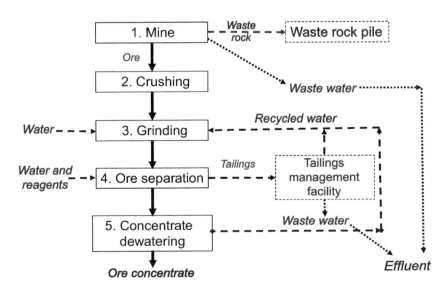

Figure 11.2 Typical treatment of an ore to produce a concentrate for metallurgical purification. Figure reproduced courtesy of Environment and Climate Change Canada (ECCC).

material may also be acid-generating (see Section 11.2.3). For any or all of these reasons, areas where mine and mill wastes are deposited and left untreated are usually barren and inhospitable to plant colonization. The physical instability that results from lack of plant cover renders potentially toxic metals more mobile (e.g., by wind erosion), and the ecological effects of the lack of plant cover are clearly inhibitory for any succession to a functioning ecosystem.

11.2.3 Substances of Concern

Having briefly described a typical mining operation, let us now consider the substances of concern that may be mobilized or formed and released during the overall process; extraction, concentration (beneficiation) and smelting are the steps that generate the greatest environmental contamination (Box 11.2). The nature of the substances released will obviously depend upon the type of ore that is being processed. Sulphidic ores are of particular concern, since they are acid-generating and the dominant global source of copper, nickel, lead and zinc. In addition to these commercially exploited metals, such ores typically contain a suite of chalcophilic (sulphur-seeking) elements as impurities (e.g., Ag, As, Cd, Co, Ge, Mo, Sb, Se, Te). It follows that the untreated effluents from mines that are extracting sulphidic ores, as well as stack gases from smelters, are likely to include significant concentrations not only of the dominant metals but also of many of these other elements.

The beneficiation step often involves the use of flotation agents such as xanthates to separate the mineral

from the host rock. Since water is involved as the solvent for the flotation agent, this step requires wastewater collection and treatment. As a result, the mine effluent will be a likely source not only of the dissolved and particulate metals described above, but also of the residual process chemicals used in the flotation process.

The milling and flotation steps yield metal-rich concentrates, but they also generate large quantities of fine-grained tailings that are considered waste material. In the case of sulphidic ores, these tailings contain large quantities of reduced sulphur in the form of pyrite (FeS_2) or pyrrhotite ($Fe_{1-x}S$, $x \leq 0.2$). These reduced sulphides, on exposure to air and water as they are brought to the surface, form sulphuric acid through a series of microbial and chemical oxidations that can be represented globally by the equation

$$4\,FeS_2 + 14\,O_2 + 4\,H_2O \rightarrow 4\,Fe^{2+} + 8\,SO_4^{2-} + 8\,H^+$$

$$(11.1)$$

The resulting sulphuric acid dissolves iron and other metals from the ore and other geological materials, generating a highly acidic metal-bearing solution that is commonly referred to as acid mine drainage (or acid rock drainage in the geological literature). Because the tailings have been ground down to a fine size, greatly increasing their **specific surface area**, they are particularly susceptible to this oxidation step. Acidophilic bacteria, notably iron and sulphur oxidizing species (e.g., *Acidithiobacillus* spp.; *Thiobacillus thiooxidans*) that use these reduced chemicals as their energy source, are found at sites affected by acid mine drainage, where they act as catalysts, accelerating the oxidation of the iron sulphides.

The generation of acid mine drainage is an autocatalytic process, meaning that it is far better to stop it from developing than it is to try to control it once it has become established. It is a huge problem worldwide and is largely unresolved in cases where reactive sulphidic tailings were simply left lying on the ground when the mine closed, or used for local road construction. One of the obvious ways of limiting oxidation is to limit contact between the tailings and the atmosphere. Subaqueous disposal of tailings is one way to achieve this state, but only if the receiving waterbody is enclosed (not a running water) and stable. Tailings ponds are a very frequent feature of mining sites, where they offer an effective strategy for limiting the oxidation of sulphide-rich reactive tailings. However, they are often engineered

---

Box 11.2 Examples of potential toxicants released from mining operations

- Waterborne dissolved and particulate metals (mine effluent; surface run-off)
- Waterborne dissolved and particulate metals + sulphuric acid (**acid mine drainage** from sulphide-rich tailings)
- Process chemicals used in the flotation beneficiation process (e.g., xanthates)
- Sulphur dioxide (SO_2) from smelting
- Airborne particulate metals from smelting
- Airborne particulate metals (dust) from wind erosion of tailings

structures, i.e., dammed enclosures, and thus there is always a finite risk of dam failure and the sudden release of the tailings downstream, particularly in seismically active areas such as Papua New Guinea or elsewhere in the Indonesian archipelago (Section 14.3.1).

The final source of undesirable chemicals is the smelting process. Smelter emissions potentially include airborne gases (e.g., SO_2 for sulphidic ores; HF for bauxite) and airborne metal-containing particulates. For centuries in the United Kingdom and Europe, and early in the twentieth century in the New World, roasting and smelting were carried out in open pits, resulting in ground-level fumigation by acids and metal particulates. The adjacent ecosystems were exposed to very high concentrations of contaminants, often with dramatic results. As recently as the 1960s, the area around the copper and nickel smelters in Sudbury, Ontario, Canada was frequently referred to as a 'moonscape', reflecting the destruction of all the local vegetation in response to the high concentrations of SO_2, which is extremely phytotoxic (Capsule 12.1). The subsequent practice of erecting tall stacks to disperse the volatile and other airborne waste products did improve local air quality, with beneficial effects for human health and for the ecosystems close to the smelting operation. Note that some of this

'greening' of Sudbury is described in Chapters 12 and 13. However, the tall stacks led to the widespread airborne dispersal of precursors to acid precipitation and of fine metal particulates, a process that occurs irrespective of provincial or national borders. As mentioned earlier, the final step now practised in modern smelters is to scrub the waste stream of stack gases to remove the SO_2 and also to capture the majority of the particulate metals before they escape into the atmosphere (Figure 11.3).

In addition to airborne emissions, smelting also produces solid waste in the form of slag, a heterogeneous mixture of metal oxides (e.g., those of Al, Si, Ca and Mg). The slag may also contain residues from any fluxing material that was used to protect the molten metal from oxidation during the smelting process. Formerly just considered a waste material, in modern times slag can be recovered as a product of the smelting process, ground and used in blended cement.

11.2.4 Ecotoxicological Impacts of Metal Mining and Smelting

Our analysis of the environmental impacts of mining and smelting activities follows an approach similar to that used earlier in our discussion of individual metals

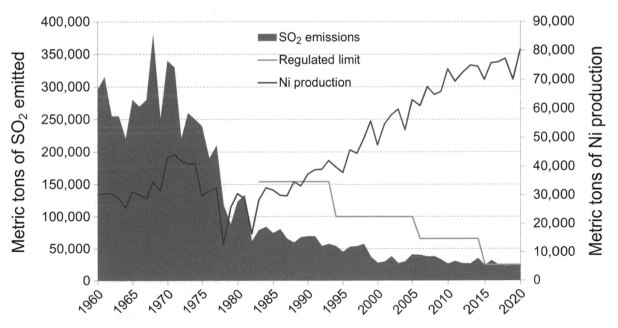

Figure 11.3 Changes in SO$_2$ emissions from one of the Sudbury smelters between 1960 and 2020. The reduction in emissions was achieved despite the steady increase in nickel production. Figure reproduced with permission from Glencore Canada, Sudbury, ON, Canada.

(Chapter 6), and we build on the science presented in that chapter. Much of the information discussed below falls in the category of what we could call 'legacy effects', namely effects that were documented before mining practices were subject to environmental regulations. For convenience, we will first address impacts on running waters and then consider lakes.

11.2.4.1 Rivers

Historically, mine tailings tended to be left lying on the ground at the mine site, exposed to alternating wet and dry periods (ideal for generating acid mine drainage). Run-off from uncontrolled acid generation is highly acidic (pH values as low as 2) and rich in dissolved metals. Downstream from the source, the dissolved ferrous iron is oxidized to ferric ion ($Fe^{2+} \rightarrow Fe^{3+}$), leading to precipitation of highly visible Fe(III) oxyhydroxides, a tell-tale sign of acid mine drainage. Many of the originally dissolved metals associate with the oxyhydroxides as coatings on rocks, pebbles and on bottom sediments. These metals associated with the oxyhydroxides (and any metals that have simply precipitated as the pH increased) are much more available than they were before the mining step, when bound to reduced sulphur, buried underground and protected from atmospheric oxygen.

There are also some situations where acid mine drainage completely overwhelms the capacity of the river to neutralize the acidic inputs. For example, the Rio Tinto drains the Iberian Peninsula in Spain, a sulphide-rich area that has been actively mined for over 2,000 years. Currently, its waters enter the Atlantic Ocean at a pH of about 2.5 with dissolved cadmium, copper and zinc concentrations in the millimolar range, more than 10,000 times higher than the concentrations in the uncontaminated coastal waters.

In some cases, mine tailings have been directly sluiced into nearby rivers or simply allowed to erode into the river during high rainfall events, leading to contamination of the downstream sediments. The fate of metals discharged into a river in particulate form, or of those that are converted to particulate form by sorption or precipitation, will depend on the size of the particles and the hydrodynamic conditions in the river. Fluvial transport of particles varies both temporally and spatially, especially in smaller rivers, meaning that the metal-bearing particles are subject to a sequence of sedimentation and resuspension events.

From an ecological perspective, tailings that end up in depositional areas along the main river channel are less problematic than those that are deposited in the floodplain. In the first case, the sediments will always be water-covered and thus relatively protected from atmospheric oxygen. However, metal-bearing tailings that are deposited in the floodplain are subject to a wetting and drying cycle that favours oxidation of the sulphide-rich particles and the local release of their metal content. For this reason, modern mining practice emphasizes the need to avoid release of tailings or other metal-bearing particles into running waters.

One of the best-studied examples of the downstream effects of metal mining is the Clark Fork system in Montana, USA. Mining of copper and silver started late in the nineteenth century, and metal contamination eventually extended 350–450 km downstream from the mine (located near Butte, MT), affecting the streambed and the floodplain (Table 11.4). Even after the inputs from the mine site were curbed, the downstream fauna still reflected continuing periodic metal stress events. These were traced to the river floodplain, where large quantities of mine tailings had been deposited early in the mining cycle. The wetting–drying–rewetting cycle led to the injection of acidic and metal-rich pulses into the mainstream of the river, which effectively prevented full recovery.

Another example of downstream effects is afforded by the Fly River in Papua New Guinea, one of the largest tropical rivers in the world. The Ok Tedi copper and gold mine is situated in the mountainous, turbulent and highly turbid headwaters of the Ok Tedi, one of the tributaries of the Fly River. Because the region is seismically active and subject to very high rainfall, tailings ponds could not be constructed, and it was decided to inject the tailings directly into the river. During mine operation, the suspended solids load of the river increased from about 60 mg/l above the mine to more than 10,000 mg/l immediately below the mine, and then declined to about 150 mg/l at the confluence of the Ok Tedi with the Fly River. Monitoring of the river fauna indicated that there had been significant reductions in fish catches in the Ok Tedi after the mine started operations in 1984. Tissue concentrations of copper, cadmium, lead and zinc were elevated in the liver and kidney of a range of fish species, and these concentrations showed a general decreasing trend as distance increased downstream from the mine, suggesting a mine-related effect (Table 11.4). Later it became evident that the effects of the tailings release extended into the

Table 11.4 **Representative studies showing the geochemical and ecotoxicological effects of metal mining and smelting.**

Location	Operation	Field effect	References
Clark Fork River, Montana, USA	Copper mine and smelter	Mine footprint detected 500 km downstream. Deposition of mine tailings in river floodplain. Periodic wetting and drying; tailings oxidation. Periodic fish kills; avoidance by fish of some former habitats.	Moore and Luoma (1990)
Ok Tedi and Fly Rivers, Papua New Guinea	Copper and gold mine	Tailings, mine waste discharged directly to river. Suspended solids increase from 65 mg/l to >10,000 mg/l in passing from above to below mine effluent. Mine footprint detected 200 km downstream. Deposition of mine tailings in river floodplain. Reduced fish captures downstream from mine. Forest dieback in areas where tailings deposited overbank.	Apte *et al.* (1995); Swales *et al.* (1998); Apte (2008); Stauber *et al.* (2008); Storey *et al.* (2008)
Rio Tinto, Iberia, Spain	Copper mines	pH close to 2.5 year-round; mean concentrations (mg/l) Cu 29, Zn 45; (μg/l) Cd 144, Ni 218, Pb 166. No plants, fish or other higher organisms survive. Some 'extremophiles' present.	Van Geen *et al.* (1997); Olías *et al.* (2020)
Sudbury, Ontario, Canada	Copper, nickel mines and smelters	Smelter airborne metal contamination 60 km downwind. Impoverished aquatic flora and fauna. Lake sediment core evidence of effects of acidification and metals on phytoplankton community. Resident metal-tolerant fish (yellow perch) show evidence of reduced aerobic capacity, impaired olfaction, stunted growth, shorter lifetime.	Freedman and Hutchinson (1980); Dixit *et al.* (1989); Keller (1992); Borgmann and Norwood (2002); Rajotte and Couture (2002); Pyle *et al.* (2005); Couture and Pyle (2008)
Rouyn-Noranda, Quebec, Canada	Copper mine and smelter	Smelter airborne metal contamination 65 km downwind. Lake sediment core evidence of effects of acidification and metals on phytoplankton community. Resident metal-tolerant fish (yellow perch) show evidence of endocrine disruption, stunted growth. Evidence for Cd being the metal responsible.	Campbell *et al.* (2003); Dixit *et al.* (2007); Campbell *et al.* (2008); Couture and Pyle (2008); Rasmussen *et al.* (2008)
Rupert Inlet, British Columbia, Canada	Copper mine 1971–1998, marine fjord	Decline in benthic fauna when [Cu] > 300 μg/g and tailings layer thickness >15–20 cm. Within 3 years of mine closure, total taxa values in near (<5 km) and mid-field (5–16 km) areas fell within the range of far-field stations (>20 km). Polychaetes dominated at tailings stations after closure; amphipods recolonized sporadically but sensitive to tailings instability.	Burd (2002)

floodplain, where large areas of the forest bordering the river were dying. There are also reports that the plume from the Fly River does reach the marine environment (Torres Strait), but there is no conclusive evidence that mine-related wastes are affecting the seagrasses or the coral reefs.

The Clark Fork and the Ok Tedi rivers are clear examples where the river remains impaired after remedial measures were undertaken, in large measure because of the tailings that had been deposited outside the main river channel, on the floodplain. Many other examples exist in more complicated settings, where the rivers are affected not only by mining and smelting activities but also by other anthropogenic inputs (e.g., Rainbow, 2018).

11.2.4.2 Lakes

To this point, we have emphasized the effects of mining and smelting on rivers, a decision that reflects the universal tendency of humans to use rivers to get rid of their wastes, whether they are of domestic or industrial origin. However, mining and smelting activities also generate airborne contamination, linked to the smelting process and, in dry climates, to wind erosion of tailings placed on land. Small metal-bearing particles are released during the smelting process and are dispersed with the stack gases. Indeed, in early attempts to calculate global budgets for metals, smelters were identified as a major source of atmospheric emissions for arsenic, copper, cadmium, antimony and zinc (Nriagu and Pacyna, 1988), with substantial contributions to the budgets for chromium, lead, selenium and nickel.

Metal-induced effects downwind from base metal smelter emissions have been identified in many studies conducted on the Canadian Precambrian Shield (Table 11.4). The following examples come from observations on lakes located around the smelters in Sudbury, Ontario (Cu, Ni) and Rouyn-Noranda, Quebec (Cu, Zn). The areas downwind from these point sources are largely undeveloped, meaning that the attribution of observed biological effects to the smelter emissions is relatively unambiguous. Sediment cores collected from the profundal zones of lakes located downwind from the smelters indicated that the smelter 'footprints' extend up to 65 km in the direction of the prevailing winds. Dated depth profiles of cadmium, copper and nickel in the sediment cores showed that the increase in metal concentrations coincided with the onset of smelting activities in the early part of the twentieth century. Palaeolimnological analysis

of sediment cores from the Sudbury lakes yielded evidence of temporal changes in composition of the diatom and chrysophyte communities, which could be explained by a combination of acidification and increased metal concentrations. Similar results were obtained from a sediment core from a Rouyn-Noranda lake located only 8 km from the smelter and were interpreted in terms of metal contamination rather than acidification.

Studies of the present-day communities in the lakes downwind from the Rouyn-Noranda smelter suggested that cadmium was the metal largely responsible for the observed effects (see Section 6.5.3.2), whereas nickel was implicated in the lakes downwind from the Sudbury mining and smelting complex. Some of the algae surviving in the highly metal-contaminated sites near Sudbury showed tolerance to high concentrations of Cu and Ni. In both the Sudbury and Rouyn-Noranda field studies, a suite of adverse biological responses was detected in yellow perch (*Perca flavescens*) collected along the metal concentration gradient and they could be related to the internalized metal dose (Table 11.4). For example, adult and juvenile perch collected from the more contaminated lakes exhibited endocrine disruption, as characterized by an attenuated cortisol stress response *in vivo* and a lower secretory capacity in response to adrenocorticotropic hormone *in vitro*. The aerobic capacity of yellow perch collected from contaminated lakes in the Sudbury area was reduced, as indicated by the activity of enzymes involved in aerobic metabolism as well as by swim performance and respiration rates. Subtle indirect effects of chronic metal exposure were also identified and attributed to food-web-mediated effects. The simplification of the benthic invertebrate communities in response to higher metal exposure led to stunted growth of the yellow perch, a bottom-feeding species during part of its normal growth cycle.

To integrate what we have described for rivers and lakes, Box 11.3 presents a guide to the biological effects that have been attributed to mining and smelting activities.

11.2.4.3 Coastal Marine Environments

To complete this summary of the impacts of mining and smelting on the environment, we should briefly consider the marine environment as a potential target. Historically, there were many examples of mine wastes being dumped into rivers that emptied into shallow coastal areas. This practice has fortunately largely disappeared (<1% of industrial mines currently discharge their

Box 11.3 Characteristics of the ecotoxicological impacts attributed to mining and smelting activities

- Often episodic events, associated with high rainfall and flooding
- Especially acute when these events occur after prolonged dry periods (droughts), which favour the oxidation of sulphides
- Normally characterized by the presence of metal mixtures and sometimes include acidification
- Secondary minerals often involved (i.e., forms that are less stable, more soluble than the original ore)
- Avoidance of metal-affected habitats, notably by fish
- Disturbed benthic communities
- Altered diets for bottom-feeding fish
- Diet-borne metals often implicated

tailings into such areas; Ramirez-Llodra et al. (2015)). However, there are examples where mine tailings have been or are being discharged into deep-sea environments, such as fjords in Norway, Greenland and western Canada (British Columbia). Fjords are long and narrow inlets of the sea, often separated from the open ocean by a sill. The sill limits the tidal flushing of the fjord, and in some cases the bottom waters are anoxic.

Many of the studies on submarine tailings disposal have been carried out in Norway, the country with the most experience in this area (Ramirez-Llodra et al., 2015). The immediate effect of such practices is the smothering of the original benthic environment and a loss of habitat heterogeneity as the tailings cover the seafloor. The most vulnerable members of the benthic community are the sessile fauna that have limited abilities to escape the greatly increased sedimentation. During active discharge, the process chemicals used in the beneficiation process may also contribute to loss of sensitive species. Sulphidic tailings deposited in such settings are relatively stable (see the discussion about tailings ponds in Section 11.2.3), and the buffering capacity of the overlying seawater limits any local acidification. However, the metal-rich tailings particles may serve as a vector for dietary metal uptake. When tailings discharge ceases and natural sedimentation begins to add organic matter to the submerged tailings, initial recolonization of the benthic environment by opportunistic species occurs quite rapidly. However, long-term studies have shown that the recovery to a 'natural' community similar to that present before tailings discharge started is a slow process, taking up to 15 years or more (Ramirez-Llodra et al., 2015).

11.2.5 Lessons Learned

What does the future hold for the mining and smelting sector? Worldwide metal usage continues to increase, notably in countries with developing economies such as China, India and Brazil, and there is no sign of a decrease in demand (UNEP, 2013; Reichl and Schatz, 2021). Recycling can help meet this demand, but it seems inevitable that metal mining and processing will also increase. These activities will remain a potential environmental stressor, not only because of the factors that we have discussed in this section, but also because of the demands that this sector puts on the global supply of energy (mining and refining currently use about 8% of the world's energy supply). In addition, there are notable changes in the marketplace, with demands for elements that used to be considered 'exotic' but now play key roles in various new 'green' technologies (Section 14.3).

Fortunately, significant progress has been made in recent decades in developing and enforcing strict mining codes of practice. New scientific research, regulatory oversight and industry actions have led to continuous improvements in the sector's environmental performance in many developed countries. These improvements have included reductions in overall energy usage, and in emissions of sulphur dioxide and greenhouse gases. 'Green' mining is no longer a pie-in-the-sky concept but rather a feasible target. At the time of writing, green mining initiatives can be found in resource-intensive countries (e.g., Australia, Brazil, Canada), as well as in China, Europe and the United States.

11.3 Engineered Nanomaterials

We have discussed particles earlier, notably in Chapters 6 and 7, where we emphasized the role that particle surfaces play in adsorbing metals and hydrophobic organic contaminants and limiting their environmental mobility. However, these natural particles generally fall into the micrometre to millimetre size range and do not remain in suspension over time. The terms **nanoparticle** and **nanomaterial** are sometimes used synonymously, but nanoparticle usually refers to an object with all three dimensions at nanoscale and nanomaterial refers to an object with at least one dimension at nanoscale, i.e., between 1 and 100 nanometres (nm) (Jeevanandam *et al.*, 2018). The adjective 'engineered' is used to distinguish manufactured nanomaterials from those that occur naturally (i.e., natural **colloids**). **Engineered nanomaterials** (ENMs) include those based on carbon (e.g., carbon nanotubes, **fullerenes**, graphene) and metal-based nanomaterials (e.g., elemental metals, metal oxides, quantum dots), as shown in Box 11.4. However, new ENMs are being produced all the time.

Engineered nanomaterials may be manufactured by changing the size and shape of existing bulk materials, or they may be synthesized at the nanoscale. Inorganic nanomaterials are often stabilized with a protective coating designed to limit dissolution or agglomeration (Section 11.3.2), but the use of such surface applications is much less frequent for carbon-based nanomaterials. Over the past 20 years, nanomaterials have evolved from

being a scientific curiosity to their current status as an emerging global technology, one with many potential environmental uses (European Commission, 2017) and medical applications (Jeevanandam *et al.*, 2018).

The principal justification for including nanoparticles as a Complex Issue in this chapter is that marked changes occur in the chemical and physical properties of materials on the nanoscale. Notably, the **specific surface area** of the particle (surface area divided by particle volume or mass) increases as the size of a particle decreases. The decreased radius of curvature imposes strains on the chemical bonds among the surface atoms, changing the particle's surface properties. In addition, nanoparticles could conceivably attain a particle size that would allow them to move through cell membranes (Chapter 3) or interact with biological molecules. These observations have led to the concept of 'nanotoxicity' (Box 11.5). The second reason for including nanomaterials in the present chapter is linked to how environmental scientists and regulators responded to the surge of interest in nanotechnology that occurred in the early 2000s. This response represents an example of how an emerging and complex environmental issue (temporarily) challenged both researchers and regulators.

We start our examination of engineered nanomaterials by examining the routes by which they may reach the receiving environment and their fate once there. We then

Box 11.4 Examples of the most used and researched engineered nanomaterials (ENMs)

ENM	Typical use
Metallic silver Ag^0	Biocide
Copper oxide CuO	Biocide; electronics
Zinc oxide ZnO	Sunscreen
Cerium dioxide CeO_2	Catalyst; cosmetics; fuel additive
Titanium dioxide TiO_2	Pigment; cosmetics; sunscreen
Iron oxide Fe_xO_y Zerovalent iron ZVI	Wastewater treatment; electronics Water treatment (remediation)
Carbon nanotubes CNT	Electronics; material science
Fullerenes C_{60}	Drug delivery; catalyst

Nanoplastics refers to fragmentation products generated by physical abrasion and UV photo-oxidation of expanded polystyrene and other plastic products (see Chapter 14).

Box 11.5 Nanotoxicity: A regulatory dilemma

Nanotoxicity refers to the possible toxicity of a nanoparticle that is attributable to its very small size, rather than to its chemical composition. For example, copper oxide (CuO) and zinc oxide (ZnO) exist as distinct chemical compounds. In their bulk form (particles in the millimetre to micrometre range), their toxicity is normally assessed by quantifying their tendency to release Cu or Zn in cationic form into solution (i.e., the 'free ion'). However, these two oxides can also exist in nanoform and are used commercially (e.g., CuO is used in antifouling paints; ZnO is used in sunscreens). Could they exert nanotoxicity because of the unique properties of nanosized particles that are not shared by the bulk material? If the answer to this question were positive, regulations for nanomaterials would have to be different from (and possibly stricter than) those for the bulk materials.

consider how particles of any dimension might exert toxicity and evaluate how nanoparticles fit into this scheme. Finally, we assess the evidence for true nanotoxicity.

11.3.1 Routes of Exposure and Environmental Fate

Nanomaterials may enter the aquatic environment with industrial discharges, especially in those associated with the manufacture of ENMs, and to a lesser extent in effluents from industries where nanomaterials are used in the manufacture of other commercial objects. After use, nanomaterials may be discharged into domestic wastewater and subsequently enter the aquatic environment (with or without prior wastewater treatment). For example, weathering of painted surfaces is known to release metallic nanoparticles (e.g., TiO_2) during rain events, and surface run-off from roads and soils may also contribute to loadings to the aquatic environment. Recognized routes for the addition of nanomaterials to terrestrial ecosystems include direct addition to soils in fertilizers or plant protection products, and indirect addition through application to land of wastewater treatment products such as sludges or biosolids.

Once they reach the receiving environment, ENMs are subject to a wide array of physical and chemical transformations (Schultz *et al.*, 2015). These processes are summarized in Box 11.6, and many of them are depicted in Figure 11.4. The most important of these processes is aggregation, which includes self-aggregation (homoaggregation) and aggregation with naturally occurring colloids or larger particles (heteroaggregation). Natural colloids include clays, hydrous iron and manganese oxides, humic acids and exopolymers exuded by algae and other microorganisms. In the receiving environment, these colloids and particles will be present in much higher concentrations than the ENMs, particularly in soils, with the result that heteroaggregation will be much more important than homoaggregation. Aggregation of ENMs is favoured under conditions of high ionic strength, so homoaggregation is favoured in surface waters with high alkalinity or in salt water. To counter their tendency to aggregate, ENMs are often manufactured with organic coatings (e.g., polyvinylpyrrolidone) to promote their dispersion in suspension, and these coatings will influence their fate in the aquatic and terrestrial environment by slowing the rate of dissolution and inhibiting aggregation.

Box 11.6 Transformations of ENMs in wastewater, receiving waters and soils

- Aggregation, especially with natural colloids (heteroaggregation)
- Dissolution (pH-sensitive; gut pH; vacuole pH)
- Oxidation/reduction (Ag(0) → Ag(I); Ce(IV) → Ce(III))
- Interaction with natural organic matter
- Sulphidation (Ag → Ag_2S; ZnO → ZnS)
- Biodegradation (carbon-based nanomaterials)

With the exception of biodegradation, these transformations are analogous to the speciation changes that occur when a dissolved metal enters the receiving environment.

Carbon-based nanomaterials such as carbon nanotubes or fullerenes are essentially insoluble in natural waters, but they may be subject to slow biodegradation. In contrast, most metal-containing nanoparticles (e.g., CuO, ZnO, Ag(0)) are subject to dissolution, which is often pH-sensitive and is favoured as the size of the particle decreases and the specific surface area increases. Dissolution reactions may involve an initial oxidation or reduction step, in which case the dissolution rate will be sensitive to the redox conditions in the receiving environment. The undissolved ENMs may interact with various solutes present in the environment, notably with natural organic matter (amino acids, proteins, carbohydrates, fulvic and humic acids). In so doing, the ENMs acquire a coating (referred to as a 'corona') which may affect their dissolution and their tendency to participate in aggregation reactions, and will also influence how the ENM is recognized at cell surfaces (Pulido-Reyes *et al.*, 2017). Finally, the ENMs may also react with anions such as sulphide, in reactions that effectively transform the ENMs and remove them from the nanoscale domain.

The nanoparticles that enter wastewater treatment facilities will tend to aggregate with other mineral and organic components of the wastewater, with the result that most of the original ENMs end up associated with other solids and do not remain as dispersed nanosized suspensions. Also, under the reducing conditions in wastewater, many ENMs will become sulphidated. Given the array of physical and chemical transformations illustrated in Figure 11.4, exposures to pristine, well-dispersed nanomaterials will occur only rarely in nature.

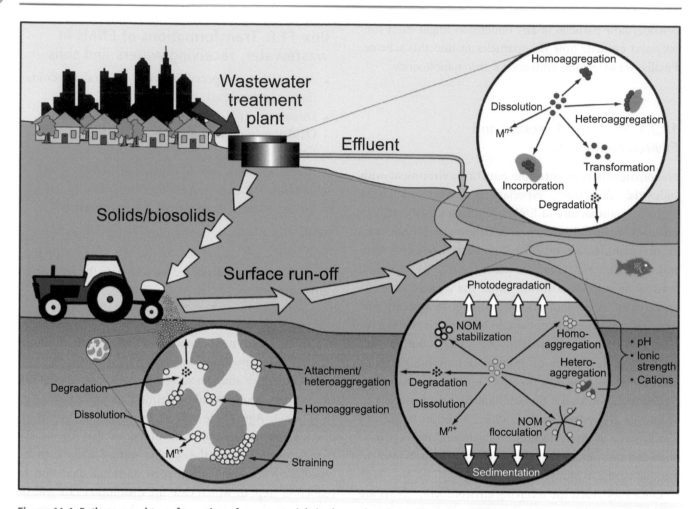

Figure 11.4 Pathways and transformation of nanomaterials in the environment. Note that the figure applies to generic nanoparticles, including carbon-based ENMs and those with organic coatings, which are thus subject to degradation. Emissions to the atmosphere are considered to be negligible. NOM = natural organic matter. Figure reproduced from Batley *et al.* (2013) with permission.

It follows that analytical methods are needed to assess environmental exposures to ENMs, to understand their behaviour in the environment, and to detect and track the assimilation and internal distribution of the modified forms described above. To quote from the review paper by Lead *et al.* (2018), "Current methods for determining ENM concentrations are limited for various reasons, including inadequate sensitivity and selectivity in relation to the complexity of both NMs and environmental conditions, lack of resolution (for imaging techniques), an inability to provide full quantification, and a lack of broad applicability." This is a major analytical challenge that demands a radical change in our ability to identify and quantify the man-made nanoparticles in the presence of similarly sized natural (colloidal) particles. As of 2022, this challenge has not yet been met, but significant progress has been made with techniques designed to detect single particles present in a matrix of similarly sized

natural colloids and to determine their chemical composition or signature. To compensate for the absence of suitable analytical techniques, exposure models have been widely used to simulate the environmental fluxes of ENMs from various sources and to predict their environmental concentrations. The environmental concentrations of ENMs in natural waters predicted by such models are typically less than 20 µg/l (Lead *et al.*, 2018). These exposure models cannot yet be routinely validated by comparison with *measured* concentrations (Nowack, 2017), but the few reported values in the literature are all well below this threshold (Azimzada *et al.*, 2021).

11.3.2 How Do Engineered Nanomaterials Enter Living Organisms?

Based on what we have discussed in Chapters 3 and 6, there are several mechanisms by which particles could be incorporated into living organisms and exert toxicity. The first of these is ingestion by animals, which involves the transfer of what is perceived as a food item from the external environment to the intermediate environment of the gut. As discussed in Section 3.6.2, the conditions in the gut lumen are largely controlled by the host organism and are often very different from those in the external environment. The fate of the particle in the lumen will depend on such factors as the pH, the oxidation–reduction potential and the transit time of the 'food' in the gut. Nanomaterials are too small to be retained by the feeding structures of aquatic filter-feeding organisms. However, in the real world such particles might well acquire a coating of natural organic matter and become more food-like or form larger particles as a result of homoaggregation or heteroaggregation. In soils, ENMs are likely to be ingested by terrestrial organisms (e.g., earthworms) in association with soil particles. As the particles are broken down within the organism, the liberated solutes are transported across the gut epithelial structures and enter the circulatory system, whereby they are transported to the hepatopancreas or liver and reach other internal organs. Negative effects of ingested particles may also occur in the digestive tract itself (e.g., effects on the intestinal membrane or on the intestinal microbiome).

Particles can also be taken up by endocytosis, a process whereby the cell membrane engulfs small areas of the plasmalemma and releases the resulting vesicles into the cytosol (Section 3.2.8). The invagination process is initiated by contact with a small particle and results in the formation of an intracellular vesicle that contains the engulfed particle. Physical contact between the particle and the cell membrane is necessary but not sufficient of itself to trigger phagocytosis, i.e., some form of 'particle recognition' is necessary. In this example, one could argue that the particle has not truly entered the cytosol, since it is still held within a membrane-bound vesicle. However, the vesicles are short-lived and typically fuse with endosomes within which the particle is subjected to digestion. As we pointed out in Sections 3.1.3 and 3.2.8, the ability to take up particles by endocytosis varies considerably from one type of epithelium to another, and is better documented in invertebrates (gills and intestines) than in vertebrates. Laboratory studies with cultured gill or intestinal cells have shown that some intact ENMs can be taken up by endocytotic mechanisms.

In principle, particles could interact with the epithelial surfaces of an exposed organism, where they might disturb the cell membrane's phospholipid bilayer (Wu *et al.*, 2013). The chemical conditions within the boundary layer at such surfaces, where solutes are being exchanged across the epithelium, may very well differ from those in the bulk solution (Section 3.2.7). In other words, the particle may be in a microenvironment that differs from the conditions prevailing in the external medium (e.g., water, pore water, sediment, soil). In the specific case of metal-based particles, these conditions might favour the release of the metal, for example by desorption or by oxidation of a **zerovalent metal**. With this release, the metal concentration in the epithelial microenvironment could exceed that in the terrestrial or aquatic environment and lead to greater toxicity than would have been predicted from concentrations in the ambient environment. We will refer to this particular mechanism as '**particle-assisted toxicity**'.

11.3.3 In Search of Nanotoxicity

Much of the early work on the toxicity of ENMs was conducted at unrealistic exposure concentrations that greatly exceeded those estimated from even the most conservative exposure models. In addition, many of these early toxicity tests were run at nominal concentrations, with no consideration of the fate of the ENMs in the toxicological test media or of the coating materials that are often used to stabilize nanoparticles (Box 11.7). Fortunately, the errors of this early phase were recognized, and there is now a much-improved appreciation of the complexities involved in testing ENMs at

> ### Box 11.7 Examples of problems encountered in the toxicity testing of ENMs
>
> - Aggregation of the test material in the exposure medium, meaning that it loses its nanoscale properties
> - Oxidation and/or dissolution of the (metal-containing) ENM, leading to concurrent exposure to both the dissolved metal and the ENM
> - Loss of the ENM by adsorption, both to the walls of the toxicity test enclosure and to the epithelial surface of the test organism(s)
> - Formation of coatings surrounding the ENMs
> - Release from ENMs of toxic organic contaminants present as a result of the manufacturing process

environmentally realistic concentrations (Handy *et al.*, 2012; Schultz *et al.*, 2014; Lead *et al.*, 2018).

When testing the toxicity of an ENM, it is important to determine whether the tested nanoparticle has intrinsic toxicity. In other words, does the nanoparticle cause toxicity because of its nanometre size, its extremely high surface area to volume ratio, its surface charge, or its ability to cause inflammation of and/or cross (epithelial) biological membranes? If the answers to all these questions were negative, the ENM would be deemed not to have any intrinsic nanotoxicity. In such cases, the assumption of 100% dissolution of the nanoparticulate form to yield its contents in dissolved form would be a conservative estimate of the potential toxicity of the material. However, if the answer to any of these questions were positive, then the ENM would have to be evaluated as a specific substance in its own right. In addition, it may be necessary to evaluate individual ENMs of different dimensions and sizes (e.g., nanoparticles, nanorods, nanofibres).

The most commonly observed mechanisms for the intrinsic or direct toxicity of ENMs are oxidative stress and inflammation. Although these mechanisms have been widely demonstrated in a variety of organisms, as well as in rodent experimental models and in humans occupationally exposed to ENMs, the steps involved in the pathway(s) from exposure to these outcomes are poorly defined. Engineered nanomaterials with larger dimensions (e.g., >50 nm) are generally recognized and ingested by phagocytic cells such as macrophages and **Kuppfer cells**, and this could generate an inflammatory response in affected tissues. ENMs with smaller dimensions may be able to cross cell membranes and interfere with enzymes and macromolecules involved in reducing the levels of reactive oxygen species (ROS) in cells, such as interfering with the cycling of glutathione between oxidized and reduced forms (Section 6.4.3). Some photoactive ENMs such as TiO_2 may generate ROS on epithelial surfaces when exposed to sunlight. There is ample evidence that exposures to ENMs result in increased expression of genes that are involved in combatting oxidative stress, such as catalase, glutathione peroxidase and glutathione reductase.

Transport of individual or aggregated ENMs across the cell membrane by endocytosis would bring the ENM into a new environment within the digestive vesicle. It has been suggested that in this environment the coating of the nanoparticles by proteins would give them new properties and make them 'recognizable' by the cellular

machinery (Nel *et al.*, 2009; Stern *et al.*, 2012). In other words, the ENMs might be directed and handled in a specific manner, eventually leading to specific modes of action. In this context, autophagy and lysosomal dysfunction have been suggested as possible mechanisms of nanoparticle toxicity.

To explore the question of nanoparticle toxicity, Notter *et al.* (2014) carried out a meta-analysis of three different types of metal-based nanomaterials (nano-Ag, nano-ZnO and nano-CuO) in which toxicity data were compiled for studies published between 2007 and 2013. The published half-maximal effective concentration (EC_{50}) values were compared for both the nanosized form and the corresponding dissolved metal (72 articles in 25 different peer-reviewed journals and 453 data pairs). In almost all cases for Ag (94%), Cu (100%) and Zn (81%), the nanosized form was *less* toxic than the dissolved metal in terms of total metal concentration. These results suggest that dominant intrinsic nanotoxicity is the exception rather than the rule, but they do not rule out the possibility that, under certain circumstances and with particular ENMs, exposure to the ENM might lead to a nanospecific effect.

Indeed, there are some examples where careful laboratory work has demonstrated that ENMs can possess intrinsic toxicity, and three such examples are shown in Figure 11.5. In the first case (A), a unicellular green alga (*Pseudokirchneriella subcapitata*) was exposed to suspensions of cerium oxide, either as nanoparticulate CeO_2 (10–20 nm) or as its microparticulate equivalent (≤5 μm) (Rogers *et al.*, 2010). Inhibition of algal growth was greater in the presence of nano-CeO_2 than in the experiment with the larger particles. Since CeO_2 is stable in aqueous suspension and underwent negligible dissolution, the experimental results suggest that the direct interaction of nano-CeO_2 with the algal surface was responsible for the toxicity. Indeed, the algal cells exposed to nano-CeO_2 showed evidence of cell membrane damage. It was also noted that this was only the case when experiments were carried out in the light, a result that was traced to the photocatalytic activity of the CeO_2 particles and the generation of hydroxyl radicals close to the cell membrane surface.

In the second example (Figure 11.5B), also carried out with a unicellular green alga (*Chlamydomonas reinhardtii*), the toxicity of nanoparticulate silver was compared to that of dissolved silver (Ag^+) (Navarro *et al.*, 2008). Based on total Ag concentrations, inhibition of algal growth was 18 times higher for dissolved Ag than for nano-Ag

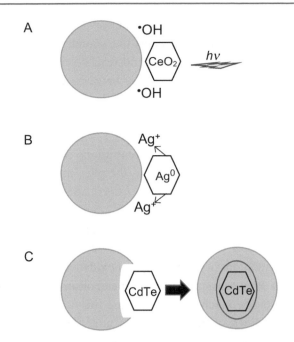

Figure 11.5 Three examples showing how a nanoparticle might exert toxicity. The green circle represents an algal cell, and the hexagons represent a nanoparticle. The symbol *hv* represents light energy. A: Association of a photoreactive cerium oxide nanoparticle with the algal surface, causing damage to the cell membrane. B: Association of nano-Ag with the algal surface where it acts as a localized source of ionic silver close to the cell membrane. C: Endocytotic uptake of a cadmium–tellurium quantum dot by the algal cell.

(25 ± 13 nm). However, when toxicity was compared on the basis of the concentrations of the free Ag^+ ion, toxicity in the presence of nano-Ag was much higher than in its absence. It was concluded that the nano-Ag particles were binding to the algal surface, where they acted as a local source of Ag^+ in the microlayer surrounding the algal cell (i.e., the 'phycosphere'). In other words, the toxicity of nano-Ag towards the alga was mediated by Ag^+ but it involved a direct interaction of the nanoparticles with the algal surface. This is an example of the particle-assisted toxicity scenario described earlier (Section 11.3.2).

The third example (Figure 11.5C) again involves an alga exposed to an ENM, but it differs from the first two in that there was evidence for uptake of the nanoparticle and for a nanospecific effect on the alga (*C. reinhardtii*). The algal cells were exposed either to a soluble Cd salt or to CdTe/CdS semiconductor quantum dot nanoparticles (<6 nm) in short-term experiments (<30 min) (Domingos *et al.*, 2011). Partial dissolution of the nanoparticles occurred during the experiments, yielding free

Cd^{2+} in solution. Bioaccumulation of Cd by the algal cells could be largely accounted for by uptake of dissolved Cd, but uptake of the quantum dots occurred in parallel, as determined by the intracellular Cd:Te ratios. More importantly, transcriptomic screening using RNA-seq (Chapter 5) demonstrated that free Cd^{2+} and nanoparticulate CdTe elicited distinctly different biological effects.

These examples demonstrate that the phenomenon of 'nanotoxicity' is real. Although the meta-analysis carried out by Notter *et al.* (2014) suggests that dissolution of metal-based ENMs is the driving force behind the toxicity to aquatic organisms, we cannot assume that all ENMs can be regulated based on the toxicity of the bulk material.

11.3.4 Lessons Learned

The inclusion of ENMs as a 'Complex Issue' in this chapter was guided in part because of the difficulties encountered when ENMs first appeared as potential environmental contaminants. These complications reflected a weakness in the regulatory framework as it existed circa 2000, in that it had not been designed to consider size-related effects at the nanoscale. Even now, 20+ years later, there is still some debate about how to regulate nanomaterials, but there is also a reassuring degree of similarity among the approaches that have been adopted in Europe, North America and Australasia.

We also felt that it was important to explore the initial 'bandwagon' effect of research on the fate and toxicity of ENMs. Regulatory agencies throughout the developed world, including the OECD, were demanding quick answers to the thorny problem of how to regulate the flood of new nanomaterials appearing on the market. Prior to 2000, there were very few products containing ENMs, but now there are several hundred in an array of commercial products and devices. Initially, scientists did not have the experimental tools to conduct appropriate toxicity tests or to detect ENMs in various environmental media. Environmental scientists were certainly aware of the importance of taking the environmental fate of a new contaminant into account. However, because of analytical limitations for detecting ENMs in water, soils and sediments at environmentally relevant concentrations, the only real considerations related to environmental fate were: (i) their solubility in water and (ii) their tendency to remain intact. As we have discussed here, ENMs are subject to an impressive array of transformations when they reach the receiving environment (Box 11.6). As a result, well-dispersed and pristine nanomaterials will rarely persist in the receiving environment or, for that matter, in laboratory toxicity tests. Our full appreciation of these complexities has developed slowly, but new analytical techniques and protocols for toxicity tests are helping to resolve the challenge of determining the fate of an ENM once it has been added to a toxicity test medium, a wastewater stream or other receiving environments.

A final lesson to be retained is that environmental scientists should take advantage of the findings of medical research on nanomaterials, especially with respect to drug delivery. We have much to learn from medical researchers about the nano–bio interface and how nanoparticles interact with biological membranes.

Case Study 11.1 Whole-lake Addition of Nanosilver

A whole-lake ecosystem experiment was conducted at the IISD Experimental Lakes Area (IISD-ELA) in northwestern Ontario, Canada, to evaluate the impacts of adding suspensions of silver nanoparticles (AgNPs) to a small oligotrophic lake (Lake 222). Additions of 9 kg of AgNPs in 2014 started in mid-June and ended in late October (18 weeks), and additions in 2015 of a further 6 kg of AgNPs started in early May and ended in late

August (14 weeks). Soon after additions of AgNP suspensions began at a point along the shoreline, silver was detected throughout the lake and reached concentrations in the water column that varied between approximately 5–10 µg/l in both field seasons. AgNPs with a size range of approximately 15–40 nm were detected suspended in the water at densities of approximately 1×10^{10} particles per litre, and dissolved silver concentrations were very low. The concentrations of silver in the gills, liver and other tissues of northern pike (*Esox lucius*) and yellow perch (*Perca flavescens*) increased rapidly after additions started and declined slowly after additions stopped (Martin *et al.*, 2018). Yellow perch monitored before, during and after AgNP additions developed signs of oxidative stress in liver tissues. Prey consumption, total metabolism and the condition of the perch declined during the AgNP additions and remained depressed one year after (Hayhurst *et al.*, 2020). As shown in Figure 11.6, perch population densities declined after AgNP was added to the lake, but perch populations were stable in a nearby reference lake.

After additions of AgNPs ended, the levels of silver in the water column of the lake declined to below limits of detection, and the silver was primarily deposited into the lake sediments. Taken together, these results show a negative impact on fish from chronic exposure to AgNPs at environmentally relevant concentrations. Enforcement of the Canadian Water Quality Guideline of 0.25 µg/l for long-term exposure to total silver should be protective for fish and other aquatic organisms exposed to AgNPs.

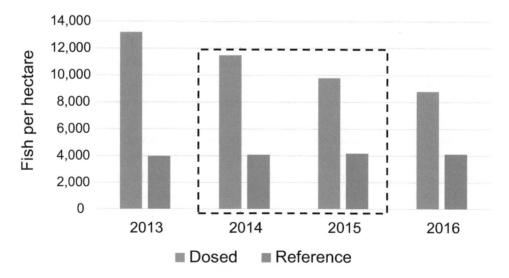

Figure 11.6 Densities of yellow perch (fish per hectare) in a lake dosed with AgNPs (Lake 222) and a reference lake (Lake 239) at the IISD-ELA. The dashed box indicates the period of additions of AgNPs to the lake in 2014 and 2015. Figure drawn from data published by Hayhurst *et al.* (2020).

11.4 Pulp and Paper Production

The production of pulp and paper from wood is considered a complex issue in ecotoxicology for several reasons. Wood itself is a complex mixture of structural components, including wood fibre that is the target of the process, polymeric glues (lignin) that bind the fibres together into a rigid structure, and many compounds that are components of plant metabolism, ranging from sugars to steroid analogues. When wood is processed to create paper, air and water emissions contain mixtures of these components plus thousands of new compounds formed when pulping and bleaching chemicals degrade the lignin that binds wood fibres together. Many are bioactive, posing a risk to aquatic and terrestrial species and to human health, and chlorination reactions can increase their persistence and toxicity. The mix of compounds in effluents also varies day-to-day with changes in the nature of the wood pulped, changes to the process, and process 'upsets'. The production of pulp and paper starts with deforestation, and includes physical, chemical and biological effects on rivers, lakes and marine coastal waters caused by hydroelectric dams, log floating and the smothering of benthic habitats by wood fibre. Many past activities, such as log driving, interacted with the fate and effects of current chemical emissions.

This section provides a brief history of papermaking and a review of the processes by which wood fibre is converted to white paper, wastes are generated and treated, and chemicals are emitted to the environment. The immediate effects of each activity are usually quite obvious, but their resolution has often uncovered previously unrecognized ecological impacts, or even created new issues. The online Further Reading list and Case Study 4.1 provide additional details about the pulp and paper issue and its complexity.

Paper is an inexpensive product that is valued primarily for what we print or paint on it (particularly if it is currency!), and for a multitude of other applications, including packaging. However, this was not always the case. Before paper, records were etched on stone tablets or written on wood or animal skins. Although these media outlasted the civilizations that created them, they were costly and inconvenient. In China, about 1,900 years ago, the first paper was made from plants by a process that is conceptually the same as modern papermaking. Materials such as bark or reeds were mashed in hot water to separate the long and strong fibres comprised of cellulose from the less fibrous components that glued the fibres together. When a slurry of plant fibres was dried in thin sheets, the result was a lightweight material with a surface suitable for painting, writing and printing. An excellent example is papyrus, paper made from the *Cyperus papyrus* reed that grows along the River Nile.

Papermaking spread via the Silk Road and Ottoman Turk invasions of China, first reaching Spain with the Crusades between 1100 and 1200 CE, and was widely adopted in Europe. Some paper was (and still is) handmade from recycled cotton or linen, which provides the finest quality. However, the process became increasingly mechanized with the development of water- and windmills, and the increased demand for paper after 1453, when Gutenberg invented the printing press. By the late 1800s, machines were separating fibre from wood in high volumes, primarily from coniferous softwood trees (e.g., pine, fir) which yielded very long and strong fibres that could be harvested in large quantities. The modern pulp and paper industry began primarily in Scandinavia, Canada and the northern United States where softwood forests and water resources are abundant. Depending on the degree of processing, paper made from wood fibre ranges in quality from boxboard (brown, little processing) to bright white (highly processed) (Table 11.5). Specialty papers are also manufactured to suit requirements for waterproofing, glossy colour printing, sanitary tissues, corrugated cardboard, packaging, etc.

In the latter half of the twentieth century, technologies were developed to produce pulp and paper from tree species other than coniferous softwoods (e.g., eucalyptus, aspen). The result was a rapid expansion of the industry in more tropical countries where trees are abundant and regenerate more quickly than northern softwoods, and where labour is less expensive. These new mills are much larger than traditional mills (e.g., 3,000 versus 300 air-dried metric tonnes (ADMT) per day of pulp). Although some older mills in North America and Scandinavia have been expanded and automated since the 1990s, others have been closed because they were too small, inefficient and expensive to retrofit, and they could not compete in a changing economy. The Internet displaced newspapers as a source of information, and the transportation costs of softwood increased when nearby forests were depleted. The rising costs of waste disposal led to an increase in paper recycling because it was a less expensive source of fibre. Wastepaper was converted to an aqueous slurry, washed to remove impurities such as ink, and mixed with virgin pulp to create new paper, reducing the market for virgin pulp. In Europe, more than 72% of all paper is currently recycled, and 35–40% of newsprint in the United States is made from recycled fibre.

11.4.1 Evolution of Pulp and Paper Environmental Issues

Since its inception, the pulp and paper industry has caused an array of significant environmental issues. As each was recognized and addressed, their resolution was followed by new or previously unrecognized problems, often as important and difficult to deal with as the first, and many interacted in unexpected ways to create new concerns.

11.4.1.1 Making Paper from Wood

Wood consists of long fibres of cellulose and hemicellulose bonded together by lignin, a glue-like polymer of phenolic compounds; in combination, fibre and lignin create the structure and rigidity of wood. Wood pulping separates the fibres from the lignin by mechanical, semi-chemical and chemical processes (Table 11.5). Pulp is produced mechanically by grinding wood chips between rotating abrasive discs. This 'groundwood' pulp retains its lignin, giving a very high yield of product and a low percentage of waste. However, the residual lignin is slowly oxidized, and products made from groundwood pulp darken and weaken and are unsuitable for long-term applications. Pulping quality is enhanced by combining mechanical grinding with chemicals such as sulphite, for mild chemical digestion of lignin (chemi-mechanical) and lighter coloured pulp (less lignin). When more corrosive chemicals are used for pulping, the pulp is progressively lighter in colour, which means that it contains less residual lignin and is more easily bleached to create white

paper. However, the yield of useful pulp declines with an increased degree of refining, and in the purely chemical 'kraft' process, 45–60% of the mass of wood chips is lost as chemical waste. This means that the wastewater from pulp washing contains progressively higher concentrations of organic compounds. In particular, the phenolic derivatives of lignin and resin acids and fatty acids are highly toxic to aquatic species, and should be removed by treating wastewaters before they are discharged.

Despite its lower efficiency, chemical digestion without mechanical grinding reduces the extent to which fibres are broken during pulping; longer fibres create stronger paper. 'Kraft' (German for strength) pulping is currently the most widely used process and produces a light brown pulp comprising 95% cellulose and hemicellulose, and 5% residual lignin. In kraft pulping, wood chips (approximately 5 cm long × 3 cm wide × 1 cm thick) are cooked in a sealed digester at high temperatures and pressures. The cooking liquid (white liquor) is a highly corrosive alkaline solution of sodium sulphide (NaS) and sodium hydroxide (NaOH). The high pressure forces this liquor into the pores and capillaries of each wood chip where the chemicals cleave the bonds between phenolic monomers of lignin, without affecting the structure and strength of cellulose and hemicellulose fibres. After several hours, the pressure in the digestion vessel is released explosively. The water inside the chips expands instantly as steam and the sudden pressure blows apart the digested wood chip like popcorn, creating a slurry of pulp. The pulp is washed repeatedly to remove spent cooking liquor (black liquor) which is a concentrated solution of digested lignin and wood extractives (Box 11.8).

When the digester is 'blown', some of the volatile components of wood are recovered by condensation as turpentine, a useful by-product. The black liquor is recycled by evaporating water to raise the solids content to about

Table 11.5 **Examples of different wood pulping processes.**

Process	Pulp colour (before bleaching)	Yield (% of wood recovered as usable fibre)	Uses
Thermo-mechanical	Brown	>95	Boxboard, brown paper bags, newsprint
Chemi-mechanical	Light brown	85–95	Newsprint, specialty papers
Semi-chemical	Beige brown	60–80	Newsprint, paper bags
Chemical – kraft and sulphite	Light brown (white after bleaching)	40–55	Newsprint, fine papers

Box 11.8 Major products and by-products of wood pulping

Products
- Cellulose
- Hemicellulose

By-products
- Lignin
- Tall oil (fatty acids)
- Turpentine (alcohols and phenolics)
- Phytosterols

Waste to effluents
- Lignin
- Resin and fatty acids
- Phenolics
- Sugars
- Biological oxygen demand (BOD)
- Chemical oxygen demand (COD)

Box 11.9 Major by-products of pulp bleaching by chlorine

Waste to effluents
- Chlorinated resin and fatty acids
- Chlorinated phenols
- Chlorinated dioxins and furans
- Mercury from Cl_2 generation

80%, when it is combusted to remove the organic matter and to recover energy (heat), NaS and NaOH for reuse in pulping. During recycling of the black liquor, wood extractives are recovered as useful by-products. These include tall oil (resin and fatty acids) used in adhesives and soaps, and β-sitosterol, an analogue of cholesterol used in medicine, and a component of some cosmetics. Although these products are recovered from black liquor, a portion is lost to wastewater when the pulp is washed before bleaching, including bioactive compounds such as juvabione, stilbenes, genistein, pinosylvin and betulin. Compounds that are vertebrate androgens (e.g., testosterone, epitestosterone and androstenedione) also occur in wood and probably play a role in plant physiology. Large quantities of methanol, and volatile and odorous sulphidic compounds are released as air pollutants.

Kraft pulp is bleached to remove the residual lignin and to create different grades of white paper. Until the 1990s, pulp was most often bleached with elemental chlorine gas (Cl_2) under acidic conditions. Multiple washing steps neutralized the pH, solubilizing many of the chlorinated by-products of lignin degradation, such as chlorinated resin and fatty acids, and chlorinated phenols (Box 11.9). Hydrophobic by-products, such as polychlorinated dibenzo-p-dioxins and dibenzofurans (PCDDs/

PCDFs), were removed bound to particulates. These by-products contained variable numbers of chlorine atoms, and the degree of chlorination determined their environmental persistence, bioaccumulation and biomagnification in food webs (Chapter 7).

Chlorine gas is an extremely hazardous material to transport. Until its use ceased, chlorine was generated on site by a mercury-cell chlor-alkali process in which carbon electrodes (anodes) were inserted into a brine solution (NaCl) over a pool of liquid Hg, which acted as a cathode. When power was applied to the cell, chlorine gas was captured at the anode. At the cathode, sodium ions were reduced to elemental sodium which formed an amalgam with the mercury (Na/Hg) that subsequently reacted with water to form hydrogen gas and NaOH for pulping. Unfortunately, inorganic Hg contaminated the chlorine gas and NaOH solutions, and thus Hg was lost to wash waters and pulp-mill effluents. Impurities in the brine solutions also reacted with the Hg electrode and reduced its efficiency. The periodic removal and disposal of contaminated Hg in crude landfills became a second source of Hg to soils and surface waters.

The conversion of wood pulp to various products, such as papers for tissues, printing and packaging, is a continuous process in which a flow of pulp is spread and dried in thin sheets and captured on giant rolls of paper. Wastewater from paper making contains new chemical agents needed for specialty products; for example, clay filler is added to create glossy magazine paper. In the past, antimicrobial and antifungal agents were commonly used within the paper mill to limit the growth of microbial and fungal slimes that stain paper products. Mercury-based slimicides, such as phenylmercuric acetate, were the most problematic because they contributed to Hg loadings in effluents and paper products; they have since been replaced by organic compounds.

11.4.1.2 Power Dams: Pulp Mills Need Water

For both pulping and bleaching, immense volumes of water are required even though wash waters are reused in counter-current systems to conserve water. For example, water used to wash almost clean pulp is reused to wash more-contaminated pulp. Nevertheless, kraft mills consume about 100 m^3 of water/ADMT of pulp. For an average-sized mill producing 1,000 ADMT/day, the total water use is 100,000 m^3/day (about 40 Olympic-sized swimming pools), all of which must be treated before release to a river, lake or ocean.

To meet their water needs, pulp mills were typically located near rivers where hydroelectric dams were built to store water and to generate power. The reservoirs ensured a constant supply of fresh water and an inexpensive way to float logs from upstream forested areas and to keep them wet and suitable for pulping during storage. However, the dams caused significant environmental impacts. They changed seasonal thermal and flow regimes of rivers and blocked the migration of fish species to critical habitat such as spawning grounds. Floating logs scoured shallow sediments and riverbanks, and they lost large amounts of bark and wood fibre. Many became waterlogged, destroying habitats for benthic invertebrates and fish by smothering sediments, which became anoxic. Habitats were further degraded by toxic resin and fatty acids that leached from freshly cut logs and from waste bark discharged directly to rivers. Despite these obvious problems, log driving and log storage were not eliminated until the 1990s.

Because sunken logs are preserved for decades in sediments, the impacts on rivers will persist until new sediments accumulate over slowly degrading fibre deposits, or until floods wash the accumulated debris downstream, where it would continue to affect benthic environments. Newer mills do not cause these issues because they draw electricity from the grid and are often located near shipping ports to facilitate the import of wood chips and the export of finished paper.

11.4.1.3 Oxygen Consuming and Toxic Wastes from Wood Pulping

Until the mid-twentieth century, organic wastes from pulp and paper mills were discharged untreated into rivers, lakes and marine coastal waters. In the view of the time, "dilution was the solution to pollution", and adverse effects were "the price of progress". The organic wastes included compounds that were readily degraded by aerobic microbes, as measured by biological oxygen demand (BOD), and compounds that spontaneously reacted with oxygen in water (chemical oxygen demand or COD). The discharge of untreated waste stimulated blooms of aerobic degrader microbes that depleted oxygen concentrations in receiving waters, creating an 'oxygen sag'. These sags were characterized by well-oxygenated water upstream and anoxia (no oxygen) at the site of discharge and for many kilometres downstream. Oxygen concentrations recovered slowly with distance downstream as the waste was 'treated' *in situ* by natural processes and dilution by tributaries, as primary producers (algae) increased in abundance, and as turbulence entrained atmospheric oxygen.

The impacts were dramatic. Healthy aquatic communities of algae, invertebrates and fish were replaced by microbial slimes on benthic substrates, and high densities of sludge worms (*Tubifex tubifex*) that tolerate hypoxia. If not killed by a lack of oxygen, mobile species, including birds and mammals, avoided the affected regions. The quality of receiving water as a source of drinking water was also impaired by foul tastes and odours. Starting in the 1950s and 60s, new or upgraded mills incorporated primary treatment of effluents, followed in the 1980s and 90s by secondary treatment. Early treatments simply mixed alkaline pulping wastes with acidic bleachery wastes to prevent extremes of pH in receiving waters. The addition of primary settling ponds removed grit, woody debris and unrecovered fibre, but did little to remove BOD and COD. However, secondary treatment added aeration and stabilization of effluent, in which oxygen was added continuously to well-stirred ponds or tanks of effluent. The oxygen removed COD and supported communities of aerobic bacteria and protozoans that degraded organic compounds associated with BOD and that caused acute lethality (Table 11.6). Secondary treatment generated CO_2 and biosolids (sludge) which contain the remnants of the microbial biomass and the solids recovered from settling ponds.

When **primary** and **secondary treatment systems** were scaled to suit the size of pulp mills and were operated properly, receiving water ecosystems recovered dramatically as oxygen depletion and acute toxicity were eliminated. Fish kills that had been common downstream of pulp mills, particularly during conditions of low river flow, were avoided. Biosolids recovered from waste treatment were also useful resources to restore the soil organic content of agricultural and forestry lands.

However, the solution of one problem uncovered several others. The restoration of downstream aquatic ecosystems soon revealed that persistent non-chlorinated and chlorinated compounds, including PCDDs and PCDFs, bioaccumulated, biomagnified and impaired fish reproduction (Case Study 4.1; Section 11.4.1.4). These compounds also threatened the health of human consumers of fish, and many commercial and recreational fisheries were closed until chlorine bleaching was eliminated. Contamination of biosolids with chlorinated compounds limited their use as soil amendments, and they were often incinerated and lost as CO_2 emissions. Residual phenolics in effluents also caused taste and odour issues in species harvested for human consumption.

One surprising result of waste treatment was the discovery of extreme concentrations of **retene** (7-isopropyl-1-methylphenanthrene) in sediments of treatment ponds and receiving waters. Retene is an alkyl polynuclear aromatic hydrocarbon produced by the aromatization of abietic acid by anaerobic microbes that flourish in anoxic, organically rich sediments (Figure 11.7). In forested watersheds, aquatic sediments contain about 0.050 mg retene per kg dry weight of sediment, whereas lake sediments affected by pulp-mill effluents may contain up to 3,300 mg/kg dry wt. Retene can partition from contaminated sediments into water at concentrations within its solubility limit (about 16 µg/l), sufficient to cause up to 100-fold induction of cytochrome P4501A (CYP1A) detoxification enzymes in fish and embryotoxicity to fish eggs. Retene mimics the toxicity of dioxin-like molecules to embryos and is phototoxic to any semi-transparent aquatic species exposed to sunlight. Similar concentrations of retene were found in anoxic downstream sediments, reinforcing the concern for the discharge of residual abietic acid and related wood extractives.

A third problem became evident at mills that used highly engineered systems to accelerate effluent treatment and to reduce the area of land needed for treatment ponds. Pulp logs and pulp-mill effluents contain low concentrations of the nitrates and phosphates that are needed by degrader organisms. To obtain rapid waste treatment, fertilizers were added to treatment ponds.

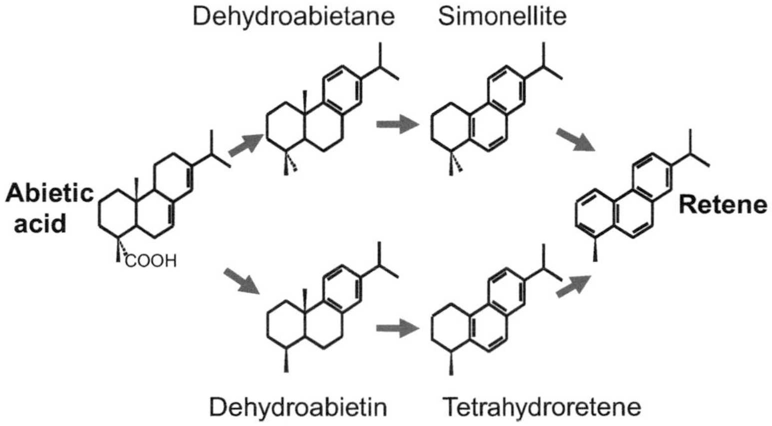

Table 11.6 **Secondary treatment of pulp-mill effluents reduces oxygen demand and acute toxicity.**

	Untreated effluent	Treated effluent
Total organic carbon (mg/l)	200–750	50–100
Total suspended solids (mg/l)	>>100	10–100
Biological oxygen demand (BOD) (mg/l)	270–900	<10
Chemical oxygen demand (COD) (mg/l)	500–2,000	<300
pH (6.8–8.5 = typical surface water)	2.0–11.0	7.0–8.0
Resin acids (mg/l)	0.2–3.5	<0.1
Fatty acids (mg/l)	1–15	0.1–10
Phenols (mg/l)	40–125	10–40
Acute lethality (LC$_{50}$ as % effluent)	<1.0	>100

Figure 11.7 The conversion of abietic acid to retene by anaerobic bacteria. Adapted from Wakeham *et al.* (1980) with permission from Elsevier.

If not recovered by the addition of **tertiary treatment** systems, these nutrients caused eutrophication and renewed oxygen depletion in receiving waters. Tertiary treatment was also needed at some mills where human waste was diverted to the treatment system. Free Cl_2 was added to sterilize the final effluent, followed by chlorine removal to prevent toxicity in the receiving waters. As might be expected, the addition of more 'technological fixes' to address emerging environmental issues increased the energy consumption, CO_2 emissions and cost of pulp production.

11.4.1.4 Toxic Chemicals from Pulp Bleaching

Mercury contamination of predatory fish is one of the best-known and severe effects of pulp bleaching (Section 11.4.1.1) and a significant risk to human and wildlife consumers of fish (Chapter 1). This was a particular issue for isolated communities that rely on local fisheries as a source of food and income, as well as for larger commercial fisheries. For example, the commercial and recreational walleye (*Stizostedion vitreum*) fisheries of Lakes St. Clair and Erie, which form part of the boundary between the United States and Canada, were closed between the 1970s and 1990s because of Hg contamination. It was not the inorganic Hg discharged from chlor-alkali plants that caused toxicity, but rather the methylmercury (CH_3Hg) that was formed by aquatic and soil microbes. As explained in Section 6.8.1.3, the transformation of hydrophilic Hg to lipophilic CH_3Hg facilitates its bioaccumulation and biomagnification through aquatic food webs. Most importantly, CH_3Hg can cross the blood–brain barrier where its neurotoxic effects are expressed. Because sediments downstream of pulp mills are rich in BOD and wood fibre, they foster the microbial communities that methylate Hg, illustrating the complex interactions among the different environmental impacts of pulp mills. Once aquatic ecosystems are contaminated, remediation is difficult, expensive and very slow (Capsule 13.1).

Chlorine bleaching also created new and often unexpected by-products. For example, chlorinated phenols were more toxic to aquatic organisms than their non-chlorinated parent compounds and contributed to taste and odour issues in fisheries products. Similarly, the release of PCDDs and PCDFs, highly toxic to human and wildlife consumers of aquatic species, caused the closure of many freshwater and coastal fisheries, particularly for marine crabs and shellfish. Less obvious effects on fish production were caused by the embryotoxicity of PCDDs and PCDFs accumulated by fish and transferred maternally to developing ova. Embryotoxicity is characterized by induction of CYP1A enzymes (Sections 7.4, 7.5) and by cardiotoxicity, yolk sac and pericardial oedema, craniofacial deformities, spinal curvatures, and a failure to develop to the free-swimming and feeding stage (Rigaud *et al.*, 2014). Adult fish downstream of bleach kraft pulp mills also exhibited CYP1A induction and signs of reproductive impairment, despite increased growth rates due to nutrient enrichment. Reproductive effects included a disrupted regulation of reproductive hormones, reduced size of gonads, feminization or masculinization, intersex condition, and an increased liver size (Case Study 4.1). Embryotoxicity and reproductive impairment reduced the recruitment and abundance of affected fish species, effects that have been observed in field studies in Scandinavia, North and South America, and Australasia. These effects can be mimicked by laboratory chronic toxicity tests of effluents, although only a few of the compounds causing these effects have been identified. In contrast, changes in the abundance and taxon richness of benthic invertebrates downstream of effluent discharges could not be predicted by acute effluent lethality tests.

Both the mercury and chlorine issues were largely resolved after 1990 by implementing alternative bleaching technologies (Table 11.7). Elemental chlorine is no longer used as a bleaching agent, eliminating the risk of Hg contamination from the production of chlorine. Alternative bleaching technologies avoid the reactions that generate chlorinated compounds. To remove residual lignin, pulp is often pre-treated by oxygen delignification under alkaline conditions to facilitate bleaching with milder agents, such as ozone and hydrogen peroxide. Chlorine dioxide is still used as a strong bleaching agent because it does not cause chlorination reactions, but it is highly toxic to workers, requiring that it be produced under strict safety conditions. Nevertheless, contamination by PCDDs and PCDFs remains a problem with recycled paper and cardboard. When stored outdoors prior to collection in large cities, used paper and cardboard accumulate dust containing PCDDs and PCDFs, likely from the combustion of fossil fuels. Fortunately, the majority are the highly chlorinated and least toxic congeners.

Table 11.7 Bleaching alternatives and the production of chlorinated waste products.

Process	Bleaching agents	Characteristics
Chlorine, hypochlorous acid	Cl_2; $HClO$	Very efficient, little fibre breakage; generated chlorinated compounds
Alternatives		
Chlorine dioxide, hypochlorous acid	ClO_2; $HClO$	Efficient; minimal chlorinated compounds; very dangerous
Peroxide	H_2O_2	Less efficient; no chlorinated compounds; can be used alone or in combination with ClO_2
Ozone	O_3	Less efficient; no chlorinated compounds; can be used alone or in combination with ClO_2

11.4.2 Lessons Learned

Many older pulp mills are still in operation. However, they are becoming smaller contributors to environmental problems as they are displaced by newer and larger mills in countries with low-cost production. Improvements to existing technologies have increased the efficiency of converting wood to paper and reduced the generation of wastes. Controls include more efficient carbon recovery in pulping, computer process control to reduce mill 'upsets', oxygen delignification to reduce chemical bleaching, increased water recycling, conversion of waste fibre to 'green' energy and focused treatments for problem waste streams. Nevertheless, applying technological fixes to effluent treatment systems is like peeling an onion: as each layer is removed, new issues are revealed, including complex interactions between past and present problems (Figure 11.8). For each new layer, research is needed to understand cause–effect relationships, to identify the source of the problem, and to develop waste treatment technologies that will resolve the issue. Research since the 1990s has demonstrated that natural components of wood and new compounds created during pulping, bleaching and papermaking are the agents affecting fish reproduction. However, most have not been identified, and toxicity varies widely with the type of wood being pulped, and with pulping, bleaching and waste treatment technologies.

For environmental protection and a competitive industry, repeatedly upgrading effluent production and treatment seems illogical in the face of effluent and ecosystem complexity. To achieve both goals, the future of the industry lies with zero-emissions technologies, such as a bleached chemi-thermo-mechanical pulp (BCTMP) mill in Meadow Lake, Saskatchewan (Canada), the world's first 'closed loop' mill (www.meadowlakepulp.com). It

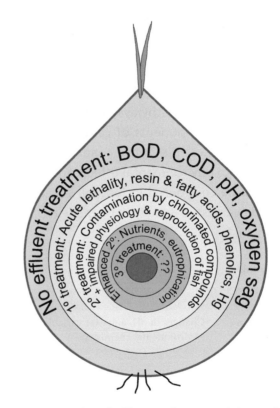

Figure 11.8 The pulp-mill effluent onion. As each layer of the cause–effect onion is peeled away by primary (1^0), secondary (2^0) and tertiary (3^0) treatment, another layer is revealed, progressing from severe ecological damage to subtle effects on fish. Is more effluent treatment the answer, or less effluent?

emits no liquid effluents. All wastewater is treated and recycled, and the carbon recovered is burned for renewable energy. This mill is independent of large supplies of water, buys its wood from suppliers that apply sustainable forest management and produces white pulp without the environmental complexities of chemical emissions and waste treatment.

Summary

In the four scenarios described in Sections 11.1 to 11.4, complexity is manifest in many different ways. For acidification, the single factor of an increase in the hydrogen ion concentration results in a wide array of geochemical and biochemical changes. The remarkable feature of acidification is the breadth and diversity of effects on aquatic and terrestrial ecosystems, and the interactions with other chemical stressors such as metals. In contrast, the complex effects of emissions from metal mining and smelting illustrate how the exploitation of a natural mineral resource can generate a range of potential stressors (gaseous, dissolved and particulate), some of which are infinitely persistent (the metals themselves) and can lead to legacy effects. The nanomaterials example serves as a useful reminder of the importance of the physical state and size of a potential stressor, and how these properties may change when a substance enters the receiving environment. In the review of engineered nanomaterials, we considered the implications of moving from the macro- to the nanoscale and explored the concept of nanotoxicity. Finally, pulp and paper processing follows a cradle-to-grave sequence, but one that reflects the evolution of the technology for producing paper. This example also illustrates our progressive understanding of the increasingly subtle effects of sublethal toxicity that are revealed when effluents are no longer acutely lethal.

Adverse outcome pathways (Section 5.9) imply linear progressions from molecular interactions between chemicals and components of cells to impairment of the health and performance of exposed species. In contrast, each example in this chapter involves non-linear networks of interactions that link chemical fate and toxicity to environmental effects and ecological impacts. Other complex issues reviewed in this textbook that could be considered include oil spills (Section 1.2.5) and endocrine disrupting chemicals (Section 8.6). Each and every issue is likely to be made more complex with additional stresses resulting from climate change (Section 14.1).

Summary table

Case	Contributors to complexity		
Acidification	Changes in pH, a master variable in environmental chemistry	Airborne transport of contaminants (across borders)	Interactions of the H^+ ion with metals, essential (Ca) and non-essential (Al, Cd, Hg), and with epithelia
Mining and smelting	Infinite lifetime of metals; legacy effects of past mining activities	Airborne transport of contaminants (across borders)	Food-web-mediated effects of metals
Engineered nanomaterials	Effect of physical size on the toxicity of particles	Analytical difficulties in detecting ENMs in environmental matrices	Extensive transformations of ENMs in receiving environment
Pulp and paper	Effects of process chemicals (Hg^0; Cl_2) and production of new chemicals (e.g., chlorinated organics; retene)	Effects of chemicals present in natural wood (resin acids, phyto-oestrogens)	Evolution from gross pollution (BOD, O_2) and acute lethality to subtle effects on sexual maturation and reproduction (onion model)

Complex issues such as those analysed in the present chapter should be subjected to environmental risk assessment (ERA) before any regulatory action is taken. Indeed, as discussed in Chapter 12, some countries' industries such as mining and smelting, and pulp and paper, are already regulated as a whole, not just on the basis of the individual chemicals that they emit.

Review Questions and Exercises

. . . .

1. What is the basis of the observation that "singly operating stresses may actually be quite rare"?

2. List the chemical changes related to acidification in freshwater aquatic and terrestrial systems. Include an explanation of the mechanism(s) for each of the changes that you list.

3. How may acidification increase water clarity?

4. What are the effects on freshwater biota of a decrease in pH? Include the effects from the population to ecosystem level, where known.

5. Describe the link(s) between terrestrial acidification and the receiving fresh waters.

6. What changes have been observed in terrestrial and aquatic systems following abatement of emissions of acid precursors?

7. Following reduction of SO_2 emissions, why does chemical recovery from acidification tend to occur faster than biological recovery?

8. Explain the mechanism underlying ocean acidification and discuss the relationship with global warming.

9. Why do we have limited understanding of the biological/ecological effects of ocean acidification?

10. Explain how mining and smelting activities can affect aquatic and terrestrial ecosystems located up to 100 km from the mine site.

11. Why does the exploitation of sulphidic ore bodies pose a particular problem, and how can this problem be minimized?

12. Why is the disposal of mine tailings into rivers (as was done in the Clark Fork River and Ok Tedi examples) particularly problematic?

13. How can lake sediments be used to uncover past missteps in the handling of mine-generated wastes?

14. Explain what is meant by the term 'legacy effects', as it applies to metal mining and smelting.

15. What is the difference between a colloid and a nanoparticle?

16. What are the major routes by which engineered nanomaterials enter the receiving environment?

17. When a nanoparticle enters the receiving environment, what transformations are likely to occur?

18. How might a nanoparticle enter a living organism? Be careful here to distinguish between simple ingestion and true bioaccumulation.

19. How might a nanoparticle affect a living organism?

20. Explain what is meant by the term 'nanotoxicity'.

21. During wood pulping and papermaking, what are the sources of chemicals that are removed to wastewater?

22. When bleached kraft pulp-mill effluents are discharged, what types of effects on aquatic ecosystems are associated with untreated effluents? Primary treated effluents? Secondary treated effluents?

23. What effects on ecosystem services might be expected following the discharge of effluents from pulp mills that bleach wood fibre with chlorine?

24. How do the natural organic components of wood extracted during wood pulping harm aquatic species? Provide three examples of sublethal effects and the compounds that might cause them.

25. What are the limitations to more sophisticated and efficient effluent treatment technologies? Given these limitations, what do you think is the most logical way to prevent the environmental impacts of pulp and paper production?

Abbreviations

ADMT	Air-dried metric tonnes	IPCC	Intergovernmental Panel on Climate Change
AgNP	Silver nanoparticle		
BCE	Before the Common Era	LRTAP	Long-range transport of air pollutants
BOD	Biological oxygen demand	NO_X	Oxides of nitrogen (NO_2 = nitrogen dioxide; $NO^\bullet$ = nitric oxide)
COD	Chemical oxygen demand		
CYP1A	Cytochrome P4501A	OA	Ocean acidification
DIC	Dissolved inorganic carbon (carbonates)	OECD	Organisation for Economic Cooperation and Development
DOM	Dissolved organic matter	PCDDs/PCDFs	Polychlorinated dibenzo(p)dioxins and dibenzofurans
ECCC	Environment and Climate Change Canada		
ENM	Engineered nanomaterial	RNA-seq	RNA-sequencing
ICMM	International Council on Mining and Minerals	ROS	Reactive oxygen species
		SO_2	Sulphur dioxide
IISD	International Institute for Sustainable Development	UV	Ultraviolet radiation
		UVB	Ultraviolet radiation, 280–315 nm

References

Allen, J., Maunoury-Danger, F., Felten, V., *et al.* (2020). Liming of acidified forests changes leaf litter traits but does not improve leaf litter decomposability in forest streams. *Forest Ecology and Management*, 475, e118431.

Apte, S. C. (2008). Biogeochemistry of copper in the Fly River. In B. Bolton (ed.) *Developments in Earth and Environmental Sciences, The Fly River, Papua New Guinea: Environmental Studies in an Impacted Tropical River System*. Elsevier, 321–373.

Apte, S. C., Benko, W. I. & Day, G. M. (1995). Partitioning and complexation of copper in the Fly River, Papua New Guinea. *Journal of Geochemical Exploration*, 52, 67–79.

Arnott, S. E. & Yan, N. D. (2002). The influence of drought and re-acidification on zooplankton emergence from resting stages. *Ecological Applications*, 12, 138–153.

Azimzada, A., Jreije, I., Hadioui, M. *et al.*, (2021). Quantification and characterization of Ti-, Ce-, and Ag-nanoparticles in global surface waters and precipitation. *Environmental Science & Technology*, 55, 9836–9844.

Batley, G. E., Kirby, J. K. & McLaughlin, M. J. (2013). Fate and risks of nanomaterials in aquatic and terrestrial environments. *Accounts of Chemical Research*, 46, 854–862.

Borgmann, U. & Norwood, W. P. (2002). Metal bioavailability and toxicity through a sediment core. *Environmental Pollution*, 116, 159–168.

Breitbarth, E., Achterberg, E. P., Ardelan, M., *et al.* (2010). Iron biogeochemistry across marine systems–progress from the past decade. *Biogeosciences*, 7, 1075–1097.

Brezonik, P. L., Eaton, J. G., Frost, T. M., *et al.* (1993). Experimental acidification of Little Rock Lake, Wisconsin: chemical and biological changes over the pH range 6.1 to 4.7. *Canadian Journal of Fisheries and Aquatic Sciences*, 50, 1101–1121.

Burd, B. J. (2002). Evaluation of mine tailings effects on a benthic marine infaunal community over 29 years. *Marine Environmental Research*, 53, 481–519.

Burns, D. A., Aherne, J., Gay, D. A. & Lehmann, C. M. B. (2016). Acid rain and its environmental effects: recent scientific advances. *Atmospheric Environment*, 146, 1–4.

Campbell, P. G. C., Hontela, A., Rasmussen, J. B., *et al.* (2003). Differentiating between direct (physiological) and food-chain mediated (bioenergetic) effects on fish in metal-impacted lakes. *Human and Ecological Risk Assessment*, 9, 847–866.

Campbell, P. G. C., Kraemer, L. D., Giguère, A., Hare, L. & Hontela, A. (2008). Subcellular distribution of cadmium and nickel in chronically exposed wild fish: inferences regarding metal detoxification strategies and implications for setting water quality guidelines for dissolved metals. *Human and Ecological Risk Assessment*, 14, 290–316.

Couture, P. & Pyle, G. (2008). Live fast and die young: metal effects on condition and physiology of wild yellow perch from along two metal contamination gradients. *Human and Ecological Risk Assessment*, 14, 73–96.

Deppeler, S. L. & Davidson, A. T. (2017). Southern Ocean phytoplankton in a changing climate. *Frontiers in Marine Science*, 4, e40.

Dixit, A. S., Alpay, S., Dixit, S. S. & Smol, J. P. (2007). Paleolimnological reconstructions of Rouyn-Noranda lakes within the zone of influence of the Horne Smelter, Quebec, Canada. *Journal of Paleolimnology*, 38, 209–226.

Dixit, S. S., Dixit, A. S. & Evans, R. D. (1989). Paleolimnological evidence for trace-metal sensitivity in scaled chrysophytes. *Environmental Science & Technology*, 23, 110–115.

Domingos, R. F., Simon, D. F., Hauser, C. & Wilkinson, K. J. (2011). Bioaccumulation and effects of CdTe/CdS quantum dots on *Chlamydomonas reinhardtii* – Nanoparticles or the free ions? *Environmental Science & Technology*, 45, 7664–7669.

Doney, S. C., Fabry, V. J., Feely, R. A. & Kleypas, J. A. (2009). Ocean acidification: the other CO_2 problem. *Annual Review of Marine Science*, 1, 169–192.

Driscoll, C. T., Driscoll, K. M., Mitchell, M. J. & Raynal, D. J. (2003). Effects of acidic deposition on forest and aquatic ecosystems in New York State. *Environmental Pollution*, 123, 327–336.

European Commission (2017). *NanoData Landscape Compilation – Environment*. Luxembourg: European Commission, Directorate-General for Research and Innovation.

Freedman, B. & Hutchinson, T. C. (1980). Long-term effects of smelter pollution at Sudbury, Ontario, on forest community composition. *Canadian Journal of Botany – Revue Canadienne de Botanique*, 58, 2123–2140.

Frost, T. M., Montz, P. K., Kratz, T. K., *et al.* (1999). Multiple stresses from a single agent: diverse responses to the experimental acidification of Little Rock Lake, Wisconsin. *Limnology and Oceanography*, 44, 784–794.

Gao, K., Beardall, J., Häder, D. P., *et al.* (2019). Effects of ocean acidification on marine photosynthetic organisms under the concurrent influences of warming, UV radiation and deoxygenation. *Frontiers in Marine Science*, 6, e322.

Gorham, E., Martin, F. B. & Litzau, J. T. (1984). Acid rain: ionic correlations in the Eastern United States, 1980-1981. *Science*, 225, 407–409.

Gunn, J. M., Snucins, E. D., Yan, N. D. & Arts, M. T. (2001). Use of water clarity to monitor the effects of climate change and other stressors on oligotrophic lakes. *Environmental Monitoring and Assessment*, 67, 69–88.

Handy, R. D., Cornelis, G., Fernandes, T., *et al.* (2012). Ecotoxicity test methods for engineered nanomaterials: Practical experiences and recommendations from the bench. *Environmental Toxicology and Chemistry*, 31, 15–31.

Hayhurst, L. D., Martin, J. D., Wallace, S. J., *et al.* (2020). Multi-level responses of yellow perch (*Perca flavescens*) to a whole-lake nanosilver addition study. *Archives of Environmental Contamination and Toxicology*, 79, 283–297.

Hessen, D. O. & Rukke, N. A. (2000). UV radiation and low calcium as mutual stressors for *Daphnia*. *Limnology and Oceanography*, 45, 1834–1838.

Hoffmann, L. J., Breitbarth, E., Boyd, P. W. & Hunter, K. A. (2012). Influence of ocean warming and acidification on trace metal biogeochemistry. *Marine Ecology Progress Series*, 470, 191–205.

Hrabik, T. R. & Watras, C. (2002). Recent declines in mercury concentration in a freshwater fishery: isolating the effects of de-acidification and decreased atmospheric mercury deposition in Little Rock Lake. *Science of the Total Environment*, 297, 229–237.

Huang, J., Zhang, W., Mo, J. M., *et al.* (2015a). Urbanization in China drives soil acidification of *Pinus massoniana* forests. *Scientific Reports*, 5, e13512.

Huang, Y. M., Kang, R. H., Mulder, J., Zhang, T. & Duan, L. (2015b). Nitrogen saturation, soil acidification, and ecological effects in a subtropical pine forest on acid soil in southwest China. *Journal of Geophysical Research – Biogeosciences*, 120, 2457–2472.

Hurd, C. L., Beardall, J., Comeau, S., *et al.* (2020). Ocean acidification as a multiple driver: how interactions between changing seawater carbonate parameters affect marine life. *Marine and Freshwater Research*, 71, 263–274.

Jeevanandam, J., Barhoum, A., Chan, Y. S., Dufresne, A. & Danquah, M. K. (2018). Review on nanoparticles and nanostructured materials: history, sources, toxicity and regulations. *Beilstein Journal of Nanotechnology*, 9, 1050–1074.

Jeziorski, A., Yan, N. D., Paterson, A. M., *et al.* (2008). The widespread threat of calcium decline in fresh waters. *Science*, 322, 1374–1377.

Keller, W. (1992). Introduction and overview to aquatic acidification studies in the Sudbury, Ontario, Canada, area. *Canadian Journal of Fisheries and Aquatic Sciences*, 49, 3–7.

Keller, W., Yan, N. D., Gunn, J. M. & Heneberry, J. (2007). Recovery of acidified lakes: lessons from Sudbury, Ontario, Canada. *Water, Air and Soil Pollution: Focus*, 7, 317–322.

Kroglund, F., Rosseland, B. O., Teien, H.-C., *et al.* (2008). Water quality limits for Atlantic salmon (*Salmo salar* L.) exposed to short term reductions in pH and increased aluminum simulating episodes. *Hydrology and Earth System Sciences*, 12, 491–507.

Lawrence, G. B., Burns, D. A. & Riva-Murray, K. (2016). A new look at liming as an approach to accelerate recovery from acidic deposition effects. *Science of the Total Environment*, 562, 35–46.

Lawrence, G. B., Hazlett, P. W., Fernandez, I. J., *et al.* (2015). Declining acidic deposition begins reversal of forest-soil acidification in the Northeastern U.S. and Eastern Canada. *Environmental Science & Technology*, 49, 13103–13111.

Lead, J. R., Batley, G. E., Alvarez, P. J. J., *et al.* (2018). Nanomaterials in the environment: behavior, fate, bioavailability, and effects – an updated review. *Environmental Toxicology and Chemistry*, 37, 2029–2063.

Mackey, K. R. M., Morris, J. J., Morel, F. M. M. & Kranz, S. A. (2015). Response of photosynthesis to ocean acidification. *Oceanography*, 28, 74–91.

Martin, J. D., Frost, P. C., Hintelmann, H., *et al.* (2018). Accumulation of silver in yellow perch (*Perca flavescens*) and northern pike (*Esox lucius*) from a lake dosed with nanosilver. *Environmental Science & Technology*, 52, 11114–11122.

Moore, J. N. & Luoma, S. N. (1990). Hazardous wastes from large-scale metal extraction – a case-study. *Environmental Science & Technology*, 24, 1278–1285.

Navarro, E., Piccapietra, F., Wagner, B., *et al.* (2008). Toxicity of silver nanoparticles to *Chlamydomonas reinhardtii*. *Environmental Science & Technology*, 42, 8959–8964.

Nel, A. E., Maedler, L., Velegol, D., *et al.* (2009). Understanding biophysicochemical interactions at the nano–bio interface. *Nature Materials*, 8, 543–557.

Notter, D. A., Mitrano, D. M. & Nowack, B. (2014). Are nanosized or dissolved metals more toxic in the environment? A meta-analysis. *Environmental Toxicology and Chemistry*, 33, 2733–2739.

Nowack, B. (2017). Evaluation of environmental exposure models for engineered nanomaterials in a regulatory context. *NanoImpact*, 8, 38–47.

Nriagu, J. O. & Pacyna, J. M. (1988). Quantitative assessment of worldwide contamination of air, water and soils by trace metals. *Nature*, 333, 134–139.

Olías, M., Cánovas, C. R., Macías, F., Basallote, M. D. & Nieto, J. M. (2020). The evolution of pollutant concentrations in a river severely affected by acid mine drainage: Río Tinto (SW Spain). *Minerals*, 10, e10070598.

Pachauri, R. K., Allen, M. R., Barros, V. R., *et al.* (2014). *Climate Change 2014: synthesis report. Contribution of Working Groups I, II and III to the Fifth Assessment Report of the Intergovernmental Panel on Climate Change*. Geneva, Switzerland: IPCC.

Pulido-Reyes, G., Leganes, F., Fernandez-Pinas, F. & Rosal, R. (2017). Bio-nano interface and environment: a critical review. *Environmental Toxicology and Chemistry*, 36, 3181–3193.

Pyle, G. G., Rajotte, J. W. & Couture, P. (2005). Effects of industrial metals on wild fish populations along a metal contamination gradient. *Ecotoxicology and Environmental Safety*, 61, 287–312.

Raddum, G. G. & Fjellheim, A. (2002). Species composition of freshwater invertebrates in relation to chemical and physical factors in high mountains in soutwestern Norway. *Water, Air and Soil Pollution: Focus*, 2, 311–328.

Rainbow, P. S. (2018). Metals and Mining. *Trace Metals in the Environment and Living Organisms – The British Isles as a Case Study*. Cambridge, UK: Cambridge University Press, pp. 22–67.

Rajotte, J. W. & Couture, P. (2002). Effects of environmental metal contamination on the condition, swimming performance, and tissue metabolic capacities of wild yellow perch (*Perca flavescens*). *Canadian Journal of Fisheries and Aquatic Sciences*, 59, 1296–1304.

Ramirez-Llodra, E., Trannum, H. C., Evenset, A. *et al.* (2015). Submarine and deep-sea mine tailing placements: a review of current practices, environmental issues, natural analogs and knowledge gaps in Norway and internationally. *Marine Pollution Bulletin*, 97, 13–35.

Rasmussen, J. B., Gunn, J. M., Iles, A., *et al.* (2008). Direct and indirect (foodweb-mediated) effects of metal exposure on the growth of yellow perch (*Perca flavescens*): implications for ecological risk assessment. *Human and Ecological Risk Assessment*, 14, 317–350.

Reichl, C. & Schatz, M. (2021). *World Mining Data 2021*. Vienna, Austria: Federal Ministry of Agriculture, Regions and Tourism.

Riebesell, U., Bellerby, R. G. J., Grossart, H.-P. & Thingstad, F. (2008). Mesocosm CO_2 perturbation studies: from organism to community level. *Biogeosciences*, 5, 1157–1164.

Rigaud, C., Couillard, C. M., Pellerin, J., Legare, B. & Hodson, P. V. (2014). Applicability of the TCDD-TEQ approach to predict sublethal embryotoxicity in *Fundulus heteroclitus*. *Aquatic Toxicology*, 149, 133–144.

Roberts, D. A., Birchenough, S. N. R., Lewis, C., *et al.* (2013). Ocean acidification increases the toxicity of contaminated sediments. *Global Change Biology*, 19, 340–351.

Rogers, N. J., Franklin, N. M., Apte, S. C., *et al.* (2010). Physico-chemical behaviour and algal toxicity of nanoparticulate CeO_2 in freshwater. *Environmental Chemistry*, 7, 50–60.

Schindler, D. W. (1988). Effects of acid rain on freshwater ecosystems. *Science*, 239, 149–157.

Schultz, A. G., Boyle, D., Chamot, D., *et al.* (2014). Aquatic toxicity of manufactured nanomaterials: challenges and recommendations for future toxicity testing. *Environmental Chemistry*, 11, 207–226.

Schultz, C., Powell, K., Crossley, A., *et al.* (2015). Analytical approaches to support current understanding of exposure, uptake and distributions of engineered nanoparticles by aquatic and terrestrial organisms. *Ecotoxicology*, 24, 239–261.

Stauber, J. L., Apte, S. C. & Rogers, N. J. (2008). Speciation, bioavailability and toxicity of copper in the Fly River System. In B. Bolton (ed.) *Developments in Earth and Environmental Sciences. The Fly River, Papua New Guinea: Environmental Studies in an Impacted Tropical River System*. Amsterdam, The Netherlands: Elsevier, 375–408.

Stern, S. T., Adiseshaiah, P. P. & Crist, R. M. (2012). Autophagy and lysosomal dysfunction as emerging mechanisms of nanomaterial toxicity. *Particle and Fibre Toxicology*, 9, e20.

Storey, A. W., Yarrao, M., Tenakanai, C., Figa, B. & Lynas, J. (2008). Use of changes in fish assemblages in the Fly River System, Papua New Guinea, to assess effects of the Ok Tedi copper mine. In B. Bolton (ed.) *Developments in Earth and Environmental Sciences. The Fly River, Papua New Guinea: Environmental Studies in an Impacted Tropical River System*. Amsterdam, The Netherlands: Elsevier, 427–462.

Swales, S., Storey, A. W., Roderick, I. D., *et al.* (1998). Biological monitoring of the impacts of the Ok Tedi copper mine on fish populations in the Fly River system, Papua New Guinea. *Science of the Total Environment*, 214, 99–111.

UNEP (2013). *Environmental Risks and Challenges of Anthropogenic Metals Flows and Cycles*. Nairobi, Kenya.

Van Geen, A., Adkins, J. F., Boyle, E. A., Nelson, C. H. & Palanques, A. (1997). A 120 yr record of widespread contamination from mining of the Iberian pyrite belt. *Geology*, 25, 291–294.

Wakeham, S. G., Schaffner, C. & Giger, W. (1980). Polycyclic aromatic-hydrocarbons in recent lake-sediments. 2. Compounds derived from biogenic precursors during early diagenesis. *Geochimica et Cosmochimica Acta*, 44, 415–429.

Wang, W. & Xu, P. (2009). Research progress in precipitation chemistry in China. *Progress in Chemistry*, 21, 266–281.

Wannicke, N., Frey, C., Law, C. S. & Voss, M. (2018). The response of the marine nitrogen cycle to ocean acidification. *Global Change Biology*, 24, 5031–5043.

Wright, R. & Schindler, D. (1995). Interaction of acid rain and global changes: effects on terrestrial and aquatic ecosystems. *Water, Air, and Soil Pollution*, 85, 89–99.

Wu, Y. L., Putcha, N., Ng, K. W., *et al.* (2013). Biophysical responses upon the interaction of nanomaterials with cellular interfaces. *Accounts of Chemical Research*, 46, 782–791.

Yang, Y. H., Ji, C. J., Ma, W. H., *et al.* (2012). Significant soil acidification across northern China's grasslands during 1980s–2000s. *Global Change Biology*, 18, 2292–2300.

Zeebe, R. E. (2012). History of seawater carbonate chemistry, atmospheric CO_2, and ocean acidification. *Annual Review of Earth and Planetary Sciences*, 40, 141–165.

IV

Management

12 Regulatory Toxicology and Ecological Risk Assessment

Peter V. Hodson, Pamela M. Welbourn and Peter G. C. Campbell

Learning Objectives

Upon completion of this chapter, the student should be able to:

- Describe how governments translate environmental concerns into regulatory actions for environmental protection, nationally and internationally.
- Understand the language and regulatory tools that governments apply to manage chemical emissions, environmental contamination and ecological effects.
- Discuss the role of ecotoxicology in regulating contaminants, including the development of environmental quality objectives, criteria, guidelines and standards, ecological risk assessment and environmental monitoring.
- Explain how standards and regulations are enforced and how environmental monitoring can assess the success of regulations.
- Discuss the obstacles to management of chemicals and describe how some industries are moving to self-regulation.

In Chapter 1, we learned that ecotoxicology includes both basic and applied science, developed in response to environmental problems caused by chemical contaminants. In Chapter 12, we explore how **ecotoxicology** is translated into environmental protection through regulatory actions by various levels of government. More broadly, international agreements foster global controls on emissions to protect individuals and nations from the transboundary movement and the global distribution of contaminants in air, water and commerce.

12.1 The Need for Chemical Management and Regulation

Throughout this textbook, we have learned about the toxicity of different compounds and their effects on natural ecosystems, mediated by their environmental behaviour and distribution. Chapter 1 traces the emergence of ecotoxicology as a science, beginning with Paracelsus' observations relating **toxicant** exposure (dose) to biological responses, and it highlights Rachel Carson's role in alerting the public to the ecological impacts of **pesticide** use. Subsequent chapters reinforce the need to understand the environmental impacts of substances associated with issues such as acid rain, **endocrine disruption**, mine wastes and oil spills, and Chapter 14 identifies emerging issues of novel contaminants such as micro- and nanoplastics.

Ideally, our understanding of ecotoxicology should be transformed into actions to reduce chemical emissions, remediate contaminated environments and restore ecosystem structure and function. Such actions, expressed as policies, laws, regulations and enforcement, are primarily the responsibility of governments, which by their very nature have the power and mandate to take initiatives on our behalf. However, this does not absolve us, as individuals or as organizations, from taking responsibility for our own use, disposal and management of chemicals. Indeed, there are many examples of citizen- and industry-led programmes to 'green' the economy (Section 12.4). As illustrated in Figure 1.3, public awareness and pressure are effective in compelling governments to develop laws and regulatory programmes to manage contamination. In the final analysis, who is responsible? The answer is: any and all of individuals, society, scientists, governments and industries.

The development of regulations requires that we translate technical knowledge about the sources, distribution and ecological effects of chemicals into:

- **environmental quality criteria** and **guidelines**. These are recommended limits to chemical emissions and concentrations in environmental media to protect ecosystem structure and function and the quality of food for humans and livestock;
- enforceable **standards** that limit emissions in solid, liquid and gaseous wastes;
- requirements for the safe storage, use and disposal of chemicals;
- controls on the use and application rates of agrochemicals such as pesticides;
- bans on the production and use of compounds for which environmental risks are unacceptable; and
- international conventions and trade agreements to limit the use, disposal and long-range transport of compounds that cause global impacts.

The development and application of these regulations, and the role of ecotoxicology in supporting regulation, are described in Sections 12.2 and 12.3, with a particular emphasis on **ecological risk assessment (ERA)**.

12.2 Legislation for Chemical Management

Environmental law is relatively new in legal history (UNEP, 2019), and is still evolving, with many **statutes** dating from the 1970s; by 2017, 176 countries had enacted framework laws for environmental protection. In some cases, existing **regulations** or **common law** are applied to a new issue or a specific case, which is often more expedient than waiting for new statutes to be enacted (Box 12.1).

An example of existing **legislation** with a broad mandate (**enabling legislation**) adapted to the management of pollution is Canada's Federal Fisheries Act. Following Canada's Confederation (1867), the Fisheries Act (1868) was created to support the new Department of Marine and

Box 12.1 Common law versus statutory law

- Common law is typically used by individual plaintiffs in response to injuries to their own property. It is largely based on precedents derived from court decisions about disputes related to negligence, **nuisance** or strict liability.
- Statutory law is a formal expression of a rule or set of rules established by a particular authority, such as a national government.

Fisheries. A key section prohibited the disposal of **fish offal** (remains of fish processed for food) from fish-packing plants into small harbours. Fish offal caused organic pollution, objectionable odours, unsightly scums, oxygen depletion and the destruction of marine life. To resolve this issue, a section of the Act prohibited "the deposit of **substances deleterious** to fish in waters frequented by fish" (www.canada.ca/en/environment-climate-change/services/managing-pollution/effluent-regulations-fisheries-act.html). Convictions only required evidence that the accused had control of the substance, that the substance was deleterious (harmful) to fish, and that fish would normally be present where the substance was deposited or discharged. When concerns about chemical pollution arose in the twentieth century, the simple wording and concepts of this prohibition provided a powerful tool for regulating effluents and spills. For more comprehensive applications, a deleterious substance is now defined as:

. . . any substance that, if added to any water, would degrade or alter or form part of a process of degradation or alteration of the quality of that water so that it is rendered or is likely to be rendered deleterious to fish or fish habitat or to the use by man of fish that frequent that water.

Enforcement of the sections of the Fisheries Act relevant to contaminants is shared among Fisheries and Oceans, Environment and Climate Change Canada, and the Canadian Food Inspection Agency.

Box 12.2 lists ways in which emissions are regulated. Many regulations are based on recommended numerical limits to chemical concentrations in environmental media (criteria, guidelines, standards; Section 12.3.1). Concentrations above the recommended limits are likely to cause ecological impacts, and the limits provide a baseline for assessing whether emissions conform to the regulations.

Box 12.2 Examples of regulatory measures to limit chemical emissions and effects

Prescribed releases of wastes (e.g., industrial effluents)

- Effluent discharge limits (concentrations, amounts, toxicity)

Accidental releases (e.g., oil spills)

- Regulations controlling the transport, storage and use of hazardous products (e.g., gasoline)
- Shipping regulations (e.g., vessel construction; navigation equipment)
- Marine traffic management (speed limits, shipping lanes, pilots, training standards)
- Prescribed spill response and clean-up measures and standards

Deliberate releases (e.g., pesticides; cleaning products)

- Mandatory assessment of environmental fate, effects and toxicity before product approval
- Regulations prescribing labelling, application rates, application methods

Legacy chemicals (e.g., PBT: persistent, bioaccumulative and toxic)

- Limits or bans on production, use or emissions (virtual elimination)
- Concentration limits in food (e.g., **methylmercury**)
- Contaminated site assessment and **remediation** standards

12.2.1 The Process of Regulation

Figure 1.3 (Chapter 1) suggests that regulations do not arise as a direct result of scientific research. Rather, they represent a progression from public concerns (political pressure) to government statements of policy and ultimately to political action expressed as laws and regulations. Translating environmental policy into laws and regulations can be a lengthy process involving risk assessors, government agencies, legislators, public officials and stakeholders.

12.2.1.1 Policy

A policy is an expression of the principles or plans agreed upon by a government or other body, related to a specific situation or subject. Environmental policy refers to any action or intent to prevent, reduce or mitigate harmful effects of human activities on the environment and natural resources. Although environmental policy broadly reflects public opinion, it is also influenced by activists and the media and by scientists when they communicate their findings to the general public. In fact, the very choice of a research topic by one or more scientists may alert the public to an environmental problem that has become serious and warrants policy development. Although policies are aspirational and do not of themselves have legal implications, they often lead to or precede new laws and influence the implementation of related laws.

12.2.1.2 Legislation

Legislation is defined as a law or set of laws proposed by local, regional or national governments and made official by a vote of elected representatives. Responsibility for environmental law is frequently predetermined for a specific level of government by the constitution of a country, with appropriate ministries or departments having jurisdiction over respective areas (Section 12.2.1.4). Legislation for environmental protection typically includes the regulation of polluting emissions, prohibition or taxation of activities that threaten the environment and human health, alternative solutions such as carbon emissions trading and environmental permitting, and requirements for environmental impact assessments.

There are variations among jurisdictions in the process of law making. In most, a governing party prepares and proposes legislation that is examined in detail by legislative committees that propose amendments, often with input from affected industries, technical experts and grassroots advocates. However, not all policy is translated into new laws. If flaws are found in a proposed law or if public support (**social licence**) is not widespread, bills to pass the law may be voted down, forcing a government either to accept further amendments, or to drop the proposed legislation and start again. Once a law has been passed, implementation is the next step, including the development of detailed regulations.

12.2.1.3 Regulations

A regulation is a rule within a law that is vital to its implementation and that specifies how the provisions of the law are to be translated into action (e.g., limits to chemical emissions). Regulations provide some flexibility because they can be modified without changing the overarching law. Governments are often required to seek feedback on new regulations through public comment, which provides us all with an opportunity to influence how laws are applied in our communities.

12.2.1.4 Departmental Responsibilities and Options for Chemical Management

Given the diversity of chemical sources and the complexity of ecological interactions with chemical fate and effects, it is not surprising that managing contaminant issues requires interventions from multiple government agencies. The number involved reflects the complexity of economies (the source), of ecosystems that are contaminated, and of responsibilities for local (municipalities), regional (states, cantons, provinces), national and international issues. Depending on the history and form of government in any country, the division of responsibilities among different government departments can vary considerably. Nevertheless, a common feature is an environment department or ministry responsible for implementing statutes to control contamination, enforce regulations on emissions, and monitor the extent of environmental contamination and effects. Ministries of health and consumer protection, natural resources and agriculture regulate the concentrations of substances in the human food chain, while ministries of transport define safe modes of transport and measures for spill response. Although environmental agencies might define the amounts and forms of compounds that can be imported or exported, border security and public safety departments administer these regulations and monitor trade. When regulations are violated, the appropriate departments collaborate with departments of justice to lay charges, collect evidence and assist in court proceedings (Section 12.3.4). It is clear that successful management requires coordination among departments and statutes framed to recognize and authorize their unique and overlapping roles.

In keeping with the principle of 'the polluter pays', governments also regulate some contaminants by assigning responsibilities to the private sector. For example, companies that own and ship crude or refined oil may be required to contribute a small fee per barrel to an oil spill response fund. The purpose is to develop and maintain a capacity to respond to an oil spill, for example an organization formed to clean up spills under the supervision of a nation's Coast Guard.

To manage existing and emerging compounds of concern, data on toxicity, environmental fate and distribution and ecological effects are needed. Thus, environment departments often include laboratories that provide **toxicity tests**, chemical analyses and research to support **monitoring** and enforcement of regulations, although some of these services can be provided by commercial laboratories (Section 12.3.4). Other units within environment departments would develop the standards and guidelines needed to manage contamination of air, water and soil (Section 12.3.1) and apply methods for ERAs (Section 12.3.2).

12.2.2 International Law and Multilateral Agreements

The term **International Law** was coined by the English philosopher Jeremy Bentham (1748–1832) to describe a system of principles and rules that govern relationships among sovereign states and other internationally recognized parties. International Environmental Law (IEL) (Section 1.4) is exemplified by treaties, conventions and agreements (Table 1.2) that respond to scientific knowledge about the current state of our global environment. Examples of relevant and important IELs include the Montreal Protocol (1985), Stockholm Convention (2001) and **Minamata Convention** (2013), all of which were preceded by major international science meetings, such as the Rio de Janeiro Earth Summit (1992). With the signing of the Paris Agreement in 2016, a new global initiative was formulated to confront our most important environmental challenge, climate change (Section 14.1).

Some IELs are binding, while others are voluntary because they are not typically administered by any single international authority. Instead, they represent collective efforts to manage the environment at a level beyond individual nations. One notable exception is the International Seabed Authority, established in 1994 by the United Nations, which has authority over all mineral-related activities in the seabed that lie outside national jurisdictions (Case Study 14.1).

International agreements by their very nature are difficult to administer and enforce, but some have proved effective. For example, the Minamata Convention, signed by 128 parties, came into force in 2017. It advocates a life-cycle approach to managing the production, use, emissions, handling and disposal of mercury. To implement the Convention, global cooperation and coordination are critical to the collection and analysis of the technical information needed to assess the effectiveness of Convention measures and to support management decisions (Evers *et al.*, 2016; Selin *et al.*, 2018). Since the recognition of anthropogenic mercury as an environmental contaminant in the 1970s, most industrialized countries have successfully reduced contamination through guidelines or standards for environmental media and legislation that limits uses and emissions. For example, the last chlor-alkali mercury cell in Canada (Section 11.4.1) was closed in 2008, and the use of mercury in measuring instruments was restricted by the US EPA in 2011. Under the Minamata Convention, limits on mercury use in dental amalgams were established on 1 July 2018. Nevertheless, the disposal of phased-out products containing mercury still represents a significant risk to the environment (Section 6.8.1).

While there are gaps in many of the laws, the growth of IEL has been dramatic. The United Nations Environment Programme (UNEP) undertook the first ever global assessment of the environmental rule of law in 2019 and reported a 38-fold increase in relevant laws established since 1972 (UNEP, 2019). In some regions such as the European Union, there are free trade and environmental regulations that apply to all member nations, and each must translate these regulations into enforcement. A similar approach was developed as part of the 1994 North American Free Trade Agreement and continues under the 2018 United States–Mexico–Canada Trade Agreement. These Agreements include a Commission for Environmental Cooperation to address common environmental issues and concerns, and to avoid trade distortions caused by different environmental protection and sustainability goals. On a broader scale, the Organisation for Economic Co-operation and Development (OECD) (37 countries representing 80% of world trade) has a mandate to protect the economic welfare of member states. The OECD sets environmental benchmarks and operating procedures that members are encouraged to follow. These include standard protocols to determine the properties, environmental behaviour and toxicity of new compounds before they enter the marketplace (Section 2.2.7).

One older and highly successful IEL began as the 1909 Boundary Waters Treaty between the United States and Canada (Hall, 2008). The original purpose was to resolve disputes over the use of waters shared by the two countries, and the Treaty was not specifically concerned with contaminants. The International Joint Commission is a bi-national organization established under the Treaty to report to both countries on the state of trans-boundary waters, progress towards resolving existing issues, and the emergence of new concerns. In 1972, its responsibilities were expanded with the signing of the Great Lakes Water Quality Agreement (GLWQA) to address Great Lakes problems of **nutrient** enrichment (eutrophication), chemical contamination and invasive species. Such a bilateral agreement between two countries that have much in common culturally, and that share hundreds of lakes and rivers along their borders, is a 'natural' one, although it has not been entirely conflict-free.

12.2.3 Regulatory Challenges and Disparities

The regulation of contaminants is not based solely on scientific information but also reflects social, economic and political factors. Therefore, it is no surprise that there are differences among jurisdictions in regulatory policies and the regulated concentrations of chemicals. For example, there is a wide disparity among countries in some environmental quality criteria even though they are all developed by the objective application of science (Section 12.3.1). The inconsistencies reflect technical, conceptual and philosophical differences among nations in the data that are selected for review, the interpretation of those data and the degree of conservatism used in applying a precautionary approach to define a criterion value.

12.2.3.1 Factors That Affect the Development and Implementation of Chemical Regulations

Many factors influence the nature and success of regulatory actions, including challenges that fall outside the scope of existing regulations for chemical management. Recent examples include environmental contamination related to the production, recycling and disposal of rare earth minerals used in batteries and electronics, and the disposal and dispersal of plastics (Section 14.2). There are also events beyond the control of individual nations, including wars, pandemics and natural disasters that divert attention and resources from environmental management and create new sources of pollution. With global climate change, the frequency of these events is likely to increase.

Despite the prolific growth of laws and agencies, the UN found weak enforcement to be a global trend that is exacerbating environmental threats (UNEP, 2019). A failure to fully implement and enforce domestic laws and IELs is one of the greatest challenges to mitigating climate change, reducing pollution and preventing widespread loss of species and habitats. Although many nations seek international recognition when enacting domestic laws and signing IELs, the political will and funds needed to implement them are often lacking, and this collective failure undermines their effectiveness.

Important barriers to implementing chemical regulations are differences among nations in culture, politics, wealth, development, infrastructure, technology and industry–government relationships. In some nations, inadequate resources for environmental regulation may reflect political pressure to address other societal needs, cultural and philosophical attitudes towards nature and environmental protection, or corruption that diverts resources to private gain. Even when governments favour environmental protection over economic growth, the relative emphasis on environment or economy can change when new governments are elected. For example, in early 2016, the United States began to issue permits to drill for oil on Alaska's North Slope, reversing policies that protected an Arctic wildlife reserve and prolonging a conservation conflict dating back to the 1980s (Docherty, 2001).

Among many nations, there are divergent positions about chemical management, as discussed by Selin (2005) for the European Union's Registration, Evaluation, Authorisation and Restriction of Chemicals (REACH) programme. The author suggests that if the European Union and the United States adopt different environmental management policies, there may be a growing polarization among other nations on the politics of chemical issues. Differences in the management and disposal of waste, including recyclables and electronics, illustrate discrepancies among jurisdictions. For example, the global trade in waste makes sense if recyclable materials are in excess in some countries and potentially valuable in others. However, the export of waste from developed to less developed countries (LDCs), so-called 'toxic colonialism', has been strongly criticized, and the 1989 Basel Convention (www.basel.int/) was designed specifically to prevent such transfers. The very definition of waste is often unclear and susceptible to fraud. The Convention does not differentiate recyclable plastic from contaminated mixed plastic waste, and to avoid fraudulent shipments, China banned imports of all types of used plastic in 2018. In response, exporters diverted these wastes to Malaysia, Vietnam, Thailand, Indonesia and the Philippines, illustrating the discrepancy among nations in standards. Malaysia and the Philippines subsequently shipped full containers of fraudulent waste back to the highly embarrassed countries of origin. In 2021, an amendment to the Basel Convention allowed only clean, homogeneous, and readily recyclable non-halogenated polymers to be freely traded globally.

One of the most controversial debates in this context is whether pollution-intensive industries are relocating to countries with less stringent environmental policies, turning these countries into 'pollution havens' (Kukenova and Monteiro, 2008). The concern is the growing role of multinational firms that exploit developing countries, combined with the progressive easing of regulations on foreign investments, which undermine efforts to end poverty and environmental degradation. Economic globalization has shifted mining from developed to less developed 'host' countries (Section 14.3), creating competition among LDCs to attract foreign investment with tax breaks and concessions in environmental regulation (Reed, 2002). An example is the Ok Tedi gold mine in Papua New Guinea that was permitted to discharge mine wastes directly to the Ok Tedi and Fly Rivers (Section 11.2.4.1). Despite a successful class action suit by landowners against the mining company, mine wastes continue to flow into the river. Downstream communities believe that damage to their fisheries, farming, forestry and economies far outweighs any local benefits of the mine. Similar issues exist in other LDCs rich in minerals but beset by poverty.

A flood of foreign investment has also meant an increased risk of corruption. In 2014, the OECD estimated that 19% of all foreign bribery cases occur in the oil, gas and mining industries, higher than in any other sector (Rauter, 2019). One solution is to regulate multinational firms by requiring the disclosure of investment details. The idea is to increase payment transparency and to reduce illicit financial outflows from developing countries (https://eur-lex.europa.eu/eli/reg/2019/2088/oj). Ayentimi et al. (2016) emphasize that linking investment incentives with local environmental regulations could potentially support and strengthen industrial development, but it is necessary to 'strike a balance'. One questions when and how such a balance can be achieved.

12.3 Applying Ecotoxicology to Support Chemical Management

To be effective, regulations to limit the use, discharge and environmental concentrations of chemicals must be technically sound. The science of ecotoxicology supports these regulations (Section 12.3.3) by employing research and testing to:

- define acceptable concentrations of compounds in different environmental media (Section 12.3.1);
- assess the site-specific ecological risks and impacts of chemical discharges, planned or accidental (Section 12.3.2);
- define and enforce limits to emissions from known sources (e.g., industrial effluents; waste disposal sites) and guidelines for remediating contaminated sites (Section 12.3.4); and
- support monitoring to define trends in chemical concentrations and effects, to detect emerging contaminant issues, and to protect **ecosystem services** (e.g., edible fish) (Section 12.3.5).

However, to apply this information to environmental management, the concepts and data generated must be synthesized to define numerical limits for chemical concentrations in different media that will protect species, ecosystem function and human health.

12.3.1 Numerical Limits: Criteria, Objectives, Standards, Guidelines[1]

Guidelines, **objectives**, criteria and standards are numerical values that reflect policy decisions taken by governments to control the releases of toxic substances and nutrients to the environment. In non-human situations, guidelines are generally set to protect populations, rather than individuals, although there are exceptions for species at risk. These numerical values may be used:

- with monitoring data to ascertain whether environmental concentrations of specific compounds exceed **thresholds** for effects;
- in the hazard or effects component of ecological risk and impact assessments;
- in the development of controls on emissions and contaminated site remediation;
- as performance measures to evaluate the success of management actions; and
- as targets in sustainable development strategies.

Definitions of guidelines, objectives, criteria and standards can vary widely among jurisdictions, depending on

language, culture and legal definitions. A 'guideline' is the actual numeric limit in one nation, or the derivation method itself in another. The term 'objectives' often, but not always, implies additional societal considerations outside of ecotoxicology, including socio-economic concerns (affordability, achievability) or available control technologies. In general, however, guidelines, criteria and objectives are non-binding recommended limits on concentrations or emissions, whereas 'standards' imply enforceable limits. However, among nations, implementation may be voluntary or prescribed in legislation.

12.3.1.1 How Numerical Limits Are Developed

For the most part, water quality guidelines (WQGs), used here synonymously with water quality criteria, are developed for single substances from laboratory acute and **chronic** toxicity data and are thus considered 'risk-based'. Mixtures are not considered during guideline development unless it is for groups of compounds with similar mechanisms of action, such as chlorinated dioxins and furans (toxic equivalents model; Section 2.3.4). There are also approaches such as the **toxic unit** concept where monitoring data may be compared to WQGs and exceedances combined into a single metric of toxicity to aid in management (Section 2.3.4.1).

In many jurisdictions, the preferred method for developing guidelines, where there are sufficient toxicity data, is to use a **species sensitivity distribution** (SSD; Section 2.3.3.2), but here too there is considerable variation. Some nations use '**no-effect concentrations**' (NECs; see also NOEC, Box 2.5) below which effects are not expected, or 'effect concentrations' (ECs) above which effects would be expected. These metrics define a guideline value as the HC_5, or **hazard concentration** protecting all but the most sensitive 5% of test species. Other percentiles may also be chosen, for example 1% for high-value resources, 5% generally, or 10% or 20% for already disturbed systems (Warne *et al.*, 2018). Prior to the widespread use of SSDs, 'application factors' (also known as 'safety', 'adjustment' or 'uncertainty' factors) were frequently applied to adjust for scientific uncertainty or to add conservatism (Section 12.3.2.6). Although largely supplanted by SSDs, **safety factors** are still retained to develop guidelines from small datasets. Other factors that may influence WQGs include exposure duration (indefinite versus fixed duration toxicity tests), exposure–effect relationships observed in field studies, effect **endpoints** ('no effect', 'predicted or probable effect'

---

[1] Douglas J. Spry, Environment and Climate Change Canada, Ottawa, Ontario, Canada contributed Section 12.3.1 on Numerical Limits.

or 'effect' data), and natural background concentrations tolerated by endemic species. It is not surprising that WQGs vary among countries, given the wide range of dates when they were developed and these myriad decisions by guideline developers based on science and policy. Yet, all are deemed to be protective of aquatic life in their respective jurisdictions.

The development of WQGs for cationic metals is increasingly sophisticated, with the recognition of toxicity modifying factors that affect toxicity and derived criteria (Adams *et al.*, 2020). The empirical observation that increased water hardness reduces the toxicity of many metals (Section 6.7.3) has been incorporated into WQGs since the 1970s. Current guidelines for some metals are derived using multiple linear regressions to relate toxicity to hardness, pH and dissolved organic carbon. Several countries now apply mechanistic models such as the **biotic ligand model (BLM)** to account for metal bioavailability when setting site-specific limits for surface waters, sediments and soils (Section 6.3). Guidelines for some metals can vary by more than 27-fold between fresh and marine waters, depending on the chemistry of each metal and on physiological differences between freshwater and marine species. The bioavailability and toxicity of organic contaminants are less affected by water chemistry so that toxicity modifying factors are seldom used.

Site-specific WQGs can also be derived that recognize how differences in toxicity modifying factors at a site can make metals more or less toxic and guidelines more or less stringent. Four methods include:

1. recognizing high natural background concentrations of the compound of concern;
2. recalculating SSDs, excluding aquatic species not endemic to the specific site;
3. determining a water-effect ratio, i.e., adjusting the guideline by testing key species in laboratory water with few toxicity modifying factors compared with site water with natural toxicity modifying factors; and
4. restricting toxicity data to those species captured at the site and tested in site water.

In other words, although numerical limits are based on toxicity data, there is room for scientific judgement with respect to what data to consider and the need for greater stringency to protect species at risk. With time, the numerical values are becoming more uniform among nations as the techniques to derive WQGs improve and toxicity databases expand.

12.3.1.2 Numerical Limits for Soils, Sediments and Biological Tissue

Unlike WQGs, sediment quality guidelines (SQGs) are not typically based on laboratory toxicity data, but rather on correlations between chemical concentrations in a particular sediment sample and the health of the associated benthic **infauna** (Section 4.3.1). The observations are separated into EC and NEC comparisons, and two or three estimates of the ECs are prepared. Limits may also be established from NECs, low ECs and severe ECs using SSDs to estimate the 5th and 95th percentile of the distribution. For organic contaminants, SQGs may be derived from WQGs via **equilibrium partitioning** (Burgess *et al.*, 2013). In this approach, the concentration of an organic compound dissolved in the sediment **interstitial or pore water** is considered a surrogate for the bioavailable concentration, and the WQG for the compound defines the permissible concentration in the sediment pore water. Using this value, the equilibrium partitioning coefficient between the sediment and the pore water, and the sediment composition (notably its organic carbon content), the regulator can calculate the guideline value for the bulk sediment. A similar approach, also focused on sediment pore water, has been used for metals (Section 6.8.3; Box 6.2).

Soil quality guidelines have a slightly different premise. They are developed to help identify sites that are contaminated relative to a desired land use, typically agricultural, residential/parkland, commercial or industrial. Sites that are identified as contaminated should be remediated to the guideline values to restore an acceptable level of ecological function and to protect human health (Section 13.4). Soil guidelines are therefore remedial rather than preventive; they are pathways-based (soil contaminants leaching to groundwater or surface water, **uptake** via food or soil contact) and typically set around the 25th percentile of the SSD for environmental protection. A key function of soil quality guidelines is to limit liability upon sale of the property or upon change to a more sensitive use.

To help protect against toxicity from persistent, bioaccumulative substances such as polychlorinated biphenyls (PCBs), a tissue residue approach can be used. The numerical limits may be based on measured or modelled **critical body burdens** (Section 2.3.4.3) to protect the organism *per se* but are more commonly used to limit concentrations in prey items to prevent secondary poisoning of fish-eating birds and mammals, including humans.

12.3.1.3 Future of Numerical Limits

Although guideline development is slow to adopt new science, it is becoming more sophisticated with the greater acceptance of models that predict the toxicity and bioaccumulation of untested compounds (Section 2.3.3). For instance, the BLM for copper was published in 2000, but its first use in guideline development was in 2007, complementing the existing multiple linear regression methods. Statistical methods have improved with the transition to SSDs, but the basic input data from acute and chronic toxicity tests with fish, invertebrates and plants have changed little since the 1970s. These indicators of toxicity should continue to be the primary metrics for setting guidelines, but the use of 'omics' and *in vitro* tests, such as 'EE2-equivalents' for total oestrogenicity, shows promise for estimating chronic 'no to low' effect concentrations from acute exposures. High-throughput testing based on adverse outcome pathways (AOPs) will also generate data quickly and reduce the need for vertebrate toxicity testing (Sections 2.2.8.4, 5.10.1, 14.4.3).

12.3.2 Ecological Risk Assessment (ERA)

The development of the science of ERAs can be traced back to the 1980s, when scientists and regulators recognized a need to evaluate the ecological effects of human activities and to incorporate this information into regulatory decisions. In this context, ERA is defined as "A process that evaluates the likelihood that adverse ecological effects are occurring or may occur as a result of exposure to one or more stressors" (US EPA, 1992). The US Environmental Protection Agency (EPA) first published a framework for ERAs in 1992, followed by the European Union in 1996 (European Commission, 1996), and the general approach has since remained relatively consistent (Suter et al., 2003).

There are key differences between a human health risk assessment (HHRA), which recognizes that industrial chemicals can affect human health outside the workplace, and an ERA. The former is concerned with protecting only one species and all individuals of that species. In contrast, the goal of an ERA, as with WQGs (Section 12.3.1), is to protect populations and communities, not individuals, except for threatened or endangered species. Achieving this goal may involve protecting the most sensitive species, whereas in other cases some level of damage may be accepted, provided that populations and communities are not unduly affected and that ecosystems are functional. For example, in some jurisdictions the goal for aquatic species is to protect 95% of species 95% of the time.

The original focus of ERAs was on highly polluted environments where contaminants caused adverse effects on endemic biota. As these areas were assessed and, in some cases, remediated, researchers and managers realized that the effects of contaminants were often more widespread than originally anticipated and that contaminants were not the only stressors. Physical stressors such as habitat loss and biological stressors such as introduced species could also impair or foster site recovery (Section 13.9.2). As a result, applications of ERAs have expanded beyond specific sites to the regional or watershed level and consider both non-chemical and chemical stressors (Bradbury et al., 2004).

12.3.2.1 The Methodology of Ecological Risk Assessment

In its simplest form, an ERA starts with a concern about existing environmental contamination (substances present where they should not be, or present above baseline levels), or the potential for contamination associated with a proposed development. However, the focus is not on the mere presence of contaminants but rather on the potential for adverse effects. As we saw in Box 1.1, a contaminant is not necessarily a pollutant. For an ecological risk to exist, two conditions must be satisfied:

1. the contaminant has the inherent ability to cause one or more adverse effects (i.e., toxicity); and
2. the contaminant co-occurs with or contacts an ecological component (**receptor**) with sufficient duration and magnitude (exposure) to elicit the adverse effect (Box 2.1).

For a specific compound, an ERA determines a predicted (or measured) environmental concentration (PEC) and compares it to a predicted (or measured) NEC (i.e., PNEC). The PEC/PNEC ratio is referred to as the hazard quotient (HQ). When the PEC and/or PNEC are based on generic rather than site-specific information, the HQ only provides an indication of hazard. Estimates of risk are improved with site- and contaminant-specific information, which also allows more complex probabilistic comparisons of exposure and effects.

The detailed procedure for carrying out an ERA involves a series of prescribed steps. The philosophy and essential components differ little among jurisdictions, although details vary, such as the level of protection required (Section 12.3.1). Certain terms have been adopted in the practice of ERA, but the reader is likely to encounter some variation in the definitions and the uses of such terms, as expected of an evolving process (Suter *et al.*, 2003). Figure 12.1 illustrates the US EPA (1998) framework, showing three boxes:

1. problem formulation;
2. analysis – exposure characterization and ecological effects characterization; and
3. risk characterization.

In Canada, the European Union and Australia, there are four components, with some differences in nomenclature, but the overall procedure is comparable; other countries may use alternative frameworks. Except for the analysis stage, the process involves the active participation of risk managers and interested parties (stakeholders), an important component of problem formulation. Data collection (literature searches) and data generation (e.g., modelling, experimentation or *in situ* measurements) occur throughout the process, typically in an iterative process, with mid-course adjustments being normal.

12.3.2.2 Problem Formulation

The problem formulation stage is critically important in structuring the risk assessment. It includes the hazard identification referred to earlier, but also a conceptual

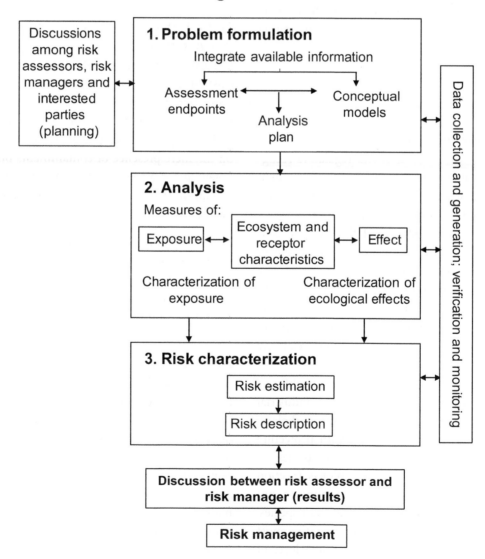

Figure 12.1 A typical ERA framework based on the scheme proposed by the US EPA (1998).

model with a non-quantitative pathway analysis and identification of receptors, geographical areas and ecosystems at particular risk. Pathway analysis considers the entry of a substance into the environment, from both point and diffuse sources, and the way it partitions among air, soil, water and sediment. Information may be available on the physical and chemical behaviour of the substance under varying environmental conditions, or the behaviour may be modelled, based for example on its molecular structure (QSARs, Sections 2.3.3.3, 4.5).

12.3.2.3 Analysis

The analysis stage involves the critical review of existing data, often complemented by modelling. The first step, exposure characterization, identifies the sources and quantities of chemical releases, and data on emissions from the problem formulation stage are verified and refined. This step may involve direct measurements or estimates of exposure concentrations, followed by evaluation of uptake by receptors. Data on uptake may be available from controlled experiments, but predicting the uptake by receptors *in situ* often requires exposure models (Section 4.5.2). When exposure is estimated from chemical concentrations in the environment, ideally the bioavailable forms of substances rather than the total concentrations are used in the calculation. For compounds that are not changed through metabolism (Section 7.5), exposure may be evaluated through indirect methods such as measurements of tissue residues in wildlife, as discussed in Chapters 2, 4 and 6.

Effects characterization should lead to a critical toxicity value (PNEC), typically based on SSDs (Section 2.3.3.2) or environmental quality guidelines (Section 12.3.1). Otherwise, information on toxicity is compiled for a wide range of trophic levels as a first indication of the population, community or ecosystem effects of a particular substance. Once PNECs are determined, they are used as inputs to the next phase, risk characterization (Figure 12.1).

12.3.2.4 Risk Characterization

Risk characterization involves two stages: risk estimation and risk description. Risk estimation determines the likelihood and magnitude of adverse effects that would result from exposure to the substance of concern. This step incorporates the results of Stage 2 (characterization of exposure and effects) to derive a worst-case risk. As an example, the first screening tier employs the worst-case HQ for each endpoint. If HQ <1.0 for all available endpoints, the substance is deemed to pose no risk, and no further analysis is required. However, if any worst-case quotients exceed 1.0, higher tiers of risk estimation are required. Finally, the preceding stages of risk characterization are summarized in a risk description that is communicated to decision makers in a comprehensive but non-technical manner. This process leads to the practical use of the assessment, i.e., risk management, which lies outside the ERA framework. Management of the ecological risks of one or more compounds can be achieved by limiting emissions, restricting use, improving methods for waste treatment or remediating and restoring contaminated sites (e.g., Case Study 13.2).

12.3.2.5 Applications of ERA: Specific Chemicals

The ERA paradigm is used internationally to predict and manage the toxicity of individual compounds. Early applications (**retrospective ERAs**) focused on substances already in commercial use, known to be present in industrial emissions, or suspected to cause ecological effects. Subsequently, the scope was enlarged to include potential new sources of compounds not yet in use but slated for introduction into the marketplace (e.g., prospective ERA for a new industry). The ERA approach was honed on PBT compounds (Box 12.2), and screening values for these criteria have been adopted in most jurisdictions to identify new chemicals that merit further consideration. For example, compounds that have half-lives greater than the prescribed values for persistence (e.g., $t_{1/2} > 2$ months in surface waters, or $t_{1/2} > 6$ months in soil) will be flagged. Similarly, a bioconcentration factor (BCF) value >2,000 identifies compounds with a potential for bioaccumulation. These screening values vary somewhat among different jurisdictions, as is the case for WQGs and SQGs (Section 12.3.1), but they coalesce around similar values.

Given that the PBT approach is relatively simple when applied to **hydrophobic** organic chemicals, there was an understandable attempt to extend its use to all chemicals, including metals, **metalloids** and their compounds. However, as discussed in Section 6.1 (Table 6.1), environmental contaminants do not all behave like hydrophobic organic molecules. For example, with the exception of some radioelements, metals cannot be created or destroyed; they are inherently persistent, and the 'P' criterion has no discriminatory power. Similarly, the internal concentrations of many essential metals are regulated physiologically (Section 6.5.1) to maintain relatively constant tissue concentrations. If ambient metal concentrations decrease, the apparent BCFs and bioaccumulation factors (BAFs) will increase (a constant numerator divided by a smaller denominator); conversely, if ambient concentrations increase, BCFs and BAFs will decrease. In other words, BCFs and BAFs determined for metals are *not* intrinsic properties of the element, but rather integrate the ambient exposure conditions, the metal's form in the exposure medium and each species' unique metabolic requirements for each metal. Clearly, the 'B' criterion also fails for metals.

The shortcomings of the PBT approach were eventually recognized in a revised ERA framework for metals (US EPA, 2007), which explains in a comprehensive way how to account for environmental factors that determine their environmental fate and bioavailability. The 2007 framework also considers the variable background concentrations of naturally occurring substances, and their long-term influence on indigenous fauna and flora. In this connection, physiological tolerance to these naturally occurring concentrations must also be taken into account. For example, a key issue for resource managers is whether metal-tolerant communities (Section 6.5) provide the same or similar ecological goods and services as metal-intolerant communities.

In contrast, the bioavailability of organic contaminants is generally considered an intrinsic property of a compound (e.g., K_{OW}, molecular size and shape; Section 2.3.4.3). However, the bioavailability of some compounds also varies with the species exposed and environmental conditions (Table 2.1; Sections 3.6.2.1, 4.5.2). Clearly, there is a need to look beyond the standard conditions of laboratory tests to understand how environmental conditions interact with chemical form, behaviour and ecological risk.

12.3.2.6 Handling Uncertainty: An Integral Part of ERA

The strength of ERA is the systematic, disciplined and prescribed approach for collecting and analysing data relevant to the ecological risks of chemical contamination, often in the absence of 'perfect' data. Box 12.3 provides specific recommendations for 'evidence-based ecotoxicology' to support reliable ERAs.

The nature of the input data is particularly important. For example, the information used to estimate a PNEC is often derived from single-species toxicity tests (Chapter 2). These tests are carefully designed and controlled to yield reproducible results when performed in different

---

Box 12.3 Recommendations to support evidence-based ecotoxicology (Martin *et al.*, 2019)

- Consider all applicable studies.
- Report all findings of experimental studies.
- Make ecotoxicity studies publicly accessible.
- Implement reporting guidelines for publishing ecotoxicity studies.
- Apply transparent and consistent evaluation criteria to all ecotoxicity studies.
- Improve the regulatory guidance for weight-of-evidence (WoE) evaluations.
- Increase collaborations among all stakeholders.
- Declare interests.
- Improve training and knowledge transfer among all stakeholders.

---

laboratories, but the results are used to predict adverse effects on whole ecosystems. As discussed in Section 4.2, extrapolating from the laboratory to the receiving environment is not a simple matter. For example, exposure conditions in laboratory toxicity tests are carefully controlled, but in natural ecosystems, exposures to a compound will occur over a range of conditions as the seasons and weather patterns change. Similarly, in most laboratory tests, food web connections are absent, whereas in the natural environment the toxicant may impinge upon predator–prey or herbivore–plant interactions. The SSD approach to derive environmental quality guidelines also has its limitations. Examples include the weak ecological relevance of some test species and endpoints and the absence of a guiding theory in ecotoxicology to justify a particular statistical distribution (Fox *et al.*, 2021). In other words, the PNEC comes with an inherent uncertainty of unknown size.

The derivation of a PEC, although conceptually simpler, also yields a value with inherent uncertainty if monitoring data are sparse, as is currently the case for many engineered nanomaterials and emerging organic contaminants. Extrapolating exposure estimates from the laboratory to the field is fraught with difficulties because ecosystems exhibit unique properties not shared with individuals or species (Section 4.1). There are also significant challenges in accounting for the potential interactions of target compounds with other contaminants, urban, industrial or agricultural developments, or natural stressors such as disease.

Clearly, uncertainties are a fact of life in ecotoxicology and risk assessment, and there are techniques to deal with them, including the application of safety factors as a precautionary approach to setting conservative guidelines for exposure concentrations or emission limits (Section 12.3.1.1). Risk assessors should ensure that uncertainties are treated in the same way for all chemicals and make statements about the *relative risk* that different chemicals represent rather than authoritative statements about the *absolute risk* posed by a specific chemical.

The uncertainty in estimated risks emphasizes the need for follow-up monitoring to determine whether risk management has succeeded in limiting environmental contamination and ecological impacts (Sections 4.3, 12.3.5). If monitoring reveals unexpected effects, ERAs can be revisited to identify needs for better site-specific data and for research to improve risk estimates and environmental protection. Each ERA can also guide assessments in related ecosystems and build a record of factors that should be considered in future assessments.

For relatively small areas such as contaminated harbours, ERAs provide useful models of environmental fate and effects, and strong support for regulation and remediation. Over larger landscapes, risk predictions become less certain, owing to environmental characteristics that vary spatially and seasonally and to the uneven distribution of receptor species and chemicals of concern. As illustrated by Capsule 12.1, this requires a **probabilistic** approach to risk assessment based on extensive and intensive assessments of contaminant distributions, environmental conditions and receptors at risk across the diversity of geographic features.

12.3.3 Regulations for Individual Chemicals and Complex Mixtures in Environmental Media

Industrial emissions of liquid, gaseous or solid wastes are obvious targets for controls, and unique regulations are often developed for specific industrial sectors, such as mines and pulp and paper mills (Sections 11.2, 11.4). Such regulations are often derived from published WQGs and specify limits for the concentrations of highly toxic components such as abietic acids extracted from wood. However, it is important to recognize that regulating effluents compound-by-compound overlooks the potential for interactive toxicity among chemicals (Section 2.3.4), and regulating all compounds individually in a complex mixture is often impractical. An alternative

is to regulate the whole effluent as a 'substance', for example by requiring that undiluted whole effluent be non-toxic to test species in routine tests.

Recognizing that effluents will be diluted in a receiving water, an important question about effluent regulations is whether they apply at the 'end-of-pipe', or at the edge of a specified mixing zone. Mixing zones are controversial because the distribution and dilution rate of effluent plumes can be highly variable, aquatic species may not actively avoid harmful chemical exposures, and a portion of the receiving environment becomes a *de facto* waste treatment facility. Effluent regulations may also specify other characteristics, such as pH, temperature, turbidity and biological oxygen demand, that potentiate or antagonize the toxicity of individual compounds and contribute to the broader environmental impacts of the whole effluent. These same concepts apply to regulating air emissions.

For specific compounds such as agricultural pesticides, regulations specify where they can be used (e.g., on soils or vegetation) and in what form (e.g., sprays, granules) and may require application by licensed operators. When applied, there are limits to the concentrations and amounts used and the environmental conditions during application (e.g., wind direction and speed). Some regulations even require **surveys** of pest densities to justify applications. Insect traps containing pheromones that attract a target pest can be deployed in orchards to determine whether pest densities warrant pesticide spraying. This strategy avoids 'precautionary spraying', reduces pesticide use and contamination, and lowers the cost of fruit production.

Regulations can also affect ecosystem services and the resource industries that depend on them. Frequently, commercial fisheries and the sale, import or export of fish products have been banned when concentrations of contaminants exceeded guidelines for human consumption. Recreational fishers are protected primarily by non-regulatory fish consumption advisories and voluntary compliance (Box 12.4; Section 4.3.5; Capsule 12.2). However, protecting subsistence fishers in remote communities presents unique challenges due to the cost and scarcity of fresh produce and the poor nutritional quality of many processed foods. In this case, the dietary and cultural benefits of consuming 'country food' may outweigh the risks of contamination. The problem is often best managed by combining education to promote balanced diets with subsidies for importing fresh food.

Limits on chemical concentrations in food also have implications for international trade. When pesticide

Box 12.4 Fish consumption advisories – mercury as an example

Fish consumption guidelines have been established in many countries where aquatic ecosystems have been contaminated with mercury. The guidelines are based on calculated safe levels of mercury, predominantly as methylmercury, in fish muscle. Advisories are published when chemical monitoring demonstrates that certain species and sizes of fish in a particular water body exceed the calculated safe concentrations of total mercury. Among jurisdictions, this value ranges from about 0.1 to 1 µg/g fresh weight. Safe concentrations are based on two key factors: (1) estimates of fish consumption by average populations, and (2) a reference dose, i.e., the dose that can be absorbed daily or weekly for a lifetime without a significant risk of adverse effects. The Joint FAO/WHO Expert Committee on Food Additives (JECFA) has established provisional tolerable weekly intakes (PTWIs) for total mercury (inorganic plus methylmercury) at 5 µg/kg body weight and for methylmercury at 1.6 µg/kg body weight (UNEP and WHO, 2008). For many countries, advisories typically show the name and location of a water body, a list of fish species and sizes, and advice about the number of allowable fish meals per week, often adjusted for age and sex.

regulations are different among nations or not enforced evenly, the trade in some foods may be banned in response to pesticide monitoring by the importing nation. Import bans on contaminated products encourage the exporting nation to regulate and remediate the sources of contamination.

12.3.4 Enforcement of Environmental Regulations

An essential component of chemical management is the enforcement of environmental laws and regulations. In Section 12.2.1.4, we learned that many government agencies regulate one or more aspects of chemical production, use and disposal. However, the primary responsibility usually falls to departments of the environment with significant roles for natural resource and human health agencies. To ensure compliance, enforcement agencies

apply various strategies, including coercive powers to demand compliance to environmental laws, generally termed 'command and control'. However, others rely increasingly on conciliatory, educational and collaborative strategies to persuade individuals, organizations and governments to comply with environmental laws and regulations (Wyeth, 2006), and industries may pre-empt enforcement through voluntary self-regulation (Section 12.2). Environmental non-government organizations (ENGOs) such as Greenpeace have also played an important role in drawing attention to polluters and enhancing compliance with international laws. Their role is particularly important where states lack the capacity or political will to enforce international agreements and is sometimes termed 'ecoterrorism' by those who are the target (Eilstrup-Sangiovanni and Bondaroff, 2014).

Enforcement is often triggered by a spill, by toxicity testing of industrial emissions, or by the spatial or temporal patterns of chemicals in environmental media demonstrated by environmental monitoring (Section 12.3.5). When a problem is detected, enforcement agencies investigate the source and extent of contamination and contributing factors such as extreme weather, the efficiency of waste treatment systems, or negligence in the management of the substance spilled. For industries that routinely monitor their effluent quality, the source and responsible party may be quite clear. However, for unknown or diffuse sources, such as groundwater leakage from old or undocumented landfills, considerable effort is needed to track down the responsible party and to understand why contamination occurred (Section 13.1). If enforcement determines that a regulation has been broken, charges are laid against the organization thought to be responsible, and departments of justice collaborate with enforcement agencies to prosecute the case in court or to negotiate a settlement.

Ecotoxicology is an important part of enforcement. Strict protocols for collecting, labelling, shipping, storing and analysing environmental media are needed to meet legal standards for quality of evidence. These **forensic** protocols are derived from methods developed by environmental chemists to investigate the fate and distribution of compounds at the ultra-low concentrations typical of natural environments. Similarly, toxicity test methods to demonstrate compliance are versions of standard protocols (Chapter 2) adapted to the needs of specific regulations. Although government laboratories may conduct toxicity tests and chemical analyses to initiate or support

Capsule 12.1 The Sudbury Soils Study: An Area-wide Ecological Risk Assessment

Christopher D. Wren

LRG Environmental, R.R. #1, Markdale, ON, Canada

Glen Watson

Vale Canada Limited, North Atlantic Operations Centre, Copper Cliff, ON, Canada

Marc Butler

Glencore Sudbury Integrated Nickel Operations, Falconbridge, ON, Canada

Background and Study Rationale

The City of Greater Sudbury, Ontario, Canada, is recognized internationally for its mining and smelting operations. The vast nickel ore body in Sudbury was discovered in 1883. Several companies began mining, but by the mid-1930s, only INCO (now Vale) and Falconbridge (now Glencore) remained. These two companies dominated the world nickel and copper market for many years with a mix of open pit and underground mining.

Annually, smelting of the sulphide-rich ores released hundreds of thousands of tonnes of sulphur dioxide (SO_2) and hundreds of tonnes of metal particles to the atmosphere. The SO_2 emissions caused significant damage to the regional landscape, and the phenomenon of acid rain was recognized in the mid-1960s (Section 11.1). In response, the companies began to reduce SO_2 emissions in the 1970s by improvements in technology, and the province applied emission limits by 1984 (Watson *et al.*, 2012). As a result, emissions of SO_2, copper and nickel declined quickly.

In 2001, the Ontario Ministry of the Environment (MOE) (in 2021, Ministry of Environment, Conservation and Parks, MECP) published a review of soil monitoring studies conducted in the Sudbury area between 1971 and 2000. The review revealed that concentrations of copper, nickel, cobalt and arsenic exceeded the 1997 MOE soil quality guidelines, particularly around the historical smelting sites. In response, the Government recommended that a more detailed soil study be undertaken to fill data gaps, and that an HHRA and an ERA be undertaken (MOE, 2001). Both INCO and Falconbridge voluntarily accepted these recommendations, and in 2001 the Sudbury Soils Study (www.Sudburysoilsstudy.com) was initiated to evaluate the ecological risks associated with soils contaminated primarily by historical smelter air emissions (Wren, 2012).

Study Organization and Process

A Technical Committee (TC) was formed in 2001 to oversee the Soils Study and to provide technical guidance. The membership was determined by the two mining companies in discussion with the Ontario MOE and included INCO, Falconbridge, Ontario MOE, the Sudbury & District Health Unit, the City of Greater Sudbury, and the First Nations and Inuit Health Branch of Health Canada.

The committee retained the services of a consortium of consulting firms called the Sudbury Area Risk Assessment Group to undertake much of the work. Although the study was funded by the two mining companies, all technical information and reports generated were shared simultaneously with all members of the TC. To ensure an objective and transparent process, a Public Advisory Committee, an independent Process Observer and an independent Science Advisor to the TC were involved, and all reports and findings were peer reviewed by an external group of experts.

The Sudbury ERA followed the regulatory guidance provided by the MOE (2009), and was considered an area-wide risk assessment, as it evaluated a large geographic area, rather than a site-specific risk assessment of an individual property. Its success depended heavily on the scientific integrity of the studies as well as the transparency of the process. To date, the Sudbury Study remains one of the largest risk assessments ever completed in Canada and lasted from 2001 to 2009.

ERA Goals and Objectives

The overall goal of the ERA was "To characterize the current and future risks to terrestrial ecosystem components due to Chemicals of Concern (COCs) from the particulate emissions of Sudbury smelters, and to provide information that will support activities related to the recovery of regionally representative, self-sustaining ecosystems in areas of Sudbury affected by the COCs".

To achieve that goal, three specific objectives were identified to:

1. evaluate the extent to which the COCs are preventing the recovery of regionally representative, self-sustaining terrestrial plant communities;
2. evaluate the risks of the COCs to terrestrial wildlife populations and communities; and
3. evaluate the risks of the COCs to individuals of threatened or endangered terrestrial species.

Although Objectives 2 and 3 appear similar, the regulatory guidance required that explicit consideration be given to threatened or endangered species. For common animal species, ERAs examine the risk to populations of organisms. In contrast, ERAs for threatened and endangered species are aimed at individual organisms, as are HHRAs. The Sudbury Study HHRA was released to the public first because the community had expressed greater concerns about human health than about ecological receptors.

Approach to the ERA

The risk assessment framework was applied to the ERA in three distinct phases: problem formulation, sampling and analyses to fill data gaps, and the detailed ERA, including exposure assessment, hazard assessment and risk characterization.

Problem Formulation

The problem formulation established the goals, scope and focus of the assessment. It was a systematic planning step that identified the major factors to be considered and was linked to the regulatory and policy context of the assessment. Once completed, the problem formulation became the blueprint for the detailed ERA that followed.

Definition of the Study Area

In response to the 2001 recommendations, the study area for the ERA was defined as the area from which soil samples had been collected in the 2001 survey. A comprehensive soil sampling and analysis programme was undertaken that collected soil samples from 1,200 sites over a 200 km × 200 km sampling grid (40,000 km^2), roughly the size of Switzerland (Figure 12C.1). The sampling density was higher close to the original smelting sites and decreased with distance away to capture background conditions at locations unaffected by smelter emissions.

Identification of Chemicals of Concern

After careful analysis of the 2001 soil survey results, seven elements were selected as the COCs: arsenic, cadmium, cobalt, copper, nickel, lead and selenium. A total of 8,148 samples was analysed for 20 inorganic parameters, and results for the seven key elements in the upper 20 cm soil layer are summarized in Table 12C.1.

Table 12C.1 **Concentrations of select elements (mg/kg) in the upper 20 cm of soil.**

Element	Range	Average	MOE (2004) soil quality guideline
Arsenic	2.5–620	16	20
Cadmium	0.4–6.7	1	12
Cobalt	1–190	9	40
Copper	2.7–5,600	260	150
Nickel	7–3,700	264	150
Lead	1–790	35	200
Selenium	0.5–49	2	10

Concentrations are from the 2001 survey (Wren *et al.*, 2012).

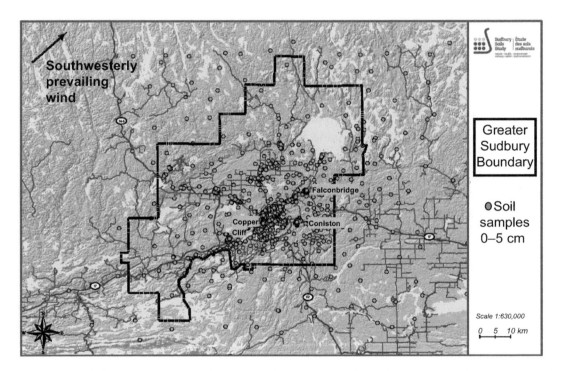

Figure 12C.1 **The area of The Sudbury Soils Study. The black line represents the Sudbury municipal boundaries and the study area extends to the limits indicated by the green dots. Each green dot is a sampling site, and the three smelters are identified as Falconbridge, Copper Cliff and Coniston. Modified and reproduced from Wren (2012) with permission of Maralte BV, Voorschoten, The Netherlands.**

Identification of Valued Ecosystem Components (VECs)

The VECs were selected based on input by stakeholders including naturalist groups and members of the public during focus workshops. The final list of nine wildlife VECs for Objectives 2 and 3 included:

- Mammals: Northern short-tailed shrew (*Blarina brevicauda*), meadow vole (*Microtus pennsylvanicus*), moose (*Alces alces*), white-tailed deer (*Odocoileus virginianus*), red fox (*Vulpes vulpes*) and beaver (*Castor canadensis*)
- Birds: American robin (*Turdus migratorius*), ruffed grouse (*Bonasa umbellus*) and peregrine falcon (*Falco peregrinus*)

Objective 1

This was a key objective of the ERA and very specific to Sudbury. As a result of smelter emissions, large tracts of land were barren or semi-barren, although there had been considerable progress towards the 're-greening' of the Sudbury region (Section 13.9.1). From 1978 to 2017, over 3,500 ha of land were limed and seeded, and over 9.8 million trees planted. However, an estimated 28,000 hectares of damaged terrestrial landscape have yet to be treated and biological recovery in these areas is very slow. Even in the reclaimed or treated areas, plant **species richness** remains low (Sudbury, 2018).

To address Objective 1, there were three primary environmental conditions that required extensive original research:

- metal mixtures present in soils;
- the low pH of soils; and
- the combined effects of the first two factors on plant species and communities.

Detailed data for four lines of evidence (LOEs) were collected from 22 study sites (18 exposed sites, three reference sites, and one historically limed and treated site). The LOEs included:

- an *in situ* plant community assessment;
- laboratory toxicity tests with plant and animal species;
- detailed analyses of soil physical and chemical characteristics; and
- an assessment of decomposition rates using *in situ* litter bags.

The LOEs were integrated using a WoE approach to determine whether the concentrations of COCs in the soil were impeding recovery of a self-sustaining forest system.

Objective 1 Findings

The study clearly showed that the levels of COCs in soil have affected plant communities and continue to impede the recovery of a self-sustaining forest community in the Sudbury region. Many plant species are not thriving because of high metal concentrations, low pH (<5.2) and low concentrations of nutrients and organic matter. Overall, the study identified harmful effects on the landscape that are still pronounced and extend farther from the smelters than previously inferred from aerial photo interpretation.

Objectives 2 and 3

The Sudbury ERA used a probabilistic modelling approach for these two objectives (Hull and Breese, 2012). The exposure assessment estimated the daily amount of each COC accumulated by each species or VEC. The effects assessment presented **toxicity reference values** (TRVs) for each VEC and COC, as well as a summary of what is known about the populations around Sudbury. Risk was characterized by comparing exposure estimates to a TRV and calculating the exposure ratio (ER), as well as the probability of exceeding the TRV.

Objectives 2 and 3 Findings

Given the **weight-of-evidence** available, the COCs in the soil are not exerting direct toxic effects on wildlife populations. However, the effects of COCs on plant communities (Objective 1) have reduced food availability for herbivores and habitat cover for many wildlife species and are, therefore, likely to be affecting birds and mammals indirectly.

Overall ERA Study Outcomes

The Sudbury ERA demonstrated that vegetation and habitat quality for wildlife remain affected over a larger area than previously considered. Effects are related to elevated metal concentrations combined with low soil pH and other soil conditions such as their shallowness and relative lack of organic matter. Existing soil quality is impeding the recovery of a healthy, self-sustaining forest ecosystem. Significantly, the results were supported by both the public and the scientific community and all members of the TC. Study findings led to the development of a local Biodiversity Action Plan that identified risk management strategies and techniques (Sudbury, 2018). The Action Plan is a concise, plain language document that recognizes community participation as being vital to its implementation. The Plan is updated regularly, and its implementation continues to be funded by both mining companies and the City of Greater Sudbury.

Capsule References

Hull, R. and Breese, K. (2012). Evaluating risks to terrestrial wildlife. In C. D. Wren (ed.) *Risk Assessment and Environmental Management. A Case Study in Sudbury, Ontario, Canada.* The Netherlands: Maralte Publishing, 343–382.

MOE (2001). *Metals in Soil and Vegetation in the Sudbury Area.* Toronto, ON, Canada: Ontario Ministry of the Environment, Government of Ontario.

MOE (2004). *Record of Site Conditions. Ontario Regulation 153/04. Soil, Ground Water and Sediment Standards for Use Under Part XV.1 of the Environmental Protection Act.* Toronto, ON, Canada: Ontario Ministry of the Environment, Government of Ontario.

MOE (2009). *Ontario Regulation 153/04. A Guide on Site Assessment.* Updated 2009. Toronto, ON, Canada: Ontario Ministry of the Environment, Government of Ontario.

Sudbury (2018). *Living Landscape: A Biodiversity Action Plan for Greater Sudbury.* City of Greater Sudbury, Ontario, Canada. https://www.greatersudbury.ca/live/environment-and-sustainability1/regreening-program/pdf-documents/biodiversity-action-plan/

Watson, G. M., Greenfield, M., Butler, M. & Wren, C. D. (2012). Historical smelter emissions and environmental impacts. In C. D. Wren (ed.) *Risk Assessment and Environmental Management. A Case Study in Sudbury, Ontario, Canada.* The Netherlands: Maralte Publishing, 23–48.

Wren, C. D. (ed.) (2012). *Risk Assessment and Environmental Management. A Case Study in Sudbury, Ontario, Canada.* The Netherlands: Maralte Publishing.

Wren, C. D., Spiers, G., Fergusson, G. & McLaughlin, D. (2012). The 2001 soil survey and selection of chemicals of concern. In C. D. Wren (ed.) *Risk Assessment and Environmental Management. A Case Study in Sudbury, Ontario, Canada.* The Netherlands: Maralte Publishing, 49–70.

court cases, many countries also rely on licensed commercial laboratories to provide similar services.

The penalties assigned (e.g., fines, control orders) for chemical spills or contamination often depend on the law that was broken, legal precedents from related cases, the amount spilled, the area affected and the severity of ecological effects. Thus, environmental chemists and ecotoxicologists are needed to design and interpret surveys of the nature and extent of contamination and effects. They might also act as expert witnesses to assist a court in understanding technical information submitted by either the prosecution or defence.

Chemical spills and discharges often create opportunities for research in ecotoxicology on scales that are not possible in the laboratory or not permitted in field experiments. These 'spills of opportunity' enable the study of:

- the environmental fate of compounds under conditions that vary spatially and temporally;
- species that cannot be studied in the laboratory;

- responses at the level of populations, communities and ecosystems;
- ecosystem services, represented by changes in the nature and productivity of natural resources (fisheries, forestry) and agriculture;
- the human health aspects of chemical exposures or the stresses associated with a loss of ecosystem services; and
- techniques for site remediation (Case Study 13.2).

The *Exxon Valdez* oil spill (1989) in Prince William Sound in the Gulf of Alaska and the Deepwater Horizon oil spill (2010) in the Gulf of Mexico provide two excellent examples of such opportunities (Section 1.2). In both cases, the oil industries responsible for the spill were required by the US government to support research as part of the Natural Resources Damage Assessment programme (https://darrp.noaa.gov/). The purpose was to measure the nature and extent of impacts of crude oil on natural resources as a starting point for programmes of ecosystem restoration. The results provided unique insights about oil distribution and effects, the unexpected persistence of oil and effects in some ecological settings, rates of ecosystem recovery and the efficacy of spill response measures (Lee *et al.*, 2015). Unfortunately, in some countries, the agencies responsible for enforcing regulations do not have the mandate or budgets to support related research, so that opportunities to learn from spills are lost.

12.3.5 Environmental Surveillance and Monitoring

From a regulatory perspective, surveillance and monitoring are akin to regulatory 'quality control'. Are regulations designed to reduce chemical emissions, environmental concentrations of specific compounds, and adverse effects on ecosystem or human health having the desired effect (Case Study 12.1)? However, their weakness is that the ecosystems we wish to protect are themselves the indicators of contamination and effects. By the time contaminants accumulate to the point that damage is measurable, ecosystem services may be lost, often for decades, despite environmental remediation and more stringent controls on emissions (Capsule 12.1). This is not to say that environmental monitoring is not needed, but rather to emphasize the importance of detecting problems as early as possible.

Some industries are required to monitor the local environment near their effluent pipes as part of their discharge permits, and the results are submitted to government agencies to ensure compliance (Box 12.5). Regional and national-scale monitoring is typically carried out by government agencies because a long-term commitment is needed for environmental sampling, the maintenance and analysis of databases, and communicating the results to regulators and the public. Sampling frequencies depend on the perceived risks and public concerns about chemical exposure, the severity of expected effects, the rates of change in contamination or effects, and the geographic extent of contamination. Cost and the need to sustain monitoring efforts for the long term are critical drivers for research to improve the power of monitoring. Examples include innovations in integrative passive samplers (Box 2.10), biomarkers of toxicity or of biodiversity (toxicogenomics; eDNA; Section 5.10.2), and ecological modelling (Section 4.5). Monitoring the contamination of food to protect public health also contributes to ecosystem protection if the data illustrate geographic and temporal trends of chemical concentrations (Section 4.3.5).

Box 12.5 Environmental Effects Monitoring (EEM): a comprehensive strategy for environmental protection

Canada's EEM programme assesses the effectiveness of regulations established to limit the environmental effects of effluents from specific industry sectors, such as pulp and paper or mining and smelting. The advantages of a sectoral approach wherein all industries within a sector are monitored with the same tools and held to the same standards are efficiency and fairness. The purpose of EEM is to limit the discharges of substances deleterious to fish, fish habitat or the use of fish by man, as defined by Canada's Fisheries Act (Section 12.2). Under the Act, fish are defined broadly as any aquatic species, and 'deleterious effects' can be assessed by literature reviews, acute or chronic toxicity tests of effluents, or monitoring chemical exposure and effects on fish in the receiving water. Effects may include biomarkers of exposure (**induction** of CYP1A enzyme activity) and changes in reproductive capacity (gonadosomatic index), disease prevalence (cancers; bacterial infections) and population **demographics** (abundance; growth rate; survival rate). The quality of 'fish habitat' is assessed from measures of sediment contamination and the abundance and diversity of plankton, invertebrate and fish communities, using approaches such as the **sediment quality triad** (Section 4.3.3.4). Deleterious effects on 'the use of fish by man' are evident from chemical analyses of fishery products (e.g., edible portions, fish oil) compared to consumption advisories. When needed, organoleptic taste-testing panels can detect substances with a disagreeable taste or odour (e.g., phenolics) that would limit the marketability of fish products.

Biological or effects monitoring supports the regulation of contamination but is not linked to enforcement to the same extent as chemical monitoring. For example, effluent monitoring that demonstrates the exceedance of a numerical chemical standard can trigger legal actions to mitigate the problem by improving waste treatment. In contrast, except for acute lethality, there are few clear 'red line' standards for biological responses, and in an **ecological context**, biological responses are not easily linked to specific emissions (Chapter 4).

Case Study 12.1 Monitoring Rivers to Assess the Adequacy of Pesticide Regulations

Chlorpyrifos is an organophosphate insecticide used in Spain and Portugal to control insect pests in orchards, and residues are commonly found in soils, surface waters and on citrus fruit. In the European Union, pesticides are approved for specific uses under the REACH programme, which relies on prospective ERAs to define limits on the amounts of pesticide used and the timing and rates of application. However, risks to non-target aquatic invertebrate species are estimated from toxicity tests of standard laboratory species exposed only once to chlorpyrifos. In contrast, in both countries, chlorpyrifos is applied repeatedly throughout a growing season characterized by seasonal droughts that affect river characteristics and the quality of aquatic habitats.

To assess the adequacy of chlorpyrifos regulations, Rico *et al.* (2021) analysed 5 years of river monitoring data using a retrospective probabilistic ERA as a quality control check on the prospective ERA. Assuming that chlorpyrifos was used as directed, would the monitoring data demonstrate that aquatic invertebrates are protected from acute toxicity? The data analysed were derived from a Water Information System for each of 19 river basins sampled between 2012 and 2017. Chlorpyrifos was measured in 14,600 surface water samples that were collected from 1,766 sites. Most sites were sampled 1–4 times per year, but 8% were sampled monthly. To assess risk, a threshold effect concentration ($HC_5 = 30$ ng/l) was calculated from an SSD for 67 taxa of aquatic invertebrates.

Clear patterns of contamination were consistent with the agricultural use of chlorpyrifos in fruit-growing areas in the south and east of Spain. The proportions of samples exceeding the HC_5 were highest in the Jucar, Andalusian, Mediterranean, Segura and Ebro River basins, with high rates but fewer data in several other basins. Concentrations were highest in the summer growing season, and lowest in January, with chlorpyrifos inputs corresponding to rainfall that washed dissolved and particle-borne pesticide from terraced orchards. Median concentrations ranged from a few ng/l to <1,000 ng/l, but were much higher, up to 96,000 ng/l, in three river basins. Overall, the risks of pesticide effects were considered unacceptable in five river basins.

From a regulatory perspective, the study demonstrated that:

1. Multiple applications of chlorpyrifos to single orchards caused unacceptable and long-term environmental exposures. Where exceedances were relatively low, only highly sensitive species of Cladocera, Ephemeroptera and Amphipoda were at risk. Where concentrations were higher, 50% or more of all species were likely affected, consistent with previous biological surveys showing a lower abundance during spray periods.

2. For relatively arid regions, there are weaknesses in the current ERA models used to set standards for pesticides in the specific climate and agricultural conditions of Mediterranean citrus crops:

   a. The models do not consider intermittent river flows and low dilution conditions during periods of low rainfall.

   b. Prospective ERAs are limited to acute lethality because there are too few data on chronic toxicity, despite evidence of chronic exposures.

3. Monitoring should include high-frequency sampling by passive samplers, autosamplers triggered by rainfall, and standardization of sensitive analytical methods.

4. Biological and chemical monitoring should be coordinated to link pesticide use to environmental effects. Both types of data should be integrated with ecological models to assess the resilience of aquatic communities to repeated and long-term pesticide exposure under conditions of intermittent water flow.

Overall, this study highlights the need for post-registration monitoring to assess the adequacy of pesticide regulations, to identify pesticide misuse and to improve the prospective risk-assessment framework.

12.4 The Future of Environmental Regulation

An important gap in the regulatory process is the lag between a rapidly expanding knowledge base in ecotoxicology and the tools and data typically incorporated into regulations. For example, using survival, growth and reproduction to define guidelines, to conduct ERAs and to link cause and effect with biomonitoring should be augmented by the expanding array of molecular and physiological biomarkers of toxicity within an AOP framework (Section 4.3.3; Case Study 5.1). Given that the toxicity and bioavailability of most chemicals in commerce have not been tested, we can expect a significant increase in the development and application of bioaccumulation and toxicity models to support the development of guidelines and ERAs. As well, the suitability of establishing guidelines from empirical safety factors or from an understanding of the mechanisms of exposure and effect should be assessed in light of the environmental and economic costs of under- or over-estimating guideline concentrations.

Two major obstacles to implementing and enforcing chemical regulations are their administrative complexity and the technological challenges of treating complex wastes to standards that change in response to emerging issues of environmental contamination (Section 12.2.1.4). Industries are sensitive to the consequences of violating laws because specific commercial activities, such as exporting waste plastic for recycling, could be banned, and the cost of liability insurance continues to rise. Industries are also pressured by consumers, investors and government procurement agencies who demand 'green' goods and services that meet all legal requirements to minimize contamination and environmental impacts.

To remain competitive, many industries have adopted voluntary self-regulation, even before being legally required to do so, and often with innovations that reduce emissions and increase efficiencies. In the past, industries responded to discharge limits with 'end-of-pipe' treatment systems that were costly to install and operate. A more profitable alternative is a switch to production processes that reduce chemical emissions as part of a 'circular economy' that maximizes the economic, environmental and social value of products (www.nature.com/collections/hpcvbjppgy/). Circular production creates durable, repairable and recyclable products, often by redesigning the production process, the product, how it is marketed, and the overall business model (Agrawal *et al.*, 2019). Increasingly, governments encourage these innovative strategies by taxing emissions rather than defining concentration or loading limits. Among the best-known examples are carbon taxes, which encourage industries to adopt low-carbon technologies that reduce CO_2 emissions and production costs and increase their competitiveness. Adopting 'green' production technologies and sustainable development also frames corporate social responsibility. For example, the mining industry now promotes sustainability and practises voluntary reporting on environmental issues (Section 11.2.5) to improve its image and demonstrate its environmental and social performance (Dashwood, 2014).

Although the future of chemical management will include more self-regulation by emitters, traditional 'command-and-control' strategies will still be needed to regulate contaminated sites, the production and release of POPs, the occurrence of spills, and emissions of complex mixtures. Hence, there is a need to develop more cost-effective methods for toxicity testing and environmental monitoring. Most importantly, the economic costs of ecosystem services degraded by chemical contamination must be better understood to provide incentives for regulation.

An emerging component of chemical regulation is the recognition of the traditional environmental knowledge of Indigenous communities derived from their connections to the land and natural resources and from their political, economic, social and cultural traditions (Capsule 12.2). Many sovereign states have adopted the 2007 United Nations Declaration on the Rights of Indigenous Peoples (UNDRIP). As a consequence, Indigenous communities globally will play a much greater role in the conservation and protection of their lands, territories and resources, including issues related to chemical storage or disposal.

Capsule 12.2 Legislation for Chemical Management – Traditional Environmental Knowledge in the Regulation of Chemical Contaminants?

F. Henry Lickers

Canadian Commissioner, International Joint Commission, Canada–United States

Shekon (Greetings)

I hope this narrative finds you and your families in good health and spirits.

Introduction

I am a citizen of the Haudenosaunee, Seneca Nation, Turtle Clan. I am not learned in law and policy, but I have been invited to discuss Traditional Environmental Knowledge and the regulation of contaminants. It may seem strange to have this type of narrative in such a learned text, but it is the nature of my people to relay information in stories.

What is Akwesasne?

Akwesasne occupies the territory between Canada and the United States near Cornwall, Ontario and Massena, New York. The territory straddles part of the borders of Quebec, Ontario and New York State. To say that Akwesasne exists in a jumble of jurisdictions is an understatement. To add to the confusion, Akwesasne has three internal forms of government: two local elected governments, the Mohawk Council of Akwesasne in Canada and the St. Regis Tribal Council in the United States; and thirdly, the Mohawk Nation Council of Chiefs, the traditional government of the Haudenosaunee, which views itself as the sovereign government of all Mohawk Peoples. Each of these governing bodies has its own rules for environmental protection. To discuss any environmental subject, a meeting in Akwesasne must be convened with relevant departments of two federal and three provinces/states governments, and the local counties and institutions in the different jurisdictions. Each has its own set of environmental standards, criteria and laws.

Some 44 years ago, as a young biologist, I came to Akwesasne with my degree clutched tightly in my hand, ready to take on all issues. I began to study my environment and all of the texts that existed about Akwesasne. I learned that conflicts in the Mohawk territories of Akwesasne over resources, governance and understanding were common: first with the French, then with the English, and later with the United States and Canada.

As I travelled across the territory, I noticed the immense variety of ecological communities and complex **ecotones**. During my university studies, I was accustomed to a much more simplified description of the environment. I realized that the complexity existed, not as a legislative or legal outcome, but as an outcome of responsibility. The critical environments of marshes, wetland and forested areas existed because the Mohawk People honoured and utilized these areas, which were protected by the people more than by any law. As I studied these environments, I noticed that certain people took responsibility for certain areas and activities. Farmers cared for the agricultural lands but worked with the trapper/hunter, who took care of the marshes and who in turn worked with the fishers who took care of the river. Each had pride and vast knowledge about their respective interests. They shared their environmental goods with each other, each depending upon the other as much as they depended upon the environments, not as a money-based economy but as a barter economy. Each group gently taught this young biologist their knowledge. However, as my knowledge began to grow, I noticed something peculiar about the environment of Akwesasne. Yes, it was important to know what was in Akwesasne, but it was more important to know *what was missing*.

The first time that I realized this, I was crossing a buckwheat field. As I crossed the field, my mind went back to my childhood and being warned not to run through my grandfather's buckwheat field because I'd get stung by a honeybee. However, in this field, I noticed that there were no honeybees. With the knowledge I had gained, I knew that the production of this field would be decreased due to the lack of pollinators, and with decreased pollination, the income for the farmer would be decreased. It would also mean no honey. I had learned that honeybees are susceptible to fluoride, which was an airborne waste product of the local aluminium industries. As I stood there, I wondered why the law did not protect these honeybees. In that moment, I knew that laws could not protect everything in the environment and as an Indigenous person, I had a responsibility to care for the honeybees and to protect them. From then on, I looked for what was *missing* from the environment of Akwesasne. Upland game birds, rabbits and other small mammals should have thrived on these protected lands, but their populations were low compared to those in my own reserve in Six Nations of the Grand River near Brantford in south-west Ontario.

The Problem: The Complexities of Governance and Environmental Protection

There are misunderstandings between Indigenous governance and the relevant government agencies in the United States and Canada, including misunderstandings about chemical management and legislation. The 'Elected Band Councils' have the responsibility to govern their lands under a complex set of rules, regulations and policies. Of the many Canadian Federal acts that deal with the environment, only a few actually mention Indigenous people. The Elected Band Councils are expected to adhere to Federal and some Provincial Acts, but with confused jurisdiction and ignorance of the Indigenous culture and philosophy, the lands and waters remain unprotected and vulnerable. The Indigenous communities and the various governments are coming from different environmental and legislative philosophies, and there is a lack of empathy concerning protection of the environment: the government does not trust Indigenous people, and Indigenous people do not trust the government.

This lack of empathy is in part because the North American public is largely ignorant of Indigenous beliefs, history and concerns. They do not understand that Indigenous people have a different set of norms when it comes to the environment, but like everyone else will exploit the environment for their personal gain. Desperate people will most definitely act to protect themselves and their communities, but they are desperate because they have been kept in the economic wilderness for the past 300 years. In Akwesasne, the Mohawk People for the past 300 years have maintained the wetland while at Walpole Island they have protected oak savanna grassland, and there are many other examples, large and small. Some plant and animal species that are in danger of becoming extinct are protected by Indigenous peoples on their small reserve lands. Lately, oil sands and pipeline developments have become a focus of concern because carbon resource development is viewed as destroying the environment. As a result, Indigenous people believe it must be stopped or more strictly controlled. The 2007 United Nations Declaration on the Rights of Indigenous Peoples (www.un.org/development/desa/indigenouspeoples/declaration-on-the-rights-of-indigenous-peoples.html) is beginning to move the world to a place where Indigenous thoughts and responsibilities will be recognized. Progress includes moves towards reconciliation in Canada, inclusion of Maori peoples in New Zealand and Aboriginal peoples in Australia, and reinvigorated discussions about dismantling the Doctrine of Discovery (https://www.afn.ca/implementing-the-united-nations-declaration-on-the-rights-of-indigenous-peoples/), but we have a long road to share with each other.

Example: Chemical Contamination of Fish and the Protection of Human Health

In Akwesasne, the Mohawk Council established a Department of the Environment (1976) to protect and enhance the natural environment of Akwesasne. One of our first tasks was to understand the numerous contaminants in Akwesasne. The Department reached out to Federal, Provincial and State governments and found a woeful lack of data concerning fish contamination. We offered to help with collecting and analysing data on fish in local waters. Our fishermen were hired and helped with the design of the study and the collection of fish samples. Since that time, many fish contaminant surveys have been carried out, each with the task of measuring contaminants and applying fish consumption guidelines (Box 12.4). Each jurisdiction has its own guidelines and regulations, depending upon the geographic area within Akwesasne. For example, the maximum amount of fish a person should consume could change by simply moving your boat 3 m!

As a result, the Mohawk Council of Akwesasne advised its people that women of childbearing age, pregnant women and children under the age of 16 years should not consume *any* fish species from the St. Lawrence River. The Council also lobbied for a major health study of its people. This study was carried out in 1983–1985 by Irving Selikoff of Mount Sinai Medical School, Brooklyn, New York. At the same time, Canada and the United States, through the International Joint Commission, had declared the St. Lawrence River near Massena, NY and Cornwall, Ontario an Area of Concern and began to develop a Remedial Action Plan (Section 13.2). These studies found unacceptable levels of many contaminants including mercury, Mirex, PCBs, dioxins and dibenzofurans in fish and other biota important to the people of Akwesasne. Snapping turtles at that time contained so many compounds and at such high levels that they were declared hazardous waste and had to be disposed of with appropriate hazardous waste measures.

One of the problems facing the people of Akwesasne was that each of the many jurisdictions based their consumption guidelines on concentrations of single contaminants, such as mercury or PCBs, but the fish that people ate contained many different compounds. Another weakness of the guidelines was the mismatch between the amount of fish consumed by Akwesasne people during a meal, week and month (metrics used in calculating consumption guidelines) and the much smaller amounts consumed by non-indigenous people. Prior to 1976, the Mohawk People of Akwesasne depended upon the fishery for 75% of their protein, but after 1976, fish consumption dropped drastically. As a result, the diet of the people included increased amounts of carbohydrates that contributed to an epidemic of diabetes.

By then, the people of Akwesasne were also divided between those who loved eating fish and those who were afraid, with arguments among fish consumers, Indigenous governments, fishers and Akwesasne mothers about the safety of St. Lawrence River fish. The scientists (biologists, nutritionists and chemists) also had different opinions based on their respective understandings. The biologist looked at the fish for tumours and any manifestation of the poisons, the chemist analysed fish samples for single poisons and the nutritionist expounded on the virtues of eating fish. Some of the studies showed that even minute amounts of certain compounds were harmful to human development. As a scientist, I believe in science combined with the traditional knowledge of my people. However, to try to understand a complex ecosystem, the scientists were not looking at the question clearly but were using a model focused on single contaminants, financial concerns and government environmental policy. A fish swims and uses the whole ecosystem – it cannot differentiate between good compounds and toxic ones but accumulates them all. We depend upon our experts for advice, but if they are also confused, we suffer.

A Solution

The development of different fish consumption guidelines by each government is understandable since legislation is to protect all people. However, special cases like Akwesasne are very hard to fit into a 'one-size-fits-all' policy. Single contaminant and a 'standard person' analysis no longer fit the complex environment we live in. To overcome these problems, algorithms are needed to integrate the complexities of the environmental fate and behaviour of multiple chemicals in different ecosystems with their effects on people who have different cultural and dietary practices.

So how do we as scientists look for what is missing? What does a normal healthy environment really look like? When do we achieve good health and stable well-being? By analogy, climate change and the COVID-19 pandemic have shown us that multiple disciplines working together can give us a better understanding of workable solutions to complex problems. For contaminants, we need the chemist to investigate models of contaminant interactions in the bodies of flora and fauna, including humans. We need the biologist to investigate the pathways and interactions in the ecosystems. We need the health professional to look at the mortality and morbidity of our communities and match them with our contaminant profiles, both in ourselves and within the environment. And we need them all to work together with Community Knowledge to integrate these complexities into *culturally appropriate solutions*. As the elders say, a cause is like dropping a pebble into a pond, the spreading ripples representing the ever-increasing effects of the cause on the associated hierarchies. This is not a new concept but one that requires greater and more vigorous application to the whole contaminants issue. Indigenous communities may be uniquely able to help the scientific world to develop these concepts.

Conclusions: The Need for Personal and Collective Responsibility

Over the past thousand years, we have seen an increase in life expectancy and improvements for human health, but we have also seen some of the greatest species extinction events and some of the greatest wars. We have seen education increase in our populations but also a stubborn adherence to harmful precepts ('fake news'). How is it we can believe that we can dump tonnes of toxic waste into the environment with no negative effects? The application of the precautionary approach, zero discharge (Section 11.4.2) and the impact of Liebig's Law of the Minimum must be integrated into our work. Past norms must be re-evaluated, established axioms challenged, and our models made understandable to lawmakers and the public. As scientists we must look to the information and media sciences to translate and integrate our science into our everyday thoughts. Research on the minuscule will always be needed, but today's problems are also found at the level of family circles, communities, nations and confederacies.

However, it is not enough to just communicate. In the current Global Digital culture, we are learning to integrate, translate and communicate within networks of people. Canada ranks around 23rd in the world with 90% of our population of about 36,000,000 people being connected electronically. A good number of these are Indigenous People, and we have tried to warn society about the environmental damage being caused. We have approached these problems with a holistic view considering all aspects of the problem. So, what does this have to do with legislation for chemical management and with Traditional Environmental Knowledge in the regulation of chemical contaminants? Legislation is one small tool to protect society and the environment. Rights and laws are the weaker half-brother of responsibility. It is time to embrace the concepts of responsibility. If we want to manage the myriad chemicals in the environment, we must accept our responsibility for them (Section 12.1). Any damage caused by these compounds must be visited on the producers and not emptied into the commons. PCBs, dioxins, carbon dioxide, fluoride and perfluorinated compounds will affect our environment for a very long time!

Skennen (Go in the way of Peace)
Henry

P.S. In the meantime, I'll be rooting for the Bees!

Summary

This chapter provides an overview of how environmental law translates public concerns about environmental contamination into regulatory action. Although the emphasis has been on the technical role of ecotoxicology in chemical management, differences among nations in regulations reflect a wide array of societal concerns and attitudes, although there is clear need to include Traditional Environmental Knowledge. Environmental protection is intimately connected to national and international priorities for economic and social development. Recognizing the interrelationships among these issues and the ecological and economic costs of contamination will be key to developing regulatory and sustainable production strategies that achieve environmental protection for a diverse community of nations, particularly Indigenous Peoples.

Review Questions and Exercises

1. Compare and contrast environmental objectives, criteria, guidelines and standards. Which are legally enforceable?

2. Who is ultimately responsible for ensuring that chemical pollutants are regulated?

3. What is the fundamental difference between human health risk assessment and ecological risk assessment?

4. What is enabling legislation and why is it an efficient means of regulating chemical emissions?

5. What is social licence? How would you measure it?

6. **Project:** For your own country, state (province, canton), and municipality (city), work in groups of three to five students to identify the government departments and agencies that regulate chemical emissions, environmental concentrations and ecological effects, and the nature of the substances or activities that they regulate.

7. Why is International Environmental Law so difficult to enforce? Provide some examples of differences among nations in regulations that create obstacles to environmental protection.

8. If you were asked to assess the ecological risks of compounds found at high concentrations in soils, what steps would you take to estimate risk? What factors would contribute to your confidence in the estimated risk, and what factors would create uncertainty?

9. What is a safety factor and where is it applied? Why may water quality criteria vary among countries, even those with similar ecosystems?

10. What does PBT refer to? What sorts of compounds might be B and T but not P? Why are metals not considered P or B?

11. How does the monitoring of chemical concentrations in the human food chain contribute to environmental protection?

12. How would you determine the effectiveness of regulations to limit the discharge of an organic compound that biomagnifies in food chains and inhibits the testosterone production and gonadal development of male fish?

13. **Project:** Choose several products from your daily life (e.g., pickles, mobile phones), investigate the substances used in their production, use and disposal, and devise alternative processes that might lead to a circular economy for those products.

Abbreviations

AOP	Adverse outcome pathway	MOE	Ministry of the Environment
BAF	Bioaccumulation factor	NEC	No-effect concentration
BCF	Bioconcentration factor	NOEC	No-observable-effect concentration
BLM	Biotic ligand model	OECD	Organisation for Economic Co-operation
BMF	Biomagnification factor		and Development
COC	Chemical of concern	PBT	Persistent, bioaccumulative and toxic
EC	Effect concentration	PCBs	Polychlorinated biphenyls
eDNA	Environmental deoxyribonucleic acid	PEC	Predicted environmental concentration
EEM	Environmental effects monitoring	PNEC	Predicted no-effect concentration
ERA	Ecological risk assessment	REACH	Registration, Evaluation, Authorisation and
FAO	Food and Agricultural Organisation		Restriction of Chemicals
GLWQA	Great Lakes Water Quality Agreement	SQG	Sediment quality guideline
HC	Hazard concentration	SSD	Species sensitivity distribution
HHRA	Human health risk assessment	TC	Technical committee
HQ	Hazard quotient	UNEP	United Nations Environment Programme
IEL	International Environmental Law	US	United States
K_{OW}	Octanol–water partition coefficient	VEC	Valued ecosystem component
LDC	Less developed country	WoE	Weight-of-evidence
LOE	Line of evidence	WHO	World Health Organisation
MECP	Ministry of Environment, Conservation and Parks	WQG	Water quality guideline

References

Adams, W., Blust, R., Dwyer, R., *et al.* (2020). Bioavailability assessment of metals in freshwater environments: a historical review. *Environmental Toxicology and Chemistry*, 39, 48–59.

Agrawal, V. V., Atasu, A. & Wassenhove, L. N. V. (2019). OM forum – new opportunities for operations management research in sustainability. *Manufacturing and Service Operations Management*, 21, 1–12.

Ayentimi, D. T., Burgess, J. & Brown, K. (2016). Developing effective local content regulations in sub-Sahara Africa: the need for more effective policy alignment. *Multinational Business Review*, 24, 354–374.

Bradbury, S. P., Feijtel, T. C. J. & Van Leeuwen, C. J. (2004). Meeting the scientific needs of ecological risk assessment in a regulatory context. *Environmental Science & Technology*, 38, 463a–470a.

Burgess, R. M., Berry, W. J., Mount, D. R. & Di Toro, D. M. (2013). Mechanistic sediment quality guidelines based on contaminant bioavailability: equilibrium partitioning sediment benchmarks. *Environmental Toxicology and Chemistry*, 32, 102–114.

Dashwood, H. S. (2014). Sustainable development and industry self-regulation: developments in the global mining sector. *Business and Society*, 53, 551–582.

Docherty, B. (2001). Challenging boundaries: the Arctic National Wildlife Refuge and international environmental law protection. *New York University Environmental Law Journal* 10, 70–116.

Eilstrup-Sangiovanni, M. & Bondaroff, T. N. P. (2014). From advocacy to confrontation: direct enforcement by environmental NGOs. *International Studies Quarterly*, 58, 348–361.

European Commission (1996). *Technical Guidance Document in Support of the Commission Directive 93/67/EEC on Risk Assessment for New Notified Substances and the Commission Regulation (EC) 1488/94 on Risk Assessment for Existing Substances*, Parts I–IV. Luxembourg: Office for Official Publications of the European Communities.

Evers, D. C., Keane, S. E., Basu, N. & Buck, D. (2016). Evaluating the effectiveness of the Minamata Convention on Mercury: principles and recommendations for next steps. *Science of the Total Environment*, 569, 888–903.

Fox, D. R., van Dam, R. A., Fisher, R., *et al.* (2021). Recent developments in SSD modeling. *Environmental Toxicology and Chemistry*, 40, 293–308.

Hall, N. (2008). The centennial of the Boundary Waters Treaty: a century of United States–Canadian transboundary water management. *Wayne Law Review*, 54, 1418–1449.

Kukenova, M. & Monteiro, J.-A. (2008). Does lax environmental regulation attract FDI when accounting for 'third-country 'effects?. *Munich University Personal Archive*. Munich, Germany; https://mpra.ub.uni-muenchen.de/11321/

Lee, K., Boufadel, M., Chen, B., *et al.* (2015). *Expert Panel Report on the Behaviour and Environmental Impacts of Crude Oil Released into Aqueous Environments*. Ottawa, ON: Royal Society of Canada.

Martin, O. V., Adams, J., Beasley, A., *et al.* (2019). Improving environmental risk assessments of chemicals: steps towards evidence-based ecotoxicology. *Environment International*, 128, 210–217.

Rauter, T. (2019). Disclosure regulation, corruption, and investment: Evidence from natural resource extraction. *New Working Paper Series No. 31*. Chicago, IL: Stigler Center for the Study of the Economy, University of Chicago Booth School of Business.

Reed, D. (2002). Resource extraction industries in developing countries. *Journal of Business Ethics*, 39, 199–226.

Rico, A., Dafouz, R., Vighi, M., Rodríguez-Gil, J. L. & Daam, M. A. (2021). Use of postregistration monitoring data to evaluate the ecotoxicological risks of pesticides to surface waters: a case study with chlorpyrifos in the Iberian Peninsula. *Environmental Toxicology and Chemistry*, 40, 500–512.

Selin, H. (2005). European over-REACH? Efforts to revise European Union chemical legislation and regulation. In *Conference on Human Dimensions of Global Environmental Change*. Berlin, Germany; http://userpage.fu-berlin.de/ffu/akumwelt/bc2005/papers/selin_bc2005.pdf

Selin, H., Keane, S. E., Wang, S. X., *et al.* (2018). Linking science and policy to support the implementation of the Minamata Convention on Mercury. *Ambio*, 47, 198–215.

Suter, G. W., Norton, S. B. & Barnthouse, L. W. (2003). The evolution of frameworks for ecological risk assessment from the Red Book ancestor. *Human and Ecological Risk Assessment*, 9, 1349–1360.

UNEP (2019). *Environmental Rule of Law: First Global Report*. Nairobi, Kenya: United Nations Environment Programme; https://wedocs.unep.org/handle/20.500.11822/27279

UNEP & WHO (2008). *Guidance for Identifying Populations at Risk from Mercury Exposure*. Geneva, Switzerland: United Nations Environment Programme (UNEP), Division of Technology, Industry and Economics (DTIE); World Health Organization (WHO), Department of Food Safety, Zoonoses and Foodborne Diseases, Cluster on Health Security and Environment.

US EPA (1992). *Framework for Ecological Risk Assessment*. Washington, DC: US Environmental Protection Agency, Risk Assessment Forum.

US EPA (1998). *Guidelines for Ecological Risk Assessment*. Washington, DC: US Environmental Protection Agency.

US EPA (2007). *Framework for Metals Risk Assessment*. Washington, DC: US Environmental Protection Agency, Office of the Science Advisor, Risk Assessment Forum.

Warne, M. S., Batley, G. E., van Dam, R. A., *et al*. (2018). *Revised Method for Deriving Australian and New Zealand Water Quality Guideline Values for Toxicants – update of 2015 version*. Prepared for the revision of the Australian and New Zealand Guidelines for Fresh and Marine Water Quality. Canberra, Australia: Australian and New Zealand Governments and Australian State and Territory Governments.

Wyeth, G. B. (2006). Standard and alternative environmental protection: The changing role of environmental agencies. *William & Mary Environmental Law and Policy Review*, 31, 5–74.

13 Recovery of Contaminated Sites

Pamela M. Welbourn and Peter V. Hodson

Learning Objectives

Upon completion of this chapter, the student should be able to:

- Appreciate the historical background that results in contaminated sites and understand what is meant by contaminated site, remediation, ecological restoration and ecosystem function.
- Show how ecotoxicology can guide an understanding of the methods and approaches that are termed: engineering, natural recovery, bioremediation, recolonization or recruitment, and phytoremediation.
- Discuss the advantages and limitations of the monitored natural recovery (MNR) approach to recovery of contaminated sites.
- Distinguish between the approaches of those who study and manage chemicals and those who study and manage ecosystems, and critically review a systems solution to this potential conflict.

This chapter deals with scientific and technical options for clean-up of contaminated sites. Typically, these sites are the legacy of past activities, often subject to minimal or no regulation when the contamination originally occurred. We describe the procedures involved in assessment, remediation and recovery for contaminated systems, drawing from our understanding of the mechanisms of toxic effects described in earlier chapters. A multi-disciplinary team is typically involved in the planning of a clean-up operation. The priority of some members will be for 'ecological restoration', while others will emphasize the removal or isolation of contaminants. The latter group focuses on risk assessment. We address physical engineering as well as biological methods; sometimes these are used in combination. Decisions about the approach to be used depend in part on the ecological significance of a particular site, but also on socio-economic considerations and the proposed land use following clean-up. In many examples, regulations play a part in guiding these approaches.

13.1 Background

Until the second half of the twentieth century, there was tacit acceptance of some degree of change or damage to the environment resulting from industrial, agricultural and other human activities. In Chapters 6, 7, 8 and 10 we address the contamination of air, soil and water by inorganic, organic and radioactive substances, and in Chapter 11 we discuss how such human activities can result in complex ecotoxicological hazards. Contaminated, damaged, impaired or degraded terrestrial or aquatic sites or ecosystems may be abandoned, and for many sites, the origins or sources of contamination are unknown. For other sites, the origins of contamination may be traceable to former owners, although perhaps their activities were not illegal when contamination occurred. There are now national and even international laws that hold responsible parties liable for restoring such contaminated sites, which is not true for other environmental concerns, such as loss of habitat and introduction of invasive species. Activities directed to the recovery of damaged sites represent part of the broader paradigm of environmental protection. With advancements in our understanding of ecosystem function, along with improvements in technology, it is now possible to anticipate restoring waters and lands that were once deemed unusable because of extremely high levels of contamination.

Not all contaminated sites are of equal size, ecological value or complexity. Commercial sites or local small industries, such as abandoned petrol stations, former glassworks or tanneries in urban or suburban settings, present very different scenarios compared with wetlands that have been damaged by oil spills, or land deforested by severe air pollution. Decisions about **remediation** of a site or an ecosystem must take into consideration and may even be directed by the proposed future use of the site, as well as by its perceived ecological value. For example, most '**brownfield**' sites (Section 1.2) are in urban areas where any sign of a functioning ecosystem has long gone. Their future use is likely to be some kind of 'redevelopment' such as residential housing or non-industrial commercial operations, or even paved surfaces. In no sense will there be any attempt to restore a functioning ecosystem, so the goal of clean-up for such sites is to remediate and reduce risk for humans and other organisms that are exposed to residual contaminants after clean-up. In contrast, for projects like the Upper Clark Fork River Basin, Montana (see Sections 11.2.4.1, 13.9.1), efforts are being made to restore functional native plant communities in the basin. For other projects, the goal may lie somewhere between these two extremes.

Regulations typically prescribe clean-up procedures starting with risk assessments, but there are many additional factors that need to be considered when planning recovery options. As pointed out by Hobbs *et al.* (2014), complicating factors beyond the actual contamination of sites include climate change, changes in land use and species invasions. These authors also show that some changes are irreversible, for example the secondary salinization of wetlands in Australia or the 'creation' of novel substrates such as shale-oil spoil heaps in Scotland. In such cases, a recovery to the previous state is no longer feasible, or indeed, possible. Furthermore, financial, technical and social obstacles may also limit recovery to the original condition, which effectively renders some changes irreversible.

13.2 Component Disciplines and Goals

To restore a contaminated or damaged site, expertise is required in a wide range of disciplines (e.g., ecotoxicology, engineering, restoration ecology, sociology, economics, risk assessment, risk communication and risk management). Typically, the procedure has been linear, first remediating (removing or treating to render 'safe') the site, and then restoring some kind of functioning ecosystem. As discussed above, some systems are not amenable to recovery to the original condition, nor in many cases is that the goal. Frequently, there are concerns over aesthetic values such as landscape conservation, or loss of valued species. Beginning in the 1980s, a body of regulations has developed in a number of jurisdictions for the restoration of contaminated sites (Section 1.2.4).

Among theoretical ecologists and practitioners, their respective emphases on the goals for restoration have given rise to some controversy. Restoration clearly includes not only the science of ecology but also societal decisions on appropriate **endpoints** for restoration. In 2014, experts from Australia, Canada, Mexico, the United Kingdom and the United States met to discuss methods to advance the practice of ecosystem restoration of sites contaminated by industrial processes. The workshop was provocatively entitled "Restoration of impaired ecosystems: An ounce of prevention or a pound of cure?" It involved ecotoxicologists, restoration ecologists and experts from related fields, and the goal was to provide a forum to discuss the best scientific practices to restore ecosystems, while simultaneously addressing contaminants of concern (Farag *et al.*, 2016).

As a result of the workshop, Wagner *et al.* (2016) proposed a framework for determining restoration goals (Figure 13.1). This framework aimed to establish 'achievable goals' for ecological restoration, to be used in combination with goals for remediation. This model illustrates the point that for each case, restoration depends first on an understanding or definition of what is meant by a 'restored' system. If defined on a case-by-case basis, then goals can be set. The framework also recognizes situations in which restoration may not be feasible and allows for "off-site compensatory restoration of an alternative site". This provision corresponds to a philosophy of 'no net loss', whereby a governing authority can issue a permit to construct habitat equivalent to that which had been 'lost' because restoration of the contaminated site was deemed unfeasible. To ensure equivalence, post-restoration monitoring is needed to

Determining restoration goals

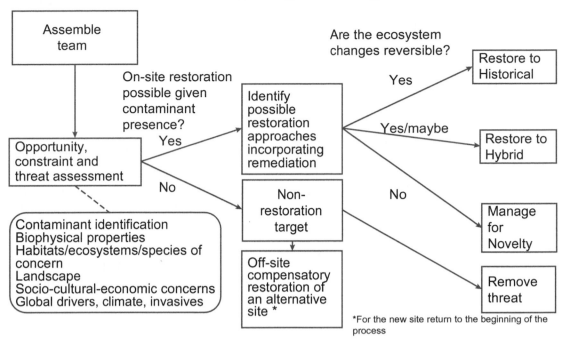

Figure 13.1 Decision tree for determining restoration goals for contaminated sites. Rectangles represent actions and the lozenge in the first column represents inputs to assessment. Arrows indicate decisions with respect to the questions (outside the rectangles) posed above them. Reproduced from Wagner *et al.* (2016), with permission from John Wiley and Sons.

confirm that the quality and productivity of the constructed habitat is equal to or greater than what was lost. Although the intent of 'no net loss' is admirable, its value can only be realized through close supervision.

For the Laurentian Great Lakes, restoration goals have been set by a somewhat more pragmatic and anthropocentric approach by the International Joint Commission (IJC) under the Canada–US Agreement on Great Lakes Water Quality (Section 12.2.2). In 1987, the IJC identified 43 Areas of Concern (AOCs), i.e., sites where environmental quality was degraded by chemical contamination or urban and industrial development. The sites were defined by the loss or degradation of one or more ecosystem services, designated as Beneficial Use Impairments (BUIs); Box 13.1).

The listing of a harbour or region as an AOC creates a challenge for municipalities seeking economic growth, tourism and a reputation as a good place to live. In response, local authorities collaborated with state and provincial governments to develop **remedial action plans (RAPs)** that included:

1. A list of local or regional BUIs and their likely causes;
2. The criteria for restoring the BUIs, taking into account local conditions and in consultation with local communities;

> ## Box 13.1 Examples of impairments of beneficial use (ecosystem services) identified within the North American Great Lakes Areas of Concern
>
> - Restrictions on Fish and Wildlife Consumption
> - Bird or Animal Deformities or Reproduction Problems
> - Fish Tumours or Other Deformities
> - Degradation of Fish and Wildlife Populations
> - Degradation of Benthos
> - Eutrophication or Undesirable Algae
> - Beach Closings
> - Degradation of Phytoplankton and Zooplankton Populations
> - Loss of Fish and Wildlife Habitat

3. Remedial measures to be taken, including identifying the parties responsible for implementing these measures;
4. A **monitoring** and evaluation programme to track remedial measures and how they lead to resolving BUIs; and
5. Annual summaries of the measures taken and their progress in removing each BUI.

The causes of BUIs and their technical solutions were not always obvious, and extensive research and surveys were often required before progress could be made on resolving each issue and ultimately delisting each AOC. Since 1987, five of the 43 sites have been delisted, several more are in recovery (Case Study 13.1), and all have made significant progress towards delisting. The strength of this approach is the focus on specific environmental issues relevant to local citizens, and their engagement in finding solutions and learning about the science behind BUIs and their resolution.

In summary, ideally restoration occurs when realistic goals and objectives are set, funding is adequate, all **stakeholders**, including Indigenous peoples, are thoroughly involved, and the costs and benefits of the selected restoration plan are carefully considered.

13.3 Definitions and Concepts

Some of the terms that are encountered in the context of the recovery of contaminated sites are also in general use, a situation that has led to misunderstandings between stakeholders and environmental scientists. For example, the term 'stakeholder' is considered to be pejorative among Indigenous peoples who consider it a term that reflects colonial claims of resources within their Traditional Lands. Even in the scientific literature, many terms have been used inconsistently. We have already defined 'contaminant' in Section 1.1, but other terms have been used in the context of site recovery. We use the term 'recovery' in a very broad sense as a concept, but a number of different terms are encountered in the related literature (Box 13.2).

13.4 Triggers for Action Towards Recovery

We should begin by asking how a contaminated site is identified and how action towards recovery is initiated. In fact, it may be a rather *ad hoc* process. The trigger may originate from public pressure – for example, the past contamination of the Love Canal site in the Niagara region of North America had not been previously recognized, but the issue became highly publicized, and action followed (Section 1.2). It may also be triggered by a land sale in which potential buyers are aware of the history of a site and insist on a site assessment of soil contamination before a sale proceeds. Restoration may have a regulatory driver, in which case there will be liability for the current owner of

> ### Box 13.2 Terms used in the context of recovery
>
> - Contaminated site: a site that has received substances that were not previously present at that site, or that are at significantly higher concentrations than were normal for the site.
> - Remediation: the removal of contaminants present in excess.
> - Ecological restoration: the process of assisting the recovery of an ecosystem that has been degraded, damaged or destroyed. Restoration includes the re-establishment of the pre-existing biotic integrity in terms of species composition and community structure.
> - Ecosystem function: the processes that control the fluxes of energy, nutrients, inorganic and organic matter through any environment.

> ### Box 13.3 Questions that can guide clean-up
>
> - Which contaminated sites are likely to result in the exposure of living organisms to chemical contaminants? Of these, which are likely to represent a real risk of toxicity and ecological impacts?
> - What is the current and future use of a given site?
> - What technical means are available to remedy any potential risk that has been identified?
> - Who determines whether a site requires clean-up?
> - Who is responsible for the site if it is determined that it requires clean-up?

the site. Restoration activity may stem from a systematic review of known former industrial areas, in locations that have current value for redevelopment. In such an example, an assessment will normally be required when there is an application for a change in land use or zoning.

It is not possible to clean up or even identify all contaminated sites. The selection of candidate sites, as well as the choice of procedures for remediation or restoration, will depend on a number of factors, not least the regulatory climate in specific jurisdictions. In an ideal world, the questions in Box 13.3 need to be addressed prior to any decision about clean-up and recovery. The first two points are essentially components of risk assessment as described in Section 12.3.2.

Once the environmental and human safety needs for clean-up are established, the actual procedure must be determined. As pointed out in Section 13.2, this is a multi-disciplinary and multi-stakeholder activity. Technical and scientific considerations are almost invariably weighed along with economic and social concerns.

13.5 Methods and Approaches for Recovery

The objectives for the clean-up of contaminated sites may include the following (which are not mutually exclusive):

- to return a site to its natural or historical condition – for conservation, safety or aesthetic reasons;
- to manage risks to human health or the ecosystem as a whole;
- to comply with regulations (which vary by jurisdiction);
- to comply with the requirements of an **Ecological Risk Assessment** for a proposed development.

The Society for Ecological Restoration (SER) defines a restored ecosystem with nine attributes (Box 13.4). The Primer from which these attributes are extracted continues with "full expression of all of these attributes is not essential to demonstrate restoration. Instead, it is only necessary for these attributes to demonstrate an appropriate trajectory of ecosystem development towards the intended goals or reference" (Society for Ecological Restoration International, 2004). In other words, the attributes can be considered as 'ideal'. In practice, a variety of options exist, not all of which would qualify under the SER definition.

The characteristics of the contaminants at a site can influence the range of appropriate remedial and restoration opportunities. Similarly, biophysical factors such as geological, ecological and climatological conditions will also influence restoration practices. Achievable goals are determined after careful consideration of the opportunities and constraints that the site may present. Once the nature, location and quantities of contaminants are known, an understanding of the environmental behaviour, toxicology and ecotoxicology of the contaminants will guide the choice of the most appropriate technology.

In contrast to the SER position on attributes, it may be argued that removal of contaminated material or making it unavailable should be the primary goal of remediation and restoration. Removal addresses the risk assessment and risk management goals, but in fact, some of the

Box 13.4 Society for Ecological Restoration (SER) attributes of a restored ecosystem

1. The restored ecosystem contains a characteristic assemblage of species that occur in the reference site.*
2. The restored ecosystem consists of indigenous species.
3. All functional groups necessary for the continued development and/or stability of the restored ecosystem are represented or, if they are not, the missing groups have the potential to colonize by natural means.
4. The physical environment of the restored ecosystem is capable of sustaining reproducing populations of the species necessary for its continued stability.
5. The restored ecosystem apparently functions normally for its ecological stage of development.
6. The restored ecosystem is suitably integrated into a larger ecological matrix or landscape, with which it interacts through abiotic and biotic flows and exchanges.
7. Potential threats to the health and integrity of the restored ecosystem from the surrounding landscape have been eliminated or reduced as much as possible, and contaminants are eliminated.
8. The restored ecosystem is sufficiently resilient to endure the normal periodic stress events in the local environment that serve to maintain the integrity of the ecosystem.
9. The restored ecosystem is self-sustaining to the same degree as its reference ecosystem and has the potential to persist indefinitely under existing environmental conditions.

*"The SER Primer emphasises that using a reference or reference ecosystem is a critical aspect of achieving restoration success as it provides a clear depiction of the goals of the restoration project and a development state to evaluate against" (Wortley et al., 2013).

options that have been developed do not necessarily involve physical removal of material. Examples of possible approaches include the following:

- engineering – physical intervention including removal ('dig and dump' or 'pump and treat'), containment on site, the addition of binding materials to immobilize contaminants *in situ*, or construction of barriers;

- removal of the source, followed by natural or 'passive' recovery. In some cases, the source is not removed but natural **attenuation** can still occur. In both cases, recovery should be monitored (e.g., MNR);
- **bioremediation** – using selected organisms that can break down and detoxify organic contaminants;
- **phytoremediation** – using (mostly vascular) plants that tolerate and accumulate or detoxify contaminants;
- recolonization/revegetation, sometimes referred to as translocation and/or reintroduction of the original flora and fauna, as well as tolerant organisms from local or non-endemic sources.

These categories are for convenience/administrative neatness and are not mutually exclusive. Ideally, long-term monitoring is a component of the process.

As will be seen in the more detailed descriptions of the respective approaches, there may be overlap or blending among them, or a sequence of approaches that evolve over time. Engineered solutions often require further action in the form of revegetation; or if natural recovery does not proceed as expected, intervention in the form of bioremediation may be necessary. For example, attenuation of organic contaminants in MNR may not proceed passively, but may require nutrient enrichment, or inoculation with degrader organisms, which would move such a project into the category of bioremediation.

13.6 Engineering

A number of physical engineering options are available to remove or contain contaminants; this approach addresses ERA and risk management and many regulatory requirements, but not ecosystem recovery.

13.6.1 Removal and Off-site Disposal of Contaminated Material

The excavation of soil, dredging of sediment or interception of groundwater discharge were some of the earliest approaches to the clean-up of contaminated sites. Indeed, excavation may be mandated if contaminants such as PCBs exceed a certain threshold of concentration (Case Study 13.2). These options often require large machinery, which may cause physical disturbance of the site. There may be occupational risks for the operators as well as ecotoxicological risks. Physical removal of contaminants of concern raises the very real possibility of their

mobilization or of changes in their bioavailability during the initial collection and removal steps. Particular concern has been expressed for dredging sediments. Sediment disturbance will result in oxidation of **anoxic** sediments and resuspended contaminants might affect aquatic biota. In a detailed report from the US Army Corps of Engineers on guidelines for dredging contaminated sediments, Palermo *et al.* (1998) concluded that "Environmental dredging removes contaminants from the aquatic environment. However, the potential for sediment resuspension may limit the effectiveness... as a remedial approach at some sites". Recent work has shown that concerns about dredging may have been overestimated, at least for metals, or at least that they represent only a short-lived problem. One of the reasons has been discussed by Jones-Lee and Lee (2005). They show that during sediment suspension into the water column, ferrous iron in the sediments and dissolved oxygen in the water column react to produce a precipitate of $Fe(OH)_3$, which rapidly scavenges many contaminants. The precipitated material is deposited back into the sediment. However, habitat disruption for benthic organisms may be of more concern than the mobilization of contaminants as a result of dredging. Fortunately, other options are available to reduce risks from contaminated sediments (Section 13.7.2).

The ecotoxicological concerns for groundwater contamination relate mainly to the discharge of groundwater into surface water and the risk of toxicity to aquatic biota, although there are also significant concerns for human health when groundwater is used as a potable water supply. Dissolved chemicals, including metals, industrial solvents and fuel oil, can enter groundwater from various sources, including improperly engineered landfills and leaking storage tanks. Particularly problematic are the so-called **dense non-aqueous phase liquids (DNAPLs)** such as chlorinated solvents (Engelmann *et al.*, 2019). Because they are denser than water, DNAPLs sink and contaminate deep groundwater. Removal is notoriously difficult and expensive and, in some cases, DNAPLs have been left *in situ* and the use of groundwater limited. Once contaminated, an aquifer is very difficult to clean up, although barriers can be placed to restrict the movement of contaminant plumes. The main approach to risk management is to pump and treat the trapped groundwater before it contacts surface ecosystems or is used for human or livestock consumption. The contaminated water is either piped to sewers or sent to an above-ground treatment system for physical or chemical treatment or bioremediation (Section 13.7). More recently, permeable reactive barriers have been designed to intercept contaminated plumes of groundwater and remove the chemicals *in situ* through reactions with the components of the barrier. This approach can be much more efficient than pump-and-treat.

One of the obvious concerns for all off-site disposal options is the long-term fate of contaminants at the disposal site. The process might be perceived in the worst case as simply moving the problem from one site to another. Disposal options and the extent to which material is removed from contact with biota vary, depending on risk. For example, dredged sediments that pose a low risk may be managed by open or unconfined open-water disposal; those with moderate to high risk require more confined or contained disposal. Landfilling is one option, but no landfill is perpetually secure. Some of the approaches, including bioremediation (Section 13.8), have the potential to clean up disposed material for later reuse. In common with all clean-up options, long-term monitoring of disposal and treatment sites is essential, and some regulations require a 'cradle-to-grave' approach to the clean-up.

The complete removal of contaminated material is sometimes not technically or economically feasible, and for future uses of a contaminated site, risk management may limit its use if remediation is incomplete. An example of the limitation of use was illustrated by the attempt to remediate the site of an old coal gasification operation in Kingston, Ontario, Canada. During the clean-up, it became evident that the extent of contamination of fractured limestone by DNAPLs was much greater than expected and that much of this material was inaccessible without extensive bedrock excavation. All future development was limited to above-ground structures so that volatile organic compounds do not contaminate basements and cause health problems. The unrecovered DNAPLs continue to migrate with groundwater through cracks in bedrock and along pathways created by sewer installations. With the proximity of this site to Lake Ontario, long-term monitoring was a critical component of risk management.

13.6.2 On-site Remediation

Wastes can be buried on the site and covered with a membrane or an impervious material such as clay to limit infiltration of precipitation, which would favour leaching or erosion. A modification of this method involves sealing off the contamination in a prepared location on the site, an option that may be preferable to simple burial without treatment. Various procedures, including entombment and vitrification (heating to high temperature to form or embed in glass) are available.

On-site containment is particularly appropriate for local 'hotspots' when the majority of a site is 'clean'. It has the advantage of limiting disturbance of the site, with less risk of re-entrainment of contaminants. It also minimizes the cost and environmental risks of transporting large volumes of contaminated material to alternative disposal sites (Case Study 13.1).

These approaches can be viewed as long-term 'storage' solutions, rather than clean-ups. Even today, burial may sometimes be approved for long-term disposal, depending on the geological and hydrogeological conditions at the site. However, as illustrated by the Love Canal situation (Section 1.2.4), long-term events may reveal that the burial solution was unsafe. In that particular case, the landfill was not leaking until municipal services were installed that breached the seal and allowed water to enter and chemicals to migrate. The lesson is that the long-term security of stored waste chemicals cannot be guaranteed, no matter how well engineered the storage site. The disposal and 'storage' of radioactive waste (Section 10.3.5) represents what is arguably the greatest long-term risk from burial of contaminants.

In aquatic systems, capping contaminated sediments is an acceptable approach, provided there are no strong currents. Sediments by their nature accumulate deposited material over time, removing contaminants from the aqueous phase, as illustrated by examples below for natural recovery (Section 13.7.2). If the sources of contaminants are regulated, the subsequent accumulation of uncontaminated sediments will create a natural cap over the contaminated materials. At such sites, activities such as the anchoring of large ships must be controlled to avoid disturbing the sediment cap and the underlying contaminants.

Case Study 13.1 Entombment

The Randle Reef Sediment Remediation Project in Hamilton Harbour, Ontario, Canada illustrates another engineered approach. The harbour is one of 43 AOCs identified in the Great Lakes Water Quality Agreement between Canada and the United States (Section 13.2). Randle Reef is believed to be the largest site contaminated by polycyclic aromatic compounds (PACs) and coal tar on the Canadian side of the Great Lakes, with an estimated 675,000 m^3 of contaminated sediments requiring clean-up. The source is a large steel mill that converts coal to metallurgical coke for high-temperature steel production (Hall and O'Connor, 2016). A by-product of coke production is the release of PACs into the atmosphere and into mill effluents. Environmental assessment began in 2003, with a great deal of exploratory research preceding any remediation. After rejecting the options of removal and incineration or removal and off-site storage, the stakeholders, representing technical, industrial and community interests, chose an *in situ* engineered containment facility (ECF). The ECF would cap about 130,000 m^3 of contaminated sediments, enclose about 500,000 m^3 of dredged contaminated sediments from the surrounding areas, and involve dredging and on-site water treatment.[1] In 2017, the most severely contaminated material was sealed in a 6.2-hectare steel container, with walls extending 24 metres into the lakebed. This ECF is expected to hold the contaminated materials without leakage for at least 200 years. This first phase concluded with the reconstruction of the pier wall next to the container to allow for dredging alongside it.

13.7 Monitored Natural Recovery (MNR)

The MNR approach involves removing the source of contamination and allowing natural physical, chemical and biological processes to aid recovery (Rohr *et al.*, 2016). This approach, if successful, would return a site to a functioning ecosystem, resembling or identical to the original one, thus satisfying all or most of the SER attributes in Box 13.4. In contrast to some other approaches, MNR involves no intervention or construction beyond the removal of the source of contamination. The apparent cost savings that are sometimes claimed for MNR may ultimately prove ephemeral if, for example, recovery does not proceed as expected. In such cases, further remedial work, such as soil removal, sediment capping or dredging, may be necessary. Therefore, it is essential that this approach be applied only to appropriate systems.

[1] www.canada.ca/en/environment-climate-change/news/2017/07/randle_reef_contaminatedsedimentremediationproject.html

13.7.1 Passive Recovery for Surface Water

The water in lakes and rivers is replaced over time, with residence/flushing times varying from 0.5 to >1 year for medium-sized lakes, to 330 years for the Russian Lake Baikal and 13,300 years for Lake Vostok, Antarctica. If specific sources of contaminants are removed, the surface water system is likely to recover in response to the natural replacement of water, degradation processes, accumulation of contaminants in sediments and in some cases volatilization. The time for recovery will depend on the relative rates of each of these processes, which vary with the nature of the contaminants of concern and the characteristics of the affected ecosystems. As well, the capacity to detect recovery may be hindered by wide variations in environmental characteristics and changes in analytical methods.

The recovery of a large lake such as Lake Superior from contamination by widely dispersed persistent organic pollutants (POPs) such as PCBs is an example of long-term and slow MNR. The production and use of PCBs in North America were banned in 1977 (Figure 1.4), but in 2016, concentrations in Lake Superior lake trout (*Salvelinus namaycush*) still exceeded consumption limits (Urban *et al.*, 2020). This slow rate of recovery may be caused by uncontrolled diffuse sources, including leaking landfills, the leaching or erosion of contaminated soils, and the long-range atmospheric transport of contaminants (the **global distillation** model; Figure 7.6). Lake Superior is a 'sink' that accumulates airborne POPs when surface waters are near 0 °C during the winter, and a source in the summer when POPs volatilize back to the atmosphere. The rate of decline of PCB concentrations in air over Lake Superior indicates a half-life of 12–17 years, with corresponding declines in concentrations in water and trout until the 1990s. After 1995, the rate of decline in contamination of trout slowed considerably, with large variations in concentrations among different regions of the lake (Urban *et al.*, 2020). Some of these changes are probably caused by spatial and temporal changes in food web structures that determine lake trout diets, growth rates and contamination, and some are simply 'noise' due to differences in analytical methods among monitoring agencies. Nevertheless, it is clear that the rate of decline of POP concentrations in fish is not a simple function of lake flushing. Instead, the decline reflects long-term changes in loadings from the atmosphere and lake watershed, and regional changes in food web structures, energy flows and biomagnification.

A somewhat different situation occurs for lake acidification. A fairly recent example of MNR for surface waters followed the abatement of emissions of sulphur oxides. As the pH of atmospheric deposition has increased in North America and Europe, a number of lakes and soils that were acidified or threatened with acidification have shown chemical recovery, with increasing pH and decreasing aluminium concentrations (Sections 11.1.3.1, 11.1.3.2).

Lakes in the Sudbury region of northeastern Ontario, Canada, provide a rather extreme example of the potential for recovery. These lakes were arguably among those most affected by acid precipitation in North America. Following major decreases in atmospheric SO_2 and metal emissions, beginning in the 1970s, chemical recovery has been documented. Biological recovery is slower, possibly because of additional environmental stressors such as climate change and declines of calcium (Section 11.1.1), and the relatively slow recovery of the terrestrial ecosystems of each lake's watershed. Furthermore, for species with limited dispersal, fish in particular, it may be necessary to recolonize or stock the lakes to return to the original ecosystem (Section 13.9.2).

Monitored natural recovery was the original management strategy for the mercury contamination in the English Wabigoon River system in northern Ontario (Capsule 13.1). After the source of mercury was removed, concentrations in fish began to decline. However, the recovery did not continue, and other management strategies were sought. This required a detailed investigation of the mechanism(s) underlying the continuing mercury contamination of the fish.

In addition to the freshwater examples described above, natural attenuation is also a possibility for contaminated coastal marine systems. Capsule 13.2 provides an excellent example of microbial species that evolved to degrade petroleum hydrocarbons in response to ongoing natural oil seeps in terrestrial and aquatic ecosystems. Some research on natural attenuation is still at the laboratory stage. For example, a study of the potential for microbially mediated natural attenuation of weathered bitumen was described by Schreiber *et al.* (2019) in the context of water bodies on the coast of British Columbia that may be affected by the transport and shipping of diluted bitumen (**dilbit**) (Section 7.6.1). In **microcosm** experiments, natural microbial communities treated with three dilbit blends decreased the concentrations of low-molecular-weight water-soluble aromatics and alkanes. There was an almost complete degradation of acutely toxic n-alkanes over the 28 days of the experiments, but only very limited degradation of the chronically toxic higher-molecular-weight multi-ringed PACs. In general, the less water-soluble compounds tend to be more chronically toxic, and the experiments did not show any degradation of those compounds.

13.7.2 Passive Recovery and Natural Attenuation for Sediments and Soils

For sediments, if the source is removed, fresh materials deposited on top of the contaminated layers will often seal off the internal source of contamination (Case Study 4.3), this being the underlying principle behind palaeo-ecotoxicology (Section 4.3.4). Several assumptions underpin this statement, including that sediments are not disturbed during this period and that there is no vertical migration of contaminants within the sediment bed; this can be an issue for arsenic and selenium, as their oxidation states respond to anoxia. Natural recovery of contaminated sediments is normally preferable to physical removal, namely dredging (see Section 13.6.1). However, in areas where there are many disturbances of sediments from spring freshets and seiches, or from boat traffic, physical burial of contaminated sediments is impossible (Buell *et al.*, 2021).

As is the case for sediments, contaminated soils may also recover naturally. However, unlike sediments, soil recovery is not a result of covering the contamination with clean material, but rather is based on changes in soil chemistry. The declining acidic deposition that has led to chemical recovery for surface waters has also been shown to reverse forest soil acidification. At some sites, decreases in aluminium concentrations in the organic upper layers of soils have also been noted. Such recovery has been described by Lawrence *et al.* (2015) for some forest soils in the northeastern United States and eastern Canada, as well as for forest soils in Sweden. Between 1980 and 1996, emissions of SO_2 in Europe have decreased by 50–60%. Sverdrup *et al.* (2005) applied the SAFE (Soil Acidification in Forest Ecosystems) model to 645 productive forest sites. The results showed that chemical recovery had begun at many sites, including chemical changes with higher pH and lower Al concentrations. The authors point out that forest harvesting will delay recovery. Driscoll *et al.* (2001) identify the second phase in

Capsule 13.1 The Enduring Legacy of Point-source Mercury Pollution

John W. M. Rudd

Chief Scientist, Experimental Lakes Area, Department of Fisheries and Oceans Canada (Ret.)

Carol A. Kelly

Department of Microbiology, University of Manitoba (Ret.)

Mercury pollution is a worldwide phenomenon. It is manifested by high methylmercury (MeHg) concentrations in fish, resulting in toxicity to human and wildlife consumers (Section 6.8.1). The Grassy Narrows First Nation in northwestern Ontario, Canada is a prime example of the very real human health effects, as well as related socio-economic effects. The English–Wabigoon River was polluted by mercury discharges from a chlor-alkali plant between 1962 and 1969. Prior to the pollution, there was an active subsistence fishery and 100% employment of the Grassy Narrows people at a fly-in fishing lodge. Immediately after the pollution was discovered, the subsistence fishery was closed, as was the fishing lodge, resulting in 100% unemployment. Fifty-nine years later, the fishery and lodge are still closed. Recently it has been found that the life expectancy of the Grassy Narrows people has been shortened by the mercury pollution (Philibert *et al.*, 2020).

There are several different types of mercury pollution. The first occurs on a global scale and is caused by mercury emissions during coal combustion. It impacts thousands of lakes, even in remote regions. In this case, atmospheric deposition of quite small amounts of inorganic mercury that is highly available to mercury methylators pollutes lakes and their watersheds. A second type involves no change in inorganic mercury sources but is caused by increased rates of the microbial activities that produce MeHg. The main example of this is reservoir construction. In the areas flooded, MeHg concentrations become remarkably high because microbial activity (and concomitant mercury methylation) is very high while bacteria are decomposing flooded vegetation (Section 6.8.1).

Here we discuss in detail a third particularly pernicious type of mercury pollution – point-source mercury pollution from chlor-alkali plants. Chlor-alkali plants produce a chlorine bleaching agent for the paper industry (Section 11.4.1.1) and sodium hydroxide, which is needed to refine aluminium (Bloom *et al.*, 1999). The chlorine and sodium hydroxide are produced by an electrolytic process that uses liquid elemental mercury, Hg^0. A single chlor-alkali plant can contain 400 metric tonnes (t) of elemental mercury. As part of the operation of chlor-alkali plants, often 10–30 t mercury are lost over a period of years during handling of the liquid mercury. This is the type of mercury pollution that has severely impacted the Grassy Narrows people for decades.

The liquid mercury that is lost from chlor-alkali plants during handling seeps through cracks in concrete floors supporting the cathode cells; spilled mercury collects in pools along with NaCl that is also used in the electrolytic process. High NaCl concentrations promote oxidation of elemental mercury to ionic mercury (Hg^{2+}), which then dissolves in groundwater as mercuric chloride complexes $(HgCl_n^{(n-2)-})$ before being transported to receiving waters.

After arriving in a water body in groundwater discharge, the concentrated $HgCl_2$ is diluted and the $HgCl_2$ dissociates into the Hg^{2+} and Cl^- ions; Hg^{2+} is very particle-reactive and binds strongly to negatively charged clay particles and to S^- moieties found in particulate organic matter. Often these particles are very small and easily transported in aquatic ecosystems, so the attached mercury can be transported long distances before being deposited in bottom sediments (Kelly and Rudd, 2018).

The waters of flowing rivers and pristine lakes are oxic, but their sediments are usually anoxic within a few millimetres of the sediment–water interface. When mercury-containing particles are deposited from the water column and incorporated into anoxic sediments, MeHg is produced by anaerobic bacteria as described in Section 6.8.1.3.

A little mercury goes a very long way! For perspective, fish MeHg concentrations increased by 35% when an 8-ha lake (5.5×10^5 m^3) and its 43-ha watershed in northwestern Ontario, Canada were experimentally contaminated by adding only 1/4 teaspoon (17 g) of mercury annually, over a period of 3 years. So it is not surprising that discharging 10–30 t of mercury to a waterway over a period of years causes a very large problem.

Chlor-alkali plants began operation in the early 1960s, but most have been phased out after the discovery of elevated concentrations of mercury in fish and wildlife in their receiving waters. Unfortunately, there are two phases of such point-source mercury pollution. A first acute phase occurs during the unregulated operation of a chlor-alkali plant when large quantities of mercury are being discharged to a waterway. During this acute phase, mercury concentrations in fish as high as 27 µg/g were recorded in Ball Lake on the English–Wabigoon River system, which is 135 times above the limits for fish mercury consumption in Ontario. When mercury concentrations are this high, they are toxic to the fish themselves (Wiener, 2013) and to fish-eating birds (Rudd *et al.*, 2018). This acute phase usually passes within about 10 years of the chlor-alkali plant closure (Figure 13C.1).

After the decommissioning of a chlor-alkali plant, a second phase usually follows during which mercury concentrations in fish are lower than at peak levels but continue to be much too high for consumption. This second less acute, but still unacceptable phase can last many decades, as it has in the English–Wabigoon system (Figure 13C.1).

There are at least two causes for these decades-long elevated mercury concentrations in fish and sediments, as described below. Distinguishing between these two causes is the first step in addressing the long-term problem.

The first cause is that in some instances, after the chlor-alkali plants have been decommissioned, small residual amounts of mercury continue to leak into receiving waters from pools of elemental mercury in contaminated soils under the former chlor-alkali cells. For example, Lavaca Bay, Texas, USA is an estuary connected to the Gulf of Mexico where aluminium is refined. Mercury concentrations in Lavaca Bay fish initially declined after 1979 when the chlor-alkali plant was decommissioned, but then remained high for three decades. Investigations found that most of the discharged mercury had been buried in the sediments of Lavaca Bay early in the pollution event at depths that were below the zone of MeHg production (i.e., more than about 4 cm below the sediment–water interface). Therefore, this deeply buried mercury was no longer

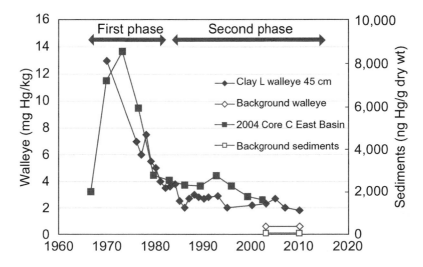

Figure 13C.1 Time courses of mercury concentrations in surface sediments of the east basin of Clay Lake and in 45-cm walleye fish (*Stizostedion vitreum*). The data show a rapid decline in fish and sediment mercury concentrations during the first decade after the chlor-alkali plant was closed in Dryden, ON. Following the initial phase, a second phase lasting several decades followed, during which concentrations in sediments and fish declined very slowly.

contributing to the problem. However, there was still a small ongoing discharge from contaminated groundwater at the chlor-alkali plant site. An engineering approach called 'pump-and-treat' (Section 13.6.1) was employed. This approach involved installing groundwater wells and pumping groundwater from the wells to intercept the mercury-contaminated groundwater before it could flow into Lavaca Bay. A treatment process was then employed to remove mercury from the groundwater before it was discharged. After this solution was installed, fish mercury concentrations in Lacava Bay resumed their decline towards background mercury concentrations.

A second case of decades-long elevations in fish mercury, which required a different solution, was illustrated by the HoltraChem chlor-alkali plant located on the upper Penobscot Estuary in Maine, USA. Mercury concentrations in fish and sediment declined quickly after the chlor-alkali plant was closed and then very slowly for the next five decades. However, at the HoltraChem site there was no significant ongoing leakage of mercury into the upper Penobscot Estuary in contrast to Lavaca Bay. Instead, the explanation for the continuing high concentrations of mercury lay in the unusual hydrodynamics of the upper Penobscot Estuary and the resulting persistence of legacy mercury in the estuary. In the upper Penobscot, there are very high tides (~4 m), which reverse flow into and out of the upper estuary twice daily. During each inflowing tide, a dense salt-wedge of seawater invades the lower depths of the upper estuary, resulting in the trapping of an enormous pool of contaminated mobile suspended sediments (estimated to be 300,000 tonnes (t); Rudd et al., 2018). This mobility prevented the burial of the mercury contamination below the zone of methylation, which would occur in a more quiescent environment such as in Lavaca Bay or in lakes. The continued presence of high mercury concentrations in surface sediments of the Penobscot Estuary sustained high methylation rates, which resulted in ongoing high concentrations of MeHg in aquatic life and toxic MeHg concentrations in aquatic and wetland birds.

A multi-disciplinary scientific team, including geochemists, hydrodynamicists, ecotoxicologists, microbiologists and ecosystem modellers, recommended an engineering solution for the Penobscot estuary in which two dredges would be installed. One was to remove suspended sediments from the contaminated mobile pool and transport them to the lower Penobscot Bay. These contaminated dredged sediments would be buried below the depth of mercury methylation in confined aquatic disposal pits (CADs). At the same time, a second dredge would move the clean fine-grained sediments obtained during the excavation of the CADs in lower Penobscot Bay to the upper estuary and use them to dilute the mercury in the contaminated mobile pool. After the scientific investigations were completed, the team handed the work on to a team of engineers who are now working on detailed plans for remediation.

A similar solution was proposed for the English–Wabigoon system in the 1980s. Two decades after mercury was first discharged to the English–Wabigoon River, a Federal/Provincial study recommended diluting the highly contaminated particulate material in the English–Wabigoon River with clean glacial clay taken from an upstream lake. *In situ* studies in Clay Lake had demonstrated that there could possibly be a 90% reduction in fish mercury concentrations if this remediation approach were applied. Unfortunately, even though this approach was recommended by the Ontario cabinet to the Premier of the day, it was decided that the system was to be left to recover naturally. This recovery has yet to occur (Figure 13C.1).

Unlike the early work on the English–Wabigoon system, more recent work has involved the Grassy Narrows people in the planning and execution of studies, and they are now also involved in the remediation. In 2016, they commissioned a report to determine whether there was a solution for the English–Wabigoon River, which in 2016 had endured 54 years of mercury pollution (Rudd et al., 2016). This report led to a proposal to clean up the river (Rudd et al., 2017) and the establishment of a Can$85M trust fund by the Ontario Government to begin the remediation. A recent study found that the perpetuation of the English-Wabigoon mercury problem had similarities to the Penobscot problem (Rudd et al., 2021) in that there was no ongoing low-level discharge of mercury. Instead, there is a continuing resuspension of mercury-contaminated particles, which had been deposited in the floodplain of the Wabigoon River between

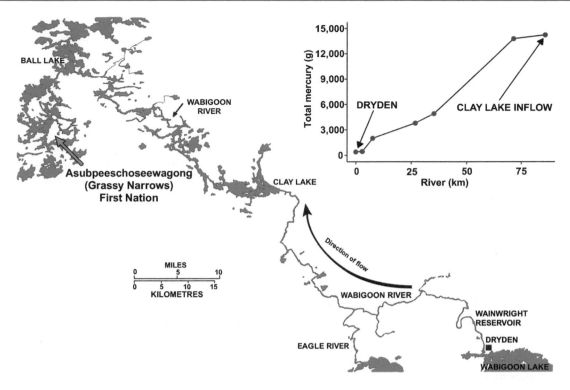

Figure 13C.2 Beginning in 1962, a 150-km reach of the English–Wabigoon River between Dryden and Ball Lake was polluted by mercury discharged from a chlor-alkali plant in Dryden. Although the pollution event occurred in 1962–1969, there was continued pollution of the river by gradual erosion and downstream transport of mercury from the floodplain of the Wabigoon River. The graph in the inset shows that loading of mercury (mercury concentration × flow) increased all the way between Dryden and Clay Lake. Figure based on Rudd *et al.* (1983), with permission from Canadian Science Publishing.

Dryden and Clay Lake, during the operation of the chlor-alkali plant. This resuspension occurs as the river banks are scoured during the formation of meanders in low-relief wetlands and agricultural land between Dryden and Clay Lake (Figure 13C.2). The clean-up proposal was the same as that put forward in 1983, namely that clean glacial clay be dredged continually from Wabigoon Lake, which is upstream of Dryden, and added to the downstream contaminated reaches of the English–Wabigoon River. Hopefully, after 59 years of procrastination, English–Wabigoon remediation will be undertaken this time, and the Grassy Narrows people can begin to recover from this pollution event.

The overall conclusion that we see from this body of work is that a whole ecosystem understanding of how mercury is being transported, buried and methylated is needed before a solution can be found. Also, the impacts of long-ago mercury contamination can last for many decades.

Capsule References

Bloom, N. S., Gill, G. A., Cappelino, S., *et al.* (1999). Speciation and cycling of mercury in Lavaca Bay, Texas, sediments. *Environmental Science & Technology*, 33, 7–13.

Kelly, C. A. & Rudd, J. W. M. (2018). Transport of mercury on the finest particles results in high sediment concentrations in the absence of significant ongoing sources. *Science of the Total Environment*, 642, 1471–1479.

Philibert, A., Fillion, M. & Mergler, D. (2020). Mercury exposure and premature mortality in the Grassy Narrows First Nation community: a retrospective longitudinal study. *The Lancet Planetary Health*, 4, e141–e148.

Rudd, J. W. M., Bodaly, R. A., Fisher, N. S., *et al.* (2018). Fifty years after its discharge, methylation of legacy mercury trapped in the Penobscot Estuary sustains high mercury in biota. *Science of the Total Environment*, 642, 1340–1352.

Rudd, J. W. M., Harris, R., Kelly, C. A., Sellers, P. & Townsend, B. (2017). Proposal to Clean Up (Remediate) Mercury Pollution in the English–Wabigoon River. Prepared for the Asubpeeschoseewagong Netum Anishinabek (Grassy Narrows First Nation) and the Government of Ontario. DOI: 10.13140/RG.2.2.28734.08004

Rudd, J. W. M., Harris, R. & Sellers, P. (2016). Wabigoon–English River System Advice on Mercury Remediation – Final Report. Report submitted to: Asubpeeschoseewagong Netum Anishinabek (Grassy Narrows First Nation) Ontario: Canada Working Group on Concerns Related to Mercury. DOI: 10.13140/RG.2.2.11914.88004

Rudd, J. W. M., Kelly, C. A., Sellers, P., Flett, R. J. & Townsend, B. (2021). Why the English–Wabigoon River is still polluted by mercury 57 years after its contamination. *FACETS*, 6, 1–25, DOI: 10.1139/facets-2021-0093

Rudd, J. W., Turner, M. A., Furutani, A., Swick, A. L. & Townsend, B. (1983). The English–Wabigoon River system: I. A synthesis of recent research with a view towards mercury amelioration. *Canadian Journal of Fisheries and Aquatic Sciences*, 40, 2206–2217.

Wiener, J. G. (2013). Mercury exposed: advances in environmental analysis and ecotoxicology of a highly toxic metal. *Environmental Toxicology and Chemistry*, 32, 2175–2178.

ecosystem recovery as biological recovery, pointing out that this can occur only if chemical recovery is sufficient to allow survival and reproduction of plants and animals. In common with aquatic systems, there are indications from some studies that depletion of base cations, notably Ca^{2+} (Section 11.1.2), may be one reason why biological recovery lags behind chemical recovery. Recovery of terrestrial ecosystems after decreases in acidic deposition is likely to require several decades after soil chemistry is restored, reflecting the long life of tree species and the complex interactions among soil, roots, microbes and soil biota. Studies of severely metal- and acid-impacted soils in Sudbury, Ontario, have shown that intervention is required to help biological recovery of forest ecosystems. The use of lime to raise the soil pH above 5.2, the addition of fertilizer and seeding greatly stimulate vegetation recovery, although full recovery is likely to be a decades-long process (see also Capsule 12.1).

A process that may aid in the recovery of terrestrial systems damaged by organic contaminants is **natural attenuation**, involving the action of physical, chemical, and existing biochemical and microbial agents on the contaminants of concern. Mulligan and Yong (2004) reviewed this procedure, referring to some alternative terms, 'intrinsic remediation', 'bio-attenuation' and 'intrinsic bioremediation', that have been used. Processes include volatilization, microbial transformation and immobilization. Most examples illustrate the recovery from organic contaminants such as benzene, toluene, ethylbenzene and xylene (BTEX), and chlorinated hydrocarbons by volatilization and microbial transformation. However, the immobilization of inorganic chemicals has also been demonstrated. Microbial products can influence the desorption of hydrocarbons and metals from the soil. Bacteria and yeasts produce a range of surfactants, which have been shown to remove metals and hydrocarbons from contaminated soil by disruption of the pollutant/soil bonds (Mulligan and Yong, 2004). Gadd (2004) discusses how fungi, yeasts and bacteria can affect the mobility of metals and metalloids in soils. In solid matrices, metals can be mobilized by leaching, chelation or methylation. The soluble or volatilized metals can then be removed from soils, sediments and solid industrial wastes. Among these processes, leaching, which is already used in nonentry mining (Section 11.2), has also been applied to the treatment of contaminated soil.

In general, these attenuation processes, while attractive in terms of being non-invasive, are for the most part very slow, and caution is required in using them for remediation. Case Study 13.1 illustrates a combination of MNR and engineering.

Case Study 13.2 Recovery of Saglek Bay, Labrador[2]

Background

The southern shore of Saglek Bay, Labrador in subarctic Canada (Figure 13.2) was used as a United States Air Force communication station from the early 1950s until 1971, at which time control of the facility was transferred to the Canadian Forces North Warning System Long

---

[2] This Case Study was researched and authored by Adrian Pang, Queen's University, Kingston, ON, Canada.

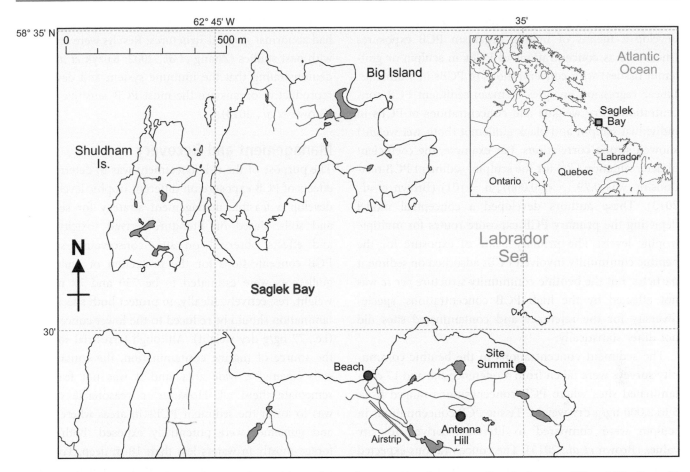

Figure 13.2 Location of Saglek Bay, showing the positions of three former PCB point sources (Beach, Antenna Hill and Site Summit) at Saglek, Labrador (LAB-2) and background sites (Big Island and Shuldham Island). Figure modified from Ficko *et al.* (2013), reproduced with permission from Elsevier.

Range Radar network (LAB-2). A preliminary assessment of the site in 1996 found PCB-contaminated soils that had eroded into the adjacent marine environment, resulting in elevated concentrations in marine sediments. The source was the power supply and communications systems that contained PCBs, which are chemically stable, do not conduct electricity, and are excellent heat transfer agents. Because of their high hydrophobicity, toxicity and tendency to persist in biological tissues (Section 7.7.3), the presence of PCBs in marine sediments was of concern.

Assessment

For the preliminary assessment, which began in 1996, soils, plants, marine sediments and aquatic biota were sampled during 1997–1999. Near the source of the contamination, PCB concentrations (total of 51 congeners) in sediment were as high as 62,000 ng/g dry weight, over 1,000 times the Canadian Environmental Protection

Act regulation of 50 ng/g dry weight. Concentrations decreased with distance from the contamination source (Ficko *et al.*, 2013) and were determined to be derived exclusively from soil erosion at the sealift landing beach area.

Ecological risk assessments (ERAs) were conducted following the 1996 investigation to provide a scientific rationale for the best management strategy for the PCB-contaminated marine sediment and terrestrial soil in the surrounding area. The risk assessments and remedial measures were designed to protect multiple upper trophic levels, as Saglek Bay is a popular hunting and fishing area, mainly for indigenous communities. In the marine environment, the primary focus was on three trophic levels consisting of mixed benthic fauna, a benthic-feeding demersal fish, the shorthorn sculpin (*Myoxocephalus scorpius*), and a secondary consumer, the black guillemot (*Cepphus grylle*). The ERA results were intended to support decision-making about possible remediation of the

PCB contamination at Saglek. Hence, it was important to develop a means of linking organism PCB exposures (measured as concentrations of PCBs in sculpin or guillemot tissues) with concentrations of PCBs in sediments. Linear regression analyses of mean sediment PCB concentrations (dry weight) and concentrations of PCBs in individual sculpin and black guillemot (both wet weight) showed strong correlations. For example, the coefficient of determination (R^2) for the sculpin–sediment PCB relationship was 0.79 ($p < 0.0001$, $n = 101$) (Brown *et al.*, 2013). These authors developed a conceptual model depicting the primary PCB exposure routes for multiple trophic levels. The primary route of exposure for the benthic community involved PCBs adsorbed on sediment particles, but the benthic community structure *per se* was not affected by the high PCB concentrations; species diversity for the reference and contaminated sites did not differ statistically.

The sediment concentrations in the benthic community surveys were taken from a reference site and 17 contaminated sites where PCB concentrations ranged from 5 to 2,000 ng/g dry weight. Tissue PCB concentrations in sculpin were compared to literature-derived toxicity values (Brown *et al.*, 2013). The concentrations expected to affect shorthorn sculpin reproduction ranged from 1,000 to 5,000 ng/g wet weight, and the average tissue concentration for sculpin at the source of contamination was within this range (2,660 ng/g wet weight). Liver ethoxyresorufin-*o*-deethylase (EROD) activity was also measured in the sculpin and was strongly correlated with tissue PCB concentrations ($R^2 = 0.60$, $p < 0.05$). Shorthorn sculpin within 4.5 km of the contamination source were found to have tissue PCB concentrations exceeding the **threshold** linked to EROD induction. Although the biological significance of elevated EROD activity in shorthorn sculpin remains unclear, results indicate that the sculpin are experiencing sublethal effects (e.g., altered metabolism, growth and reproductive capabilities) in response to the PCB exposure.

Black guillemot were chosen to represent seabirds in the area. These birds do not migrate during the winter months and create nests in cliff faces near the site. They forage on benthic invertebrate and fish species, making them susceptible to bioaccumulation of PCBs. Hatchling success and chick growth showed no relationship to PCB concentrations in chicks. However, adult black guillemots had enlarged livers, increased thymus mass and elevated steroid hormone concentrations. Additionally, many male guillemots collected from the contaminated sites had abnormal gonadal structures. Results were consistent with past studies (Zhang *et al.*, 2002; Kuzyk *et al.*, 2003) demonstrating that the immune system and developing reproductive organs are the most PCB-sensitive systems (Brown *et al.*, 2013).

Management and Recovery

The purpose of the risk assessment was to determine the effects of PCB exposure on different trophic levels and to develop a feasible management strategy for sediments and soils. Based on literature-derived toxicity values and effects observed on site, conservative sediment PCB concentrations for the protection of sculpin and guillemots were estimated to be 750 and 77 ng/g dry weight, respectively. Ideally, to protect both species, contamination should be reduced to the lower concentration (i.e., 77 ng/g dry weight). Although terrestrial soils were the source of marine contamination, the contaminated soils covered a wide area, and it was not feasible to remediate them all. However, a reasonable objective was to lower the sediment PCBs in areas where sculpin and guillemot were potentially exposed. Both species forage mostly in waters less than 18 m deep. To protect both sculpins and guillemots, contaminated terrestrial soils that exceeded 50 µg/g PCBs were removed to ensure that concentrations in shallow marine sediments would decrease to <77 ng/g. Between 1999 and 2004, approximately 20,000 m^3 of PCB-contaminated soils exceeding 50 µg/g were removed from the site and/or isolated to reduce marine sediment PCB concentrations in the affected area, and to protect terrestrial biota (Ficko *et al.*, 2013).

Saglek Bay is a highly energetic environment with many parts of the seabed disturbed by wave action, resulting in large amounts of resuspension and redistribution of sediments to the west and southwest. In view of this, any attempt to physically remove contaminated sediment would probably be ineffective. Furthermore, in the short term, dredging sediments could resuspend material and lead to a wider distribution of contaminants. A broad stakeholder group, including government regulators and the Labrador Inuit Association, was consulted to decide on a final management strategy. They decided not to remove any more soil, nor to dredge any sediment, but to allow the ecosystem's sediments to recover naturally, after the removal of the point-source contamination.

As part of the long-term monitoring programme, the marine environment at Saglek Bay was revisited in 2006–2007 to evaluate recovery (Ficko *et al.*, 2013). PCB concentrations were measured for a second time in sediments, sculpin and guillemots. The PCB concentrations in sediments closest to the contamination source were lower than previously recorded, with a maximum of 800 ng/g dry weight). Sediments to the north and northeast also contained significantly lower PCB concentrations. However, owing to the direction of the wave energy, disturbed PCB-contaminated sediments settled in areas west of the site. Nevertheless, these western areas also contained lower PCB concentrations, albeit not statistically different from the previous values. Sculpin and guillemot samples were taken from the contaminated site ("Beach") and an island 6 km north of the Beach. Concentrations in sculpin livers decreased significantly between 1998 and 2006. At the beach, liver PCB concentrations decreased 19-fold over the 8-year period, whereas liver concentrations only decreased 3-fold in samples taken from the island (Brown *et al.*, 2009). Similarly, in guillemot sampled at the beach and island, PCB concentrations decreased 20- and 6-fold, respectively (Brown *et al.*, 2009).

After the initial removal of contaminated soil with concentrations greater than 50 ng/g dry weight, from 1999 to 2004, representative species from the terrestrial ecosystem were monitored annually or biannually. Between 2001 and 2011, PCB concentrations in dwarf birch (*Betula glandulosa*), white willow (*Salix* spp.), roseroot (*Rhodiola rosea*), mountain sorrel (*Oxyria digyna*) and deer mice (*Peromyscus maniculatus*) decreased 3- to 4-fold, except in localized rocky outcrops where soil replacement was difficult.

Although these results indicate a downward trend in PCB concentrations in soils, marine sediments and organisms, PCB contamination is expected to persist for many more years, owing to its long half-life. Ecological resilience is often used to describe the capacity of an environment to maintain its normal patterns (e.g., diversity, nutrient cycling and productivity) after damages caused by disturbances. However, an ecosystem's resilience to disturbance is difficult to measure and quantify, and is still an active area of research. This study shows that for Saglek Bay, Labrador, the ecosystems showed resilience in their capacity to recover from PCB contamination. Concentrations declined to acceptable levels after the original source of contamination was removed, although elevated concentrations remain in some areas.

13.8 Bioremediation

Natural attenuation, as discussed above, includes the action of microorganisms in breaking down, degrading or otherwise detoxifying contaminants. It harnesses the metabolic activities of native communities of microorganisms, which use the contaminant as a substrate in the environmental medium of concern. The same concept has long been in use for waste management, for example in **secondary sewage treatment** and composting. We already know that organisms exist in the field that are capable of detoxifying various contaminants, including some xenobiotics. A logical extension of this is to bring the organisms into culture, where they can be grown in quantity, and then introduce (inoculate) them into the environment to degrade and detoxify target compounds. The organisms may be isolated, enhanced and reintroduced into the original environment, or isolated from elsewhere and brought to the site requiring treatment.

Bioremediation, defined as the breakdown of contaminants by biological mechanisms, is an increasingly popular alternative to conventional engineering methods for cleaning up contaminated sites. It is particularly successful for organic contaminants. In common with natural attenuation, bioremediation may be applied with the objective of ecosystem restoration, or the goal may be solely remediation to limit the migration and ecotoxicity of the contaminants. For remote sites such as those in the Far North or South, it has the added advantage of not requiring elaborate engineering equipment. The basis of bioremediation is the application of microorganisms capable of utilizing organic contaminants as a carbon source, e.g., by degrading or breaking down the contaminant (catabolism) by oxidation or reduction reactions mediated by enzymes or cofactors. Naturally occurring bacteria and fungi, sometimes including communities of such, are isolated and cultured, and then applied to degrade and detoxify target compounds. Green plants have been included in the category of bioremediation but to date, microorganisms have been applied much more frequently and successfully; in this text we are limiting examples of bioremediation mainly to bacteria, yeasts and fungi. The microbes added to wastes can be anaerobic or aerobic, and the process can be *in situ* or *ex situ*.

Many studies on bioremediation have been reported and the recent scientific literature reflects the progressive emergence of various advances in bioremediation techniques. Genetic engineering techniques offer the

Capsule 13.2 Bioremediation of Oil Spills

Charles W. Greer

National Research Council Canada, Montreal, QC

Introduction

An extensive body of literature has been produced over the past several decades on the natural attenuation of oil spills in the marine environment by processes such as **biodegradation**, bioremediation and photo-oxidation (Prince *et al.*, 2017; Ward *et al.*, 2018). The two largest spills in the United States, the *Exxon Valdez* in Alaska and the Deepwater Horizon in the Gulf of Mexico, provided numerous opportunities to study the environmental fate and effects of oil (Atlas and Hazen, 2011). These field studies and 'spills of opportunity' (Section 12.3.4) are critically important for studying the impact and fate of oil in the natural environment, because laboratory studies are not considered to be true environmental surrogates. There remain important questions related to the long-term ecological effects of oil and the ultimate fate of the introduced oil. Can the rate and effectiveness of oil spill remediation be increased, especially in the context of protecting coastlines and susceptible biota?

Oil as a Natural Product

Oil is a natural product that has been in our biosphere for millions of years. Crude oils comprise thousands of individual compounds in four major categories: **saturates, aromatics, resins and asphaltenes**, referred to as SARA (Section 7.6.1.2). Ultra-high-resolution mass spectrometry has already identified more than 17,000 individual compounds in crude oil. The compounds in this mix have a broad range of physicochemical properties that influence their biodegradability, toxicity and ultimate fate in the environment. Hydrocarbons originate from three basic sources, depending on how they are formed: petrogenic (geologically formed from plant and animal remains), pyrogenic (combustion of organic matter) and biogenic (synthesized by a variety of organisms) (Lee *et al.*, 2015). The saturates and aromatics are hydrocarbons that are very energy-rich compounds, serving as food sources for prokaryotic (bacteria, archeae) and eukaryotic (fungi, protozoa, algae) microorganisms. The resins and asphaltenes are structurally more complex hetero-compounds (containing carbon, hydrogen, oxygen, nitrogen, sulphur and metals) and through condensation reactions between different simple structures can become increasingly complex, contributing to their resistance to biodegradation in the environment.

Natural Attenuation and the Role of Biodegradation in Bioremediation

Natural attenuation is a combination of physical, chemical and biological processes that all play a role in determining the fate of oil in the environment, whether introduced naturally or through human activity. These processes include evaporation, dispersion, dissolution, photo-oxidation, emulsification, oil particle aggregation and biodegradation, all of which are part of natural weathering of the spilled oil. As oil weathers, the more soluble compounds dissipate into the water column, the more volatile components evaporate and the more biodegradable components are consumed, gradually increasing the relative concentration of the more recalcitrant components of the oil. These residual components may adhere to other materials in the water to form oil–particle aggregates (aggregates of oil with organic and inorganic matter) which, if conditions are appropriate, will sink.

The biodegradation of the alkane and aromatic fractions of hydrocarbons has been well documented (Peng *et al.*, 2008; Das and Chandran, 2011). In many instances, the biochemical pathways as well as the genes involved have been elucidated, and the microorganisms that have this capacity have been identified and characterized. All three major domains of life have demonstrated the ability to metabolize hydrocarbons to various degrees, and there are currently several hundred bacterial, algal and fungal genera that are known to use hydrocarbons as carbon and energy sources (Prince *et al.*, 2019).

Hydrocarbons are biodegraded by both aerobic and anaerobic microorganisms, although the rates of degradation are typically much faster under aerobic conditions. Rates of natural biodegradation vary considerably, depending on the hydrocarbon compounds as well as environmental conditions, with half-lives ranging from days to months or years. Simple alkanes are the most rapidly biodegraded oil components, with rates depending on chain length and degree of branching. Small aromatics are also degraded quite rapidly, with rates decreasing as the number of rings and the extent of ring substitution increase (e.g., position and number of alkyl substitutions). The resin and asphaltene fractions are considered to be more recalcitrant to biodegradation, although some of the simpler structures may be slowly biodegraded.

Oil Spills in Marine Versus Freshwater Environments

Hydrocarbon-degrading bacteria are ubiquitous in the marine environment (Head *et al.*, 2006). The presence of natural oil seeps in the ocean means that marine microorganisms have been exposed to oil over millions of years. In fact, of the approximately 8 million barrels of oil that enter our oceans every year, half comes from natural seeps (Lee *et al.*, 2015). These natural sources of hydrocarbons have played a major role in the evolution of a specialized group of bacteria, known as hydrocarbonoclastic (obligate hydrocarbon-degrading) bacteria, which have become so specialized that hydrocarbons serve as their sole carbon and energy source. These bacteria are probably the reason we do not see a persistent hydrocarbon sheen on the surface of the ocean, especially in locations with natural seeps. More than 0.5 million barrels of oil enter the Gulf of Mexico every year from natural seeps. During the Deepwater Horizon blowout in 2010, the natural abundance of oil degraders is believed to have contributed greatly to the rapid onset of biodegradation of the underwater oil plume at a depth of more than 1,000 m. Despite the low water temperature of ≤ 5 °C, microbes living in this environment were already primed to degrade oil, and their numbers were probably high enough to reduce any lag times before biodegradation rates could be measured accurately (Hazen *et al.*, 2010). Monitored rates of oil biodegradation in the deepwater plume following the Deepwater Horizon blowout showed half-lives of only a few days, which was considered quite unusual at the time. It is also important to recognize that bacteria can effectively degrade oil components at temperatures below 0 °C, temperatures that are common in the oceans at both poles of the globe.

In terrestrial environments where oil in natural formations is near to or at the surface, freshwater microorganisms can access these compounds as food sources. This is the case for the Athabasca River of northern Alberta, Canada, where oil-bearing formations are in direct contact with the river.

Factors Affecting Bioremediation

Microbial contact can be increased substantially by increasing the surface area of the oil. The organisms need to have direct contact with the substrate, which is accomplished through physical mixing (waves and currents) or chemical dispersion (Section 7.6.1.2). To date, the largest and perhaps the most successful application of chemical dispersants was in the Deepwater Horizon deep-sea oil well blowout. Dispersant was added both to the surface oil slicks by aerial spraying and to the escaping oil, when injected directly at the well-head at a depth of 1,500 m. In this case, the application of dispersant was shown to have a positive effect on the dispersion of oil as fine droplets and the rate of oil biodegradation.

In addition to the hydrocarbons in oil, there are other essential nutrients that are required by oil degraders to function effectively. For the microbes to assimilate the carbon contained in the oil, nitrogen and phosphorus are the key nutritional elements. The biodegradation of hydrocarbons in most environments would probably be stimulated by ensuring that these essential nutrients are present in sufficient quantity. Although the ocean surface typically has very low concentrations of available nitrogen, adding nutrients is not currently an accepted practice for oil spills, owing to concerns about causing eutrophication and **harmful algal blooms**.

Other factors that influence the rate of biodegradation of spilled oil include temperature, concentrations of oxygen or other electron acceptors (e.g., nitrate, iron, sulphate), and sometimes salinity, because hydrocarbon accessibility is reduced by high salinities (Abou Khalil et al., 2021). In addition, the overall rate of oil biodegradation depends on oil interactions with other materials (e.g., minerals, biota and their macromolecules) as well as the level of predation and parasitism (viruses) on degrader microorganisms.

Importance of Field Studies to Support Feasibility and Natural Rates of Biodegradation

A comprehensive report on the behaviour and ultimate fate of oil in aquatic environments provided extensive documentation of the various mechanisms that affect the fate of oil components and their ecological effects (Lee et al., 2015). This report served as a wake-up call by identifying key knowledge gaps and outlining critical research needs to better address oil spills and their remediation in different aquatic environments. Controlled field experiments were emphasized as the best way to close some of these knowledge gaps. Other research needs included:

- environmental impacts in poorly documented environments, including the Arctic, deep ocean, and inland and coastal areas as well as wetlands;
- impacts on wildlife and aquatic organisms at the population, community and ecosystem levels;
- baseline studies to characterize environments that could be impacted in the future and potentially sensitive areas;
- evaluating the efficiency of spill response strategies, making use of 'spills of opportunity';
- improving spill prevention and response strategies, and incorporating them into decision-making; and
- updating and refining risk assessment protocols and procedures.

This report also provides a detailed list of previous oil spills, the volumes involved and their impacts on the environment, and it is a valuable reference document for the history of oil spills.

In addition to spills of opportunity and controlled field studies (Case Study 4.4), another type of experimental system is now being employed. The natural biodegradation potential of different marine environments can be evaluated with *in situ* microcosms that permit the incubation of oil-coated substrates in the presence of the natural microbial flora in their native environments. These experimental systems, while not as robust as controlled field studies, can nevertheless provide valuable data on environmental conditions that influence the rates of oil biodegradation *in situ* (Vergeynst et al. 2019). The advent of 'omics' technologies (Chapter 5) has also improved our ability to study natural microbial ecosystems, to decipher the role of individual species in many ecosystem functions, to understand the effects of oil spills and to monitor ecosystem recovery.

Conclusions: Importance of Biodegradation in Oil Spill Remediation

Most saturate and aromatic compounds in oil, including some of the most toxic, typically disappear rapidly following an oil spill through natural attenuation. Biodegradation is one of the most important processes in natural attenuation. Over millennia, microbes in all environments have developed the capacity to utilize petroleum hydrocarbons as nutrient sources because they are present constantly in our biosphere. Since biodegradation is difficult to quantify in open environments, controlled field studies are essential to understand the fate of spilled oil, particularly how factors such as nutrients, environmental conditions and predation on microbes affect the kinetics of biodegradation. Developing baseline data for potentially sensitive and remote environments is needed for comparative analyses and to monitor long-term impacts and ecosystem recovery. Having access to these data will be invaluable for assessing the effectiveness of microbial oil treatment strategies, and to evaluate the potential recovery time and extent of recovery of affected ecosystems, should a spill occur.

Capsule References

Abou Khalil, C., Prince, V. L., Prince, R. C., *et al.*, 2021. Occurrence and biodegradation of hydrocarbons at high salinities. *Science of the Total Environment*, 762: 143165.

Atlas, R. M. & Hazen, T. C., 2011. Oil biodegradation and bioremediation: a tale of the two worst spills in U.S. history. *Environmental Science & Technology* 45: 6709–6715.

Das, N. & Chandran, P., 2011. Microbial degradation of petroleum hydrocarbon contaminants: an overview. *Biotechnology Research International* 2011: 941810.

Hazen, T. C., Dubinsky, E. A., DeSantis, T. Z., *et al.*, 2010. Deep-sea oil plume enriches indigenous oil-degrading bacteria. *Science* 330: 204–208.

Head, I. M., Martin-Jones, D. & Röling, W. F. M., 2006. Marine microorganisms make a meal of oil. *Nature Reviews Microbiology* 4: 173–182.

Lee, K., Boufadel, M., Chen, B., *et al.*, 2015. *Expert Panel Report on the Behaviour and Environmental Impacts of Crude Oil Released into Aqueous Environments*. Ottawa, ON: Royal Society of Canada.

Peng, R.-H., Xiong, A.-S., Xue, Y., *et al.*, 2008. Microbial biodegradation of polyaromatic hydrocarbons. *FEMS Microbiology Reviews* 32: 927–955.

Prince, R. C., Butler, J. D. & Redman, A. D., 2017. The rate of crude oil biodegradation in the sea. *Environmental Science & Technology* 51: 1278–1284.

Prince, R. C., Amande, T. J. & McGenity, T. J., 2019. Prokaryotic hydrocarbon degraders. In T. J. McGenity (ed.) *Taxonomy, Genomics and Ecophysiology of Hydrocarbon-Degrading Microbes, Handbook of Hydrocarbon and Lipid Microbiology*. New York, NY: Springer, 1–41.

Vergeynst, L., Christensen, J. H., Kasper, U. K., *et al.*, 2019. In situ biodegradation, photooxidation and dissolution of petroleum compounds in Arctic seawater and sea ice. *Water Research* 148: 459–468.

Ward, C. P., Sharpless, C. M., Valentine, D. L., *et al.*, 2018. Partial photochemical oxidation was a dominant fate of *Deepwater Horizon* surface oil. *Environmental Science & Technology* 52: 1797–1805.

possibility to equip and enhance the catabolic potential of bioremedial organisms, particularly for the breakdown of xenobiotics. Bioremediation has even been applied, in combination with phytoremediation (Section 13.9), for dechlorination of PCBs (Sharma *et al.*, 2018), which are notoriously persistent. Transgenic plants and genetically modified microorganisms have proved to be effective in the bioremediation of PCBs. LaRoe *et al.* (2014) isolated a strain of the anaerobic bacterium *Dehalococcoides mccartyi* that dechlorinated PCBs, and they identified 85 distinct PCB dechlorination reactions and 56 different PCB dechlorination pathways.

13.9 Recolonization and Phytoremediation

We have mentioned earlier that biological recovery is normally much slower than chemical recovery. One of the reasons for this lag is the limited recruitment of some flora and fauna following chemical recovery. Indeed, some organisms may never return through natural processes, or if they do, it may take many years. This inherent lag raises the option of actively promoting recolonization or recruitment.

13.9.1 Recolonization by Plants

Not all recolonization has the goal of restoring a contaminated site to a state resembling the original or a natural ecosystem. In some examples the goal is simply to attain some vegetation cover on contaminated land or wetland that is barren or sparsely populated, for reasons of soil or aquatic conservation, or aesthetics. The use of metal-tolerant species is one important application that has shown promise in the pursuit of this more limited goal. The potential of metal-tolerant herbaceous plants to reclaim metalliferous waste sites was recognized and demonstrated in the second half of the twentieth century, notably by Bradshaw and his group at Liverpool University, UK. They successfully used native grasses, isolated from mine sites, to recolonize sparsely vegetated sites. With initial fertilizer amendments, the plants became established and persisted for at least 9 years. One of the outcomes was the commercialization of seed of several metal-tolerant cultivars. The potential for using metal-tolerant plants, in some case genetically engineered, has been addressed further since Bradshaw's early work.

Sites in the Upper Clark Fork River basin, Montana, USA (Section 11.2) have received considerable attention since 1990, with several demonstration projects completed. In fact, these sites have served as an educational and research resource for several years, during which time the riparian zone, the floodplain and upland areas have been addressed, but long-term results are still to be determined. In the late twentieth century and into the twenty-first, efforts were made to restore functional native plant communities in the upper basin. The development of acid-tolerant cultivars began in 1995, and the study continued in 2004 with the development of acid- and metal-tolerant plants; the average soil pH was 4.5. Most soil samples in the project area contained arsenic and copper at concentrations exceeding the US Environmental Protection Agency (US EPA)'s upper range for phytotoxicity, and some samples accumulated phytotoxic concentrations of zinc. The study involved seed collection of native indigenous plants already growing in the Upper Clark Fork River Basin. The seeds were planted and evaluated at various parts of the Anaconda metal smelter site within the overall Clark Fork Superfund area (USDA, 2014). Revegetation was not successful at all sites; for some, additional soil amendments were required.

Successful establishment in experimental situations does not guarantee success in large-scale revegetation projects. A reliable seed source of the plants is a critical need for large-scale revegetation. Another consideration is the long-term need to support grazers to maintain a functioning ecosystem, so plants that provide food for wildlife will also be valuable. In 2018, a draft plan to restore degraded fisheries and to conserve wildlife habitat for this Superfund site was published. It has taken many years of research as well as industrial and political wrangling to reach this point, and still most of the clean-up remains to be done. This is not atypical for a large contaminated site.

The revegetation of the roadsides in Sudbury, Ontario illustrates another approach to recolonization. As a result of the metal contamination and acidification of soil, roadsides in the Sudbury area were very sparsely vegetated or bare (see Capsule 12.1). Since 1978, roadsides in the Regional Municipality of Sudbury have been revegetated, in what has been termed 'the greening of Sudbury'. The area was rapidly colonized with relatively little expense and no complex machinery, just surface application of ground limestone, with or without fertilizer and/ or native seed application. In fact, even when no seed was

applied, a combination of the existing seed bank and wind-dispersed seeds soon resulted in a vegetation cover. This 'ecosystem' is self-perpetuating and has undergone **succession**, with birch, aspen and willows very soon becoming established. Both rhizobial (nitrogen-fixing symbiotic bacteria associated with leguminous plant roots) and actinorhizal (symbiotic nitrogen-fixing actinomycetes) species played an important role in the restoration of vegetation, and the role of ectomycorrhizae (symbiotic fungal association with plant roots) was probably critical.

In actinorhizal species of plants, there is a promising system for revegetation and reclamation of lands contaminated with both metals and organic chemicals. Alders (*Alnus* sp.) are resilient pioneers that are already used in nurse cropping (interplanting) for forestry and mine spoil reclamation. The recovery of previously forested landscapes is a much longer-term prospect than for herbaceous vegetation, for the simple reason that the life cycle of trees spans many years. A certain amount of manipulation can accelerate the revegetation of such sites. The techniques are similar to those used in any reforestation programmes but also require appropriate selection of species that can tolerate contaminated soils. Roy *et al.* (2007) discuss the potential for alder with its ectomycorrhizal fungus and the nodule-forming nitrogen-fixing actinomycete *Frankia* sp. The authors make a strong case that "long-term studies are essential to conduct fundamental research into the impact of anthropogenic stress on these organisms, and to develop efficient, sustainable and economically viable phytotechnologies".

13.9.2 Recolonization by Fish and Other Animals

Some recolonization of previously contaminated sites by animals will occur, as it does for plants and microorganisms, by the natural spread of species. Stocking may be necessary for species with limited mobility e.g., fish confined by barriers to migration, compared with planktonic species and ambulatory or flying animals. As functioning communities of 'pioneer' microbes, plants and animals develop, habitat characteristics will change to favour colonization by larger animal species at higher trophic levels. Stocking would accelerate this progress only if integrated with other restoration techniques that foster succession, such as revegetation (Section 13.9.1). Important factors in recolonization will be the proximity of 'seed' populations in unaffected areas, and physical barriers to migration.

Natural recovery and recolonization of terrestrial sites will not always be uniform over an entire area; 'patches' of suitable habitat surrounded by large areas of unsuitable habitat may not support viable populations of animals. In contrast, for aquatic ecosystems, the movement and mixing of waters may ameliorate the effects of patchy contamination, so that recolonization depends more on lake-wide declines in contaminant concentrations than is the case for large terrestrial sites.

Recovery due to recolonization can sometimes be more apparent than real. For example, the recovery of reproduction by Lake Ontario Lake trout followed the slow decline in lake-wide sediment concentrations of dioxin-like compounds, taking about 20–25 years (Case Study 4.3). Populations of the bald eagle (*Haliaeetus leucocephalus*) throughout the Great Lakes began to recover as contaminants in fish declined from the 1980s onwards. However, recovery was uneven. The reproduction of eagles that nested and hunted along the Great Lakes shorelines, or along tributary rivers downstream of dams, was insufficient to maintain stable populations. Poor reproduction was associated with tissue contaminants derived from consuming contaminated Great Lakes fish and fish that migrated up rivers as far as the first dams (Giesy *et al.*, 1995). In contrast, eagles that nested upstream of dams consumed less contaminated fish, and their reproductive rate was sufficient to sustain both upstream populations as well as downstream and lakeshore populations through migration of offspring.

13.9.3 Phytoremediation

The logical progression from applying tolerant plants to revegetate damaged sites is to use them to clean up their environment. In phytoremediation, plants remove, transfer, stabilize and/or degrade contaminants in soil, sediment and water. The idea of this type of environmental remediation is not particularly new. However, recent research has refined and extended the use of this approach/technology. Padmavathiamma and Li (2007) point out that phytoremediation is simply a concept, but several distinct methods of clean-up are recognized.

Phytoextraction refers to the use of plants that absorb metals from terrestrial or aquatic systems and translocate them to tissues where they accumulate and can subsequently be harvested. The ability of plants (wild types, selected **ecotypes** and transgenics) to **bioaccumulate** high concentrations of elements without suffering toxicity has been described in Section 6.5.4. Aquatic macrophytes are particularly suited for phytoremediation of aquatic systems, because the rooted species can take up contaminants from sediments as well as from the water column. The leaves can accumulate metals from water or air, as well as from their roots. Nguyen *et al.* (2021) discuss the mechanisms by which macrophytes accumulate metals.

The need to clean up mine sites and restore viable ecosystems and landscapes drives some phytoextraction projects. Indeed, **phytomining**, sometimes called **agromining**, utilizing the capacity of hyperaccumulators to take up metals from their substrate, is a real possibility. In this connection, harvesting hyperaccumulator plants could potentially become an economically feasible component of non-entry mining (Section 11.2.2). The idea of plant-mining was first described 500 years ago by German scientist/metallurgist Georgius Agricola (1494–1555), who smelted leaves of metallophytes. However, the true potential of phytomining is yet to be established; research in this area is still mainly at the laboratory scale (Sheoran *et al.*, 2009; Wang *et al.*, 2020).

Another practical factor directing the development of phytoextraction research is the urgent need to restore contaminated agricultural land to a condition where crops can be raised. An estimated 20% of China's agricultural land is already contaminated but is needed for food production (Ye-Tao *et al.*, 2012). The agricultural produce must be safe for human consumption and as feed for cattle and other domesticated animals. The use of hyperaccumulators is a valid option for clean-up of agricultural soils, particularly for metal contamination (Neilson and Rajakaruna, 2015). Several naturally occurring plant species, including some in the genus *Brassica*, have this capacity. Wang *et al.* (2020) showed that amendment of soil with chelating agents such as ethylenediamine tetraacetic acid (EDTA) or ethylenediamine-N,N'-disuccinic acid (EDDS) favours the **apoplastic** absorption of some metals by hyperaccumulators (Figure 3.11).

Hyperaccumulators are of course not edible, and the resulting biomass is in effect toxic waste, or may be suitable for phytomining. An alternative to phytoremediation involves growing tolerant but non-accumulating plants, or plants that do not accumulate contaminants in their edible portions. Legumes and grasses have low transfer factors, meaning the contaminants are not passed on to consumers in large amounts (Neilson and Rajakaruna, 2015). Most of these ideas are still at the research stage, but practical applications seem feasible. Further, genetic modification has successfully increased the potential for metal accumulation in several plants, including *Brassica* spp. (Gall and Rajakaruna, 2013).

Phytostabilization is accomplished by plants retaining metals in roots, rather than removing and transporting contaminants. The requirements for this approach include a **tolerance** to high concentrations of the contaminant(s) of concern, and large root biomass to immobilize and retain the contaminants, preventing transport into shoots. There is no contaminated above-soil biomass to be handled and disposed of. The use of plants for stabilization illustrates the overlap between soil clean-up by the phytostabilization process and the use of plants for revegetation. Two other categories often listed as phytoremediation are **phytofiltration**, involving plants to clean various aquatic environments, and **phytovolatilization**, the use of plants to extract certain contaminants from soil and then release them into the atmosphere by volatilization. In contrast to the three previous approaches to phytoremediation, phytovolatilization does not retain the contaminants in the plant; rather, they are moved from one medium (soil or water) to another (the atmosphere).

The use of plants as groundwater pumps is another approach to environmental clean-up and can be seen as a type of phytoremediation. **Phreatophytes** are plants that depend on groundwater that lies within reach of their roots for their water supply. These are frequently trees, and they 'waste' water by transpiring large amounts that they draw from the ground. There has been considerable interest in these plants because of their influence on water supplies, particularly in arid regions where their effect can be more damaging than in more humid and subhumid regions. This spectacular capacity to pump and transpire groundwater has been harnessed for a system of remediation. Groundwater is carried in the xylem from soil to roots to leaves, to remove dissolved contaminants in the transpiration stream. Removal of some organic chemicals as well as water-soluble inorganic substances can be accomplished in this way (Landmeyer, 2011). There is an assumption of no harm to the plant from the dissolved contaminants, but the contaminants will gradually accumulate, probably in the leaves. Some may be lost in leaf fall, potentially transferring the problem to another medium. The choice of approach for phytoremediation will be guided by the nature of the contaminants and their concentration, the site conditions, the level of clean-up required and the types of plants available.

13.10 Conclusions

A number of options are available for assisting the recovery of contaminated sites. Some are well established, but some types of bioremediation and phytoremediation, while attractive, have rarely been field-tested.

Recovery, aided by appropriate intervention, is not only necessary to reduce current risks for human health and ecosystem integrity, but is also part of our responsibility towards future generations. For many jurisdictions, it is also driven by legal regulations. Without intervention, recovery of contaminated sites will involve **primary, secondary** and **tertiary succession**, these being ecological processes that would occur naturally at sites that had been affected by geological events such as volcanic eruptions. The rates of natural recovery will vary widely, depending on climate, the proximity of ecosystems that provide 'seed' populations of microbes, plants and animals, and the size of the area affected. Or, if unassisted, recovery may never happen.

Risk assessment and an understanding of ecotoxicological processes aid decisions about the type of intervention to pursue. These decisions are also guided by the goal of the clean-up, which ideally is determined by the future use of the site and its perceived ecological value. However, the rate of recovery is not always predictable, and even with intervention, success is not guaranteed. Therefore, following risk management, long-term monitoring should be mandatory.

The costs of site remediation and restoration, and the loss of ecosystem services when a site is contaminated or access to the site is denied during clean-up and recovery, emphasize the importance of chemical regulations. Uncontrolled chemical emissions represent the externalization of the costs of production, costs that are borne by the public and frequently by specific groups that lose access to natural resources or whose health and livelihood are affected. In short, as quoted in Section 13.2, we should remember the adage "An Ounce of Prevention or a Pound of Cure?" It is much less expensive to avoid chemical contamination by applying stringent emission controls than to decontaminate, restore and monitor affected sites.

Summary

. . . .

Chapter 13 has presented a number of approaches for cleaning up contaminated sites with the overall goal of preventing or minimizing exposure and effects of the contaminants on plants and animals. Although most of the examples in the chapter come from North American work, the scientific principles, philosophy and many of the regulatory principles apply across most of the developed world. Using physical, chemical and biological approaches, the challenge is to remove the chemical contaminants (excavate and off-site treatment), to destroy or detoxify them (e.g., biodegradation), to clean them up by biological processes (e.g., phytoremediation) or to contain them in perpetuity (e.g., on-site containment). Restoring ecosystem structure and function (ecological restoration) is more challenging than remediation and risk management, particularly if site decontamination involved major physical disruption of the site, e.g., the removal of soils or the installation of caps or chambers to isolate contaminated materials.

Review Questions and Exercises

. . . .

1. Describe three sets of circumstances that can lead to the legacy of contaminated terrestrial or aquatic systems.

2. What are the goals of ecosystem restoration, according to the Society for Ecological Restoration?

3. Using examples, explain the concepts of: Areas of Concern, Beneficial Use Impairments, and remedial action plans.

4. Explain three triggers that can initiate a clean-up operation.

5. The primary goal of some clean-up operations is to remove or contain contaminants. Why is this not always compatible with ecosystem restoration?

6. List three objectives for the clean-up of contaminated sites. Are they mutually exclusive?

7. Describe, with examples, three on-site methods for remediation. What are the respective advantages and limitations of these methods?

8. What is monitored natural recovery? What are its possible limitations/disadvantages?

9. What is meant by bioremediation? Give examples of this process and identify contaminants that have been cleaned up by this approach.

10. **Group project.** Using role-playing, set up a multi-disciplinary panel faced with planning the clean-up and long-term risk management of a site that is known to be contaminated with dense non-aqueous phase liquids (DNAPLs).

11. Outline the principles of phytoremediation, including the properties of plants that have potential for this approach. Provide examples of its application in the recovery of a contaminated site. List other possible applications of plants with such properties.

12. In the Introduction, we quoted: "... An Ounce of Prevention or a Pound of Cure?" Discuss this in the context of the clean-up of contaminated sites.

Abbreviations

AOC	Area of concern	MNR	Monitored Natural Recovery
BUI	Beneficial use impairment	RAP	Remedial Action Plan
DNAPL	Dense non-aqueous phase liquid	SER	Society for Ecological Restoration
ERA	Ecological risk assessment	US EPA	United States Environmental Protection
IJC	International Joint Commission		Agency
LAB-2	North Warning System Long-Range Radar Network		

References

Brown, T. M., Kuzyk, Z. Z. A., Stow, J. P., *et al.* (2013). Effects-based marine ecological risk assessment at a polychlorinated biphenyl-contaminated site in Saglek, Labrador, Canada. *Environmental Toxicology and Chemistry*, 32, 453–467.

Brown, T. M., Sheldon, T. A., Burgess, N. M. & Reimer, K. J. (2009). Reduction of PCB contamination in an Arctic coastal environment: a first step in assessing ecosystem recovery after the removal of a point source. *Environmental Science & Technology*, 43, 7635–7642.

Buell, M.-C., Johannessen, C., Drouillard, K. & Metcalfe, C. (2021). Concentrations and source identification of PAHs, alkyl-PAHs and other organic contaminants in sediments from a contaminated harbor in the Laurentian Great Lakes. *Environmental Pollution*, 270, e116058.

Driscoll, C. T., Lawrence, G. B., Bulger, A. J., *et al.* (2001). Acidic deposition in the northeastern United States: sources and inputs, ecosystem effects, and management strategies. *BioScience*, 51, 180–198.

Engelmann, C., Händel, F., Binder, M., *et al.* (2019). The fate of DNAPL contaminants in non-consolidated subsurface systems – discussion on the relevance of effective source zone geometries for plume propagation. *Journal of Hazardous Materials*, 375, 233–240.

Farag, A. M., Hull, R. N., Clements, W. H., *et al.* (2016). Restoration of impaired ecosystems: an ounce of prevention or a pound of cure? Introduction, overview, and key messages from a SETAC-SER workshop. *Integrated Environmental Assessment and Management*, 12, 247–252.

Ficko, S. A., Luttmer, C., Zeeb, B. A. & Reimer, K. (2013). Terrestrial ecosystem recovery following removal of a PCB point source at a former pole vault line radar station in Northern Labrador. *Science of the Total Environment*, 461, 81–87.

Gadd, G. M. (2004). Microbial influence on metal mobility and application for bioremediation. *Geoderma*, 122, 109–119.

Gall, J. E. & Rajakaruna, N. (2013). The physiology, functional genomics, and applied ecology of heavy metal-tolerant Brassicaceae. In M. Lang (ed.) *Brassicaceae: Characterization, Functional Genomics and Health Benefits*. New York: Nova Science, 121–148.

Giesy, J. P., Bowerman, W. W., Mora, M. A., *et al.* (1995). Contaminants in fishes from Great Lakes-influenced sections and above dams of three Michigan Rivers: III. Implications for health of bald eagles. *Archives of Environmental Contamination and Toxicology*, 29, 309–321.

Hall, J. D. & O'Connor, K. M. (2016). Hamilton Harbour remedial action plan process: connecting science to management decisions. *Aquatic Ecosystem Health and Management*, 19, 107–113.

Hobbs, R. J., Higgs, E., Hall, C. M., *et al.* (2014). Managing the whole landscape: historical, hybrid, and novel ecosystems. *Frontiers in Ecology and the Environment*, 12, 557–564.

Jones-Lee, A. & Lee, G. F. (2005). Water quality aspects of dredged sediment management. *Water Encyclopedia*, 2, 122–127.

Kuzyk, Z. Z. A., Burgess, N. M., Stow, J. P. & Fox, G. A. (2003). Biological effects of marine PCB contamination on black guillemot nestlings at Saglek, Labrador: liver biomarkers. *Ecotoxicology*, 12, 183–197.

Landmeyer, J. E. (2011). *Introduction to Phytoremediation of Contaminated Groundwater: Historical Foundation, Hydrologic Control, and Contaminant Remediation*. Dordrecht, The Netherlands: Springer Science & Business Media.

LaRoe, S. L., Fricker, A. D. & Bedard, D. L. (2014). *Dehalococcoides mccartyi* strain JNA in pure culture extensively dechlorinates Aroclor 1260 according to polychlorinated biphenyl (PCB) dechlorination Process N. *Environmental Science & Technology*, 48, 9187–9196.

Lawrence, G. B., Hazlett, P. W., Fernandez, I. J., *et al*. (2015). Declining acidic deposition begins reversal of forest-soil acidification in the northeastern U.S. and Eastern Canada. *Environmental Science & Technology*, 49, 13103–13111.

Mulligan, C. N. & Yong, R. N. (2004). Natural attenuation of contaminated soils. *Environment International*, 30, 587–601.

Neilson, S. & Rajakaruna, N. (2015). Phytoremediation of agricultural soils: using plants to clean metal-contaminated arable land. *Phytoremediation*. Springer, 159–168.

Nguyen, T. Q., Sesin, V., Kisiala, A. & Emery, R. N. (2021). Phytohormonal roles in plant responses to heavy metal stress: implications for using macrophytes in phytoremediation of aquatic ecosystems. *Environmental Toxicology and Chemistry*, 40, 7–22.

Padmavathiamma, P. K. & Li, L. Y. (2007). Phytoremediation technology: hyper-accumulation metals in plants. *Water, Air, and Soil Pollution*, 184, 105–126.

Palermo, M. R., Clausner, J. E., Rollings, M. P., Williams, G. L. & Myers, T. E. (1998). *Guidance for Subaqueous Dredged Material Capping*. Vicksburg, MS: Army Engineer Waterways Experiment Station Vicksburg.

Rohr, J. R., Farag, A. M., Cadotte, M. W., *et al*. (2016). Transforming ecosystems: when, where, and how to restore contaminated sites. *Integrated Environmental Assessment and Management*, 12, 273–283.

Roy, S., Khasa, D. P. & Greer, C. W. (2007). Combining alders, frankiae, and mycorrhizae for the revegetation and remediation of contaminated ecosystems. *Botany*, 85, 237–251.

Rudd, J. W., Turner, M. A., Furutani, A., Swick, A. L. & Townsend, B. (1983). The English–Wabigoon River system: I. A synthesis of recent research with a view towards mercury amelioration. *Canadian Journal of Fisheries and Aquatic Sciences*, 40, 2206–2217.

Schreiber, L., Fortin, N., Tremblay, J., *et al*. (2019). Potential for microbially mediated natural attenuation of diluted bitumen on the coast of British Columbia (Canada). *Applied and Environmental Microbiology*, 85, e00086–19.

Sharma, J. K., Gautam, R. K., Nanekar, S. V., *et al*. (2018). Advances and perspective in bioremediation of polychlorinated biphenyl-contaminated soils. *Environmental Science and Pollution Research*, 25, 16355–16375.

Sheoran, V., Sheoran, A. & Poonia, P. (2009). Phytomining: a review. *Minerals Engineering*, 22, 1007–1019.

Society for Ecological Restoration International (2004). *The SER International Primer on Ecological Restoration*. Tucson, AZ: Society for Ecological Restoration, International Science & Policy Working Group.

Sverdrup, H., Martinson, L., Alveteg, M., *et al*. (2005). Modeling recovery of Swedish ecosystems from acidification. *AMBIO: A Journal of the Human Environment*, 34, 25–31.

Urban, N. R., Lin, H. & Perlinger, J. A. (2020). Temporal and spatial variability of PCB concentrations in lake trout (*Salvelinus namaycush*) in Lake Superior from 1995 to 2016. *Journal of Great Lakes Research*, 46, 391–401.

USDA (2014). *Acid- and Heavy-Metal Tolerant Plants for Restoring Plant Communities in the Upper Clark Fork River Basin*. Plant Materials Technical Note No. 97. Bridger, MT: Natural Resources Conservation Service.

Wagner, A. M., Larson, D. L., DalSoglio, J. A., *et al*. (2016). A framework for establishing restoration goals for contaminated ecosystems. *Integrated Environmental Assessment and Management*, 12, 264–272.

Wang, L., Hou, D., Shen, Z., *et al*. (2020). Field trials of phytomining and phytoremediation: a critical review of influencing factors and effects of additives. *Critical Reviews in Environmental Science & Technology*, 50, 2724–2774.

Wortley, L., Hero, J. M. & Howes, M. (2013). Evaluating ecological restoration success: a review of the literature. *Restoration Ecology*, 21, 537–543.

Ye-Tao, T., Teng-Hao-Bo, D., Qi-Hang, W., *et al*. (2012). Designing cropping systems for metal-contaminated sites: a review. *Pedosphere*, 22, 470–488.

Zhang, C., Fang, C., Liu, L., Xia, G. & Qiao, H. (2002). Disrupting effects of polychlorinated biphenyls on gonadal development and reproductive functions in chickens. *Journal of Environmental Science and Health, Part A*, 37, 509–519.

CHAPTER

14 Emerging Concerns and Future Visions

David A. Wright and Peter G. C. Campbell

Learning Objectives

Upon completion of this chapter, the student should be able to:

- Comprehend the complexity of interactions between the abiotic factors affected by climate change (e.g., temperature; oceanic pH; sunlight intensity; salinity; precipitation patterns) and the toxicity of contaminants.
- Appreciate the distinction between *toxicant-induced climate susceptibility* and *climate-induced toxicant sensitivity*, and recognize the importance of interspecific effects in understanding the adverse ecotoxicological effects of climate change.
- Discuss the importance of food webs in understanding the effects of micro- and nanoplastics.
- Describe what is meant by 'technology-critical elements' and discuss their role in the greening of our economies.
- Understand how forensic approaches to monitoring such as non-targeted screening differ from traditional environmental monitoring.
- Communicate the role of predictive ecotoxicology in environmental risk assessment and understand how the development of the adverse outcome pathways (AOPs) can contribute to risk assessment.
- Link the subject areas discussed in this chapter with those presented elsewhere in the book.

As indicated by its title, this final chapter offers some ideas about the future of the science of ecotoxicology and considers issues that do not fit neatly into traditional ecotoxicological frameworks. Pre-eminent among these is the complex interaction between climate change and ecotoxicology. We then review current thinking about the possible effects of synthetic polymer (plastic) waste materials of different dimensions (macro-, micro- and nanoplastics) on living organisms. The final two sections of the chapter are devoted to what are often termed 'emerging contaminants', with examples taken from both the inorganic and organic realms. The term 'emerging' is used first to describe technology-critical elements, which occur naturally but are beginning to play key roles in the greening of the world's economy, and then to describe synthetic organic molecules that have only recently been detected in the environment.

14.1 Climate Change and Its Role in Ecotoxicology

Climate change, in particular global warming resulting from human activity, is now unequivocal and has accelerated in recent decades. Each of the last four decades has been successively warmer than any decade that preceded it, since 1850. Global surface temperature in the first two decades of the 21st century (2001–2020) was 0.99 °C higher than 1850–1900. Physical manifestations of this phenomenon include the retreat of glaciers since the 1960s, significant net loss of Arctic sea ice since 1979, and increasing loss of Greenland sea ice and Antarctic shelf ice since 1993. The rate of sea level rise since the beginning of the nineteenth century has been larger than the mean rate during the previous 2,000 years, and the rate and extent of global warming projected for the twenty-first century have been estimated by Pecl *et al.* (2017) to be comparable to the largest global changes seen in the past 60 million years.

Climate change has also contributed to the desertification of drylands (Huang *et al.*, 2016), which comprise 40% of the global land area. Effects include changes in soil quality, vegetation diversity and cover, species composition and altered hydrologic cycles. The combination of these effects has had many adverse consequences such as sandstorms that threaten ecosystems and areas of human habitation including the western United States, Australia, China and Africa. These effects have been exacerbated by unsustainable land use practices such as excessive grazing, logging and over-exploitation of groundwater. Such activities can eliminate the vegetation cover that protects the soil against erosion by wind and water. Other influences include shifts in water usage in agriculture (irrigation from deepwater wells), municipalities and industry (reduced water use, more highly concentrated effluents).

Over the past 40 years, many papers have dealt with the scientific basis for climate change, and the fundamental reasons for this warming trend are now well established. Of paramount importance has been the release of so-called **greenhouse gases** during fossil fuel burning and some industrial and agricultural practices (Box 14.1). Currently, fossil fuels provide about 80% of global energy needs. The most important greenhouse gases that humans have influenced in the past century are carbon dioxide, methane, nitrous oxide, fluorine-containing halogenated compounds and ozone.

Box 14.1 Causes of the greenhouse effect

Approximately half of the solar energy entering the Earth's atmosphere reaches the Earth's surface, and about 5% of this energy is reflected as infrared radiation. Some gases that are emitted as a result of anthropogenic activity warm the Earth's surface by trapping and reflecting some of this energy back into the Earth's atmosphere, a phenomenon known as the **greenhouse effect**.

- **Carbon dioxide (CO_2):** It is estimated that average atmospheric concentrations have increased from 280 parts per million (ppm) at the beginning of the Industrial Revolution to 405 ppm today, an increase of approximately 45%. CO_2 levels are higher than at any time in the past 2 million years. Over the next 100 years, the relative contribution of CO_2 to the greenhouse effect is estimated to be approximately 55%.
- **Methane (CH_4):** Anthropogenic sources of methane, primarily from livestock farming, agriculture, natural gas usage and decomposition of waste, are estimated to be responsible for at least half of the total atmospheric methane, which is now higher than at any time in the last 800,000 years. The relative contribution of methane to the global warming effect over the next 100 years is estimated to be approximately 15%.
- **Nitrous oxide (N_2O):** The primary source of nitrous oxide is fossil fuel combustion, although it is also a product of some industrial processes and bacterial denitrification reactions in soils. It has an estimated potential contribution to the global warming effect of 7% over the next 100 years.
- **Chlorofluorocarbons (CFCs):** Chlorofluorocarbons were developed beginning in the 1930s as refrigerants, aerosol propellants and for the manufacture of polystyrene foam. As a group, fluorine-containing halogenated compounds have been estimated to contribute an estimated 11% to the global warming effect over the next 100 years, taking into account the mitigating cooling effect of some CFCs through their interaction with ozone (see below and Chapter 1, Montreal Protocol).
- **Ozone (O_3):** Ozone is a gas that has been both increased and depleted in different regions of the atmosphere as a result of anthropogenic activity. Ozone is formed through the reaction between the gases carbon monoxide, nitrogen dioxide and volatile organic carbon compounds such as hydrocarbons from fossil fuels. However, fossil fuel consumption has contributed to a new tropospheric ozone layer, which blocks heat from escaping the planet's surface. The potential contribution made by ozone to the greenhouse effect is approximately 11%.

Table 14.1 **Global warming potential (GWP) and atmospheric lifetime for major greenhouse gases (IPCC, 2007).**

Greenhouse gas	Chemical formula	GWP ranking	Atmospheric lifetime (yr)
Carbon dioxide	CO_2	1	100
Methane	CH_4	25	12
Nitrous oxide	N_2O	298	114
Chlorofluorocarbon-12 (CFC-12)	CCl_2F_2	10,900	100
Hydrofluorocarbon-23 (HFC-23)	CHF_3	14,800	270

The term '**global warming potential**' has been adopted as a means of comparing the relative contributions of different compounds to the global warming phenomenon (Table 14.1). The resulting temperature change is often referred to as radiative forcing (RF), which is expressed as $W\ m^{-2}$ and can be positive in the case of a net warming effect but may be negative in the event of net cooling. Cooling may occur where some gases affect the concentrations of other compounds, which may themselves be greenhouse gases. However, such a comparative index remains subject to changes in interpretation and presentation, depending on the availability of new data on sources and sinks of greenhouse gases. Another variable affecting calculation of the greenhouse effect is the time period projected for the global warming potential. The estimated percentage temperature change over a 100-year period is commonly stated.

Increasing concern over the consequences of global warming prompted the 1997 Kyoto Agreement, wherein several developed countries gave assurances to cut their level of emissions of greenhouse gases by agreed percentages, relative to 1990 emission levels. This was superseded by the Paris Agreement, which entered into effect by 2016 and, by May 2021, 190 countries had ratified the agreement (Section 12.2).

The subject of global climate change is of vital importance to the future biodiversity of both terrestrial and aquatic ecosystems. Yet, despite the critical role of anthropogenic pollutants as causative factors, much of this subject is beyond the scope of this book, and readers are directed to the many reviews of the subject (e.g., Rosenzweig et al., 2008; Pecl et al., 2017). For those interested in the atmospheric science perspective, the Intergovernmental Panel on Climate Change continues to provide a forum for global research on the physical science involved. The IPCC provides data on overall climate trends and updates on sources and sinks of related compounds, and on models of their interactions including feedback loops related to energy transfer (IPCC, 2021).

The overlap between climate change and ecotoxicology is a critical emerging concern and represents a major contemporary challenge for environmental scientists. Both subjects examine empirical data to interpret ongoing changes and formulate predictive models. A major challenge will be to integrate these models to show how various aspects of climate change might influence the behaviour and toxicity of specific chemicals at the cellular/individual organism level. Altered exposure patterns may lead to detrimental effects at higher levels of biological organization. In the following section, we discuss some of the primary climate-related influences on chemical contaminants and the implications for ecotoxicology.

14.1.1 Interactions Between Climate Change and Ecotoxicology

Climate change affects ecosystems in many different ways. Temperature shifts and physical changes resulting from factors such as altered precipitation patterns and sea-level rise will in turn lead to range shifts of individual species according to their physicochemical tolerances. For example, both terrestrial and marine animals are driven towards the poles in response to global warming trends (Poloczanska et al., 2013). Latitudinal shifts in fish distribution have been observed frequently on the west coast of North America due to shifting ocean currents caused by climate-related El Nino events, such that observations of tropical sunfish in coastal waters of Canada and Alaska are more frequent than in the past.

The migration of marine fish to deeper waters may result from an increase in temperature of surface waters (Dulvy et al., 2008), although the abundance of some species may decline if they require shallow water habitats. Increased short-term temperature variations have also prompted concern. Verheyen et al. (2019) have suggested that daily temperature fluctuations (DTFs) in combination with pesticide exposure could pose a greater threat to aquatic insect populations than do longer-term mean temperature changes.

14.1.1.1 Ecotoxicological Effects of Climate Change on Individual Species

In Chapters 6 and 7, we discuss extrinsic factors affecting the bioavailability and toxicity of both inorganic and organic chemicals. Although temperature change is the predominant issue related to climate change, there are interactive effects with other factors that may be directly or indirectly related to climatic shifts and subsequent biotic reactions.

Temperature. Much of our knowledge of the effect of temperature and other abiotic factors on chemical **toxicokinetics** results from controlled laboratory experiments on individual test species, particularly **poikilotherms** (e.g., fish, amphibians, invertebrates) that do not regulate their internal temperatures. Many of these experiments have indicated that chemical toxicity increases with increasing temperature. A review of temperature effects on the toxicity of a range of chemicals (largely metals) to representatives of five different phyla showed that in 80% of cases examined chemical toxicity increased at higher temperatures (Sokolova and Lannig, 2008). A widespread interpretation of this effect is that a heat-induced increase in metabolism may result in an increase in toxicant uptake beyond a toxic threshold. A parallel explanation suggests that heat and chemical stresses may exact a joint metabolic cost in terms of the fitness of the organism. However, a study of temperature effects on **retene** toxicity to trout embryos showed that toxicity did not change with temperature when exposure time was expressed as thermal units (degree-days of exposure) (Honkanen et al., 2020). Temperature determined the rate of development to the embryonic stage sensitive to retene toxicity, but did not change the concentrations of waterborne retene that were toxic or the dose accumulated. Retene appeared more toxic in warm water because the embryos reached the retene-sensitive developmental stage in much less time than in cold water. This acceleration of exposure and the onset of toxic effects increases the likelihood that short-term fluctuations or pulses of chemical concentrations will cause toxicity and ecological effects.

In several studies where temperature-related chemical toxicity and tissue or whole-body chemical concentrations were both measured, it was found that higher mortality was often associated with greater accumulation of the chemical. Sokolova and Lannig (2008) record several instances of temperature-enhanced chemical uptake associated with temperature-related chemical toxicity for a broad spectrum of taxa. However, for several other examples there is no such relationship.

Even from studies going back to the 1970s, it was clear that the role of temperature in modifying chemical toxicity could be highly complex, with interactions among the effects of biotic and abiotic factors. In a review of early work on this subject, focusing primarily on acute exposures, Cairns et al. (1975) reported that, although increased metal toxicity was usually associated with higher acclimation temperatures, a few studies showed little or no temperature effect over the normal temperature range of the test organism. Other studies demonstrated a negative relationship between temperature and chemical toxicity. Figure 14.1 shows examples of the principal relationships between pollutant toxicity and ambient temperature that have been demonstrated for a variety of taxa, notably aquatic **ectotherms**.

In Figure 14.1, Type I and Type II responses indicate an overall 'positive' relationship with chemical toxicity increasing as the temperature rises, which may or may not include a threshold temperature for the toxic effect. As such, this relationship may be seen as analogous to a dose–response. However, interpretation of temperature–toxicity interactions requires some caution depending on the rate of temperature acclimation relative to the duration of toxic exposure. A Type III response (lower toxicity with increasing temperature) is more common for some pesticides (e.g., DDT). The Type IV response characterizes a situation where chemical toxicity is lowest in the temperature range to which the organism is best adapted but may increase at higher or lower temperatures. The final Type V response indicates constant toxicity irrespective of the exposure temperature, which is more typical of **homoeothermic** animals than poikilotherms.

At the single-species level, we need to consider the balance and possible degree of interaction among chemical stress and the various homeostatic mechanisms whereby organisms adjust to their physical environment. For example, we might consider the degree to which chemical stress could affect the ability to withstand climate-related stresses, e.g., oxygen availability, ionic/osmotic stress, temperature stress. This effect has been given the name **toxicant-induced climate susceptibility** (TICS) (Hooper et al., 2013). Alternatively, temperature

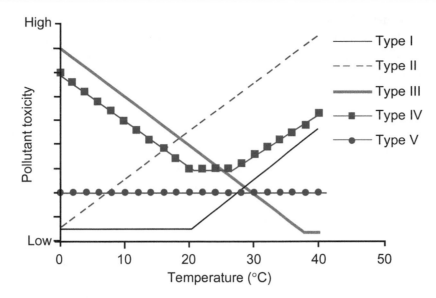

Figure 14.1 Patterns of contaminant-induced toxicity as a function of ambient temperature. Reproduced from Sokolova & Lannig (2008), with permission from Inter-Research.

or other abiotic stresses could affect the ability of organisms to withstand chemical stress. As we have discussed, climate-related stress may affect the general fitness of the organism through effects on aerobic metabolism, or may directly affect mechanisms involving the uptake, metabolism, detoxification and excretion of specific chemicals, an effect called **climate-induced toxicant susceptibility** (CITS) (Hooper *et al.*, 2013). Consequent metabolic changes, with influences on energy budgets, may also affect behavioural trends and reproductive performance of some species and will inevitably alter their relationship with other species such as consumers or competitors.

While temperature may affect the bioavailability of environmental pollutants by altering their chemical or physical speciation, most temperature effects on poikilotherms result from biological factors. Therefore, where increased temperature results in greater chemical bioaccumulation associated with increased toxicity, we usually need to look to biological factors for an explanation. One example is the temperature-related increase in the activity of transmembrane carrier molecules. This may include the facilitated transport of essential trace metals or the hijacking of ATPase-driven transporters responsible for ionic homeostasis. Contributory factors also include the effect of temperature on the permeability of key membranes involved with chemical uptake and regulation. Changes in membrane permeability are likely to result from polyunsaturated fatty acids being replaced by saturated fatty acids at higher temperatures as an acclimation response to stabilize membrane fluidity and

function (Box 2.8). Chemical uptake may also be affected by accelerated rates of ventilation of epithelia resulting from low oxygen and increased temperature, e.g., increased circulation and respiratory activity.

However, it is important to bear in mind that, in addition to chemical bioaccumulation, detoxification and elimination processes are also affected by temperature, and they may influence any of the plots illustrated in Figure 14.1. As we have noted, temperature-related patterns of chemical toxicity do not always result directly from increased toxicant bioaccumulation. In Section 6.5, we describe several metal detoxification mechanisms, some of which involve the isolation of metals within the organism in a non-toxic form. Such an outcome would not necessarily be reflected in tissue or whole-body chemical analyses. Additionally, only a small proportion of the chemical body burden may be responsible for the toxic effect.

The detoxification of chemicals and their elimination from critical receptor sites also involve energy-sensitive processes that can be either inhibited or enhanced at higher temperatures, depending on the organism and chemical involved. Table 14.2 shows examples of the effect of temperature on chemical toxicity over different taxonomic groups. In the majority of cases, toxic effects are enhanced at higher temperatures. However, the underlying causes of these temperature effects may vary.

Salinity. Changes in salinity may result from altered run-off associated with climate change. Shifts in precipitation and evaporation patterns caused by climate change have resulted in higher salinities in tropical and subtropical oceans and a decrease in the salinity of mid- and high-latitude waters. Sea-level rise linked to climate change is projected to lead to saltwater intrusion into previously freshwater habitats (IPCC, 2007). However, a decrease in salinity will occur in some areas receiving higher inputs of fresh water from glacial melting, while glaciers exist. In brackish water ecosystems, salinity reductions are predicted to result from increased run-off. In contrast, regional drought may result in salinity increases in some areas. These changes in salinity patterns will lead to shifts in species distributions with consequent changes in food web interactions (Section 14.1.1.2).

A change in salinity can influence the toxicokinetics of chemicals directly or may alter the physiological functioning of species in a way that enhances the toxicity of the chemical (Fortin *et al.*, 2008). Organic compounds are generally less soluble at higher salinities, a phenomenon commonly referred to as 'salting out'. This trend would be expected to increase the tendency for organics to partition into lipids, thereby potentially increasing their bioavailability, persistence and toxicity in water bodies experiencing increased salinity through evaporation under drought conditions. Where saltwater intrusion occurs, interactive effects of salinity and temperature are common. For example, increased temperature and salinity

had an additive effect on pesticide toxicity to grass shrimp, *Palaemonetes pugio* (DeLorenzo *et al.*, 2009).

Aridity. In terrestrial ecosystems, climate change can affect the degree of aridity, the species present, and their relative susceptibilities to chemical exposure (e.g., plants with thicker cuticles in drier areas). Where rainfall increases, the relative importance of biomagnification may decrease owing to biomass dilution, but the potential for flood-induced mobilization of chemicals from waste disposal sites may increase.

14.1.1.2 Interspecific Effects of Climate Change on Ecotoxicology

In Section 4.2 we consider the often subtle effects of toxicants on interspecific relationships and resultant changes at the population and community level. These same issues confront researchers attempting to understand the effects of climate change, and in many respects the subjects are inextricably linked. Differential adaptation to climate-driven environmental changes can significantly modify the responses of organisms to toxicants at the ecosystem level. Range shifts driven by climate change will result in the introduction of species into different environments and new species–species interactions. As we have seen in Section 4.3.3.3, toxic effects of chemicals on populations and communities take many forms. For

Table 14.2 **Effects of temperature on toxic effects of chemicals to a range of organisms.**

Organism	Chemical	Toxic effect and biomarkers	Effect of temperature on toxicity	Reference
Juvenile crayfish (*Orconectes immunis*)	Cd, Cu, Pb, Zn	Mortality	Increased	Khan *et al.* (2006)
Oyster (*Crassostrea virginica*)	Cd	Mortality, lipid peroxidation, GSH diminished, condition index	Increased	Lannig *et al.* (2006)
Mussels (*Mytilus galloprovincialis*)	Benzo[a]pyrene (BaP)	Multinucleated cells, AChE, BaP hydroxylase diminished	Increased	Kamel *et al.* (2012)
Rainbow trout (*Oncorhynchus mykiss*)	Organophosphate, endosulfan	Mortality	Increased	Capkin *et al.* (2006)
Lizard (*Anolis carolinensis*)	Pyrethrin (with piperonyl butoxide)	Mortality	Decreased	Talent (2005)
Water flea (*Daphnia magna*)	Pyrethrins: fenvalerate, cypermethrin, deltamethrin	Mortality, fecundity	Increased	Ratushnyak *et al.* (2005)
Periphytic algal community	Herbicide mixture: diuron, isoproturon, atrazine, terbutryn	Chlorophyll content	Decreased	Larras *et al.* (2013)

example, a decrease in more sensitive species attributable to chemical toxicity (direct effect) can lead to displacement by a more resistant species and a cascade of indirect effects on the food web. Contributory factors include altered competition for food resources, which may reduce the energy budget available for reproduction, resistance to toxic stress and escape from predators. The effect of temperature on aerobic metabolism has an important impact on physiological performance, especially at the boundaries of an organism's distribution range (Pörtner *et al.*, 2017). Shifts in temperature may result in decreased performance of organisms and changes in the geographical distribution of species, with resulting impacts on the entire ecosystem structure and functions.

Abiotic stress factors associated with climate change can influence all of these interactions, either directly or in combination with contaminants. Shifts in food web dynamics can result in changes at the community-level that can, in turn, alter exposure patterns for different species. Dietary changes may result in quantitative and qualitative effects on chemical bioaccumulation (McKinney *et al.*, 2011). This may have detrimental effects, particularly for animals at the limits of their natural range. A decline in numbers and condition of polar bears (*Ursus maritimus*) in the western Hudson Bay has been shown to result from several contributory factors associated with climate change. These include a decline in sea ice, leading to a shortening of the hunting season, prolonged fasting and a decline in numbers of optimal prey species. These stresses have collectively resulted in increased tissue (and milk) concentrations of PCBs and pesticides such as chlordane, including an increase in the mobilization of these compounds from adipose tissue due to starvation and fat depletion. This has led to the decline in the body condition, survival and reproductive rate of this population over the past 25 years.

14.1.2 Regional Considerations

The Arctic and Antarctic are predicted to be among the Earth's regions that will be most affected by climate change (Box *et al.*, 2019). Polar regions are characterized by low temperatures, reduced rates of weathering and dilute surface waters with low hardness. Permafrost, a subsurface layer of soil that normally remains frozen throughout the year, is often present and constitutes a very large and inherently unstable reservoir of water, organic carbon and other trapped solutes. The

hydrological cycle in polar regions features marked variations in stream flows, and there are pronounced seasonal fluctuations in the day–night cycle, i.e., in the incident light.

Effects of climate change that have been identified so far at polar latitudes include (Box *et al.*, 2019):

- 'intensification' of the hydrological cycle, with increased rates of precipitation and peak flows leading to changes in the dynamics of water residence in lakes and ponds;
- changes in the duration of ice-covered and ice-free periods;
- increased delivery of organic matter and nutrients to fresh waters and to near-coastal zones; and
- permafrost thawing, which has the potential to release massive quantities of methane and other volatiles, some of which can potentiate climate change.

Niinemets *et al.* (2017) have suggested that because of direct effects of temperature on primary productivity and microbial decomposition, the rising temperatures in the Arctic will alter the balance in aquatic ecosystems between carbon sequestration and greenhouse gas release. They imply that there is a positive feedback loop between environmental effects on polar ecosystems and climate change, which would result in accelerated climate change in these regions.

Nutrient run-off. As is the case with salinity, it has been demonstrated that changes in patterns of nutrient run-off have been exacerbated by several aspects of climate change, including a greater frequency and severity of storms (IPCC, 2007). This in turn can lead to the enrichment of receiving waters and proliferation of phytoplankton. In Section 9.3.1 we discuss the part played by climate in the development of harmful algal blooms. In particular, the massive expansion of cyanobacteria blooms now poses a serious threat to the biological integrity of several global water bodies, including some of the world's largest lakes (Paerl *et al.*, 2020). Toxins produced by cyanobacteria affect a broad range of aquatic and terrestrial animals, including humans. Increased water temperature and enhanced stratification of water bodies can create conditions that favour the formation of dense surface blooms. Cyanobacteria possess several properties that enable their rapid spread and give them a competitive advantage over other species. Their rapid growth and buoyancy cause the formation of opaque surface mats capable of taking advantage of increasing atmospheric

CO_2 by direct absorption from the atmosphere. The shading effect caused by these mats has an inhibitory effect on the growth of subsurface algae through light deprivation. Run-off-induced stratification of the water column can contribute to oxygen depletion at the sediment/water interface, which can promote the release of phosphorus from sediments. Several cyanobacteria are capable of high-affinity phosphate uptake as a nutrient source, which can put them at a competitive advantage over other species. This, again, illustrates one of the most complex, yet understudied aspects of climate change: its effect at the multi-species level.

14.1.3 Future Considerations

Our current knowledge of the joint detrimental effects of climate-related factors and chemical contaminants comes almost exclusively from controlled laboratory experiments involving individual species, a single chemical and an element of climate change, usually temperature or salinity but sometimes both. Few multifactorial studies have been performed. In one example, Van de Perre et al. (2018) employed a factorial design to investigate zinc toxicity to juvenile and adult *Daphnia longispina*. The exposure matrix consisted of three Zn concentrations, two temperatures and the presence/absence of competitor species *Brachionus calysiflorus*. Results indicated a temperature-related competitive effect wherein competition for food reduced the energy available to *Daphnia* for reproduction and counteracting Zn stress. Studies of multifactorial interactions, even those with relatively few variables, serve to illustrate the complexities involved with making such risk assessments. As is the case with risk assessments associated with chemical contaminants, it is particularly challenging to forecast the influence of climate change on chemical toxicity at levels of biological organization from molecular to population levels, based on data from single-species experiments with few variables.

In acknowledging this problem, Hooper et al. (2013) have advocated the deployment of an adverse outcome pathway (AOP) approach to define linkages across different levels of biological organization. The AOP approach was originally conceived as a means of defining the relationship between adverse chemical effects at the cellular level and at higher orders of biological organization (Chapters 4 and 5). Expansion of this concept to include the influence of climate change would incorporate the 'reciprocal' effects of TICS and CITS at the intraspecific

level (Section 14.1.1.1) and projected population effects. A generalized schematic of the principal components of an AOP as applied to global warming is shown in Figure 14.2. The pathways and links portrayed here represent a 'template' for more focused studies of specific taxa, and defined chemicals and environment changes. Using this framework, Hooper et al. (2013) presented several case studies on subjects as diverse as climate-change influence on ultraviolet-induced photoactivation of polycyclic aromatic hydrocarbons (PAHs), organohalogen profiles in polar bears, mercury exposure to nestling passerines, and the effects of endocrine disrupting chemicals (EDCs) on fish reproduction and amphibian metamorphosis. The stated objective of this approach has been to develop an understanding of toxic pathways that can be expanded to more predictive models that can be applied in a more generalized manner. As discussed in Section 4.5, an important role of such models is to identify gaps in our knowledge that may be narrowed by more mechanistic solutions.

An important deficiency in our knowledge of climate change/contaminant interactions concerns interspecific reactions. Although food-chain transfer of organohalogens in bears has been shown to alter with dietary shifts (McKinney et al., 2011), interspecific pathways (e.g., competition, predator–prey relationships) are not explicitly addressed in Figure 14.2. It is anticipated that these relationships would feature in future generations of these models. The interactions of climate with exposure and effects of contaminant mixtures are another needed area of study.

14.2 Microplastics and Nanoplastics

Micro- and nanoplastics are ubiquitous throughout the terrestrial, freshwater and marine environments, including polar ice caps and deep ocean trenches. The potential threat from these contaminants continues to grow, spurred by the versatility of the products, their convenience and their low production cost.

The contamination of terrestrial habitats and, most noticeably, waterways and oceans by synthetic polymer (plastic) waste is a subject of growing concern worldwide. Since the introduction of plastic products in the 1950s, global production has increased dramatically and currently exceeds 330 million tonnes annually, an amount that is more than the production of almost all other manufactured materials except steel and concrete. Over

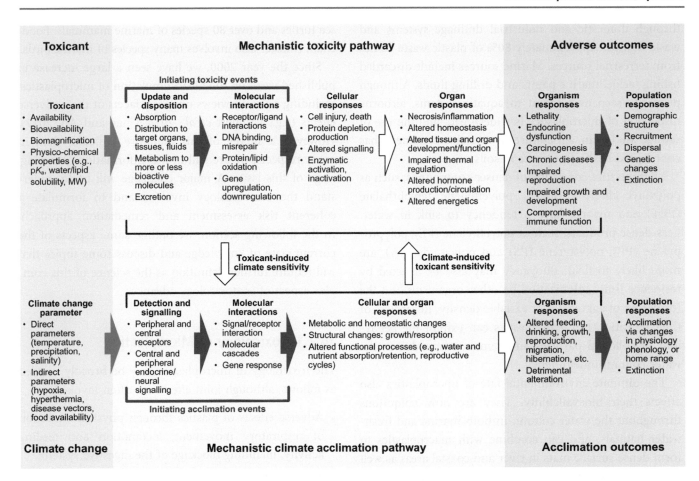

Figure 14.2 Relationships among climate acclimation and adverse outcome pathways (AOPs) for mechanistic assessments of contaminants and global climate change interactions. Adapted from Hooper *et al.* (2013), with permission from John Wiley and Sons Inc.

30% of all manufactured plastic is still in use. By the end of 2015, the sum total of plastic waste generated from primary plastics had reached approximately 6,000 million tonnes. A significant proportion of this, perhaps 2–5%, reaches the aquatic environment, although this amount has been difficult to determine accurately. From human populations living within 50 km of a coastline, Jambeck *et al.* (2015) give a mid-range estimate of 150 million tonnes of cumulative waste plastic reaching the oceans by the year 2025.

From the monitoring and regulatory standpoint, it has been difficult to define the nature of the threat posed by plastic contamination, in large part because of the complexity of the products involved and the varying types of exposure that result in adverse effects on biota. Over 4,000 different polymer types have been found in the environment in a huge variety of shapes and sizes. The term 'microplastics' is given to particles less than 5 mm in diameter, whereas 'nanoplastics' are defined as particulate plastics less than 100 nm in at least one dimension. Other

complexities relate to whether plastic fragments are filamentous (1D), flat (2D) or three-dimensional 'spherules' (3D). Further definitions refer to how such particles are formed. For example, the name 'primary microplastics', is applied to particles that are specifically manufactured for industrial or domestic use. Industrial applications include additives to paints and drilling fluids, resin pellets, which are the raw materials of plastics manufacture, and nano-fillers that are added to low-density polyethylene to increase its resistance to photo-oxidation. Domestic products containing microplastics include cosmetics, gels, facial scrubs, toothpastes, lotions and sunscreens. 'Secondary microplastics' are formed from the fragmentation of bulk plastic products through exposure to ultraviolet radiation or mechanical abrasion. The major sources are breakdown products from vehicle tyres, household appliances, plastic containers and sheeting from domestic and agricultural use, and disintegration products from clothing and footwear manufactured from synthetic compounds. Many domestic products are channelled through domestic and industrial drainage systems and waste dumps. Approximately 80% of plastic waste comes from terrestrial sources. Marine sources include discarded fishing tackle, marine paints and drilling fluids. Although primarily seen as a threat to aquatic systems, airborne dispersion of microplastics is widespread, and the use of sewage sludge in agriculture as a soil amendment has contributed to their presence in soils.

Microplastics consisting of denser compounds, such as polyvinyl chloride (PVC), polyethylene terephthalate (PET) and nylons, have a tendency to sink in water. Less-dense products, such as polyethylene (PE), polypropylene (PP), polystyrene (PS) and polyamide (PA), are more likely to float. Buoyancy may also be affected by residence time, colonization by other organisms and the formation of aggregates of variable density. Regardless of their densities, tiny nanoplastics can exhibit the behaviour of colloidal particles and remain suspended in the water column indefinitely.

The ultimate environmental fate of microplastics also affects their bioavailability. They are now ubiquitous throughout the water column in both marine and freshwater habitats, and can combine with macroplastics to form dense surface mats in river and coastal areas as well as in some open ocean locations. High concentrations of plastics have also been found in coastal sediments close to heavily populated areas (Wright *et al.*, 2013; Auta *et al.*, 2017). Particle numbers in coastal waters may reach $100,000/m^3$ near urban areas, although remote areas including mid-ocean gyres have also shown dramatic increases in recent decades. Between 1972 and 2010, densities of plastic pellets, including microplastics, increased in the North Pacific subtropical gyre by a factor of 100. The increase in the amount of plastic material found in the aquatic environment has also been associated with decreases in mean particulate size, reflecting a weathering effect. However, plastics remain extremely persistent in the aquatic ecosystems, owing to a high degree of resistance to degradation by microorganisms.

The tiny size of microplastics makes them available for ingestion by a wide range of aquatic species, which mistake them for food items. Invertebrate consumers include marine bivalves, crustaceans, echinoderms, coelenterates, annelids and a variety of zooplankton from several different phyla. Ingested microplastics may affect the consuming organism, and in addition they may be transferred through food webs. Accumulation of microplastics has been demonstrated in numerous secondary consumers such as fish,

sea turtles and over 80 species of marine mammals. Food-chain transfer also involves many species of marine birds.

Since the year 2000, we have seen a large increase in published work on the bioaccumulation of microplastics, including several reviews on many facets of their adverse effects (e.g., Lambert *et al.*, 2014; Chae and An, 2017). However, a survey of this literature indicates that, in many respects, we are still in the 'information gathering' stage of this issue. A major challenge will be to understand the ecotoxicology involved and to formulate a coherent risk assessment and remediation approach. In the following section, we outline some aspects of the current state of knowledge and discuss some topics that will require further definition as the science of this complex aspect of ecotoxicology advances.

14.2.1 Toxicology of Microplastics

The toxicology of microplastics may be broadly divided as follows, although joint effects are often invoked:

- Adverse effects of plastics through physical inhibition of respiratory movement, locomotion and feeding activity, including blockage of the digestive system;
- Tissue invasion by smaller particles irrespective of their chemistry causing cellular stress reactions, i.e., the 'nanoeffect' (see Section 11.3);
- Toxicity of the structural chemical constituents of polymers following ingestion. Effects may also include leaching of additives through the digestive process or cellular intrusion of particles;
- Toxicity of environmental chemicals that become adsorbed to the surface of microplastics and which may enter the plastic matrix prior to ingestion by biota. As with chemical components of plastics, bioaccumulation may occur via digestion or direct cellular uptake.

14.2.1.1 Adverse Physical Effects Through Tissue Damage and Inhibition of Movement

The most conspicuous extracellular effects on wildlife result from exposure to a variety of shapes and sizes of plastics, and include physical impairment of feeding, respiratory and locomotory activity through entanglement and physical damage to digestive and other epithelial tissue. Blockage of the digestive system may lead to nutritional stress and loss of condition in a variety of aquatic organisms, and has proven to be a serious threat to some seabirds (Roman *et al.*, 2020).

14.2.1.2 Cellular Invasion by Small Particles (Nanospecific Effect)

Particles at the lower end of the size range may enter cells, primarily through endocytosis, where they may cause stress reactions independently of their chemical constituents. As stated previously, the term 'nanoplastics' is applied to a subset of microplastics defined only by size (<100 nm). They are assumed to be spherical (or at least granular) in shape. As such, in many reviews of the subject they are often included with 'engineered nanoparticles' of varying chemical composition. In Section 11.3 we discuss the nanospecific effect of very small, engineered particles, i.e., toxicity relating only to size. Although nanoplastics are not specifically mentioned in that section, it would be expected that any nanospecific effect might be shared by nanoplastics. For example, Figure 11.5 illustrates the association of a photoreactive cerium oxide nanoparticle with a unicellular alga, which resulted in cell membrane damage and growth inhibition. Similarly, Bhattacharya et al. (2010) demonstrated that adsorption of 20-nm polystyrene particles hindered photosynthesis in the algae *Scenedesmus* sp. and *Chlorella* sp., possibly through reduction of light intensity and through increased production of reactive oxygen species (ROS). Von Moos *et al.* (2012) demonstrated the internalization via endocytosis of high-density polyethylene (HDPE) particles, described as '0–80 μm', into digestive cells of the mussel *Mytilus edulis* and their incorporation into lysosomes. This process resulted in the formation of granulocytomas, an inflammatory cellular response. Several studies have demonstrated the passage of nanoplastics into a variety of tissues, indicating their ability to cross cell membranes including the blood–brain barrier (e.g., Browne *et al.*, 2008), but their mode(s) of action remains unclear.

14.2.1.3 Toxicity of Chemical Constituents of Microplastics

In addition to demonstrating the damaging physical effects of microplastics, laboratory studies have also implicated the potential toxicity of their chemical constituents as an important aspect of their risk to wildlife. Chemical toxicity may be associated with component compounds including plasticizers and other additives. Some of the most widely used plasticizers are di(2-ethylhexyl)phthalate, dimethyl phthalate, dibutyl phthalate, butyl benzyl phthalate and bisphenol A

(BPA) (Section 7.11). Other additives include organotin compounds and nonylphenols, which are used as stabilizers. The dramatic increase in plastic waste over the past few decades has provided the impetus for numerous studies of these contaminants. Although they tend not to be acutely toxic, the identification of several phthalates and **organotins** as endocrine disruptors (Section 8.5) has prompted renewed interest in this subject. Much of the toxicological literature associated with these compounds deals with controlled doses of pure compounds with a primary focus on their modes of action. Although some reviews discuss factors affecting the release of polymer components and additives through ingestion (e.g., Oehlmann *et al.*, 2009), there remain gaps in our understanding of the role of ingested microplastics as a concentrated source of these compounds. A related issue concerns the dynamics of bio-transfer of microplastics and associated chemicals through the food chain.

14.2.1.4 Toxicity of Adsorbed Chemicals

Several studies have demonstrated that microplastics may act as a dietary source for a concoction of inorganic and organic chemicals that become associated with micro- and nanoplastics in the course of environmental exposure. These include persistent organic pollutants (POPs) such as chlordanes, DDT, DDE, DDD, toluene, PAHs and PCBs, as well as metals such as mercury, zinc and lead. Small microplastic particles offer a large surface area for adsorbed chemicals and may provide a concentrated source of potentially toxic chemicals through the ingestion of contaminated particles. The strength of adsorption of organic compounds to microplastic surfaces is dependent on polymer type and density (Fries and Zarfl, 2012). Adsorption of polar compounds may also be influenced by the electric charge at the microplastic surface. The importance of this source of chemicals relative to other routes of uptake remains speculative, and the subject of debate and further refinement (Koelmans *et al.*, 2016).

In some cases, it may be difficult to differentiate between adverse effects due to micro- and nanoplastics themselves and those related to adsorbed and constituent compounds. Cellular uptake of microplastics can have a direct effect on detoxification mechanisms by enhancing ROS production. Such reactions may result in toxic effects of microplastic exposure added to those of adsorbed contaminants. Mussels (*Mytilus* spp.) exposed to micro-polystyrene (micro-PS) particles and fluoranthene both separately and individually showed the highest levels of antioxidant biomarkers and greatest histopathological damage following joint micro-PS and fluoranthene exposure. Higher tissue concentrations of fluoranthene in the presence of micro-PS were attributed to inhibition of depuration resulting from inhibition of the detoxification process (Paul-Pont *et al.*, 2016).

Based on the currently available data, a variety of potentially toxic effects may result from wildlife exposure to micro- and nanoplastics, an outcome no doubt heavily influenced by the size of these particles, their varied chemistry and the degree of contamination by adsorbed chemicals. The association of contaminants with plastics can have both positive and negative effects on their bioavailability. For example, plastics may act as a transport mechanism to move contaminants from the aqueous phase to sediment sinks, where ultimately the compounds become unavailable to aquatic food webs. Where ingestion does occur, microplastics may, in some cases, have a scavenging or dilution effect that may mitigate chemical uptake from the digestive system. On the other hand, ingestion of microplastics may also have a dilution effect on the nutritive quality of ingested material, leading to a reduction in energy assimilative efficiency (Besseling *et al.*, 2013).

14.2.2 Future Considerations

Since the initial recognition of micro- and nanoplastics as a truly global pollution problem in the 1970s, perhaps the biggest single issue facing researchers and regulators has been to evaluate their risk of toxicity to biota. To assess this threat, it is necessary to understand and quantify the various ways organisms become exposed to particulate plastics and their chemical constituents. Only through a knowledge of the physical and chemical processes involved will we be able to establish remedial measures.

14.2.2.1 Establishing Cause and Effect

To assess the risk from realistic environmental exposures, and to establish regulatory measures to reduce this threat, a particular difficulty will be to apply data obtained from laboratory experiments to the complex array of plastics and associated chemicals encountered in the field. One aspect of this will be to establish environmentally realistic dose–response relationships. Complicating factors include partitioning of constituent chemicals between various plastics and surrounding water, a process that is highly dependent on the type of plastic, its period of immersion

and the physicochemical characteristics of the water body. In this respect they share properties with other, naturally occurring particulates, both organic and inorganic. Approaches with potential for investigating the bioavailability and toxicity of plastic-related chemicals are discussed in Chapter 2. These include the passive sampling of aqueous solutions via semipermeable membranes and through the partitioning of dissolved compounds into plastics (Section 2.2.7), and passive dosing, wherein compounds are partitioned from high concentrations in plastics to low concentrations in tissue (Section 2.2.8).

In laboratory and field studies, toxic effects of micro- and nanoplastics have been shown to occur in all size classes of biota from zoo- (and phyto-) plankton to large vertebrates. Size-related effects of micro- and nanoplastics irrespective of their chemical toxicity include mortality, impaired feeding, inhibited growth, development and reproduction. Cellular effects include decreased immune response, oxidative stress, liver toxicity, neurotransmission dysfunction and even genotoxicity. However, in most cases the mechanisms of toxic effects are largely unknown and require further study. An important aspect of this research will be to differentiate between adverse effects due to particle size (the 'nano-effect'), and toxic effects due to associated chemicals such as structural polymers and adsorbed contaminants that are ingested along with the particulate plastics. Related questions include the identification of polymers that are most susceptible to leaching of additives and those that are most prone to adsorption and surface contamination by other contaminants.

14.2.2.2 Mitigation

Because of the perceived risk of microplastics to ecosystems, it is inevitable that measures for mitigation will need to be implemented in parallel with the development and refinement of a risk assessment strategy. Such measures can be applied during both manufacturing and waste disposal phases and thereby provide another example of the 'life-cycle' (cradle-to-grave) approach discussed in Chapter 11. The mitigation process has already been initiated, beginning with goals for more efficient waste management. However, such measures have an economic component, which will result in uneven implementation, at least in the short term. The European Union has set a target of only 10% of plastic waste to landfill by 2030, compared with approximately 30% at present.

Currently, the two major options for limiting the release of waste plastic into the environment are recycling and incineration, measures that were introduced in the 1980s. Of the approximately 6,000 million tonnes of plastic waste accumulated between 1950 and 2015, about 800 million tonnes have been incinerated and 600 million tonnes recycled. It should be noted that recycling delays rather than eliminates final disposal, yet offers significant energy savings in addition to reducing the mass of waste plastic. Bearing in mind that 90% of raw plastic is currently manufactured from fossil fuels, current life-cycle analyses show that the use of recycled plastic in place of raw plastic can cut greenhouse gas (GHG) emissions associated with plastic manufacture by 20–50%. As part of the 2018 legislation, the European Union introduced a mandatory stipulation for beverage containers to contain at least 35% recycled plastic by 2025. Incineration rates as a proportion of total waste plastic have increased steadily in most countries, reaching 40% in Europe and 30% in China. In most other countries the incineration rate is less than 20%. The overall impact of waste plastic incineration will depend strongly on both emission control and energy recovery.

Apart from recycling and thermal destruction, a large proportion of waste plastic continues to be consigned to waste disposal sites. These vary from containment in managed systems such as sanitary landfills to uncontrolled dump sites including the natural environment. Managed systems usually offer little more than a means of isolating plastic products from potentially sensitive ecosystems, but might play a remedial role where containment and composting methods can enhance both physical and biodegradation. Plastics vary widely in their biodegradability. Polyhydroxyalkanoates (PHAs) such as polyhydroxybutyrate (PHB) and polylactic acid (PLA) are known to be highly biodegradable, whereas polyethylene (PE), polyurethane (PU) and polystyrene (PS) have relatively low biodegradability. The terms 'bio-based plastics' and 'bioplastics' are often encountered in the literature, which can lead to some confusion regarding production estimates, particularly with respect to degradable polymers. For example, readers are cautioned that the term 'bioplastics' is not synonymous with 'biodegradable'. Although figures vary, the global manufacture of bioplastics as a percentage of worldwide plastic production is currently estimated at only 2%, with the biodegradable polymers accounting for about half of that output (Box 14.2). This percentage is projected to

Box 14.2 Bioplastics and biodegradable plastics

Bioplastics derive some or all of their carbon content from living biota but may nevertheless include non-biodegradable polymers (e.g., biopolyethylene). Bio-based non-biodegradable polymers are estimated to comprise ~45% of total bioplastics.

Conversely, *biodegradable plastic* may be manufactured abiotically and not bio-based (e.g., polycaprolactone and polybutylene adipate-co-terephthalate). Biodegradable bioplastic is estimated to comprise ~55% of total bioplastics.

increase substantially over the next decade, with a higher proportion of biodegradable polymers, along with a significant reduction in CO_2 output on a life-cycle basis compared with plastic manufacture from petroleum-based products.

Maximizing the impact of plastic degradation as a remedial measure involves decisions at both ends of the life-cycle process, including the choice of plastics to be manufactured, and a waste management strategy that facilitates both abiotic and biotic degradation. Mitigation strategies during the manufacture of plastics involve the development of polymers that are more easily biodegraded. The enzymatic degradation of polymers involves the action of both extracellular and intracellular enzymes. Extracellular enzymes secreted by microorganisms cause the breakdown of complex polymers into molecules such as oligomers, monomers and dimers, which are small enough to be taken up by bacteria, a process known as depolymerization. These short-chain molecules are mineralized into the end products H_2O, CO_2 and CH_4, which can be utilized as a carbon and energy source by microorganisms. In this way, several polymers can be degraded by bacteria and fungi. Both of these groups of organisms have been identified as sources of extracellular enzymes capable of degrading PE, and biodegradation of polyurethane (PU) has been demonstrated as a characteristic of the fungus *Curvularia senegalensis* through the action of an extracellular polyurethanase produced by this species (Shah *et al.*, 2008). Development of biodegradable plastics and strategies for enhancing the degradation of waste products will play an increasingly important role in reducing the contamination of the environment by plastic materials of all dimensions.

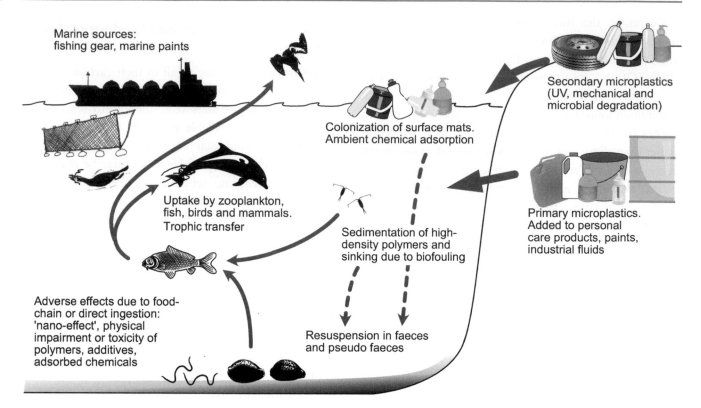

Marine sources:
fishing gear, marine paints

Secondary microplastics
(UV, mechanical and
microbial degradation)

Colonization of surface mats.
Ambient chemical adsorption

Uptake by zooplankton,
fish, birds and mammals.
Trophic transfer

Primary microplastics.
Added to personal
care products, paints,
industrial fluids

Sedimentation of high-
density polymers and
sinking due to biofouling

Adverse effects due to food-
chain or direct ingestion:
'nano-effect', physical
impairment or toxicity of
polymers, additives,
adsorbed chemicals

Resuspension in faeces
and pseudo faeces

Figure 14.3 Fate of plastics in the aquatic environment.

Figure 14.3 illustrates the fate of plastics in the aquatic environment and the toxic threat posed by these contaminants to aquatic organisms. While plastic debris in lakes, rivers and oceans is the most conspicuous form of this class of pollutants, it should be noted that contamination of aquatic media represents only a small proportion of global plastic waste. The toxic threat from aerial and terrestrial plastic contamination remains largely undefined.

14.3 Emerging Inorganic Contaminants

In discussing inorganic contaminants, we are necessarily restricted to the Periodic Table of the elements (Figure 6.1). In other words, in contrast to the situation with organic contaminants, we generally do not have to deal with new inorganic compounds. The only obvious exceptions to this statement are the engineered metal-based nanomaterials (Section 11.3), where the shift from the micrometre scale to the nanometre scale may lead to differences in the behaviour of solid particles that are otherwise of the same chemical composition.

At this point, the astute reader may be asking why we have included inorganic contaminants in this chapter on

emerging concerns. As developed below, two observations led to the inclusion of inorganic contaminants in this final chapter: (i) the marked changes that are taking place in the geographical distribution of new mining ventures around the world, and (ii) the trend towards the exploitation of elements that previously were rarely used.

14.3.1 Trends in Mining Activities

According to world mining data published annually by the International Organising Committee for the World Mining Congresses, global mining production was quite stable between 1990 and 2000 ($\sim$10–11 $\times$ 10^9 metric tonnes/yr), but has almost doubled since 2000 ($\sim$18 $\times$ 10^9 metric tonnes/yr in 2018) (Reichl and Schatz, 2020). In addition to this documented increase in global mining activity, there has also been a shift in the location of the mining activity, with Asia having become a major player (notably China and Indonesia), accounting for almost half of the worldwide production. In the 2000–2018 period, mining activity in China and Australia increased by over 100%, whereas in Africa, South America and North America the increases were more modest (28%, 22% and 17%, respectively). Only in Europe did mining activity decrease (−19%).

From an ecotoxicological perspective, it is important to note that mining activity in areas such as Indonesia, Papua New Guinea and New Caledonia often takes place in near-coastal environments, in regions with very high seasonal rainfall and sometimes in zones of recurrent seismic activity (which make dam building problematic). In other words, environmental risk assessments for new mines will often have to deal with tropical marine environments, where the tailings management techniques currently used in temperate inland mining locations may not be applicable. For example, there is often a lack of physical space that can be set aside for safe terrestrial disposal of tailings; in such cases, deep-sea tailings placement is attracting a good deal of attention (Kwong *et al.*, 2019). Currently, there is also some debate about the use of toxicological data derived from temperate test organisms to derive environmental protection thresholds for tropical environments. However, to date, comparisons between tropical and temperate organisms have not revealed any consistent trends indicating that native tropical organisms are more (or less) sensitive to metals than those found in temperate environments (Wang *et al.*, 2019). As mentioned in Section 14.1.1.1, for these comparisons to be valid, the actual toxicity tests must be carried out at the temperatures typical of those in the local climates.

Another factor to consider is the future expansion of mining activities to regions formerly considered to be 'off limits', either because of logistical considerations (deep-sea mining) or climatic restrictions (Arctic and Antarctic). In the first example, there is increasing pressure to develop methods to exploit the mineral content of deep-sea polymetallic nodules. These nodules are metal-rich and offer the enticing possibility of extracting metals such as manganese, cobalt, copper and nickel without generating appreciable tailings (see Case Study 14.1 on deep-sea mining). However, we still know very little about the nature of the communities living at great depth in our oceans. These communities are subject to intense pressure, complete darkness and limited nutritional sources, and their inherent sensitivity to chemical and physical stressors is virtually unknown. As for the Earth's polar regions, we do know more about the native fauna than is the case for the deep ocean, but here too the native fauna and flora have evolved with minimal past contact with humans. In addition, these areas are inherently very sensitive to climate change; any expansion of mining activities into the Arctic or Antarctic will have to contend with a receiving environment that is rapidly changing (Section 14.1).

14.3.2 Trends in Metal Use

Until the end of the twentieth century, most industrial uses of nonferrous metals involved elements found in the first four rows of the Periodic Table, i.e., elements such as copper, lead, nickel or zinc. However, recent developments in areas such as computing, electrical engineering and renewable energy technologies have expanded this range of metals. To quote Gulley *et al.* (2018), "While previous ages of human history can broadly be defined by a single metal or alloy (i.e., Iron Age or Bronze Age), the material compositions of today's emerging technologies encompass almost the entire Periodic Table and are constantly evolving."

A new term has been defined, **technology-critical elements (TCEs)** or simply **critical elements**, to describe these elements (Box 14.3) (Cobelo-García *et al.*, 2015). The term 'critical' refers to the increasing technological importance of these elements and, in some cases, to the limited number of countries with known reserves of the elements. For example, from 1995 to 2005, China was responsible for more than 90% of the world production of **rare earth elements** or **REEs** (elements 21, 39 and 57–71 in the Periodic Table, including several examples in Box 14.3: cerium, Ce, and dysprosium, Dy). This monopoly stimulated intensive research for alternative sources of some of

Box 14.3 Examples of technology-critical elements and their uses

Cerium (Ce)	Catalysts
Cobalt (Co), lithium (Li)	Rechargeable batteries
Dysprosium (Dy)	Wind turbines
Gallium (Ga)	Smart phones
Germanium (Ge)	Photovoltaic technologies
Indium (In)	Flat screen displays
Niobium (Nb)	High-strength steel alloys
Platinum, palladium, rhodium (Pt, Pd, Rh)	Catalysts
Rhenium (Re)	Jet engines, turbines
Tellerium (Te)	Photovoltaic technologies

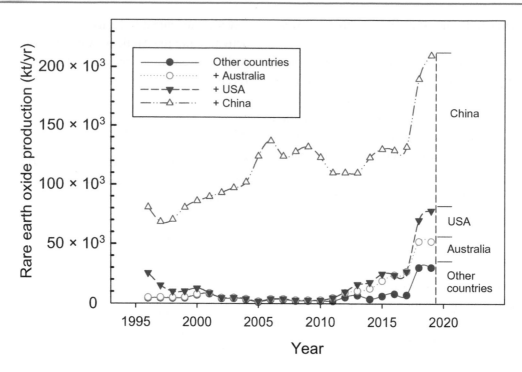

Figure 14.4 Global production of rare earth oxides (kt/yr), plotted from data published annually by the US Geological Survey in its Mineral Commodity Summaries.

these elements, and the latest production figures show a somewhat more balanced inventory of sources (Figure 14.4). As a second example, the demands for lithium and cobalt increased more than 6-fold and almost 3-fold, respectively, in the period from 2000 to 2018, largely because of their use in rechargeable batteries. In 2020, ~80% of the lithium came from just two countries, Australia and Chile (see Capsule 14.1), and >60% of the cobalt was mined in the Democratic Republic of the Congo (Reichl and Schatz, 2021).

One of the important characteristics of the TCEs is that many of them are data-poor with respect to the thermodynamic constants needed to calculate their speciation in solution (see Section 6.2.1). For the monovalent and divalent metals that dominate the ecotoxicological literature (e.g., Ag, Cd, Cu, Ni, Pb, Zn), most of the data needed to calculate their speciation in aqueous solutions are readily available in publicly available databases. However, for many of the TCEs, the thermodynamic data are incomplete, in part because the generation of such data is no longer considered to be a priority (for example, the US National Institute of Standards and Technology database of Critical Stability Constants was unceremoniously discontinued in 2004). This means that for many of the TCEs, we cannot calculate their speciation in solution with confidence.

To compound the problem, the TCEs are also data-poor with respect to the geochemical and ecotoxicological data needed to predict their behaviour, fate and effects in the receiving environment. Extensive ecotoxicological databases exist for many of the monovalent and divalent metals found in the first four rows of the Periodic Table. Examples of such databases include the US-ECOTOX database (US EPA, 2020) and the European ECHA database (ECHA, 2020), which are used in ecological risk assessments around the world. In the case of most TCEs, however, very few toxicity tests have been performed and the ecotoxicological data are still very scarce, limiting our ability to properly assess the possible environmental impacts of these 'new' elements.

In the absence of robust toxicological data, regulators could in principle estimate the potential toxicity of a given TCE using quantitative ion character–activity relationships (QICARs). This approach is analogous to the use of quantitative structure–activity relationships (QSARs) to predict the bioaccumulation and toxicity of novel organic contaminants (Section 2.3.3). This methodology has been used successfully to predict the toxicity of data-rich elements (Walker et al., 2013), but few attempts to predict the toxicity of TCEs have yet appeared in the literature. One of the potential complications is that the metals for which the QICAR approach has been shown to work are

Capsule 14.1 Lithium – A Critical Mineral Element: Sources, Extraction and Ecotoxicology

Heather Jamieson

Queen's University, Kingston, ON, Canada

Lithium (Li) is the lightest metal in the Periodic Table (atomic weight 3), has unique properties and is required for specific technological applications, notably energy storage. The need for Li in rechargeable batteries has increased in recent years, owing to the increased demand for portable electronic devices, electric vehicles and grid storage applications (USGS, 2020). In 2018, 58% of Li produced was used for batteries, increasing from 20% in 2012. As an indication of the importance of these batteries to society, the 2019 Nobel Prize for Chemistry was shared by three scientists who did pioneering work on Li-ion batteries. The second most important use of Li is in the ceramics industry, with lesser proportions used for industrial and medical applications. Figure 14C.1 illustrates increase in both total production and the proportion of Li used for batteries over a recent 15-yr period.

Recently, because of its importance in modern technology and the risk of supply interruption, Li has been designated as a *technology-critical element* by both the European Union and the United States (USGS, 2018; European Commission, 2020). In previous analyses, it fell just outside the criticality envelope. Policies that encourage the conversion of personal and commercial vehicles to electric models are likely to drive the demand and maintain the *critical* designation.

Sources and Extraction

Lithium is produced from two very different types of ore deposits: 'hard-rock' deposits with Li dominantly hosted in the mineral spodumene ($LiAlSi_2O_6$), and brine deposits, where Li is extracted by favouring the precipitation of Li salts from groundwater. Lithium is also concentrated in some clay minerals. This presents a third type of potential source, not currently exploited. The hard-rock deposits have a higher concentration of Li per ton of material than do the brines, and they can be exploited using conventional mining methods. High-quality spodumene deposits are found in Australia, Brazil, Canada and Zimbabwe, with ore grades of 0.6 to 1.2% Li. Brine deposits have lower grades (0.05 to 0.2% Li) but are much larger in size. The largest are found in the Lithium Triangle in the arid region of the joint borders of Chile, Argentina and Bolivia, with

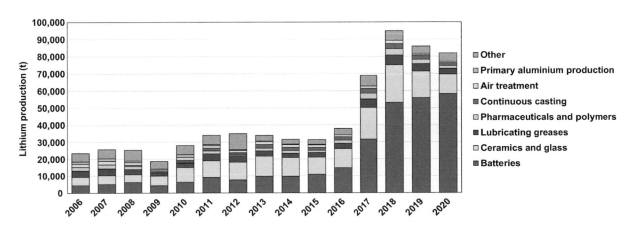

Figure 14C.1 Production and uses of lithium since 2006. The figure shows global production, which is dominated by Australia and Chile, but excludes US production which originates from one brine operation in Nevada. The information for 2020 is estimated. Source: USGS Mineral Commodity Summaries.

smaller ones located in China and the United States. Lithium-clay deposits with grades of 0.2% Li have been identified in the United States, Mexico and Peru (Bowell *et al.*, 2020). Other potential sources of Li are deep geothermal brines associated with oil fields (e.g., Texas), granitic batholiths (Cornwall, England) and seafloor rifting (Reykjanes, Iceland).

Extraction Processes

As is the case for other minerals, the extraction and potential environmental impact of mining lithium depend on the type of ore deposit. The hard-rock mines generate land disturbance, noise and dust associated with all operations of this nature, particularly when open-pit mining is used. Processing involves multiple steps of crushing, roasting and chemical extraction (Flexer *et al.*, 2018). Waste from the ore itself, whether from spodumene or clay minerals, is not likely to generate acid drainage from sulphide oxidation (Section 11.2), but other materials extracted during mining may do so. This potential concern will be site-specific, since the type of rock surrounding the deposit, some of which is likely to be excavated during mining, will vary from place to place. Waste rock and tailings need to be tested for acid-generating potential, and the metals or metalloids associated with any type of leachate need to be monitored. Pegmatite and granitic rocks that host Li in hard-rock deposits may contain other valuable elements such as tin (Sn), tantalum (Ta), beryllium (Be), rubidium (Rb) and caesium (Cs), such that Li is only one of several commodities produced, as is the case for the Tanco deposit in Manitoba, Canada (Bowell *et al.*, 2020). The host rocks may also contain potentially hazardous trace elements or radionuclides that are released in the form of fine particles in dust, or dissolved species in water, and monitoring is needed to test for their presence. Turbidity caused by suspended sediment in streams and lakes can affect aquatic flora and fauna. Many jurisdictions have regulations controlling these types of potential environmental damage (Section 12.3.1) and, if enforced, should limit environmental impact.

Brine lithium deposits require a very different sort of extraction process, and its environmental impacts differ from those of conventional mining. Brines are 2 to 24 times more saline than seawater (Bowell *et al.*, 2020). They can be available at the surface in very arid regions but are usually pumped from underground aquifers and concentrated by evaporation. Figure 14C.2 shows an oblique air photograph of a Li mine in northern Chile.

Extraction from brine deposits involves both evaporation and chemical separation to remove other ions such as magnesium and borate. Sun and wind are used for evaporation, and the process takes 12 to 24 months (Flexer *et al.*, 2018). For successful extraction of lithium, up to 95% of the water needs to be evaporated. Brine water would not be suitable for drinking water or irrigation. However, producing Li from brines also involves using freshwater resources, and there is the potential for mining operations to compete with community freshwater needs, as well as concern that the brine and freshwater subsurface aquifers may

Figure 14C.2 Lithium evaporation pond, Atacama Desert, Chile. Reproduced from Bustos-Gallardo *et al.* (2021) with permission from Elsevier.

mix (Flexer *et al.*, 2018). The extraction of brines for lithium recovery may also affect hydrological balance, including the stability and biodiversity of the nearby saline lakes that provide habitat and migration staging areas for multiple bird species (Gajardo and Redón, 2019).

As is the case with all mining, brine extraction produces a solid waste that must be managed, mainly the economically undesirable sodium, magnesium and potassium salts. In the long term, the land used for the evaporation ponds will need to be remediated after the lithium extraction process has ceased operation.

Hard-rock mines accounted for approximately 60% of the global production of lithium in 2019 (Sterba *et al.*, 2019). Brine deposits represent the balance of current production but are described as containing the large majority of the global Li resources, and many analysts predict that this source of Li will eclipse the hard-rock mines in the future. Mineral resources available for future exploitation are normally defined based on drilling and excavating of rock, so the estimation of Li hosted in brine deposits is challenging to assess (Bowell *et al.*, 2020). The total global resources from both brines and hard-rock Li deposits are expected to be adequate to the year 2100, even with the expectation of sharply increased demand associated with widespread electrification of vehicles (Ambrose and Kendall, 2020a; Bowell *et al.*, 2020).

Life-cycle Analysis

Ambrose and Kendall (2020a, 2020b) have quantified lithium resource use and environmental impact by combining resource production and life-cycle analysis models. They predict that both hard-rock and brine mining will be used to meet demand to 2100. The environmental impact component of the life-cycle analysis includes a wide range of possible effects, including CO_2 emissions, impacts of acid and nutrient drainage (eutrophication) on water quality, as well as human health and ecotoxicological effects. Although demand for Li may eventually require exploitation of lower-grade hard-rock and brines, which generally involves greater environmental effects, the authors predict that the overall impacts will be modest (Ambrose and Kendall, 2020b).

Recycling of batteries is a potential source of lithium production, although the current collection rate is very low and many of the batteries used for small electronic devices go to landfills or are stored in households (Velázquez-Martinez *et al.*, 2019; Ambrose and Kendall, 2020a). The recovery of other elements present in the batteries, such as cobalt, copper, iron and aluminium, may be more profitable than for Li, as the cost of the leaching chemicals needed to extract Li from spent batteries may result in the recycled Li being more expensive than Li from primary sources (Ambrose and Kendall, 2020b). Methods of recovering Li indicate that it can be accomplished at a rate of approximately 50% (Zeng *et al.*, 2014). Ideally, as many metals as possible should be recovered from the batteries, requiring a combination of mechanical, hydro-metallurgical and pyrometallurgical processes (Velázquez-Martinez *et al.*, 2019). A standardized Li-battery design would increase Li recycling efficiency.

Ecotoxicology

Although little work has been done on the ecotoxicity of lithium, it clearly is biologically active, as illustrated by its known effects on humans, for whom it has been used as a successful mood stabilizer since 1949 (Bibienne *et al.*, 2020). However, given the properties of the lithium ion, notably its marked tendency to form ionic rather than covalent bonds and its weak affinity for reduced sulphur, its predicted toxicity using QICAR models is low. The scattered results in the toxicological literature agree with this prediction and suggest that its toxicity is similar to that of other hard cations such as Na^+, K^+, Ca^{2+} and Mg^{2+} (Kszos and Stewart, 2003; Aral and Vecchio-Sadus, 2008). In other words, current concern about the environmental impacts of the surge in Li demand is related to the mining and extraction techniques, not to releases of Li to the receiving environment during its use or after its disposal.

Capsule References

Ambrose, H. & Kendall, A. (2020a). Understanding the future of lithium: Part 1, resource model. *Journal of Industrial Ecology*, 24, 80–89.

Ambrose, H. & Kendall, A. (2020b). Understanding the future of lithium: Part 2, temporally and spatially resolved life-cycle assessment modeling. *Journal of Industrial Ecology*, 24, 90–100.

Aral, H. & Vecchio-Sadus, A. (2008). Toxicity of lithium to humans and the environment – a literature review. *Ecotoxicology and Environmental Safety*, 70, 349–356.

Bibienne, T., Magnan, J. F., Rupp, A. & Laroche, N. (2020). From mine to mind and mobiles: society's increasing dependence on lithium. *Elements*, 16, 265–270.

Bowell, R. J., Lagos, L., de los Hoyos, C. R. & Declercq, J. (2020). Classification and characteristics of natural lithium resources. *Elements*, 16, 259–264.

Bustos-Gallardo, B., Bridge, G. & Prieto, M. (2021). Harvesting lithium: water, brine and the industrial dynamics of production in the Salar de Atacama. *Geoforum*, 119, 177–189.

European Commission (2020). *Critical Raw Materials Resilience: Charting a Path Towards Greater Security and Sustainability*. COM/2020/474 final.

Flexer, V., Baspineiro, C. F. & Galli, C. I. (2018). Lithium recovery from brines: a vital raw material for green energies with a potential environmental impact in its mining and processing. *Science of the Total Environment*, 639, 1188–1204.

Gajardo, G. & Redón, S. (2019). Andean hypersaline lakes in the Atacama Desert, northern Chile: between lithium exploitation and unique biodiversity conservation. *Conservation Science and Practice*, 1, e94.

Kszos, L. A. & Stewart, A. J. (2003). Review of lithium in the aquatic environment: distribution in the United States, toxicity and case example of groundwater contamination. *Ecotoxicology*, 12, 439–447.

Sterba, J., Krzemien, A., Fernandez, P. R., Garcia-Miranda, C. E. & Valverde, G. F. (2019). Lithium mining: accelerating the transition to sustainable energy. *Resources Policy*, 62, 416–426.

USGS (2018). *Final List of Critical Minerals 2018*. Reston, VA: US Department of the Interior, US Geological Survey.

USGS (2020). *Mineral Commodity Summaries 2020 – Lithium Fact Sheet*. Reston, VA: US Department of the Interior, US Geological Survey.

Velázquez-Martínez, O., Valio, J., Santasalo-Aarnio, A., Reuter, M. & Serna-Guerrero, R. (2019). A critical review of lithium-ion battery recycling processes from a circular economy perspective. *Batteries*, 5, e5040068.

Zeng, X. L., Li, J. H. & Singh, N. (2014). Recycling of spent lithium-ion battery: a critical review. *Critical Reviews in Environmental Science & Technology*, 44, 1129–1165.

dominated by monovalent and divalent cations. For these metals, the free metal ion is a consistently good predictor of metal bioaccumulation and toxicity (Section 6.3.2). However, many of the TCEs exist in oxidation states of +III or higher, and we do not yet know enough about their speciation or their interactions with living organisms to be able to conclude that the free metal ion will be a reliable predictor of their toxicity too.

14.3.3 Future Considerations

In conclusion, obvious challenges remain in the inorganic contaminants sector, linked both to the predicted increase in the intensity of mining activities and to the use of previously little used and untested elements, many of which will end up in landfills. However, unlike the challenge of engineered nanomaterials (ENMs), which

required a quantum change in our ability to detect and manipulate ENMs at environmentally realistic concentrations, most of the challenges posed by the TCEs should be resolvable with the approaches that were developed and refined for the more common metals and metalloids that are discussed in Chapter 6.

Case Study 14.1 Deep-sea Mining[1]

The anticipated need for high-technology metals continues to fuel interest in launching commercial deep-sea mining ventures, both for these metals and for more common ones such as copper and nickel. Technological advances

---

[1] This Case Study was co-authored with Corinna Dally-Starna, Queen's University, Kingston, ON, Canada.

combined with a perception of increased commercial viability have brought into closer reach areas of the Earth that had previously remained largely protected from physical intrusion. However, the desire to mine these areas has a relatively long history. Mineral exploitation of the ocean floor was recognized as an option as early as 1876 (Heim, 1990). However, not until the 1970s was its feasibility as a complement to land-based mining considered in earnest. Given the unfavourable balance between costs and economic value at the time, researchers predicted in the 1980s that commercial exploitation of the deep seas most likely would not occur until the twenty-first century. About 40 years later, areas off the shores of Papua New Guinea and inside what is referred to as the South Pacific Clarion–Clipperton Zone (CCZ) may soon see their first mining activities. The CCZ covers an area of about 6×10^6 km^2 (roughly the size of Europe) with water depths of 4 to 6 km. The amounts of manganese, cobalt, nickel, tellurium, thallium and yttrium in the CCZ are currently estimated to exceed total known terrestrial reserves of these elements. A test mining operation in the CCZ was conducted in the early 1970s, but at the time of writing no large-scale commercial mining has yet taken place.

Legal Framework

To frame an analysis of the issue, we need first to ask who owns the oceans, and who is responsible for protecting them? The short answer is: no one, and everyone. The legislative groundwork for the exploitation of what is deemed *res communis*, or the 'territory of no nation', was formally laid in the latter part of the last century when the area of the ocean beyond 200 miles from the shores of coastal nations was deemed 'common heritage of mankind' (Heim, 1990). With that understanding came the requirement for multinational involvement in decision-making about activities such as seabed mining. In 1982, the UN Convention of the Law of the Sea (UNCLOS) established the International Seabed Authority (ISA). This was followed by the signing of the 1994 Agreement Related to the Implementation of Part XI of UNCLOS, which was to govern the mining of the seabed beyond the limits of national jurisdictions. To quote Levin *et al.* (2016):

Under UNCLOS, where mining activities may cause serious harm, the ISA has the power to: (i) set-aside areas where mining will not be permitted; (ii) deny a new application for a contract to conduct seabed mineral activities; (iii) suspend, alter or even terminate operations, and (iv) hold the contractor and its sponsoring state liable for any environmental harm if it ensues.

Another important milestone was the establishment of the Legal and Technical Commission under ISA. The mandate of ISA is to formulate environmental rules and regulations, as well as to prepare environmental impact assessments and implement monitoring programmes, among other responsibilities (Lodge *et al.*, 2014). Consistent with the 'common heritage' principle, exploration contracts have been extended to entities representing nations from around the world, including developing countries. Under these agreements, contractors are required to undertake and submit to the ISA environmental baseline studies. Exploitation permits are only granted once the ISA has judged the comprehensiveness and quality of the data and determined it to be adequate (Lodge *et al.*, 2014; Levin *et al.*, 2016).

Target Metal Deposits and Anticipated Mining Practices

Mineral deposits on the seabed are found primarily in three distinctive topographical environments, for which exploration permits have been issued: abyssal plains, hydrothermal vents and **seamounts** (Levin *et al.*, 2016). The types of deposits found in these environments include (1) manganese (polymetallic) nodules, (2) sea-floor massive (polymetallic) sulphides created by submarine volcanic events, and (3) cobalt-rich (polymetallic) crusts (Table 14.3).

Manganese Nodules

The discovery of these spherical nodules dates from the pioneering voyages of the HMS *Challenger* in the 1870s. They are formed by the accretion of iron and manganese oxyhydroxides around an original tiny nucleus (e.g., a sand grain or a shell fragment). They form on top of the seabed and in commercially exploitable deposits they usually range in size from 5 to 10 cm in diameter. Because of their location, the nodules are readily accessible, and their distribution and abundance are easy to estimate by remote sensing. Based on this dispersal and the results of small-scale feasibility tests, it is anticipated that the mining of manganese nodules will involve the collection of a thin surface layer of the sediments in reasonably flat areas (e.g., the abyssal plains). In test mining experiments, a hydraulic lift has been used to raise the

Table 14.3 **Mineral deposits found in deep-sea environments.**

Type of deposit	Location and physical appearance	Typical metals
Manganese nodules	Abyssal plains; patchy distribution at the seabed surface	Co, Cu, Mn, Ni $\pm$ Ge, Li, Mo, Pt, Ti, Y, Zn, Zr, rare earth elements (REEs)
Massive sulphide	Consolidated mounds around hydrothermal vents	Co, Cu, Pb, Zn $\pm$ Ag, As, Au
Polymetallic crusts	Crusts on flanks of seamounts, ridges and plateaus where currents keep seabed sediment-free	Co, Cu, Mn, Ni $\pm$ Bi, Mo, Pt, Te, Ti, Tl, W, Zr + REEs

suspended nodules to the ocean surface where they were separated from the slurry and retained. In such an approach, the sediment accompanying the nodules would then be released back into the ocean, at a depth to be determined by the regulator. In contrast to a land-based operation, this mining system would be portable (ship-based) and could be moved from site to site.

Worldwide demand for cobalt, lithium and nickel has increased markedly in response to their use in the rechargeable batteries that will be needed to achieve the 'green' transition to a carbon-free economy. Because of this trend, it is currently predicted that manganese polymetallic nodules will be the first type of deep-sea deposit to be subjected to commercial mining (Cuyvers *et al.*, 2018).

Massive Polymetallic Sulphides

The massive sulphide deposits found on the seafloor and those on the land are products of similar geological and geochemical processes. The marine deposits are associated with seafloor spreading, the penetration of seawater into the ocean's crust and the subsequent ejection of hydrothermal fluids. As it enters the crust and approaches the **magma**, the penetrating water is heated and interacts with the crustal rocks and, in some cases, with magmatic fluids. The resulting hydrothermal fluids are metal-laden and acidic, with pH values as low as 2. The hot hydrothermal fluids mix with the cold seawater, and a wide variety of minerals precipitate. As their name suggests, the seafloor massive sulphides are formed from reduced sulphur.

Current plans for the mining of massive sulphide deposits are not as far advanced as are those for nodules, but mining will probably be limited to the surface (strip mining) and shallow subsurface (open-cast mining) (Section 11.2). Hydraulic pumps are possible systems to move the ore from the seabed to the ocean surface for processing. It seems likely that the deposits targeted

initially will be those that are highly enriched in gold and base metals, close to land (i.e., in territorial waters) and located at depths <2,000 m. In 2017, Japan announced that it had successfully retrieved a small quantity of polymetallic sulphides from a depth of 1,600 m in waters close to their shore, using a continuous ore retrieval and lifting procedure. This announcement was hailed as the first successful demonstration of such an approach.

Polymetallic Crusts

Found at depths from 400 to 4,000 m, cobalt-rich ferromanganese crusts occur throughout the global ocean on seamounts (extinct volcanoes), ridges and plateaus, in areas where currents have kept the rocks swept clean of sediments for millions of years. Crusts and nodules are both rich in manganese and iron oxyhydroxides, but they differ mineralogically and in chemical composition (e.g., high copper in nodules and high cobalt in crusts). Crusts precipitate out of cold ambient seawater onto hard-rock substrates, growing very slowly (e.g., 1 to 6 mm over 10^6 years), and eventually form pavements up to 250 mm thick. The action of the crusts has been compared to those of a 'sponge', removing metals and other elements from seawater over these long periods of time.

Despite their enticingly high metal contents, relatively little effort has been expended for the development of techniques adapted to the mining of polymetallic crusts. The crusts are cemented to hard substrates and their removal would be destructive, probably requiring their physical fragmentation and collection by a remote operating vehicle (ROV) followed by transport to the surface vessel.

Nature of the Habitats and Their Existing Communities

As is the case for terrestrial mining activities (Section 11.2), the biological communities most affected by deep-sea mining will be those living near the mining sites. In other words, the habitats of primary interest are those close to polymetallic nodule fields in the ocean's abyssal plains, on seamounts that host polymetallic crusts, and near the hydrothermal vents and seeps that feed the massive sulphide deposits. Understandably, given their remoteness and the difficulties of working at great depths in the ocean, these ecosystems remain poorly understood. This lack of knowledge severely hampers any efforts to evaluate the potential environmental impacts of mining activities and invites a precautionary approach to deep-sea mining.

Taking these habitats in the order mentioned above, the abyssal plains are the deepest part of the ocean where the resident biota live in complete darkness (no photosynthesis), at cold temperatures (1–2 °C) and very high pressures (300–380 atm) (Box 14.4). Organic matter produced by photosynthesis in the upper ocean degrades as it settles, so that virtually none reaches this benthic habitat. Estimates of the total biomass are very low, but the community is surprisingly diverse. It has been suggested that nodules create habitat heterogeneity in these plains and that the fauna associated with them is more abundant and more diverse than in nodule-free areas.

In contrast to the relative homogeneity of the abyssal plains, seamounts offer a wide variety of physical characteristics such as size, shape and local hydrodynamics. As a result, they are considered to be biodiversity hotspots (e.g., some 800 species of fish have been specifically associated with seamounts (Cuyvers et al., 2018)). The seamount hard rock surface is typically dominated by filter-feeders such as corals and sponges. These organisms influence the existing ecosystem structure by forming reefs and introducing additional microhabitats that attract more organisms, including crustaceans, molluscs and echinoderms.

Hydrothermal vents, the final habitat to be considered, exhibit marked chemical and temperature gradients, and they host unique ecosystems. In the absence of light, specialized bacteria utilize inorganic nutrients and energy derived from the hydrothermal fluids to fix carbon into organic molecules, supporting a rich array of life. More than 500 species, many of them previously unknown to science, have been identified in these ecosystems, including bacteria, polychaetes, gastropods, crustaceans and fish (Cuyvers et al., 2018).

Anticipated Impacts of Deep-sea Mining Activities

Even under the best of circumstances, assessment of environmental risks and potential effects of industrial activities on the aquatic environment can present a conundrum to ecotoxicologists and risk assessors. Challenges are elevated to another level when considering the deep-sea environment. Chief among these is the lack of data about and access to deep-sea ecosystems. In addition, the delineation of spatial and temporal boundaries within the deep-sea environment is confounded by hydrodynamics. Contaminants travelling with water do not adhere to political, geographic or ecosystem boundaries; their dispersion occurs in response to the forces of currents present at various depths.

Beyond limited knowledge about life at depths of up to 6 km, other challenges include a paucity of information about process chemicals (if any) that might be used in separation of the target mineral from the accompanying sediment. There are also uncertainties associated with extrapolation from laboratory results to distant biotic communities or understudied species. Knowledge acquired from other resource exploitation activities, such as oil and gas extraction, may prove to be useful, although these activities are usually carried out at much shallower depths.

Potential environmental impacts are expected to vary as a function of individual mining activities and geographic location, as well as with the type of target mineral to be exploited (Levin et al., 2016). The physical destruction of seafloor habitat, the generation of contaminant plumes from physical excavation activities at depth, and dewatering processes at the sea surface are widely

Box 14.4 Effects of temperature and pressure on deep-ocean chemistry

Temperature and pressure are the dominant variables that define the deep-sea environment. From a physical chemistry perspective, these two factors often counterbalance one another. For example, at constant temperature, an increase in pressure will favour compression and ordering of molecules. On the other hand, at constant pressure, an increase in temperature will have the opposite effect, favouring an increase in molecular disorder. Among different biomolecular structures, lipid membranes are very sensitive to pressure changes, whereas protein primary structures are relatively insensitive.

acknowledged as inevitable stressors. The fate of the contaminants present in these plumes will be determined by the depth at which the plumes are generated and by the water temperature at this depth (in relation to temperature of the plume itself) as well as by the strength of the local ocean currents. When the suspended solids resettle, they could affect sessile fauna, especially filter-feeders.

To address knowledge gaps and the technological challenges of assessing ecotoxicological impacts on deep-sea species, some scientists have employed creative approaches and methods (e.g., the use of pressure chambers to mimic the effect of varying sea depths). In a limited number of cases, 'proxy' shallow-water species have been selected to investigate the effects of changes in temperature and hydrostatic pressure on organism responses to contaminant exposure. The responses of the proxy species are then compared to those of analogous deep-sea species (Brown *et al.*, 2017). However, it is still unclear whether this approach will yield useful predictions.

Recovery

Given the extremely slow rates at which sediment accumulates in the ocean depths, areas disturbed by deep-sea mining are unlikely to recover at what would be considered a 'reasonable' rate on the Earth's surface. To evaluate the resilience of benthic deep-sea fauna to mining activities associated with the three main topographical terrains, Gollner *et al.* (2017) studied community recovery at sites that have witnessed major disturbances. Examples of such sites included areas affected by volcanic eruptions at vents, fisheries on seamounts and experiments that mimicked nodule mining on abyssal plains. They reported a wide range of recovery rates among taxa, varying notably with the size and mobility of the fauna. Although taxa in some habitats were able to recover and even exceed pre-disturbance conditions, others reportedly remain affected after decades.

The abyssal seafloor is an obvious example of an inherently stable habitat, where the ability of native fauna to cope with physical disturbance is likely to be limited. Indeed, the nematode communities that were disturbed by experimental nodule mining in the CCZ in 1978 had not recovered when the tracks left by the self-propelled test miner were revisited in 2004, 26 years after the original disturbance (Miljutin *et al.*, 2011).

Mitigation

The ISA has recommended that 'impact reference zones' (IRZs) and 'preservation reference zones' (PRZs) be established (ISA, 2017).

- IRZs: areas set aside within the area to be mined, representative of the whole area and located close to an area to be exploited early in the mining contract;
- PRZs: areas in which no mining will occur, to ensure representative and stable biota on the seabed in order to assess any background changes in the flora and fauna of the marine environment.

Explicit mention of these zones has been incorporated into ISA's exploration regulations as an integral part of programmes for monitoring and evaluating the impacts of deep seabed mining.

Other potential mitigation strategies that have been identified include:

- the design of mining patterns that leave large, adjoining areas undisturbed by direct mining;
- the improvement of remotely operated mining equipment to reduce sediment plume dispersion and seafloor compaction;
- the development of shipboard waste disposal techniques to reduce their impact;
- the introduction of artificial substrates to help with the recolonization process; and
- the development and implementation of emergency response procedures (remediation).

To quote from Cuyvers *et al.* (2018), "Spatial management through the designation of no-mining areas will undoubtedly contribute to ensuring site-specific and regional-scale conservation of ecosystems and biodiversity."

Conclusions

The deep seas are home to a diversity of flora and fauna that has yet to be fully studied, catalogued and their interdependencies within the larger ecosystem understood. There are particular concerns over potentially irreversible environmental impacts of deep-sea mining on these unique and fragile biological communities. Significant challenges include data collection for the development of analytical frameworks and models to aid impact assessments and environmental protection planning ahead of resource exploitation. In the final analysis, those as-of-yet missing data will be critical for the development of analytical frameworks and models. These, in turn, are essential for developing policies, regulations and environmental management plans (Brady et al., 2017) to protect the 'common heritage of humankind,' or 'Earth's final frontier,' against adverse effects.

14.4 Emerging Concerns about Organic Contaminants[2]

Issues with emerging organic contaminants differ markedly from those that we identified in the preceding section for emerging inorganic contaminants. Emerging organic contaminants greatly outnumber their inorganic counterparts, and, thanks in part to the inventiveness of organic chemists, they are inherently more variable in composition. The term 'emerging' is used here to describe organic molecules that have only recently been detected in the environment and for which regulatory guidelines have not yet been developed. It also applies to products that have been commercialized recently and have not yet been detected in environmental matrices, but have properties or characteristics that suggest they may cause toxicity and ecological harm.

With approximately 200,000 chemicals in commerce (among which >80,000 are produced in Europe in quantities exceeding 1 tonne/year), the European Union has established 'watch lists' for organic chemicals considered to be contaminants of emerging concern that should be monitored as part of water policy (European Commission, 2020). The substances listed in Box 14.5 correspond to a snapshot in time, with individual chemicals being removed from or added to the list as the monitoring results are compiled and examined.

The examples in Box 14.5 include fungicides and insecticides used in agriculture that are released directly into the environment. Alternatively, some products may be used in a manner that leads to their delayed but known release to the environment. Examples of such products include chemicals found in discharges of municipal wastewater, such as antibiotics and antidepressants, as well as fungicides used in antidandruff shampoos (Clotrimazole) and in antifungal ointments (Fluconazole). However, other contaminants may 'leak' into the environment,

> ### Box 14.5 Examples of 'contaminants of emerging concern' (CECs) on the European Union watch lists (2020)
>
> *Fungicides (azole compounds)*: Clotrimazole; fluconazole; imazalil; ipconazole; metconazole; miconazole; penconazole; prochloraz; tebuconazole; tetraconazole
> *Fungicide*: Dimoxystrobin; famoxadone
> *Insecticide*: Metaflumizone
> *Antibiotics*: Amoxicillin; ciprofloxacin; sulfamethoxazole; trimethoprim
> *Antidepressant*: Venlafaxine, and its metabolite O-desmethylvenlafaxine

sometimes from unanticipated sources. Substances such as these, which have not yet been detected and identified, represent the hidden part of the chemical 'iceberg'.

The scientific challenge here is threefold: (i) how to monitor the environment to determine if these compounds or their metabolites are present, and if so, at what concentrations; (ii) how to evaluate the toxicity and environmental behaviour of these new compounds, and (iii) how to apply lessons learned in the laboratory to predict the toxicity of a new contaminant in the natural environment, as has been stressed in Chapter 4. Possible ways of meeting these challenges are discussed in the following sections.

14.4.1 Monitoring

Part of the answer to the challenge of looking for and identifying chemicals of emerging concern lies in 'non-targeted screening', which involves the analysis of a variety of carefully chosen media that may contain new compounds. Examples of such media include industrial and municipal wastewater effluents, sediments, soils, passive samplers or tissues collected from **biomonitor** species. This line of attack is not designed to detect specific known compounds (i.e., 'targeted' analysis), but rather to separate candidate compounds, referred to as 'features', that warrant further investigation. For example, one might use a particular solvent or a specific adsorbent to concentrate substances from water, sediments or biota; the components of the resulting mixture would be separated chromatographically. In the first step in the subsequent workflow, high-resolution mass spectrometry (HRMS) is used

---

[2] Christopher Metcalfe also contributed to this section.

to determine the exact mass of these candidate compounds so that they can be identified using one or several techniques. As is described in Section 14.4.2, this approach is not new. There are several examples from the past where unexpected chromatographic peaks have been seen during the analysis of environmental samples and the compounds have subsequently been identified. What is new is the improvement in mass spectrometry instrumentation and data analysis software that allows rapid identification of the compounds responsible for these unexpected peaks.

14.4.2 Non-targeted Screening

Until recently, monitoring of organic contaminants in the environment typically focused on a relatively small number of compounds that were of regulatory interest or had been previously identified as contaminants of concern. Chemicals that are targeted for analysis typically number in the hundreds, but this is only a fraction of the many thousands that are present in the environment. In many cases, new classes of organic contaminants were discovered by accident, and this often occurred when their concentrations in environmental matrices were so high that they interfered with the analysis of other priority contaminants.

For example, while analysing the muscle of sea eagles in Sweden in 1966, a Swedish researcher named Søren Jensen fortuitously detected unknown compounds that were interfering with his analysis of DDT. After two further years of study, Jensen and his co-workers were able to demonstrate that these compounds were polychlorinated biphenyls (PCBs), and that they were present in remarkably high concentrations in Baltic Sea fauna. These PCBs had been discharged into the environment undetected for more than 37 years (Jensen et al., 1969). Similarly, Robert Huggett made the fortuitous discovery in the late 1970s that the James River in Virginia, USA, was contaminated by Kepone (chlordecone), a pesticide that had been discharged into the river by the manufacturer for at least 10 years (Huggett and Bender, 1980).

Obviously, we cannot protect the environment by depending on serendipitous discoveries by researchers that occur years after the contamination begins. Fortunately, new analytical tools are being applied to identify new classes of contaminants, in principle before they reach levels that are a hazard to the environment.

High-resolution mass spectrometry using full-spectrum acquisition techniques is a powerful analytical tool for identifying unknown compounds in the environment. Advanced software systems that are now available for identifying candidate compounds from matches with extensive mass spectral databases can be applied to screen environmental samples for potentially hazardous organic contaminants. Figure 14.5 illustrates an example of the workflow that can lead to the identification of chemicals of emerging concern. The steps are numbered sequentially: (1) Extracts from environmental samples are separated by liquid or gas chromatography and the exact masses determine by HRMS. (2) Using various databases that

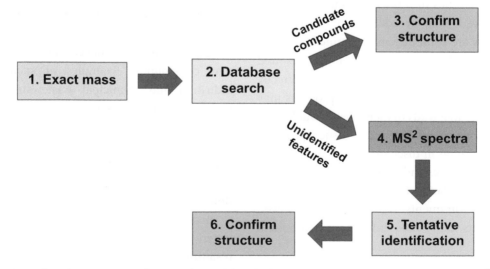

Figure 14.5 Typical workflow for non-targeted approaches to identify chemicals of emerging concern. The term MS² in box 4 refers to tandem mass spectrometry.

provide data on exact masses, the compounds may be tentatively identified and then (3) the structures confirmed by targeted analysis. (4) For those compounds that cannot be identified from their exact mass, it may be necessary to fragment the compounds in a collision cell in the mass spectrometer to generate tandem mass spectra, often abbreviated as MS^2 spectra. (5) Using software designed for elucidating chemical structures, it may be possible to use these spectra to tentatively identify the mystery compound. Commercially available analytical standards for these compounds, if available, can then be used to confirm the identity of some of the candidate compounds that are tentatively identified (Gosetti *et al.*, 2016). Several recent examples of how this workflow has been used to identify contaminants previously unsuspected to be present in the aquatic environment have been compiled in Table 14.4.

In an interesting variation of this approach, Seiwert *et al.* (2020) proposed what they refer to as 'source-related smart suspect screening'. In this approach, non-targeted screening is performed (i) on samples that represent a potential source of contaminants, as well as (ii) on environmental samples collected close to the suspected source, and (iii) on blank samples. The features present in three samples are compared, and all subsequent work is focused on the features that appear in both samples (i) and (ii) but do not occur in the blank. The stated advantage of this approach is that it "introduces a rational criterion for selecting features of interest, rather than criteria like signal intensity, which may be unfounded and misleading." This approach was used in the urban streams example summarized in Table 14.4 and led to the positive identification of toxic tyre-wear

compounds that are washed into surface waters from roads during rain events. By comparing tyre particle leachate as the source sample (i) with a 'close to source' sample (ii), the authors were able to eliminate almost 90% of the features in the source sample and to concentrate on those that were common to both the potential source and the creek environment.

14.4.3 Toxicity Evaluation

Given the number of organic chemicals in commerce, it would be completely unrealistic to subject each and every compound to a battery of toxicity tests with living organisms. There are also ethical considerations, such as the need to minimize animal testing, particularly for vertebrates. How then can we aspire to evaluate their toxicity? For compounds that are presumed to act as narcotics, toxicity can be estimated from the target lipid model (Section 2.3.4.3). For molecules that are not narcotics, QSARs can be developed to predict toxicity from the nature, size and location of molecular substituents associated with a specific mode of action (MoA). Both approaches can identify priority compounds for toxicity testing.

In other cases, where the mode of action of the compound can be predicted (e.g., for new insecticides or herbicides, or for pharmaceuticals), *in vitro* testing can be used to determine the relative toxicity of the new compounds in comparison with existing compounds. As described in Section 5.10.1, this approach involves the use of high-throughput *in vitro* toxicity tests to predict which chemicals may be most harmful to wildlife and humans.

Table 14.4 **Examples of non-targeted analysis as applied to organic contaminants.**

Location	Sampling and analysis	Examples of chemicals identified	Reference
Groundwater, France	Passive sampling (POCIS); LC-QTOF	Dechlorinated degradation product of metolachlor (herbicide)	Soulier *et al.* (2016)
Urban creek, Leipzig, Germany Road dust, tyre rubber crumb	Aqueous extraction; UHPLC; HRMS	Hexamethoxymethyl melamine and other methoxymethyl melamines; four benzothiazoles	Seiwert *et al.* (2020)
Urban streams, Seattle, WA, USA	Tyre particle aqueous leachate and solid phase extract from ambient water; UHPLC; QTOF-HRMS	Transformation product of N-(N-1,3-dimethylbutyl)-N'-phenyl-p-phenylenediamine (6PPD-quinone)	Tian *et al.* (2021)
Urban estuary, Puget Sound USA	Solid phase extraction; UHPLC; QTOF-HRMS	Two novel transformation products of tebuthiuron (herbicide)	Tian *et al.* (2020)

HRMS = High-resolution mass spectrometry; LC = liquid chromatography; POCIS = polar organic chemical integrative sampler; QTOF = quadrupole time-of-flight mass spectrometer; UHPLC = ultra-high-performance liquid chromatography.

Some of these tests use animal cell lines and involve the measurement of whole-cell responses to the test chemical. Other tests focus on signalling pathways and evaluate the capacity of the test chemical to bind to proteins, enzymes, DNA, RNA, or specific biomolecules such as oestrogen, thyroid, androgen or aryl hydrocarbon receptors. The response to the test chemical is compared with the responses obtained with a suite of chemicals known to affect the tested pathway, and test chemicals that rank highly can then be prioritized for standard laboratory testing. For example, the ToxCast database established by the US EPA contains information about the biologically relevant activity of ~1,800 chemicals, as obtained *in vitro* from more than 700 high-throughput assays (Kavlock *et al.*, 2012).

In a pioneering example of this approach, Corsi *et al.* (2019) considered chemical monitoring data from 57 rivers that drain into the Laurentian Great Lakes from the United States. In an earlier study, a total of 67 synthetic organic chemicals had been identified and quantified, but traditional water quality guidelines were only available for 37 of these chemicals (Baldwin *et al.*, 2016). To expand the evaluation, Corsi and co-workers performed a parallel comparison with data from the ToxCast high-throughput *in vitro* assays. To do so, they compared the measured concentration of a given chemical with the effect concentration of the chemical in the ToxCast database, i.e., an exposure–activity ratio (EAR). The chemicals that were identified in this manner to be of potential concern were: 4-nonylphenol; bisphenol A; metolachlor; atrazine; *N,N*-diethyl-*m*-toluamide (DEET); caffeine; tris(2-butoxyethyl) phosphate; tributyl phosphate; triphenyl phosphate; benzo(a)pyrene; fluoranthene; benzophenone. These qualify as well-known chemicals for which there is an emerging recognition of their occurrence in environmental media.

The EAR is analogous to a hazard quotient (HQ) as defined in Chapter 12, but of course the 'effect' here is a response in an *in vitro* assay rather than a traditional response in an *in vivo* toxicity test such as growth inhibition, mortality or reproductive impairment. Despite this difference, for chemicals for which both water quality benchmarks and ToxCast EARs were available, the chemical classes that were identified to be of concern were consistent. The use of the ToxCast data doubled the number of chemicals that could be evaluated, and several additional molecules were identified as potential concerns. Note that the EAR approach is just the start of a long process of research to establish whether the compounds identified have any environmental impacts at environmental concentrations.

The *in vitro* testing approach originally focused largely on human health impacts, but as illustrated by the foregoing study, it also has potential for predicting adverse impacts on natural fauna and flora. The further development of this link between *in vitro* testing and ecological risk assessment will undoubtedly benefit from the ongoing development of the AOP framework (Section 5.9, Figure 14.2), which emphasizes linkages among the types of molecular or biochemical activities that are monitored in ToxCast assays and adverse outcomes. Chemicals that are identified with the *in vitro* approach can be further tested in animal models to better discern the relationship between exposure and adverse effects.

In the European Union, the concept of 'trigger values' for toxicological endpoints generated from *in vitro* assays has entered into the regulatory framework. Under this scenario, if the toxic effects of extracts prepared from wastewater, surface waters or drinking water exceed a certain level in an *in vitro* assay, then further work is mandated to identify the compounds that are inducing the response. For instance, if the *in vitro* oestrogenic activity in a drinking water sample exceeds a trigger value of 0.1 µg/l oestradiol-equivalents, further studies are required to identify which compounds are present in the drinking water that could cause this response.

14.4.4 Predictive Toxicology

The remaining challenge is to improve our ability to predict the environmental effects of these chemicals of emerging concern. By environmental effects, we mean effects in natural environments such as rivers, lakes, pastures, forests and oceans, but also in urban and industrial locations. The challenge is how to improve our ability to extrapolate from empirical data obtained for a limited number of test species and to predict impacts on the much larger number of species from different phyla that are found in the target ecosystem. We expect to find marked differences in physiology, diet and life-history traits among these native species, and there are also likely to be differences between them and the cultured species that are used in toxicity tests (e.g., Box 2.6).

With the increased recognition of the limitations of traditional toxicity testing, there has been considerable interest in the AOP approach as a means to link chemical exposure to a molecular initiating event (MIE) and then to apical endpoints across various levels of biological organization (Sections 5.9 and 14.1.2, Figure 14.2). For this approach to work in predicting effects in untested species, it would be necessary that the AOP apply across a diversity of organisms, i.e., it requires that the AOP pathway be conserved in different phyla. This may be a reasonable anticipation for a substance with a very specific mode of action (e.g., an insecticide that inhibits acetylcholine esterase), but pathways other than the central nervous system in vertebrates and invertebrates are often less tightly conserved. Note too that this approach assumes that there is only one dominant AOP within a species or phylum.

Consideration of the AOP approach leads naturally to a consideration of the causes of the differences in species sensitivity. Mechanisms that contribute to resistance, i.e., reduced sensitivity, include: (i) a decreased uptake of the toxicant; (ii) an increased rate of metabolic detoxification; and (iii) decreased sensitivity of the target site with which the toxicant (or its metabolites) interacts. In other words, if we are to predict the toxicity of compound 'X' on an untested species, we need to know whether the species is likely to exhibit reduced or enhanced sensitivity. How might we do this? There are two dominant approaches for predicting the sensitivity of different phyla to a specific toxicant: (a) *phylogenetic comparison of species* and (b) *target ortholog assessments*. These two terms are described in the following paragraphs.

Phylogenetic comparisons are designed to assess the likelihood that related species, which share metabolic systems and target structures, show similar sensitivity to a given chemical. In other words, we are asking if their response to exposure to a particular toxicant bears a **phylogenetic signature**. Research in this area suggests that for some chemical categories (e.g., organochlorines and metals) the phylogenetic signature is weak, but other classes of contaminants do have a strong signature (e.g., organophosphates for fish and other vertebrates – cf. Section 2.1.1.2, Box 2.3). For chemicals in this latter group, with a strong phylogenetic signature, this approach could help predict species sensitivity when toxicity data are unavailable (Hylton et al., 2018).

Target ortholog assessments are a natural by-product of the promise of the AOP approach and its emphasis on the MIE, the subsequent key events and the eventual adverse outcome. Given the success of the AOP approach for individual species, research in the field of predictive toxicology has focused on critical receptors that are involved in the MIEs and their presence or absence (abundance) in other species (a concept known as 'target conservation as an indicator of sensitivity'). This approach was initially proposed by Gunnarsson et al. (2008) and enlarged upon by Verbruggen et al. (2018). Working with specific human drug molecules, many of which have been detected in the environment, these researchers demonstrated that the presence of well-conserved receptors in a given species was linked to an increased sensitivity to the drug. In a comparison across different phyla, the presence of orthologs to the drug targets decreased in the order zebrafish (86%) > arthropod (61%) > green alga (35%). An analogous example is provided by the CYP1A system and its link to chemical metabolism and toxicity, both for toxicity enhancement (e.g., forming carcinogenic metabolites) and detoxification (enhanced degradation and excretion of toxicants) (Section 7.5.1). The CYP system is highly developed and diverse among vertebrate species, less well developed in invertebrates, and absent in plants.

From a practical perspective, this approach involves the identification of the gene that encodes the receptor involved in the MIE, cataloguing receptor **orthologs** in different phyla and organisms, and determining if the orthologs are functional in these other organisms. A potential limitation of this approach, where one interprets the presence of orthologs as a predictor of sensitivity, occurs in cases where the orthologs have retained sequence homology but in the course of evolution have assumed roles that lie outside the target function or tissue. For example, appreciable concentrations of cholinesterases and carboxylesterases have been found in non-neural tissues in earthworms. Predictions of functionality might involve examination of homology in ligand-binding domains and in other areas that can affect protein conformation and thus affect ligand–receptor interactions. In other words, determination of the presence or absence of orthologs will be an important first step in predicting sensitivity, but confirmation that binding to the receptor leads to the predicted downstream events is also needed.

14.4.5 Applications of Predictive Toxicology in Ecological Risk Assessment

Predictive toxicology is an essential feature of ecological risk assessments. As described in Section 2.3.3.2, species sensitivity distributions (SSDs) are currently widely used to extrapolate from the laboratory to the field. However, SSDs are widely acknowledged to be imperfect tools, in part because of the small number of species that are used to develop the statistical distributions (often less than 15). In principle, *phylogenetic comparisons* and *target ortholog assessments* could be used to identify phyla and species likely to be affected by emerging contaminants. In their thought-provoking article on future developments in ecotoxicology, Spurgeon *et al.* (2020) suggest that "It may be possible to evaluate the proportion of species present in an SSD that possesses one or more traits (e.g., absence of relevant detoxification pathway or presence of a receptor ortholog) linked to sensitivity." To minimize the possibility of underestimating toxicity in the real-world environment, these authors also suggest that "an SSD may be derived only for species likely to bioaccumulate a chemical, that possess a receptor ortholog with the relevant ligand binding domain, or which lack competing off-target receptors for the chemical ligand that may play a protective role."

14.4.6 Future Considerations

Emerging organic contaminants obviously pose a much greater numerical challenge than do the inorganic contaminants considered in Section 14.3. However, considerable progress has been achieved in the design of **forensic** monitoring programmes to detect novel organic contaminants and link them to possible sources. This use of non-targeted screening should be encouraged. Once identified, these compounds could then be subjected to high-throughput *in vitro* toxicity screening to provide an initial estimate of their toxicity. In addition, the predictive modelling of the toxicity of 'new' organic contaminants is poised to improve radically. The AOP methodology shows promise as a means of predicting links between chemical exposure and apical endpoints in untested species, provided that the AOP pathway is conserved in different phyla. Similarly, the use of phylogenetic comparisons and target ortholog assessments should help to identify species that are potentially sensitive to the new chemicals and thus are candidates for targeted toxicity testing.

Summary

In this chapter we have discussed four quite different topics that have assumed increasing urgency over the past few decades. In view of the disparate nature of these topics, each of the above sections contains a brief outline of 'Future Considerations'. However, here we provide a recap of some of the main points of emphasis in this chapter as a whole. We provide a summary of the progress made, subsequent shifts in emphasis, and deficiencies that we hope that the next generation of ecotoxicologists can address. This summary is followed by a brief final perspective on the science of ecotoxicology, past, present and future.

Climate change. It is particularly challenging to forecast the influence of climate change on chemical toxicity and chemical exposure patterns at levels of biological organization from molecular to population levels; to date, we mainly have data from single-species experiments with few variables. Solutions include the deployment of an AOP approach to define linkages across different levels of biological organization. Other important deficiencies in our knowledge of climate change/contaminant interactions concern interspecific reactions and the impact of climate stress on mixtures of contaminants.

Microplastics and nanoplastics. Readers will appreciate that we are, again, confronted by the difficulty of extrapolating from laboratory experiments to multiple external stressors in an ecosystem. In understanding 'dose–response', we need to differentiate between toxicological effects caused by plastic particle ingestion, and those resulting from the release and assimilation of associated chemicals. Answers to these questions will have a direct bearing on regulatory and mitigation strategies, including plastic manufacturing and disposal (i.e., 'cradle-to-grave') choices.

Emerging inorganic contaminants. The challenges related to inorganic chemicals should be simpler to address than those for organic contaminants, in part because of the much smaller number of emerging inorganic contaminants. In addition, the approaches that have been developed and refined for the more common elements in the Periodic Table should also prove useful for the newly defined technology-critical elements. However, many of these TCEs exist in oxidation states of +III or higher in aqueous solution, and the applicability of such tools as the biotic ligand model for such elements remains uncertain. The challenges posed by inorganic contaminants also have a distinct regional component, in that increased mining activities are projected for ecologically sensitive or unknown areas, such the deep sea and the Arctic/Antarctic.

Emerging organic contaminants. A forensic approach to monitoring organic contaminants, including the use of non-targeted screening, should be encouraged. Predictive modelling of the toxicity of 'new' organic contaminants is poised to improve radically, with a combination of the AOP methodology, high-throughput *in vitro* toxicity screening and the use of phylogenetic comparisons and target ortholog assessments to identify potentially sensitive species. These approaches address the need to reduce the lag time between the results of the above steps and the appropriate regulatory response.

Review Questions and Exercises

1. Give a summary of the gases that are the primary drivers of climate change, including their sources and their relative contributions to global warming. What remedial efforts are in development and projected for the future?

2. What evidence is there for the effects of climate change on existing ecosystems? Give examples of regional aspects involved.

3. Give an account of the interactive effects on biota of physical factors resulting from climate change and stresses due to chemical contaminants.

4. What interspecific effects involving chemical stress and climate change have been demonstrated or projected? Give an account of the strengths and weaknesses of current investigative methods and discuss the merits of future approaches.

5. Write an essay on the history of plastic contamination of the environment, giving examples of the types of plastics involved.

6. Describe the ways in which nano- and microplastics pose a threat to wildlife. Give examples of physical and chemical factors involved, including interactive effects.

7. What strategies are being considered to mitigate the threat posed by plastic pollution worldwide? How do these strategies intersect with other forms of environmental regulation?

8. Global mining production has almost doubled since 2000. What factors are driving this expansion?

9. Is metal recycling a viable approach for attenuating the increased demand for metals?

10. How would you define a 'critical' metal?

11. How should humans manage the minerals that are present in the deep ocean? Should they be left undisturbed, or can they be recovered with acceptable environmental impacts?

12. Explain the links between rare earth elements and the greening of the world's economy.

13. How would you explain non-targeted screening to someone with minimal knowledge of analytical chemistry? How does it differ from traditional environmental monitoring?

14. How can high-throughput *in vitro* testing help us deal with the profusion of new organic chemicals in the world's commerce?

15. Explain how target ortholog assessments can be used to predict the sensitivity of certain phyla to a new chemical contaminant.

16. How could you determine if a particular chemical has a 'phylogenetic signature'? How might this be useful in identifying untested species that might be particularly sensitive to the chemical?

17. How could the phylogenetic signature of a novel compound help uncover its mode of toxic action?

18. **Group project**. Set up a team to enact a task force/public enquiry into the use and disposal of plastics, including use of nanoplastics. Design your Terms of Reference and include ENGOs, professional scientists and technologists, municipal and other levels of government and members of the general public on the team.

19. **Group project**. Design and enact a television documentary on technology-critical elements (TCEs), for an audience that is well-educated but non-specialized in technology.

Abbreviations

AOP	Adverse outcome pathway		PE	Polyethylene
BPA	Bisphenol A		PET	Polyethylene terephthalate
CCZ	Clarion–Clipperton zone		PHAs	Polyhydroxyalkanoates
CEC	Contaminants of emerging concern		PHB	Polyhydroxybutyrate
CFC	Chlorofluorocarbon		PLA	Polylactic acid
CITS	Climate-induced toxicant susceptibility		POCIS	Polar organic chemical integrative sampler
CYP1A	Cytochrome P4501A		PP	Polypropylene
DDD	Dichlorodiphenyldichloroethane		PRZ	Preservation reference zone
DDE	Dichlorodiphenyldichloroethylene		PS	Polystyrene
DDT	Dichlorodiphenyltrichloroethane		PU	Polyurethane
DNA	Deoxyribonucleic acid		PVC	Polyvinyl chloride
DTFs	Daily temperature fluctuations		QICAR	Quantitative ion character–activity relationships
EAR	Exposure–activity ratio			
EDC	Endocrine disrupting chemical		QSAR	Quantitative structure–activity relationships
ENM	Engineered nanomaterial		QTOF	Quadrupole time-of-flight mass spectrometer
GHG	Greenhouse gas			
GWP	Global warming potential		REE	Rare earth element
HDPE	High-density polyethylene		RF	Radiative forcing
HQ	Hazard quotient		RNA	Ribonucleic acid
HRMS	High-resolution mass spectrometry		ROS	Reactive oxygen species
IPCC	Intergovernmental Panel on Climate Change		ROV	Remote operating vehicle
IRZ	Impact reference zone		SSD	Species sensitivity distribution
ISA	International Seabed Authority		TCE	Technology-critical element
LC	Liquid chromatography		TICS	Toxicant-induced climate susceptibility
MIE	Molecular initiating event		UHPLC	Ultra-high-performance liquid chromatography
MoA	Mode of action			
MS^2	Tandem mass spectrometry		UNCLOS	United Nations Convention of the Law of the Sea
PA	Polyamide			
PAH	Polycyclic aromatic hydrocarbon		USGS	United States Geological Survey
PCB	Polychlorinated biphenyl		UV	Ultraviolet radiation

References

Auta, H. S., Emenike, C. U. & Fauziah, S. H. (2017). Distribution and importance of microplastics in the marine environment: a review of the sources, fate, effects, and potential solutions. *Environment International*, 102, 165–176.

Baldwin, A. K., Corsi, S. R., De Cicco, L. A., *et al.* (2016). Organic contaminants in Great Lakes tributaries: prevalence and potential aquatic toxicity. *Science of the Total Environment*, 554, 42–52.

Besseling, E., Wegner, A., Foekema, E. M., van den Heuvel-Greve, M. J. & Koelmans, A. A. (2013). Effects of microplastic on fitness and PCB bioaccumulation by the lugworm *Arenicola marina* (l.). *Environmental Science & Technology*, 47, 593–600.

Bhattacharya, P., Lin, S. J., Turner, J. P. & Ke, P. C. (2010). Physical adsorption of charged plastic nanoparticles affects algal photosynthesis. *Journal of Physical Chemistry C*, 114, 16556–16561.

Box, J. E., Colgan, W. T., Christensen, T. R., *et al.* (2019). Key indicators of Arctic climate change: 1971–2017. *Environmental Research Letters*, 14, e045010.

Brady, S. P., Monosson, E., Matson, C. W. & Bickham, J. W. (2017). Evolutionary toxicology: toward a unified understanding of life's response to toxic chemicals. *Evolutionary Applications*, 10, 745–751.

Brown, A., Wright, R., Mevenkamp, L. & Hauton, C. (2017). A comparative experimental approach to ecotoxicology in shallow-water and deep-sea holothurians suggests similar behavioural responses. *Aquatic Toxicology*, 191, 10–16.

Browne, M. A., Dissanayake, A., Galloway, T. S., Lowe, D. M. & Thompson, R. C. (2008). Ingested microscopic plastic translocates to the circulatory system of the mussel, *Mytilus edulis* (L.). *Environmental Science & Technology*, 42, 5026–5031.

Cairns, J., Heath, A. G. & Parker, B. C. (1975). Effects of temperature upon toxicity of chemicals to aquatic organisms. *Hydrobiologia*, 47, 135–171.

Capkin, E., Altinok, I. & Karahan, S. (2006). Water quality and fish size affect toxicity of endosulfan, an organochlorine pesticide, to rainbow trout. *Chemosphere*, 64, 1793–1800.

Chae, Y. & An, Y. J. (2017). Effects of micro- and nanoplastics on aquatic ecosystems: current research trends and perspectives. *Marine Pollution Bulletin*, 124, 624–632.

Cobelo-García, A., Filella, M., Croot, P., *et al.* (2015). COST action TD1407: network on technology-critical elements (NOTICE) – from environmental processes to human health threats. *Environmental Science and Pollution Research*, 22, 15188–15194.

Corsi, S. R., De Cicco, L. A., Villeneuve, D. L., *et al.* (2019). Prioritizing chemicals of ecological concern in Great Lakes tributaries using high-throughput screening data and adverse outcome pathways. *Science of the Total Environment*, 686, 995–1009.

Cuyvers, L., Berry, W., Gjerde, K., Thiele, T. & Wilhem, C. (2018). *Deep Seabed Mining: A Rising Environmental Challenge*. Gland, Switzerland: International Union for Conservation of Nature (IUCN), Gallifrey Foundation.

DeLorenzo, M. E., Wallace, S. C., Danese, L. E. & Baird, T. D. (2009). Temperature and salinity effects on the toxicity of common pesticides to the grass shrimp, *Palaemonetes pugio*. *Journal of Environmental Science and Health Part B: Pesticides Food Contaminants and Agricultural Wastes*, 44, 455–460.

Dulvy, N. K., Rogers, S. I., Jennings, S., *et al.* (2008). Climate change and deepening of the North Sea fish assemblage: a biotic indicator of warming seas. *Journal of Applied Ecology*, 45, 1029–1039.

ECHA. (2020). *ECHA Information on Chemicals – Registered Substances* [Online]. Helsinki, Finland: European Chemicals Agency. Available at https://echa.europa.eu/information-on-chemicals/registered-substances [Accessed 17 Nov 2020].

European Commission (2020). Directive 2020/1161 of the European Parliament and of the Council, establishing a watch list of substances for Union-wide monitoring in the field of water policy pursuant to Directive 2008/105/EC of the European Parliament and of the Council.

Fortin, M. G., Couillard, C. M., Pellerin, J. & Lebeuf, M. (2008). Effects of salinity on sublethal toxicity of atrazine to mummichog (*Fundulus heteroclitus*) larvae. *Marine Environmental Research*, 65, 158–170.

Fries, E. & Zarfl, C. (2012). Sorption of polycyclic aromatic hydrocarbons (PAHs) to low and high density polyethylene (PE). *Environmental Science and Pollution Research,* 19, 1296–1304.

Gollner, S., Kaiser, S., Menzel, L., *et al.* (2017). Resilience of benthic deep-sea fauna to mining activities. *Marine Environmental Research,* 129, 76–101.

Gosetti, F., Mazzucco, E., Gennaro, M. C. & Marengo, E. (2016). Contaminants in water: non-target UHPLC/MS analysis. *Environmental Chemistry Letters,* 14, 51–65.

Gulley, A. L., Nassar, N. T. & Xun, S. (2018). China, the United States, and competition for resources that enable emerging technologies. *Proceedings of the National Academy of Sciences USA,* 115, 4111–4115.

Gunnarsson, L., Jauhiainen, A., Kristiansson, E., Nerman, O. & Larsson, D. G. J. (2008). Evolutionary conservation of human drug targets in organisms used for environmental risk assessments. *Environmental Science & Technology,* 42, 5807–5813.

Heim, B. E. (1990). Exploring the last frontiers for mineral resources: a comparison of international law regarding the deep seabed, outer space, and Antarctica. *Vanderbilt Journal of Transnational Law,* 23, 819–850.

Honkanen, J. O., Rees, C. B., Kukkonen, J. V. K. & Hodson, P. V. (2020). Temperature determines the rate at which retene affects trout embryos, not the concentration that is toxic. *Aquatic Toxicology,* 222, e105471.

Hooper, M. J., Ankley, G. T., Cristol, D. A., *et al.* (2013). Interactions between chemical and climate stressors: a role for mechanistic toxicology in assessing climate change risks. *Environmental Toxicology and Chemistry,* 32, 32–48.

Huang, J. P., Yu, H. P., Guan, X. D., Wang, G. Y. & Guo, R. X. (2016). Accelerated dryland expansion under climate change. *Nature Climate Change,* 6, 166–171.

Huggett, R. J. & Bender, M. E. (1980). Kepone in the James River. *Environmental Science & Technology,* 14, 918–923.

Hylton, A., Chiari, Y., Capellini, I., Barron, M. G. & Glaberman, S. (2018). Mixed phylogenetic signal in fish toxicity data across chemical classes. *Ecological Applications,* 28, 605–611.

IPCC (2007). *Climate Change 2007: Impacts, Adaptation, and Vulnerability. Contribution of Working Group II to the Fourth Assessment Report of the Intergovernmental Panel on Climate Change.* Cambridge, UK: Intergovernmental Panel on Climate Change.

IPCC (2021). *Climate Change 2021: The Physical Science Basis. Contribution of Working Group I to the Sixth Assessment Report of the Intergovernmental Panel on Climate Change.* Cambridge, UK and New York, NY: Intergovernmental Panel on Climate Change.

ISA (2017). *Report of ISA Workshop on the Design of "Impact Reference Zones" and "Preservation Reference Zones" in Deep-Sea Mining Contract Areas.* Kingston, Jamaica: International Seabed Authority.

Jambeck, J. R., Geyer, R., Wilcox, C., *et al.* (2015). Plastic waste inputs from land into the ocean. *Science,* 347, 768–771.

Jensen, S., Johnels, A. G., Olsson, M. & Otterlind, G. (1969). DDT and PCB in marine animals from Swedish waters. *Nature,* 224, 247–250.

Kamel, N., Attig, H., Dagnino, A., Boussetta, H. & Banni, M. (2012). Increased temperatures affect oxidative stress markers and detoxification response to benzo[a]pyrene exposure in mussel *Mytilus galloprovincialis. Archives of Environmental Contamination and Toxicology,* 63, 534–543.

Kavlock, R., Chandler, K., Houck, K., *et al.* (2012). Update on EPA's ToxCast program: providing high throughput decision support tools for chemical risk management. *Chemical Research in Toxicology,* 25, 1287–1302.

Khan, M. A. Q., Ahmed, S. A., Catalin, B., *et al.* (2006). Effect of temperature on heavy metal toxicity to juvenile crayfish, *Orconectes immunis* (Hagen). *Environmental Toxicology,* 21, 513–520.

Koelmans, A. A., Bakir, A., Burton, G. A. & Janssen, C. R. (2016). Microplastic as a vector for chemicals in the aquatic environment: critical review and model-supported reinterpretation of empirical studies. *Environmental Science & Technology,* 50, 3315–3326.

Kwong, Y. T. J., Apte, S. C., Asmund, G., Haywood, M. D. E. & Morello, E. B. (2019). Comparison of environmental impacts of deep-sea tailings placement versus on-land disposal. *Water Air and Soil Pollution,* 230, e287.

Lambert, S., Sinclair, C. & Boxall, A. (2014). Occurrence, degradation, and effect of polymer-based materials in the environment. *Reviews of Environmental Contamination and Toxicology,* 227, 1–53.

Lannig, G., Flores, J. F. & Sokolova, I. M. (2006). Temperature-dependent stress response in oysters, *Crassostrea virginica*: pollution reduces temperature tolerance in oysters. *Aquatic Toxicology, 79,* 278–287.

Larras, F., Lambert, A. S., Pesce, S., *et al.* (2013). The effect of temperature and a herbicide periphytic algae mixture on freshwater. *Ecotoxicology and Environmental Safety, 98,* 162–170.

Levin, L. A., Mengerink, K., Gjerde, K. M., *et al.* (2016). Defining "serious harm" to the marine environment in the context of deep-seabed mining. *Marine Policy, 74,* 245–259.

Lodge, M., Johnson, D., Le Gurun, G., *et al.* (2014). Seabed mining: International Seabed Authority environmental management plan for the Clarion-Clipperton Zone. A partnership approach. *Marine Policy, 49,* 66–72.

McKinney, M. A., Letcher, R. J., Aars, J., *et al.* (2011). Regional contamination versus regional dietary differences: understanding geographic variation in brominated and chlorinated contaminant levels in polar bears. *Environmental Science & Technology, 45,* 896–902.

Miljutin, D. M., Miljutina, M. A., Arbizu, P. M. & Galéron, J. (2011). Deep-sea nematode assemblage has not recovered 26 years after experimental mining of polymetallic nodules (Clarion–Clipperton Fracture Zone, Tropical Eastern Pacific). *Deep Sea Research Part I: Oceanographic Research Papers, 58,* 885–897.

Niinemets, U., Kahru, A., Noges, P., *et al.* (2017). Environmental feedbacks in temperate aquatic ecosystems under global change: why do we need to consider chemical stressors? *Regional Environmental Change, 17,* 2079–2096.

Oehlmann, J., Schulte-Oehlmann, U., Kloas, W., *et al.* (2009). A critical analysis of the biological impacts of plasticizers on wildlife. *Philosophical Transactions of the Royal Society B: Biological Sciences, 364,* 2047–2062.

Paerl, H. W., Havens, K. E., Hall, N. S., *et al.* (2020). Mitigating a global expansion of toxic cyanobacterial blooms: confounding effects and challenges posed by climate change. *Marine and Freshwater Research, 71,* 579–592.

Paul-Pont, I., Lacroix, C., Fernandez, C. G., *et al.* (2016). Exposure of marine mussels *Mytilus* spp. to polystyrene microplastics: toxicity and influence on fluoranthene bioaccumulation. *Environmental Pollution, 216,* 724–737.

Pecl, G. T., Araujo, M. B., Bell, J. D., *et al.* (2017). Biodiversity redistribution under climate change: impacts on ecosystems and human well-being. *Science, 355,* eaai9214.

Poloczanska, E. S., Brown, C. J., Sydeman, W. J., *et al.* (2013). Global imprint of climate change on marine life. *Nature Climate Change, 3,* 919–925.

Pörtner, H. O., Bock, C. & Mark, F. C. (2017). Oxygen- and capacity-limited thermal tolerance: bridging ecology and physiology. *Journal of Experimental Biology, 220,* 2685–2696.

Ratushnyak, A. A., Andreeva, M. G. & Trushin, M. V. (2005). Effects of type II pyrethroids on *Daphnia magna*: dose and temperature dependences. *Rivista Di Biologia Biology Forum, 98,* 349–357.

Reichl, C. & Schatz, M. (2021). *World Mining Data 2021*. Vienna, Austria: Federal Ministry of Agriculture, Regions and Tourism.

Roman, L., Butcher, R. G., Stewart, D., *et al.* (2020). Plastic ingestion is an underestimated cause of death for southern hemisphere albatrosses. *Conservation Letters,* e12785.

Rosenzweig, C., Karoly, D., Vicarelli, M., *et al.* (2008). Attributing physical and biological impacts to anthropogenic climate change. *Nature, 453,* 353–U20.

Seiwert, B., Klockner, P., Wagner, S. & Reemtsma, T. (2020). Source-related smart suspect screening in the aqueous environment: search for tire-derived persistent and mobile trace organic contaminants in surface waters. *Analytical and Bioanalytical Chemistry, 412,* 4909–4919.

Shah, A. A., Hasan, F., Hameed, A. & Ahmed, S. (2008). Biological degradation of plastics: a comprehensive review. *Biotechnology Advances, 26,* 246–265.

Sokolova, I. M. & Lannig, G. (2008). Interactive effects of metal pollution and temperature on metabolism in aquatic ectotherms: implications of global climate change. *Climate Research, 37,* 181–201.

Soulier, C., Coureau, C. & Togola, A. (2016). Environmental forensics in groundwater coupling passive sampling and high resolution mass spectrometry for screening. *Science of the Total Environment, 563,* 845–854.

Spurgeon, D., Lahive, E., Robinson, A., Short, S. & Kille, P. (2020). Species sensitivity to toxic substances: evolution, ecology and applications. *Frontiers in Environmental Science, 8,* e588380.

Talent, L. G. (2005). Effect of temperature on toxicity of a natural pyrethrin pesticide to green anole lizards (*Anolis carolinensis*). *Environmental Toxicology and Chemistry*, 24, 3113–3116.

Tian, Z., Peter, K. T., Gipe, A. D., *et al.* (2020). Suspect and nontarget screening for contaminants of emerging concern in an urban estuary. *Environmental Science & Technology*, 54, 889–901.

Tian, Z., Zhao, H., Peter, K. T., *et al.* (2021). A ubiquitous tire rubber-derived chemical induces acute mortality in coho salmon. *Science*, 371, 185–189.

US EPA (2020). ECOTOX User Guide: ECOTOXicology Knowledgebase System. Version 5.3 (available at https://cfpub.epa.gov/ecotox/). Washington, DC: US Environmental Protection Agency.

Van de Perre, D., Janssen, C. R. & De Schamphelaere, K. A. C. (2018). Combined effects of interspecies interaction, temperature, and zinc on *Daphnia longispina* population dynamics. *Environmental Toxicology and Chemistry*, 37, 1668–1678.

Verbruggen, B., Gunnarsson, L., Kristiansson, E., *et al.* (2018). ECOdrug: a database connecting drugs and conservation of their targets across species. *Nucleic Acids Research*, 46, D930–D936.

Verheyen, J., Delnat, V. & Stoks, R. (2019). Increased daily temperature fluctuations overrule the ability of gradual thermal evolution to offset the increased pesticide toxicity under global warming. *Environmental Science & Technology*, 53, 4600–4608.

Von Moos, N., Burkhardt-Holm, P. & Kohler, A. (2012). Uptake and effects of microplastics on cells and tissue of the blue mussel *Mytilus edulis* l. after an experimental exposure. *Environmental Science & Technology*, 46, 11327–11335.

Walker, J. D., Newman, M. C. & Enache, M. (2013). *Fundamental QSARS for Metal Ions*, Boca Raton, FL: CRC Press, Taylor & Francis Group.

Wang, Z., Kwok, K. W. H. & Leung, K. M. Y. (2019). Comparison of temperate and tropical freshwater species' acute sensitivities to chemicals: an update. *Integrated Environmental Assessment and Management*, 15, 352–363.

Wright, S. L., Thompson, R. C. & Galloway, T. S. (2013). The physical impacts of microplastics on marine organisms: a review. *Environmental Pollution*, 178, 483–492.

Epilogue: A Final Perspective

Updating Ecotoxicology

The science of ecotoxicology was defined by Truhaut about 50 years ago, so it can be considered a 'young' branch of science. However, very soon there was a perceived need for ecotoxicology to escape the confines of traditional toxicology and embrace a more holistic view of stress from environmental toxicants, including how these stresses might be measured at the ecosystem level. Quite independently, this initiative coincided with a period of rapid advances in theoretical ecology that incorporated concepts such as connectivity, species richness, stability (resistance to change) and energy flow. Estimates of chemical flow through ecosystems were also enlightened by the application of fugacity and other models, which provided a means of tracking, in a predictive capacity, chemical partitioning and fluxes through various environmental compartments under specified temporal and spatial conditions.

Many aspects summarized above were discussed by Wright and Welbourn (2002). In revisiting the question of future trends 20 years later, we consider here how issues identified then have evolved, and identify emerging issues that have assumed new prominence. Since 2002, some of the most significant advances have been made with studies at the molecular level. For example, molecular biomarkers help to identify the nature of the toxic agent and its action. They also contribute to the development of adverse outcome pathways (AOPs) that link exposure to fitness, reproduction and survival. Nevertheless, assessment of toxic chemical stress at the ecosystem level remains one of the most challenging issues for ecotoxicologists. Assessment approaches will continue to evolve through the integration of single- and multi-species laboratory toxicity tests with ecological frameworks, as defined by models of ecosystem and community structure and function. Multi-species tests are subject to a high degree of complexity. For example, altered abundance of one trophic level may result in changes at other trophic levels through top-down or bottom-up regulation. Data to differentiate between these two mechanisms can only be acquired through large-scale field experiments. At the other extreme, recent advances in analytical techniques such as environmental DNA screening offer novel ways of identifying taxonomic shifts in populations resulting from various environmental stressors, including toxicants. As pointed out many times in this 2022 textbook, a major challenge for ecotoxicology will be addressing the balance between laboratory-based repeatability and environmental realism.

Ecological Risk Assessments; Environmental Decision-making and Indigenous Rights

The transition in terminology from environmental toxicology to ecotoxicology is also reflected by the increasing adoption by regulatory agencies of the term ecological risk assessment (ERA), in place of environmental risk assessment. However, there is some variation in its application among different countries. For example, in the United States, in common with several other countries, ERA is often associated with a regional activity, where there is a clearly defined source of contamination. However, the REACH programme adopted by the European Union involves a much broader application of the term ERA that covers over 100,000 chemicals manufactured and marketed within the European Union. Hazard assessment under REACH also makes provision for predictive toxicological approaches designed to address an increasingly large and complex inventory of toxicants. As a result, there has been significant global effort towards developing computational chemistry, high-throughput screening, and omics technologies for prioritization of chemical testing.

Although ERAs and resulting regulations will require a global approach, regional and cultural considerations will also have a part to play. Many jurisdictions now require that indigenous peoples and their knowledge systems be included in environmental decision-making, including the optimization of the goals and strategies for restoration programmes. To varying degrees, this requirement is included in the UN Declaration on the Rights of Indigenous Peoples (UNDRIP), the policies of the US Environmental Protection Agency ("Environmental Justice for Working with Federally Recognized Tribes and Indigenous Peoples"), the Great Lakes Water Quality Agreement between the United States and Canada, the "Vision Mātauranga" science policy framework in New Zealand, and several pieces of federal legislation in Canada. This requirement is a necessary step to protect Indigenous rights, sovereignty and relationships with the environment, and to achieve the free, prior and informed consent that is so heavily emphasized in the UNDRIP. While there has been a shift in policy at international and national levels, there remains a need to incorporate traditional environmental knowledge in the regulatory process by engaging Indigenous communities in environmental management.

Reliance on Environmental Modelling in Evaluating New Chemicals

Regulatory agencies around the developed world rely on ERAs to evaluate whether new chemicals present a potential risk to the environment, before they are approved for entry into the marketplace. Typically, these risk assessments are conducted in a 'tiered approach' in which ever more complex laboratory studies are required to yield larger datasets that can more accurately predict the impacts of exposure to chemicals. These ERAs rely heavily on environmental modelling to predict how such exposures are influenced by the altered fate of chemicals, including capacity for long-range transport, movement

from the terrestrial to aquatic environments and potential for bioaccumulation. Various modelling approaches can also be used to predict toxicity, including quantitative structure–activity relationships (QSARs) and dynamic energy budget models. Such models are particularly useful in assessing the combined toxicity of chemical mixtures. For endocrine disrupting chemicals (EDCs), it is anticipated that we will increase the assessment and regulation of EDC mixtures, instead of studying individual responses.

Interactions Between Ecotoxicology and Human-induced Environmental Changes

External influences such as a changing physical environment and altered exposure patterns resulting from climate change will play an increasingly important part in ecotoxicology over the next several decades. In reflecting on the complexity and global nature of many of the issues confronting us, it is useful to consider some of the interconnections among seemingly unrelated subject areas. For example, the extraction of several of the metals described in this book as 'emerging inorganics' is specifically related to their potential application in the energy sector, with implications for the reduction of fossil fuel combustion and greenhouse gas emissions. Greenhouse gases are also by-products of plastic manufacture involving petroleum hydrocarbons.

While 'linear' concepts such as AOPs remain vital to the establishment of cause-and-effect links related to chemical toxicity in the biosphere, a major challenge for the science of ecotoxicology will involve addressing the influence of indirect effects. Such effects take various forms, including changes in ecosystems resulting from shifts in regional patterns of chemical contamination and environmental conditions. Changes in population demographics due to differential tolerance of chemical and physical circumstances will inevitably result in territorial shifts of animal and plant populations and communities. These in turn will generate new species–species interactions with resultant cascade effects on food-web demographics. At the organism level, a useful perspective in improving our understanding of the joint effects of climate and toxic chemicals has been the distinction between the terms toxicant-induced climate susceptibility (TICS) and climate-induced toxicant susceptibility (CITS). More complex, multifactorial models may consider such disparate factors as the mode of action (MoA) of a particular group of chemicals, or the 'ecological function' of various species to explain complex ecotoxicological effects.

Other complexities could include, for example, pesticide impacts in systems already affected by other industrial emissions or changes associated with agriculture or urban development. Despite the continuing improvement of multivariate models, it is important to bear in mind the interconnected nature of contributing factors. As an example, it would not be unreasonable to define a major portion of climate change as a toxicity-enhancing event brought about by the release of certain atmospheric contaminants.

Looking to the Future

Since the 1800s, we have entered an unprecedented period of human influence on the global environment. In what some have called the Anthropocene epoch, we are now faced with several global environmental issues requiring urgent global solutions. These issues will involve universal behavioural changes in our relationship with natural resources, including not only commodities but the wildlife that are an integral part of our ecosystem. Laboratory and field research conducted over several years or even decades have often been required to generate sufficient data to justify regulation or a complete ban on problematic chemicals. Past mistakes have given us the opportunity to learn from retrospective studies, and history has shown that we are making improvements in responding to the risks associated with exposure to xenobiotics – we can learn from our past mistakes! It took several decades of data gathering before PCBs were banned from commerce, and by that time, the concentrations of these contaminants had reached mg/kg concentrations in biota from all over the world, including in humans. In contrast, the banning of BDEs was relatively rapid, once research indicated that the levels of these compounds were doubling in biota every 5 to 8 years. As a result, the maximum concentrations of BDEs observed now in biota typically do not exceed μg/kg levels.

Many subject areas discussed in this book illustrate major advances that have already been made in recent decades in assessing risk from chemical contaminants at different levels of biological organization. Improved effectiveness of risk assessment has many aspects. In the laboratory the need to curtail the use of live animals, particularly vertebrates, for toxicity testing is driven by both ethical concerns and the need to reduce the time and high costs of *in vivo* toxicity testing. These constraints will inevitably lead to the development of specific cell lines and *in vitro* culture techniques that facilitate the study of chemical pathology at the tissue level. However, this will not eliminate the need to account for the differences between experiments *in vivo* with whole organisms and tests *in vitro* with cells in culture. Improved molecular techniques will inform the establishment of chemical-specific MoAs. While experiments to understand MoAs are not always designed to predict thresholds of toxicity, they are important in providing the basis for regulatory models such as quantitative structure–activity relationships (QSARs). In future we will also see the increased phasing out of universal toxicity endpoints such as LC_{50}s and their replacement by the molecular initiating event (MIE), which is the initial key event defining the course of an AOP.

We anticipate the increasing importance of molecular endpoints in the establishment of models used for toxicity assessment and associated chemical regulation. Molecular techniques such as high-throughput screening will also play an increasingly important role in tracking relationships among microbiomes and their chemical environment. It is to be expected that future toxicity assessment will adopt a more integrated approach using data from a variety of different sources including many established models and endpoints. This strategy, known as the Integrated Approach to Testing and Assessment (IATA), is already being adopted in several countries and will diminish the reliance on standard whole-organism testing. It is anticipated that this will result in a more flexible approach, allowing toxicity estimates to be reassessed as new data are acquired.

Constraints on data collection, plus the large numbers of new chemicals that are entering the marketplace, are significant challenges for regulatory agencies. Of paramount importance will be the need to develop guidelines and standards that are consistent among nations. Global mechanisms will be required to oversee and enforce these regulations and to limit the international trade in wastes such as plastics. A corollary of this will be the need for more environmental monitoring to assess the success of chemical regulations in limiting chemical emissions and ecological effects. Such assessments will need to consider the benefits of ecosystem services as well as the costs associated with their loss and the effort needed to restore them. In the future, it is anticipated that part of the regulatory burden will be borne by chemical developers themselves. The obligatory involvement of manufacturers in potential clean-up or mitigation procedures will hopefully foster the careful design of new chemicals and products. This approach is embodied in the 'circular economy' concept, which is gaining traction in the European Union and several other countries. The circular economy approach seeks to integrate all aspects of manufacture, marketing and product maintenance as well as disposal in ways that minimize the ecological impacts of product development and use. In common with the notion of 'sustainability', this strategy has the potential to play a major role in reshaping chemical risk assessment and management.

However, progress does not always follow a smooth 'linear' pathway. We note that, with respect to defining the toxicological effects of plastics, we are still in the information gathering phase. In some cases, this situation leads to intense but unfocused research activity prompted by the appearance of new but incompletely understood issues. Earlier in this book, we refer to the initial 'bandwagon' effect of research on the fate and toxicity of engineered nanomaterials (ENMs). This phenomenon of false starts is, of course, not unique to the science of ENM ecotoxicology. It often reflects the evolution of the broader science, as exemplified by the earliest experiments with pesticides, PCBs, EDCs and many other chemicals. It

seems that we have to crawl, stumble and walk before we can run.

Overall, vigilance is required to ensure that we do not repeat the mistakes of the past and that we continue to protect our ecosystems and the organisms that live in them. The process will involve a greater role for environmental scientists in public debates and in educating the public on environmental management and the consequences of personal choices in this process. This will be critically important for the development of the informed public opinion needed to drive effective ecosystem management. As stewards of the environment, you, the reader, can play an important role in protecting our environment by recognizing the implications of your purchases (e.g., country of origin, material and fate) and your activities, and by supporting ecologically sound decisions. Educate your peers and your family. You can create the world in which you would like to live – you just need to be actively involved in that dream of yours!

Reference

Wright, D. A. & Welbourn, P. M. (2002). *Environmental Toxicology*, Cambridge, UK: Cambridge University Press.

INDEX

Page numbers in italics refer to figures, boxes and tables.

17α-ethinyloestradiol (EE2), 142, *155*, 327, 341,
    See pharmaceuticals
  oestrogen receptor, activation, 153
  transcription, alteration, 142, 158
2,3,7,8-tetrachlorodibenzo-p-dioxin (TCDD), 281, *283*, *295*,
    296–297. *See also* halogenated aromatic hydrocarbons
3,3′,4,4′,5,5′-hexachlorobiphenyl. *See* PCBs
3,3′,4,4′,5,5′-pentachlorobiphenyl, 297
4-nonylphenol, 542
5α-dihydrotestosterone (5α-DHT), 333. *See also* steroidogenesis

acetylcholinesterase
  organophosphate pesticides, 106, 109
acid mine drainage, 426–427
  causes, 426
acid rock drainage. *See* acid mine drainage
acidic deposition, 10–11. *See also* acidification
  acid precipitation, 10, 411, 419, 427
  airborne contaminants, 10
  forest dieback, 10
  global contamination, 10
  precursors, 10
  recovery from acidification, 10
acidification, 411–450
  freshwater
    biological effects, 413–415
    causes, 412
    episodes, 412
    physicochemical effects, 412–413
  mobilization, metals, 178, 186, 235, 412
  ocean, 11
    causes, 417
    effects, *417*
  regulation, abatement, 415–417
  speciation, metals, 207, 413
  terrestrial, chemical biological effects, 412–414
  transparency, water, 413
  treatment, freshwater and terrestrial, 417
adrenocorticotropic hormone (ACTH), 329. *See also* hypothalamic–
    pituitary–adrenal axis (HPA)
adverse outcome pathway (AOP), 24, 54, 100, *101*, *130*, 140,
    152–153, *523*
  key event (KE), 152, *153*

linkages, molecular interactions, ecosystem change, 100
modes of action, 100, 106, 129
molecular initiating event (MIE), 152, *153*, 154
pharmacodynamics, pharmacokinetics, 100
reductionist versus holistic studies, *101*
sequence, exposure to effects, 100
aflatoxins, 286, 364–365, 377. *See also* mycotoxins
  Aspergillus flavus, 364
  Aspergillus parasiticus, 364
  Aspergillus sp., 377
African clawed frog (*Xenopus laevis*), 333, 338, *338*, 344, 347
aliphatic compounds, 276. *See also* alkanes, alkenes, alkynes
alkanes, 276. *See also* pentane
  cycloalkanes, 276
alkenes
  ethylene ($H_2C=CH_2$), 276
alkynes
  acetylene ($HC\equiv CH$), 276
allelopathy, 356, 366, 371, 376
amatoxins, 364–365
  Amanita ocreata, 355
  Amanita phalloides, 364, 377
amino acid conjugation, 288. *See also* Phase II reactions
anatoxins, 361
  Anabaena lemmermannii, 361
  Anabaena sp., 361
androgens, 281. *See also* cellular receptors
anthrax, 363–364, 370
  Bacillus anthracis, 363, 377
  Bwabwata National Park, Namibia, 364
  toxins, 377
antibiotics, 364
  cephalosporin, 364
  penicillin, 364
aquaporins, 65–68, 252
aromatase, 333, *See* steroidogenesis
aromatic compounds, 276–277, *277*. *See also* benzene,
    benzo(a)pyrene, BTEX compounds, naphthalene,
    phenanthrene
  anilines (phenylamines), 277
  aromatic alcohol (phenol), 276
  benzene, *277*
  benzoic acid, 276

aromatic compounds (cont.)
  phenanthrene, 277, 286, 291
  phenol, benzoic acid, nitrobenzene, toluene,
      methylphenanthrene, *277*
  phthalic acid, 314
  toluene, 276. *See also* BTEX compounds
arsenic, 248–255
  bioaccumulation, uptake, 251–252
    anion transport, 252
    aquaporins, 251
  biochemistry, 251–252
    biomethylation, 252
  biogeochemistry, 249–251
    arsenate, 249–250
    arsenite, 249–250
    arsenosugars, 249
    forms-speciation, 249–250
    groundwater, 248
    methylated species, 249
    mobility, pH effect, *251*
    mobility, role Fe(III), *250*, 251
    oxyanions, 249, 251
    redox reactions, kinetics, 249–250
    sediments, diagenesis, 250
    thioarsenates, 249
  detoxification
    depuration (aquaporins), 252
  ecotoxicity, 252–254
    diet-borne, 253–254
    signs of As toxicity, *253*
  occurrence, sources, uses, 248–249, 426
  organoarsenicals
    arsenobetaine, 252
    dimethylarsinic acid, 249
    monomethylarsonic acid, 249
  trophic transfer, 252–253
aryl hydrocarbon receptor (AhR), 281, *282*, 283, 297
aryl hydrocarbon receptor nuclear translocating (ARNT) protein, 281
atrazine, 343, 542
autocrine hormones, 329

Basidiomycetes, 364–365, 377
behaviour, avoidance, 109
benzene
  hydroxylation, *285*
benzo(a)pyrene, 277, *286*, 542
benzo(a)pyrene-7,8-dihydrodiol-9,10-epoxide, *286*
benzophenone, 542
  sunscreen agents, *317*
bioaccumulation
  growth dilution, 87
  metals, 87–88
  organic molecules, 88–91
  perfluoroalkyl contaminants, 91
  polychlorinated biphenyls, 88
bioaccumulation models. *See* modelling, bioaccumulation
bioassays versus toxicity tests, *51*

bio-attenuation. *See* recovery:natural:attenuation
bioindicators, ecosystem effects, 107–111
  chemical contamination, 101
  definition, 101
  ecological change, 101
  ecosystem services, 101
  ecosystem structure, 101, *104*
  energy flow, 101
  indicator species, 110, *110*, 114
  keystone species, 100–101
  natural resources, 101
  population demographics, *101*, 101, 109–110
  predator–prey relationships, 101
  primary, secondary production, 101
  species abundance, distribution, diversity, 101–104, *104*,
      110–111, 114, 121
  trends, spatial, temporal, 101–102
bioinformatics, 150–152
Biological and Toxin Weapons Convention, 370
biomagnification, 81, 87
  lipophilic molecules, 88
  proteinophilic molecules, 91
  toxins, 366, *371*
biomarkers, ecotoxicogenomics, 141, 154. *See also* omics, adverse
      outcome pathways (AOPs)
  enzyme-linked immunosorbent assays (ELISA), 148
  lipid, 147, *159*
  metabolite, 140–141, 146–147, 152, 157–159, *159*, 161
  proteins, 144, *146*, *159*
  receptor, 141–142, 147, 150, 152–153, 155
  vitellogenin, 142, 147, 153, 156, 158, 160
biomarkers, exposure, effects, 108–109
  definition, 107
  examples, 108–109
  induction, gene expression, enzyme activity, 108
  metabolite concentrations, 108
  mode of action (MoA), 109, *130*
bioremediation, 492–493, 503–507, 510
  aerobic, anaerobic microorganisms, 505
  chemical dispersants, 505
  controlled field studies, 506
  crude oil
    saturates, aromatics, resins, asphaltenes (SARA), 504
  definition, 503
  hydrocarbonoclastic (obligate hydrocarbon-degrading)
      bacteria, 505
  natural oil seeps, 495
  oil spills, 504–507
    Deepwater Horizon, 504–505
    Exxon Valdez, 12, 504
    spills of opportunity, 506
biotic ligand model, 50–51
  metal mixtures, 51
  potency factor, 51
biotransformation, 85–87
  metals, 85
  organic molecules, 86

bisphenol A (BPA), 332, *339–340*, 525, 542
botulinum, 377
   Clostridium botulinum, 377
brevetoxins, 357, 359, 375
   Gymnodinium sp., 359, 375
   Karenia brevis, 375
brominated flame retardants, 294, 305–307, *305–307*, 316, 343
   brominated diphenyl ethers (BDEs), 305
   decabromodiphenylethane (DBDPE), 306
   hexabromocyclododecane (HBCD), 306
   tetrabromobisphenol A (TBBPA), 306
Brookhaven National Laboratory, 399
brownfields, 11–12
   definition, 11
   remediation, 11, 13
Bt insecticide, 369–370, 377
   Bacillus thuringiensis, *369*, 369
   Bt gene, 369
   Bt toxin, 369, *369*, 377
   Cry toxin, *369*, 369, 377
BTEX compounds
   benzene, ethylbenzene, toluene, xylene, 276
bystander effect, 395

cadmium, 226–228
   bioaccumulation, uptake, 228
      cation mimicry, 228
      role hardness, 228
      role pH, 228
   biochemistry, 226–227
   biogeochemistry, 226
      forms-speciation, 226
   detoxification
      granules, 227
      metallothionein, 227
   ecotoxicity, *143*, 199–202, 227–228
      diet-borne, 228
      DNA methylation, 148
      protein alteration, 146
      signs of Cd toxicity, *227*
   hyperaccumulation
      ecological advantage, 205
   occurrence, sources, uses, 226, 426
carbamate insecticides
   carbaryl, *299*
   carbofuran, *299*
carbamazepine, 315
carbon tetrachloride, 286
Carson, Rachel
   organochlorines, 7
   pesticides, 7, 16
   Silent Spring, 5, 7, 11, 16, 355
CAS Registry, American Chemical Society, 289
Casparian strip, *77*, 77
cause–effect relationships, ecosystems
   challenges, 102
   chemicals in commerce, diversity, 100

correlation, causation, 101–102, 106
   descriptive surveys versus experiments, 102
   multiple stressors, 102, 105
   reference systems, controls, 102–103, 107, *111–112*, *117*
   spatial, temporal patterns, contamination, ecological change, 101, 103, 112, 114, *116*
cell death
   apoptosis, 393
   necrosis, 393
cell membrane, 61–64
   apical surface, 63, 71, 75
   basolateral surface, 63, 71, 79
   composition, 62
   endocytosis, 71
   exocytosis, 71, 79, 89
   extrinsic protein, 63
   integral proteins, 63
   phagocytosis, 71
   pinocytosis, 71
   role, 72–73
cellular receptors, 281–282, *282*, 283
chemical half-lives ($t_{1/2}$), 280
chemical markers, exposure, effect, 108
   age structures, plants, animals, *108*, 108
   definition, 107
   exposure history, *108*
   tissue concentrations, 107
chemical monitoring, human food supply, 115–117, 127
   bioaccumulation, biomagnification, 116
   chemicals of concern, 116
   ecological interactions, food web contamination, 116
   edible portion, whole fish, 116
   methylmercury, 116
   persistent organic pollutants (POPs), 116
   public awareness, 116
   risk management, 116
   temporal trends, 116
chemical warfare, 356, 365
chlorofluorocarbons (CFCs), *516*
chlorophenoxy herbicides, *303*
   2,4,5-trichlorophenoxy acetic acid (2,4,5-T), 302
   2,4-dichlorophenoxy acetic acid (2,4-D), 302
chorionic gonadotropin (CG), 332
climate change
   aridity, 520
   contaminated sites, 488, 495
   effects, interspecific, 520–521
   Global Warming Potential, 517
   greenhouse gases, *516–517*, 516–517, 521, 527, 552
   interactions, ecotoxicology, 515–521
   Intergovernmental Panel on Climate Change (IPCC), 417, 517
   natural toxins, 356, *357*
   polar bears (*Ursus maritimus*), 521–522
   precipitation, 517, 519, 521
   regional considerations, 521–522
   salinity, 519–520

climate change (cont.)
   temperature, 518–519
     acclimation, 518, 523
     bioaccumulation, 519
     toxicity, *519–520*
climate-induced toxicant susceptibility, 514, 519, 553
coke, 291, 319. *See also* polycyclic aromatic compounds (PACs),
     pyrogenic
Colborn, Theo, 12–13
   diethylstilboestrol (DES), 13
   endocrine disruption, 12–13
   Our Stolen Future, 327
community indices, ecological effects, *110*, 110–111
   benthic assessment of sediment (BEAST) model, 110
   biomass size spectra, 110
   diversity indices, *110*
   ecological function, 111
   European Water Framework Directive, 111
   growth, reproduction, mortality, 110
   index of biotic integrity (IBI), 111
   indicator species, 110–111
   invasive species, 110
   role of species, 111
   species richness, evenness, 110
comparative endocrinology, 330–331
competition, interspecific, 366, 370, *See* venoms
   role of toxins, *371, 375. See also* allelopathy
complex issue
   cradle-to-grave, 386, 411, 493, 526, 544
   definition, 411, 446
Comprehensive Environmental Response, Compensation and
     Liability Act (CERCLA), 11
computational toxicology, 154–157
   case study, glyphosate, 155–157
   ToxCast, 154–155
constitutive androstane receptor (CAR), *282*, 338
contaminant
   inorganic, emerging, 528
   organic, emerging, 539, *539*, 542
   organic, monitoring, 539
   organic, non-targeted screening, 540, *540–541*, 544
   plastics, adsorbed chemicals, 525
contaminated site, 487–488, 490–492, 502–503, 508, 510
   abandoned petrol stations, 488
   contaminants of concern, 488, 492, 500
   damaged sites, 488, 509
   degradation organic contaminants, 505–506
   degraded sites, 489
   groundwater, 492, 496, 510
   liability, 490
   mobilization of contaminants, 492
   sediments, 492–495, 502–503
copper, 234–239
   bioaccumulation, uptake
     diet-borne, 238
     role Cu(I), 236

     role DOM, 237
     role hardness, 236–237
     role intracellular chaperones, 236
     role pH, 237
   biochemistry, 236–237
   biogeochemistry, 235–236
     complexation (DOM), 237
     forms-speciation, 235
     redox reactions, 235
   detoxification, 236–237
     granules, 236
   ecotoxicity, 237–239
     behavioural effects, 194, 238
     diet-borne, 237
     fungicide, 234–235
     signs of Cu toxicity, *237*
   occurrence, sources, uses, 234–235, 426
corticosteroids, 281, 329. *See also* hypothalamic–pituitary–adrenal
     axis (HPA), cellular receptors
corticotropin-releasing hormone (CRH), 329. *See also*
     hypothalamic–pituitary–adrenal axis
criteria, objectives, guidelines, standards, 463–465
   contaminated site remediation, 463
   definitions, 463
   ecological impact assessment, 463
   ecological risk assessment, 463
   emission limits, 463
   in vitro tests, 465
   oestrogenicity, 465
   omics, 465
   success, chemical management, 463
   targets, sustainable development, 463
current use pesticides (CUPs)
   bipyridilium herbicides, 303
     diquat, 303
     paraquat, 303
   carbamate insecticides, 299–301
   chlorophenoxy herbicides, 302–303
   fungicides, 304–305, *305*
     carbendazim, embryotoxicity, teratogenicity, endocrine
       disruption, 304
     quintozene, teratogenicity, 304
     vinclozolin, anti-androgenic activity, 304
   glyphosate, 147, 155–157, 303–304
     transcriptomics, 156
   introduction, 297–298
   neonicotenoid insecticides, 302
   organophosphate insecticides, 298–299
     activation chlorpyrifos, 299
     activation parathion, 298
     dichlorvos, *298*, 299
     malathion, 299
     monocrotophos, *298*
     parathion, 285, *298*
   phenylpyrazole insecticides, 301
   pyrethroid insecticides, 301–302

triazine herbicides, 304
usage, 297
cyanobacteria, 358, 361, 368, 370–371, 375
cypriniform fish *Catla catla*, 332
cytochrome P450 enzyme system, 104, *105*, 281–282, 284, *285*, *289*.
      See also Phase I reactions, Phase I enzymes
  CYP4501A (CYP1A), 281, 286
  cytochrome b5, 284
  hydrocarbon excretion, 108
  hydrocarbon oxygenation, 108
  induction, gene expression, enzyme activity, 108
  inhibition
    carbon monoxide, 286
    cyanide, 286
    naphthoflavone, 286
    piperonyl butoxide, 286, 301

deleterious substance, definition, 458
dichlorodiphenyltrichloroethane (DDT), 279, 294
  aquatic, terrestrial food webs, 7
  biomagnification, 7
  decline in usage, 294
  regulation, 294, *295*
  toxicity, 294
diclofenac, 317. *See also* pharmaceutically active compounds
di(2-ethylhexyl)phthalate (DEHP), 314
digestion
  contaminants, fate, 88–90
  intracellular, 71
dinitrophenol (2,4-DNP), 277
dinoflagellates, 358–361, 370–371, 375
dinoseb (2,4-dinitro-2-sec-butylphenol), 277
dioxin-like compounds (DLCs), 104, *105*, 108, 114–115, *115*
dipole moment, 279
  dipole–dipole interactions, *279*
DNA adducts, 109
domoic acid, 357–358, *358*, 375
  Pseudo-nitzschia australis, 357, 375
dopamine, 330, 332, 362

ecdysteroid receptor (invertebrates), *282*
ecoepidemiology, 102, 105–107, *106*
  cause–effect relationships, 106
  consistency of association, 106
  definition, *106*
  fish cancers, 107
    aetiology, 107
    benthic fish species, 107
    benzo[a]pyrene, mode of action (MoA), 107
    polycyclic aromatic compounds, carcinogenic, 107
    sediment contamination, steel mills, 107
    skin, liver, 107
    weight-of-evidence (WoE), 107
  relative risk, 106
  specificity, 106
  strength of association, 106

time order (before, after), 106
  weight-of-evidence (WoE), *107*
ecological context, 104, 106, 111
ecological modelling, integrating lab and field studies, 102
ecological risk assessment (ERA), 5, *14*, 103, 465–469, 490–492, 506
  before, after study, 103
  chemical emissions, sources, timing, fate, distribution, 102
  comparison, human health risk assessment (HHRA), 465
  developments, new, existing, 103
  ecological context, 104
  framework, *466*
  geographical scope, 465
  risk management, 488, 492–493, 510–511
  stressors – biological, physical, 465
ecological risk assessment (ERA), area wide. *See* Sudbury Soils Study
ecological risk assessment (ERA), methodology, 465–467, *466*
  analysis, exposure, effects, 467
  effects characterization, 467
  exposure characterization, 467
  exposure modelling, 467
  focus, 465
  framework, metals, 468
  framework, organics, 467
  hazard quotient (HQ), 465, 467
  pathway analysis, 467
  predicted environmental concentration (PEC), 465
  predicted no-effect concentration (PNEC), 465, 467
  problem formulation, conceptual model, 466
  prospective, retrospective, 477
  receptors, 467
  risk characterization, 467
  risk, exposure, hazard, 465
  toxicity, critical value, 467
  variations among countries, 466
ecological risk assessment (ERA), specific chemicals, 467–468
  bioavailability, environmental conditions, 468
  commercial, industrial, 467
  metals, metalloids, 467–468
  naturally occurring chemicals
    background concentrations, 468
    tolerance, 468
  new chemicals, sources, 467
  PBT approach, shortcomings, 468
  screening value
    bioconcentration, 467
    half-life, 467
ecological risk assessment (ERA), uncertainty, 468–469
  evidence-based ecotoxicology, *478*
  exposure
    data limitations, 468
    site-specific data, 469
  extrapolation, lab to field, 468
  monitoring, validation, 468
  probabilistic risk assessment, 469
  receptors, spatial, seasonal variation, 469
  risk, relative, absolute, 469

ecological risk assessment (ERA), uncertainty (cont.)
  safety factors. *See* precautionary approach
  test species, site-specific relevance, 468
economic, social concerns, 491
  ecosystem services, *489*, 510
  perceived ecological value, 488, 510
  social obstacles, 488
ecosystem change, toxicity, environmental conditions, 100
ecosystem diversity, complexity, 100
  ecological niches, 100
  environmental conditions, 100
  habitat, 100
  response to chemicals, 100
  species, traits, 100, 102
ecotoxicology
  definition, comparison with environmental toxicology, 5–7, 99
  historical landmarks, *9*
  major features, *6*
  recent methodological advances, *15*
  unanticipated benefits for ecological research, 4
ecotoxicology, role, chemical management, 462–463
  criteria, guidelines, 462
  ecological impact assessment (EIA), 462
  ecological risk assessment (ERA), 462
  ecosystem services, 462
  emission limits, 462
  monitoring, 462
  remediation, contaminated sites, 462
  research, testing, 462
EE2. *See* 17α-ethinyloestradiol
electrochemical gradient, 64
elimination, contaminants, 79
  kinetics, 81–85
  routes, 80
endocrine disrupting chemicals (EDCs), 327
  biomarkers, 337, 342. *See also* vitellogenin
  examples, *340*
  field evidence
    decline in fathead minnow populations, 341
    effects on steroid homeostasis in polar bears, 345
    feminization of fish in UK rivers, 341
    imposex in gastropod molluscs, 345
    intersex in rainbow darters, *341*, 341–342
  legacy contaminants, 344–345. *See also* organochlorine pesticides
    dichlorodiphenyldichloroethylene (DDE), *340*
    dichlorodiphenyltrichloroethane (DDT), *340*
    organotins, 345–346
    polychlorinated biphenyls (PCBs), *340*, 344–345
    polychlorinated dibenzodioxins (PCDDs), *340*
  mode of action (MoA), 327, 337–338
  organochlorine pesticides, 332
  organophosphate pesticides, 332
endogenous compounds, 276, 281–282, 284, 287–288
endoplasmic reticulum. *See* cytochrome P450 enzyme system

energy dispersive X-ray analysis (EDXA), 108
engineering, remedial, 488
  capping contaminated sediments, 493–494
  containment, 491, 511
    entombment, 493–494
    vitrification, 493
  dig and dump, 491
  disposal options, site, 493
  dredging, 492–495
  genetic engineering, 507
  immobilization contaminants *in situ*, 491
  intervention, *489*, 491–492, 500, 510
  pump and treat, 491–492, 498
  removal of contaminated material, 491, 493
  Saglek Bay, 500
enterocyte, 72, 79, 89–90
environmental contamination
  contaminant versus pollutant, definition, 6
  early history, 5, 7, 12–13
    endocrine disrupting chemicals, 12
    mercury, 10
    oil spills, 12
    pesticides, 7
environmental law, 458–462
  common versus statutory, 458
  development, process, 459
environmental law, enforcement, 470–476
  coercive, 470
  departments, ministries, 470
  environmental non-government organizations (ENGOs), 470
  natural resources damage assessment (NRDA), 475
  persuasive, 470
  protocols, sample collection, analysis, 470
  role of ecotoxicology, 470–475
  spills of opportunity, 475–476
  toxicity test methods, 470
  triggers, spills, toxicity tests, monitoring, 470
  voluntary self-regulation, 470, 478
environmental movement, 16
  Carson, Rachel, 7, 11
  Colborn, Theo, 12
environmental non-government organizations (ENGOs), 7, 14, 18
environmental surveillance, monitoring, *476*, 477
  deleterious effects, 476, *476*
  discharge permits, 476
  early detection, 476
  Fisheries Act, Canada, *476*
  food contamination, spatial, temporal trends, *476*
  habitat quality, *476*
  innovations, 476
  long-term commitment, cost, 476
  organoleptic tests, *476*
  pesticides, 477
  quality control, regulations, 476

sampling frequency, 476
scale, regional, national, 476
environmental toxicology, *15*
comparison with ecotoxicology, 5–7
definition, principles, 5–7, *24*
enzyme-linked immunosorbent assays (ELISA), 337
epidemiology
definition, *106*
epigenetics, 147–148
epithelium, 71
boundary layer, 78
gill, 74
intestine, 75
mucus, 73–74
epizootiology
definition, *106*
ERICA tool, 400–402
essential element, 172
arsenic (?), 253
copper, 236
dose–response relationship, *174*
nickel, 240
selenium, 256
zinc, 245
exogenous compounds, 86, 281–282, 284
experimental ecosystems, 119–122, *120*, 122
acidification, whole-lake experiment, 121
endocrine disruptors, whole-lake experiment, 122
legal, ethical issues, 121
limnocorrals, 121
microcosms, mesocosms, macrocosms, characteristics, 119
model ecosystems, types, *119*
obstacles, 121
oversampling, 121
site availability, 121
size, complexity, realism, 121
biogeochemical processes, 121
long-lived species, 121
natural communities, 121
nutrient recycling, 121
replication, statistical power, 121
stability, 121
exposure–response, 27–28, 33
exposure gradients, 27, 33, *34*
monotonic relationship, 28, 33
response gradient, 33
threshold, 28

farnesoid receptor (invertebrates), *282*
fathead minnow (*Pimephales promelas*), 341
field experiments, 100–101, 117–122
biodegradation, spilled oil
bioavailability, hydrocarbons, 118
confinement stress, snails, 118
experimental treatments, 118
freshwater wetland, 118
plant growth, sediment oxygen, 118
sediment microbes, oil degradation, 118
sediment toxicity, snails, 118
snail sensitivity, herbivore, detritivore, 118
temporal trends, sediment oil concentration, 118
definition, 117
experimental ecosystems, *119*, 119–122, *122*
experimental plots, 118–119
agricultural research, 118
gradients, ecosystem characteristics, 118
randomization, treatment within blocks, 118
replication, statistical power, 118
in situ toxicity tests, 117–118
caging stress, 117–118
effects measured, 117–118
exposure gradients, 117–118
exposure time, 117–118
integrated chemical samplers, 117–118
monitoring chemical exposure, 117–118
test communities, algae, zooplankton, 117–118
test communities, colonized substrates, 117–118
test species, 117–118
upstream–downstream experiments, 117
lab versus field, *101*
flavoprotein monooxygenases (FMOs), amine oxygenases, 284. *See also* monooxygenases
fluoranthene, 542
fluoxetine (Prozac®), 147, 332
follicle-stimulating hormone (FSH), 330, 332
free radicals, 391–394
fugacity, 123–126
capacity, definition, *123*
definition, *123*
equilibrium, definition, *123*
example, dissolved oxygen, 124
Quantitative Water, Air, Sediment Interaction (QWASI) model, 125
scalability, 124
steady state, definition, *123*

gamma-amino butyric acid (GABA), 332
gene expression, 144
CRISPR-Cas9, 160
qPCR, 141
RNA-seq, 142
global distillation hypothesis, grasshopper effect, 124, *278*, 278
glucuronic acid, 287, *287*
glutamate, 332
glutathione (GSH) conjugation, 287–288, *288*. *See also* Phase II reactions
glutathione transferase, 287
gonadotropin-releasing hormone (GnRH), 329. *See also* hypothalamic–pituitary–gonadal axis (HPG)

gonadotropins, 332–333. *See also* hypothalamic–pituitary–gonadal axis (HPG)

government agencies, chemical management, 460, 470
   Canada, 458
   local, national, international, 460
   management options, 460
   monitoring, testing, enforcement, 460
   responsibilities, 460

groundwater ubiquity score (GUS), 280

growth hormone (GH), 330

guidelines, soil, sediment, tissue, 464
   adverse outcome pathway (AOP), *465*
   bioavailability, 464
   contaminated site remediation, 464
   correlations, cause, effect, 464
   critical body burden (CBB), 464
   partitioning, sediment, water, tissue, 464
   persistent bioaccumulative substances, 464
   sediment organic carbon, 464

guidelines, water quality, 463–464
   application factor, 463
   biotic ligand model (BLM), 464
   development, 463–464
   hazard concentration (HC), 463
   metals, 464
   no-effect concentration (NEC), 463
   no-observable-effect concentration (NOEC), 463
   risk-based, 463
   scientific judgement, 464
   site-specific, 464
   species sensitivity distribution (SSD), 463
   test species, endemic, standard, 464
   toxic unit (TU), 463
   variations among countries, 464
   water-effect ratio (WER), 464

halogenated aromatic hydrocarbons (HAHs), 277, 285. *See also* PCBs

Hanford Site, 396, 398

harmful algal blooms (HABs), 357
   climate change, *357*

hazard, definition, 24
   hazardous concentration, 47

heat-shock protein, *283*

Henry's Law constant, 123–124, 279

homeostasis
   copper, 236
   zinc, 245

hormesis, *26*
   adaptive cellular stress response pathways, *27*
   biphasic exposure-response, *27*
   cancer risk, ionizing radiation, arsenic, *27*
   conditioning reaction, *27*
   cytoprotective proteins, *27*

human health risk assessment (HHRA), 465

hypothalamic–pituitary–adrenal axis (HPA), 329–330. *See also* corticosteroids; adrenocorticotropic hormone (ACTH)

hypothalamic–pituitary–gonadal axis (HPG), 329. *See also* gonadotropins; gonadotropin-releasing hormone

hypothalamic–pituitary–thyroid axis (HPT), 329. *See also* thyroid hormones; thyroid-stimulating hormone (TSH); thyroid-releasing hormone (TRH)

IISD Experimental Lakes Area (ELA), 41, 220
   17α-ethinyloestradiol (EE2), 122, 341
   acidification, 121, 413
   cadmium, 226
   nanosilver, 438

indicator species, 110
   invasive, 110
   K (carrying capacity)-selected species, *110*
   r (reproduction)-selected species, *110*

Indonesia, 427, 528

International Atomic Energy Agency (IAEA), 400

International Commission on Radiological Protection (ICRP), 395–402

International Environmental Law (IEL), 460
   enforcement, 460

international treaties, conventions, agreements, 460–461
   Basel Convention, 462
   Great Lakes Water Quality Agreement, 461, 489
   International Joint Commission (Canada–USA), 461, 489
   International Seabed Authority, 460
   Joint FAO/WHO Expert Committee on Food Additives (JECFA), 470
   Minamata Convention, 460–461
   Montreal Protocol, 460
   Paris Agreement, 460
   Registration, Evaluation, Authorisation and Restriction of Chemicals (REACH), 462, 477
   Stockholm Convention, 460
   trade agreements, 458
   United Nations Declaration on the Rights of Indigenous Peoples (UNDRIP), 478, 552

intrinsic bioremediation. *See* recovery:natural:attenuation

ionizing radiation
   epigenetic effects, 394–395
   hormesis, 395

ionizing radiation, carcinogen, 385–386, 394–396
   one-hit model, 396

ionizing radiation, dose–response, 396
   committed equivalent dose, 397
   effective dose, 382, 395–396
   linear non-threshold dose–response, 396, *405*

ionizing radiation, isotopes, 380–381, 385–388, 390–391

ionizing radiation, risk assessment, 395, 398, 402

ionizing radiation, sources of radiation, 383
   alpha particle, *381–382*, 381, 383, 385–386
   beta particle, 381, *382*, 386
   electromagnetic photons, 381
   neutrons, 380–381, *382*, 384, 386–387, 389
   X-rays, 381, *382*, 385

ionizing radiation, units of measurement, 379, 382, *382*
  becquerel, *382*, 389
  coulomb, *382*
  curie, *382*
  gray, *382*
  linear transfer energy, 383
  roentgen, *382*
  sievert, *382*, 383
ionizing radiation, weighting factors, 383, 395
isotocin, 331

juglone, 362, 370, 376
  7-methyl juglone, 362, 376
  Juglans nigra, 362, 376

karlotoxins, 357, 360–361, 368
  Gymnodiniuim veneficum. See karlotoxins, *Karlodinium*
    veneficum
  Karlodinium veneficum, 374–375
  Pfiesteria, 360
Kelch-like ECH-associated protein 1 (Keap-1), 288
killer whales (*Orcinus orca*), 281, 306
kisspeptin, 331
Kyshtym, 402

Lake Superior, 494–495
largemouth bass (*Micropterus salmoides*), 332
lead, 239–243
  bioaccumulation, uptake, 232–233
    role hardness, 231
  biochemistry, 231–232
  biogeochemistry, 230–231
    forms-speciation, 230
    isotopes, 230
    sediment cores, 231
  detoxification, 232
  ecotoxicity, 232–234, *233*
    diet-borne, 233
    elemental Pb(0), 233
    fish and wildlife, 233
    signs of Pb toxicity, *232*
  occurrence, sources, uses, 228–230, *230*
    tetraethyllead, 228, *231*
lectins, 363, 376
legacy contaminants, 277, 294
  regulation, 294, *295*
levonorgestrel (LNG) (synthetic gestagen), 333
Love Canal, 11, 490, 493
luteinizing hormone (LH), 330, 332

Manhattan Project, 398
markers, indicators, 107–117
  bioindicators, *101*
  biomarkers, *108–109*
  chemical markers, *108*
  community indicators, *110*

linking chemical cause, ecological effects, 107–111
metal-binding proteins
  metallothionein, 108
  phytochelatin, 108
  sediment quality triad, *112*
menadione, 362. *See also* naphthoquinones
mercury, 210–225
  amalgamation, *212*, 421
  bioaccumulation, uptake, 219–221
    flooding effect, 220
    monitoring, 221
    pH effect, 219, *220*
    selenium effect, 220
  biogeochemistry, 212–217
    Atmospheric Mercury Depletion Events (AMDEs), 213
    biogeochemical cycle, *215–217, 217*
    complexation (DOM), 209
    forms-speciation, 212
    redox reactions, 212
  biomagnification, 187, 217–219
    definition, *81*
    examples, *218*
  detoxification, 224–225
    demethylation, 225
    metallothionein, 224
    selenium, 224
  ecotoxicity, 221–224
    fish and wildlife, 222–224
    Minamata, 221
    signs of Hg toxicity, *222*
  elemental Hg(0), 10, *76*, 211–213, *212*, 215, 423, 496
  methylation, 214–215
    biochemistry, 214
    microbiology, 214–215
  methylmercury, *67, 80*, 89, 112, 150, 175, 212, 214, 217, 221, 364,
    366, *459, 470*, 496
  occurrence, sources, uses, 210–212, *211*
metalloids
  definition, 175
metals
  analytical challenges, *179*
    filtration, 178
  comparison Cd, Cu, Hg, Ni, Pb, Zn, As, Se
    covalent index, *210*
    crustal abundance, *210*
    mining production, *210*
    mobilization factor, *210*
    relative toxicity, *210*
  comparison, organic contaminants, *174*
  dose–response relationships, 172, *174*
  Periodic Table of the elements, 172, *173*
metals, bioavailability, 179–187
  biotic ligand model (BLM), 181–184, *182*
    assumptions, 183
    exceptions, 183
  diet-borne, 186

metals, bioavailability (cont.)
    biomagnification (definition), 187
    digestive tract conditions, 187
    trophic transfer, trophic transfer factor (TTF), 187
  dissolved metals, 180–184
  metal mixtures, 50, 184
  particulate metals, 184–185
  sediments
    acid-volatile sulphide (AVS), 184
    AVS-SEM model, 184
    pore water, 184, *185*
  soils, 185–186
    extractions, 185, *186*
    pore water, 185
    rhizosphere, 185
  uptake
    biological membranes, 180
    ion mimicry, 180
    mechanisms, *180*
metals, biogeochemistry, 174–178
  sedimentation, 178, 230
  sediments, 184–185
    acid-volatile sulphide (AVS), 184, 247
    interstitial water. *See* metals, biogeochemistry:sediments:
      pore water
    oxic, anoxic, 184
    pore water, 184
  soils, 185–186
    serpentine, 202
    smelter-affected, 186
  sources, mobilization, 178–179
    mobilization factor, 178, *210*
metals, detoxification, 195–202, *197*
  field observations, 199–202, *201*
    Chaoborus sp. (phantom midge), 200
    Perca flavescens (yellow perch), 200
    Pyganodon grandis (floater mussel), 200
  intracellular metal speciation, 195–196, *196–197*
    apoenzyme, 195
    chaperone, 195
    finger protein, 195
    metallothionein, *197*
    phytochelatin, *197*
  laboratory observations, 198–199
  mechanisms, 195
  metabolic cost, 202
  subcellular partitioning, 196–197
    protocol, *198*
    spillover hypothesis, 197–199, *199*
metals, hard, borderline, soft, 178
  cadmium, 227
  copper, 236
  mercury, 224
  nickel, 241, 243
metals, organometallic molecules, 206
  arsenobetaine, 206

methylmercury, 206
  regulation, 206
  tetraethyllead, 206
  tributyltin (TBT), 206, 345–346
  triphenyltin (TPT), 346
metals, speciation, 174–178, *175–176*
  chemical equilibrium modelling, 176
  complexation, 175, 207–209
  coordinate bond, 175
  covalent bond (organometals), 175, 206
  hard, soft cations, 178
  ligands, 175
  oxidation, reduction reactions (redox), 174
metals, tolerance, 202–206
  acclimation, 183, 205, 207
  adaptation, 202, 254
  co-tolerance, 205
  definition, 202
  ecological advantage, 205, 243
  ecotoxicological implications, 205
  examples, *204*
  mechanisms, 203–205
  metallothionein, 203
  phytochelatin, 203
  plants, hyperaccumulation, 202–203, 509
    Arabidopsis sp., 203
    Thlaspi sp., 204, 247
  pollution-induced community tolerance (PICT), 205
  soils, serpentine flora, 202
metals, toxicity
  biotic ligand model (BLM)
    cadmium, 227
    copper, 237–238
    lead, 232
    nickel, 241
    regulation, 464
    zinc, 246–247
  intrinsic toxicity, 177, *210*, 225, 241–242, 246
  quantitative ion character–activity relationships (QICARs), 178
  toxicity modifying factors, 206–209
    dissolved organic matter (DOM), 209, 464
    hardness, 208, 464
    pH, 207, 464
    salinity, *177*, 208
    temperature, 206
metals, toxicity mechanisms, 172, 187–190, *188–190*
  adverse outcome pathway (AOP), 188
  altered enzyme conformation, 188
  behavioural effects, 190
  displacement essential cations, 188
  gene transcription, 190
  molecular initiating event (MIE), 188
  neurotoxicity, 190
  olfaction, fish, 191
  oxidative stress, 189
methane, 516, *517*, 521

metolachlor, 542
microcystins, 361, 371, 374
  Anabaena sp., 361
  Microcystis aeruginosa, 361
  Microcystis sp., 361
  Oscillatoria sp., 361
Minamata
  disease, 8–10
  mercury, methylmercury, 8–10, *17*
Minamata Convention, 10, *17*, 460–461
  mercury uses, emissions, 460
  mercury, fish consumption advisory, 469
mining
  beneficiation, 425–426
  Clark Fork River, MT, USA, 428, 508
  effects
    aquatic, 427–431
    floodplain, 428
  Fly River, Papua New Guinea, 428
  history, 419
  mine cycle, 424
  non-entry, 425
  open-pit, 424–425, 532
  underground, 424–425
mining and smelting
  effects
    aquatic, 200, *447*
    coastal marine, 430
    ecotoxicological, *430*
    lakes, 430
  emissions, 426, *427*, 430
  history, silver and mercury, 420–424
  mercury, 420
  patio process, 422
  regional considerations, 528
  silver, 420
  temporal trends, *427*, 430–431, 528, *530–531*
mining, deep-sea, 534–539
  abyssal plains, 535, 537–538
  anticipated impacts, 537–538
  Clarion–Clipperton Zone (CCZ), 535, 538
  deep-ocean chemistry (T, P), *537*
  hydrothermal vents, 537
  International Seabed Authority (ISA), 535, 538
  manganese nodules, 535
  massive sulphides, 536
  mineral deposits, *536*
  mitigation, 538
  polymetallic crusts, 536
  recovery, 538
  regulation, 535
  seamounts, 536–538
  tailings, 529
  typical metals (Co, Cu, Mn, Ni, Te, Tl, Y), 536
mixotrophy, mixotroph, 360, 375
mode of action (MoA), *101*

model types, ecotoxicology, 123
modelling, bioaccumulation, 126–129, *127*
  agricultural, 127
  bioamplification, biodilution, biotransformation, 126
  complexities, long-lived species, 128–129
  dynamic models, 123–124, 129
  equilibrium, definition, 123
  equilibrium, non-equilibrium tissue distribution, 126
  life-history events, 127
  linked models, trophic levels, 127
  maternal transfer, *128*
  routes, uptake, elimination, 126
  single species, multi-compartment, 126
  steady-state, 87, *125*, *127*, 129
  time-varying exposures, 127
  wildlife models, 128
modelling, chemical fate, *123*, 123–126, *126*
  behaviour, transport, distribution, 123–124
  chemical loading, 124
  fate, definition, *123*
  food web biomagnification, 127, 129
  multisegmented models, 125
  remediation options, 123
  steady-state, 123–124, *126*
modelling, integrated effects, 129–130
  aggregated exposure pathways (AEPs), 130, *130*
  dynamic energy budget, 130
  quantitative ion character–activity relationships (QICARs), 129
  quantitative structure–activity relationships (QSARs), 129
  toxic effects models, 129
modelling, PCB contamination, killer whales, 128–129
  contamination, age, sex, reproduction, *128*
  individual-based model, 129
  killer whale habitat, *128*
  maternal transfer, *128*
  population-based model, 129
  sediment dredging, contamination, 129
  threats to viability, 128–129
modelling, rationale, 122–123
  chemical regulation, management, 122–123
  diversity
    developmental stages, species, 122–123
    environmental conditions, 122–123
    environmental contaminants, 122–123
  Stockholm Convention, global-scale models, 125
  test hypotheses, fate and effects, 123
monitoring, orchard pesticide, *477*
  coordination, biological, chemical monitoring, 477
  distribution, spatial, temporal, 477
  ERA, prospective, retrospective, 477
  exceedances, HC_5, 477
  inputs, rainfall, 477
  monitoring, post-registration, 477
  regulatory weakness, 477
  river dilution, intermittent flow, 477
  rivers, Spain, Portugal, 477

monitoring, orchard pesticide (cont.)
   samplers, passive, 477
   sampling, high frequency, 477
   species at risk, 477
monocrotophos, 299. *See also* Swainson's hawk
monooxygenases
   cytochrome P450 enzymes, 284. *See also* Phase I
      reactions
   flavoprotein monooxygenases (FMO). *See* Phase I reactions
multidrug resistance proteins (MRPs), 288
   MRP2, MRP3, MRP4, 288
mummichog (*Fundulus heteroclitus*), 286
Murray rainbow fish (*Melanotaenia fluviatilis*), 342
mycotoxins, 282, 286, 364–365. *See also* aflatoxins
myzocytosis, 360

N,N-diethyl-*m*-toluamide (DEET), 542
nanomaterial, engineered
   definition, 432
   examples, *432*
nanoparticle, engineered
   definition, 432
   detection, 434
   fate, transformation, *433–434*
   intrinsic toxicity. *See* nanotoxicity
   metal-based
      cadmium, 437
      cerium, 436
      copper, 192
      silver, 438
   nanotoxicity, 435–437, *437*
      regulation, *432*
   toxicity, 432
      experimental challenges, *435*
      oxidative stress, 436
      particle-assisted, 435, 437
   uptake, 434–435
      endocytosis, 436
nanoparticle, natural
   colloid, 433–434
naphthalene, 290
naphthazarin, 362
naphthoflavone, 286
naphthoquinones, 362–363, 370, 376
   7-methyl juglone (7-methyl-5-hydroxy-1,4-
      naphthoquinone), 362
   juglone (5-hydroxy-1,4-naphthoquinone), 362, 370
   menadione (2-methyl-1,4-naphthoquinone), 362
   naphthazarin (5,8-dihydroxy-1,4-naphthoquinone), 362
   plumbagin (2-methyl-5-hydroxy-1,4-naphthoquinone),
      362, 370, 376
narcosis, 276
neonicotinoid insecticides, *302*
neuroendocrine disruption, 331–332. *See also* oxytocin; isotocin;
      serotonin

nickel, 239–243
   bioaccumulation, uptake
      role DOM, 241
      role hardness, 240–241
      role intracellular chaperones, 241
      role pH, 240–241
      role salinity, 242
   biochemistry, 240–241
   biogeochemistry, 240
      forms-speciation, 240
   detoxification, 241
   ecotoxicity, 241–243
      fish and wildlife, 242–243
      signs of Ni toxicity, *242*
   hyperaccumulation
      ecological advantage, 205, 243
   occurrence, sources, uses, 239–240, 426
      serpentine soils, 239
nicotinamide adenine dinucleotide (NAD), *285*
nicotinamide adenine dinucleotide hydride (NADH), *285*
nicotinamide adenine dinucleotide phosphate (NADP),
      284, *285*
nicotinic acetylcholine receptor (nAChR), 302
non-ionizing radiation, *379–380*, 379–380
nonylphenol, *340*
nuclear factor erythroid 2-related factor 2 (Nrf2), 288
nuclear power, 381, 386–388, 390–391, 403
   Chernobyl, 389–390, 398, 402
   enrichment, 386–387
   Fukushima Daiichi, 390
   mine tailings, 386
   nuclear reactor, 381, 386, 388
   spent fuel, 387
nuclear receptors, 281. *See also* aryl hydrocarbon receptor
   amino terminal transactivation domain (NTD), 281
   central DNA binding domain (DBD), 281
   ligand-binding domain (LBD), 281
nuclear waste, 387–389
   United States Nuclear Waste Policy Act, 388
   waste fissile, 386
   Waste Isolation Pilot Plant, 388
   Yucca Mountain Nuclear Waste Repository, 388
nuclear weapons, 380, 385–387, 398
   Hiroshima, 380, 385, 389
   Nagasaki, 380, 385, 389
nucleic acid sequencing, 139, 148–150, *151*, 161. *See also* omics;
      metagenomics; microbiome
   RNA-Seq, 142
nutrients versus toxicants, 25, *26*

objectives, guidelines, standards, purpose, 102, 111
octanol, 278, *318*. *See also* partition coefficient
oestrogens, 281. *See also* cellular receptors
oil spills, 12
   Atlantic Empress, 12

Deepwater Horizon, 12, 475, 504
Exxon Valdez, 12, 475, 504
Natural Resources Damage Assessment (NRDA), 12
Oil Pollution Act (USA), 12
public's perspective, public education, 12
okadaic acid, 357, 360, 375
    Dinophysis sp., 375
    Phalocroma rotundatum, 375
    Prorocentrum sp., 375
omics, 139–142, 150–151
    adverse outcome pathways (AOPs), 152–153
    DNA methylation, 147–148
    ecological risk assessment (ERA), 140, 142, 152, 154, 157
    environmental DNA (eDNA), 149
    environmental effects monitoring, 157
    exposome, 160–161
    lipidomics, 147, *159*
    metabolomics, 140, 146–147, *151*, *159*, 159
    metagenomics, 149–150
    microbiome, 149–150
    microRNA (miRNA), 142, 148
    multi-omics. *See* omics, exposome
    proteomics, 144–146, *151*, *153*, 156, *159*, 159
    regulatory toxicology, 152, 153–154, 157
    transcriptomics, 141–142, 151–152, *151*, *153*, 156–159, *159*
        applications, 142–143
        applications, fish, *143*
opioids
    codeine, 362
    heroin, 362
    morphine, 362
organic compounds
    bioaccumulation, 279–281, 289, 291, 298, 308, 314, 318
    bioaccumulation factor (BAF), 281
    bioavailability, 280, 303, 309
    bioconcentration factor (BCF), 281, 291
    biomagnification, 281, 306
    food chain, 281, 294, 306, 309–313. *See also* biomagnification
    heterocyclic, 277. *See also* polycyclic aromatic hydrocarbons
    hydrophilic, 279
        products Phase I, Phase II metabolism, 283
    hydrophobic (lipophilic), 279
        sorption to soil, sediment, 279
    metabolism, 282–283. *See also* Phase I reactions, Phase II reactions, Phase III reactions
    perfluoroalkyl, 281, 307–313
        mobility, soils, 309
        perfluorooctane sulphonic acid (PFOS), 307
        perfluorooctanoic acid (PFOA), 307
        surfactant properties, 307
    polyfluorinated, 307
    proteinophilic, 281, 318
    routes of uptake, 280
        gut endothelium, 280
        plant leaves, 280, 304
        plant roots, 280
        respiratory epithelia, 280
organic contaminants, 284, 286, 294, 310
    effect CYP450 enzyme system on toxicity, 286
    partitioning in the environment, 278, *278*, 311
    persistent organic pollutants (POPs), 294, *295*, 494–495
    regulation, 294
organochlorine pesticides, *340*. *See also* legacy contaminants
organophosphate flame retardants, *307*, 307
organotins
    tributyltin (TBT), *340*
    triphenyltin (TPT), *340*
osmotrophy, 360
Our Stolen Future, 331. *See also* Colborn, Theo
oxidative stress, 283, 285, 289
oxytocin, 331–332. *See also* neuroendocrine disruption

P450 NADPH reductase, 284. *See also* monooxygenases
palaeo-ecology, definition, 113
palaeo-ecotoxicology, 112, *113*, 114–115, *115–116*
    bioindicators, 113–114
    community structure, 114
    core dating, ^{210}Pb, ^{137}Cs, stable lead, 113
    definition, 114
    degradable organic compounds, 114
    ecosystem change, recovery, 114
    metals, 430
    methods, 113, *113*
    microfossils, plankton, 113
    persistent substances, 113–114
    presence, absence, predatory fish, 114
    redox-sensitive elements, 114
    reproductive failure, lake trout, 114–115
        biota–sediment accumulation factors (BSAFs), 114
        embryotoxicity, 114–115
        sediment cores, contamination history, *115*
        toxic equivalent concentration (TEC), eggs, 114
        toxic equivalent factor (TEF), 114
        toxicity threshold, 115
    retrospective assessment, 112
    species composition, abundance, 113
    terrestrial plant spores, seeds, 113
    time order, exposure, effects, 112
palaeolimnology
    definition, 113
palynology
    definition, 113
Papua New Guinea, *218*, 427–428, 529, 535
Paracelsus, *4*
    dose, 4–5
    poison, 4–5
paracrine hormones, 329
partition coefficient, 90
    air–water (K_{AW}), *126*, *278*
    biota–water, *126*

partition coefficient (cont.)
  octanol–air (K_{OA}), 90, 124, 126, *278*
  octanol–water (K_{OW}), 45, 81, 124, 126, *278*
  sediment–water, *126*, 280
  soil–water, 280
passive dosing, 39
passive samplers, *38*
pentane, 276
perfluorooctanesulphonic acid (PFOS), *340*
perfluorooctanoic acid (PFOA), *340*
peroxisome proliferator-activated receptor (PPAR), *282*, *340*, 342.
      See also phthalates
personal care products, 316, *316*. *See also* benzophenone; triclosan
  benzophenones, *340*
  parabens, *340*
  synthetic musks, 316
  triclosan, *340*
P-glycoproteins, 288
phagocytosis, 393, 395
pharmaceutically active compounds (PhACs), 315–318
  antimicrobial resistance (AMR), 317
  carbamazepine, 315
  diclofenac, 317
pharmaceuticals
  17α-ethinyloestradiol, *340*
  carbamazepine, *340*
  fluoxetine, *340*
  ibuprofen, *340*
Phase I reactions, 283–287
  CYP450 enzymes in plants, 284
  epoxide formation, 286
  location CYP450 enzyme system in eukaryotes, 284, *285*
  Phase I enzymes, 282–285
  reactive functional groups, 284
Phase II reactions, 283–284, 287–288, *289*
  amino acid conjugation (amidation), 288
  glutathione-mediated conjugation, 287, *288*
  sulphate conjugation, 288
  sulphotransferases, 288
Phase III reactions, 283, 288
phthalates, 277, *282*, *314*, 314, *340*, 342
  butylbenzyl, 525
  dibutyl, *339*, 525
  diethylhexyl, 314, *339*, 525
  dimethyl, 525
phytochemicals, 361–362
  plant alkaloids, 362
  plant steroids, 362
  polyphenols, 362
  quinones, 361
phyto-oestrogens, 281
phytoremediation, 492, 507–508
  agromining, 509
  hyperaccumulators, 509–510
  phreatophytes, 510

phytoextraction, 509
phytofiltration, 510
phytomining, 509
phytostabilization, 510
phytovolatilization, 510
plasma membrane, 61, 71. *See also* cell membrane
plasmodesmata, 77
plasticizers. *See* bisphenol A; phthalates
  microbiome, modification, 150
  oestrogen receptor, activation, 153
plastics
  bioplastics, *527*
  degradation, 524, 527
  fate, aquatic environment, *528*
  global production, 522
  microplastics, 522–526, 544
    adsorbed chemicals, 525
    endocytosis, 525
    mitigation, 526–528
    physical effects, 524
    plasticizers, 525
    primary, 523
    secondary, 523
    toxicology, 524–526
    trophic transfer, 524, 526
  nanoplastics, 432, 523, 525–526, 544
plumbagin, 362, 376
  Bignoniaceae, 370
  Plumbaginaceae, 370
plutonium, 381, 385–389, 396, 398
policy, 459
  basis for legislation, 459
  definition, 459
  development, 459
'polluter pays' principle, 460
polyamide (PA), 524
polybrominated diphenyl ethers (PBDEs), *340*
polychlorinated biphenyls (PCBs), 296–297, 494–495. *See also*
      halogenated aromatic hydrocarbons
  Aroclor 1242, 1254, 1260, 296
  coplanar PCBs, 297
polychlorinated dibenzodioxins (PCDDs). *See* halogenated
      aromatic hydrocarbons
polychlorinated dibenzofurans (PCDFs), 296. *See also* halogenated
      aromatic hydrocarbons
polycyclic aromatic compounds (PACs), 276, 290–292
  biogenic, 291
  petrogenic, 291
  pyrogenic, 290–291
polycyclic aromatic hydrocarbons (PAHs), 277, 282, 290, *290*, 291,
      See benzo(a)pyrene; coke; retene
polyethylene (PE), 524, 527
polyethylene terephthalate (PET), 524
polyhydroxyalkanoates (PHAs), 527
polyhydroxybutyrate (PHB), 527

polylactic acid (PLA), 527
polypropylene (PP), 524
polystyrene (PS), 524, 527
polyurethane (PU), 527
polyvinyl chloride (PVC), 277, 286, 313, 524
population demographics, *101*, 109–110
   Leslie matrix model, 109
precautionary approach, 461, 469
   application factor, guidelines, 463
predation, 360, 365–366, *367*, *371*, 375
predictive models, mixture toxicity, 27, 48–51
   bioconcentration factor, 50
   critical body burden (CBB), narcosis, 50
   dioxin-like compounds, 49
   error limits, 50
   mixtures as 'substances', 48
   molar, gravimetric units, 48
   multiple modes of action, 50
   partition coefficient, octanol–water (K_{OW}), 50
   risk assessment, 48
   target lipid model (TLM), 49–50
   toxic equivalent factor (TEF), *49*, 49
   toxic unit (TU) model, 48
   water–lipid partitioning, 50
pregnane X receptor (PXR), *282*
prolactin (PRL), 330, 332. *See also* neuroendocrine disruption
protein expression
   gel electrophoresis, 144–146
   tandem mass spectrometry (LC-MS-MS), 145–146
pulp and paper industry, 439–446
   'closed-loop' mill, zero-effluent, 445
   complex environmental interactions, 444–446
   fibre sources, 439–440
   history, global distribution, 439–440
   physical environmental effects, 439
   production volumes, 439
   products, byproducts, *441*
   pulp-mill effluent 'onion', *443*, 445
   pulp production technologies, *440*
pulp-mill effluent treatment, 442–444
   biosolids, *442*
   oxygen demand, biological and chemical, 442, *443*
   technologies, *443*, *445*
pulp-mill effluent, environmental impacts, 442–445
   benthic habitat damage, 439, 442, 444
   fisheries closures, 444
   oxygen depletion, 442–443
   toxicity, 439, 442–444
pulp-mill emissions
   chemicals, 439–442, *441*, *444*
   chlorinated compounds
     biomagnification, 443
     risks to fish consumers, 444
     sources, effects, 110, 443–445
     taste and odour, 443

mercury, 441, 444
   methylation, biomagnification, 444
   slimicide, 441
   resin and fatty acids, 442
   retene, biogenic production, *443*
pulp-mill production technologies, 440–442
   chemicals used, 440–441
   water use, 442
pulp mill, bleaching, 441
   chlorine and alternatives, *444*
   chlorine, byproducts, *444*
pyrethrins, 301, 361, *369*
   Chrysanthemum cinerariaefolium, *369*
   Tanacetum coccineum, *369*

quantitative ion character–activity relationships (QICARs),
   45
quantitative structure–activity relationships (QSARs), 25, *41*, 45,
   47–48
   mode of action, 48
   physicochemical properties, 47–48
   principal component analysis, 48
   regression, linear, multiple, 48
   substructures, chemical properties, 47

radioactive decay, 381
radioactive half-life, 381
radioisotope, 381, 385, 387–388, 399
radionuclide, 381, 385, 387, 389, 397–398, 400–402
radon, 382–384, 386
rainbow darter (*Etheostoma caeruleum*), 341
rainbow trout (*Oncorhynchus mykiss*), 332
recolonization, 495, 507–509
   acid-tolerant cultivars, 508
   actinorhizal species, 508
   ectomycorrhizae, 508
   fish, other animals, 508–509
   metal-tolerant cultivars, plants, species, 507
   reclamation, 508
   reintroduction, 492
   revegetation, 492, 508, 510
   rhizobial species, 508
recovery
   biological, 495–500, 507
   chemical, 495–500
   ecosystem function, 488, *490*
   natural, 493, 495, 509–510
     attenuation, 492, 500, 503–504, 506
     lake flushing, 494
     microbially mediated attenuation, 495
     monitored natural recovery (MNR), 494–495
     passive, 494–495
   soil, 495–500
   terrestrial ecosystem, 495, 500
   to the original condition, 488

recovery (cont.)
  to the previous state, 488
  triggers for action, 490
regulations
  development, *8*
  early history, 15–16
  international agreements, 16, *17*
  public awareness, 458
  purpose, 5, 457
regulations, complex mixtures, 469
  effluents as substances, 469
  end-of-pipe, 469
  industry sectors, 469
  toxicity interactions, mixtures, 469
  toxicity interactions, water quality, 469
regulations, contaminated sites
  change to land use or zoning, 490
  clean-up procedures, 488
  management, 495, 501–503. *See also* ecological risk assessment
     (ERA):risk management
  regulatory climate, 490
  regulatory principles, 511
regulations, fish consumption advisories, 469, *470*
  consumption estimates, *470*
  contaminated fish, sale, import, export, 469
  country food, contamination versus nutrition, 469
  Joint FAO/WHO Expert Committee on Food Additives
     (JECFA), 470
  mercury, *470*
  reference dose, *470*
regulations, future, 477, 478–482
  circular economy, 478
  corporate social responsibility, 478
  economic costs, loss of ecosystem services, 478
  emission taxes, 478
  lags, regulatory practice, 477
  models, bioaccumulation, toxicity, 477
  traditional ecological knowledge. *See* traditional
     environmental knowledge
  traditional environmental knowledge, 478–482
  United Nations Declaration on the Rights of Indigenous
     Peoples (UNDRIP), 478
  voluntary self-regulation, 478
regulations, pesticide applications, 469
  amounts, concentrations applied, 469
  licensed operators, 469
  location, 469
  precautionary spraying, 469
  surveys, pest density, 469
regulatory challenges, 461–462
  corruption, 462
  culture, 462
  enforcement, effectiveness, 461
  global trade, waste, 462
  political will, 462, 470
  politics, 462

pollution havens, 462
poverty, 462
toxic colonialism, 462
war, natural disaster, 461
regulatory options, *459*
  bans, production, use, *459*
  criteria, guidelines, 458
  spill response, *459*
  standards, environment, food, 458
  trade agreements, 461
remediation, 488, 490–491, 493, 498–499
  definition, *490*
  on-site, 493
restoration, 488–490, 508
  definition, *490*
  functional native plant communities, 488, 508
  functioning ecosystem, 488, 494, 508
  goals, *489*
  landscape conservation, 488
  loss of valued species, 488
  no net loss philosophy, 488
  off-site compensatory action, 488
restoration, ecological, 488–489, *511*
  definition, *490*
  ecological function, *490*
  ecosystem function, 506
  reference ecosystem, *491*
  restored ecosystem, *491*
  theoretical ecologists, 488–489
retene, 286, 291, 518
retinoic acid receptor (RAR), *282*
retinoid X receptor (RXR), *282*
ricin, 363, 370
  Ricinus communis (castor oil plant), 363, 376
  Sambucus ebulus (dwarf elder), 363
  Viscum album (mistletoe), 363
risk
  ecological risk assessment, 5, 13–14
  risk benefit analysis, 5
  risk perception, 5
river continuum, 104

Saglek Bay, Labrador, Canada, 500–507, *501*
  ethoxyresorufin-*o*-deethylase (EROD), 502
  marine sediment PCBs, 501–503
  removal contaminated soil, 503
  soil PCBs, 501–502
  stakeholder group, 502
saxitoxin, 358–359, 361, *375*
  Alexandrium sp., 358, *375*
  Anabaena sp., 358
  Aphanizomenon sp., 358, *375*
  Gymnodinium sp., *375*
  puffer fish, 358
sediment quality triad, 40, 111–112, *112*
  bioaccumulation, 111

community structure, 40, 111
confounding stressors, 111
purpose, 111
sediment characteristics, 40, 111
sediment contamination, 111
strengths and weaknesses, 111
toxicity, 111
weight-of-evidence (WoE), 111
selective serotonin reuptake inhibitors (SSRIs), 316. *See also*
    pharmaceutically active compounds
selenium, 255–259
    bioaccumulation, uptake
        anion transport, 257
    biochemistry, 256–257
        antioxidant, 256
        detoxification Cd Hg As, 256
        methylation, 255
    biogeochemistry, 255–256
        biogeochemical cycle, *256*
        forms-speciation, 255–256
        methylated species, 255
        mobility, 256
            pH effect, 256
            role Fe(III), 256
        mobilization factor, 255
        redox reactions, 255
        selenate, 255
        selenide, 255
        selenite, 255
    ecotoxicity, 256–259
        Belews Lake, Hyco Reservoir, Kesterson Reservoir, 258
        diet-borne, 253–257
        egg-laying vertebrates, 258
        embryo toxicity, 258
        field observations, 258–259
        maternal transfer, 258
        mechanisms, 258
        signs of Se toxicity, *258*
        teratogenesis, 258
    hyperaccumulation
        ecological advantage, 257
    occurrence, sources, uses, 255
    organoselenium
        dimethyl diselenide, 256
        dimethyl selenide, 256
        glutathione peroxidase, 257
        methylselenol, 255
        selenocysteine, 257
        selenomethionine, 257
        selenoprotein P, 257
    trophic transfer, 257–258
serotonin, 328, 331–332
shellfish poisoning
    amnesic shellfish poisoning (ASP), 357
    diarrhoeal shellfish poisoning (DSP), 357
    neurotoxic shellfish poisoning (NSP), 357

smelting history
    silver and mercury, 420–424
social concerns, 8, 16
    public awareness, 5, 7, 12
    public perception, 5
Society for Ecological Restoration (SER), *491*, 491
species at risk, sensitivity
    iconic, 100
    keystone, 100
    test species, 100
species evenness, definition, *110*
species richness, definition, *110*
species sensitivity distributions (SSDs), 28, 47, *47*
    error limits, 47
    hazardous concentration, threshold, 47
    method, 47
    response measured, 47
    sample size, statistical power, 47
    species protected, proportion, 47
    species selected, 47
steady state, definition, *123*
steroid and xenobiotic receptor/pregnane X receptor
        (SXR/PXR), 338
steroid hormone nuclear receptors, 336
    effect di(2-ethylhexyl)phthalate on 5α-reductase, 333
    testosterone synthesis, role 5α-reductases, 333
steroidogenesis, 333, 337, 342
    CYP11, 333. *See also* cytochrome P450 enzyme system
    CYP17, 333. *See also* cytochrome P450 enzyme system
steroidogenic acute regulatory (StAR) protein, *334*
Sudbury Soils Study, 471–475
    chemicals of concern, metals, metalloids, *472*, *473*
    ecological risk assessment, area wide, 472
    ecosystem recovery
        lines of evidence, 474
        soil pH, metal mixtures, 474
    goal, objectives, 472
    history, metal mines, 471
    organization, progress, 471–472
    problem formulation, 472
    recovery goal, 472
    risk management, 475
    risks, probabilistic modelling, 474
    risks, toxicity reference values (TRVs), 474
    soil quality guidelines, 471
    study area, definition, map, *473*
    threatened or endangered species, 472
    valued ecosystem components (VECs), 473–474
    weight-of-evidence (WoE), 474
surveys
    design, 102–105
        assimilation capacity, definition, 104
        baseline, reference sites, 103
        gradient study, *103*
        statistical power, 104
        upstream–downstream, *104–105*

surveys (cont.)
  pulp-mill effluents, 104–105
  monitoring, 102
    cause, effect, 103
    definition, purpose, *102*
surveys, monitoring. *See* environmental surveillance, monitoring
Swainson's hawk (*Buteo swainsoni*), 299. *See also* monocrotophos

tailings
  erosion, wind, 186, 244, 386
  leaching, 244, 254, 425
  management, 426–427, 529, 532
target lipid model, 73
  bioaccumulation, 50
  bioconcentration factor, 50
  correction factor, 50
  critical body burden
    species, 50
    species differences, bioaccumulation, toxicity, 50
  critical body burden (dose), 50
  error limits, 50
  membrane function, 50
  mixture toxicity, 49–50
  modes of action, multiple, 50
  narcosis, baseline toxicity, 49–50
  non-linearity, 50
  non-polar versus polar organic chemicals, 50
  octanol–water partition coefficient (K_{OW}), 50
  toxic unit (TU), 50
  water–lipid partitioning, 50
technology-critical elements (TCEs), 529–534, *529*
  cobalt, 530, 536
  data-poor elements, 530
  lithium, uses deposits ecotoxicology, 531–534, 536
  rare earth elements, 529, *536*
tetrabromobisphenol A (TBBPA), 337, 343
tetrodotoxin, 4, 357, 361, *367*, *375*
  puffer fish, 358, 375
thyroid hormones, 281, 333–336. *See also* cellular receptors;
    hypothalamic–pituitary–thyroid axis (HPT)
  thyroxine (T4), 333, 335
  triiodothyronine (T3), 333, 335
thyroid-releasing hormone (TRH), 329. *See also* hypothalamic–
    pituitary–thyroid axis (HPT)
thyroid-stimulating hormone (TSH), 329. *See also* hypothalamic–
    pituitary–thyroid axis (HPT)
tolerance, resistance, pesticides, 31
toxic equivalent factor (TEF)
  dioxin-like compound, 49
  mixture toxicity, *49*, 49
  mode of action (MoA), 49
  toxic equivalent quantity (TEQ), 49
toxic equivalent factors (TEFs), 297
toxic equivalent quantity (TEQ), 297
toxicant-induced climate susceptibility, 518, 553

toxicity, acute versus chronic, 28
  delayed effects, 33
  effect measured, 28
  relative to life span, 28
toxicity, definition, *24*
  chemical structure, properties, *24*, 24–25
  effects, populations, communities, *24*
  exposure–response relationship, *24*
  molecular receptors, 24
toxicity, measures, *24*, 26–28
  dose, concentration, 26
  exposure time, 27, 33–35
  lethal versus sublethal effects, 26
toxicity, mixture dissection, 51–52
  bioassays versus toxicity tests, *51*
  chemical composition, *51*
  effects-driven chemical fractionation (EDCF), crude oil,
    51–52, *52*
  effluent treatment, toxicity, 51
  effluent, toxicity source, 51
  toxicity identification and evaluation (TIE), 51
  toxicity reduction and evaluation (TRE), 51
toxicity, mixture interactions, 48–50, *49*
  additivity, strict, more than, less than, 48
  independent, joint action, 48
  synergism (potentiation), antagonism, 48–49
toxicity, mode of action (MOA), *25*
  acetylcholinesterase, 25
  metabolism and toxicity, 25
  organophosphate pesticide, *25*
  pentachlorophenol, 29
toxicity, predictive models, metal mixtures, 50–51
  additivity, 50
  additivity, independent action, 51
  bioavailability, 50
  biotic ligand model (BLM), 50
  free-ion activity, 50
  metal mixture modelling exercise, 50
  metal–metal interactions, 50
  toxicity-response functions, 51
toxicity, predictive models, single compounds, 45–48
  acute-to-chronic ratio (ACR), 46
  data quality, uncertainty, 46
  error limits, 46
  mode of action (MOA), 46, 48
  octanol–water partition coefficient (K_{OW}), 45
  quantitative ion character–activity relationships (QICARs), 45
  quantitative structure–activity relationships (QSARs), 25, 45,
    47–48
  regulatory tools, 45
  species sensitivity distributions (SSDs), 28, *47*
  target lipid model (TLM), 45
toxicity test
  sentience, definition, 42
toxicity test designs, 28

in vitro, 29
omics, emerging technologies, 29
population metric, 29
purpose, 29
range-finding versus definitive, *29*
recommendations, valid tests, 29
responses measured, mode of action, 29
test conditions, 29, 36–38
toxicity test statistics, 26–28, 42–44
  Abbott's formula, control responses, 44
  confidence limits, 42
  continuous data, binomial data, 44
  linear regressions, 44
  logistic regressions, 44
  median responses, 27
  multiple regressions, 44–45
    multiple variables, effects, 44
    response surfaces, 45, *45*
  non-parametric tests, skewed distributions, 44
  parametric tests, normal distributions, 44
  polynomial regressions, 44
  probit transformation, analysis, *43*
  repetition, replication, 42
  response rate, 28
  sample size
    acute, chronic tests, 44
    precision, 42
  sigmoidal exposure–response curve, 43
  subsampling
    multiple responses, 44
    selection bias, 44
    time series, 44
  threshold toxicity, estimating, 28, *35*, 46, *46*
  thresholds, legal, 42
toxicity test, alternative organisms, *41*
toxicity test, animal welfare, *41*, 41
toxicity test, biomarkers, *41*
toxicity test, characterizing test conditions, 38–39
  bioavailable fractions, 39
  exposure gradients, 38
  exposure, tissue dose, 38
  modelling, exposure, toxicity, 39, *41*
  nominal versus measured concentrations, 36
  non-steady-state concentrations, 38
  passive samplers, *38*
  quality assurance, quality control (QA/QC), 39
  sediment pore water, *38*
  specific ion electrode, metals, 39
  variance, 38
  water quality, 38
toxicity test, chemical exposures, *24*
  bioaccumulation, 25
  bioavailability, 32
  concentration, dose, 26
  exposure gradients, 33, *34*

exposure monitoring, 38
exposure route, 32
oxygen concentration, 32
partitioning, 32
pulse exposure, 33
solution renewal, 32
static, semi-static, continuous, 33
water quality, 32
water solubility, *32*
toxicity test, definition, 24, 28
  bioassays versus toxicity tests, *51*
toxicity test, exposure route
  diet, 33, 36
  test media, 26, 32
toxicity test, exposure time, 33–35
  acute, chronic, 34
  observation, intervals, 34
  thresholds, 28, *35*, 35, 46, *46*
  time–toxicity relationship, 33, *35*
toxicity test, modelling, *41*
toxicity test, novel, 52–54
  adverse outcome pathway (AOP), 54
  assimilated energy allocation, 52
  behaviour, 53, *53*
  biomarkers, 54
  chemoreception, 53
  developmental toxicity, 52
  dynamic energy budget (DEB), 52
  energetics, 29
  omics, 53–54
  reproduction, 52
  scope for growth (SFG), 52
  sensitivity, relevance, 52
toxicity test, organism sensitivity, 26
  confinement, 37–38
  cover, 37–38
  crowding, competition, dominance, 37–38
  handling stress, 37–38
  light intensity, 37–38
  movement, vibration, 37–38
toxicity test, organisms, 29–32
  acclimation, acclimatization, adaptation, *31*
  animal welfare, 41
  confinement stress, 30
  culture conditions, 29
  development rate, embryo, 31
  disturbance, 30, 38
  embryos, 31
  full life-cycle tests, 31
  genetic diversity, 30
  growth, reproduction, mortality, 31
  juveniles, 31
  lethal, sublethal responses, 32
  life stages, 26, 29
  sentience, 42

toxicity test, organisms (cont.)
  sexually mature, 31
  social interactions, 30
  species, 30–31, *30–31*, *41*
  strains, sensitivity, 30
  surrogate species, *41*, 41
  tolerance, 31
toxicity test, quality assurance, quality control (QA/QC),
      32–36
  chemical analyses, 39
  controls, negative, positive, solvent, 36, 39
  expectation bias, 36
  observer effect, 36
  precision, accuracy, 33
  randomization, treatments, observations, 33, 38
  repeatability, 33
  responses, false negative, false positive, 36
  solution renewal, 32
toxicity test, realism, 41
  complexity, extrapolation, 41
  design, 41
  experimental ecosystems, 41
toxicity test, role in chemical management
  guidelines, standards, 42
  research, hypothesis testing, 42
  risk, predictions, 42
toxicity test, sediments, 38–41
  bioavailability, 40
  chemical disequilibrium, 40
  chemical dosing, 40
  ecosystem integrity, indicators, 40
  feeding strategies, 40
  interactions, chemical, spiked sediments, 40
  leachates, extracts, overlying water, 40
  mixing, 40
  oxidation state, anoxic versus oxic, 40
  physical, chemical characteristics, 40
  pore water, *38*
  porewater peepers, *40*
  sediment quality triad, *111*
  spiked sediment tests, *40*
  standard protocols, 38, 40
  storage time, 40
  surface, deep, 40
  test species
    benthos, infauna, *30*, 40
    sediment characteristics, 40
  weight-of-evidence, 111
toxicity test, soils
  bioavailability, equilibration time, 40
  control soil, 41
  physical, chemical characteristics, 40–41
  species
    optimal conditions, 40

  responses, 40
  selection criteria, *30*, 40
toxicity test, sparingly soluble compounds, 36, 39
  bioavailable fraction, 39
  continuous flow, 39
  equilibrium partitioning, 36, 38–39
  non-steady-state exposures, 36, 38–39
  passive dosing, 39
  solubility limit, 39
  solution stability, 39
  solvent controls, 39
  solvents, 39
toxicity test, standard protocols
  consistency, comparability, 38
  risk assessment, 38
  test conditions, realistic, 41
toxicity test, test conditions, 36–38, *37*
  bioaccumulation, 36, *37*
  bioavailability, *37*
  biomass loading, *37*
  chamber materials, *36*
  chemical absorption, photolysis, hydrolysis, 39
  chemical form, 36
  confinement, wild organisms, 37
  dilution water characteristics, 36
  disturbance, *37*
  feeding, diets, *37*
  handling stress, 37
  lighting, *37*
  nominal versus measured concentrations, 36
  organic matter, biofilms, 36
  organism sensitivity, 36
  static, semi-static, continuous flow, *36*, 39
  temperature, *36*
toxicity, threshold definition, 28
toxicokinetics, 83–85
toxicology, predictive, 542–545
  adverse outcome pathway (AOP), 152–153, 543–544
  ecological risk assessment (ERA), 543
  in vitro testing, 541–542
  phylogenetic comparisons, 543–544
  quantitative ion–character activity relationships
      (QICARs), 530
  quantitative structure–activity relationships (QSARs), 541
  target lipid model, 541
  target ortholog assessments, 543–544
  ToxCast, 154–155, *155*, 542
toxins, natural, 355–356, *368*, *371*
  biological warfare, 363–364, 368, 370, 377
  climate change, 356, *357*
  collateral damage, collateral effects, 356, 366
  defence mechanisms, 368
  ecological advantages, 356, 366–368, 371, 375
  herbal remedies, 370

metabolic cost, 365–366
pesticides, 355–356, 368, 377
trade, environmental agreements
    Commission for Environmental Cooperation, USMCTA, 461
    European Union, 461
    North America Free Trade Agreement, 461
    Organisation for Economic Co-operation and Development
        (OECD), 461
    United States-Mexico-Canada Trade Agreement
        (USMCTA), 461
traditional environmental knowledge (TEK), 478–482
    Akwesasne territory, 479
    culturally appropriate solutions, 481
    environmental goods, sharing, 479
    fish consumption guidelines, weaknesses, 480–481
    governance
        environmental protection, complexity, 479–480
        federal, provincial, state, county, 479
        Haudenosaunee, Seneca Nation, Turtle Clan, 479
        jurisdictions, 479
        Mohawk Council, Akwesasne, Canada, 479
        St.-Regis Tribal Council, USA, 479
    indigenous culture, philosophy, diet, 480–481
    missing ecosystem components, 479
    networks, integrate, translate, communicate, 482
    pipeline development, 480
    responsibility
        personal, collective, 482
    St. Lawrence River Area of Concern (AOC), 480
    trust, empathy, 480
    United Nations Declaration on the Rights of Indigenous Peoples
        (UNDRIP), 480
transgenic crops, 369
transport, transcellular
    gill, intestine, lung, 71
transport, transmembrane, 64–71
    active transport, 65, 70
    antiporter, 70
    carrier-mediated, 69
    channel, gated, 68, 301, 365, 369
    cotransport, 70
    diffusion, 65
    metals and metalloids, 65, 72
    multidrug resistance proteins (MRPs), 89, 288
    passive transport, 65, 69
    polychlorinated biphenyls, 65, 80
    pores and channels, 65–69
    pump, 70
    role, ecotoxicology, 72
    symporter, 70
    transported substances (examples), 67, 76
    uniporter, 70
treatment, point-source Hg pollution
    English–Wabigoon River system

chlor-alkali plant, 496–497, 497, 499, 499
Clay Lake, 497, 498–499
Grassy Narrows, 498
mercury concentrations, fish, 497–498
mercury pollution, 496–498
Penobscot Estuary, Bay, 498
tributyl phosphate, 542
trichloroethyl phosphate, 307. See also organophosphate flame
    retardants
triclosan, 280
    microbiome, alteration, 150
    protein, alteration, 146
triphenyl phosphate, 542
tris(2-butoxyethyl) phosphate, 542
Truhaut, René
    ecotoxicology, 5

UDP glucuronosyltransferase (UGT), 285. See also Phase II reactions
uptake, contaminants, 73–79
    boundary layer, 78, 435, 437
    digestive system, 75
    endocytosis, 78
    gills, 74
    kinetics, 81–85
    lungs, 73
    olfactory system, 76
    plant foliage, roots, 76
    skin, 73
uranium, 381–382, 381, 384–385, 387, 397

vasopressin (antidiuretic hormone), 331–332
venoms, 365–368, 367, 371, 377
    bristles, 365
    coevolution, rattlesnake venom, squirrel resistance, 366
    defence, 365–366
    fangs, 365
    nematocysts, 365
vinyl chloride, 286, See polyvinyl chloride
vitellogenin (VTG), 336–337, 342
    oestrogenic compounds, response, 109
vultures, Gyps spp., population decline, 317, 317

weight-of-evidence (WoE), 107, 112
Windscale Piles Facility, 398
World Health Organisation (WHO), 364

xenobiotic regulatory element (XRE), 281
xenobiotics, 275
    definition, 276
    recognition by receptors, 281, 282

zinc, 243–248
    bioaccumulation, uptake
        bioaccumulation factor (BCF), 247

zinc (cont.)
    role DOM, 247
    role hardness, 247
    role pH, 247
  biochemistry, 245–246
    enzyme cofactor, 245
    enzyme stabilization, 245
    finger proteins, 245
    homeostasis, 245–246

    metabolic roles, *245*
    role metallothionein, 246
  biogeochemistry, 244–245
  ecotoxicity, 246–248
    sediments (AVS-SEM model), 247
    signs of Zn deficiency or toxicity, *172*
  hyperaccumulation, 248
  occurrence, sources, uses, 243–244
    galvanization, 244